Network Programming

Katta G. Murty

Department of Industrial and Operations Engineering
University of Michigan, Ann Arbor

Prentice Hall
Englewood Cliffs, New Jersey 07632

Library of Congress Cataloging-in-Publication Data

Murty, Katta G.
   Network Programming
      p. cm.
   Includes bibliographical references and index.
   ISBN 0-13-615493-X
   1. Systems programming (Computer science) 2. Computer networks.
   I. Title.
   QA76.66.M87 1992
   658.4'032--dc20   91-45726
                                CIP

Acquisitions editor: *Marcia Horton*
Production editor: *Irwin Zucker*
Cover design: *Lundgren Graphics, Ltd.*
Prepress buyer: *Linda Behrens*
Manufacturing buyer: *Dave Dickey*
Supplements editor: *Alice Dworkin*
Editorial assistant: *Diana Penha*

The authors and publisher of this book have used their best efforts in preparing this book. These efforts include the research, development, and testing of the theory and programs in the book to determine their effectiveness. The authors and publisher make no warranty of any kind, expressed or implied, with regard to these programs or the documentation contained in this book. The authors and publisher shall not be liable in any event for incidental or consequential damages in connection with, or arising out of, the furnishing, performance, or use of these programs.

Printed in the United States of America

10 9 8 7 6 5 4 3 2 1

0-13-615493-X

Prentice-Hall International (UK) Limited, *London*
Prentice-Hall of Australia Pty. Limited, *Sydney*
Prentice-Hall Canada Inc., *Toronto*
Prentice-Hall Hispanoamericana, S.A., *Mexico*
Prentice-Hall of India Private Limited, *New Delhi*
Prentice-Hall of Japan, Inc., *Tokyo*
Simon & Schuster of Asia Pte. Ltd., *Singapore*
Editora Prentice-Hall do Brasil, Ltda., *Rio de Janeiro*

Contents

iv

Preface

The earliest instance of a network model for a routing problem seems to have been constructed by L. Euler in his solution of the Königsberg bridges problem in the early 18th century. Initially, theoretical aspects of networks were developed by mathematicians under the name of graph theory. It was not until the 1940's that network models were used to study transportation and distribution problems by F. L. Hitchcock, L. V. Kantorovitch, and T. C. Koopmans. The study of network models got a big boost with the development of linear programming, and particularly after G. B. Dantzig showed that the computational version of the simplex method developed by him in the late 1940s simplifies considerably for the special case of the transportation problem. But the publication in 1962 of the book *Flows in Networks* by L. R. Ford and D. R. Fulkerson is the landmark event after which the study of network flows really took off.

Since then, network optimization has become an established branch of operations research that attracted a lot of talented researchers. It has grown rapidly in its sheer theoretical elegance, in its scope, and in its range of application. Beginning with the 1960s, work on data structures developed by computer scientists has been adopted for improving the performance of implementations of network algorithms. For several network optimization problems, the best implementations available today can solve very large-scale problems with reasonable running times and memory requirements. Because of this, network models and algorithms for solving them, have come into widespread use in many areas.

Flows (of goods, vehicles, people, services, messages, etc.) are an essential component of modern society. In fact, as the Telugu poet Venkateswara Rao Dhulipala states, life itself is a manifestation of the flow of blood and of nerve signals.

సెట్‌వర్కింగ్ కోసం ఫోను లైన్లు, కంప్యూటర్లు చూడాలా ; పుస్తకాలు తిరగేయాలా?
గుండెల్లో మండి రక్త నాళాల ద్వారా నిరంతరం ప్రవహించే రుధిరం
మెదడు నుండి నాడీ మండలం ద్వారా అనుక్షణం నడిచే యార్తాయాహనం
ప్రతి ప్రాణి ఒక పెద్ద సెట్‌వర్క్ సుమా, ఇ లలో ఇంతకు మింటి వేరొకటి లేదు సుమా !

(For flows and networks, does one have to look at modern innovations like phone lines and computer systems?, nay, these things are as old as life itself; for, the essence of life is nothing but the flow of the red fluid through blood vessels, and of nerve signals through the nervous system). Network models find many applications in controlling the various flows in our society. The application of network algorithms to solve these models benefits humanity enormously by optimizing the distribution, communication, construction, and transportation costs.

Thus, network flow models are a special class of mathematical programs for which very efficient algorithms exist. This special network methodology constitutes a fundamental pillar of optimization. New developments are occurring at a fast pace due to the dramatic increase in research efforts devoted to network problems. This makes the study of network flows very interesting and exciting.

The primary aim of this book is to cover the significant advances in network flow methods ranging across modeling, applications, algorithms, their implementations and computational complexity. While almost all real world problems are essentially nonlinear, the importance of linear models stems from their ability to provide a reasonable approximation to many problems. This point is brought home by the well-known engineering proverb which exhorts researchers to *"be wise, linearize."* This book deals with problems on network structures that can be handled by linear programming techniques or their adoptations.

Some familiarity with matrix algebra (which can be gained from an undergraduate course), especially the concept of linear independence and the pivotal methods for solving systems of linear equations is necessary. Also some knowledge of linear programming (LP) is needed. The basics of LP are reviewed in Chapter 1.

Chapter 1 presents definitions of all the important network concepts which form the basis for the material developed in the rest of the book, and several formulation examples of various types of network models. At the end of this chapter, we discuss a formulation of the Chinese postman problem (an important routing problem with many applications) as a minimum cost perfect matching problem. We also discuss algorithms for finding an Euler route in an Eulerian network, which are needed to solve the Chinese postman problem, once the solution of the corresponding minimum cost perfect matching problem is given. This chapter concludes with several formulation exercises from a variety of areas of application of network models.

Chapter 2 presents the maximum value flow problem in pure single commodity flow networks. Various algorithms for this problem are discussed, and their computational complexity is analyzed. Methods for doing sensitivity analysis in this problem are also discussed.

Chapter 3 covers the Hungarian and signature methods for assignment problems, and the classical primal-dual method for the transportation problem together with the polynomially bounded version of it obtained by scaling.

Chapter 4 discusses various algorithms for several types of shortest chain problems and their applications.

Chapter 5 presents algorithms for minimum cost flow or circulation problems in pure networks, based on a variety of approaches. We also present a recently developed strongly polynomial algorithm for minimum cost circulation problems. Chapter 6 covers an efficient structured programming implementation of the primal simplex algorithm applied to a pure single commodity network flow problem subject to additional linear constraints.

Chapter 7 deals with applications of network algorithms in scheduling the various jobs in a project according to the precedence constraints. Discussed here are critical path methods (CPM), and methods for finding the project shortening cost curve.

In Chapter 8 we discuss efficient implementations of the primal simplex method for solving generalized network flow problems using rooted loop labelings.

Chapter 9 presents various algorithms for finding minimum cost spanning trees in undirected networks.

And, in Chapter 10, we present blossom algorithms for matching, and matching/edge covering problems.

Most chapters contain many exercises including formulation problems from a variety of application areas.

This book is ideally suited for a second semester course in the first year of graduate programs dealing with mathematical programming, following an introductory course on linear programming. Such courses are offered in disciplines such as operations research, industrial engineering, other branches of engineering, or business schools. It could also serve as a text for a course on algorithmic graph theory in mathematics or computer science programs or as a supplemental book in the study of computational complexity and algorithms. Portions of this book can be used to supplement the material in other courses (such as Chapters 1 and 7 in a course on CPM in civil engineering).

I am deeply indebted to several friends and colleagues for their help in preparing this book. The one person whose explanations and insights helped me enormously is Santosh Kabadi. He reviewed several revisions of this manuscript, pointed out errors and ways of correcting them, and suggested many improvements. I also learnt a great deal about network flows and matchings through many discussions with R. Chandrasekaran and Clovis Perin, and from the subtle questions raised by enquiring students in my classes. The comments of many people including Abdo Alfakih, Horst Hamacher, Mohammed Partovi, Tamas Solymosi, Subbarao Ghanta, Vasant Tikekar, and Paul Tseng have been very helpful. I would also like to thank Jeffery K. Cochran, R. Gary Parker, and Horst W. Hamacher for reviewing the manuscript. The intellectually stimulating environment in our department at the University of Michigan in Ann Arbor has motivated me to think and work harder. I am grateful to Bala Guthy, Teresa Lam, Jolene Glaspie, and Ruby Sowards for help and advice with word processing and typesetting. Finally, my thanks to Vijaya Katta for everything.

Katta Gopalakrishna Murty
Ann Arbor, Michigan.

Other books by Katta G. Murty:

→ *Linear and Combinatorial Programming*, first published in 1976, available from R. E. Krieger Publishing Co., Inc., P. O. Box 9542, Melbourne, FL 32901.

→ *Linear Programming*, published in 1983, available from John Wiley & Sons, Inc. Publishers, 605 Third Avenue, New York, NY 10158.

→ *Linear Complementarity, Linear and Nonlinear Programming*, published in 1988, available from Heldermann Verlag, Nassauische Str. 26, D-1000 Berlin 31, Germany.

Glossary

$(\mathcal{N}, \mathcal{A}, \ell, k, c, \check{s}, \check{t}, \bar{v}$ or $V)$ denotes a network with all the data. See below for the meanings of these symbols.

Bold face capital letters such as $\mathbf{X}, \mathbf{Y}, \mathbf{A}, \mathbf{M}, \mathbf{E}, \Gamma, \Delta,$ etc. denote sets. In citing references, we give the authors' names in capital letters, year of publication, title of the article or book, journal name or its abbreviation as defined below, volume, number in the volume if available, and inclusive pages within parentheses, in that order. Figures, equations, exercises, tableaus, arrays, comments, etc. are numbered serially within each chapter. So, for each of these, the entity $i.j$ refers to the jth entity in Chapter i.

The *size* of an optimization problem is a parameter that measures how large the problem is. Usually it is the number of digits in the data in the problem when it is encoded in binary form. When n is some measure of how large a problem is (either the size, or some qualtity which determines the number of data elements in the problem), a finitely terminating algorithm for solving it is said to be of order n^r or $O(n^r)$, if the worst case computational effort required by the algorithm grows as αn^r, where α is a number that is independent of the size and the data in the problem. The n and m that appear in the computational complexity measures for algorithms for network problems are usually the number of nodes and lines in the network. An algorithm is said to be *polynomially bounded* if the computational effort required by it to solve an instance of the problem is bounded above by a fixed polynomial in the size of the problem. $\mathbf{P}$ is the class of all problems for which there exists a polynomially bounded algorithm. *NP-complete* is a class of decision problems in discrete optimization, satisfying the property that if a polynomially bounded algorithm exists for one problem in the class, then polynomially bounded algorithms exist for every problem in the class. So far no polynomially bounded algorithm is known for any problem in the NP-complete class, and it is believed that all these problems are hard problems (in the worst case, the computational effort required for solving an instance of any of these problems by any known algorithm grows faster, asymptotically, than any polynomial in the size of the instance). See Garey and Johnson [1980 of Chapter 1] for definitions of NP-complete and NP-hard classes of problems.

wrt	With respect to
iff	If and only if
$\mathcal{N}$	The finite set of nodes in a network
$\mathcal{A}$	The set of lines (arcs or edges) in a network
$G = (\mathcal{N}, \mathcal{A})$	A network with node set $\mathcal{N}$ and line set $\mathcal{A}$
(i, j)	An arc joining node i to node j
$(i; j)$	An edge joining nodes i and j
Head(e), tail(e)	The head and tail nodes on an arc e
$(i, j)_t, (i; j)_t$	The t-th arc (edge) among a set of parallel arcs (edges) joining i to j(i and j)
e, e_t	An arc or an edge in a network
e	The column vector of all 1s in $\mathbb{R}^n$
ℓ_{ij}, ℓ	The lower bound for flow on arc (i, j). ℓ is the vector of ℓ_{ij}
k_{ij}, k	The upper bound or capacity for flow on arc (i, j). k is the vector of k_{ij}
κ_{ij}, κ	The residual (or remaining) capacity for flow on arc (i, j). κ is the vector of κ_{ij}
ϵ	Usually, the residual capacity of an FAP or FAC
f_{ij}, f_t, f	The flow amount on arc (i, j) (or arc e_t) in a node-arc flow vector f
c_{ij}, c_t, c	The unit cost coefficient or length or weight of arc (i, j) or edge $(i; j)$ (or line e_t). c is the vector of c_{ij}.
p_{ij}, p	The multiplier associated with arc (i, j) in a generalized network. p is the vector of p_{ij}.
$\check{s}, \check{t}$	The source and sink nodes in a network
V_i, V	The exogenous flow amount at node i. V is the vector of V_i.
v	Value of a flow vector in Chapters 2, 5; or a vector of dual variables associated with columns of a transportation array in Chapter 3, or a node in Chapter 10
u	A vector of dual variables associated with rows of a transportation array in Chapter 3, or a node in other chapters
π_i, π	The dual variable or node price associated with node i, π is the vector of π_i

μ_σ, μ	The dual variable associated with the σth blossom constraint in Chapter 10. μ is the vector of μ_σ
$\bar{c}_{ij}, \bar{c}$	The reduced or relative cost coefficient on arc (i,j),wrt a node price vector π, it is $c_{ij} - (\pi_j - \pi_i)$. $\bar{c}$ is the vector of $\bar{c}_{ij}$.
$\mathbf{X}$	A set of nodes, e.g., the labeled nodes, or those defining the partition for a cut, etc.
$\overline{\mathbf{X}}$	$\mathcal{N} \backslash \mathbf{X}$
$[\mathbf{X}, \overline{\mathbf{X}}]$	A cut in a directed network
$(\mathbf{X}, \overline{\mathbf{X}})$	Forward arcs of the cut $[\mathbf{X}, \overline{\mathbf{X}}]$ in a directed network
$(\overline{\mathbf{X}}, \mathbf{X})$	Reverse arcs of the cut $[\mathbf{X}, \overline{\mathbf{X}}]$ in a directed network
$(\mathbf{X}; \overline{\mathbf{X}})$	A cut in an undirected network
$\mathbb{T}$	A tree
$P(i), S(i)$	The predecessor and successor indices of node i
$EB(i), YB(i)$	The elder brother and younger brother indices of node i
$H(\mathbb{T}, j)$	The family of a node j in the rooted tree $\mathbb{T}$, the set consisting of node j and all its descendents in $\mathbb{T}$
$(\mathcal{N}_1, \mathcal{N}_2; \mathcal{A})$	A bipartite network with node bipartition $(\mathcal{N}_1, \mathcal{N}_2)$
$\mathcal{P}$	A path
$\mathcal{C}$	A chain
$\mathbb{C}$	A cycle
$\overrightarrow{\mathbb{C}}$	A circuit
$\mathbf{A}_i$	After i, set of head nodes on arcs incident out of i
$\mathbf{B}_i$	Before i, set of tail nodes on arcs incident into i
$G(f)$	Residual network of G wrt bound feasible flow f
$\mathcal{A}(f)$	Set of residual arcs wrt a bound feasible flow f
$AN(f) = (N(f), A(f))$	Auxiliary network wrt feasible flow f. $N(f)$, $A(f)$ are nodes, arcs in it
$A^+(f), A^-(f)$	The $+, -$ labeled arcs in $A(f)$ in an auxiliary network
$\mathcal{L}$	A layered network
$A_r^+(f), A_r^-(f)$	The sets of $+, -$ labeled arcs in the rth layer in the layered network wrt f
$f(\mathbf{X}, \mathbf{Y}), \ell(\mathbf{X}, \mathbf{Y}), k(\mathbf{X}, \mathbf{Y})$	Sum of f_{ij}, ℓ_{ij}, k_{ij} respectively, over arcs (i,j) with $i \in \mathbf{X}, j \in \mathbf{Y}$.

n, m	Usually, number of nodes, lines respectively in a network.
R_i, C_j	Row i, column j of an array, or the associated nodes.
$\underline{\tau}_{ij}, \overline{\tau}_{ij}$	The crash and normal durations for a job (i, j) in a project network.
M	A matching
$B = (\mathcal{N}_B, \mathcal{A}_B)$	A blossom defined by the subset of nodes $\mathcal{N}_B$ in Chapter 10. $\mathcal{A}_B$ is the set of edges in it.
$\mathcal{A}_*(\pi, \mu)$	Set of equality edges wrt the dual solution (π, μ) in Chapter 10
$G_*(\pi, \mu)$	The equality subnetwork wrt the dual solution (π, μ) in Chapter 10
$G^1 = (\mathcal{N}^1, \mathcal{A}^1)$	The current network in Chapter 10. $\mathcal{N}^1, \mathcal{A}^1$ are the sets of current nodes, and edges
$\mathcal{A}_*^1(\pi, \mu)$	Set of current equality edges wrt the dual solution (π, μ) in Chapter 10
$G_*^1(\pi, \mu)$	Current equality subnetwork wrt the dual solution (π, μ) in Chapter 10
$d_{ij}(\pi, \mu)$	A quantity used in defining the duals of matching, edge covering problems in Chapter 10. Its definition depends on the type of the primal problem.
$\mathbf{Y}^-(x), \mathbf{Y}^+(x)$	Quantities associated with a solution vector x and an odd set of nodes $\mathbf{Y}$, in matching, edge covering problems, used in defining the blossom constraint corresponding to $\mathbf{Y}$ in that problem in Chapter 10.
MB, CB	The index sets corresponding to matching and covering blossom constraints in a 1M/EC problem in Chapter 10.
$(\mathcal{N}^{\leq}, \mathcal{N}^{=}, \mathcal{N}^{\geq}, \mathcal{N}^0)$	Subsets of nodes in a partition of $\mathcal{N}$ in a 1-M/EC problem
AP	A path in which edges are alternately matching, non-matching edges in Chapter 10
FAC	Flow augmenting chain
FAP	Flow augmenting path
ECR	Edge covering route
LP	Linear program
BFS	Basic feasible solution
POS	Partially ordered set

$\succ$	The relationship between elements in a POS
$\blacksquare$	Symbol indicating end of a proof
s. t.	Such that
c.s. conditions	The complementary slackness optimality conditions
Parent(e), son(e)	The parent and son nodes on an in-tree line e in a rooted tree

$\mathbb{R}^n$ Real Euclidean n-dimentional vector space

$|\alpha|$ Absolute value of real number α

$|\mathbf{F}|$ Cardinality of the set $\mathbf{F}$

$\lceil \alpha \rceil$ Ceiling of real number α, smallest integer $\overset{\geq}{=} \alpha$. e.g., $\lceil -4.3 \rceil = -4, \lceil 4.3 \rceil = 5$

$\lfloor \alpha \rfloor$ Floor of real number α, largest integer $\overset{\leq}{=} \alpha$. e.g., $\lfloor -4.3 \rfloor = -5, \lfloor 4.3 \rfloor = 4$

n^{r} n to the power of r. Exponents are set in this type style to distinguish them from ordinary superscripts

$O(n^{\mathrm{r}})$ A positive valued function $g(n)$ of the nonnegative variable n is said to be $O(n^{\mathrm{r}})$ if there exists a constant α s. t. $g(n) \overset{\leq}{=} \alpha n^{\mathrm{r}}$ for all $n \overset{\geq}{=} 0$. For meaning in the context of computational complexity see the beginning of this glossary.

$A_{i.}$ The ith row vector of the matrix A

$A_{.j}$ The jth column vector of the matrix A

$n!$ n factorial

$||x||$ Euclidean norm of the vector x. For $x = (x_1, \ldots, x_n)$ it is $+\sqrt{x_1^2 + \ldots + x_n^2}$

∞ Infinity

$\in$ Set inclusion symbol. $a \in \mathbf{D}$ means that a is an element of $\mathbf{D}$. $b \notin \mathbf{D}$ means that b is not an element of $\mathbf{D}$

$\subset$ Subset symbol. $\mathbf{E} \subset \mathbf{F}$ means that set $\mathbf{E}$ is a subset of $\mathbf{F}$, i.e., every element of $\mathbf{E}$ is an element of $\mathbf{F}$

$\cup$ Set union symbol

$\cap$ Set intersection symbol

$\emptyset$ The empty set

$\backslash$ Set difference symbol. $\mathbf{D} \backslash \mathbf{H}$ is the set of all elements of $\mathbf{D}$ that are not in $\mathbf{H}$.

$\geqq$ Greater than or equal to

$\leqq$ Less than or equal to

$\sum$ Summation symbol

$\sum(x_j : \text{over } j \in \mathbf{J})$ Sum of x_j over j from the set $\mathbf{J}$.

Abbreviations for Journal Names

AOR	*Annals of Operations Research*
BAMS	*Bulletin of the American Mathematical Society*
CACM	*Communications of the Association for Computing Machinery*
COR	*Computers and Operations Research*
DAM	*Discrete Applied Mathematics*
EJOR	*European Journal of Operations Research*
IPL	*Information Processing Letters*
JACM	*Journal of the Association for Computing Machinery*
JORS	*Journal of the Operational Research Society*
MOR	*Mathematics of Operations Research*
MP	*Mathematical Programming*
MPS	*Mathematical Programming Study*
MS	*Management Science*
NRLQ	*Naval Research Logistics Quarterly*
OR	*Operations Research*
ORQ	*Operations Research Quarterly*
QAM	*Quarterly of Applied Mathematics*
TS	*Transportation Science*
OR Letters	*Operations Research Letters*

Chapter 1

Network Definitions and Formulations

1.1 Introduction

Wanting to optimize is a basic human trait. We begin with an incident which illustrates the fact that the urge to optimize provides universal motivation.

When I joined the University of Michigan, we stayed in an apartment close to campus. Our daughter was three years old then, and we had a small poodle to which she was very much attached. One day the poodle was missing. We searched the entire neighborhood for it, but had no luck. Another day went by, but it did not return. Unable to console my tearful daughter, I sought the advice of a colleague. He suggested that I put an advertisement in the campus newspaper. They recommended the offer of a reward for the poodle's return. Everything was agreed and they began running the advertisement offering a reward of $150 for anyone who finds and returns the poodle. We were very confident that this would bring quick results. A week passed by, and I called the newspaper office to check the status. The girl who received the call recognized my voice and asked whether I was the professor with the missing poodle. I said yes, and asked to be connected to the advertising manager. She said that he was out. I then asked to be connected to his assistant. She said that he was out too. Then I wanted to speak with the editor. She replied that he was out too. I remarked "Goodness! Is everyone out?" She replied, "Yes Professor, they are all out looking for your dog!"

In this story, all the people have an optimizing attitude which is almost universal. With many successful optimization algorithms, and modern digital computers for implementing them, many organizations are using them routinely to optimize their operations.

While constructing an optimization model for a system these days, the emphasis is on keeping it *computable* or *tractable*. The network models discussed in this book

1

are among the most tractable. We have very efficient algorithms for solving them. Taking advantage of the special structure, they are able to solve very large problems many times faster than general purpose linear programming algorithms. Excellent computer implementations of these algorithms are widely available. In this book we discuss network algorithms in all their variety and depth. In this chapter we discuss the basic network definitions and present some formulation examples.

A *network* is a pair of sets $(\mathcal{N}, \mathcal{A})$, where $\mathcal{N}$ is a set of *points* (also called *vertices* or *nodes*) and $\mathcal{A}$ is a set of *lines*, each line joining a pair of points, together with some associated data. Nodes i, j are said to be *adjacent* if there is a line joining them.

A line joining i, j that can only be used in the direction from i to j is called an *arc*, and denoted by the *ordered pair* (i, j) with a comma, it is *incident into j* and *out of i* , node i is its *tail* , and j is its *head*. If (i, j) is denoted by e, then $i =$ tail(e), $j =$ head(e). For example $(2, 1)$ is an arc in the network in Figure 1.1 with tail 2, and head 1.

A line joining two points x, y that can be used either from x to y, or from y to x, is called an *edge* and denoted by the *unordered pair* $(x; y)$ with a semicolon, it is *incident* at x and y. For example $(4; 2)$ is an edge in the network in Figure 1.1 incident at nodes 4 and 2.

There can be more than one arc with the same orientation, or more than one edge joining points i, j; such lines are called *parallel lines*. They will then be denoted by $(i, j)_1, (i, j)_2$, etc. When the network has such parallel lines, each of them is considered a separate distinct line by itself. However, we assume, except in Chapter 8, that there are no *self loops*, which are lines joining a point with itself. Figure 1.1 is a network with 8 points, 16 arcs, and 7 edges. Arcs $(3, 5)$ and $(2, 6)$ intersect in it, but their point of intersection is not a point in the network. The network is a *directed network* if all the lines in it are arcs, an *undirected network* if all the lines are edges, and a *mixed network* if it has both arcs and edges.

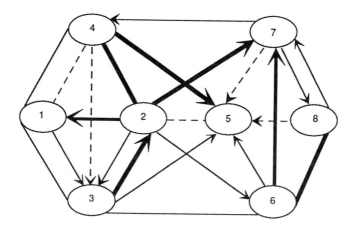

Figure 1.1

The *degree* of a point in a network is the number of lines incident at it. In a directed network, the *indegree (outdegree)* of a node is the number of arcs incident into (out of) it.

A *subnetwork* of G = $(\mathcal{N}, \mathcal{A})$ is a network F = $(\mathcal{N}, \bar{\mathcal{A}})$ with the same set of points, but with $\bar{\mathcal{A}} \subset \mathcal{A}$.

A *partial network* of G is a network $(\hat{\mathcal{N}}, \hat{\mathcal{A}})$ in which the set of points is $\hat{\mathcal{N}} \subset \mathcal{N}$, and the set of lines is $\hat{\mathcal{A}}$ which is the set of all the lines in G that have both their incident points in $\hat{\mathcal{N}}$. This partial network is also called the *partial network* or *subnetwork of G induced by* $\hat{\mathcal{N}}$. A *partial subnetwork* of G is a partial network of a subnetwork of G. For example, by omitting all the thick and dashed lines and all the lines incident at node 8 in Figure 1.1, we get the subnetwork in Figure 1.2. The partial network of the network in Figure 1.1 induced by the subset of nodes {1, 2, 6, 8 } is given in Figure 1.3. See Figure 1.4 for a partial subnetwork.

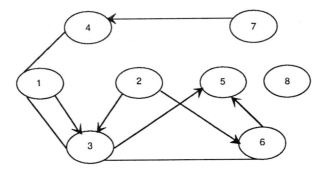

Figure 1.2 A subnetwork of the network in Figure 1.1.

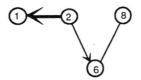
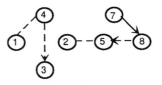

Figure 1.3 A partial network of the network in Figure 1.1

Figure 1.4 A partial subnetwork of the network in Figure 1.1

In a directed network, the *forward star (reverse star)* at a node i is the set of all arcs incident out of (incident into) i. It is sometimes convenient to represent directed networks by storing the forward stars of the nodes, or the reverse stars, or both. Searching through forward (or reverse) stars of nodes is a common operation in many network algorithms, this representation is convenient for carrying out this search. In a directed network G = $(\mathcal{N}, \mathcal{A})$, for $i \in \mathcal{N}$, the set $\mathbf{A}(i) = \{j : (i,j) \in \mathcal{A}\}$ is called the *after i* set, and $\mathbf{B}(i) = \{j : (j,i) \in \mathcal{A}\}$ is called the *before i* set.

Mathematically, an undirected network which has no self loops, is called a graph if it has no parallel edges, or a *multigraph* otherwise. A directed network which has no self loops is called a *digraph* if it has no parallel arcs, or a *multidigraph* otherwise. What distinguishes these graphs from networks is the fact that networks usually have some data such as arc lengths, or arc capacities associated with points and/or lines. Our problems come with data; hence we will use the term *network* to refer to our structures.

Network models are used extensively to analyze and optimize the operation of many systems. For example, vehicular traffic in highway systems is studied using the network in which points are traffic centers and lines are roadway segments joining pairs of points. Similarly, airline systems, railroad systems, shipping systems, and all transportation systems can be analyzed using appropriately defined networks. Natural gas, crude oil, or other fluid flows are analyzed using the appropriate pipeline and/or ship or truck route networks. The telephone network is the network for studying the flow of calls. Economic models can be treated as network models by representing factories, warehouses, and markets as points and by treating highways, railroads, waterways, and other transportation channels as lines. Even software and computer systems are analyzed using network models. A software system is often a collection of many components such as program modules, command procedures, data files, etc. The execution-time relationships and data communications between them form a directed network called the *call graph* of the program, nodes in it are procedures, and each arc represents one or more invocations of a procedure by another. Similarly, distributed computer systems are analyzed using appropriate network models that represent their working modes.

Network models find applications in the design and analysis of many other types of systems, as most systems have to transmit goods or messages etc., through an appropriate transportation or communication medium. The application of network algorithms to solve these models benefits society by optimizing distribution, communication, and transportation costs, and improving productivity. Also, being pictorial, a network model is visually informative and easy to construct and explain to top management.

The origin of network theory can be traced to the work of the famous Swiss mathematician Leonhard Euler on the *Königsberg bridges problem* in the year 1736. The problem originated as a children's game in Königsberg on the banks of the river Pregel. There were two islands (denoted as land areas 1, 4; 2, 3 denote the two banks of the river) and seven bridges each joining a pair of land areas as in Figure 1.5. The question is: Is there a route beginning in one of the land areas, passing through each of the bridges exactly once, and returning to the initial land area at the end? The town's children had long amused themselves running across the various bridges endlessly trying to find such a route, but no one succeeded. It intrigued Euler, and when he grew up, he constructed a network model and developed an elegant method to answer this question. He represented each land area by a node, and each bridge by an edge joining the corresponding pair of nodes as in Figure 1.6. The desired route corresponds to one that begins at a node, traverses through each edge exactly once, and terminates at the starting node. Such a route is nowadays called an *Euler route* or *Euler circuit*. He proved that such a route exists in an undirected network iff it is connected (i.e., it is possible to pass from any node to any other node using the edges of the network), and every node has even degree (Theorem 1.11 of Section 1.3.8). Since the network in Figure 1.6 contains nodes which are not of even degree, the answer to the Königsberg bridges problem is no. Euler provided an elegant answer to an existence question. Section 1.3.8 discusses efficient methods for actually computing an Euler route when it exists, and their application to solve important routing problems.

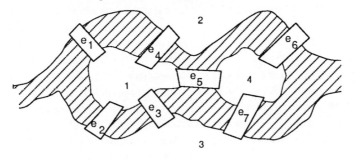

Figure 1.5 The Königsberg bridges.

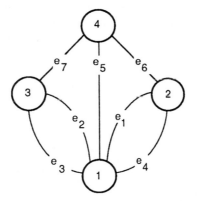

Figure 1.6 Network for the Königsberg bridges problem model.

Notes On Computational Complexity, $O(n^r)$

All the problems that we discuss in this book come with input data. The word *problem* normally refers to the general question to be answered, stated using mathematical symbols to represent the input data. An *instance* of this problem is obtained by providing specific values for all the input data symbols. For example, *the maximum value flow problem* is the problem of finding a maximum value flow in the single commodity flow network G $= (\mathcal{N}, \mathcal{A}, \ell, k, \v{s}, \v{t})$ ($\ell, k, \v{s}, \v{t}$ are data, their meanings are explained later). Finding the maximum value flow in the specific network with data given in Figure 2.8 in Chapter 2, is an instance of this problem.

Measures of the *largeness* of a problem instance are provided by its parameters such as its *dimension* (the total number of data elements in it), or its *size* when all the data is rational (the total number of digits in the data when it is encoded in binary form).

Algorithms are step-by-step procedures for solving problems. We will measure the *computational effort* of an algorithm to solve a problem instance by the number of basic operations (additions, subtractions, multiplications, divisions, comparisons, lookups, etc.) that it takes when it is applied to solve that instance.

We are interested in finding the most efficient (or the fastest, in the sense of requiring the least computational effort in general) algorithms to solve problems. We would expect the relative difficulty of problem instances to increase in general with their largeness measure, so the computational effort of an algorithm should be expressed in terms of this largeness measure. But, even among problem instances with the same largeness measure, the computational effort of an algorithm may vary considerably depending on the actual values for the input data; this makes it very difficult to develop efficiency measures for algorithms.

One possible measure of the efficiency of an algorithm, called the *average computational complexity* or the *average time complexity* of the algorithm, is the average computational effort (i.e., the mathematical expectation of the computational effort) expressed as a function of the largeness of the instance, under the assumption that all the input data are random variables with known distributions. Deriving this average complexity requires complete knowledge of the probability distributions of the data in the class of problem instances on which the algorithm will be applied, this kind of detailed information is not available in general. Therefore, we will not discuss this measure.

Another measure of efficiency of an algorithm is the *empirical average computational complexity* (i.e., the observed average computational effort of the algorithm in computational experiments with randomly generated data). This measure is used in an informal way; it is not theoretically satisfactory, since the observed performance of the algorithm may depend critically on the probability distributions used to generate the data in the experiments.

A third measure of the efficiency of an algorithm, known as the *worst case computational complexity*, is a mathematical upperbound (i.e., the maximum) for the computational effort needed by the algorithm, expressed as a function of the largeness measure of the instance, as the values of the input data elements take all possible values in their range. This is the measure used commonly in computer science to classify algorithms as good or bad; this is the measure that we will use. So, in the text, *computational complexity* of an algorithm refers to this worst case computational complexity function.

When n is some measure of how large a problem instance is (either the size, or the dimension), an algorithm for solving it is said to be of order n^r or $O(n^r)$ if its worst case computational effort grows as αn^r, where α and r are numbers that are independent of the largeness measure and the data in the instance. The n and m that appear in our computational complexity measures are usually the number of nodes and lines in the network.

As an example, consider the following problem that came up in the previous discussion: Given a connected undirected network $G = (\mathcal{N}, \mathcal{A})$ with $|\mathcal{N}| = n \geq 2$, $|\mathcal{A}| = m$, check whether all its nodes have even degree. We now state a simple algorithm for this problem formally in the style that we will use and determine its computational complexity. The algorithm scans each edge once. It maintains numbers called degree indices for the nodes, which become the degrees at termination. The list is the set of unscanned edges.

Step 1 Initialization   For each $i \in \mathcal{N}$, set d_i, its degree index, to 0. List = $\mathcal{A}$.

Step 2 Select an Edge to Scan   If list = $\emptyset$, go to Step 4. Otherwise, select an edge e from the list to scan. Delete e from the list.

Step 3 Scanning   Let e be the edge to be scanned. Add 1 to the degree indices of each of the two nodes on e. Go to Step 2.

Step 4 Termination   The vector of degree indices, $d = (d_i : i \in \mathcal{N})$ at this stage is the vector of degrees of the nodes in G. Check whether all the d_i in it are even and find the answer to the problem. Terminate.

The work in this algorithm consists of $2m$ additions, and checking the evenness of n integers (i.e., a total of $2m + n$ operations, in terms of these operations). It will be shown later that $m \geq n - 1$ in this problem, so the computational effort of this algorithm in terms of these operations is $\leq 4m$; hence the computational complexity of this algorithm is $O(m)$. Therefore, given a connected undirected network with m edges, the existence of an Euler route in it can be checked with an effort of $O(m)$, by the previous discussion.

In this simple problem n, m are parameters describing the dimension of the problem, there is no numerical data, and the computational effort of the algorithm for it is $2m + n$. In the algorithms that follow, there will be numerical data, and the computational effort in them usually depends not just on the dimension or size of the problem but on the actual values of these numbers and possibly on the manner in which the algorithm is executed (specific rules used to select edges or nodes to scan, which may have been left open in the statement of the algorithm, etc.). Therefore, to determine the computational complexity of these algorithms it will be necessary to determine the maximum possible effort (or a reasonably close upper bound for it) as these data elements take all possible values in their range.

Any algorithm whose time complexity function cannot be bounded above by a polynomial function in the size of the problem instance is classified as an *exponential time algorithm* in the worst case.

A Network Model We will now provide a network model for a typical production distribution problem.

EXAMPLE 1.1

A company makes chairs at 3 plants with wood, for which there are 2 suppliers, and sells them through 3 wholesalers. Relevant data is given in the following tables.

The whole process here can be seen as a flow of wood from the suppliers, through the plants where it is converted into chairs, to the wholesalers. There are limits, both lower and upper, on the amounts of flow through the various points, and either costs or revenues associated with these flows.

On an average a chair requires 20 lbs. of wood. It is convenient to define a *chair unit* of wood to be 20 lbs. and measure all flows in terms of chairs/day. The revenue obtained by selling chairs is treated as negative cost.

Plant	Production cost $/*chair* at plant	Production capacity in chairs/day	Lower bound on chairs made/day
1	8	1200	0
2	4	600	200
3	5	450	300

Supplier	Cost of shipping wood ($/lb) to plant			Minimum quantity to be purchased from supplier	Selling price of wood ($/lb)
	1	2	3		
1	.02	.03	.04	8 tons	.10
2	.05	.03	.03	10 tons	.075

Plant	Cost of shipping ($/chair) to wholesaler		
	1	2	3
1	2	1	1
2	1.5	2	1
3	1	1.5	2
Selling price ($/chair)	25	20	22
Max. chairs wanted/day	2100	1600	1700
Min. chairs wanted/day	500	400	300

Nodes in the network arc S_1, S_2 (representing the two suppliers for wood), P_1, P_2, P_3 (the three plants), and W_1, W_2, W_3 (the wholesalers). Arcs in the network represent channels along which material is shipped between pairs of nodes. The flow amount on an arc (x, y) represents the amount of material (chair units/day) shipped from x to y.

Figure 1.7 shows the network model. The 3 numbers on a node are the lower and upper bounds for total flow either originating at that node (for S_1, S_2) or passing through that node (for P_1, P_2, P_3), or being shipped to that node (for W_1, W_2, W_3), and the unit cost associated with that flow. The number on each arc is the cost/unit flow on it. Remember that a ton of wood is 100 chair units. The problem is to find a flow vector in this network that minimizes the total cost subject to the bounds specified.

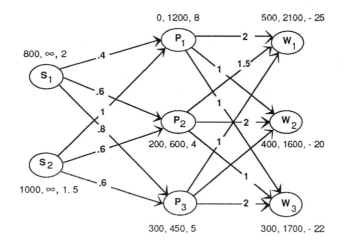

Figure 1.7 Network for the chairmaker's problem. Lower and upper bounds for flow on each arc are $0, \infty$, respectively.

In real world applications, the networks usually involve very large numbers of nodes and arcs, but the mathematical structure of the problem remains very similar to that in this example.

Neglecting parallel lines, the total number of lines in a directed network $G = (\mathcal{N}, \mathcal{A})$ with $|\mathcal{N}| = n$, $|\mathcal{A}| = m$, can be at most $n(n-1)$. If m is close to this number, the network is said to be *dense*. If m is much less than $n(n-1)$, the network is said to be *sparse*. The *sparsity* of G is measured by how small m is in comparison with $n(n-1)$. In very sparse networks, the number of nonparallel lines incident at any node will be much smaller than the possible $n-1$. The networks occurring in most practical applications tend to be very sparse.

1.2   Notation and Preliminaries

In this section, G will denote the network $(\mathcal{N}, \mathcal{A})$. Let **X**, **Y** be two subsets of nodes, not necessarily disjoint. If G is undirected, the symbol (**X**; **Y**) or (**Y**;**X**) denotes the set of all edges in G with one node in **X** and another node in **Y**. For example, in the network in Figure 1.6, ({1, 2} ; {3, 4}) = $\{e_3, e_2, e_5, e_6\}$ and ({1, 2, 3} ; {2, 4}) = $\{e_1, e_4, e_5, e_6, e_7\}$. If G is directed, the symbol (**X**, **Y**) denotes the set of arcs in G with tail in **X** and head in **Y**. So, in the directed case (**X**, **Y**), (**Y**, **X**) may not be the same. For example, in the network in Figure 1.7, ({ P_1, S_1, S_2}, { S_1, P_3, W_2}) = {$(P_1, W_2), (S_1, P_3), (S_2, P_3)$}.

1.2.1   Linear Programming Background

The *standard form* for an LP (we will use this abbreviation for *Linear Program*) is (1.1), where A is a given real matrix of order $m \times n$ and $c = (c_j) \in \mathbb{R}^n$. Without any loss of generality, we assume that the rank of A is m; otherwise, if (1.1) is feasible, some of the equality constraints in it must be redundant, and they can be eliminated one at a time until in the remaining system the matrix of coefficients is of full row rank. We also assume that each column vector of A is nonzero. We denote by $A_{i.}, A_{.j}$ the ith row vector, jth column vector of A.

$$\begin{array}{rlcl}
\text{Minimize} & z(x) = cx & & \\
\text{Subject to} & Ax & = & b \\
& x & \geqq & 0
\end{array} \qquad (1.1)$$

In (1.1) $A_{.j}$ is said to be the *original column*, and c_j the *original cost coefficient* of $x_j, j = 1$ to n; $b = (b_i)$ is called *the vector of original right hand side constants* in (1.1).

A *solution* for (1.1) is a vector x satisfying $Ax = b$, whether or not it satisfies $x \geqq 0$. A *feasible solution* is an x satisfying both $Ax = b$ and $x \geqq 0$. A solution x for (1.1) is said to be a *basic solution* if the set $\{A_{.j} : j \text{ such that } x_j \neq 0\}$ is linearly independent. Assuming that A is of full row rank, a *basis for (1.1)* is a square nonsingular submatrix of A of order m. If $B = (A_{.j_1}, \ldots, A_{.j_m})$ is a basis for (1.1), the vector of variables associated with the columns in B, namely $x_B = (x_{j_1}, \ldots, x_{j_m})$, is known as the *basic vector*, and x_D, the vector of all the variables not in x_B is known as the *nonbasic vector*, associated with it. Let D denote the submatrix of A consisting of the columns of nonbasic variables. Then, after rearranging the variables, the constraints in (1.1) can be written as:

$$\begin{array}{rcl}
Bx_B + Dx_D & = & b \\
& & \\
x_B \geqq 0, x_D & \geqq & 0
\end{array} \qquad (1.2)$$

Then the basic solution of (1.1) corresponding to this basic vector is obtained by setting $x_D = 0$ in (1.2) (i.e., fixing all the nonbasic variables at their lower bound, which is the only finite bound on them) and then solving (1.2) for the values of x_B. This is given by:

$$x_D = 0, x_B = B^{-1}b \qquad (1.3)$$

The basic solution is said to be *degenerate* if at least one of the components in $B^{-1}b$ is zero, *nondegenerate* otherwise. Thus every basis for (1.1) leads to a basic solution; and conversely, every basic solution of (1.1) corresponds to at least one basis for (1.1).

The basis B and the basic vector x_B for (1.1) are said to be *primal feasible* if the basic solution in (1.3) is feasible to (1.1) (i.e., if $B^{-1}b \geqq 0$); otherwise they are *primal infeasible*. In the former case, the solution in (1.3) is said to be a *basic feasible solution* abbreviated as BFS for (1.1). Each BFS corresponds to an *extreme point* of the set of feasible solutions for (1.1), and vice versa. See Murty [1983].

Canonical Tableaus

The *canonical tableau* of (1.1) wrt the basis B or the basic vector x_B is obtained by multiplying the system of equality constraints in it on the left by B^{-1}. It is given below.

Basic variables	x	
x_B	$B^{-1}A$	$B^{-1}b$

When the basic and nonbasic columns are rearranged in proper order as in (1.2), the canonical tableau becomes:

Basic variables	x_B	x_D	
x_B	I	$B^{-1}D$	$B^{-1}b = \bar{b}$

The vector $\bar{b}$ is known as the *updated right hand side constants vector* in the canonical tableau. The column of x_j in the canonical tableau, $B^{-1}A_{.j} = \bar{A}_{.j}$ is called *the updated column of x_j wrt the basis B, or the associated basic vector x_B.* This can be computed from the original column of this variable, and the basis inverse.

Interpretation of the Updated Column of a Nonbasic Variable as the Representation of Its Original Column

The updated column $\bar{A}_{\cdot j} = (\bar{a}_{1j}, \ldots, \bar{a}_{mj})^T$ of x_j in the canonical tableau of (1.1) wrt the basis $B = (A_{\cdot 1}, \ldots, A_{\cdot m})$ say, where $A_{\cdot 1}, \ldots, A_{\cdot m}$ are the original columns of the basic variables in x_B, is $B^{-1} A_{\cdot j}$. So, $A_{\cdot j} = B \bar{A}_{\cdot j} = \bar{a}_{1j} A_{\cdot 1} + \ldots + \bar{a}_{mj} A_{\cdot m}$ (i.e., *the updated column $\bar{A}_{\cdot j}$ is the vector of coefficients in the representation of the original column $A_{\cdot j}$ as a linear combination of the basic columns*, see Murty[1983]). This is an important result relating updated columns in canonical tableaus for (1.1) with the corresponding original columns and the columns of the basis. This result is used later in deriving the canonical tableau for the system of flow conservation equations of a single commodity flow problem combinatorially without performing any pivot steps.

EXAMPLE 1.2

Consider the following system of constraints for an LP.

x_1	x_2	x_3	x_4	x_5	x_6	x_7	b
1	−1	0	1	1	0	0	1
0	−3	1	1	1	1	0	2
1	0	1	0	1	1	1	7

$$x_j \gtreqless 0 \text{ for all } j$$

Consider the basic vector $x_{B_1} = (x_1, x_3, x_4)$ for this problem. The corresponding basis is B_1.

$$B_1 = (A_{\cdot 1}, A_{\cdot 3}, A_{\cdot 4}) = \begin{pmatrix} 1 & 0 & 1 \\ 0 & 1 & 1 \\ 1 & 1 & 0 \end{pmatrix}, \quad B_1^{-1} = \begin{pmatrix} 1/2 & -1/2 & 1/2 \\ -1/2 & 1/2 & 1/2 \\ 1/2 & 1/2 & -1/2 \end{pmatrix}$$

Canonical Tableau wrt The Basic Vector (x_1, x_3, x_4)

Basic variables	x_1	x_2	x_3	x_4	x_5	x_6	x_7	b
x_1	1	1	0	0	1/2	0	1/2	3
x_3	0	−1	1	0	1/2	1	1/2	4
x_4	0	−2	0	1	1/2	0	−1/2	−2

We denote by $A_{\cdot j}$ the original column of x_j, and by $\bar{A}_{\cdot j}$ its updated column in the above canonical tableau. So, $\bar{A}_{\cdot 2} = (1, -1, -2)^T$ and it can be verified that $A_{\cdot 2} = A_{\cdot 1} - A_{\cdot 3} - 2A_{\cdot 4}$, as discussed above.

The basic solution of this system wrt x_{B_1} is $\bar{x} = (3, 0, 4, -2, 0, 0, 0)^T$, and since $x_4 = -2 < 0$ in this solution, it is not a feasible basic solution for this system.

The Dual Problem

Associate the dual variable π_i with the ith equality constraint in (1.1), $i = 1$ to m. Let $\pi = (\pi_1, \ldots, \pi_m)$ be the row vector of dual variables. The dual of (1.1) is:

$$\text{Maximize} \quad \pi b$$
$$\text{Subject to} \quad \pi A_{\cdot j} \overset{\leq}{=} c_j, j = 1 \text{ to } n. \tag{1.4}$$

In matrix notation the constraints in (1.4) are $\pi A \overset{\leq}{=} c$. Given the basic vector $x_B = (x_{j_1}, \ldots, x_{j_m})$ and the associated basis B, the row vector $c_B = (c_{j_1}, \ldots, c_{j_m})$ is the associated *original basic cost vector*. The *dual basic solution* corresponding to the basis B is obtained by solving the system of dual constraints corresponding to the basic variables in x_B as equations, i.e.,

$$\pi A_{\cdot j} = c_j, \text{ for each } j = j_1, \ldots, j_m \tag{1.5}$$

or equivalently, $\pi B = c_B$ yielding $\pi = c_B B^{-1}$. The *dual slack vector* corresponding to the basis B is $\bar{c} = (c - \pi A) = (c - c_B B^{-1} A)$, and it is also known as the vector of *reduced* or *relative cost coefficients* in (1.1) wrt the basis B. The dual basic solution obtained from (1.5) is said to be *dual feasible* if it satisfies all the dual constraints in (1.4) (i.e., if $\bar{c} \overset{\geq}{=} 0$). If this is satisfied, the basis B is said to be *dual feasible* for (1.1). The BFS corresponding to a primal and dual feasible basis is an *optimum feasible solution* for (1.1).

A feasible solution $\bar{x}$ for (1.1) is optimal iff there exists a dual vector $\bar{\pi}$, such that $\bar{x}, \bar{\pi}$ together satisfy all the following conditions.

$$
\begin{array}{rrcl}
\text{Primal feasibility} & A\bar{x} & = & b, \bar{x} \overset{\geq}{=} 0 \\
\text{Dual feasibility} & \bar{\pi} A & \overset{\leq}{=} & c \\
\text{Complementary slackness} & \bar{x}_j(c_j - \bar{\pi} A_{\cdot j}) & = & 0, \text{ for all } j = 1 \text{ to } n
\end{array} \tag{1.6}
$$

Bounded Variable LPs

A *bounded variable LP* is a problem of the following form:

$$
\begin{array}{rrcl}
\text{Minimize} & cx & & \\
\text{Subject to} & Ax & = & b \\
& \ell_j \overset{\leq}{=} x_j & \overset{\leq}{=} & k_j, \text{ for each } j
\end{array} \tag{1.7}
$$

where A is a matrix of order $m \times n$, and $\ell = (\ell_1, \ldots, \ell_n)^T, k = (k_1, \ldots, k_n)^T$ are given lower bound and upper bound vectors satisfying $\ell \overset{\leq}{=} k$. Some of the k_j may be $+\infty$, but we assume that ℓ is finite. As before, we assume that the rank of A is m, and that each column of A contains at least one nonzero entry. Let $B = (A_{.j_1}, \ldots, A_{.j_m})$ be a nonsingular square submatrix of A of order m, and $x_B = (x_{j_1}, \ldots, x_{j_m})$. Then x_B is the *basic vector* corresponding to the basis B for (1.7). Basic solutions of (1.7) correspond to *partitions of the variables*, (x_B, x_L, x_U), where x_B is a basic vector and x_L, x_U are the vectors of nonbasic variables made equal to their lower, upper bounds respectively in the basic solution. Notice that x_U should be such, that for every x_j in it k_j must be finite. The basic solution corresponding to this partition is obtained by solving for the values of the basic variables in x_B from the system of equations in (1.7) after fixing the nonbasic variables in x_L, x_U at their respective bounds. It is:

$$
\begin{aligned}
x_j &= \ell_j \text{ if } x_j \text{ is in } x_L, \text{ or } k_j \text{ if } x_j \text{ is in } x_U \\
x_B &= B^{-1}\left(b - \sum(\ell_j A_{.j} : \text{ over } j \text{ s. t. } x_j \in x_L) \right. \\
&\qquad\qquad \left. - \sum(k_j A_{.j} : \text{ over } j \text{ s. t. } x_j \in x_U)\right)
\end{aligned}
\tag{1.8}
$$

The basic solution in (1.8) is said to be *degenerate* if the value of at least one basic variable is equal to either its lower or upper bound; *nondegenerate* otherwise. The solution in (1.8) is feasible to (1.7) if the value of every basic variable satisfies the bounds on it, and in this case the partition (x_B, x_L, x_U) is said to be a *primal feasible partition*, and the solution itself is called a *basic feasible solution (BFS)*. Thus every BFS for (1.7) is associated with a primal feasible partition for it and vice versa.

Associate a dual variable π_i to the ith equality constraint in (1.7), $i = 1$ to m; dual variables μ_j, γ_j with the bound restrictions on x_j (γ_j is defined only if k_j is finite), $j = 1$ to n. Let $\pi = (\pi_i), \mu = (\mu_j), \gamma = (\gamma_j)$ denote the row vectors of these dual variables. The dual is:

$$
\begin{aligned}
\text{Maximize} \quad & \pi b + \mu \ell - \sum(\gamma_j k_j \quad : \quad \text{over } j \text{ s. t. } k_j \text{ is finite}) \\
\text{Subject to} \quad & \pi A_{.j} + \mu_j - \gamma_j = c_j, \text{ for each } j \text{ s. t. } k_j \text{ is finite} \quad (1.9) \\
& \pi A_{.j} + \mu_j = c_j, \text{ for each } j \text{ s. t. } k_j \text{ is } \infty \\
& \mu \geqq 0, \gamma \geqq 0
\end{aligned}
$$

The dual basic solution corresponding to the partition (x_B, x_L, x_U) is defined to be $\pi = c_B B^{-1}$, where c_B is the row vector of original basic cost coefficients, and B is the associated basis. This partition is said to be *dual feasible* if $\bar{c} = c - \pi A = c - c_B B^{-1} A$ satisfies:

$$
\bar{c}_j \begin{cases} \geqq 0 & \text{for each } j \text{ such that } x_j \in x_L \\ \leqq 0 & \text{for each } j \text{ such that } x_j \in x_U \end{cases}
$$

If these are satisfied, define for each $j = 1$ to n, $\mu_j = \bar{c}_j$ if $\bar{c}_j > 0, \mu_j = 0$ if $\bar{c}_j \leqq 0$; and $\gamma_j = -\bar{c}_j$ if $\bar{c}_j < 0, \gamma_j = 0$ if $\bar{c}_j \geqq 0$. Then verify that $(\pi = c_B B^{-1}, \mu, \gamma)$ satisfy (1.9).

A feasible solution $\bar{x}$ for (1.7) is optimal iff there exists a dual vector $\bar{\pi}$, such that $\bar{x}, \bar{\pi}$ together satisfy *primal feasibility* $(A\bar{x} = b, \ell \leqq \bar{x} \leqq k)$, *dual feasibility* $((c_j - \bar{\pi}A_{.j}) \geqq 0$ for all j such that $k_j = +\infty)$, and the following *complementary slackness conditions for optimality*

$$
\begin{aligned}
c_j - \bar{\pi}A_{.j} > 0 & \quad \text{implies} \quad \bar{x}_j = \ell_j \\
c_j - \bar{\pi}A_{.j} < 0 & \quad \text{implies} \quad \bar{x}_j = k_j \\
\ell_j < \bar{x}_j < k_j & \quad \text{implies} \quad c_j - \bar{\pi}A_{.j} = 0
\end{aligned}
\tag{1.10}
$$

1.2.2 Paths, Chains, Trees, and Other Network Objects

A *path* $\mathcal{P}$ from x_1 (*origin* or *initial node*) to x_k (*destination* or *terminal node*) in G is a sequence of points and lines alternately, $x_1, e_1, x_2, e_2, \ldots, e_{k-1}, x_k$, such that for each $r = 1$ to $k - 1$, e_r is either the arc (x_r, x_{r+1}) or the arc (x_{r+1}, x_r) or the edge $(x_r; x_{r+1})$ with some orientation selected for it, which we will again treat as an arc. It is a sequence of lines connecting the points x_1 and x_k, but the lines need not all be directed towards x_k. An arc whose orientation coincides with (is opposite to) the direction of travel from origin to destination is called a *forward (reverse) arc* of the path. A *chain* is a path in which all the arcs are forward arcs. A path (chain) is said to be a *simple path (simple chain)* if no point or line is repeated on it. See Figures 1.8, 1.9, 1.10. A path (chain) is said to be an *elementary path (elementary chain)* if it does not pass through any line more than once, but it may pass through nodes more than once. The chain in Figure 1.9 going through arcs e_1 to e_5 in that order is an elementary, but not simple chain.

The network $(\mathbf{N}, \mathbf{A})$ where $\mathbf{N}$ is the set of distinct points and $\mathbf{A}$ is the set of distinct lines on a path $\mathcal{P}$, is called the *underlying partial subnetwork of G corresponding to $\mathcal{P}$*. A *cycle (circuit)* is a path (chain) from a node back to itself satisfying the property that in the underlying partial subnetwork $(\mathbf{N}, \mathbf{A})$ corresponding to it, each node in $\mathbf{N}$ is incident to an even number of lines in $\mathbf{A}$. A *simple cycle (simple circuit)* is a cycle (circuit) in which no node or line is repeated, except of course for the initial and terminal nodes which are the same. So, a simple cycle is a cycle that does not contain another cycle as a proper subsequence. Also, every cycle must either be simple itself or contain a simple cycle as a subsequence. And every point has degree 2 in the underlying partial subnetwork of a simple cycle. A cycle that does not pass through any line more than once but may pass through nodes more than once is called an *elementary cycle*.

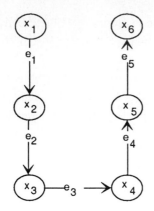

Figure 1.8 A simple chain

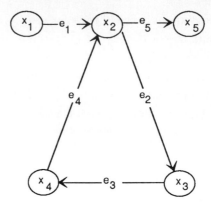

Figure 1.9 An elementary but not a simple chain

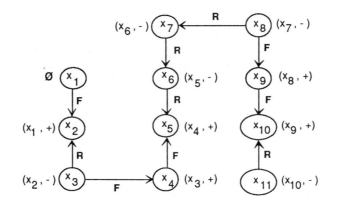

Figure 1.10 A path from x_1 to x_{11}. "F" indicates a forward arc, and "R" indicates a reverse arc. This path is simple. Node labels for storing it are entered by the side of the nodes.

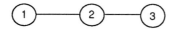

1, (1, 2), 2, (2, 3), 3, (3, 2), 2, (2, 1), 1 which uses these two edges in forward and backwards directions is not a cycle.

1, $(1,2)_1$, 2, $(2,1)_2$, 1 consisting of one of these parallel edges in the forward and the other in the reverse direction is a simple cycle.

Figure 1.11

Let $x_1, e_1, x_2, e_2, \ldots, e_{u-1}, x_u$ be a simple path $\mathcal{P}$ from x_1 to $x_u \neq x_1$ in G. We can store $\mathcal{P}$ using *node labels*. The origin x_1 has no *predecessors*, so its *predecessor index* (or *predecessor label*) is $\emptyset$. For $2 \overset{\leq}{=} r \overset{\leq}{=} u, x_{r-1}$ is the immediate predecessor of x_r on $\mathcal{P}$, so the predecessor index of x_r is x_{r-1}. To remember the orientation of the arc incident into it, we define the label on x_r to be $(x_{r-1}, +)$ $[(x_{r-1}, -)]$ if e_{r-1} is (x_{r-1}, x_r) $[(x_r, x_{r-1})]$ i.e., a forward [reverse] arc on $\mathcal{P}$. See Figure 1.10. The simple path itself can be traced by a *backwards trace of these predecessor labels* beginning at the terminal node. The last arc on $\mathcal{P}$ is the forward arc (x_{u-1}, x_u) if the label on x_u is $(x_{u-1}, +)$, or the reverse arc (x_u, x_{u-1}) if that label is $(x_{u-1}, -)$. Now go back to the predecessor node x_{u-1}, look up the label on it and continue in the same manner. The trace stops when the node with the $\emptyset$ label, the origin, is reached. In simple chains all the arcs are forward, so for storing them, only the predecessor indices are used without the $+, -$ signs. In all network algorithms, labels of this type are used; that's why they are called *labeling algorithms*. This labeling scheme is only useful for representing simple paths; verify that it is not adequate to represent nonsimple paths.

A simple cycle in a directed network G can be oriented in one of two possible ways. An *oriented cycle* in a directed network is a simple cycle for which an orientation has been selected. Arcs on an oriented cycle are classified into forward arcs (those whose orientation coincides with that of the cycle), and reverse arcs (those whose orientation is opposite to that of the cycle). Changing the orientation of an oriented cycle interchanges its sets of forward and reverse arcs. As an example, consider the simple cycle 2, (2, 6), 6, (6, 8), 8, (7, 8), 7, (3, 7), 3, (2, 3), 2 in the network in Figure 1.16 (see later on). As it is written, this simple cycle is oriented in the clockwise direction, with $\{(2, 6), (6, 8)\}$ as forward arcs, and $\{(2, 3), (3, 7), (7, 8)\}$ as reverse arcs. If this simple cycle is oriented in the anticlockwise direction, these sets switch their roles.

The network G is said to be *connected* if there exists a path between every pair of points in it. A network that is not connected is just two or more independent connected networks put together. The connected networks in it are called its *connected components*.

A directed network G $= (\mathcal{N}, \mathcal{A})$ is said to be *strongly connected* if there exists a chain from each node to every other node in it. If the directed network G is not strongly connected, it can be separated into its *strongly connected components*, where, each strongly connected component of G is a maximal partial network of G that is strongly connected.

Exercise

1.1 Suppose G $= (\mathcal{N}, \mathcal{A})$ is a network with $\mathcal{N} = \{1, \ldots, n\}$, and $\mathcal{A}$ is given as a list, listing the nodes on each line. Develop an efficient algorithm to check whether G is connected, and, if it is not, to identify each connected component in it. Determine its computational complexity. Similarly, if G is a directed network, develop an efficient algorithm to check whether it is strongly connected, and, if it is not, to identify its strongly connected components.

Lower Bounds, Capacities and Node-Arc Flow Variables

Let G $= (\mathcal{N}, \mathcal{A})$ be a directed connected network. In single commodity flow models on G, the flow amount on arc $(i, j) \in \mathcal{A}$ (the amount of commodity transported from node i to node j along this arc, in units per unit time) is denoted by $f_{i,j}$, or $f(i, j)$, or just f_{ij}, and called the *node-arc flow amount* or *flow variable* corresponding to arc (i, j). In applications there are usually lower and upper bounds specified on f_{ij}; these are the *lower bound*, ℓ_{ij} and *capacity*, k_{ij}, of arc $(i, j) \in \mathcal{A}$. The arcs in $\mathcal{A}$ are arranged in some order, and $f = (f_{ij}), \ell = (\ell_{ij}), k = (k_{ij})$ denote the flow, lower bound, and capacity vectors in which these quantities are ordered in the same order as arcs are in $\mathcal{A}$. Often $\ell = 0$, but in some applications it may be nonzero. We will always have $\ell \overset{\le}{=} k$.

If $\mathbf{X}, \mathbf{Y}$ are two subsets of $\mathcal{N}$, not necessarily disjoint, we define $f(\mathbf{X}, \mathbf{Y}) = \sum (f_{ij} : \text{over}(i, j) \in \mathcal{A} \text{ with } i \in \mathbf{X}, j \in \mathbf{Y})$, i.e., it is the sum of f_{ij} over arcs (i, j) in the set $(\mathbf{X}, \mathbf{Y})$ defined earlier. The symbols $\ell(\mathbf{X}, \mathbf{Y}), k(\mathbf{X}, \mathbf{Y})$ carry similar meanings. When $\mathbf{X}$ is a singleton set containing only one node, i say, we denote $f(\mathbf{X}, \mathbf{Y}), \ell(\mathbf{X}, \mathbf{Y}), k(\mathbf{X}, \mathbf{Y})$ by $f(i, \mathbf{Y}), \ell(i, \mathbf{Y})$, and $k(i, \mathbf{Y})$ respectively.

As an example, for the network in Figure 1.7, $f(\{P_1, S_1, S_2\}, \{S_1, P_3, W_2\}) = f_{P_1 W_2} + f_{S_1 P_3} + f_{S_2 P_3}$. In this notation it can be verified that $f(\mathcal{N}, \mathcal{N})$ is the sum of the flow amounts on all the arcs in G; and $f(i, \mathcal{N}), f(\mathcal{N}, i)$ are the sums of the flow amounts on arcs in the forward star, reverse star of i respectively. And if $\mathbf{X}_1, \mathbf{X}_2$ is a partition of $\mathbf{X}$ (i.e., $\mathbf{X}_1 \cup \mathbf{X}_2 = \mathbf{X}, \mathbf{X}_1 \cap \mathbf{X}_2 = \emptyset$), and $\mathbf{Y}_1, \mathbf{Y}_2$ is a partition of $\mathbf{Y}$, then $f(\mathbf{X}, \mathbf{Y}) = f(\mathbf{X}, \mathbf{Y}_1) + f(\mathbf{X}, \mathbf{Y}_2) = f(\mathbf{X}_1, \mathbf{Y}) + f(\mathbf{X}_2, \mathbf{Y}) = f(\mathbf{X}_1, \mathbf{Y}_1) + f(\mathbf{X}_1, \mathbf{Y}_2) + f(\mathbf{X}_2, \mathbf{Y}_1) + f(\mathbf{X}_2, \mathbf{Y}_2)$.

In many single commodity flow models, two special nodes, one called the *source node* (which we denote by $\check{s}$), and another called the *sink node* (which we denote by $\check{t}$) are specified, and the commodity is required to be shipped from $\check{s}$ to $\check{t}$ in G. All the other points are called *intermediate points* or *transit nodes* in these models.

In some models, a unit cost coefficient c_{ij}, the cost per unit flow on arc (i, j), is given for each arc, $c = (c_{ij})$ is the vector of these cost coefficients.

With all this data, the network itself is denoted by the symbol G $= (\mathcal{N}, \mathcal{A}, \ell, k, c, \check{s}, \check{t})$. In some models there may be even more data elements, in others less, then the network is denoted by a corresponding symbol consisting of all the data elements in it.

Now consider the case where the network G $= (\mathcal{N}, \mathcal{A})$ is undirected. The edge $(i; j)$ can be treated as the pair of arcs $(i, j), (j, i)$ as in Figure 1.12. In single commodity flow models, a flow of 10 units in the direction i to j and 6 units in the direction j to i as in Figure 1.13, is equivalent to a net flow of 4 units from i to j as in Figure 1.14 (this argument is not valid if there are two are more distinct commodities and the flows from i to j and j to i are of different commodities). So, in single commodity flow models we can assume that each edge in the network will only be used in one of the two possible directions. We assume that lower bounds for flows along all the edges are 0, and that the capacity restriction applies in the direction in which it is used. Under this assumption, each edge in the network can be replaced by a pair of arcs as in Figure 1.12 with the same data holding for both arcs in the pair. Hence, in the study of single commodity flow problems, we assume without any loss of generality that the network is a directed network.

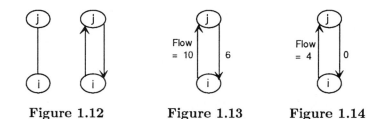

| Figure 1.12 | Figure 1.13 | Figure 1.14 |

In network models, nodes usually represent processing centers, warehouses etc. In some of these models *node capacities* may be specified. The node capacity of a node represents the maximum amount of material that can either enter or leave the node. It represents the maximum amount of flow that the node can process per unit time. As an illustration, the network model for the chair making company problem in Example 1.1 had node capacities specified. In Section 2.1, we show how to transform the model so as to modify node capacities into arc capacities.

Cuts, Cutsets

Let $G = (\mathcal{N}, \mathcal{A})$ be a connected network. A *cut* in G is a subset of lines, the deletion of all of which disconnects the network.

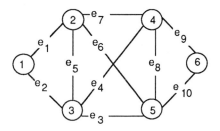

Figure 1.15

First consider the case where G is undirected. Cuts in undirected networks are commonly used in the graph-theoretic study of electrical networks. Let $\mathbf{X} \subset \mathcal{N}, \bar{\mathbf{X}} = \mathcal{N} \backslash \mathbf{X}$ where $\mathbf{X}, \bar{\mathbf{X}}$ are both nonempty. This partition of $\mathcal{N}$ generates the cut $(\mathbf{X}; \bar{\mathbf{X}})$ which is the set of all edges with one node in $\mathbf{X}$ and the other in $\bar{\mathbf{X}}$. $(\mathbf{X};$

$\bar{\mathbf{X}}$) is a *disconnecting set* because it has the *path blocking property* (after deleting all the edges in this set there exists no path in the remaining network from any node in $\mathbf{X}$ to any node in $\bar{\mathbf{X}}$). The *cut vector* of this cut is its 0-1 incidence vector over the set of edges $\mathcal{A}$. For example, in the network in Figure 1.15 the cut ({ 1, 2, 3 }; { 4, 5, 6 }) = { e_3, e_4, e_6, e_7 }, and hence its cut vector is $(0,0,1,1,0,1,1,0,0,0)$ when the edges are arranged in the order e_1 to e_{10}. A *cutset* in G is a minimal set of edges whose removal disconnects G (i.e., it is a cut satisfying the property that no proper subset of it is a cut). Equivalently, a cutset in the connected undirected network G, is a minimal set of edges whose removal disconnects G into exactly two connected components. As an example, in Figure 1.15 the cut ({ 1, 2, 3 }; { 4, 5, 6 }) is a cutset. But the cut ({ 1, 6 }; { 2, 3, 4, 5}) is not a cutset.

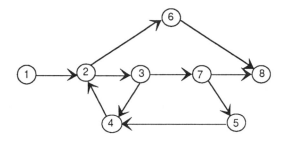

Figure 1.16

Now consider the case where the connected network G is directed. Let G = $(\mathcal{N}, \mathcal{A}, \ell, k)$. Let $\mathbf{X}$, $\bar{\mathbf{X}}$ be a partition of $\mathcal{N}$ with both the sets nonempty. This partition generates a cut denoted by $[\mathbf{X}, \bar{\mathbf{X}}]$ it consists of $(\mathbf{X}, \bar{\mathbf{X}})$, called the *set of forward arcs of this cut*, and $(\bar{\mathbf{X}}, \mathbf{X})$ called the *set of reverse arcs of this cut*. Notice that the order in which the sets $\mathbf{X}$, $\bar{\mathbf{X}}$ are recorded in $[\mathbf{X}, \bar{\mathbf{X}}]$ is important, switching them exchanges the forward and reverse arc sets in the cut. In the network remaining after all the forward arcs of the cut $[\mathbf{X}, \bar{\mathbf{X}}]$ are deleted, there exists no chain from any node in $\mathbf{X}$ to any node in $\bar{\mathbf{X}}$, and if both forward and reverse arcs are deleted there exists no path from any node in $\mathbf{X}$ to any node in $\bar{\mathbf{X}}$. Thus in directed networks while a cut has the path blocking property, the set of forward arcs of a cut by itself has the *chain blocking property*. In network algorithms the concept of a cut plays a role dual to that of a path.

In the single commodity maximum value flow problem on the directed connected network G = $(\mathcal{N}, \mathcal{A}, \ell, k, \check{s}, \check{t})$, special cuts called *cuts separating the source $\check{s}$ and the sink $\check{t}$* play a significant role. These are cuts $[\mathbf{X}, \bar{\mathbf{X}}]$ satisfying the property $\check{s} \in \mathbf{X}$ and $\check{t} \in \bar{\mathbf{X}}$. So deletion of all the forward arcs of a cut separating $\check{s}$ and $\check{t}$ destroys all chains from $\check{s}$ to $\check{t}$ even though there may be a path connecting them consisting of reverse arcs, thus no flow is possible from $\check{s}$ to $\check{t}$ in the remaining network.

The *capacity* of the cut $[\mathbf{X}, \bar{\mathbf{X}}]$ in G = $(\mathcal{N}, \mathcal{A}, \ell, k)$ is defined to be $k(\mathbf{X}, \bar{\mathbf{X}}) - \ell(\bar{\mathbf{X}}, \mathbf{X})$.

Here again, a cut $[\mathbf{X}, \bar{\mathbf{X}}]$ is called a cutset if it is a minimal cut (i.e., if no proper subset of it is a cut). As an example, consider the single commodity flow network in Figure 1.16. The cut [{ 1, 2, 7 }, { 3, 4, 5, 6, 8 }] has (2, 3), (2, 6), (7, 5), (7, 8) as forward arcs, and (3, 7), (4, 2) as reverse arcs. It is not a cutset since the cut [{ 1, 2, 3, 7, 4, 5 }, { 6, 8 }] with no reverse arcs and only (2, 6), (7, 8) as forward arcs is a proper subset of it.

Forests and Trees

A *forest* in a network G is a partial subnetwork that contains no cycles. In Figure 1.1 the dashed partial subnetwork, which is redrawn in Figure 1.4, is a forest. A *tree* in G is a connected partial subnetwork that contains no cycles. Each connected component of a forest is a tree. The forest in Figure 1.4 contains two trees. A *spanning tree* in a network is a subnetwork that is a tree. Hence a spanning tree in $G = (\mathcal{N}, \mathcal{A})$ is a subnetwork $(\mathcal{N}, \hat{\mathcal{A}})$ which is connected and contains no cycles. In Figure 1.1 the subnetwork consisting of the thick lines is a spanning tree.

A single isolated node by itself constitutes a tree, which we call the *trivial tree.* It is the only tree which has no lines. Unless it is mentioned otherwise, in the sequel the word tree refers to a nontrivial tree.

A node in a tree $\mathbb{T}$ is called a *terminal node, end node, pendant node,* or *leaf node* if its degree in $\mathbb{T}$ is 1 (i.e., if it is incident to exactly one line in $\mathbb{T}$). For example, in the tree consisting of the thick lines in Figure 1.1, 1 is a terminal node, but 2 is not. In the tree in Figure 1.18 nodes 1, 2, 3, 4, 8, 10 are the terminal nodes. The terminal nodes of a tree are also called its *leaves.* An arc or edge in the tree incident at a leaf node is called a *leaf arc* or *leaf edge* of the tree.

Exercises

1.2 Prove that every nontrivial tree has at least two terminal nodes.

1.3 Prove that if a line e_1 is deleted from a tree $\mathbb{T}$, what is left is a forest. Also, if e_1 is a line containing a terminal node i of $\mathbb{T}$, then what is left after deleting e_1 from $\mathbb{T}$ is another tree and the trivial tree containing only node i.

1.4 Let G be a connected network with n nodes. Prove that every spanning tree in G contains $n - 1$ lines (Hint: use induction on n and the results in previous exercises).

1.5 Let G be a network with n nodes, and $\mathbb{T}$ a subnetwork of G containing $n - 1$ lines and no cycles. Prove that $\mathbb{T}$ must be a spanning tree in G. (You have to prove that $\mathbb{T}$ is connected. Prove that any subnetwork like $\mathbb{T}$ must have a point whose degree in $\mathbb{T}$ is one. Use induction.)

1.6 Prove that a connected network in which the number of lines is equal to the number of nodes -1 must be a tree.

1.7 If $\{e_1, \ldots, e_r\}$ is a set of r lines containing no cycles in a connected network, prove that there is at least one spanning tree containing all of these lines.

1.8 Prove that there exists a unique simple path between every pair of distinct nodes in a tree.

1.9 Prove that the number of lines is one half of the sum of the degrees of the points in a network.

1.10 Prove that the number of odd degree nodes in a network must be an even number.

1.11 In a network in which no point has a degree greater than 2, prove that the number of nodes of degree one is an even number.

1.12 Prove that every chain from the source to the sink must contain at least one forward arc of any cut separating them.

1.13 For any nonempty proper subset of nodes $\mathbf{X} \subset \mathcal{N}$ in a connected undirected network $G = (\mathcal{N}, \mathcal{A})$, define $\mathcal{A}_{\mathbf{X}}$ to be the set of edges in $\mathcal{A}$ with both their nodes from $\mathbf{X}$. Let $\bar{\mathbf{X}}$ be the complement of $\mathbf{X}$.

(i)   Prove that $(\mathbf{X}; \bar{\mathbf{X}})$ is a cutset of G iff $(\mathbf{X}, \mathcal{A}_{\mathbf{X}})$ and $(\bar{\mathbf{X}}, \mathcal{A}_{\bar{\mathbf{X}}})$ are both connected networks.

(ii)   If $\mathbf{S}$ is a cutset of G, and $\mathbf{V}_1, \mathbf{V}_2$ are the node sets of the two connected components of $(\mathcal{N}, \mathcal{A} \backslash \mathbf{S})$, then show that $\mathbf{S} = (\mathbf{V}_1; \mathbf{V}_2)$.

(iii)   Prove that a cutset of G contains at least one in-tree edge of every spanning tree in G. Further, show that a subset of edges $\mathbf{S}$ is a cutset in G iff it is a minimal set of edges containing at least one in-tree edge of every spanning tree in G.

(iv)   Prove that a cycle and a cutset of G have an even number of common edges.

Let $G = (\mathcal{N}, \mathcal{A})$ be a connected network with $|\mathcal{N}| = n, |\mathcal{A}| = m$. A subset of lines $\mathbf{A} \subset \mathcal{A}$ is said to be a *cotree* in G iff its complement $\mathcal{A} \backslash \mathbf{A}$ is the set of lines in a spanning tree for G. The word "cotree" is an abbreviation for *complement of a spanning tree*. So, for $\mathbf{A}$ to be a cotree in G, a necessary condition is that $|\mathbf{A}| = m - n + 1$.

When considering a spanning tree $\mathbb{T}$ in a network G, a line in G is said to be an *in-tree arc* or *edge* if it lies in $\mathbb{T}$; otherwise it is said to be an *out-of-tree arc* or *edge*. The set of out-of-tree lines defines the cotree corresponding to $\mathbb{T}$.

Let i, j be the nodes on an out-of-tree line e in a spanning tree $\mathbb{T}$. By Exercise 1.8, there exists a unique simple path $\mathcal{P}$ in $\mathbb{T}$ between i and j. Hence, when e is included in $\mathbb{T}$ a unique simple cycle containing e is created (it consists of e and the path $\mathcal{P}$), this is known as the *fundamental cycle of e wrt* $\mathbb{T}$. As an example, in the spanning tree consisting of the thick lines in Figure 1.1, (6, 5) is an out-of-tree arc. The fundamental cycle associated with it is 6, (6, 5), 5, (4, 5), 4, (4, 2), 2, (2, 7), 7, (6, 7), 6. All the arcs on this cycle are in-tree arcs except (6, 5).

By replacing an in-tree arc in the fundamental cycle associated with the out-of-tree arc (i, j), by (i, j), a new spanning tree is obtained. As we will see later, every basis for a single commodity pure network flow problem in a connected directed network corresponds to a spanning tree. And when this problem is solved by the simplex algorithm, every pivot step is exactly the operation of obtaining a new spanning tree by adding an out-of-tree arc and dropping an in-tree arc in its fundamental cycle.

Trees are stored using node labels. These *tree labels* make it possible to store in the computer all the information necessary to manipulate a tree. Their use has led to enormous improvements in the efficiency of computer implementations of network algorithms. Some of the node labels are: *predecessor index* (P), *successor index* (S), *elder brother index* (EB), *younger brother index* (YB), *and thread label* (TH). Besides these, the distance of the node, the number of successors of the node, and others are sometimes used in network codes. The most commonly used of these labels are defined below. Each label used in the data structure requires an array of length $n = |\mathcal{N}|$, and it imposes the work of updating this label whenever the tree changes in the algorithm.

We will first discuss tree labels for storing a spanning tree $\mathbb{T}$ in a connected directed network $G = (\mathcal{N}, \mathcal{A})$ with $n = |\mathcal{N}|$. Modifications to be made if G is undirected are mentioned later. Select any node and designate it as the *root node*. Once the root node is selected, the tree is called a *rooted tree*. A rooted tree is a tree with one of its nodes identified as the root. The labels are generated by the following procedure while drawing the tree with the root node at the top and the other nodes below it level by level. In this procedure, nodes may be in three possible states: unlabeled, labeled and unscanned, or labeled and scanned. The present sets of unlabeled, labeled and unscanned nodes are denoted by $\mathbf{Y}, \mathbf{X}$ respectively.

Initialization   Make the P, YB, EB indices of the root node all $\emptyset$, and $\mathbf{X} = \{\text{root node}\}$, $\mathbf{Y} = $ set of all non-root nodes.

Step 1 Select a node to be scanned   Terminate if $\mathbf{X} = \emptyset$. Otherwise, select a node from $\mathbf{X}$ to scan.

Step 2 Scanning a node   Let i be the node to be scanned, delete it from $\mathbf{X}$. Find $\mathbf{J} = \{j : j \in \mathbf{Y}$ and j is joined to i by an in-tree arc $\}$. Nodes in $\mathbf{J}$ are the *sons* or *children* or *immediate successors* of i, and i is their *parent* or *immediate predecessor*.

If $\mathbf{J} = \emptyset$, i has no children, define its successor index $S(i) = \emptyset$.

If $\mathbf{J} \neq \emptyset$, arrange the nodes in $\mathbf{J}$ in some order, say $j_1, \ldots, j_r$. Then j_1 is the eldest child of i. j_p is an elder brother of j_q (and j_q is an younger brother of j_p) if $p < q$. The successor index of i, S(i) is defined to be $-j_1[+j_1]$ if the in-tree arc joining i and j_1 is $(i, j_1)[(j_1, i)]$. For each $u = 1$ to r, define the predecessor index of j_u, P(j_u), to be $+i[-i]$ if the in-tree arc joining i and j_u is $(i, j_u)[(j_u, i)]$. For each $u = 1$ to r, define the elder brother index, EB(j_u) to be $\emptyset$ if $u = 1, j_{u-1}$ if $u > 1$ and the younger brother index YB(j_u) to be $\emptyset$ if $u = r, j_{u+1}$ if $u \overset{\leq}{=} r - 1$. Nodes $j_1, \ldots, j_r$ are now labeled and unscanned; transfer them from $\mathbf{Y}$ to $\mathbf{X}$. Go to Step 1.

Thus, the elder brother index of any node is the youngest among its elder brothers, and its younger brother index is the eldest among its younger brothers, when these brothers exist. Figure 1.17 explains our convention for the signs of the predecessor and successor indices. They indicate the orientation of the in-tree arc joining a node and its parent.

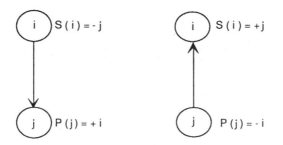

Figure 1.17 Signs of indices when i is parent of j and the in-tree arc joining them is (i, j) (on left), or (j, i) (on right). The successor index of i, S(i) is $\pm j$ only if j is the eldest child of i.

All nonroot terminal nodes have no successor. Conversely, if S(i) = $\emptyset$, i must be a nonroot terminal node.

The $\pm$ signs on the successor and predecessor indices are used only when dealing with a directed network. For a spanning tree in an undirected network, all the indices are defined in exactly the same way with the exception that the successor and predecessor indices carry no signs.

EXAMPLE 1.3

For the purpose of this illustration, only the in-tree arcs are given in Figure 1.18; all the out-of-tree arcs and the remaining data on the network is omitted. We assume that nodes which are brothers of each other are arranged from left to right in Figure 1.18 for determining the elder, younger brother relationships. This leads to the predecessor, successor, and brother labels in the table given below, for the spanning tree in Figure 1.18.

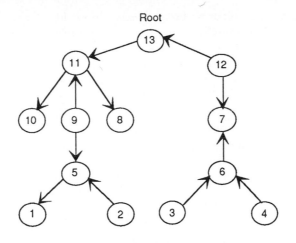

Figure 1.18 A spanning tree.

Predecessor, successor and brother labels for the spanning tree in Figure 1.18

Node i	1	2	3	4	5	6	7	8	9	10	11	12	13 root
$P(i)$	+5	-5	-6	-6	+9	-7	+12	+11	-11	+11	+13	-13	$\emptyset$
$S(i)$	$\emptyset$	$\emptyset$	$\emptyset$	$\emptyset$	-1	+3	+6	$\emptyset$	-5	$\emptyset$	-10	-7	-11
$YB(i)$	2	$\emptyset$	4	$\emptyset$	$\emptyset$	$\emptyset$	$\emptyset$	$\emptyset$	8	9	12	$\emptyset$	$\emptyset$
$EB(i)$	$\emptyset$	1	$\emptyset$	3	$\emptyset$	$\emptyset$	$\emptyset$	9	10	$\emptyset$	$\emptyset$	11	$\emptyset$

The unique path in a rooted tree $\mathbb{T}$ from a node j to the root node is called the *predecessor path of j* in $\mathbb{T}$. It can be found by a backward trace of the predecessor indices beginning with j recursively. If $P(j) = +i\ [-i]$, $(i,j)\ [(j,i)]$ is the first arc in this path. Now look up the node $P(i)$ and continue in the same manner until the root node is reached. As an example, the predecessor path of node 1 in the spanning tree in Figure 1.18 is 1, (5, 1), 5, (9, 5), 9, (9, 11), 11, (13, 11), 13.

A node i is said to be an *ancestor* or *predecessor* of another node j in the rooted tree $\mathbb{T}$ if i appears on the predecessor path of j in $\mathbb{T}$, in this case j is a *descendent* or *successor* of i.

The *family* of a node j in the rooted tree $\mathbb{T}$ is the set consisting of j and all the descendents of j, it is denoted by the symbol $\mathbf{H}(\mathbb{T}, j)$.

The *level* of a node in a rooted tree is defined to be the number of lines on its predecessor path. Thus, the root node is the only level 0 node in a rooted tree. For any r, given the set of level r nodes, level $r+1$ is empty if none of the level r nodes have a successor, otherwise level $r+1$ consists of the set all immediate successors of nodes in level r.

Let (i,j) be an in-tree arc in a rooted tree $\mathbb{T}$. Then, one of the nodes among i, j must be a parent and the other its child. If i is the parent and j the child (i.e.,

$P(j) = +i$) this arc is said to be *directed away from the root node*. On the other hand, if j is the parent and i the child, this arc is said to be *directed towards the root node*. See Figure 1.19.

For each in-tree line e in a rooted tree $\mathbb{T}$, *son(e)*, *parent(e)* refer to the son, parent nodes on it. Thus, if $\mathbb{T}$ is a directed network and the in-tree arc (i, j) is directed away from [towards the] root node, son$(i, j) = j$, parent$(i, j) = i$ [son(i, j) $= i$, parent$(i, j) = j$].

The set of younger brothers of a node j is empty if YB$(j) = \emptyset$, or is the union of {YB(j)} and the set of younger brothers of YB(j) otherwise. Using this recursively, the set of younger brothers of any node can be found efficiently. In a similar manner, the set of elder brothers of any node can be found recursively using only the EB indices. The set of brothers of a node is the union of its sets of younger and elder brothers.

The set of immediate successors of a node j is empty if S$(j) = \emptyset$, or is the union of {S(j)} and the set of younger brothers of S(j) otherwise.

The *set of descendents* of a node j is empty if S$(j) = \emptyset$. Otherwise it is the union of the set of immediate successors of j and the sets of descendents of each of the immediate successors of j. The thread label defined later, is designed to obtain this set very efficiently.

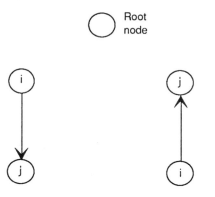

Figure 1.19 If son$(i, j) = j$ as on the left [son$(i, j) = i$ as on the right] arc (i, j) is directed away from [towards the] root node.

Let i, j be two nonroot nodes in a rooted tree $\mathbb{T}$. The first common node on the predecessor paths of i and j is known as the *apex* on the simple path between i and j in $\mathbb{T}$. The simple path between i and j in $\mathbb{T}$ is obtained by putting the predecessor paths of i and j together and eliminating the common lines on them. As an example, the simple path between 1 and 10 in the spanning tree in Figure 1.18 is 1, (5, 1), 5, (9, 5), 9, (9, 11), 11, (11, 10), 10. Node 11 is the apex on this path.

If $\mathbb{T}$ is a spanning tree in G, and (i, j) is an out-of-tree arc, the fundamental cycle of (i, j) wrt $\mathbb{T}$ consists of arc (i, j) and the simple path in $\mathbb{T}$ from j to i. As an example, the fundamental cycle of arc (10, 1) (not in the figure) wrt the spanning tree in Figure 1.18 is 10, (10, 1), 1, (5, 1), 5, (9, 5), 9, (9, 11), 11, (11, 10), 10.

Let $\mathbb{T}$ be a rooted spanning tree in a connected network $G = (\mathcal{N}, \mathcal{A})$ with $|\mathcal{N}| = n$. Let e be an in-tree line. If G is undirected let $\bar{\mathbf{X}} = \mathbf{H}(\mathbb{T}, \text{son}(e))$. If G is directed let $\bar{\mathbf{X}}$ be the set among $\mathbf{H}(\mathbb{T}, \text{son}(e))$ and its complement, which contains head(e). Let $\mathbf{X} = \mathcal{N} \backslash \bar{\mathbf{X}}$. Let $G_{\mathbf{X}}, G_{\bar{\mathbf{X}}}$ denote the partial networks of G, and $\mathbb{T}_{\mathbf{X}}, \mathbb{T}_{\bar{\mathbf{X}}}$ the partial networks of $\mathbb{T}$, induced by the sets of nodes $\mathbf{X}, \bar{\mathbf{X}}$. $\mathbb{T}_{\mathbf{X}}, \mathbb{T}_{\bar{\mathbf{X}}}$ are themselves spanning trees in $G_{\mathbf{X}}, G_{\bar{\mathbf{X}}}$; one or both of them may be trivial trees. So, both $G_{\mathbf{X}}, G_{\bar{\mathbf{X}}}$ are connected networks.$\mathbf{X}, \bar{\mathbf{X}}$ is a partition of the node set in G, this partition generates a cut in G. If G is undirected and $(\mathbf{X}; \bar{\mathbf{X}})$ is the cut, it is a cutset called the *fundamental cutset corresponding to the in-tree edge* e in $\mathbb{T}$. If G is directed, the cut is $[\mathbf{X}, \bar{\mathbf{X}}]$; it is also a cutset and is called the fundamental cutset corresponding to the in-tree arc e in $\mathbb{T}$. The only in-tree line in this fundamental cutset is e. Each in-tree line leads to a different fundamental cutset; together they form the *set of fundamental cutsets* wrt $\mathbb{T}$.

Thus, given a spanning tree $\mathbb{T}$ in a network G, each line in $\mathbb{T}$ defines a fundamental cutset in G, and each line in the corresponding cotree defines a fundamental cycle.

We now present some of the other tree labels used in network codes. One is the *number of successors* denoted by NS(i) for node i. Another is the *distance* or *depth label* which is the level of the node in the tree.

Another commonly used label is the *thread label*. Let $\mathcal{N} = \{1, \ldots, n\}$ be the set of nodes in a rooted tree $\mathbb{T}$. A thread label for $\mathbb{T}$ is a one-to-one correspondence from $\mathcal{N}$ onto $\mathcal{N}$ satisfying certain properties. Given such a correspondence $t_i, i \in \mathcal{N}$, define other maps $t_i^r, i \in \mathcal{N}$ by recursion: $t^1(i) = t_i$, for each $i \in \mathcal{N}$; $t^r(i) = t^{r-1}(t^1(i))$, for each $i \in \mathcal{N}, r \geq 2$. Then the correspondence $t_i, i \in \mathcal{N}$ is said to be a *thread label* for $\mathbb{T}$ if for each i such that NS$(i) \neq 0$ the set $\{t^r(i) : r = 1, \ldots, \text{NS}(i)\}$ is the set of all descendents of i. Many types of thread labels satisfying these conditions can be defined, but the most commonly used one is defined by:

$$
t_i = \begin{cases} \text{eldest son, i.e., } S(i), \text{ if } S(i) \neq \emptyset \\[2mm] \text{YB}(i), \text{ if } S(i) = \emptyset \text{ and } \text{YB}(i) \neq \emptyset \\[2mm] \text{YB index of first ancestor with a YB, if } i \text{ has no son or YB} \\[2mm] \text{root node, otherwise} \end{cases}
$$

Given the thread labels (t_i), the set of all descendents of node i is the largest set of the form $\{t_i, t^2(i), \ldots, t^k(i)\}$ such that the parent of $t^k(i)$ is one of the nodes $i, t_i, \ldots, t^{k-1}(i)$. So, we can find the set of all descendents of any node i efficiently by recursive application of the maps $t^r(i)$ for all r up to NS(i). As this is a commonly used operation in network algorithms, maintaining the thread label can be very advantageous. Also, the result in Exercise 1.19 states that the number of $\emptyset$ entries in S(.) and YB(.) indices together is $n + 1$. This indicates that the information in S(.) and YB(.) can be stored more economically in the single thread index.

Given the thread label $t_i, i \in \mathcal{N}$, the *preorder distance label*, PD(i), and *the last successor label*, LS(i), corresponding to it, can be defined for each $i \in \mathcal{N}$ as below.

$$PD(i) = \begin{cases} 1, & \text{if } i \text{ is the root} \\ r + 1, & \text{if } i \neq \text{root, where } r \text{ is s.t. } i = t^r(\text{root}) \end{cases}$$

$$LS(i) = \begin{cases} i, & \text{if } i \text{ has no successor} \\ t^r(i) & \text{otherwise, for } r \text{ s.t. } t^r(i) \text{ is a descendent of } i, \text{ but } t^{r+1}(i) \text{ is not.} \end{cases}$$

For the rooted tree in Figure 1.18, we provide these labels in the following table. Figure 1.20 illustrates the thread labels for a rooted tree with pointers.

NS(i), distance (d_i), thread index (t_i), PD(i), and LS(i) node labels for the rooted tree in Figure 1.18

Node i	1	2	3	4	5	6	7	8	9	10	11	12	13
NS(i)	0	0	0	0	2	2	3	0	3	0	6	4	12
d_i	4	4	4	4	3	3	2	2	2	2	1	1	0
t_i	2	8	4	13	1	3	6	12	5	9	10	7	11
PD(i)	6	7	12	13	5	11	10	8	4	3	2	9	1
LS(i)	1	2	3	4	2	4	4	8	2	10	8	4	4

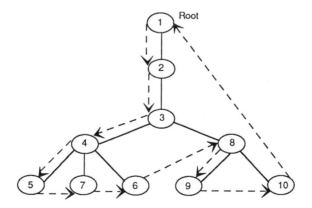

Figure 1.20 The thread labels. The tree consists of solid lines. The dotted arc (i, j) indicates that the thread label of i is j.

Thread labels facilitate the forward traversal of the tree, an operation performed many times in simplex based network codes. It can be viewed as a connecting link or thread which passes through each node exactly once in a top to bottom, left to right sequence starting from the root node.

These tree labels are mainly used in simplex based algorithms for network flow problems discussed in Chapter 5. Each basic vector for the problem corresponds to a spanning tree, and a pivot step in the algorithm consists of changing the spanning tree by adding an out-of-tree arc to replace an in-tree arc in its fundamental cycle. The tree changes by an arc in each step, and the tree labels are updated by very efficient updating schemes. While the predecessor indices are enough to trace the predecessor paths, the other indices make it possible for updating the tree labels and the node price vector efficiently.

An *outtree* or a *branching* is a directed network which is a rooted tree such that every arc on it is directed away from the root node. It has exactly one arc incident into every node other than the root, which has no arcs incident into it. See Figure 1.21. Similarly, an *intree* or *arborescence* is a directed network which is a rooted tree such that every arc in it is directed towards the root node (i.e., the predecessor path of every node is a chain from that node to the root). It has exactly one arc incident out of every node other than the root, which has no arcs incident out of it.

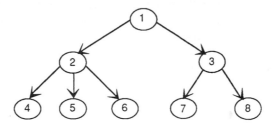

Figure 1.21 An outtree or branching with node 1 as the root.

Tree Growth Subroutines

Several of the algorithms for maximum value flow problems (Chapter 2), shortest chain problems (Chapter 4), minimum cost spanning tree problems in undirected networks (Chapter 9) and blossom algorithms for matching/edge covering problems (Chapter 10) use a scheme that begins by selecting a node in the network, say p, and labeling it with $\emptyset$. At this initial stage, p is the only labeled node, all the other nodes are unlabeled.

General step A labeled node, say i, and one of its adjacent unlabeled nodes, say j, are selected by some rule (this rule is problem dependent, the line joining i, j has to satisfy certain properties for the pair i, j to be selected) and j is labeled with i as its predecessor index. The line joining i, j is known as the *line used in labeling node j*. Now the scheme either terminates or moves on to the next step.

At any stage of this scheme, let $\mathbf{X}$ denote the set of labeled nodes and let $\mathcal{A}(\mathbf{X})$ denote the set of all lines used so far in labeling the nodes in $\mathbf{X}$. We have the following theorem.

THEOREM 1.1 *At every stage of this scheme, the partial subnetwork* ($\mathbf{X}$, $\mathcal{A}(\mathbf{X})$) *will be a tree spanning the nodes in the set* $\mathbf{X}$. *It is a rooted tree with its root at node p.*

Proof   Initially, the partial subnetwork $(\mathbf{X},\ \mathcal{A}(\mathbf{X}))$ is $(\{\ p\ \},\ \emptyset)$, the trivial tree consisting of node p. In each step, one new node is added to the set $\mathbf{X}$, if it is j, its immediate predecessor is an adjacent node which is already in $\mathbf{X}$, and the line joining it to j is added as a new arc to the set $\mathcal{A}(\mathbf{X})$. This implies that $(\mathbf{X}, \mathcal{A}(\mathbf{X}))$ is always connected and that $|\mathcal{A}(\mathbf{X})| = |\mathbf{X}| - 1$. So, by the result in Exercise 1.6, $(\mathbf{X}, \mathcal{A}(\mathbf{X}))$ is always a tree spanning the nodes in $\mathbf{X}$. Also, every node in $\mathbf{X}$ has a unique immediate predecessor, except the node p which has no predecessor, so, these predecessor labels make $(\mathbf{X}, \mathcal{A}(\mathbf{X}))$ a rooted tree with node p as the root node. ∎

So, this scheme is actually a *tree growth subroutine*, growing a rooted tree with its root node at p. Each step of the scheme is known as a *tree growth step*. It adds one new node and a line connecting it to an earlier in-tree node, to the tree. At any stage of this scheme, the in-tree nodes are the labelled nodes, and the in-tree lines are the lines joining each in-tree node and its immediate predecessor. In these schemes, sometimes several trees may be grown simultaneously in the network. The predecessor indices serve all the functions needed in these schemes, and no successor or brother indices are maintained.

Methods for Selecting a Spanning Tree in a Network

Here we discuss two algorithms for selecting a spanning tree in a connected network $G = (\mathcal{N}, \mathcal{A})$, to initiate algorithms such as the primal network simplex method discussed in Section 5.5.

ALGORITHM 1 : Initialization   Here nodes may be in 3 possible states: unlabeled, labeled and unscanned, or labeled and scanned. *List* always refers to the set of labeled and unscanned nodes. Select the root node, say n, and label it with $\emptyset$, and put it in the list. All the other nodes are unlabeled initially.

General Step   In a general step, select a node, say i, from the list to scan. Delete i from the list. Find all unlabeled nodes j such that either (i,j), or (j,i), or edge $(i;j)$ is in $\mathcal{A}$. For each such node j include one of the lines joining it to i as an in-tree line. All these nodes are children of i, give predecessor, EB, YB indices to them and include them in the list, and give the successor index to i as discussed above. If there are no unlabeled nodes left, we have a spanning tree, terminate. If there are some unlabeled nodes but the list is empty, G is not connected, terminate. Otherwise go to the next stage.

In this algorithm, if nodes in the list are maintained in the order in which they are labeled and in each stage the node for scanning is selected by the FIFO rule

(First In First Out, or first labeled first scanned rule), then it is called the *breadth-first search method*, and the spanning tree generated a *breadth-first search spanning tree* in G.

ALGORITHM 2 : Initialization This algorithm introduces one node and line into the tree per step. $\mathbf{A}$ denotes the set of lines in $\mathcal{A}$ joining an in-tree and an out-of-tree node at the present stage. This set is maintained in the algorithm. If at some stage there are out-of-tree nodes, but $\mathbf{A} = \emptyset$, G is not connected and the algorithm terminates. Select the root node, say n and declare it as an in-tree node. Make $\mathbf{A} =$ set of all lines incident at n.

General Step In a general step, select a line from $\mathbf{A}$. Introduce this line and the unlabeled node on it, say j, into the tree. If there are no out-of-tree nodes, we have a spanning tree, terminate. Otherwise, delete from $\mathbf{A}$ all lines incident at j that are currently in it. Include in $\mathbf{A}$ all lines joining j to an out-of-tree node. Go to the next step.

In this algorithm, if lines in the set $\mathbf{A}$ are maintained in the order in which they are introduced into this set, and in each step the line selected from it is chosen by the LIFO rule (Last In First Out rule, i.e., the line chosen is always the one put into $\mathbf{A}$ most recently), then the algorithm is called the *depth-first search method, or DFS*, and the spanning tree generated a *depth-first search, or DFS spanning tree* in G. A numbering of the nodes in the order of becoming in-tree nodes is called a *DFS numbering*.

Exercises

1.14 Let $\mathbb{T}_1$ denote a breadth-first search spanning tree in G with root n. For each node $i \neq n$, prove that the predecessor path of i in $\mathbb{T}_1$ is a shortest path between i and n (i.e., it contains the smallest number of lines). Also prove that there cannot be an out-of-tree arc wrt $\mathbb{T}_1$ which joins a node and one of its descendents.

1.15 Let $\mathbb{T}_2$ be a DFS spanning tree in G with root node n. Prove that every line in G connects two nodes, one of which is an ancestor of the other. Also prove that if i, j are leaf nodes in $\mathbb{T}_2$, then there is no line joining i and j in G.

1.16 Prove that every connected network has at least one spanning tree.

1.17 Let G be a connected network. Prove that every cotree $\mathbf{A}$ in G is a cutset-free subset of $\mathcal{A}$. And for any cycle-free subset $\mathbf{B} \subset \mathcal{A}$, prove that there exists a spanning tree in G including it. For any cutset-free subset $\mathbf{A} \subset \mathcal{A}$, prove that there exists a cotree in G including it.

1.18 Let $G = (\mathcal{N}, \mathcal{A})$ be a connected undirected network and $\mathbb{T}$ a spanning tree in it.

(i) If e is an out-of-tree edge, prove that the fundamental cycle of e consists of exactly those in-tree edges of $\mathbb{T}$ whose fundamental cutsets contain e.

(ii) Prove that the fundamental cutset of the in-tree edge e_1 consists of exactly those out-of-tree edges whose fundamental cycles contain e_1.

1.19 Let $\mathbb{T}$ be a rooted tree with n nodes. Let n_1, n_2 be the number of nodes in it for which the S(.), YB(.) indices are respectively $\emptyset$. Prove that the number of nodes for which the EB(.) is $\emptyset$ is n_2. Also prove that $n_1 + n_2 = n + 1$.

Bipartite Networks

A simple cycle is said to be an *odd cycle (even cycle)* if it contains an odd (even) number of lines.

A network $G = (\mathcal{N}, \mathcal{A})$ is said to be a *bipartite network* if $\mathcal{N}$ can be partitioned into two nonempty subsets $\mathcal{N}_1$ and $\mathcal{N}_2$ such that every line in $\mathcal{A}$ joins a point in $\mathcal{N}_1$ with a point in $\mathcal{N}_2$ (i.e., there are no lines in $\mathcal{A}$ joining a pair of points both of which are either in $\mathcal{N}_1$ or in $\mathcal{N}_2$). The partition $(\mathcal{N}_1, \mathcal{N}_2)$ is then called a *bipartition* of G, and G itself denoted by $(\mathcal{N}_1, \mathcal{N}_2 ; \mathcal{A})$. See Figure 1.22. As it is drawn, the bipartiteness of the network in Figure 1.7 may not be apparent, but taking the partition $\mathcal{N}_1 = \{S_1, S_2, W_1, W_2, W_3\}$, $\mathcal{N}_2 = \{P_1, P_2, P_3\}$ it can easily be verified to be so.

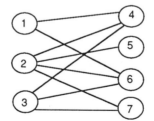

Figure 1.22 A bipartite network. Here ({1,2,3}, {4,5,6,7}) is the bipartition.

THEOREM 1.2 *A network $G = (\mathcal{N}, \mathcal{A})$ is bipartite iff it contains no odd cycles.*

Proof   Clearly a network is bipartite iff each of its connected components is. Without any loss of generality we assume that G is connected, because otherwise the proof can be repeated for each connected component separately.

Suppose G is bipartite, and $(\mathcal{N}_1, \mathcal{N}_2)$ is a bipartition for it. While traversing any cycle in G we move alternately to points in the sets $\mathcal{N}_1, \mathcal{N}_2$, and hence every cycle in G must be an even cycle.

Now suppose G is a network which contains no odd cycles. Let 1 be an arbitrary point in $\mathcal{N}$. Define $\mathcal{N}_1 = \{i :$ either $i = 1$, or there exists a simple path from 1 to i with an even number of lines$\}$, $\mathcal{N}_2 = \mathcal{N}\backslash\mathcal{N}_1$. Suppose two points $j, p \in \mathcal{N}_1$ are joined by a line $(j, p) \in \mathcal{A}$. By definition of $\mathcal{N}_1$, there are simple paths $\mathcal{P}_1$ between 1 and j, and $\mathcal{P}_2$ between 1 and p, both with an even number of lines. Let $r =$ number of common lines on $\mathcal{P}_1$, $\mathcal{P}_2$; $r_1 =$ number of lines on $\mathcal{P}_1$ not on $\mathcal{P}_2$; and $r_2 =$ number of lines on $\mathcal{P}_2$ not on $\mathcal{P}_1$. Since $r + r_1, r + r_2$ are both even, $r_1 + r_2$ is also even and it is > 0 since $j \neq p$. So, combining (j, p) with the lines on $\mathcal{P}_1$, $\mathcal{P}_2$ and then eliminating all the common lines on $\mathcal{P}_1$ and $\mathcal{P}_2$ leaves a cycle with $r_1 + r_2 + 1$ = odd number of lines. Since $\mathcal{P}_1$ and $\mathcal{P}_2$ are simple paths, this cycle decomposes into a collection of line-disjoint simple cycles, at least one of which must be an odd cycle, contradicting the hypothesis.

Suppose two points j, p in $\mathcal{N}_2$ are joined by a line $(j, p) \in \mathcal{A}$. Since G is connected, there are simple paths in G between 1 and j, or p, and by the definition of $\mathcal{N}_1$ all these paths must traverse through an odd number of lines. Either there exists a simple path between 1 and j not containing p, or there exists a simple path between 1 and p not containing j; suppose the first possibility holds. Take any simple path from 1 to j not passing through p, and add the line (j, p) at its end. This leads to a simple path from 1 to p traversing through an even number of lines, contradicting $p \in \mathcal{N}_2$. Hence there cannot be any line in G joining a pair of points in $\mathcal{N}_2$. Hence $(\mathcal{N}_1, \mathcal{N}_2)$ is a bipartition for G, and it is bipartite. ∎

The following algorithm can be used to check whether a network $G = (\mathcal{N}, \mathcal{A})$ is bipartite and generate a bipartition for it, if it is. We assume that G is connected, otherwise apply the algorithm on each connected component of G and put the results together. **X** denotes the set of included but unscanned nodes, and **Y** denotes the set of unincluded nodes at any stage.

Step 1 Initialization Select a node, say $1 \in \mathcal{N}$. Make $\mathcal{N}_1 = \{ 1 \}$, $\mathcal{N}_2 = \emptyset$, **X** $= \{ 1 \}$, **Y** $= \mathcal{N}\backslash \{ 1 \}$.

Step 2 Select a node to scan If $\mathbf{X} = \emptyset$, go to Step 4. Otherwise, select a node from **X** to scan.

Step 3 Scanning Let i be the node to be scanned, delete it from **X**. Let $\mathcal{N}'$ denote the set among the pair $\mathcal{N}_1$, $\mathcal{N}_2$ containing node i, and $\mathcal{N}''$ denote the other set in this pair. Let **J** be the set of all nodes in **Y** which are adjacent to i. If $\mathbf{J} \neq \emptyset$, check whether there is a line in $\mathcal{A}$ joining a node in **J** with a node in $\mathcal{N}''$, if so go to Step 5. Otherwise, delete all nodes in **J** from **Y** and include them in both **X** and $\mathcal{N}''$, and return to Step 2.

Step 4 Termination The subsets $\mathcal{N}_1$, $\mathcal{N}_2$ at this stage form a bipartition for G. Terminate.

Step 5 Proof of non-bipartiteness G is not bipartite, because two nodes which should belong to the same set in the partition are adjacent. Terminate.

Acyclic Networks

A directed network is said to be an *acyclic network* if it contains no circuits. It may contain cycles, but not circuits. See Figure 1.23.

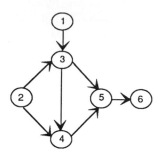
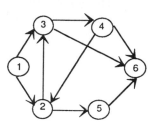

An acyclic network with acyclic numbering of nodes.

A non-acyclic network. It has the circuit with arcs (2, 3), (3, 4), (4, 2).

Figure 1.23

THEOREM 1.3 *A directed network $G = (\mathcal{N}, \mathcal{A})$ is acyclic iff its nodes can be numbered so that the number of tail(e) < the number of head(e) for each $e \in \mathcal{A}$.*

Proof If G is acyclic, there must be at least one point i in it whose before set $\mathbf{B}(i) = \emptyset$ (otherwise, by tracing the arcs backwards one can construct a circuit). Suppose there are r_1 points like this, number them 1, 2, ..., r_1 in any order, and delete them and all the arcs incident at them from G. Search the resulting network for points whose before set in it is empty, and number them starting with $r_1 + 1$. Repetition of this process leads to the desired numbering.

If the nodes in G are numbered so that $i < j$ for each $(i, j) \in \mathcal{A}$, clearly there cannot be any circuit in G; hence it is acyclic. ∎

A numbering of the nodes in a directed network G in serial order so that the number of tail(e) is < the number of head(e) for all arcs e is called an *acyclic numbering* or *topological ordering*. The acyclic numbering of the nodes in the network on the left in Figure 1.23 is obtained by applying the procedure in the proof of Theorem 1.3.

Incidence Matrices

Let G be a directed network $(\mathcal{N}, \mathcal{A})$ with n points and m arcs, and no self-loops. Let E be the $n \times m$ matrix that has a row associated with each point in G and a column associated with each arc in G, where the column associated with $(i, j) \in \mathcal{A}$ has only two nonzero entries, a "1" entry in the row associated with point i and a "−1" entry in the row associated with point j. E, defined only for directed networks with no self-loops, is known as the *node-arc incidence matrix* of G. As an example, the node-arc incidence matrix for the network in Figure 1.24 is given below.

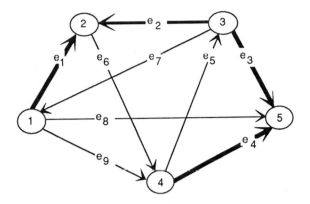

Figure 1.24

Arc →	e_1	e_2	e_3	e_4	e_5	e_6	e_7	e_8	e_9
Node 1	1	0	0	0	0	0	−1	1	1
2	−1	−1	0	0	0	1	0	0	0
3	0	1	1	0	−1	0	1	0	0
4	0	0	0	1	1	−1	0	0	−1
5	0	0	−1	−1	0	0	0	−1	0

Verify that the set of row vectors of a node-arc incidence matrix is linearly dependent since their sum is 0. Also verify that the sum of node-arc incidence vectors of arcs in a circuit is 0. Verify this for the circuit in Figure 1.24 consisting of arcs e_9, e_5, e_7. The general property is stated in the following Exercise 1.22.

Exercises

1.20 Check whether the network in Figure 1.25 is bipartite.

1.21 Check whether the network in Figure 1.26 is acyclic. If so, provide an acyclic numbering of its nodes.

1.22 Let $e_1, \ldots, e_r$ be the sequence of arcs in a cycle in a directed network G with node-arc incidence matrix E. Orient this cycle some way. Multiply the column vector of E associated with e_t by +1 if e_t is a forward arc, or by −1 if it is a reverse arc of the cycle, and add over $t = 1$ to r, show that this gives 0. Thus show that the set of column vectors of E associated with arcs in a set containing a cycle form a linearly dependent set for which there is a linear dependence relation with all the coefficients 0, or ±1.

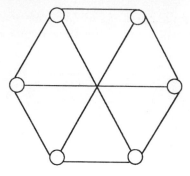

Figure 1.25

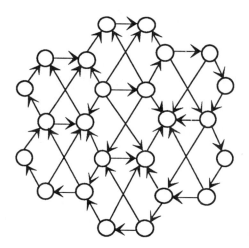

Figure 1.26

1.23 Prove that the set of columns of the node-arc incidence matrix E of a directed network, associated with the arcs in a set that contains no cycles, must be linearly independent. (This set forms a forest. So, there must be at least one terminal node. Use this repeatedly.)

1.24 Prove that the set of columns in E associated with arcs in a spanning tree is linearly independent. (Use the fact that a tree has at least one terminal node repeatedly.)

1.25 If E is the node-arc incidence matrix of a connected directed network with n points, prove that its rank is $n - 1$. Also, prove that any one of the row vectors of E could be deleted as a dependent row, and the remaining matrix will be of full row rank.

1.26 If the directed network G has n points and consists of p connected components, prove that its node-arc incidence matrix has rank $n - p$.

A nonsingular square matrix $D = (d_{ij})$ is said to be an *upper triangular matrix* if $d_{ij} = 0$ for all $i > j$. It is *lower triangular* if D^T is upper triangular, i.e., if $d_{ij} = 0$ for all $j > i$. A *triangular matrix* is a nonsingular square matrix that becomes a lower triangular matrix after a permutation of its columns and/or rows. A square matrix is triangular iff it satisfies the following properties.

1. The matrix has a row (or column) that contains a single nonzero entry.

2. The submatrix obtained from the matrix by striking off the row (or column) containing a single nonzero entry and the column (row) in which that entry lies, also satisfies Property 1. The same process can be repeated until all the rows and columns in the matrix are struck off.

For example, the following matrix A can be verified to be triangular, while B is not.

$$A = \begin{pmatrix} 1 & 1 & 0 & 0 & 0 & 0 \\ 0 & 0 & 1 & 1 & 1 & 0 \\ 0 & 0 & 0 & 0 & 0 & 1 \\ 0 & 0 & 1 & 0 & 0 & 1 \\ 1 & 0 & 0 & 0 & 0 & 0 \\ 0 & 0 & 0 & 1 & 0 & 0 \end{pmatrix}, B = \begin{pmatrix} 1 & 1 & 0 & 0 & 0 & 1 \\ 0 & 0 & 1 & 1 & 0 & 0 \\ 0 & 0 & 0 & 0 & 1 & 1 \\ 1 & 0 & 0 & 0 & 1 & 0 \\ 0 & 0 & 1 & 0 & 0 & 0 \\ 0 & 1 & 0 & 1 & 0 & 0 \end{pmatrix}$$

If D is a triangular matrix, the system of equations $Dy = d$ can be solved efficiently by back substitution. Identify the equation containing a single nonzero entry on the left hand side; solve that equation for the value of the variable associated with the nonzero coefficient in that equation; substitute the value of this variable in all the remaining equations and continue in the same manner with the remaining system. The same back substitution method can be applied to solve the system of equations $\pi D = c$ when D is triangular.

THEOREM 1.4 *Every nonsingular square submatrix of E, the node-arc incidence matrix of a directed network G, is triangular.*

Proof Let D be a nonsingular square submatrix of order r of E. Since every column vector of E contains only two nonzero entries, a $+1$ and a -1, the total number of nonzero entries in D is at most $2r$. If it is $2r$, each column of D contains a $+1$ and a -1, and the sum of all the rows of D is 0, contradicting its nonsingularity. So the total number of nonzero entries in D is at most $2r - 1$. Since D is nonsingular each row of D must contain at least one nonzero entry. As D has r rows, these facts imply that there must be at least one row of D with a single nonzero entry, and that entry is either $+1$ or -1. The same argument applies to the submatrix of D obtained by striking off a row containing a nonzero entry, and the column of that entry. So D satisfies properties 1 and 2 mentioned above for triangularity, and hence is triangular. ∎

COROLLARY 1.1 *Let $\bar{E}$ be the matrix obtained by deleting any row vector from E, the node-arc incidence matrix of a connected directed network G with n nodes and m arcs. $\bar{E}$ of order $(n-1) \times m$ is of full row rank. Let B be a basis for $\bar{E}$, and $d = (d_1, \ldots, d_{n-1})^T, c = (c_1, \ldots, c_{n-1})$. Consider the system of equations, $By = d$. Since B is triangular, and all the entries in it are 0, or ± 1, when this system is solved by the back substitution method discussed above, its solution will be of the form $y = (y_j)$ with each $y_j = \sum_{i=1}^{n-1} \alpha_i d_i$, where all the α_i are 0, or ± 1. similarly, the solution to the system of equations, $\pi B = c$, is of the form, $\pi = (\pi_i)$ with each $\pi_i = \sum \beta_j c_j$, where all the β_j are 0, or ± 1. Hence, if d, c are integer vectors, the solutions to these systems are also integer vectors.*

A real matrix $A = (a_{ij})$ is said to be *totally unimodular* if the determinant of every square submatrix of it is either 0, or ± 1. Since each element a_{ij} can be looked at as the entry in a square submatrix of A of order 1, if A is totally unimodular, all a_{ij} have to be 0, or ± 1. As examples, consider the following matrices:

$$A = \begin{pmatrix} 1 & 0 \\ 1 & -1 \end{pmatrix}, B = \begin{pmatrix} 1 & 1 \\ 1 & -1 \end{pmatrix}$$

The determinant of A is -1, and clearly A is totally unimodular. The determinant of B is -2, and so B is not totally unimodular.

THEOREM 1.5 *TOTAL UNIMODULARITY PROPERTY The node-arc incidence matrix E of a directed network is totally unimodular.*

Proof We have shown in Theorem 1.4 that every nonsingular square submatrix of E is triangular. This, combined with the fact that all the entries in it are 0 or ± 1, implies that the determinant of every nonsingular square submatrix of E is ± 1, proving the theorem. ∎

We will now discuss a result relating total unimodularity to integrality of extreme points of polyhedra, due to Hoffman and Kruskal [1958].

THEOREM 1.6 *Let A be an integer matrix of order $m \times n$, and let $\mathbf{K}(b)$ be the set of feasible solutions of*

$$\begin{aligned} Ax &\leqq b \\ x &\geqq 0 \end{aligned}$$

The following three conditions are equivalent.

(i)   *A is totally unimodular.*

(ii) *For all integral b such that* $\mathbf{K}(b) \neq \emptyset$, *all the extreme points of* $\mathbf{K}(b)$ *are integer.*

(iii) *Every nonsingular square submatrix of A has an integer inverse.*

Proof The fact that (i) implies (ii) follows by applying Cramer's rule. The fact that (i) implies (iii) follows from the definitions.

Introduce slack variables and write down the system of constraints defining $\mathbf{K}(b)$ as

$$Ax + Is \;=\; b \qquad\qquad (1.11)$$
$$x, s \;\geqq\; 0$$

Let D be a square submatrix of A of order r which is nonsingular. We will now show that (ii) implies that D^{-1} is an integer matrix. After rearranging the variables and the constraints if necessary, we can assume that D is the submatrix of A contained in rows 1 to r and columns 1 to r. Let B denote the basis for (1.11) consisting of columns $A_{\cdot 1}, \ldots, A_{\cdot r}$ and $I_{\cdot r+1}, \ldots, I_{\cdot m}$, and let the associated basic vector be denoted by y. So, B has the form given below, where I_{m-r} is the unit matrix of order $m - r$, and hence B^{-1} has the form given.

$$B = \left(\begin{array}{c|c} D & 0 \\ \hline F & I_{m-r} \end{array} \right), B^{-1} = \left(\begin{array}{c|c} D^{-1} & 0 \\ \hline -FD^{-1} & I_{m-r} \end{array} \right)$$

Let I denote the unit matrix of order m, and let i be an integer between 1 to r. Select an integer column vector $\xi \in \mathbb{R}^m$ such that $\xi + B^{-1}I_{\cdot i} \geqq 0$. Take b to be $B(\xi + B^{-1}I_{\cdot i}) = B\xi + I_{\cdot i}$, which is an integer vector since $B\xi, I_{\cdot i}$ are integer vectors. With this b vector, the BFS of (1.11) corresponding to the basis B is

$$\text{all nonbasic variables } = 0, \; y = \xi + B^{-1}I_{\cdot i} \geqq 0 \qquad (1.12)$$

Since this b vector is integer, by (ii) the y in (1.12) must be integer, i.e., since ξ is an integer vector, $B^{-1}I_{\cdot i}$ must be integer, and hence $(D^{-1})_{\cdot i}$ must be integer. Since this is true for all $i = 1$ to r, D^{-1} must be an integer matrix. That is, (ii) implies (iii).

We will now show that (iii) implies (i). Let D be any square nonsingular submatrix of A. Since D^{-1} is integer by (iii), the determinant of (D^{-1}) is integer, and since this is 1/determinant(D), determinant(D) must be ± 1. Since this applies to all square nonsingular submatrices of A, A must be totally unimodular. So, (iii) implies (i).

Hence (i), (ii), (iii) are equivalent. ∎

As a generalization of total unimodularity, an integer matrix A of order $m \times n$ and rank r is said to be *unimodular* if every one of its square submatrices of order r has determinant 0, or ± 1. There is no condition on the determinants of A of order $r - 1$ or less. Clearly every totally unimodular matrix is unimodular, but the converse may not be true. As an example, the matrix given below is unimodular but not totally. This is followed by a theorem relating unimodularity to integrality of extreme points of a polyhedron defined by constraints different from those in the previous theorem.

$$\begin{pmatrix} 1 & 0 & 0 & -1 & 0 & 0 \\ 2 & 1 & 0 & -2 & -1 & 0 \\ 2 & 2 & 1 & -2 & -2 & -1 \end{pmatrix}$$

THEOREM 1.7 *Let A be a given integer matrix of order $m \times n$. Consider the system*

$$Ax = b, x \geq 0 \tag{1.13}$$

Without any loss of generality we assume that rank(A) is m. The following statements are equivalent.

(i)   *A is unimodular*

(ii) *Whenever b is an integer vector, every BFS of (1.13) is an integer vector.*

(iii) *Every basis for (1.13) has an integer inverse.*

Proof   The proof of this theorem is very similar to that of the previous. We will first show that (ii) implies (iii). Let B be a basis for (1.13) associated with the basic vector x_B. Select an i between 1 to m, and choose an integer column vector $\xi \in \mathbb{R}^m$ such that $\xi + B^{-1}I_{.i} \geq 0$. Take b in (1.13) to be $B(\xi + B^{-1}I_{.i}) = B\xi + I_{.i}$, which is integer. With this integer b-vector, the BFS of (1.13) corresponding to the basis B is $x_B = B^{-1}b = \xi + B^{-1}I_{.i} \geq 0$, and all nonbasic variables $= 0$. Since ξ is an integer vector, by (ii), $B^{-1}I_{.i} = (B^{-1})_{.i}$ must be an integer vector. Since this is true for all i, B^{-1} must be an integer matrix. So, (ii) implies (iii).

To show that (iii) implies (i), let B be a basis for (1.13). Since determinant(B^{-1}) $= 1/(\text{determinant}(B))$, if B^{-1} is an integer matrix, determinant(B^{-1}) is integer, and hence determinant(B) is ± 1. Since this must hold for all bases for (1.13), (i) must hold.

The fact that (i) implies (ii) follows by using Cramer's rule. Hence (i), (ii), (iii) are equivalent. ∎

Let G $= (\mathcal{N}, \mathcal{A})$ be a connected directed network with n nodes, and m arcs, and let T be a spanning tree in G. Let $e_1, \ldots, e_{n-1}$ be the arcs in T, and $e_n, e_{n+1}, \ldots, e_m$ be the out-of-tree arcs. Orient each fundamental cycle so that the out-of-tree arc on it is a forward arc. For $p = n$ to m and $t = 1$ to m define

$$\lambda_{tp} = \begin{cases} 0 & \text{if } e_t \text{ not on fundamental cycle of } e_p \\ +1 & \text{if } e_t \text{ is a reverse arc on this cycle} \\ -1 & \text{if } e_t \text{ is a forward arc on this cycle} \end{cases}$$

For each $p = n$ to m, the row vector $(\lambda_{1p}, \ldots, \lambda_{mp})$ is the *incidence vector of the fundamental cycle of the out-of-tree arc* e_p, and the matrix L of order $(m-n+1) \times m$ consisting of these row vectors is known as the *fundamental cycle-arc incidence matrix* of G wrt $\mathbb{T}$. And the matrix $\lambda = (\lambda_{tp} : t = 1$ to $n-1, p = n$ to $m)$ of order $(n-1) \times (m-n+1)$ is known as the *in-tree arc-fundamental cycle incidence matrix* of G wrt $\mathbb{T}$. The matrix λ is the transpose of the submatrix of L consisting of its columns corresponding to in-tree arcs.

As an example, for the network in Figure 1.25 and the spanning tree consisting of the thick arcs, the in-tree arc-fundamental cycle incidence matrix is given below.

	Fundamental cycle of out-of-tree arc				
	e_5	e_6	e_7	e_8	e_9
In-tree arc e_1	0	0	−1	1	1
e_2	0	−1	1	−1	−1
e_3	−1	1	0	1	1
e_4	1	−1	0	0	−1

For $t = 1$ to m let $E_{.t}$ be the column vector of the node-arc incidence matrix E of G corresponding to the arc e_t. Then from the result in Exercise 1.22 we have

$$E_{.p} = \sum_{t=1}^{n-1} \lambda_{tp} E_{.t} \tag{1.14}$$

for $p = n$ to m. Thus $(\lambda_{1p}, \ldots, \lambda_{n-1,p})$ are the coefficients in the representation of the node-arc incidence vector of out-of-tree arc e_p as a linear combination of the node-arc incidence vectors of in-tree arcs.

For $t = 1$ to $n-1$, let $[\mathbf{X}_t, \bar{\mathbf{X}}_t]$ be the fundamental cutset of the in-tree arc e_t in $\mathbb{T}$. For $t = 1$ to $n-1, p = 1$ to m, define

$$g_{tp} = \begin{cases} 0 & \text{if } e_p \text{ is not in } [\mathbf{X}_t, \bar{\mathbf{X}}_t] \\ +1 & \text{if } e_p \text{ is in } (\mathbf{X}_t, \bar{\mathbf{X}}_t) \\ -1 & \text{if } e_p \text{ is in } (\bar{\mathbf{X}}_t, \mathbf{X}_t) \end{cases}$$

Then $(g_{t1}, \ldots, g_{tm})$ is known as the *fundamental cutset vector corresponding to the in-tree arc* e_t. The $(n-1) \times m$ matrix Q consisting of these rows is known as *fundamental cutset-arc incidence matrix* of G wrt $\mathbb{T}$.

Now consider an undirected network $G = (\mathcal{N}, \mathcal{A})$ which has no self loops. The *node-edge* or *vertex-edge incidence matrix* of G has a row corresponding to each node and a column corresponding to each edge in G. The entry in the row corresponding to node i and column corresponding to edge e is 1 if e contains i, 0 otherwise. As an example, the node-edge incidence matrix of the network in Figure 1.6 is given below.

Edge →	e_1	e_2	e_3	e_4	e_5	e_6	e_7
Node 1	1	1	1	1	1	0	0
2	1	0	0	1	0	1	0
3	0	1	1	0	0	0	1
4	0	0	0	0	1	1	1

The determinant of the submatrix of this matrix given by rows 1, 2, 4 and columns 4, 5, 6 is -2. So, the node-edge incidence matrix may not be totally unimodular in general.

Exercises

1.27 Let $A = (a_{ij})$, be a matrix with $a_{ij} \in \{-1, 0, 1\}$ for all i, j, in which each column has at most two nonzero entries. Prove that A is totally unimodular iff its rows can be partitioned into two sets so that

(a) If two nonzero elements of a column have the same sign, they are in different sets.

(b) If two nonzero elements of a column have different signs, they are in the same set.
(Heller and Tompkins [1958])

1.28 Show that the determinant of the node-edge incidence matrix of an odd cycle is not in $\{-1, 0, 1\}$. Prove that the node-edge incidence matrix of an undirected network is totally unimodular iff the network is bipartite.

1.29 Let H be an integer matrix of order $m \times n$. Prove that the matrix $A = (H \vdots I)$ where I is the unit matrix of order m, is unimodular iff H is totally unimodular. Thus show that any matrix of order $m \times n$ which contains the unit matrix of order m as a submatrix is unimodular iff it is totally unimodular.

1.30 Let A be an integer matrix of order $m \times n$ and rank r. If $r = m$ and B is a basis for A, prove that A is unimodular iff determinant$(B) = \pm 1$ and the matrix $\bar{A} = B^{-1}A$ is totally unimodular. If $r < m$ let B be a square nonsingular submatrix of A of order r. Rearrange the rows and columns of A so that this submatrix comes to the top left corner, as in A'.

$$A' = \left(\begin{array}{c|c} B & D \\ \hline F & H \end{array} \right)$$

Prove that A is unimodular iff $(B \vdots D)$ and $(B^T \vdots F^T)$ are unimodular. Since both these matrices are of full row rank, their unimodularity can be characterized in terms of total unimodularity as in the case above. (K. Truemper)

1.31 Let $G = (\mathcal{N}, \mathcal{A})$ be a directed connected network with n nodes and m arcs, and let $\mathbb{T}$ be a spanning tree in G. Let E be the node-arc incidence matrix, and L the fundamental cycle-arc incidence matrix of G wrt $\mathbb{T}$. Do the following: (i) Prove L is totally unimodular and its rank is $m - (n - 1)$ (i.e., its rows form a linearly independent set). (ii) Prove $LE^T = 0$. (iii) Prove that the incidence vector (written as a row vector) of any elementary cycle in G can be expressed as a linear combination of the rows of L, with all the coefficients in the expression being either 0, or ± 1. (iv) Prove that a $(m - n + 1) \times (m - n + 1)$ square submatrix of L is nonsingular iff the columns of this submatrix correspond to a cotree. (v) Let $\mathbb{T}_2$ be a spanning tree in G. Prove that the fundamental cycle-arc incidence matrix of G wrt $\mathbb{T}_2$ is $(L(\bar{\mathbb{T}}_2))^{-1}L$, where $L(\bar{\mathbb{T}}_2)$ is the square submatrix of L consisting of columns in L corresponding to the cotree wrt $\mathbb{T}_2$. (W. Mayeda [1972])

1.32 Let L, Q be the fundamental cycle-arc incidence matrix, and the fundamental cutset-arc incidence matrix respectively, wrt a spanning tree $\mathbb{T}$ in a connected directed network $G = (\mathcal{N}, \mathcal{A})$ with $|\mathcal{N}| = n, |\mathcal{A}| = m$. Do the following: (i) Prove that Q is of full row rank. (ii) Prove that $QL^T = 0, LQ^T = 0$. (iii) Rearrange the arcs in $\mathcal{A}$ so that all the $m - n + 1$ out-of-tree arcs wrt $\mathbb{T}$ appear first, and then the $n - 1$ in-tree arcs appear. Also, rearrange the columns of Q, L according to this order. Let $Q = (Q_1 \vdots Q_2), L = (L_1 \vdots L_2)$, after this rearrangement. Columns in Q_1, L_1 correspond to out-of-tree arcs wrt $\mathbb{T}$; and columns in Q_2, L_2 correspond to in-tree arcs. Then prove that $Q_2 = I_{n-1}, L_1 = I_{m-n+1}$, and $Q_1 = -L_2^T$. (iv) Prove that Q is totally unimodular (use the result in the previous bit) (v) Let $\mathbf{A} \subset \mathcal{A}, |\mathbf{A}| = n - 1$, and let $Q(\mathbf{A})$ be the submatrix of Q consisting of columns in Q corresponding to arcs in the set $\mathbf{A}$. Prove that $Q(\mathbf{A})$ is nonsingular iff $\mathbf{A}$ is the set of arcs in a spanning tree for G. (vi) Let $\mathbb{T}_2$ be any spanning tree in G and $Q(\mathbb{T}_2)$ be the submatrix of Q consisting of columns in Q corresponding to in-tree arcs in $\mathbb{T}_2$. Prove that the fundamental cutset-arc incidence matrix wrt $\mathbb{T}_2$ is $(Q(\mathbb{T}_2))^{-1}Q$. (vii) Prove L is totally unimodular. Let $\mathbf{A} \subset \mathcal{A}, |\mathbf{A}| = m - n + 1$, and let $L(\mathbf{A})$ be the submatrix of L corresponding to arcs in the set $\mathbf{A}$. Prove that $L(\mathbf{A})$ is nonsingular iff $\mathbf{A}$ is a cotree in G.

Matchings and Assignments

Let $G = (\mathcal{N}, \mathcal{A}, c)$ be an undirected network with c as the vector of edge cost coefficients. A *matching* in G is a subset of edges $\mathbf{M} \subset \mathcal{A}$ containing at most one edge incident at any node. For example, in the network in Figure 1.27, the set of thick edges forms a matching. The cost of a matching is defined to be the sum of the cost coefficients of edges in it. A matching is said to be a *perfect matching* if it contains exactly one edge incident at each node. In Figure 1.27, the set of wavy edges is a perfect matching.

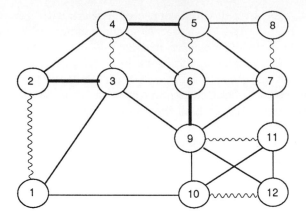

Figure 1.27 The thick edges form a matching in this network.
The set of wavy edges is a perfect matching.

Consider an undirected bipartite network with its bipartition $\mathcal{N}_1 = \{R_1, \ldots, R_n\}$, $\mathcal{N}_2 = \{C_1, \ldots, C_n\}$. A perfect matching in this network is called an *assignment* or an *assignment of order n*, it can be represented by a 0-1 square matrix $x = (x_{ij})$ of order n where $x_{ij} = 1$ if the edge $(R_i; C_j)$ is in the assignment, or 0 otherwise. We will also call such matrices as assignments. The x_{ij} in an assignment satisfy the constraints (1.15), (1.16).

$$\sum_{i=1}^{n} x_{ij} = 1, \text{ for all } j = 1 \text{ to } n$$

$$\sum_{j=1}^{n} x_{ij} = 1, \text{ for all } i = 1 \text{ to } n \qquad (1.15)$$

$$x_{ij} \gtreqless 0, \text{ for all } i, j$$

$$\text{and } x_{ij} = 0 \text{ or } 1, \text{ for all } i, j \qquad (1.16)$$

Any feasible solution $x = (x_{ij})$ to (1.15) is called a *doubly stochastic matrix*, and an assignment is a 0-1 doubly stochastic matrix (it is also a *permutation matrix*, or a 0-1 square matrix containing a single "1" entry in each row and column). Here is an assignment $\bar{x}$ of order 4.

$$\begin{pmatrix} 0 & 1 & 0 & 0 \\ 0 & 0 & 0 & 1 \\ 1 & 0 & 0 & 0 \\ 0 & 0 & 1 & 0 \end{pmatrix} \qquad (1.17)$$

Let j_r be the column which has the unique "1" entry in row $r, r = 1$ to n, in an assignment x of order n. Then the only nonzero variables in x are x_{r,j_r}, for r

= 1 to n. In this case we will denote the assignment x by the set of its unit cells, namely $\{(1, j_1), \ldots, (n, j_n)\}$. In this notation, the assignment $\bar{x}$ in (1.17) will be denoted by $\{(1, 2), (2, 4), (3, 1), (4, 3)\}$.

Assignments are useful to model situations in which there are two distinct sets of objects of equal numbers, and we need to form them into pairs, each pair consisting of one object of each set. For example, $\mathcal{N}_1$ might be a set of n boys, and $\mathcal{N}_2$ a set of an equal number of girls. If each boy in $\mathcal{N}_1$ marries a girl in $\mathcal{N}_2$, the resulting set of couples will be an assignment. Sometimes we will take both the sets $\mathcal{N}_1$, $\mathcal{N}_2$, to be $\{1, \ldots, n\}$. It should be understood that they refer to the serial numbers of distinct sets of objects.

1.2.3 Single Commodity Node-Arc Flow Models

In these models, it is assumed that the flow of the material is carried out independently along each line of the network. Consider the directed flow network G $= (\mathcal{N}, \mathcal{A}, \ell, k, \check{s}, \check{t})$. The variable f_{ij} represents the amount of material transported from i to j along the arc $(i, j) \in \mathcal{A}$, and the vector $f = (f_{ij})$ is known as the *node-arc flow vector,* or just the *flow vector.*

The quantity $\sum_{j \in \mathbf{B}(i)} f_{ji} = f(\mathcal{N}, i)$ is the total amount of material flowing into node i, and $\sum_{j \in \mathbf{A}(i)} f_{ij} = f(i, \mathcal{N})$ is that flowing out of node i, these two quantities must be equal if i is an intermediate node. The net amount of material leaving the source node $\check{s}$, $f(\check{s}, \mathcal{N}) - f(\mathcal{N}, \check{s})$, is known as the *value of the flow vector* f, and denoted by $v(f)$ or v. Since all the material leaving the source has to eventually reach the sink, the net amount reaching the sink, $f(\mathcal{N}, \check{t}) - f(\check{t}, \mathcal{N})$ should be equal to v too. So, a flow vector f is a *feasible flow vector* in G if it satisfies the following constraints:

$$f(i, \mathcal{N}) - f(\mathcal{N}, i) = \begin{cases} v & \text{if } i \text{ is source} \\ -v & \text{if } i \text{ is sink} \\ 0 & \text{if } i \text{ is an intermediate node} \end{cases} \tag{1.18}$$

$$\text{and } \ell \overset{\leq}{=} f \overset{\leq}{=} k \tag{1.19}$$

Unless otherwise specified, we will assume that the source node has an unlimited amount to be shipped out and that the sink can receive an unlimited amount. The constraints in (1.18) are known as *flow conservation equations.* For any arc (i, j), the associated variable f_{ij} appears in exactly two equations in (1.18); once with a coefficient of $+1$ in the equation corresponding to node i, and another time with a coefficient of -1 in the equation corresponding to node j. Hence, the coefficient matrix of the flow variables f_{ij} in (1.18) is the node-arc incidence matrix of G. As an example, consider the network in Figure 1.28. The flow conservation equations for this network are shown in the table given below. Verify that the flow vector marked in Figure 1.28 is feasible (e.g.,total amount reaching node 5 is 6 units along (3, 5). Total amount leaving node 5 is 6 units, 2 along (5, 2) and 4 along (5, 4). So conservation holds at node 5, etc.) This flow vector has a value of 7 units.

	f_{12}	f_{13}	f_{23}	f_{24}	f_{52}	f_{34}	f_{35}	f_{54}	f_{46}	f_{56}	$-v$	
Node 1	1	1									1	0
2	-1		1	1	-1							0
3		-1	-1			1	1					0
4				-1		-1		-1	1			0
5					1		-1	1		1		0
6									-1	-1	-1	0

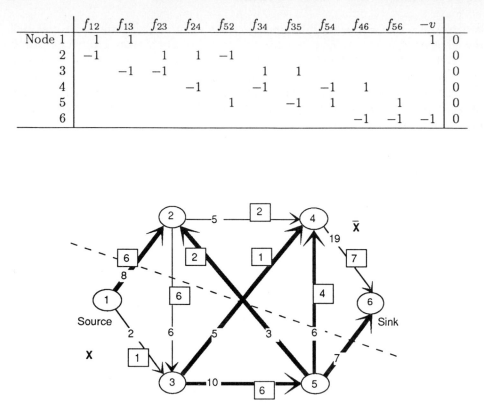

Figure 1.28 Capacities are entered in the gaps on the arcs. Lower bounds are all 0. If flow on an arc is nonzero, it is entered inside a little square by its side.

Verify that the coefficient matrix of the flow variables in the conservation equations is the node-arc incidence matrix E of G. If q denotes the node-arc incidence vector of the hypothetical $(\check{s}, \check{t})$ arc, the constraints on f for feasibility are: $Ef - qv = 0$, and $\ell \leqq f \leqq k$. The *maximum value flow problem* in G is to maximize v subject to these constraints.

If $f = (f_{ij})$ is a feasible flow vector in G, for each $(i, j) \in \mathcal{A}$, $k_{ij} - f_{ij}$ is known as the *residual capacity of arc* (i, j) wrt f. The arc (i, j) is said to be *saturated* in f if its residual capacity is 0, i.e., if $f_{ij} = k_{ij}$.

In general, there may be several source nodes where material is available, and several sink nodes where it is required. The *exogenous flow* at a node, or flow that is external to the network, refers to the quantity of such material at these nodes. At node i the exogenous flow will be denoted by V_i, the convention is that if $V_i > 0$, then i is a source with V_i units available; and if $V_i < 0$, i is a sink with a requirement of $|V_i|$ units. If $V_i = 0$, there is no exogenous flow at i, and it is a transit node.

Sometimes $-V_i$ is called the *requirement* at node i. The vector $V = (V_i)$ leads to the right-hand side constants vector in LP formulations of network flow problems.

When there are several source and sink nodes in a single commodity flow problem, we assume that the material from any source can be shipped to any sink. Otherwise, if some sources can only ship to certain sinks, it becomes necessary to keep track of the flows from each source separately, and the problem becomes a *multicommodity flow problem.*

The general LP type problem in the connected directed single commodity flow network $G = (\mathcal{N}, \mathcal{A}, \ell, k, V)$, with node-arc incidence matrix E is

$$\begin{aligned} \text{Minimize } & cf \\ \text{Subject to } Ef \ =\ & V \\ \ell \overset{\le}{=} f \ \overset{\le}{=}\ & k \end{aligned} \qquad (1.20)$$

Since the sum of all the row vectors of E is 0, a necessary condition for the existence of a feasible flow vector in G is:

$$\sum_{i \in \mathcal{N}} V_i = 0 \qquad (1.21)$$

When G is connected, from the results in the previous section it is clear that the system (1.20) contains exactly one redundant equality constraint. Any one of the equality constraints in (1.20) may be treated as a redundant constraint and eliminated, making the remaining system nonredundant.

Let $|\mathcal{N}| = n, |\mathcal{A}| = m$. Number the arcs in G as $e_t, t = 1$ to m, and denote the flow amount on e_t by f_t. Select any node, say node n, and delete the equality constraint corresponding to it from (1.20). Let the remaining system of equations there be

$$\bar{E}f = \bar{V} \qquad (1.22)$$

Since G is connected, $\bar{E}$ is of full row rank. From Theorem 1.4 we know that every basis for $\bar{E}$ is triangular. The result in Corollary 1.1 implies that every basic solution of (1.22) is an integer vector if ℓ, k, V are integer vectors. If an LP has an optimum solution it has one which is a BFS. So, we conclude that if the data in an LP type single commodity pure network flow problem is integer, and it has an optimum solution, then it has one in which all the flow amounts are integer. This includes the maximum value flow problem, and the minimum cost flow problem (1.20).

From the results in Section 1.2.2, we know that the set of columns of $\bar{E}$ associated with the arcs in a cycle is linearly dependent and that associated with a forest in G is linearly independent. These facts imply that every basic vector for (1.22) consists of node-arc flow variables associated with arcs in a spanning tree in G and vice versa. Let B denote a basis for (1.22) with its basic columns corresponding to the in-tree arcs, $e_1, \ldots, e_{n-1}$ of a spanning tree $\mathbb{T}$ in G. Let D be the matrix consisting of the nonbasic columns. When partitioned into basic, nonbasic parts this way, (1.22) becomes

$f_1, \ldots, f_{n-1}$	$f_n, \ldots, f_m$	
B	D	$\bar{V}$

Select node n as the root node for $\mathbb{T}$. Let $(\lambda_{1p}, \ldots, \lambda_{mp})$ be the incidence vector of the fundamental cycle of the out-of-tree arc $e_p, p = n$ to m, and let λ be the in-tree arc-fundamental cycle incidence matrix. Let $\bar{E}_{.t}$ denote the column of $\bar{E}$ associated with the arc $e_t, t = 1$ to m. Then from (1.14) we have $\bar{E}_{.p} = \sum_{t=1}^{n-1} \lambda_{tp} \bar{E}_{.t}$, for $p = n$ to m. This implies that $D = B\lambda$, or $B^{-1}D = \lambda$. Hence the canonical tableau of (1.22) wrt the basis B is:

$f_1, \ldots, f_{n-1}$	$f_n, \ldots, f_m$	
I	λ	$B^{-1}\bar{V}$

So, the updated nonbasic part in the canonical tableau of (1.22) is always the in-tree arc-fundamental cycle incidence matrix wrt the spanning tree consisting of the basic arcs, and hence can be constructed combinatorially without the need to perform any pivot steps. This also leads to the following property.

DANTZIG PROPERTY   In every canonical tableau for the system of conservation equations in a single commodity flow problem in a pure network, all the entries are always 0, or ± 1.

As an example consider the network in Figure 1.24. Select node 5 as the root node and eliminate the equation corresponding to it from the system of conservation equations for this network. This leads to the next tableau.

f_1	f_2	f_3	f_4	f_5	f_6	f_7	f_8	f_9	
1	0	0	0	0	0	-1	1	1	V_1
-1	-1	0	0	0	1	0	0	0	V_2
0	1	1	0	-1	0	1	0	0	V_3
0	0	0	1	1	-1	0	0	-1	V_4

For this system (f_1, f_2, f_3, f_4) is a basic vector. It corresponds to the spanning tree with thick arcs in Figure 1.24. The canonical tableau of this system wrt this basic vector is:

Basic variable	f_1	f_2	f_3	f_4	f_5	f_6	f_7	f_8	f_9	
f_1	1	0	0	0	0	0	-1	1	1	$\hat{V}_1$
f_2	0	1	0	0	0	-1	1	-1	-1	$\hat{V}_2$
f_3	0	0	1	0	-1	1	0	1	1	$\hat{V}_3$
f_4	0	0	0	1	1	-1	0	0	-1	$\hat{V}_4$

It can be verified that the updated nonbasic part under the nonbasic variables f_5 to f_9 is exactly the in-tree arc-fundamental cycle incidence matrix derived earlier in Section 1.2.2.

Computing the Inverse of a Basis for $\bar{E}$ Combinatorially

Let B be a basis for (1.22). Suppose it corresponds to the spanning tree T with in-tree arcs $e_1, \ldots, e_{n-1}$. Here we discuss how to compute B^{-1} combinatorially without the need to perform any pivot steps. Introduce artificial arcs $e_{m+t} = (t, n), t = 1$ to $n - 1$, joining each nonroot node to the root node n. With these artificial arcs, the node-arc incidence matrix of the augmented network has the following form:

Row corresponding to node	In-tree arcs $e_1, \ldots, e_{n-1}$	Out-of-tree arcs	Artificial arcs $e_{m+1}, \ldots, e_{m+n+1}$
1			
2			
	B	D	I_{n-1}
$\vdots$			
Root node n	$\ldots$	$\ldots$	$-1 \ldots -1$

From this table, and the above discussion, it is clear that the in-tree arc-fundamental cycle incidence matrix of the artificial arcs is B^{-1}, and B^{-1} can therefore be computed directly from the network.

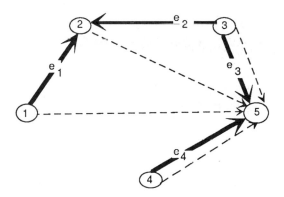

Figure 1.29

As an example, consider the basis B given by the submatrix of the node-arc incidence matrix of the network in Figure 1.24, corresponding to nodes 1 to 4 and arcs $e_1, \ldots, e_4$. To find B^{-1}, include artificial arcs $(i, 5), i = 1$ to 4. Figure 1.29 just depicts the in-tree arcs and the artificial arcs which are dashed.

The remaining matrix, after eliminating the row corresponding to root node 5 from the node-arc incidence matrix of the network in Figure 1.29, is $(B{:}I)$ given below. The in-tree arc-fundamental cycle incidence matrix for the network in Figure 1.29 is given next and it can be verified that it is B^{-1}. This is followed by Theorem 1.8 stating a fundamental property of basis inverses in pure network flow problems. This result is used in Chapter 5 to develop purely combinatorial techniques for resolving the problem of cycling under degeneracy in the primal simplex algorithm for single commodity pure network flow problems, without the need for maintaining B^{-1}.

Basic arcs				Artificial arcs			
e_1	e_2	e_3	e_4	$(1,\,5)$	$(2,\,5)$	$(3,\,5)$	$(4,\,5)$
1	0	0	0	1	0	0	0
-1	-1	0	0	0	1	0	0
0	1	1	0	0	0	1	0
0	0	0	1	0	0	0	1

	Fundamental cycle of artificial arc			
In-tree arc e_1	1	0	0	0
e_2	-1	-1	0	0
e_3	1	1	1	0
e_4	0	0	0	1

Unisign Property of Rows of Basis Inverses in Pure Network Flow Problems

THEOREM 1.8 *Let $\bar{E}$ be the matrix remaining after the row corresponding to the root node n is deleted from the node-arc incidence matrix E of a directed connected network $G = (\mathcal{N}, \mathcal{A})$. Let B be a basis for $\bar{E}$ associated with the spanning tree $\mathbb{T}$ with its columns corresponding to in-tree arcs $e_1, \ldots, e_{n-1}$, say, in that order. Then all the nonzero entries in $(B^{-1})_{t\cdot}$ are -1 if e_t is directed away from the root node, or $+1$ if e_t is directed towards the root node, in $\mathbb{T}$.*

Proof   From the procedure for computing B^{-1} discussed above, we see that the jth entry in $(B^{-1})_{t\cdot}$ is: 0 if e_t is not on the fundamental cycle of (j, n); $+1$ $[-1]$ if e_t is on this fundamental cycle with an orientation opposite to [same as] that of (j, n) (i.e., if e_t is directed towards [away from] the root node n). This clearly implies the result in the theorem. ∎

Optimality Conditions

Consider the minimum cost flow problem (1.20) in the directed single commodity flow network $G = (\mathcal{N}, \mathcal{A}, \ell, k, c, V)$. In the dual problem there are dual variables

π_i associated with the flow conservation equation at node i, they are called *dual variables*, or *node prices*, or *node potentials*. The complementary slackness optimality conditions for a feasible flow vector f are the existence of a dual vector $\pi = (\pi_i)$ satisfying: for each $(i,j) \in \mathcal{A}$, if

$$
\begin{aligned}
\pi_j - \pi_i > c_{ij} \quad &\text{then} \quad f_{ij} = k_{ij} \\
\pi_j - \pi_i < c_{ij} \quad &\text{then} \quad f_{ij} = \ell_{ij} \\
\pi_j - \pi_i = c_{ij} \quad &\text{then} \quad \ell_{ij} \leqq f_{ij} \leqq k_{ij}
\end{aligned}
$$

The potential difference $\pi_j - \pi_i$ is known as the *tension* across the arc (i,j) in the node potential vector π, the optimality conditions depend only on these tensions, and not directly on the potentials themselves. Two different node potential vectors that differ by a constant give rise to the same tension vector. The tension vector is actually $-\pi E$ where E is the node-arc incidence matrix. The quantity $\bar{c}_{ij} = c_{ij} - (\pi_j - \pi_i)$ is known as the *reduced cost coefficient* of arc (i,j) wrt π.

Flow Augmenting Paths

Let $f = (f_{ij})$ be a feasible flow vector of value v in the directed single commodity flow network $G = (\mathcal{N}, \mathcal{A}, \ell, k, \check{s}, \check{t})$. A path $\mathcal{P}$ from $\check{s}$ to $\check{t}$ is said to be a *flow augmenting path* (FAP) wrt f if it satisfies

$$
f_{ij} \begin{cases} < k_{ij} & \text{for forward arcs } (i,j) \text{ on } \mathcal{P} \\ > \ell_{ij} & \text{for reverse arcs } (i,j) \text{ on } \mathcal{P} \end{cases}
$$

The reason for this name can be easily explained. Let $\epsilon = \min\{\epsilon_1, \epsilon_2\}$ where $\epsilon_1 = \min\{(k_{ij} - f_{ij}) : (i,j) \text{ a forward arc on } \mathcal{P}\}$, $\epsilon_2 = \min\{(f_{ij} - \ell_{ij}) : (i,j) \text{ a reverse arc on } \mathcal{P}\}$. $\epsilon_1(\epsilon_2)$ is defined to be $+\infty$ if there are no forward (reverse) arcs on $\mathcal{P}$. ϵ is called the *residual capacity* of the FAP $\mathcal{P}$, it is > 0. Define a new flow vector $\hat{f} = (\hat{f}_{ij})$ by

$$
\hat{f}_{ij} = \begin{cases} f_{ij} & \text{if } (i,j) \text{ is not on } \mathcal{P} \\ f_{ij} + \epsilon & \text{if } (i,j) \text{ is a forward arc on } \mathcal{P} \\ f_{ij} - \epsilon & \text{if } (i,j) \text{ is a reverse arc on } \mathcal{P} \end{cases}
$$

Then, $\hat{f}$ is a feasible flow vector of value $\hat{v} = v + \epsilon$. This operation of computing $\hat{f}$ from f is called the *flow augmentation step* using the FAP $\mathcal{P}$. After this step, $\mathcal{P}$ is no longer an FAP wrt the new flow vector $\hat{f}$.

As an example, consider the feasible flow vector f in Figure 1.28. The thick path from source to sink with forward arcs $(1,2)$, $(5,4)$, $(3,5)$, $(5,6)$, and reverse arcs $(5,2)$, $(3,4)$ is an FAP wrt f. $\epsilon_1 = \min\{8 - 6, 6 - 4, 10 - 6, 7 - 0\} = 2$, $\epsilon_2 = \min\{2 - 0, 1 - 0\} = 1$. $\epsilon = \min\{2, 1\} = 1$. The new flow vector obtained after flow augmentation is $\hat{f} = (\hat{f}_{12}, \hat{f}_{13}, \hat{f}_{23}, \hat{f}_{24}, \hat{f}_{52}, \hat{f}_{34}, \hat{f}_{35}, \hat{f}_{54}, \hat{f}_{46}, \hat{f}_{56}) = (7, 1, 6, 2,$

1, 0, 7, 5, 7, 1). It has value 8. This example indicates that an FAP need not be a simple path. However, in the labeling algorithms discussed in Chapter 2, all FAPs identified will be simple paths.

Several of the algorithms discussed in later chapters use subroutines which generate FAPs, and hence they are called *augmenting path methods*.

An FAP is said to be a *flow augmenting chain*, FAC, if it is a chain (i.e., if all the arcs on it are forward arcs).

A feasible flow vector of value v in G is said to be a *maximum value feasible flow vector* if v is the maximum value attainable in G, or a *maximal* or *blocking feasible flow vector* if there exists no FAC wrt it.

EXAMPLE 1.4

Every maximum value flow vector is maximal, but the converse may not be true. Figure 1.30 illustrates this point.

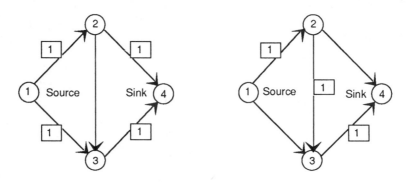

Figure 1.30 Two copies of a network. Each arc has lower bound 0 and capacity 1. The flow on an arc is entered in a box by its side if it is nonzero. The one on the left is a maximum value feasible flow vector of value 2. The one on the right has value 1. It is a maximal flow, but not of maximum value.

Residual Cycles and the Residual Networks G(f), G(f, π)

Let $f = (f_{ij})$ be a flow vector in a directed, connected, single commodity flow network $G = (\mathcal{N}, \mathcal{A}, \ell, k, c, \check{s}, \check{t})$ satisfying the bound conditions on all the flow variables (i.e., $\ell \leqq f \leqq k$), but may or may not satisfy flow conservation at the nodes.

If an arc $(i, j) \in \mathcal{A}$ satisfies $f_{ij} < k_{ij}$, the flow amount on it can be increased up to its residual capacity of $\kappa_{ij} = k_{ij} - f_{ij}$ without violating the upper bound, hence it is called a *residual arc* wrt f. The cost coefficient of this residual arc is naturally c_{ij}, since it is the unit cost of increasing the flow on (i, j).

Likewise, if $(p, q) \in \mathcal{A}$ satisfies $f_{pq} > \ell_{pq}$, the flow on it can be decreased up to $f_{pq} - \ell_{pq}$ without violating the lower bound. This is equivalent to creating a new arc (q, p), which may not be an arc in G, and increasing the flow on it from 0 by the same amount. So, in this case, we call (q, p) a residual arc wrt f, and associate the cost coefficient $-c_{pq}$ with it, since increasing the flow on it is the same as decreasing the flow on the original arc (p, q). Construct the set of arcs $\mathcal{A}(f)$ by the following rules.

1. For each $(i, j) \in \mathcal{A}$ satisfying $f_{ij} < k_{ij}$ include the arc (i, j) in $\mathcal{A}(f)$ with a $+$ label, lower bound 0, capacity $\kappa_{ij} = k_{ij} - f_{ij}$, and cost coefficient $c'_{ij} = c_{ij}$.

2. For each $(i, j) \in \mathcal{A}$ satisfying $f_{ij} > \ell_{ij}$ include the arc (j, i) in $\mathcal{A}(f)$ with a $-$ label, lower bound 0, capacity $\kappa_{ij} = f_{ij} - \ell_{ij}$, and cost coefficient $c'_{ij} = -c_{ij}$.

$\mathcal{A}(f)$ is the set of *residual arcs* and $G(f) = (\mathcal{N}, \mathcal{A}(f), 0, \kappa, c')$ is the *residual network* wrt f, it is very useful for determining flow vectors that we can move to from f while maintaining bound feasibility. Each arc in $G(f)$ corresponds to an arc in G. $(p, q) \in \mathcal{A}(f)$ corresponds to (p, q) in G if its label is $+$, or to (q, p) in G if its label is $-$. Under this correspondence, every FAP from $\check{s}$ to $\check{t}$ wrt f corresponds to a chain from $\check{s}$ to $\check{t}$ in $G(f)$ and vice versa. Let $\mathcal{C}$ be any chain in $G(f)$ and $\mathcal{P}$ the path corresponding to it in G. If $\epsilon = \min \{\kappa_{ij} : (i, j)$ is an arc on $\mathcal{C}\}$, then $\epsilon > 0$. In f, increase (decrease) the flow on all forward (reverse) arcs of $\mathcal{P}$ by ϵ, this leads to a new flow vector in G which also satisfies the bounds.

When the residual network is used in the study of the maximum value flow problem, no cost coefficients are used since there are no costs in this problem.

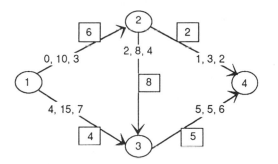

Figure 1.31 The network G and a bound feasible flow vector f in it.

As an example, consider the network G in Figure 1.31. On each arc, the lower bound, capacity, and cost coefficient are entered in that order, and a bound feasible (but conservation violating) flow vector f is marked in little squares. The residual network $G(f)$ is given in Figure 1.32 with the arc labels $+$ or $-$ entered.

Suppose we are given a bound feasible flow vector f, and a node price vector $\pi = (\pi_i)$ in G. $\bar{c}_{ij} = c_{ij} - (\pi_j - \pi_i)$ is the reduced cost coefficient of arc $(i, j) \in \mathcal{A}$.

The *residual network* wrt f, π, $G(f, \pi) = (\mathcal{N}, \mathcal{A}(f), 0, \kappa, \bar{c}')$ is the same as $G(f)$, with the exception that the arc cost coefficients in it are determined using $\bar{c}$ as the cost vector in G instead of c.

An oriented cycle $\mathbb{C}$ in G is said to be a *residual cycle* wrt the flow vector $f = (f_{ij})$ satisfying the bound conditions on all the flow variables, if $f_{ij} < k_{ij}$ on all forward arcs on $\mathbb{C}$, and $f_{ij} > \ell_{ij}$ on all reverse arcs in $\mathbb{C}$. The *capacity of this residual cycle* is defined to be min. { $k_{ij} - f_{ij}$: (i, j) a forward arc on $\mathbb{C}$ } $\cup$ {$f_{ij} - \ell_{ij}$: (i, j) a reverse arc on $\mathbb{C}$}. For example, in the network in Figure 1.31, the oriented cycle 1, (1, 3), 3, (2, 3), 2, (1, 2),1 oriented in the anticlockwise direction, is a residual cycle wrt the flow vector marked there, of capacity 6. Notice that every residual cycle wrt f corresponds to a simple circuit in the residual network and vice versa.

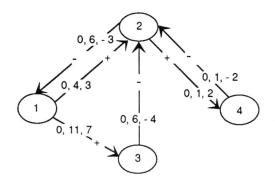

Figure 1.32 The residual network $G(f)$.

Feasible Circulations

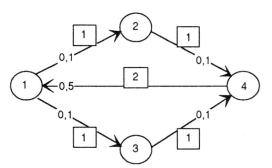

Figure 1.33 Data on the arcs is lower bound, capacity in that order; and flow amount in a box if it is nonzero.

A flow vector $f = (f_{ij})$ in a directed single commodity flow network $G = (\mathcal{N}, \mathcal{A}, \ell, k)$ is said to be a *feasible circulation* if it satisfies the bounds, $\ell \stackrel{\leq}{=} f \stackrel{\leq}{=} k$, and $f(i, \mathcal{N}) - f(\mathcal{N}, i) = 0$ for all $i \in \mathcal{N}$ (this is the same as $Ef = 0$, where E is the node-arc incidence matrix of G). See Figure 1.33 for an illustration.

1.2.4   The Arc-Chain Flow Model

Let G $= (\mathcal{N}, \mathcal{A}, \ell = 0, k, \breve{s}, \breve{t})$ be a directed single commodity flow network. In this model we determine various chains from $\breve{s}$ to $\breve{t}$, and specify how much to ship directly from $\breve{s}$ to $\breve{t}$ across each of them. The chains considered may have common arcs. Let $\mathcal{C}_1, \ldots, \mathcal{C}_P$ be all the distinct chains from $\breve{s}$ to $\breve{t}$ in G, and let x_h denote the amount of material shipped along $\mathcal{C}_h$, $h = 1$ to P. The x_h are the decision variables in this model, and $x = (x_h)$ is the arc-chain flow vector. Here, each x_h has to be clearly associated with its chain $\mathcal{C}_h$. P is likely to be large, and in specifying this vector x, it is only necessary to list the positive x_h in it, and the chains associated with them.

The *value of the arc-chain flow vector* $x = (x_h)$ (i.e., the amount of material reaching the sink in it), is clearly $\sum_h x_h$. Let $\mathbf{C}(i, j)$ denote the set of all the chains from $\breve{s}$ to $\breve{t}$ which contain the arc (i, j). We will use the same symbol to represent the set of their indices. If the arc-chain flow vector $x = (x_h)$ is implemented, the total amount of this flow passing through arc (i, j) will be $\sum(x_h : \text{over } h \in \mathbf{C}(i, j))$. This has to be $\stackrel{\leq}{=} k_{ij}$ for feasibility. So, the arc-chain formulation of the maximum value flow problem in G is

$$\text{Maximize} \sum_h x_h$$

$$\text{Subject to} \sum_{h \in \mathbf{C}(i,j)} x_h \stackrel{\leq}{=} k_{ij} \quad \text{for each } (i, j) \in \mathcal{A} \qquad (1.23)$$

$$x_h \stackrel{\geq}{=} 0 \quad \text{for all } h$$

Given an arc-chain flow vector $x = (x_h)$, in G of value v, define for each $(i, j) \in \mathcal{A}$, $f_{ij}(x) = \sum(x_h : \text{over } h \in \mathbf{C}(i, j))$, and $f(x) = (f_{ij}(x) : (i, j) \in \mathcal{A})$. Then it can be verified that $f(x)$ is a feasible node-arc flow vector of value v It is the natural and unique node-arc flow vector corresponding to the arc-chain flow vector x. Both have the same value.

The reverse question is: Given a node-arc feasible flow vector f in G, is there an arc-chain flow vector corresponding to it? We now study this question and the relationship between the two flow models.

THEOREM 1.9 *Let* $\tilde{f} = (\tilde{f}_{ij})$ *be a feasible node-arc flow vector in G of value* $\tilde{v}$. *If* $\tilde{v} > 0$, *there exists a chain from* $\breve{s}$ *to* $\breve{t}$ *in G such that* $\tilde{f}_{ij} > 0$ *on all arcs* (i, j) *along this chain.*

Proof   Assume that $\tilde{v} > 0$. Call an arc $(i, j) \in \mathcal{A}$ a P-arc if $\tilde{f}_{ij} > 0$. We need to show that there is a chain from $\breve{s}$ to $\breve{t}$ consisting of P-arcs only. Define $\mathbf{X} = \{$

$x : x \in \mathcal{N}$, there exists a chain from $\breve{s}$ to x with P-arcs only}. So, we need to show that $\breve{t} \in \mathbf{X}$.

If $\mathcal{C}_1$ is a chain from $\breve{s}$ to x with P-arcs only, and (x, y) is a P-arc, by including (x, y) at the end of $\mathcal{C}_1$ we get a chain from $\breve{s}$ to y with P-arcs only. This observation leads to the following tree growth scheme to determine the set $\mathbf{X}$. $\breve{s}$ is the root node. The list is always the present set of labeled and unscanned nodes. Each labeled node is in the set $\mathbf{X}$. Since we are only interested in showing that $\breve{t}$ is in the set $\mathbf{X}$, we will terminate the scheme whenever $\breve{t}$ is labeled.

Step 1 Label $\breve{s}$ with $\emptyset$. All other nodes are unlabeled. List $= \{ \breve{s} \}$.

Step 2 If list $= \emptyset$, terminate. Otherwise select a node from it, say i to scan as below. Delete i from list. Find $\mathbf{J} = \{ j$: j unlabeled so far and (i, j) is a P-arc}. Label each node in $\mathbf{J}$ with its predecessor index i, and include all of them in the list.

Step 3 If $\breve{t}$ is labeled, find its predecessor path by a backwards trace. This path, written in reverse order beginning with $\breve{s}$ is a chain to $\breve{t}$ with P-arcs only. Terminate.

If $\breve{t}$ is unlabeled, go back to Step 2.

Suppose this scheme terminates without $\breve{t}$ ever getting labeled. Then, $\mathbf{X} =$ set of labeled nodes at this stage. Let $\bar{\mathbf{X}} = \mathcal{N} \backslash \mathbf{X}$. Since the scheme has terminated, from the labeling rules used, we have $\tilde{f}_{ij} = 0$ for all $(i, j) \in \mathcal{A}$ with $i \in \mathbf{X}$, $j \in \bar{\mathbf{X}}$, so $\tilde{f}(\mathbf{X}, \bar{\mathbf{X}}) = 0$. From the conservation equations, we have

$$\tilde{f}(i, \mathcal{N}) - \tilde{f}(\mathcal{N}, i) = \begin{cases} \tilde{v} & \text{for } i = \breve{s} \\ 0 & \text{for } i \neq \breve{s} \text{ or } \breve{t} \end{cases}$$

Summing these over $i \in \mathbf{X}$, we get $\tilde{v} = \tilde{f}(\mathbf{X}, \mathcal{N}) - \tilde{f}(\mathcal{N}, \mathbf{X}) = \tilde{f}(\mathbf{X}, \bar{\mathbf{X}})$ - $\tilde{f}(\bar{\mathbf{X}}, \mathbf{X}) = -\tilde{f}(\bar{\mathbf{X}}, \mathbf{X})$, since $\tilde{f}(\mathbf{X}, \bar{\mathbf{X}}) = 0$. Since all $\tilde{f}_{ij} \geqq 0$, this implies that $\tilde{v} = -\tilde{f}(\bar{\mathbf{X}}, \mathbf{X}) \leqq 0$, a contradiction to the hypothesis that $\tilde{v} > 0$. Hence it is impossible for the above scheme to terminate without $\breve{t}$ getting labeled. So, it would terminate in Step 3 by producing a chain from $\breve{s}$ to $\breve{t}$ with all arcs on it satisfying $\tilde{f}_{ij} > 0$. ∎

As an example consider the network in Figure 1.34, and the feasible node-arc flow vector of value 24 marked in it. The above scheme is applied on it leading to the predecessor indices entered by the side of the nodes. The sink, node 4 is labeled. Its predecessor path written in reverse order, yields the chain $\mathcal{C}_1 = 1$, $(1, 2)$, 2, $(2, 4)$, 4 in which all the arcs are carrying positive flow.

If the above scheme is operated without Step 3, the set of labeled nodes at termination will be the set $\mathbf{X}$ defined above.

We can use the following procedure based on the scheme in the proof of Theorem 1.9 to obtain an arc-chain flow corresponding to a given node-arc flow $\tilde{f}$ of value $\tilde{v}$ in G. If $\tilde{v} < 0$ (this can happen if there are some arcs incident into $\breve{s}$ in G, and they carry positive flow) it would imply that in the flow vector $\tilde{f}$, actually $|\tilde{v}|$ units of material is flowing back from $\check{t}$ to $\check{s}$. So, if $\tilde{v} \leqq 0$, define the arc-chain flow vector corresponding to $\tilde{f}$ to be $\tilde{x} = 0$, and terminate. On the other hand, if $\tilde{v} > 0$, find a chain from $\check{s}$ to $\check{t}$, $\mathcal{C}_1$ say, with positive flows in $\tilde{f}$ on all its arcs. Define the arc-chain flow amount on $\mathcal{C}_1$ to be $\tilde{x}_1 = \min \{ \tilde{f}_{ij} : (i,j) \text{ an arc on } \mathcal{C}_1 \}$. In $\tilde{f}$ subtract $\tilde{x}_1$ from the flow amounts on all the arcs of $\mathcal{C}_1$ leading to the new node-arc flow vector $\hat{f}$ say. $\hat{f}$ is a feasible node-arc flow vector in G of value $\hat{v} = \tilde{v} - \tilde{x}_1$. If $\hat{v} \leqq 0$ terminate, otherwise repeat this step with $\hat{f}$ and continue in the same way. The arc-chain flow amounts on the chains obtained until termination define an arc-chain vector whose value is $\geqq \tilde{v}$.

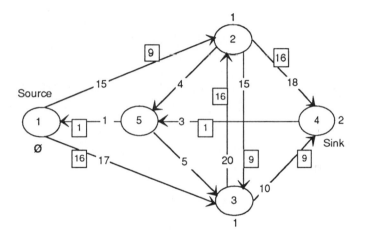

Figure 1.34 Capacities are marked on the arcs and node-arc flow amounts entered inside a box by the side of the arc. Predecessor labels are entered by the side of the nodes. Scheme terminated when the sink is labeled with 2.

For an example, we consider the node-arc flow given in Figure 1.34 of value 24. We take the first chain to be $\mathcal{C}_1$ [consisting of arcs (1, 2) and (2, 4)] obtained above, and the arc-chain flow amount on it is min $\{9, 16\} = 9 = \tilde{x}_1$. We modify the node-arc flow vector by subtracting 9 from the flows on arcs (1, 2), (2, 4) and continue the same way. We obtain the chains $\mathcal{C}_2$ [consisting of arcs (1, 3), (3, 4)] with an arc-chain flow amount $\tilde{x}_2 = 9$, $\mathcal{C}_3$ [consisting of arcs (1, 3), (3, 2), (2, 4)] with an arc-chain flow amount $\tilde{x}_3 = 7$, and the procedure then terminates. So, the arc-chain flow vector obtained has positive flows on three chains only, and its value is $9+9+7 = 25$.

The arc-chain flow vector obtained by this procedure depends on the chains obtained and the order in which they are obtained in the procedure.

THEOREM 1.10 *The maximum flow value from $\check{s}$ to $\check{t}$ in the directed single commodity flow network* G $= (\mathcal{N}, \mathcal{A}, \ell = 0, k, \check{s}, \check{t})$ *is the same, irrespective of whether it is modeled using the node-arc or arc-chain flow models.*

Proof   Since $\ell = 0$, $f = 0$ is a feasible flow vector of value 0. Therefore, the maximum flow value in G is $\geqq 0$. Given a maximum value feasible flow in G in either model, the methods discussed above can be used to construct a corresponding flow of at least the same value in the other model. This proves the theorem. ∎

The arc-chain formulation typically has too many variables. It is practical only under a method which operates by maintaining a small set of chains on which the arc-chain flow amount is positive, and generates new chains to introduce into it one by one as necessary. Since each chain corresponds to a column in the model, such approaches are called *column generation approaches*; they are used together with the revised simplex method. However, for single commodity flow problems, the node-arc formulation leads to algorithms which are much more efficient, and this model is therefore used commonly. For multicommodity flow problems the arc-chain formulation leads to a reasonable solution approach. This is discussed in Section 5.11.

Exercises

1.33 Consider the single commodity flow network in Figure 1.35 with source node

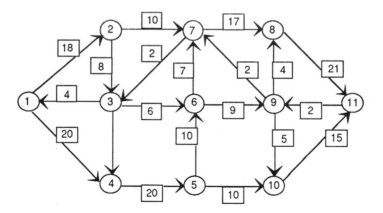

Figure 1.35

1, sink node 11, 0 lower bound on all the arcs; capacity of 20 on all the horizontal arcs, 10 on all the vertical arcs, and 25 on all the diagonal arcs. A node-arc flow vector is entered in the boxes on the arcs. Check it for feasibility and find its value. Construct an arc-chain flow vector from it. What is its value? Explain.

1.34 If the arc-chain formulation of the maximum value flow problem in the directed single commodity flow network $G = (\mathcal{N}, \mathcal{A}, 0, k, \check{s}, \check{t})$ has an optimum solution, prove that it has one in which the arc-chain flow amounts are nonzero on at most m chains, where $m = |\mathcal{A}|$.

1.3 Formulation Examples and Applications

1.3.1 The Transportation Problem

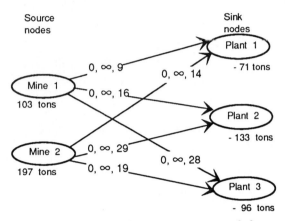

Figure 1.36 Bipartite network representation of the transportation problem. Data on each arc is its lower bound, capacity, and cost per unit flow. Exogenous flow amounts are entered by the side of the nodes.

A minimum cost flow problem on a directed network is called a transportation problem if: (1) every node is either a source node or a sink node (so, there are no intermediate or transit nodes), and (2) every arc in the network joins a source node to a sink node. So it is a minimum cost flow problem on a bipartite network. We give an example of an uncapacitated (i.e., $k_{ij} = \infty$ for all the arcs) transportation problem. A steel company wants to minimize the total shipping bill for transporting iron ore from two mines to three steel mills subject to the data given below. A bipartite minimum cost flow model for it is shown in Figure 1.36.

	Shipping cost(cents/ton)			Availability
Plant →	1	2	3	
Mine 1	9	16	28	103
2	14	29	19	197
Requirement	71	133	96	

1.3.2   The Assignment Problem

An assignment problem is a transportation problem in which the number of source nodes is equal to the number of sink nodes, and all the availabilities and requirements are equal to one. Here is an example: A large corporation is introducing a new product. There are 4 marketing zones, each requires a marketing director. Four candidates have been selected for these positions. Company estimates of the sales generated in the various zones are given in the following table, depending on which candidate is appointed in each zone. Assign candidates to zones to maximize total annual sales. This is a special bipartite minimum cost flow problem.

	Expected annual sales in zone, if candidate assigned to zone.			
Zone →	1	2	3	4
Candidate 1	90	85	139	73
2	60	130	200	112
3	60	130	200	112
4	111	88	128	94

1.3.3   The Transshipment Problem

The transshipment problem is a minimum cost flow problem on a directed network which may not be bipartite, and in which there may be intermediate nodes, arcs joining a pair of sources, or a pair of sinks. Here is an example: A company manufacturing steel shelving cabinets has 3 plants, 2 packing units and 3 sales outlets. Plants 1 and 2 make shelves, and plant 3 makes the bars, screws and all the other components. Production of every item is measured in units of the item needed for one cabinet. Plant 1 has a production capacity of 20,000 cabinets/day, but the paint shop in it is small and can handle only 10,000 cabinets/day. Shelves made in plant 1 can be shipped in any quantity to plant 2 for painting as it has a very large paint shop. Plant 2 has a production capacity of 50,000 cabinets/day. The shelves move from plants 1 and 2 to plant 3, where they are combined with bars, screws, etc. and shipped in bulk to either packing unit 1 or 2. Each packing unit combines the shelves, bars, screws etc. and packs them in cartons of one cabinet each. Shipping from the plants and packing unit 2 is by truck. The routes have the capacities (in cabinets/day) and costs (cents/cabinet) shown in Figure 1.37. Shipping from packing unit 1 to the retail outlets is carried out on water and is therefore cheaper.

Each packing unit can process at most 40,000 cabinets/day. The daily demand at sales outlets 1; 2; 3 is 25,00; 15,000; 22,000 cabinets respectively. We represent each plant, packing unit, sales outlet, by a separate node. The lines joining nodes represent the shipping channel between them. Data on each arc in Figure 1.37 is the lower bound (in cabinets/day shipped), capacity, and cost/cabinet, in that order. The problem is to determine an optimum production and shipping plan to meet the requirements at minimum shipping cost.

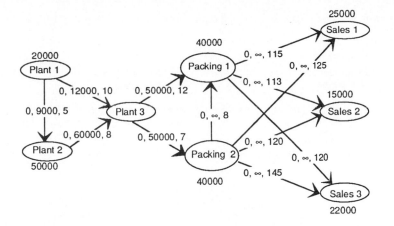

Figure 1.37 Numbers on nodes corresponding to plants and packing units are node capacities per day; and those on sales nodes are requirements per day.

Clearly, this is a minimum cost flow problem on a directed network which is not bipartite. The node corresponding to plant 2 is a source node, but it can also receive flow from other source nodes like plant 1. That's why problems like this are called *transshipment problems*.

In the same manner, problems involving production, in-process inventory, assembly, warehousing, or distribution can be represented as minimum cost flow problems on directed networks. In this model, the flow on each arc represents the amount of material (either finished or semi-finished) transferred from one manufacturing unit to another. For such problems, clearly the node-arc flow model is the most direct. Since many network flow applications are of this type, the node-arc formulation is the one most commonly discussed.

1.3.4 An Application In Short Term Investments

A corporation receives money (income) from its dealers, and has expenses to be paid out (expenditures) every week. Expenses are light in the first half of the year, and tend to be high in the second half. Income, on the other hand, is normally high at the beginning and tends to level off towards the end. The planning horizon is 52 weeks (a year), and the corporation has surplus income at the beginning which it does not need for its expenditures until later. There are several securities in which the corporation can invest this surplus income on a short term basis for an integer number of weeks from 1 to 52. The yield from the pth security as a fraction of the amount invested is c_{pq}, if the maturity period is q weeks. The corporations expected income and expenditures for each week of the year are given, as also a lower and upper bound on the amount that can be invested in each security over the year by company policy. The problem is to determine an optimal short term investment plan to maximize the yield subject to the constraints. This is a problem

of analyzing various cash flow alternatives and can be modeled as a network flow
problem.

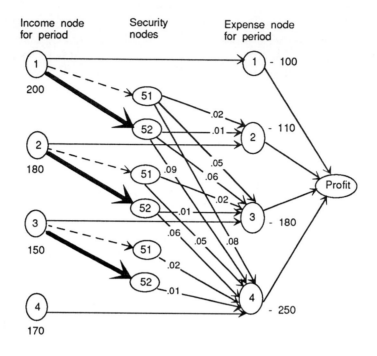

Figure 1.38

We provide a numerical example. To keep it small we consider a planning horizon of a year divided into four equal periods and only two securities, and we assume
that all investments during a year have to be cashed that year itself. The yield from
these investments is not treated as income for the company, but considered as if it
were held in a separate account which is to be maximized. The expected income
(in $1000 units) in the four periods is 200, 180, 150, 170. The corresponding ex-
penditures are 100, 110, 180, 250. The fractional returns from security 1 are 0.02,
0.05, and 0.08 if held for 1, 2, and 3 periods respectively. The corresponding figures
from security 2 are 0.01, 0.06, 0.09. No more than 150 units can be invested in any
one security over a year by company policy.

    The network model is in Figure 1.38. It contains an income node and an expense
node for each period, with the exogenous flow at them in money units entered by
the side. Flow of money directly from an income to an expense node represents
its use to meet expenses without being invested. Money flows through security
nodes represent investments, and the data on the arcs joining a security node to an
expense node represents the fractional yield on the flow through that arc. There is
also a node called profit which accumulates the excess income of this company over

its expenses (remember that yield from investments is only counted in the objective function, and does not enter as a flow anywhere in this model).

In addition to the flow conservation equations, we have two other constraints. The sum of the flows on thick arcs has to be $\leqq 150$. Likewise, the sum of the flows on the dashed arcs has to be $\leqq 150$. These constraints are not of the network flow conservation type. So, this is a minimum cost network flow problem with additional linear constraints. An algorithm for such models is discussed in Chapter 6.

In constructing this model we treated the yield from investment as the objective function, and because of this it did not enter the flow at all. So flows across the investment arcs only consist of the principal amount, this guarantees that if a certain amount of money enters an arc (i, j) at node i, the same amount comes out at node j. Networks in which this property holds are called *pure networks*. All the problems we discussed so far involved only such networks. If the return on investment is also treated as income, then an amount of 100 units entering an investment arc (i, j) carrying 2% interest at node i, would become 102 units by the time it comes out at node j. Networks with this property are called *networks with gains or losses*, or *generalized networks*. An example of a generalized network flow problem is given in Section 1.3.7, and we present algorithms for solving them in Chapter 8.

1.3.5 Shortest Chain Problem

This is the problem of finding a shortest chain from an origin to a destination in a network $G = (\mathcal{N}, \mathcal{A}, c)$ with c as the vector of line lengths. This problem appears as a subproblem in many network applications. Shortest chain algorithms are discussed in Chapter 4.

1.3.6 Project Planning Problems

Large projects usually involve many individual jobs. There may be some interdependence among the jobs, some of them cannot possibly be started until others have been completed. This defines a precedence ordering among the jobs, which can be represented by an acyclic network. Given the time needed for completing each job, the minimum time necessary to complete the project can be computed by solving a longest chain problem. Also, it may be possible to shorten the time needed to complete some jobs by spending extra money. In that case, the problem of determining how much of this extra money to spend on each job in order to complete the overall project within a specified duration at minimum extra cost can be posed as a minimum cost flow problem on an augmented network. This is the problem of computing the project cost curve as a function of project duration (see Chapter 7). This class of problems find many applications for network methods.

1.3.7 Generalized Network Flow Problems

Consider a directed network in which flow entering an arc (i,j) at node i gets multiplied by a factor p_{ij} before it reaches node j. This may happen, for example, if losses occur during transit. If at least one of these multipliers is not equal to 1, such networks are called *generalized networks*, and flow problems on them are called *generalized network flow problems*. In contrast, the networks discussed so far are called *pure networks*, and the multiplier associated with every line is 1 in them.

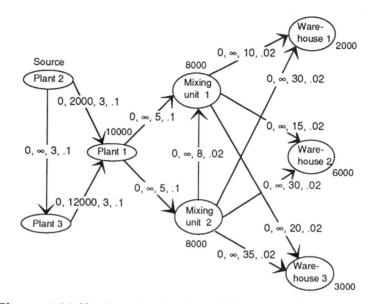

Figure 1.39 Numbers by the side of Plant 1, and Mixing units 1, 2 are the associated node capacities. Numbers by the side of warehouse nodes are requirements. Data on each arc is its lower bound, capacity, cost per unit flow, and the multiplier, in that order.

Here is an example of a generalized network flow problem: A company manufactures 20-lb. bags of fertilizer containing N (nitrogen) and Ph (phosphorous) in fixed proportion. Their plant 1 manufactures urea containing N. Their plant 2, located near their phosphate mine, can only process a limited quantity of phosphate rock into a chemical compound containing Ph. The rest of phosphate rock is shipped to plant 3 for processing. Each plant ships its output to either of two mixing units where the chemicals are blended into fertilizer containing N and Ph in specified proportion and packed into 20-lb. bags. These bags are then shipped to 3 warehouses from which they are sold to retail outlets. From mixing unit 1, material is sent to warehouses by rail, but mixing unit 2 can only ship by truck. Rail transportation is cheaper, so sometimes fertilizer bags are sent from mixing unit 2 to mixing unit 1 for shipping by rail.

There is usually a 10% loss in transit due to spillage in the rock and chemicals shipped from the plants. There is also a loss of approximately 2% due to spoilage in the fertilizer bags shipped from mixing units to warehouses.

We measure each compound in units of it needed per 20-lb. bag of fertilizer. Plant 1 has a capacity of 10,000 bags/day. Plants 2 and 3 can process 2,000, and 12,000 bags/day respectively. Each mixing unit can process and ship 8,000 bags/day. It is assumed that the lower bound, capacity and shipping cost on arc (i, j) apply to f_{ij}, the flow entering this arc at its tail node i. The remaining data is given in Figure 1.39. The problem is to find a feasible flow vector that minimizes the shipping cost.

In the same manner, distribution problems in water, electric power, etc., where losses occur in transmission, can be modeled as generalized network flow problems. Algorithms for them are discussed in Chapter 8.

1.3.8 Applications In Routing

In this section, G denotes a connected undirected network $(\mathcal{N}, \mathcal{A}, c)$ where $\mathcal{N} = \{1, \ldots, n\}$, $\mathcal{A} = \{e_1, \ldots, e_m\}$, c_t is the length of $e_t, t = 1$ to m, and $c = (c_t) \geqq 0$. For instance G might represent the street network of a town. There may be parallel edges in G. d_i denotes the degree of node i in G.

An *edge covering route (ECR)* or *postman's route* in G is an elementary cycle that begins at a node, travels along every edge at least once in some sequence, and returns to the starting node at the end. If an ECR passes l_t times through $e_t, t = 1$ to m, its length is $\sum_t l_t c_t$. The problem of finding a minimum length ECR is known as the *chinese postman problem*.

An *Euler route* is an ECR that passes through each edge of the network exactly once. Since $c \geqq 0$, if an Euler route exists in G, it must be a minimum length ECR.

THEOREM 1.11 *(Euler, 1736) There exists an Euler route in the connected undirected network G iff the degree of every node is even in it.*

Proof If Euler routes exist in G, select one and orient each edge in the direction in which it travels along that edge. An Euler route is a cycle. So, among those edges incident at a node, if the number with orientation leading into the node is r, then the number with orientation leading away from the node must also be equal to r. This implies that the degree of that node is $2r$, even. Hence, if an Euler route exists in G, the degree of every node must be even.

If every node in G has even degree, we discuss below an algorithm and prove that it will produce an Euler route in G. ∎

An undirected network is said to be an *Eulerian network* if it is connected and every node has an even degree in it.

Assume that G is Eulerian. A convenient way to represent an Euler route, is the *edge pairing representation* which we will describe now. Let Υ denote the Euler route $j_1, g_1, j_2, g_2, j_3, \ldots, g_m, j_{m+1} = j_1$ in G. So, $\{g_1, \ldots, g_m\}$ is a permutation of

$\mathcal{A}$ and $g_t = (j_t; j_{t+1})$, for each $t = 1$ to m. In Υ, edge g_t is followed by g_{t+1}, hence we say that the edges (g_t, g_{t+1}) are paired in it, for $t = 1$ to m. Also g_1, g_m are the *first and last edges*. We indicate this by including $(0, g_1), (g_m, \infty)$ as pairs. These are called the *starting and finishing pairs*. This leads to the edge pairing representation for Υ as a list of ordered pairs

$$(0, g_1), (g_1, g_2), \ldots, (g_{m-1}, g_m), (g_m, \infty)$$

The number of pairs in this representation is $1 + |\mathcal{A}|$. The initial and final edges can be retrieved from the starting and finishing pairs. The initial node is the common node on these edges. The initial edge is travelled in the direction away from the initial node. If (g_t, g_{t+1}) is a pair in the representation, these edges must have a common node, say j; the route arrives at j by travelling through g_t and leaves through g_{t+1}. Each edge in G appears in exactly two pairs in an edge pairing representation, as the left hand member in one and the right hand member in the other. It is not necessary to record the various pairs in the representation in any particular order, but an order convenient to the driver is the order of travel.

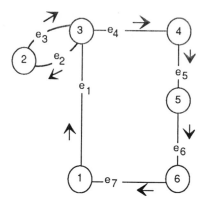

Figure 1.40 An Euler route with node 1 as the initial node.

As an example consider the Euler route in Figure 1.40 that begins at node 1, and travels through the edges in the orientation marked by their side. Its edge pairing representation is $(0, e_1), (e_1, e_2), (e_2, e_3), (e_3, e_4), (e_4, e_5), (e_5, e_6), (e_6, e_7), (e_7, \infty)$.

The *parity* of an integer is *even* if that integer is even, *odd* otherwise.

THEOREM 1.12 *Let $\mathbb{C}$ be an elementary cycle, and $\mathcal{P}_{ij}$ an elementary path between nodes i and j in G. The operation of deleting all the edges in $\mathbb{C}$ from G, or duplicating all the edges in $\mathbb{C}$ leaves the parity of the degree of every node unchanged. The operation of duplicating all the edges on the path $\mathcal{P}_{ij}$ changes the parity of the degrees of nodes i and j, but leaves that of all the other nodes unchanged.*

Proof $\mathbb{C}$ contains an even number of edges incident at every node. $\mathcal{P}_{ij}$ contains an odd number of edges incident at i and j, but an even number at every other node. The results follow from these and the fact that the parity of an integer is unchanged by subtracting or adding an even number to it, but changes when an odd number is added to it. ∎

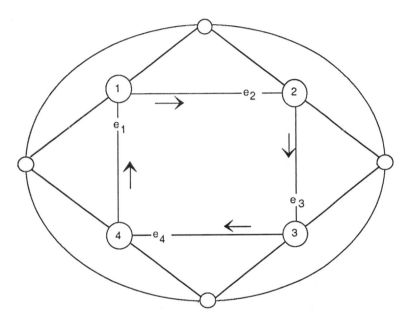

Figure 1.41

The algorithm for finding an Euler route in an Eulerian network G maintains a route which is always an elementary cycle, and grows it until it becomes an Euler route. At some stage, let $\hat{G}$ be the network consisting of the edges in G not traversed in the present route. $\hat{G}$ may not be connected, but each connected component in it is Eulerian by Theorem 1.12. Select a node i that is incident to some edges in the present route and to some not in it. Let (g_0, g_1) be an edge pair in the present route containing node i. Form an elementary cycle, $\mathbb{C}$ say, beginning and ending at i in $\hat{G}$, and record it in edge pairing representation. Let h_0, h_1 be the first and last edges travelled on this cycle, so i is the common node on them. Delete the edge pair (g_0, g_1) from the route, and include into it the edge pairs $(g_0, h_0), (h_1, g_1)$ and all the edge pairs other than the starting and finishing pairs of $\mathbb{C}$. This has the effect of inserting $\mathbb{C}$ into the route. The new route follows the old one until the edge g_0 is traversed to reach node i, then follows $\mathbb{C}$ until it is completed, and then follows the old route again. The new route contains more edges than the old. Repeat the same procedure until every edge is included in the route.

As an example, consider the Eulerian network in Figure 1.41. Start with the route $\{(0, e_1), (e_1, e_2), (e_2, e_3), (e_3, e_4), (e_4, \infty)\}$ consisting of the cycle $\mathbb{C}_1$ marked

with arrows in Figure 1.41. When all the edges e_1 to e_4 traversed in the route are
deleted from this network we get the network in Figure 1.42. This is connected, but
in general such remaining networks may not be connected. We select node 3 on the
route which is incident to some edges in the remaining network, and we find the
cycle $\mathbb{C}_2 = \{(0, e_5), (e_5, e_6), (e_6, e_7), (e_7, \infty)\}$ beginning at node 3 in the remaining
network, marked with arrows in Figure 1.42. Inserting $\mathbb{C}_2$ into the route leads to
the new route $\{(0, e_1), (e_1, e_2), (e_2, e_5), (e_5, e_6), (e_6, e_7), (e_7, e_3), (e_3, e_4), (e_4, \infty)\}$.
Now the edges e_5, e_6, e_7 on $\mathbb{C}_2$ are deleted from the remaining network in Figure
1.42, and the method is continued in the same manner.

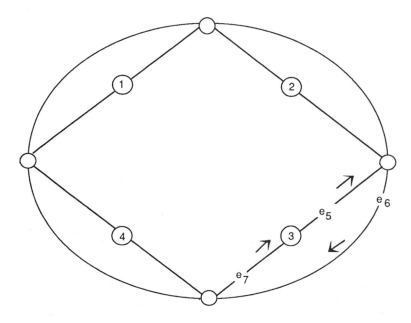

Figure 1.42

During the algorithm, edges belong to two sets, the included (in the present
route) and unincluded. The following symbols will be used; p denotes the initial
node of the new elementary cycle being formed; g_0, g_1 denote either $0, \infty$, or the
left and right edges in a pair in the present route incident at p; a denotes the first
edge of the new elementary cycle being formed incident at p; x denotes the present
node on the new elementary cycle being formed; e denotes the unincluded edge
incident at x selected for the new elementary cycle, and y the other node on it; and
e_1 denotes an unincluded edge incident at y.

Algorithm for Finding an Euler Route

Step 1 Initialization   Select any node, say 1, and an edge e_0 incident at 1. Set
$g_0 = 0, g_1 = \infty, p = 1, x = 1, e = e_0, a = e_0$. All edges are unincluded.

Step 2 Select an unincluded edge   Let y be the node $\neq x$ on e. Select an unincluded edge $\neq e$ incident at y, make it e_1 and go to Step 3. If none, we will have $y = p$. Go to Step 4.

Step 3 Make an edge pair   Include the edge pair (e, e_1) in the list for the new cycle. Now e is included. Make e_1 into the new e, y into the new x, and go to Step 2 to select new y and e_1.

Step 4 Merging the cycle into the route   Form the pairs (g_0, a) and (e, g_1) and insert these and all the other pairs generated in various occurrences of Step 3 in this cycle generation effort, into the route. Now e is included. Go to Step 5.

Step 5 Setup for new cycle   If there are no unincluded edges, terminate. Otherwise, find a node incident to both included and unincluded edges, and make it the new p. Choose an edge pair containing p from the route, delete it from the list for the route, and make the right and left members in it the new g_0, g_1 respectively. Select an unincluded edge incident at p and make it the new e. Make $a = e, x = p$, and go to Step 2.

The reason for $y = p$ when no unincluded edges are found incident at y in Step 2 is the following. At this stage $e = (x; y)$ and all edges other than it incident at y are included. Suppose $y \neq p$. What we have traced so far is a path from p to y. All the edges on this path from p up to x are 'included'. The number of unincluded edges incident at y is an even integer, and the present edge $e = (x; y)$ is one of them, hence there must be at least one more unincluded edge incident at y, a contradiction to our hypothesis. Hence when the algorithm arrives at Step 4 we must have $y = p$, and the elementary cycle being traced must be complete, and e is the last edge on it.

Each time Step 4 is completed, a new elementary cycle is formed and inserted into the route, and the set of remaining (i.e., unincluded) edges forms one or more Eulerian networks. Each time Step 3 or 4 is carried out, one new edge is 'included', so together they occur $m = |\mathcal{A}|$ times in the algorithm. If $|\mathcal{N}| = n$, Step 2 takes at most O(n) effort, and it is automatically followed by Step 3 or 4. So, with an effort of at most O(nm), the algorithm is guaranteed to find an Euler route in G.

Comment 1.1   This algorithm for finding an Euler route in an Eulerian network is due to J. Edmonds and E.L. Johnson [1973]. Their paper discusses other ways of representing Euler routes, and several other algorithms for finding Euler routes.

Now consider the case where G is connected, but not Eulerian. So, the total number of odd degree nodes in G is a positive even number. Suppose these are $1, \ldots, 2p$. Let Φ be an ECR that passes l_t times through edge $e_t, t = 1$ to m. Obtain the network G_Φ by copying the edge e_t exactly l_t times, for $t = 1$ to m. Then Φ must be an Euler route in G_Φ, and hence every node in G_Φ must be an even degree node. For $t = 1$ to m, define $b_t = 1$ if l_t is odd, 2 otherwise. Let $\tilde{G}$ be the network in which the edge set contains exactly b_t copies of e_t for $t = 1$ to m. Since $l_t - b_t \geq 0$ and even for each t, $\tilde{G}$ is Eulerian too. Let $\tilde{\Phi}$ be an Euler route in $\tilde{G}$. The length of $\tilde{\Phi}$ is $\sum_{t=1}^{m} b_t c_t \leq \sum_{t=1}^{m} l_t c_t = $ length of Φ. This clearly implies that there exists an optimum (i.e., minimum length) ECR in G which passes through each edge of G at most twice. Hence we consider only such ECRs in the sequel. A minimum length ECR of this type must minimize the sum of lengths of edges in $\bar{A}$, the subset of edges that it passes through twice.

Every odd degree node in G becomes an even degree node, and every even degree node remains an even degree node when edges in the repeated set $\bar{A}$ are duplicated in G. By Theorem 1.12, this implies that the odd degree nodes $1, \ldots, 2p$ can be partitioned into p pairs, say $(i_{11}, i_{12}), \ldots, (i_{p1}, i_{p2})$, such that there is a path from i_{r1} to i_{r2} among the set of edges $\bar{A}$, and $\bar{A}$ is the set of edges on these paths. As an example, consider the ECR $1, e_1, 4, e_6, 3, e_5, 2, e_3, 1, e_2, 3, e_2, 1, e_1, 4, e_4, 2, e_3, 1$, in the network in Figure 1.43. In this ECR, the thick edges $\{e_1, e_2, e_3\}$ have been traversed twice. This set is the union of two paths $1, e_2, 3$ and $2, e_3, 1, e_1, 4$ between the pairs of odd degree nodes 1, 3 and 2, 4 respectively.

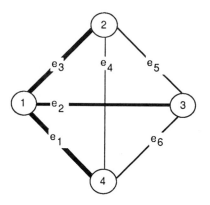

Figure 1.43 The thick edges are those traversed
twice by the ECR under consideration.

In order to minimize the sum of the lengths of the edges traversed twice, the path which has been duplicated between a pair of odd degree nodes should be a shortest path. Also the partitioning of the odd degree nodes in G into pairs should be done in such a way that the sum of the lengths of the shortest paths between the various pairs in the partition is minimized. Hence an optimum ECR in G can be obtained by the following procedure.

1. Find the shortest paths between all pairs of odd degree nodes in G. Let $\mathcal{P}(i, j)$ be a shortest path of length d_{ij}, for $i \neq j = 1$, to $2p$. Let $D = (d_{ij})$ be the $2p \times 2p$ shortest path distance matrix with $d_{ii} = \infty$ for all i.

2. Let $\hat{\mathcal{N}} = \{1, \ldots, 2p\}, \hat{A} = \{(i, j) : i, j \in \hat{\mathcal{N}}, i \neq j\}, H = (\hat{\mathcal{N}}, \hat{A})$. H is the complete undirected network on the set of odd degree nodes in G. The arc

$(i, j) \in \hat{\mathcal{A}}$ corresponds to the shortest path $\mathcal{P}(i, j)$ between the odd degree nodes i, j in G, and we define its weight to be d_{ij}. Find a minimum weight perfect matching in H. The blossom algorithm discussed in Chapter 10 can be used for this.

3. If $\{(i_{11}; i_{12}), \ldots, (i_{p1}; i_{p2})\}$ is a minimum weight perfect matching in H, an optimum pairing for the odd degree nodes in G is (i_{r1}, i_{r2}), $r = 1$ to p. Duplicate all the edges along the shortest paths $\mathcal{P}(i_{r1}, i_{r2})$, $r = 1$ to p in G, obtaining an Eulerian network, $\bar{G}$. Find an Euler route in $\bar{G}$. It is an optimum ECR in G.

The chinese postman problem provides an important application for the blossom algorithms discussed in Chapter 10. Many routing problems in trash collection, road sweeping, school bus route planning etc. can be modeled using the chinese postman problem.

1.4 Exercises

1.35 Let G $= (\mathcal{N}, \mathcal{A})$ be a non-Eulerian undirected connected network with c as the vector of edge lengths. A subset of edges **F** in G is said to be a *feasible subset* if their deletion from G makes each connected component in the remaining network Eulerian. Discuss an efficient algorithm for finding a minimum length feasible subset. (S. Biswas)

1.36 Prove that every doubly stochastic matrix, i.e., a feasible solution of (1.15), can be expressed as a convex combination of assignments.

1.37 Capacity Acquisition Problem A company requires a minimum of d_h units of warehouse capacity in period $h = 1$ to n. Capacity acquired at the beginning of period h and relinquished at the beginning of period t is said to be acquired for the interval $[h, t]$. This costs c_{ht} per unit. Let x_{ht} denote the units of capacity acquired for the interval $[h, t]$. Let y_j denote the units of acquired but unused capacity during period $j = 1$ to n. Formulate the problem of finding a minimum cost capacity acquisition to meet the requirements as an LP. Show that this problem can be transformed into a network flow problem by simple linear transformations. Do it for $n = 4$, and draw the corresponding network. (Veinott and Wagner [1962])

1.38 A company making a single product has a production capacity of 25,000 tons per period. They have to ship out respectively 15,000, 18,000, 30,000, and 8,000 tons in periods 1 to 4. The expected production cost ($/ton) during periods 1 to 4 is 50, 60, 40, and 70 respectively. The production during a period can either be shipped out in the same period or stored for later shipment at a storage cost of $2/ton/period, the charge being imposed on the quantity in storage at the end of the period. Initial inventory is zero; final inventory at the end of period 4 should be zero too. Formulate the problem of determining an optimum production storage plan as a transportation problem.

1.39 Application in Marketing   Here is a multibrand, multiattribute marketing model. Marketing is replete with examples where two brands may have the same product attribute values but enjoy very different market shares, so we include an additional component called 'brand specific effect' which measures the overall preference not explained by the attributes used in the model. This may depend on the levels or the strategy of the brand's marketing effort, etc. Let $j = 1$ to n be the different brands, $i = 1$ to m a representative sample of consumers, $p = 1$ to t the relevant product attributes, $y_{jp} = $ brand j's value on the pth attribute, $w_{ip} = $ estimated importance weight of the ith customer to the pth attribute, $b_{ij} = \sum_{p=1}^{t} w_{ip} y_{jp} = $ preference measure of customer i for brand j, $v_j = $ brand specific effect of brand j, and c_i the brand chosen by the ith consumer.

Given the v_j, $\prime b_{ij} = b_{ij} + v_j$ defines the overall preference of consumer i to brand j. It is reasonable to assume that consumer i would choose that brand j with the largest $\prime b_{ij}$, and this choice is not altered by adding the same additive constant to all the v_j. Thus, the v_js can only be determined up to an additive constant. Given the v_j, define $s_i = \max \{ \prime b_{ij} : j = 1 \text{ to } n \}$. The problem is to determine the v_j to get the best fit. Clearly, this requires minimizing $\sum_{i=1}^{m} (s_i - \prime b_{ic_i})$. Show that this can be done using algorithms for the transportation problem.

Consider the numerical problem in which $n = 3$, $m = 6$, $c = (c_i) = (1, 1, 2, 2, 2, 3)$ and b_{ij}s given below. Obtain the best estimates for (v_1, v_2, v_3) from this data.

	b_{ij}		
$j \rightarrow$	1	2	3
$i = 1$	9	8	12
2	13	11	15
3	18	13	15
4	16	10	4
5	19	5	3
6	8	6	7

(Srinivasan [1979]).

1.40 Mold Allocation Problem in a Tire Plant   To make a tire, one has to set up a "mold" into a general purpose machine called a "cavity" which is carried out by highly skilled personnel with special equipment. Every cavity can produce any type of tire, given the appropriate mold. Molds are very expensive, as they take 6 months to prepare, and they usually outlast the product for which they are designed, and hence are normally converted to another type. Assume that setups are carried out only at the beginning of each period, and consider the problem of determining the assignment of molds to cavities. The following data is given: $n = $ number of periods in the planning horizon, $m = $ number of mold types, $C = $ number of cavities available, $T_{it} = $ number of type i molds available during period t, k_t = max. possible number of setups at the beginning of period t, l_{it} = lower bound on type i molds in cavities period t, and α_{it} = cost (\$) of setup of a type i mold in a cavity at the beginning of period t. x_{io} are predetermined nonnegative integer constants that give the initial distribution of molds in cavities. The decision variables (nonnegative integer variables) are: x_{it} = number of type i molds in cavities in period t, y_{it} = number of setups of type i molds performed at the beginning of period t, and z_{it} = number of takedowns (removal of molds from cavities) of type i molds performed at the beginning of period t.

Formulate the problem of determining the optimum values of x_{it}, y_{it}, z_{it} subject to the stated constraints as a linear integer program. Develop a transformation that permits the reduction of this problem into a minimum cost flow problem. Carry out this transformation in the numerical problem with $m = 2$ and $n = 3$ $C = 15, x_{10} = 6, x_{20} = 5$, and the following data.

	T_{it}		l_{it}		α_{it}		k_t
	$i = 1$	2	$i = 1$	2	$i = 1$	2	
period 1	7	7	5	6	100	150	3
2	7	9	5	8	100	175	4
3	6	11	4	10	90	140	5

(Love and Vemuganti [1978])

1.41 Sometimes an activity can only be carried out if other activities are also carried out. Here is the data on such a situation in which there are 7 projects each of which can either be completely carried out or not at all, and only if other specified projects are also carried out. Formulate the problem of determining which subset of projects should be carried out to maximize the total net profit, as a 0-1 integer program, and show how it can be solved using a network flow approach.

Project No.	Net Return from project (million $)	Project can be carried out only if these other projects are also carried out
1	10	2
2	−8	
3	2	1,5
4	4	2,6
5	−5	
6	3	
7	2	3

(Williams [1982], and Baker [1984])

1.42 Allocating Oil Wells to Platforms The variable p is the number of platforms to be built, and w is the total number of production wells to be drilled in an oilfield, all with known locations. Platforms have to be built first, the size and cost of each depends on the number of wells to be drilled from it, and its location. The decision variable t_{ij} is 1 if production well i is assigned to platform j, 0 otherwise, for $i = 1$ to $w, j = 1$ to p, and $m_j = \sum_{i=1}^{w} t_{ij}$ is the number of production wells assigned to platform j. We are given $g_j(m_j) = $ cost of building platform j as a function of m_j, a piecewise linear convex function, and $c_{ij} = $ cost to drill well i from platform j once it is built, $i = 1$ to w, andj $= 1$ to p. A minimum cost allocation is obtained from the following problem. Show that this problem can be transformed into a minimum cost pure network flow problem.

$$\text{minimize} \quad \sum_{i=1}^{w}\sum_{j=1}^{p} c_{ij}t_{ij} + \sum_{j=1}^{p} g_j\left(\sum_{i=1}^{w} t_{ij}\right)$$

$$\text{subject to} \quad \sum_{j=1}^{p} t_{ij} = 1, i = 1 \;\; \text{to} \;\; w$$

$$t_{ij} = 0 \;\; \text{or} \;\; 1 \;\; \text{for all} \;\; i,j.$$

(Divine and Lesso [1972])

1.43 A department in a university has admitted n students in a term. They have a_1 graduate assistantships (GA's, full tuition + stipend) and a_2 tuition fellowships (TF's, tuition only) to offer. The variables p_i^1, p_i^2, p_i^3 denote the probabilities that the ith student accepts the admission if he is awarded GA, TF, or is not awarded any of these, respectively, estimated by the admissions office based on their background. W_i denotes a desirability rating given by the department to the ith student for $i = 1$ to n. Formulate the problem of selecting the set of students to be offered GA's and TF's, so as to maximize the total expected desirability rating of the incoming batch (assume that if a student is offered some aid and does not accept, then he is lost, and this aid cannot be offered to someone else). (Chandrasekaran and Subba Rao [1977])

1.44 Chromosome Classification   Karyotying is a process by which chromosomes are classified into groups by observing features like size, shape, band structure induced by staining, etc. Suppose there are n chromosomes to be assigned to m groups, and it is known that the jth group has b_j chromosomes. After observations, it has been estimated that the probability of the ith chromosome belonging to the jth group is p_{ij}, these p_{ij} are given. It is required to assign chromosomes to groups so as to maximize the product of posterior probabilities subject to achieving correct group totals. Formulate this as a transportation problem.
(Tso [1986])

1.45 The Single Depot, Unconstrained Number of Vehicles Bus Scheduling Problem   There are n trips to be operated in a planning interval T, each characterized by its starting and ending time, starting and ending places, and by its line (company may operate several lines, and typically incurs a penalty p_1\$ whenever trips of two lines are combined into a bus schedule). Assume that trips are ordered $1, \ldots, n$, by increasing value of starting time. Trip j can follow trip i in a bus schedule only if the starting time for trip j exceeds the ending time for trip i plus the driving time from ending place of i to starting place j computed with a fixed safety margin. All such pairs are specified, as well as the dead-heading cost, q_{ij}, from ending place of i to starting place of j for each such pair (i, j). Also, D, the cost incurred by each bus used in the schedule is given. Formulate the problem of forming minimum cost bus schedules as an assignment or transportation problem. Discuss how this formulation changes if there is a bound on the maximum number of vehicles to be used. (Gavish and Schweitzer [1974], Gavish, Schweitzer and Shlifer [1978], Pinto Paixão and Branco [1987], Bertossi, Carraresi and Gallo [1987])

1.46 The Caterer Problem   A caterer has to supply clean napkins each day over a period of n days. Soiled napkins can be laundered by a slow process that takes p days at a cost of $r \geq 0$ per napkin, or by a fast process that takes $0 < q < p$ days and costs $d > r$ per napkin. Also new napkins can be bought, each at a cost of $b > d$ any day. The demand, given to be a_i napkins on the ith day, $i = 1$ to n, is to be met at least cost. For the first q periods, napkins must be purchased since soiled napkins cannot be laundered quickly enough for reuse. So, initially $a_1 + \ldots + a_q$ purchased napkins are needed. Denote the total number of new napkins purchased by $a_0 + a_1 + \ldots + a_q$ where $a_0 \geq 0$ is treated as a parameter. Prove that a feasible solution exists iff $a_{\min} \leq a_0 \leq a_{\max}$, where

$$a_{\min} = \max\left\{ 0; \sum_{j=q+1}^{q+h} a_j - \sum_{i=1}^{h} a_i, \quad \text{for} \quad h = 1, 2, \ldots, n-q \right\}$$

$$a_{\max} = \sum_{j=q+1}^{n} a_j.$$

For given a_0 satisfying these feasibility conditions, show that the problem can be formulated as a $(1+n-q) \times (n-q+1)$ balanced transportation problem with rows of the array corresponding to 0, day $1, \ldots,$ day $n-q$, and columns corresponding to day $q+1, \ldots,$ day n, slack; and variables x_{0j} = number of purchased napkins used on day j, x_{ij} = number of napkins soiled on day i and reused on day j, for $i \neq 0, j \neq n+1$, $x_{i,n+1}$ = slack variable. Develop a special direct method of $O(n)$ computational effort to find an optimum solution of this transportation problem, based on the concept that slow laundered napkins are reused at the earliest possible moment, while fast laundering is delayed as much as possible. Solve the numerical problem corresponding to data, $n = 10, q = 2, p = 5, r = 2, d = 4, b = 10, (a_0$ to $a_{10}) = (3, 7, 12, 2, 6, 9, 6, 13, 8, 14, 6)$.

Develop a special direct method to obtain optimum solutions for all integer values of the parameter a_0 in its feasibility range, and for finding the best value for a_0. Apply this method to find the optimum a_0 for the numerical problem given above. (Szwarc and Posner [1985]).

1.47 Application in School Planning   The variable m is the number of school districts in a region, which has n schools in the public school system. For $i = 1$ to m, a_i is the number of students who will attend the public schools from district i. For $j = 1$ to n, b_j is the maximum number of students that school j can accommodate (this is typically the number of classrooms multiplied by the maximum number of students allowed per class, which is normally set at 25 or so.) For $i = 1$ to m, $j = 1$ to n, c_{ij} is the distance between district i and school j (this is usually the bird's flight distance between the school and the demographic center of gravity of the district). It is required to determine the number of students in each district to be assigned to each school so that no school is filled beyond its capacity, every student gets assigned to a school, and the total distance between home and school for all students is minimized (it is OK to split a district between schools, as the districts could be subdivided and renumbered). Formulate this problem.

1.48   There are m insurance agents in a region divided into n small localities called blocks. We are given the following information: w_j = expected workload (man-hours per year) in block $j, j = 1$ to n; d_{ij} = distance of block j to the location of agent $i = 1$ to m, $j = 1$ to n; a_i = ideal fraction of workload in region to be assigned to agent $i(a_i > 0$ for all i and $\sum a_i = 1)$; $a_i(1 - f_i), a_i(1 + f_i)$ = lower, upper bounds on fraction of region's workload to be assigned to agent $i(0 < f_i < 1)$.

Let x_{ij} = amount of workload in the block j assigned to agent i and assume that travel cost of agent i to block j is $x_{ij}d_{ij}$. Give a network flow formulation for assigning workloads to the agents, to minimize the travel costs of all the agents put together. (Marlin [1981])

1.49   There are r distinct groups of people planning to vacation on the beach together one night. The ith group has n_i people in it, $i = 1$ to r. There are p cars available for the drive, the jth car can seat d_j people, for $j = 1$ to p. It is required to find a seating arrangement so that no two members of the same group are in the same car. Formulate this as a network flow problem.

1.50 Natural Gas Distribution   The gas pipeline network consists of three supply systems, 1, 2 and 3. Supply system 1 consists of two source nodes with supplies of 500 ft^3 of gas each. Supply system 2 has a single source node with a supply of 1,000 ft^3, and supply system 3 has two source nodes with supplies of 2,000 and 6,000 ft^3. In each supply system all the gas flows out through a transfer node to which each source node in that system is connected by a pipeline. There are two natural gas users in the network. The delivery to user 1 has to be between 3,000 to 3,500 ft^3, and the delivery to user 2 has to be between 6,000 to 7,000 ft^3. User 1 is connected by a direct pipeline to supply systems 1 and 2, and user 2 is connected likewise to supply systems 1 and 3. There is also a redistribution node which is connected by pipeline to each supply system and to each user. Every pipeline through the redistribution node has a capacity of 3,000 ft^3. All other pipelines in the network have a capacity of 2,000 ft^3. Formulate the problem of minimizing the total flow through the redistribution node while meeting the delivery obligations.

1.51 Allocation of Contractors to Public Works   A region is geographically divided into r districts. In each district there is public work to be carried out which has to be contracted out. There are s_1 experienced and s_2 inexperienced contractors available. The work is actually carried out by teams provided by contractors and sent to the districts for this purpose. For $j = 1$ to $s_1 + s_2, n_j$ is the maximum number of teams that the jth contractor can provide. For $i = 1$ to r and $j = 1$ to $s_1 + s_2, c_{ij}$ is the price quoted by the jth contractor to send one team to district i for doing the work there. N_i is the minimum number of contractors to be allocated to district i. By policy, each district must get at least one team from an experienced contractor. Give a network formulation for the problem of allocating teams from the contractors to the districts, subject to the above constraints, at minimum cost. Construct this network formulation for the following data (c_{ij} are in units of $10,000$): $r = 5, s_1 = 2, s_2 = 4, n = (n_j) = (3, 4, 6, 8, 10, 5), N = (N_i) = (2, 3, 4, 2, 3)$.

$$c = (c_{ij}) = \begin{pmatrix} 35 & 48 & 21 & 33 & 41 & 28 \\ 56 & 29 & 19 & 22 & 38 & 50 \\ 45 & 48 & 43 & 41 & 46 & 43 \\ 65 & 58 & 54 & 59 & 52 & 51 \\ 76 & 81 & 79 & 80 & 69 & 68 \end{pmatrix}$$

(Cheshire, McKinnon, and Williams [1984])

1.52 Budget Allocation   There is a four-level hierarchy in the allocation of a state's educational budget. At the top is the state with its educational budget. The next level consists of the various universities, or campuses which are separate budget entities. The next level corresponds to colleges within each university. Finally, the lowest level corresponds to the departments within each college. Formulate this as a network flow problem, where the flow represents the budget allocation to the various units in each level. Each arc in the model will have lower and upper bounds, where lower bound = minimal requirements, and upper bound = budget request submitted by the unit administrator. The total budget is constrained by availability of state funds for education. An objective function which is composed of a weighted average of the allocations is to be optimized. The weights represent the relative importance given to the unit by the decision makers (for example, the weights may be proportional to the corresponding enrollment projections).

   Discuss the changes to be made in the formulation for determining the annual budget allocations in a multiyear planning horizon, if transfers are allowed from one year to the next.

1.53 Industrial Estate Development   A country is making a 15-year plan for industrial land development. There are 17 sites where land is available to be developed. The parameter a_i denotes the maximum amount of land (acres) available for industrial development at site i; c_i (in thousand \$/acre) is the present cost (at the beginning of the planning horizon) of developing industrial land at site i; and r_i (in thousand \$/acre) is the discounted revenue collectable over the lease period for an acre of industrial land leased out at site i. This data is tabulated below.

i	1	2	3	4	5	6	7	8
c_i	130	130	35	31	31	18	87	26
r_i	28	28	45	45	28	28	28	28
a_i	250	350	32	532	350	60	74	30

i	9	10	11	12	13	14	15	16
c_i	17	23	30	22	31	131	65	22
r_i	28	28	28	102	113	142	85	113
a_i	45	30	102	25	164	1593	321	2133

Sites 1 to 5 are considered highest priority sites. At these sites there is a requirement that a minimum of 40, 50, 30, 50, and 50 acres must be developed and leased out in the first 10 years of the planning horizon. Sites 6 to 11 are at the next priority level. At these sites there is a requirement that a minimum of 30 acres in each site must be developed and leased out during the planning horizon. The projected demand for industrial land development at all sites put together in year t of the planning horizon is d_t acres where $d_t = 300, 300, 300, 350, 350, 400, 400,$ 400, 400, 400, 450, 450, 450, 450, and 450, for $t = 1$ to 15 respectively. Formulate the problem of allocating land at the various sites for industrial development, over the years of the planning horizon, subject to the constraints mentioned above, so as to minimize the total net discounted cost (cost of developing minus the discounted revenue collected), as a minimum cost network flow problem. (Fong [1980])

1.54 A Minimum Cost Supply-Demand Problem   Over an $n-$period planning horizon a business person can buy, sell, or hold the commodity for later sale, subject to the following constraints. In the ith period, $k_i \geqq 0$ is an upper bound on the amount of commodity he can buy, $d_i \geqq 0$ is an upper bound on the amount of commodity he can hold till next period, and $\ell_i \geqq 0$ is a lower bound (because of commitments made already) on the amount he sells. The buying, selling and storage costs are $a_i \geqq 0, b_i \geqq 0, c_i \geqq 0$ respectively in the ith period. It is required to determine his optimum buying, selling, holding plan over the planning horizon, in order to maximize the net total profit. Formulate this as a minimum cost flow problem on an acyclic network. (Ford and Fulkerson [1962])

1.55 A Transshipment Model for Leveling a Road Bed When building a road through mountainous terrain, earth has to be redistributed from high points to low points to produce a relatively level road bed. The engineer must determine the number of truckloads of earth to move between various locations along the proposed road for leveling the route. Thus, high points along the proposed road bed are viewed as sources of earth, while the low points are correspondingly sinks. A terrain graph is an undirected network with nodes on it representing locations at which there are deficits (negative exogenous flow w_i) or surpluses (positive exogenous flow w_i), and edges on it representing the available routes for redistribution of earth, with the cost coefficient $c(e)$ of edge e representing the traversal cost of that edge. $c(e) > 0$ and $\sum w_i = 0$. A leveling plan is a nonnegative flow vector in this network that fulfills the requirements at all the nodes. Formulate the problem of finding a minimum cost levelling plan as a transshipment problem. Construct this model for the road construction situation described in Figure 1.44

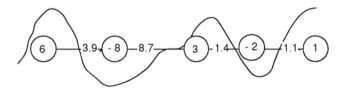

Figure 1.44 Locations on road bed are marked with circles, with the exogenous flow amount at that location entered inside the circle. Lengths of edges are marked on them.

(Farley [1980])

1.56 Machine Loading Problem There are m products to be produced on n machines. Each product can be produced on any machine, but it takes p_{ij} units of machine time and costs $C_{ij}\$$ to produce one unit of product $i = 1$ to m on machine $j = 1$ to n. At most b_j units of machine time is available on machine $j = 1$ to n, and it is required to produce a_i units item $i = 1$ to m during a period. Formulate the problem of finding an optimum production plan as a network flow problem. (Iri, Amari and Takata [1968])

1.57 $G = (\mathcal{N}, \mathcal{A})$ is a given directed network. For each $i \in \mathcal{N}$, we are given nonnegative integers a_i', a_i'' and b_i', b_i'' satisfying $a_i' \overset{\leq}{=} a_i''$ and $b_i' \overset{\leq}{=} b_i''$. It is required to find a subnetwork $\overline{G}$ of G, satisfying the property that for each $i \in \mathcal{N}$ the indegree of i in $\overline{G}$ is between a_i' and a_i'', and the outdegree of i in between b_i' and b_i''. Formulate this as a flow problem.

Suppose we are also given a vector c of arc cost coefficients in G. Define the cost of a subnetwork to be the sum of the cost coefficients of arcs in it. Discuss a formulation for the problem of finding a minimum cost subnetwork of G subject to the in- and outdegree constraints at the nodes described above, as a flow problem.

Note 1.1 The corresponding degree constrained subnetwork problem in undirected networks cannot be reduced to a max-flow problem, but it can be posed as a matching problem. See Exercise 10.19.

1.58 A medical college has 57 students to be assigned to internships at facilities over a period of 3 terms. There are 3 classes of facilities, 1) speciality, 2) rehabilitation, and 3) general/acute. Each student must intern for one term in a facility of each class. Each facility specifies the maximum number of interns it can take each term. Each student is allowed to specify three choices for each class of facility and when (which of three terms) he or she desires to intern there. The problem is to assign the students to the facilities for internship in a manner that maximizes the number of students receiving their preferred facility in the term they asked for it. Develop an efficient approach for solving this problem.

1.59 Airline Fueling Problem   Consider an $n-$leg flight for an airline with S_1 as the origin, S_{n+1} as the destination, and the ith leg consisting of a nonstop flight from S_i to $S_{i+1}, i = 1$ to n. We have the following data: d_i = normal fuel requirement (tons) for ith leg; t_i = fuel capacity (tons) on this leg based on known load; s_i = maximum fuel available (tons) for this plane at S_i; c_i = \$/ton of fuel at S_i; g_i = fuel remaining when S_{i+1} is reached per ton of extra fuel over d_i carried at S_i. Denoting by y_i the tons of fuel purchased at S_i; and by x_i the excess fuel in the plane tanks at the point of take-off from S_i, $i = 1$ to n, formulate the problem of minimizing the cost of fuel for this entire flight, as a generalized network flow problem. (Queyranne [1982])

1.60 Steel Slab Cutting With A Flame Torch   On a rectangular steel slab a cutting pattern is laid out which is a planar multinetwork. It is required to execute this cutting pattern using a flame torch. Cutting always begins at a node. To begin, the torch has to be held at that node for some time until the flame reaches from the top to the bottom (called a blowthrough) and then the torch can be moved easily along any simple path until it is lifted again. The torch cannot travel an edge a second time. If it is necessary to return to a node after passing through it once, it is necessary to make another blowthrough at it. Discuss an algorithm for determining the flame torch route to minimize the number of blowthroughs needed. What is the optimum objective value? (Manber and Israni [1984])

1.61 There are n students in a projects course. There are m available projects, with project i having a capacity of b_i students, $i = 1$ to m. Each student has to work on precisely one project. If there are no students to work on a project, it can be dropped. A total of r supervisors are available. Corresponding to each project a subset of one or more supervisors who can supervise students working on that project is specified. The parameter k_p is the maximum number of students that supervisor p can handle, $p = 1$ to r. Each student specifies a subset of projects arranged in descending order of preference. The objective is to assign students to projects and supervisors so that each student gets to work on a project that has his most preferred ranking, as far as possible. Taking as an objective function the sum of the rankings of the projects the students work on, formulate the problem of doing these assignments as a minimum cost flow problem.

1.62 In a tournament there are n players. Every pair of players play against each other precisely once, and the rules of the game exclude draws. A vector of nonnegative integers $(s_1, \ldots, s_n)$ is called a score vector for this tournament, if s_i, is the number of wins recorded by the ith player in this tournament, for $i = 1$ to n. Given a nonnegative integer vector $b = (b_1, \ldots, b_n)$, it is required to check whether b can be the score vector in such a tournament. Formulate this as the problem of finding a feasible flow vector in a capacitated bipartite network.

1.63 G $=(\mathcal{N}, \mathcal{A}, 0, k, V)$ is a directed connected single commodity flow network with V as the vector of exogenous flow amounts at the nodes. For each $(i, j) \in \mathcal{A}, r_{ij}\$$ is the cost of augmenting the flow capacity on arc (i, j) by 1 unit. It is required to find the minimum budget necessary for arc capacity augmentations in G in order to find a feasible flow vector. Formulate this as a network flow problem.

1.64 Let G$= (\mathcal{N}, \mathcal{A})$ be a directed network. $\mathbf{X}, \mathbf{Y}$ are nonempty disjoint subsets of $\mathcal{N}$ in G. $\mathbf{A} \subset \mathcal{A}$ is said to be an $\mathbf{X}, \mathbf{Y}$- separating arc set in G, if the deletion of the arcs in $\mathbf{A}$ leaves no chain from any node in $\mathbf{X}$ to any node in $\mathbf{Y}$ in the remaining network. Likewise a subset of nodes $\mathbf{Z} \subset \mathcal{N} \backslash (\mathbf{X} \bigcup \mathbf{Y})$ is said to be an $\mathbf{X}, \mathbf{Y}$- separating node set in G, if the deletion of the nodes in $\mathbf{Z}$ together with all the arcs incident at them leaves no chain from any node in $\mathbf{X}$ to any node in $\mathbf{Y}$ in the remaining network. Formulate the problem of finding a minimum cardinality $\mathbf{X}, \mathbf{Y}$- separating arc and node sets in G as network flow problems. Solve these problems for the network in Figure 1.35 with $\mathbf{X}=\{1,2,3\}$, $\mathbf{Y}=\{10,11\}$ using the algorithms discussed in the following chapters.

1.65 There are n modules each of which has to be assigned to one of two available processors. For $j = 1$ to n, c_{j1}, c_{j2} are the costs of executing module j on processors $1, 2$ respectively. In addition to these processor costs, there is a cost of communication between modules assigned to different processors. So, the communication cost between modules i, j is 0 if both i, j are assigned to the same processor, d_{ij} otherwise. Given c_{j1}, c_{j2}, and d_{ij} for all i, j, formulate the problem of assigning the modules to the two available processors so as to minimize the total cost (processor costs + communication costs) as a minimum capacity cut problem in an undirected network. Construct this model for the problem with $n = 4$, and the other data given below, and obtain an optimum solution.

		c_{j1}	c_{j2}	$i = 1$	2	3	4
						d_{ij} for	
$j =$	1	50	20	x	5	20	0
	2	60	30	5	x	34	18
	3	11	15	20	34	x	10
	4	15	10	0	18	10	x

(Dutta, Koehler, and Whinston [1982], Stone [1977])

1.66 Eulerian Trails in Directed Networks   Given a directed network, an
Eulerian trail in it is a circuit that passes through each arc exactly once. A directed
network is said to be Eulerian if there exists an Eulerian trail in it. Prove that
a connected directed network is Eulerian iff for every node in the network, its in-
degree is equal to its out-degree. Develop a version of the cycle tracing and inserting
algorithm of Section 1·3·8, to find an Eulerian trail in an Eulerian directed network.
(Ebert [1988])

1.67 Dairy Model   A co-operative dairy region has n milk processing factories,
with factory j having a capacity of b_j KL (Kilolitre)/day of milk input, $j = 1$ to
n. It has m milk suppliers with supplier i producing a_i KL milk/day, $i = 1$ to m.
Γ_i is the subset of factories to which supplier i can take his milk, $i = 1$ to m. For
$i = 1$ to $m, j \in \Gamma_i$, c_{ij}^1 is the cost ($/KL) of transporting milk from supplier i to
factory j.

$\triangle_j$ is the subset of other factories that factory j can ship oversupply to, or get
supplies from in case of undersupply; $j = 1$ to n, and c_{j_1,j_2}^2 is the cost ($/KL) of
shipping between factories $j_1, j_2 :$ $j_1 = 1$ to $n, j_2 \in \triangle_{j_1}$.

The parameter u_j is the number of different process lines at factory j with
l_{jt}, k_{jt} as the lower and upper bounds (KL/day) on input and c_{jt}^3 as the cost ($/KL
input) of processing at the tth process line $t = 1$ to u_j, $j = 1$ to n. r is the total
number of different dairy products produced in the region, and g_{jtw} is the yield of
product w in product units/KL milk input into process line t as factory j, $j = 1$ to
$n, t = 1$ to $u_j, w = 1$ to r.

It is required to measure the output of each product from each process line in
units of KL input milk, according to the average yield from all process lines at
all factories put together, rather than in product units, so that all flows can be
measured in KL milk units. In terms of these KL milk equivalent units, d_w^1, d_w^2 are
the lower and upper bounds for daily production and v_w is the return ($/unit) of
product $w, w = 1$ to r.

All milk supply has to be processed on the day of its production. Formulate
the problem of determining an optimal allocation plan so as to maximize the net
return as a generalized network flow problem. (Mellalieu and Hall [1983])

1.68 School Assignment   In a region, there are m school districts with a school
in each. For $i = 1$ to m, p_i is the number of excess pupils (if the school there has
inadequate capacity, $p_i = 0$ otherwise), and c_i is the excess school capacity (if it
does, otherwise $c_i = 0$), in district i. The distance matrix (d_{ij}) between school
districts, is given. As far as possible, all pupils will attend the school in their
district. All excess pupils from a district with inadequate capacity have to be
assigned to a district with excess capacity, but all of them to one school only. (i.e.,
there should be no splitting of excess pupils from a district between two or more
schools).

Formulate the problem of assigning the excess pupils to schools subject to the no split-constraint, so as to minimize the total distance travelled by all the students as a 0-1 integer program. Show that when the integer restrictions or the variables are relaxed, this problem can be solved as a generalized network flow problem. If the "no-splitting" constraint can be ignored, show that the problem becomes a straightforward transportation problem. Construct both these models for the problem with the following numerical data: $m = 7, p = (p_i) = (23, 18, 30, 20, 0, 0, 0), c = (c_i) = (0, 0, 0, 0, 45, 50, 30)$, and develop an efficient heuristic method based on network models to obtain a reasonable solution to this problem.

$$d = (d_{ij}) =$$

to j =	5	6	7
from i = 1	14	23	12
2	30	22	10
3	25	14	37
4	20	26	27

(Bovet [1982])

1.69 Operator Scheduling The day is divided into 48 half-hour intervals and a_i is the number of operators required on duty at the telephone company switchboard during the ith half-hour of the day $i = 1$ to 48. Operators work in shifts called tours. Assume that each tour has to be a continuous stretch of 6 or 8 half-hour intervals beginning and ending with any of the half-hours. Let c_{i1}, c_{i2} be the cost of the tour per operator beginning with the ith half-hour, of length 3 or 4 hours respectively, $i = 1$ to 48.

Formulate the problem of determining the number of operators to hire for each possible tour, so as to minimize the cost of meeting the requirements, as a minimum cost flow problem. Construct the model for the following numerical data. Here, the half-hour 1 is 7:00 a.m. to 7:30 a.m.; a_i = 12, 20, 28, 46, 86, 158, 186, 200, 200, 200, 200, 200, 198, 198, 198,192, 186, 184, 178, 158, 128, 114, 92, 102, 124, 118, 117, 116, 104, 104, 104, 104, 108, 50, 50, 16, 4, 4, 4, 4, 4, 4, 4, 4, 4, 4, 4, 4, for $i = 1$ to 48; and c_{i1} = \$17, c_{i2} = \$22 for tours beginning with 3rd to 23rd half-hours, c_{i1} = \$20, c_{i2} = \$26 for tours beginning at 24th to 27th half-hours, c_{i1} = \$25, c_{i2} = \$33 for tours beginning at 28th to 48th half-hours.

Assume now that the total duration of a tour may vary from 2 to 9 hours, and always consists of an integer number of half-hours. Tours of duration 8 to 9 hours should include two half-hour breaks. Tours of duration 5 to $7\frac{1}{2}$ hours should include one half-hour break. Tours of length $4\frac{1}{2}$ hours or less should consist of a continuous stretch. The actual timing of the break periods defines a trick. An ideal trick is one in which the breaks appear between work intervals of equal length. Alternative tricks may be built from the ideal ones by allowing the break periods to be shifted by one half-hour. Associated with each trick is a cost, which is given. Given the demand for operators by half-hours, the problem is to determine the assignment of operators to tricks at minimum cost. Formulate this problem as a linear integer programming problem. Discuss a heuristic approach for solving this problem by solving the network flow model (as in (i) above) for an auxiliary problem where the break periods are ignored temporarily. Then adjust the demands to accommodate break periods in the generated tours, which may create additional demand for operators that needs to be filled again. (Segal [1974])

1.70 Transportation Scheduling   In transportation applications, the more difficult problem is to route or schedule vehicles or aircraft to carry out the shipping, a combinatorial optimization problem. In this application there are eight locations, the travel time between location pairs and number of units to be shipped between them is given below.

(Travel Time in Airflying Minutes) and
Units to be Shipped

to →	1	2	3	4	5	6	7	8	total units leaving
from									
1	x	(52)	(71)	(75)	(70)	(115)	(115)	(56)	
		19052	8244	4209	11970	0	0	0	43475
2	(49)	x	(38)	(45)	(56)	(80)	(80)	(85)	
	25729	x	3637	1871	0	0	0	0	31237
3	(68)	(38)	x	(31)	(41)	(43)	(45)	(77)	
	10044	6703	x	5456	0	5021	9264	0	36488
4	(72)	(45)	(31)	x	(70)	(40)	(37)	(90)	
	2641	2234	667	x	0	6807	3836	0	16185
5	(70)	(56)	(41)	(70)	x	(85)	(60)	(36)	
	0	0	0	0	x	0	0	7860	7860
6	(115)	(80)	(43)	(40)	(85)	x	(24)	(120)	
	229	672	3483	2494	0	x	0	0	6878
7	(115)	(80)	(45)	(37)	(60)	(24)	x	(95)	
	1581	0	7908	0	0	0	x	0	9489
8	(56)	(85)	(77)	(90)	(36)	(120)	(95)	x	
	8010	0	0	0	0	0	0	x	8010
total units arriving	48234	28661	23939	14030	11970	11828	13100	7860	

Each aircraft has a capacity of 9,000 units. All shipping has to be completed in one night, starting at 10:00 p.m. and finishing at 5:00 a.m. Each aircraft used may start at any location at 10:00 p.m., and can finish at any location. The stoptime between arrival and departure at each location is a fixed 20 minutes independent of the quantities to be loaded or unloaded or both.

(i) A total of 43,475 units have to be shipped out of location 1. Since the capacity of a plane is 9,000 units, this implies that the number of departures from location 1 has to be $\geqq 5$. Similarly, lower bounds on the total number of departures from each location, and the total number of arrivals at each location, can be obtained. Treating these as the right-hand side constants and the travel times as the costs, formulate the problem of determining a lower bound on the total flying minutes needed in this problem (and hence a lower bound for the number of planes needed for the task) as a classical transportation problem.

(ii) Discuss an approach (based on heuristics or integer programming formulations) of taking the optimum solution of the transportation model in (i); and converting it into actual routes for the planes, to complete the shipping task subject to the constraints stated above, using the smallest number of planes.

(Wolters [1979])

1.71 Vehicle Scheduling

(i) A central office (CO) has to make u round trips. Trip r begins at clock time t_r at the CO and returns back at clock time $T_r = t_r + a_r$, where a_r is the time duration needed to complete trip r and return to the CO, $r = 1$ to u. The vehicle assigned to make trip r will therefore be available for reassignment to another trip after clock time T_r, if necessary. It is required to find the minimum number of vehicles needed to carry out all these trips. Formulate this as a minimum cost network flow problem.

(ii) Consider a generalization of the above problem in which there are l types of vehicles (small, medium, large, etc.). $V_r = \{i : \text{type } i \text{ vehicle is capable of making trip } r\}, r = 1$ to u. $a_{ir}, \text{and} c_{ir}$ are the time duration needed for a type i vehicle if it makes the rth trip, and the corresponding cost, $i \in V_r, r = 1$ to u. It is required to find an assignment of vehicles to trips, so as to minimize the total cost of making all the trips. Discuss how the formulation in (i) can be extended into a "modified" network flow problem, to provide a practical approach for solving this problem.

(Dantzig and Fulkerson [1954], Diez-Canedo and Escalante [1977])

1.72 Matrix with the Consecutive 1's Property   Let A be a 0-1 matrix with the property that in each column all the 1's are contiguous. Prove that A is a totally unimodular matrix.

1.73 Let A be an $m \times n$ integer matrix and $b \in \mathbb{R}^n$. Let $\mathbf{P}(b) = \{x : Ax \overset{\leq}{=} b, x \overset{\geq}{=} 0\}$. Prove that the following are equivalent:(i) A is totally unimodular,(ii) for each integer vector b and integer $r \overset{\geq}{=} 1$, every integer vector in $\mathbf{P}(rb)$ can be expressed on the sum of r integer vectors in $\mathbf{P}(b)$.
(Baum and Trotter [1978])

1.74 Let matrix D have rank r.

(i)   If D is a unimodular matrix, prove that every extreme point of

$$Dx = d$$

$$g \geq x \geq 0$$

is an integer vector whenever d, g are integer vectors of appropriate dimension (some or all of the components of g could be $+\infty$).

(ii)   Let A be a matrix of order $m \times n$, and $\mathbf{Q}(b) = \{x : Ax = b, x \geq 0\}$. Prove that A is unimodular iff for any integer vector b, integer $h \geq 1$, and integer vector $\bar{x} \in \mathbf{Q}(hb)$, there exist integer vectors $\bar{x}^t \in \mathbf{Q}(b)$ for $t = 1$ to h such that $\bar{x} = \sum_{t=1}^{h} \bar{x}^t$.

(iii) Let b be an integer vector. If Γ is a subset of $\mathbb{R}^n$, a point $\hat{x} \in \Gamma$ is said to be a *minimal vector in* Γ if there does not exist an $x \in \Gamma$ satisfying $x \leq \hat{x}$. Let $D(b)$ be the matrix whose row vectors constitute the set of all minimal vectors among the set of integral vectors in $\mathbf{Q}(b)$. Consider the following problems.

(1.24)			(1.25)	
maximize	$\pi \mathbf{e}$		maximize	$\pi \mathbf{e}$
subject to	$\pi D(b) \leq w$		subject to	$\pi D(b) \leq w$
	$\pi \geq 0$			$\pi \geq 0$, and integral

where $\mathbf{e}$ is the vector of all 1's. Prove that for every integer $r \geq 1$ and integer vector $\bar{x} \in \mathbf{Q}(rb)$, there exist integer vectors $\bar{x}^t \in \mathbf{Q}(b)$ for $t = 1$ to r, such that $\bar{x} = \sum_{t=1}^{r} \bar{x}^t$ iff for every integer vector w; the optimum objective values α, β in (1.24), (1.25) satisfy $\beta = \lfloor \alpha \rfloor$.
(Baum and Trotter [1977])

1.75 T is a rooted spanning tree in $\mathrm{G} = (\mathcal{N}, \mathcal{A})$. We are given the following: e is an in-tree arc, i, j, are the parent and son nodes on e. $\mathbf{X} = \mathbf{H}(\mathsf{T}, j)$ is the family of node j in T. $\bar{\mathbf{X}} = \mathcal{N} \backslash \mathbf{X}$. Give rigorous proofs of the following.

(i)   For each $p \in \mathbf{X}$, the predecessor path of p must pass through node j and actually must include e.

(ii)   For each $v \in \bar{\mathbf{X}}$, the predecessor path of v does not pass through node j.

(iii) For any out-of-tree arc in the cut $[\mathbf{X}, \bar{\mathbf{X}}]$, its fundamental cycle with respect to T must include e.

(iv)   For any out-of-tree arc both of whose nodes are either in $\mathbf{X}$ or in $\bar{\mathbf{X}}$, its fundamental cycle with respect to T does not include e.

1.76 Let A be a 0-1 matrix. It is said to have unique subsequence (or precedence) if its rows can be ordered so that all columns with a one in a particular row have their subsequent, that is next or precedent (i.e., previous) one, if it exists, in a unique common row. Notice that in a matrix with unique subsequence or precedence property, there could be more than two ones in each column. The following is a 0-1 matrix with unique subsequence property with the rows in natural order. Prove that a 0-1 matrix A with unique subsequence or precedence property is totally unimodular. (Ryan and Falkner [1988])

$$\begin{pmatrix} 1\ 1\ 1\ 1\ 0\ 0\ 0\ 0\ 0\ 0\ 0\ 0\ 0\ 0 \\ 0\ 1\ 1\ 1\ 1\ 1\ 1\ 0\ 0\ 0\ 0\ 0\ 0\ 0 \\ 0\ 0\ 0\ 0\ 0\ 0\ 0\ 1\ 1\ 0\ 0\ 0\ 0\ 0 \\ 0\ 0\ 1\ 1\ 0\ 1\ 1\ 0\ 0\ 1\ 1\ 0\ 0\ 0 \\ 0\ 0\ 0\ 0\ 0\ 0\ 0\ 0\ 0\ 0\ 0\ 1\ 1\ 0 \\ 0\ 0\ 0\ 1\ 0\ 0\ 1\ 0\ 1\ 0\ 1\ 0\ 1\ 1 \end{pmatrix}$$

1.77 Let G$= (\mathcal{N}, \mathcal{A}, \check{s}, \check{t})$ be a connected directed network. Suppose we are given a cut separating $\check{s}$ and $\check{t}$ in G as a subset of arcs. Discuss a procedure for identifying a subset of nodes $\mathbf{X} \subset \mathcal{N}$ such that this cut is $[\mathbf{X}, \bar{\mathbf{X}}]$. Is the choice of $\mathbf{X}$ unique? Also, prove that every cut separating $\check{s}$ and $\check{t}$ in G is an arc disjoint union of cutsets separating $\check{s}$ and $\check{t}$.

1.78 Let G$= (\mathcal{N}, \mathcal{A}, 0, k, V)$ be a connected directed single commodity flow network with $V = (V_1, V_2, \ldots, V_{r+1}, 0, \ldots, 0)^T$ as the vector of exogenous flow values at the nodes, where $V_1 > 0$, and V_2 to V_{r+1} are all < 0. Let $|\mathcal{N}| = n, |\mathcal{A}| = m$; and let $\mathbf{K}$ denote the set of all feasible node-arc flow vectors in G.

Prove that a flow vector $\bar{f} \in \mathbf{K}$ is an extreme point of $\mathbf{K}$ iff the set of arcs $A(\bar{f}) = \{(i,j) : (i,j \in \mathcal{A} \text{ and } 0 < \bar{f}_{ij} < k_{ij}\}$ constitutes a forest in G. If $\bar{f}$ is an extreme point of $\mathbf{K}$, prove that it is a nondegenerate extreme point if the set of arcs $A(\bar{f})$ constitutes a spanning tree in G, degenerate extreme point otherwise.

Let $\bar{f} = (\bar{f}_{ij})$ be an extreme point of $\mathbf{K}$, and $\mathbb{C}$ a simple cycle in G with an orientation. Define $\varepsilon^+(\mathbb{C}, \bar{f}), = \min\{k_{ij} - \bar{f}_{ij} : (i,j) \text{ is a forward arc on } \mathbb{C}\}, \varepsilon^-(\mathbb{C}, \bar{f}) = \min\{\bar{f}_{ij} : (i,j) \text{ a reverse arc on } \mathbb{C}\}, \varepsilon(\mathbb{C}, \bar{f}) = \min\{\varepsilon^+(\mathbb{C}, \bar{f}), \varepsilon^-(\mathbb{C}, \bar{f})\}$. Let $\mu(\mathbb{C}) = (\mu_{ij}(\mathbb{C}))$ be the incidence vector of $\mathbb{C}$ given by $\mu_{ij}(\mathbb{C}) = 0$, if (i,j) is not on $\mathbb{C}, +1$, if (i,j) is a forward arc on $\mathbb{C}$, and -1, if (i,j) is a reverse arc on $\mathbb{C}$. If $0 < \varepsilon(\mathbb{C}, \bar{f}) < \infty$, define the new flow vector $f' = \bar{f} + \varepsilon(\mathbb{C}, \bar{f})\mu(\mathbb{C})$. Prove that f' is an adjacent extreme point of $\bar{f}$ on $\mathbf{K}$ iff the following condition 1 holds.

Condition 1: The only cycle in the set of arcs in $A(\bar{f})$ and $\mathbb{C}$ put together, is either $\mathbb{C}$, or $\mathbb{C}$ with its orientation reversed.

Conversely, every adjacent extreme point of $\bar{f}$ on $\mathbf{K}$ is obtained as $\bar{f} + \varepsilon(\mathbb{C}, \bar{f})\mu(\mathbb{C})$ for some simple cycle $\mathbb{C}$ in G satisfying condition 1 and $0 < \varepsilon(\mathbb{C}, \bar{f}) < \infty$.

If $\mathbb{C}$ is a simple cycle satisfying condition 1 and $\varepsilon(\mathbb{C}, \bar{f}) = \infty$, show that $\{\bar{f} + \lambda\mu(\mathbb{C}) : \lambda \geqq 0\}$ is an extreme half-line of $\mathbf{K}$ through $\bar{f}$, and conversely every extreme half-line of $\mathbf{K}$ through $\bar{f}$ is obtained in this manner from some simple cycle $\mathbb{C}$ in G satisfying condition 1 and $\varepsilon(\mathbb{C}, \bar{f}) = +\infty$.

If $k = \infty$, prove that $\bar{f} \in \mathbf{K}$ is an extreme point of $\mathbf{K}$ iff $A(\bar{f})$ is a tree with the source node 1 as the root, and the sink nodes 2 to $r + 1$ as terminal nodes.

Let $k = \infty$ and $\bar{f}, f'$ be two extreme points of $\mathbf{K}$. Prove that $\bar{f}, f'$ are adjacent iff the arcs in $A(f')\backslash A(\bar{f})$ constitute a path which connects two nodes in $\mathbf{N}$ and does not contain any other nodes of $\mathbf{N}$, where $\mathbf{N}$ is the set of nodes on arcs in $A(\bar{f})$. (Gallo and Sodini [1979])

1.79 Let f be a feasible flow vector of value $\bar{v}$ in the directed single commodity flow network G= $(\mathcal{N}, \mathcal{A}, 0, k, \breve{s}, \breve{t}, \bar{v})$ with $k > 0$ and finite. Prove that f is an extreme flow (i.e., a basic feasible flow vector) iff in every simple cycle $\mathbb{C}$ satisfying the property that all the arcs in it carry a positive flow amount in f, there is at least one saturated arc wrt f.

1.80 Prove that a directed network G= $(\mathcal{N}, \mathcal{A})$ is strongly connected iff for every $\emptyset \neq \mathbf{X} \subset \mathcal{N}, \overline{\mathbf{X}} = \mathcal{N}\backslash\mathbf{X}$ with $\overline{\mathbf{X}} \neq \emptyset$, there exists an arc (i, j) in $\mathcal{A}$ with $i \in \mathbf{X}$ and $j \in \overline{\mathbf{X}}$.

1.81 Let G= $(\mathcal{N}, \mathcal{A})$ be a directed connected network. Develop an O($|\mathcal{A}|$) algorithm for finding all strongly connected components of G. (Tarjan [1972])

1.82 Consider a finite Markov chain with transition probability matrix P. It is required to identify all the transient states and classify the remaining states into the various closed recurrent classes. Formulate this problem as one of identifying all the strongly connected components in a directed network, and develop an efficient algorithm for it.

Comment 1.2   The first paper using a network or graph model is that of Euler [1736] on the Königsberg bridges problem. The pioneering book on network flows is that of Ford and Fulkerson [1962], it played a significant role in stimulating research and finding applications for network flow models in many areas. Other books devoted to network flows are Adel'son-Vel'ski, Dinic and Karzanov [1975], Busacker and Saaty [1965], Christofides [1975], Deo [1974], Derigs [1988], Even [1979], Gondran and Minoux [1984], Hu [1969], Iri [1969], Jensen and Barnes [1980], Kennington and Helgasson[1980], Lawler [1976], Minieka [1978], Papadimitriou and Steiglitz [1982], Rockafellar [1984], Swamy and Thulasiraman [1981], and Tarjan [1983]. The book by Bodin, Golden, Assad and Ball [1983] deals with applications in routing. The book by Burkard and Derigs [1980] provides Fortran programs for the special class of matching and assignment problems.

There are many texts in the related areas of graph theory and its applications. Among them we list Berge [1962], Bondy and Murthy [1976], Mayeda [1972], and Wilson [1972]. The book by Lovasz and Plummer [1986] specializes in matching theory; it contains a nice section describing the history of graph theory and network flow theory.

Most of the network problems that we discuss in this book are special cases of linear programming problems, and some of the algorithms discussed are specializations of variants of the simplex method of linear programming. Among the many books on linear programming, we list Bazaraa and Jarvis [1977], Chvatal [1983], Dantzig [1963], Gale [1960], and Murty [1983].

The references listed in this chapter are classified into two parts, books and research publications. The research publications in the second part deal with network models for problems in a variety of areas, algorithms for computing Euler trails, and the chinese postman problem. Several of the exercises given above are taken from these publications.

1.5   References

Books in Network flows and related areas

G. M. ADEL'SON-VEL'SKI, E. A. DINIC, and A. V. KARZANOV, 1975, *Flow Algorithms* (in Russian), Science, Moscow.

A. V. AHO, J. E. HOPCROFT and J. D. ULLMAN, 1974, *The Design and Analysis of Computer Algorithms*, Addison-Wesley, Reading, MA.

A. BACHEM, M. GROTSCHEL and B. KORTE (Eds.), 1982, *Bonn Workshop on Combinatorial Optimization*, North-Holland, Amsterdam.

M. S. BAZARAA and J. J. JARVIS, 1977, *Linear Programming and Network Flows*, Wiley, NY.

C. BERGE, 1962, *The Theory of Graphs*, Methuen, London.

L. BODIN, B. GOLDEN, A. ASSAD, and M. BALL, 1983, *Routing and Scheduling of Vehicles and Crews: State of the Art*, Special issue of *COR*, 10, no. 2, Pergamon Press, NY.

F. BOESCH, 1976, *Large Scale Networks: Theory and Design*, IEEE Press, NY.

J. A. BONDY and U. S. R. MURTHY, 1976, *Graph Theory with Applications*, American Elsevier, NY.

R. E. BURKARD and U. DERIGS, 1980, *Assignment and Matching Problems: Solution Methods with Fortran Programs*, Springer-Verlag, NY.

R. G. BUSACKER and T. L. SAATY, 1965, *Finite Graphs and Networks*, McGraw-Hill, NY.

N. CHRISTOFIDES, 1975, *Graph Theory; an Algorithmic Approach*, Academic Press, NY.

V. CHVATAL, 1983, *Linear Programming*, W. H. Freeman & Co., NY.

G. B. DANTZIG, 1963, *Linear Programming and Extensions*, Princeton University Press, Princeton, NJ.

N. DEO, 1974, *Graph Theory with Applications to Engineering and Computer Science*, Prentice-Hall, Englewood Cliffs, NJ.

U. DERIGS, 1988, *Programming in Networks and Graphs*, Lecture notes in Economics and Mathematical Systems 300, Springer -Verlag, NY.

S. E. ELMAGHRABY, 1970, *Some Network Models in Management Science*, Springer-Verlag, NY.

S. EVEN, 1979, *Graph Algorithms*, Computer Science press, Potomac, MD.

L. R. FORD and D. R. FULKERSON, 1962, *Flows in Networks*, Princeton University Press, Princeton, NJ.

D. GALE, 1960, *The Theory of Linear Economic Models*, McGraw-Hill, NY.

G. GALLO and C. SANDI (Eds.), 1986, *Netflow at Pisa, MPS*, 26.

M. R. GAREY and D. S. JOHNSON, 1980, *Computers and Intractability: A Guide to the Theory of NP-Completeness*, W. H. Freeman & Co., NY, 2nd printing.

M. GONDRAN and M. MINOUX, 1984, *Graphs and Algorithms*, Wiley-Interscience, NY.

E. HOROWITZ and S. J. SAHNI, 1978, *Fundamentals of Computer Algorithms*, Computer Science Press, Rockville, MD.

T. C. HU, 1969, *Integer Programming and Network Flows*, Addison-Wesley, Reading, MA.

M. IRI, 1969, *Network Flow, Transportation and Scheduling*, Academic Press, NY.

P. A. JENSEN and J. W. BARNES, 1980, *Network Flow Programming*, Wiley, NY.

J. KENNINGTON and R. HELGASON, 1980, *Algorithms for Network Programming*, Wiley, NY.

D. KLINGMAN and J. M. MULVEY, (Eds.), 1981, *Network Models and Associated Applications*, *MPS*,15.

E. L. LAWLER, 1976, *Combinatorial Optimization: Networks and Matroids*, Holt, Rinehart, and Winston, NY.

L. LOVASZ and M. D. PLUMMER, 1980, *Matching Theory*, North-Holland, Amsterdam.

W. MAYEDA, 1972, *Graph Theory*, Wiley-Interscience, NY.

E. MINIEKA, 1978, *Optimization Algorithms for Networks and Graphs*, Marcel Dekker, NY.

K. G. MURTY, 1976, *Linear and Combinatorial Programming*, Krieger, Malabar, FL.

K. G. MURTY, 1983, *Linear Programming*, Wiley, NY.

C. H. PAPADIMITRIOU and K. STEIGLITZ, 1982, *Combinatorial Optimization: Algorithms and Complexity*, Prentice-Hall, Englewood Cliffs, NJ.

R. T. ROCKAFELLAR, 1984, *Network Flows and Monotropic Optimization*, Wiley-Interscience, NY.

M. N. S. SWAMY and K. THULASIRAMAN, 1981, *Graphs, Networks, and Algorithms*, Wiley-Interscience, NY.

R. E. TARJAN, 1983, *Data Structures and Network Algorithms*, CBMS-NSF Regional Conference Series in Applied Math. SIAM, 44.

R. J. WILSON, 1972, *Introduction to Graph Theory*, Oliver and Boyd, Edinburgh.

Other References

B. M. BAKER, Sept. 1984, "A Network Flow Algorithm for Project Selection," *JORS*, 35, no. 9 (847-852).

K. R. BAKER, 1976, " Work Force Allocation in Cyclical Scheduling Problems: A Survey," *ORQ*, 27, no. 1,ii (155-167).

M. L. BALINSKI, 1970, "On a Selection Problem," *MS*, 17(230-231).

S. BAUM and L. E. TROTTER, Jr., 1978, "Integer Rounding and Polyhedral Decomposition for Totally Unimodular Systems," (15-23) in R. Henn, B. Korte, and W. Oettli (Eds.), *Arbeitstagung-über Operations Research und Optimierung*, Springer-Verlag, Berlin.

A. A. BERTOSSI, P. CARRARESI, and G. GALLO, 1987, "On Some Matching Problems Arising in Vehicle Scheduling Models," *Networks*, 17, no. 3(271-281).

J. BOVET, Aug. 1982, "Simple Heuristics for the School Assignment Problem," *JORS*, 33, no. 8(695-703).

R. CHANDRASEKARAN and S. SUBBA RAO, May-June 1977, "A Special Case of The Transportation Problem," *OR*, 25, no. 3(525-528).

M. CHESHIRE, K. I. M. McKINNON, and H. P. WILLIAMS, Aug. 1984, " The Efficient Allocation of Private Contractors to Public Works," *JORS*, 35, no. 8(705-709).

G. B. DANTZIG and D. R. FULKERSON, 1954, "Minimizing the Number of Tankers to Meet a Fixed Schedule," *NRLQ*, 1(217-222).

M. D. DIVINE and W. G. LESSO, April 1972, "Models for the Minimum Cost Development of Oil Fields," *MS*, 18, no. 8(B-378-387).

J. M. DIEZ-CANEDO and O. M-M. ESCALANTE, 1977, "A Network Solution to a General Vehicle Scheduling Problem," *EJOR*, 1(255-261).

R. C. DORSEY, T. J. HODGSON, and H. D. RATLIFF, 1975, "A Network Approach to a Multi-facility Multi-product Production Scheduling Problem Without Back Ordering," *MS*, 21(813-822).

A. DUTTA, G. KOEHLER, and A. WHINSTON, Aug.1982, "On Optimal Allocation in a Distributed Processing Environment," *MS*, 28, no. 8(839-853).

J. EBERT, June 1988, "Computing Eulerian Trails," *IPL*, 28, no. 2(93-97).

J. EDMONDS and E. JOHNSON, 1973, "Matching, Euler Tours and the Chinese Postman's Problem," *MP*, 5(88-124).

L. EULER, 1736, "Solutio Problematis ad Geometriam Situs Pertinentis," *Commun. Acad. Sci. Imp. Petropol.*, 8(128-140): Opera Omnia(1), Vol. 7.

A. M. FARLEY, July 1980, "Levelling Terrain Trees: A Transshipment Problem," *IPL*, 10, nos. 4/5(189-192).

C. O. FONG, Oct. 1980, "Planning for Industrial Estate Development in a Developing Economy," *MS*, 26, no. 10(1061-1067).

A. GALLO and C. SODINI, 1979, "Adjacent Extreme Flows and Application to Minimum Concave Cost Flow Problems," *Networks*, 9(95-121).

B. GAVISH and P. SCHWEITZER, 1974, "An Algorithm for Combining Truck Trips," *TS*, 8(13-23).

B. GAVISH, P. SCHWEITZER and E. SHLIFER, 1978, "Assigning Buses to Schedules in a Metropolitan Area," *COR*, 5(129-138).

P. R. HALMOS and H. E. VAUGHAN,1950, "The Marriage Problem," *American Journal of Mathematics*, 72(214-215).

I. HELLER and C. B. TOMPKINS, 1958, "Integral Boundary Points of Convex Polyhedra," (247-254) in H. W. Kuhn and A. W. Tucker (Eds.), *Linear Inequalities and Related Systems*, Princeton University Press, Princeton, NJ.

A. J. HOFFMAN and J. B. KRUSKAL, 1958, "Integral Boundary Points of Convex Polyhedra," (223-246)in H. W. Kuhn and A. W. Tucker (Eds.), *Linear Inequalities and Related Systems*, Princeton University Press, Princeton, NJ.

M. IRI, S. AMARI, and M. TAKATA, 1968, "Algebraical and Topological Theory and Methods in Linear Programming with Weak Graphical Representation," (421-464) in K. Kondo (Ed.), *RAAG Memoirs of the Unifying Study of Basic Problems in Engineering and Physical Sciences by Means of Geometry*, 4, G-iX, Gakujutsu Bunken Fukyukai, Tokyo.

W. JACOBS, 1954, "The Caterer Problem," *NRLQ*, 1(154-165).

R. R. LOVE, Jr., and R. R. VEMUGANTI, Jan.-Feb. 1978, "The Single Plant Mold Allocation Problem with Capacity and Changeover Restrictions," *OR*, 26, no. 1(159-165).

T. L. MAGNANTI and R. T. WONG, 1984, "Network Design and Transportation Planning: Models and Algorithms," *TS*, 18(1-56).

U. MANBER and S. ISRANI, 1984, "Pierce Point Minimization and Optimal Torch Path Determination in Flame Cutting," *Journal of Manufacturing Systems*, 3, no. 1(81-89).

P. G. MARLIN, 1981, "Application of the Transportation Model to a Large Scale Districting Problem," *COR*, 8, no. 2(83-96).

P. J. MELLALIEU and K. R. HALL, June 1983, "An Interactive Planning Model for the New Zealand Dairy Industry," *JORS*, 34, no. 6(521-532).

E. MINIEKA, July 1979, "The Chinese Postman Problem for Mixed Networks," *MS*, 25 no. 7(643-648).

J. PINTO PAIXÃO and I. M. BRANCO, 1987, "A Quasi-Assignment Algorithm for Bus Scheduling," *Networks*, 17, no. 3(249-269).

M. QUEYRANNE, 1982, "The Tankering Problem," CBA Working Paper Series, University of Houston, Houston.

D. M. RYAN and J. C. FALKNER, June 1988, "On the Integer Properties of Scheduling Set Partitioning Models," *EJOR*, 35, no. 3(442-456).

J. RHYS, Nov. 1970, "A Selection Problem of Shared Fixed Costs and Network Flows," *MS*, 17, no. 3(200-207).

M. SEGAL, July-Aug. 1974, "The Operator Scheduling Problem: A Network Flow Approach," *OR*, 22, no. 4(808-823).

V. SRINIVASAN, Jan.1979, "Network Models for Estimating Brand-Specific Effects in Multi-Attribute Marketing Models," *MS*, 25, no. 1(11-21).

H. S. STONE, Jan. 1977, "Multiprocessor Scheduling with the Aid of Network Flow Algorithms," *IEEE Transactions on Software Engineering*, SE-3(85-93).

W. SZWARC and M. E. POSNER, Nov.-Dec. 1985, "The Caterer Problem," *OR*, 33, no. 6(1215-1224).

F. B. TALBOT and J. H. PATTERSON, July 1978, " An Efficient Integer Programming Algorithm with Network Cuts for Solving Resource Constrained Scheduling Problems," *MS*, 24, no. 11(1163-1174).

R. E. TARJAN, 1972, "Depth-First Search and Linear Graph Algorithms," *SIAM Journal of Computing*, 1(146-160).

M. TSO, 1986, "Network Flow Models in Image Processing," *JORS*, 37, no. 1(31-34).

A. F. VEINOTT, Jr., and H. M. WAGNER, 1962, "Optimal Capacity Scheduling- I and II," *OR*, 10, no. 4(518-546).

H. P. WILLIAMS, 1982, "Models with Network Duals," *JORS*, 33(161-169).

J. A. M. WOLTERS, 1979, "Minimizing the Number of Aircraft for a Transportation Network," *EJOR*, 3(394-402).

Chapter 2

Single Commodity Maximum Value Flow Problems in Pure Networks

The problem of finding a maximum value flow in the directed single commodity flow network $G = (\mathcal{N}, \mathcal{A}, \ell, k, \check{s}, \check{t})$ can be solved by several methods, one of which is the bounded variable simplex method. In this chapter we discuss efficient network algorithms for this problem, the *augmenting path methods*, and the recently developed preflow-push algorithms.

If $\ell = 0$, one may be tempted to believe that this problem can be solved by the following simple scheme. Since $\ell = 0$, one can begin with the feasible flow vector $f^0 = 0$. Find a chain from $\check{s}$ to $\check{t}$ consisting of unsaturated arcs only, and increase the flow on each arc on it by the residual capacity of this chain; repeat this process until a stage is reached where there is no chain from $\check{s}$ to $\check{t}$ consisting only of unsaturated arcs. At that stage, terminate the process. We now have a feasible flow vector $\tilde{f}$ wrt which there exists no FAC from $\check{s}$ to $\check{t}$. This $\tilde{f}$ is a maximal or blocking feasible flow vector, but unfortunately, it may not be of maximum value, as illustrated in Figure 1.30 in Example 1.4. In this scheme, the flow on each arc either stayed the same or increased, but never decreased. To reach higher value flows, we may have to decrease the flow on some of the arcs. This leads to the possibility of flow augmentation using FAPs rather than FACs. For example, there is no FAC from 1 to 4 in the network on the right of Figure 1.30. However the path 1, (1, 3), 3, (2, 3), 2, (2, 4), 4 containing the reverse arc (2, 3) is an FAP, and augmentation using it leads to an increase in flow value by 1 unit. The first class of methods that we will discuss are based on flow augmentation using FAPs. These methods are also called *labeling algorithms* or *label tree methods*, since they grow a tree rooted at $\check{s}$ in each step to look for an FAP, and the tree itself is stored using predecessor labels on the nodes.

2.1   Simple Transformations

Supersource (Supersink) to Replace Several Source (Sink) Nodes

If there are $r(> 1)$ source nodes in the problem, i with a_i units available, for $i = 1$ to r, then introduce a new supersource $\breve{s}$ with unlimited availability and new arcs $(\breve{s}, i)$ with lower bound 0 and capacity $a_i, for\, i = 1$ to r. See Figure 2.1.

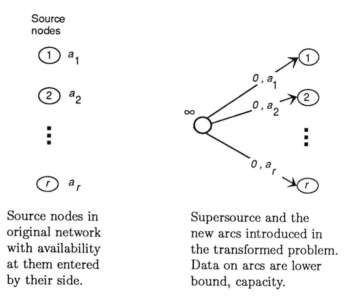

Source nodes in
original network
with availability
at them entered
by their side.

Supersource and the
new arcs introduced in
the transformed problem.
Data on arcs are lower
bound, capacity.

Figure 2.1

If there are $p(> 1)$ sink nodes, with sink node $n-j$ having a requirement of b_{n-j} units, for $j = p - 1$ to 0, introduce a new supersink $\breve{t}$ with unlimited requirement and new arcs $(n - j, \breve{t})$ for $j = p-1$ to 0. If b_{n-j} is a minimal requirement at node $n - j$, the lower bound and capacity on the arc $(n - j, \breve{t})$ can be set at b_{n-j} and ∞ respectively. If it is required to supply exactly b_{n-j} units to node $n - j$, one of the following strategies can be used: (i) Set both lower bound and capacity on the arc $(n - j, \breve{t})$ equal to b_{n-j}, or (ii) set lower bound $= 0$, capacity $= b_{n-j}$ on arc $(n - j, \breve{t})$, and look for a flow vector in which this arc is saturated.

Transformation of Node Capacities

If node i has a node transit capacity of c_i in the original network, replace it by an arc (i_1, i_2) with an arc flow capacity of c_i. Nodes i_1, i_2 represent the *receiving and departing ends* of i. See Figure 2.2.

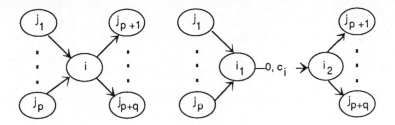

Figure 2.2 On the left is node i with transit capacity c_i in the original network. On right is the corresponding portion in the transformed network in which i is replaced by arc (i_1, i_2) with lower bound, capacity entered on it.

Combining Parallel Arcs Into A Single Arc

If there are r parallel arcs joining i to j with lower bounds $\ell_1, \ldots, \ell_r$ and capacities $k_1, \ldots, k_r$ respectively, replace them by a single arc (i, j) with lower bound $\ell_1 + \ldots + \ell_r$ and capacity $k_1 + \ldots + k_r$. In the sequel we will assume that there is at most one arc from a node to any other node.

2.2 Some Results

The conditions for the feasibility of a node-arc flow vector $f = (f_{ij})$ in the single commodity flow network $G = (\mathcal{N}, \mathcal{A}, \ell, k, \check{s}, \check{t})$ are:

$$f(i, \mathcal{N}) - f(\mathcal{N}, i) = \begin{cases} 0 & \text{if } i \neq \check{s}, \check{t} \\ v & \text{if } i = \check{s} \\ -v & \text{if } i = \check{t} \end{cases} \qquad (2.1)$$

$$\ell \overset{\leq}{=} f \overset{\leq}{=} k \qquad (2.2)$$

THEOREM 2.1 *Let $[X, \bar{X}]$ be a cut separating $\check{s}$ and $\check{t}$, and let f be a feasible flow vector in G. The quantity $f(X, \bar{X}) - f(\bar{X}, X)$, called the net flow across the cut $[X, \bar{X}]$ in f, is equal to v.*

Proof The result follows by adding the conservation equations in (2.1) over $i \in X$. ∎

THEOREM 2.2 *If a feasible flow vector exists in G, the maximum flow value is $\overset{\leq}{=}$ the capacity of any cut separating $\check{s}$ and $\check{t}$. Also, if $\hat{f}$ is a feasible flow vector of value $\hat{v}$, $[Y, \bar{Y}]$ is a cut separating $\check{s}$ and $\check{t}$ in G, and $\hat{v} =$ the capacity of the cut $[Y, \bar{Y}]$, then $\hat{f}$ is a maximum value feasible flow vector and $[Y, \bar{Y}]$ is a minimum capacity cut separating $\check{s}$ and $\check{t}$ in G.*

Proof Let f be any feasible flow vector of value v, and let $[\mathbf{X}, \bar{\mathbf{X}}]$ be any cut separating $\check{s}$ and $\check{t}$ in G. By Theorem 2.1 we have $v = f(\mathbf{X}, \bar{\mathbf{X}}) - f(\bar{\mathbf{X}}, \mathbf{X}) \leqq f(\mathbf{X}, \bar{\mathbf{X}}) - \ell(\bar{\mathbf{X}}, \mathbf{X})$ (since $f \geqq \ell$) $\leqq k(\mathbf{X}, \bar{\mathbf{X}}) - \ell(\bar{\mathbf{X}}, \mathbf{X})$ (since $f \leqq k$) = capacity of the cut $[\mathbf{X}, \bar{\mathbf{X}}]$. The second result now follows directly. ∎

THEOREM 2.3 *A feasible flow vector in G has maximum value iff there exists no FAP from $\check{s}$ to $\check{t}$ wrt it.*

Proof Let f be a feasible flow vector in G. If there exists an FAP from $\check{s}$ to $\check{t}$ wrt f, we can get a flow vector of higher value by augmentation, therefore f does not have maximum value.

Suppose there exists no FAP from $\check{s}$ to $\check{t}$ wrt f. We will now show that f has maximum value. Define $\mathbf{X} = \{i : i \in \mathcal{N},$ either $i = \check{s}$, or there exists an FAP from $\check{s}$ to i wrt $f\}$, $\bar{\mathbf{X}} = \mathcal{N} \backslash \mathbf{X}$.

If $\mathcal{P}$ is an FAP from $\check{s}$ to i wrt f (define $\mathcal{P}$ to be the empty path if $i = \check{s}$), and $j \in \mathcal{N}$ is such that

$$\text{either} \quad (i) \;\; (i, j) \in \mathcal{A} \text{ and } f_{ij} < k_{ij}, \quad \text{or} \;\; (ii) \;\; (j, i) \in \mathcal{A} \text{ and } f_{ji} > \ell_{ji}$$

then by including (i, j) under case (i) or (j, i) under case (ii) at the end of $\mathcal{P}$, we extend it to j. This implies that if $i \in \mathbf{X}$ and j satisfies (i) or (ii), then $j \in \mathbf{X}$ also. Hence, for $(i, j) \in \mathcal{A}$, we have $f_{ij} = k_{ij}$ if $i \in \mathbf{X}$ and $j \in \bar{\mathbf{X}}$, or $f_{ij} = \ell_{ij}$ if $i \in \bar{\mathbf{X}}$ and $j \in \mathbf{X}$. So, $f(\mathbf{X}, \bar{\mathbf{X}}) - f(\bar{\mathbf{X}}, \mathbf{X}) = k(\mathbf{X}, \bar{\mathbf{X}}) - \ell(\bar{\mathbf{X}}, \mathbf{X})$. And since there is no FAP from $\check{s}$ to $\check{t}$ wrt f, $\check{t} \in \bar{\mathbf{X}}$. Therefore, $[\mathbf{X}, \bar{\mathbf{X}}]$ is a cut separating $\check{s}$ and $\check{t}$. By Theorems 2.1 and 2.2, these facts imply that f is a maximum value flow vector, and $[\mathbf{X}, \bar{\mathbf{X}}]$ is a minimum capacity cut separating $\check{s}$ and $\check{t}$ in G. ∎

Given a feasible flow vector f in G, the argument in the proof of Theorem 2.3 suggests the following *labeling* or *tree growth scheme* to determine the set $\mathbf{X}$ of nodes defined there.

TREE GROWTH SUBROUTINE TO FIND $\mathbf{X}$

Step 1 Plant a tree with root at $\check{s}$ Label $\check{s}$ with $\emptyset$.

Step 2 Tree growth step   Look for a labeled node i and an unlabeled node j satisfying (i) or (ii) stated above. If (i) holds, label j with $(i, +)$; if (ii) holds label j with $(i, -)$. In either case, i is the immediate predecessor of j and the arc (i, j) in case (i) or the arc (j, i) in case (ii) is known as the *arc used in labeling node j*. It becomes an in-tree arc in this step.

Repeat Step 2 as often as possible. Terminate when no further tree growth is possible.

The set $\mathbf{X}$ defined in the proof of Theorem 2.3 is the set of all labeled (i.e., in-tree) nodes at termination. For each $i \in \mathbf{X}$, its predecessor path written in reverse order beginning with $\check{s}$ is an FAP from $\check{s}$ to i wrt f.

As an example consider the flow vector in the network in Figure 1.28. The rooted tree obtained when this scheme is applied on it is given in Figure 2.3. Nodes labeled in successive executions of Step 2 are 2, 3, 5, 4, 6, in that order. So, in this example **X** is $\mathcal{N}$.

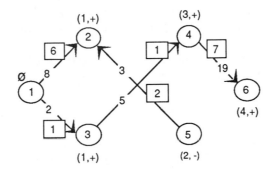

Figure 2.3 Only in-tree arcs are shown here. Node labels are entered by the side of the nodes. All lower bounds are 0. Arc capacities are entered on the arcs. If flow on an arc is nonzero, it is entered in a box by its side.

The rooted tree grown is usually called the *label tree*. At some stage of the tree growth process, if $\breve{t}$ gets labeled, it is an indication that an FAP from $\breve{s}$ to $\breve{t}$ has been identified. This event is called a *breakthrough*, and the label tree is said to have become an *augmenting tree* when it occurs. It can be interpreted as the tree bearing fruit. On the other hand if the tree growth stops without $\breve{t}$ ever getting labeled, it is an indication that there exists no FAP from $\breve{s}$ to $\breve{t}$ wrt the present flow vector f; this event is known as a *nonbreakthrough*. By Theorem 2.3, this implies that f has maximum value in G.

THEOREM 2.4 *THE MAXIMUM FLOW MINIMUM CUT THEOREM   If a feasible flow vector exists in the single commodity flow network* $G = (\mathcal{N}, \mathcal{A}, \ell, k, \breve{s}, \breve{t})$, *the maximum value among feasible flow vectors is equal to the minimum capacity of cuts separating* $\breve{s}$ *and* $\breve{t}$.

Proof   The maximum flow value in G is infinite iff there exists a chain from $\breve{s}$ to $\breve{t}$ consisting only of arcs of infinite capacity. If such a chain exists, by the result in Exercise 1.12, every cut separating $\breve{s}$ and $\breve{t}$ must contain an arc of this chain as a forward arc; hence the capacity of every cut separating $\breve{s}$ and $\breve{t}$ is also infinite, and hence the theorem holds.

If the maximum flow value in G is finite, this theorem follows from Theorems 2.2 and 2.3. ∎

Theorem 2.4 points out the connection between the maximum value flow problem and the minimum capacity cut problem. Since a cut separating $\breve{s}$ and $\breve{t}$ has the property of blocking all the paths between $\breve{s}$ and $\breve{t}$, the minimum capacity cut problem arises in disconnecting the network for interrupting the communication between $\breve{s}$ and $\breve{t}$ (i.e., for the interdiction of the physical transportation of supplies at minimum expense). A minimum capacity cut can be viewed as a minimum cost subset of arcs that intersects every path from $\breve{s}$ to $\breve{t}$. In fact, it is a project to evaluate the capacity of the Eastern European rail network to support a large scale conventional war, and the effort required for interdiction, formulated in 1956 by General F. S. Ross and T. E. Harris, that motivated L. R. Ford and D. R. Fulkerson to study the maximum value flow problem and led to their discovery of the maximum flow minimum cut theorem. See Picard and Queyranne [1982], Billera and Lucas [1978], and Hoffman [1978].

Unfortunately, the corresponding result does not hold for multicommodity flow problems (i.e., those dealing with the simultaneous shipping of several commodities). Consider a $p \;(\overset{>}{=} 2)$ commodity flow problem on the directed network $\bar{G} = (\mathcal{N}, \mathcal{A}, \ell = 0, k)$. Assume that each arc can be used for the flow of any combination of commodities, that all commodities are measured in a common unit (e.g., a truckload) and that the capacity on each arc applies to the sum of the flows of all the commodities. Clearly, this problem can be transformed into one in which there is a specified source and sink pair for each commodity. Let s_r, t_r be the source and sink for the rth commodity, $r = 1$ to p. Here a *disconnecting set of arcs* can be defined to be a subset of arcs whose removal disconnects all the chains from s_r to t_r for each $r = 1$ to p. We define the capacity of such a disconnecting set to be the sum of their capacities. In contrast to single commodity flows, the maximum value flow which maximizes the sum of the flow values of all the commodities, may be strictly less than the minimum disconnecting set capacity. We now present an example from Ford and Fulkerson [1962 of Chapter 1] to illustrate this point.

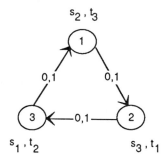

Figure 2.4 A three commodity flow network. Lower bound, capacity for total arc flow entered on the arcs; s_r, t_r are the source, sink nodes for the rth commodity, $r = 1$ to 3.

In the network in Figure 2.4, the maximum flow value is $\frac{3}{2}$ obtained by shipping $\frac{1}{2}$ unit of the rth commodity across the unique chain from s_r to t_r, $r = 1$ to 3. Any pair of arcs in this network, e.g., $\{(1, 2), (1, 3)\}$, is a disconnecting set, since removal of this pair disconnects the chains from s_r to t_r, for all $r = 1, 2, 3$. So, the minimum disconnecting set capacity is 2, strictly greater than the maximum flow value of $\frac{3}{2}$.

The Duality Interpretation of the Maximum Flow Minimum Cut Theorem

Consider the maximum value flow problem in the directed connected single commodity flow network $\bar{G} = (\mathcal{N}, \mathcal{A}, \ell = 0, k, \check{s}, \check{t})$. This is the problem of maximizing v subject to (2.1), (2.2) with $\ell = 0$. Associate the dual variable π_i to the conservation equation corresponding to node i, and the dual variable u_{ij} to the capacity constraint on arc (i, j). From Chapter 1 we know that any one of the equality constraints in (2.1) can be eliminated because of redundancy. We eliminate the equation corresponding to $\check{t}$. This has the effect of setting $\pi_{\check{t}} = 0$ in the dual problem. Hence the dual problem is equivalent to

$$\text{Minimize } z(\pi, u) = \sum (k_{ij} u_{ij} \quad : \quad \text{over } (i, j) \in \mathcal{A})$$

$$\text{Subject to } \pi_j - \pi_i + u_{ij} \overset{\geq}{} 0, \text{ for each } (i, j) \in \mathcal{A} \qquad (2.3)$$

$$\pi_{\check{s}} - \pi_{\check{t}} = 1$$

$$\pi_{\check{t}} = 0, \text{ and } u_{ij} \overset{\geq}{} 0, \text{ for each } (i, j) \in \mathcal{A}$$

Let $\Gamma = \{(\pi, u) : (\pi, u) \text{ *is feasible to* } (2.3)\}$. From the structure of the constraints in (2.3), it is clear that the values of any of the u_{ij} variables can be increased arbitrarily in any feasible solution without affecting its feasibility. Thus Γ is an unbounded set in the π, u-space. Let $(\tilde{\pi}, \tilde{u}) \in \Gamma$ be an extreme point of it. Then the results in Exercise 2.1 state that all $\tilde{\pi}$ and $\tilde{u}_{ij}$ are 0 or 1, that $[\mathbf{X}, \bar{\mathbf{X}}]$ (where $\mathbf{X} = \{i : i \in \mathcal{N} \text{ is s. t. } \tilde{\pi}_i = 1\}$, and $\bar{\mathbf{X}}$ is its complement) is a cut separating $\check{s}$ and $\check{t}$ in $\bar{G}$, and that the set $\{(i, j) : (i, j) \in \mathcal{A} \text{ is s. t. } \tilde{u}_{ij} = 1\}$ is $(\mathbf{X}, \bar{\mathbf{X}})$. Thus, $z(\tilde{\pi}, \tilde{u})$ is the capacity of the cut $[\mathbf{X}, \bar{\mathbf{X}}]$. Thus, every BFS (π, u) of (2.3) corresponds to a cut separating $\check{s}$ and $\check{t}$ in $\bar{G}$, such that $z(\pi, u) = $ the capacity of this cut. Hence, by the duality theorem of LP, when feasible flows exist, the maximum flow value is equal to the minimum capacity of cuts separating $\check{s}$ and $\check{t}$ in $\bar{G}$; this is the result in the maximum flow minimum cut theorem (Theorem 2.4). Also, the weak duality theorem of LP implies that the value of any feasible flow vector in $\bar{G}$ is $\overset{\leq}{}$ the capacity of any cut separating $\check{s}$ and $\check{t}$ in $\bar{G}$. This is the first result in Theorem 2.2.

Assume that k is a finite positive vector. Consider the problem of finding a maximum capacity cut separating $\check{s}$ and $\check{t}$ in $\bar{G}$. Since there are only a finite number of cuts separating $\check{s}$ and $\check{t}$ in $\bar{G}$, and each has finite capacity, this is a combinatorial optimization problem with a finite optimum objective value. The minimum capacity among cuts separating $\check{s}$ and $\check{t}$ is the optimum objective value in the LP (2.3). By analogy, one is tempted to look at the LP: maximize $\{z(\pi, u) : \text{over } (\pi, u) \in \Gamma\}$, for the maximum capacity cut problem. However, since Γ

is unbounded, and $k > 0$, $z(\pi, u)$ is unbounded above on Γ. So, the maximum capacity cut cannot be found directly from this LP; $z(\pi, u)$ has a finite minimum over Γ, but is unbounded above. See Figure 2.5.

Since only extreme points of Γ correspond to cuts, the maximum capacity cut problem is exactly the discrete optimization problem: maximize $\{z(\pi, u) :$ over the finite set of extreme points of $\Gamma\}$. And the minimum capacity cut problem is to minimize $z(\pi, u)$ over the finite set of extreme points of Γ. Both are discrete optimization problems requiring the optimization of a linear function over the finite set of extreme points of Γ. So, on the surface both problems seem very comparable. The main difference between them is that there exists a solution to the minimization problem which solves the LP (2.3) without any extreme point condition. Better still, given an extreme point of Γ, there are necessary and sufficient optimality conditions to check efficiently whether that point solves the minimization problem. In contrast, given an extreme point of Γ, no nontrivial optimality conditions (short of total enumeration, i.e., comparing this point with every other extreme point of Γ) are known to check whether it solves the maximization problem.

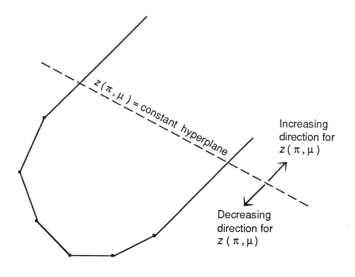

Figure 2.5

The maximum capacity cut problem is a special case of the following:

GENERAL PROBLEM   Given a finite system of linear constraints, for which the set of feasible solutions is an unbounded convex polyhedron $\mathbf{K}$, and a linear function $z(x)$ unbounded above over $\mathbf{K}$, find an extreme point of $\mathbf{K}$ which maximizes $z(x)$ over the finite set of extreme points of $\mathbf{K}$.

It is a difficult discrete optimization problem for which no efficient algorithms are known. When all the data is rational, this problem is NP-hard. Several other hard combinatorial optimization problems such as the traveling salesman problem are special cases of this problem. At the moment, the only methods known for this problem are enumerative methods with exponential growth in computational effort in the worst case.

Exercises

2.1 Prove the following. If $(\tilde{\pi} = (\tilde{\pi}_i), \tilde{u} = (\tilde{u_{ij}}))$ is a BFS of (2.3), then (i) $\tilde{\pi}_i, \tilde{u_{ij}}$ are all equal to either 0 or 1 for all i, j, (ii) for each $(i, j) \in \mathcal{A}, \tilde{u_{ij}} = 1$ iff $\tilde{\pi}_i = 1$ and $\tilde{\pi}_j = 0$; $\tilde{u_{ij}} = 0$ otherwise. Conversely, if $(\tilde{\pi}, \tilde{u})$ is a feasible solution of (2.3) satisfying (i) and (ii), then it is a BFS. Hence show that a feasible solution $(\tilde{\pi}, \tilde{u})$ for (2.3) is an extreme point of Γ iff it satisfies (i) and (ii). Then show that every BFS (π, u) of (2.3) corresponds to a cut separating $\check{s}$ and $\check{t}$ in $\bar{G}$, such that $z(\pi, u)$ = the capacity of that cut, and vice versa.

2.2 Let $A = (a_{ij})$ be a given positive square matrix of order n. For any $\mathbf{X} \subset \{1, \ldots, n\}$, define $\bar{\mathbf{X}} = \{1, \ldots, n\} \setminus \mathbf{X}$. Consider the four problems of minimizing or maximizing $a(\mathbf{X}, \bar{\mathbf{X}})$, or $a(\mathbf{X}, \bar{\mathbf{X}}) + a(\bar{\mathbf{X}}, \mathbf{X})$, over the class of subsets of $\{1, \ldots, n\}$. Which of these problems can be solved by network flow methods? Why?

2.3 Research Problem : Develop a system of linear constraints in π, u, which, when combined with those in (2.3), describes the convex hull of the set of extreme points of Γ.

2.3 Single Path Labeling Methods Beginning With A Feasible Flow Vector

We consider the problem of finding a maximum value feasible flow vector in the directed single commodity flow network $G = (\mathcal{N}, \mathcal{A}, \ell, k, \check{s}, \check{t})$. It involves two phases. They are:

PHASE 1 This phase finds an initial feasible flow vector in G if one exists. If $\ell = 0$, $f = 0$ is a feasible flow vector. Hence this phase is not needed, and we go directly to Phase 2 with $f = 0$. Methods for carrying out Phase 1 when $\ell \neq 0$ are discussed in Section 2.6.

PHASE 2 This phase requires an initial feasible flow vector, say f^0 in G, as an input. This phase constructs a sequence of feasible flow vectors of strictly increasing values, and terminates only when a maximum value feasible flow vector is obtained.

Here we discuss a class of algorithms for Phase 2. In each stage, they try to find an FAP from $\check{s}$ to $\check{t}$ wrt the present feasible flow vector $\bar{f}$, say. If none are found, $\bar{f}$ is a maximum value flow vector, and the method terminates. Otherwise an FAP is found, flow augmentation is carried out using it, and the whole process is repeated with the new flow vector. Since the methods deal with one FAP at a time, they are called *single path methods*. The methods differ in the manner in which the search for an FAP is carried out in each step.

2.3.1   An Initial Version of the Labeling Method

This version uses the tree growth scheme discussed earlier to search for FAPs. Tree growth occurs one arc at a time.

INITIAL VERSION

Step 1 Plant a tree with root at $\breve{s}$ Let $\bar{f} = (\bar{f}_{ij})$ be the present feasible flow vector. Label $\breve{s}$ with $\emptyset$.

Step 2 Tree growth   Look for a pair of nodes i, j satisfying one of the following.

  (i) **Forward labeling rule**   Node i is labeled; j is unlabeled; $(i, j) \in \mathcal{A}$ and $\bar{f}_{ij} < k_{ij}$

  (ii) **Reverse labeling rule**   Node i is labeled; j is unlabeled; $(j, i) \in \mathcal{A}$ and $\bar{f}_{ij} > \ell_{ij}$

If such a pair does not exist, there is a nonbreakthrough. Go to Step 4. If a pair of nodes i, j satisfying one of the above rules is found, label j with $(i, +)$ under rule (i), or with $(i, -)$ under rule (ii). Node i is the immediate predecessor of j. If $j = \breve{t}$, there is a breakthrough. Go to Step 3. Otherwise, repeat this Step 2.

Step 3 Flow augmentation   Since $\breve{t}$ is labeled, the predecessor path $\mathcal{P}$ of $\breve{t}$, written in reverse order beginning with $\breve{s}$ is an FAP. Compute ϵ, the residual capacity of $\mathcal{P}$, and $\hat{f} = (\hat{f}_{ij})$ where

$$
\hat{f}_{ij} = \begin{cases}
\bar{f}_{ij} + \epsilon & \text{if } (i, j) \text{ is a forward arc on } \mathcal{P} \\
\bar{f}_{ij} - \epsilon & \text{if } (i, j) \text{ is a reverse arc on } \mathcal{P} \\
\bar{f}_{ij} & \text{if } (i, j) \text{ is not on } \mathcal{P}
\end{cases}
$$

Erase the labels on all the nodes (this operation is called *chopping down the present tree*) and go back to Step 1 with $\hat{f}$ as the new flow vector.

Step 4 Termination   The present flow vector $\bar{f}$ is a maximum value flow vector in G. Let $\mathbf{X}$ = set of in-tree (i.e., labeled) nodes, and $\bar{\mathbf{X}}$ its complement. $(\mathbf{X}, \bar{\mathbf{X}})$ is a minimum capacity cut separating $\breve{s}$ and $\breve{t}$ in G. Terminate.

Discussion

In executing the algorithm, each time check whether the sink can be labeled. If so, label it and move over to Step 3. If the sink cannot be labeled, label some other unlabeled node if possible and continue. Each time Step 2 is carried out, one new node and arc join the tree. So, Step 2 can be carried out consecutively at most $n-1$ times ($n = |\mathcal{N}|$) before going to Step 3 or 4. The work between two consecutive occurrences of Steps 3 and 4 represents a *tree growth routine*, also called a *labeling routine*. It terminates either with a breakthrough or a nonbreakthrough. The total number of trees grown in this algorithm is $1 +$ the number of flow augmentations carried out.

As an example, consider the network in Figure 1.28 with a feasible flow vector of value 7 entered there. All lower bounds are zero, and capacities are entered on the arcs. Nonzero flow amounts are entered in little squares by the side of the arcs. We label the source, node 1 with $\emptyset$, then node 2 with $(1, +)$, then node 5 with $(2, -)$, and then the sink, node 6 with $(5, +)$, in that order. So, there is a breakthrough. The FAP has forward arcs $(5, 6)$, $(1, 2)$ and reverse arc $(5, 2)$; $\epsilon_1 = \min \{7-0, 8-6\}$ $= 2$, $\epsilon_2 = \min. \{2\} = 2$, the residual capacity $\epsilon = \min \{\epsilon_1, \epsilon_2\} = 2$. The new flow vector is shown in Figure 2.6.

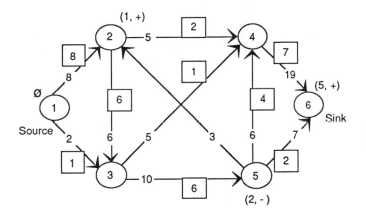

Figure 2.6

We chop down the old tree, root another tree in Figure 2.6, and grow it. There is a breakthrough again. The new FAP has forward arcs $(4, 6)$, $(2, 4)$, and $(1, 3)$ and reverse arc $(2, 3)$. Flow augmentation is carried out and the new flow vector is shown in Figure 2.7. A new tree grown in Figure 2.7 ended in nonbreakthrough. So, the flow vector in Figure 2.7, of value 10, is a maximum value flow vector in this network. A minimum capacity cut $[\mathbf{X}, \bar{\mathbf{X}}]$ separating the source and the sink is marked in Figure 2.7 with a dashed line.

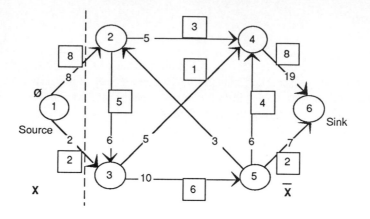

Figure 2.7

Exercises

2.4 Discuss how to find a feasible flow vector of value equal to a specified number v^* in G.

2.5 Prove that there exists r arc disjoint chains from $\check{s}$ to $\check{t}$ in the directed network $G = (\mathcal{N}, \mathcal{A}, \check{s}, \check{t})$ iff there is no cut separating $\check{s}$ and $\check{t}$ with less than r forward arcs.

2.6 In a club of m men and n women, *compatibility* is a mutual relationship in a man-woman pair. The subset of women compatible with the ith man, $i = 1$ to m is given. Formulate the problem of forming the maximum number of compatible couples as a maximum value flow problem.

2.7 Find a maximum value flow vector from source to sink in the networks in Figures 2.8 to 2.10.

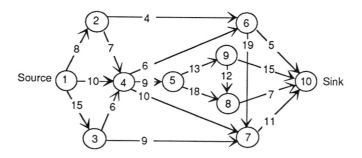

Figure 2.8 All lower bounds are 0, and capacities are entered on the arcs.

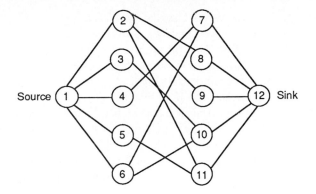

Figure 2.9 All the arcs are directed from left to right, all lower bounds are 0, and capacities are 1.

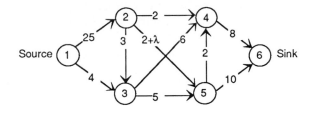

Figure 2.10 Here λ is a parameter. All lower bounds are 0, and capacities are entered on the arcs. Obtain the solution to the problem as a function of λ, for $\lambda \geqq 0$.

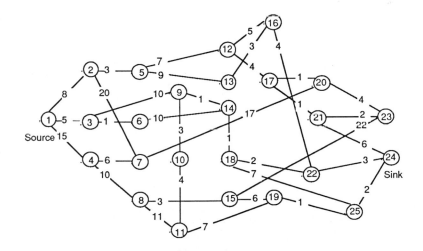

Figure 2.11

2.8 Maximum Value Flows in Undirected Networks   Consider the maximum value flow problem in a connected undirected network $G = (\mathcal{N}, \mathcal{A}, \ell = 0, k, \breve{s}, \breve{t})$. One way of handling this problem is to replace each edge $(i;j)$ by the pair of arcs $(i, j), (j, i)$ both with lower bound 0 and capacity $k_{i;j}$, but this is undesirable because it doubles the number of lines. Another way is to orient each edge arbitrarily. For example, make edge $(i;j)$ into arc (i, j) say, with capacity $k_{i;j}$ and lower bound $-k_{i;j}$. Adopt the convention that if the flow amount on this arc is a negative number, say $-\alpha$, it implies that the actual flow on the corresponding edge is $+\alpha$ in the direction opposite to the selected orientation for this edge. Use this approach to solve the maximum value flow problem on the network in Figure 2.11. All lower bounds are zero, and capacities are entered on the edges.

If there exists an FAP from $\breve{s}$ to $\breve{t}$ wrt a feasible flow vector $\bar{f}$, all the nodes along this path can be labeled and hence $\breve{t}$ can be labeled. Hence the sink will be labeled after at most $(n-1)$ tree growth steps, where $n = |\mathcal{N}|$, and a breakthrough will occur.

If ℓ, k, and the initial feasible flow vector f^0 are all integer vectors, the residual capacity of all the FAPs identified will always be a positive integer. Hence, in this case, either this version terminates with a maximum value feasible flow vector after a finite number of trees are grown, or it constructs an infinite sequence of flow vectors whose values diverge to $+\infty$.

Given an FAP which is a simple path, by labeling the nodes in the order in which they appear on it, we can guarantee that the tree growth routine will terminate by discovering that particular FAP. So, if $\{f^0, f^1, \ldots\}$ is a sequence of feasible flow vectors of increasing value, satisfying the following Property 1, then we are guaranteed that this version can be executed, beginning with f^0, so that it produces exactly this sequence of flow vectors. Hence, in order to establish the worst case computational complexity of this version, it is enough to study the following question: "What is the maximum number of feasible flow vectors in a sequence $\{f^0, f^1, \ldots\}$ satisfying Property 1 ?" Edmonds and Karp [1972] have constructed the network in Figure 2.12 to answer this question.

PROPERTY 1   For each r, f^{r+1} is obtained by augmenting the flow vector f^r using some FAP (which is a simple path) from $\breve{s}$ to $\breve{t}$ wrt f^r.

Let f_t denote the flow amount on arc e_t, $t = 1$ to 5 in the network in Figure 2.12. $\bar{f} = (\bar{f}_t : t = 1 \text{ to } 5)^T = (M, M, 0, M, M)^T$ of value $2M$ is a maximum value feasible flow vector. Define the sequence of feasible flow vectors $\{f^s : s = 0, 1, \ldots, 2M\}$ in this network, with $f^0 = 0$, $f^{2r-1} = (r, r-1, 1, r-1, r)^T$, $f^{2r} = (r, r, 0, r, r)^T$, for $r = 1$ to M. In this sequence f^s has value s for each $s = 0$ to $2M$. For $r = 1$ to M, f^{2r-1} is obtained by augmenting f^{2r-2} using the FAP 1, (1, 2), 2, (2, 3), 3, (3, 4), 4 wrt it; and f^{2r} is obtained by augmenting f^{2r-1} using the FAP 1, (1, 3), 3, (2, 3), 2, (2, 4), 4 wrt it. So this sequence of flow vectors satisfies Property 1, and it has $2M + 1$ vectors in it. Assuming that M is a positive integer, the size of the maximum value flow problem on the network in Figure 2.12 (i.e., the number of binary digits needed to store all the data in it) is $\log(1+M)$ plus a constant, and the computational effort required by this version to solve this problem, if it follows this sequence, is that for $2M$ tree growths, which grows exponentially with the size. Thus even though the initial version is a finite algorithm for problems with finite integer data, it is not a polynomially bounded algorithm. Even on networks with few nodes and arcs, it may require an unduly large amount of computational effort depending on the order in which the nodes are selected for labeling.

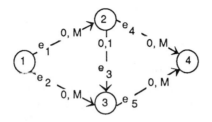

Figure 2.12 Network displaying the worst case behavior of the initial version of the labeling method. Lower bounds, capacities entered on the arcs. M is a positive integer.

EXAMPLE 2.1

Ford and Fulkerson [1962 of Chapter 1], have provided the following example to illustrate the fact that if the arc capacities are irrational numbers, and the starting feasible flow vector is an integer vector, the initial version of the labeling method may not terminate and may even converge in the limit to a flow vector whose value is strictly less than the true maximum flow value. Let $\alpha = \frac{-1+\sqrt{5}}{2}$, an irrational number satisfying $\alpha^{r+2} = \alpha^r - \alpha^{r+1}$, for integer $r \geqq 0$. Since $0 < \alpha < 1$, the infinite sum $\sum_{p=0}^{\infty} \alpha^p$ converges to a quantity which we denote by β. The network G $= (\mathcal{N}, \mathcal{A})$ is given by the following.

$$\mathcal{N} = \{\check{s}, \check{t}, x_i, y_i, i = 1 \text{ to } 4\}, |\mathcal{N}| = 10$$

$$\mathcal{A} = \{(\check{s}, x_i), (y_i, \check{t}) : i = 1 \text{ to } 4\} \cup \{(y_i, y_j), (x_i, y_j), (y_i, x_j) : \text{ for } i \neq j = 1 \text{ to } 4\} \cup \{e_i = (x_i, y_i) : i = 1 \text{ to } 4\}$$

with $\ell = 0$, capacities of e_1, e_2, e_3, e_4 to be $1, \alpha, \alpha^2, \alpha^2$ respectively, and the capacities of all other arcs to be β. In the initial feasible flow vector, f^0, there is one unit of flow on arcs $(\check{s}, x_1), (x_1, y_1)$, and $(y_1, \check{t})$, and zero flow on all the other arcs. Hence, $v(f^0)$, the value of f^0, is 1. In f^0, the residual capacities of the arcs

e_1, e_2, e_3, e_4 are $0, \alpha, \alpha^2, \alpha^2$ respectively. The following construction yields two feasible flow vectors per step and generates a sequence of feasible flow vectors in this network satisfying Property 1.

GENERAL STEP   At the beginning of this step, we have a flow vector, f^p, say, with value v_p, such that there exists a permutation of the arcs e_1, e_2, e_3, e_4, which we will denote by e_1', e_2', e_3', e_4' with residual capacities $0, \alpha^r, \alpha^{r+1}, \alpha^{r+1}$ respectively, in f^p, for some r. Denote the tail and head nodes on e_i' by $x_i', y_i', i = 1$ to 4. The chain $\breve{s}, (\breve{s}, x_2'), x_2', (x_2', y_2'), y_2', (y_2', x_3'), x_3', (x_3', y_3'), y_3', (y_3', \breve{t}), \breve{t}$, is an FAC wrt f^p whose residual capacity is α^{r+1}. Denote the flow vector obtained after augmenting f^p using this FAC by f^{p+1}. The arcs e_1', e_2', e_3', e_4' have residual capacities $0, \alpha^r - \alpha^{r+1} = \alpha^{r+2}, 0, \alpha^{r+1}$ respectively in f^{p+1}. The path consisting of arcs $(\breve{s}, x_2'), (x_2', y_2'), (y_2', y_1'), (x_1', y_1'), (x_1', y_3'), (x_3', y_3'), (x_3', y_4'), (y_4', \breve{t})$, in that order is an FAP wrt f^{p+1} with residual capacity α^{r+2}. Let f^{p+2} be the feasible flow vector obtained after augmenting f^{p+1} using this FAP, and v_{p+2} its value. $v_{p+2} - v_p = \alpha^r$. In f^{p+2}, the arcs e_1', e_2', e_3', e_4' have residual capacities $\alpha^{r+2}, 0, \alpha^{r+2}, \alpha^{r+1}$ respectively. So, rearrange them in the order e_2', e_4', e_1', e_3', and go to the next step.

This procedure leads to an infinite sequence of feasible flow vectors satisfying Property 1, whose value is strictly increasing and converges to $\sum_{r=0}^{\infty} \alpha^r = \beta$, whereas the maximum flow value in this network can be verified to be 4β. So, even though the sequence $\{f^p : p = 0, 1, \ldots\}$ converges to a limit, its limit is not a maximum value flow vector in this example.

2.3.2   The Scanning Version of the Labeling Method

In the initial version, when $\bar{f}$ is the present feasible flow vector, we search for a labeled node i and an unlabeled node j satisfying either (i) or (ii) of Step 2. If such a labeled node i is found, we can label not only this node j, but all other unlabeled nodes j that satisfy this condition with i. This operation of labeling all the unlabeled nodes j satisfying this condition is called *scanning the labeled node i*. The use of scanning leads to an improved version. This version is often called the *Ford-Fulkerson Labeling Method*. Here, nodes may be in three possible states, *unlabeled, labeled and unscanned, labeled and scanned*. *List* always refers to the present set of labeled and unscanned nodes.

SCANNING VERSION

Step 1 Plant a tree with root at $\breve{s}$ Let $\bar{f} = (\bar{f}_{ij})$ be the present feasible flow vector. Label $\breve{s}$ with $\emptyset$; $\breve{s}$ is now labeled and unscanned. List $= \{\breve{s}\}$.

Step 2 Select a node from list for scanning   If list $= \emptyset$, go to Step 5. Otherwise select a node from it to scan.

Step 3 Scanning   Let i be the node to be scanned.

(i) **Forward labeling** Identify all unlabeled nodes j satisfying $(i, j) \in \mathcal{A}$ and $\bar{f}_{ij} < k_{ij}$, and label all of them with $(i, +)$.

(ii) **Reverse labeling** Identify all unlabeled nodes j satisfying $(j, i) \in \mathcal{A}$ and $\bar{f}_{ji} > \ell_{ji}$, and label all of them with $(i, -)$.

Each newly labeled node in this step is now labeled and unscanned. Include all of them in the list. Node i is now labeled and scanned. Delete it from the list. If $\breve{t}$ is now labeled, there is a breakthrough, go to Step 4. If $\breve{t}$ is not yet labeled. Go to Step 2.

Step 4 Flow augmentation Same as Step 3 in the initial version.

Step 5 Termination Same as Step 4 in the initial version.

Discussion

As an example consider the network in Figure 2.13 with an initial feasible flow vector of value 1 marked there. When the source node 1 is scanned, both nodes 2 and 3 get labeled. At this stage the list is {2, 3}, and we select 2 from it for scanning next. This leads to a breakthrough. Node labels are shown in Figure 2.13. Notice that even though 1, (1, 3), 3, (2, 3), 2, (2, 4), 4 is an FAP, which is a simple path, we cannot obtain this FAP under the scanning version, because, when node 1 is scanned, both nodes 3 and 2 get labeled with (1, +).

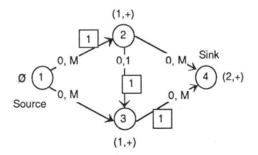

Figure 2.13 Data on the arcs is lower bound, capacity. Here $M > 2$. Nonzero flow amounts are entered in boxes by the side of the arcs.

Flow augmentation can be carried out using the FAP identified in Figure 2.13, and the method continued. It can be verified that the method terminates with the maximum value flow vector after growing three trees. This compares with $2M$ trees that the initial version may require.

Since it may not be possible to obtain some FAPs under the scanning version, we cannot use the technique of constructing a sequence of flow vectors satisfying Property 1 of Section 2.3.1 to study its computational complexity, at least not in all the networks. However, we will now show that given any network G, we can obtain a modified network $\hat{G}$ such that the behavior of the initial version, when applied on G, is exactly duplicated by the scanning version, when applied on $\hat{G}$, by choosing an appropriate order for scanning the nodes from the list. The modification consists of adding an artificial node in the middle of each arc in G (i.e., replacing each arc (i,j) in G by the pair of arcs $(i,p), (p,j)$ both with the same data and the same flow amount as the original arc (i,j) in G) where p is the artificial node introduced on (i,j).

Let $G = (\mathcal{N}, \mathcal{A})$ be the original network with $|\mathcal{N}| = n$, $|\mathcal{A}| = m$, and $\hat{G} = (\hat{\mathcal{N}}, \hat{\mathcal{A}})$ the corresponding modified network. So, $|\hat{\mathcal{N}}| = n + m$, $|\hat{\mathcal{A}}| = 2m$. Each arc in G corresponds to a unique pair of arcs in $\hat{G}$. Under this correspondence every path in G corresponds to a unique path in $\hat{G}$ with twice as many arcs. As an example, the modified network corresponding to the one given in Figure 2.13 is shown in Figure 2.14.

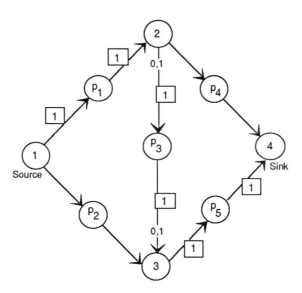

Figure 2.14 Modified network corresponding to the one in Figure 2.13. Lower bound, capacity on nonvertical arcs are 0, M; p_1 to p_5 are the artificial nodes inserted.

Let $e_1, \ldots, e_m$ be the arcs in G, and p_r the artificial node introduced in the middle of $e_r, r = 1$ to m. Let e'_r, e''_r be the pair of nodes into which e_r is split when p_r is introduced in its middle. Let f_r be the flow variable associated with e_r in G, and f'_r, f''_r the flow variables associated with e'_r, e''_r in $\hat{G}$, $r = 1$ to m. In applying the scanning version on $\hat{G}$, the following facts can be verified to hold.

In every feasible flow vector $(f'_r, f''_r : r = 1 \text{ to } m)$ in $\hat{G}$, $f'_r = f''_r$ for all r, and by making f_r equal to this common quantity, $r = 1$ to m, we get a feasible flow vector in G, and vice versa. In $\hat{G}$ an artificial node p_r gets labeled only when tail(e_r') or head(e_r'') is scanned. When an original node $i \in \mathcal{N}$ is scanned in $\hat{G}$, the only nodes that get labeled are artificial nodes. When an artificial node p_r is scanned in $\hat{G}$, at most one node gets labeled; it is either tail(e_r'), or head(e_r'').

From these facts we conclude that if $\hat{\mathcal{P}}$ is an FAP from $\check{s}$ to $\check{t}$ in $\hat{G}$ wrt a feasible flow vector $\hat{f}$, when the scanning version is applied on $\hat{G}$ with $\hat{f}$ as the flow vector, it is possible to scan the nodes on $\hat{\mathcal{P}}$ in the order in which they appear on it, and if this is done, the tree growth routine will terminate by discovering the FAP $\hat{\mathcal{P}}$.

As an example, let $\mathcal{P}$ be the FAP 1, (1, 3), 3, (2, 3), 2, (2, 4), 4 in the network G in Figure 2.13. We have seen earlier that this FAP cannot be obtained under the scanning version. The corresponding FAP, $\hat{\mathcal{P}}$, in the modified network $\hat{G}$ drawn in Figure 2.14 is 1, (1, p_2), p_2, (p_2, 3), 3, (p_3, 3), p_3, (2, p_3), 2, (2, p_4), p_4, (p_4, 4), 4. It can be verified that the nodes 1, p_2, 3, p_3, 2, p_4 can be scanned in this order when the scanning version is applied in $\hat{G}$, leading to the FAP $\hat{\mathcal{P}}$ at termination.

From these facts and the earlier results on the initial version, we conclude that if all the data and the initial feasible flow vector are all integral, then the scanning version will either terminate after a finite number of steps with a maximum value flow vector or obtain an infinite sequence of flow vectors whose value diverges monotonically to $+\infty$. And if some capacities are irrational and the initial feasible flow vector is an integer, the scanning version may produce an infinite sequence of flow vectors whose values converge to a quantity strictly less than the maximum flow value (i.e., it may not work). However, surprisingly, a simple selection rule fixes this problem.

A general selection rule, known as a *consistent labeling procedure*, is one in which the choice for the node to be scanned from the list is determined uniquely by the set of nodes currently in the list (i.e., whenever the list is a particular subset of nodes, it always selects the same node from that subset to scan). We have the following theorem by A. Tucker [1977] on the finite convergence of the scanning version operated with a consistent labeling procedure.

THEOREM 2.5 *When the scanning version is applied on the directed single commodity flow network* G $= (\mathcal{N}, \mathcal{A}, \ell, k, \check{s}, \check{t})$ *beginning with an initial feasible flow vector* f^0, *using a consistent labeling procedure, the scanning version finds a maximum value flow vector after growing at most a finite number of trees.*

Proof Let $|\mathcal{N}| = n$. We do not assume that ℓ, k, f^0 are integer vectors; the proof is valid in general. The proof is based on induction on n. Number the nodes serially, beginning with 1 for the source and ending with n for the sink. The statement of the theorem is obviously true for $n = 2$.

Induction Hypothesis The statement of the theorem is true in networks with number of nodes $\stackrel{\leq}{=} n - 1$.

Now we will prove that the theorem must also hold for G $= (\mathcal{N}, \mathcal{A})$ with $|\mathcal{N}| = n$, under the induction hypothesis. Proof is by contradiction. Suppose the theorem does not hold in G. So, there exists a consistent labeling procedure such that when the scanning version is applied on G with it, the algorithm continues indefinitely. 1 is the source node and the flow on arcs of the form $(1, i)$ is never decreased during the algorithm. Let $\mathbf{S} = \{i : i \in \mathbf{A}(1), \text{and } (1, i) \text{ is never saturated in the algorithm}\}$. $\mathbf{S} \neq \emptyset$ by the hypothesis. Suppose the consistent labeling procedure requires the selection of the node $p \in \mathbf{S}$ for scanning whenever the list is the set $\mathbf{S}$.

Let $\{f^0, f^1, \ldots\}$ be the sequence of flow vectors generated by the algorithm. From the definition of $\mathbf{S}$, there exists a finite positive integer r such that for all $i \in \mathbf{A}(1)\backslash\mathbf{S}$, $f^u_{1i} = k_{1i}$ for all $u \stackrel{\geq}{=} r$. Hence, for all $u \stackrel{\geq}{=} r$, during the uth tree growth in the algorithm, node p is scanned immediately after 1. Also since $(1, p)$ is never saturated in the algorithm, the results obtained will remain unchanged if k_{1p} is changed to $+\infty$. Make this change.

Form a new network $\bar{\mathrm{G}} = (\bar{\mathcal{N}}, \bar{\mathcal{A}})$ by just coalescing node p into node 1 in G, leaving all the data unchanged. For $u \stackrel{\geq}{=} r$, let $\bar{f}^u$ be the flow vector obtained by deleting f^u_{1p} from f^u. Now apply the same algorithm on $\bar{\mathrm{G}}$ beginning with the initial feasible flow vector $\bar{f}^r$ in it. Coalescing node p into 1 contracts the arc $(1, p)$ into node 1; this has no effect on the results of labeling in G, since k_{1p} was changed to ∞. The consistency of the labeling procedure for all nodes other than p is kept unaffected while applying the algorithm on $\bar{\mathrm{G}}$. The sequence of scanning remains the same, since node p was being scanned immediately after node 1 in G. Thus applying the algorithm on $\bar{\mathrm{G}}$ with the initial flow $\bar{f}^r$ is equivalent to applying it in G beginning with f^r. However, since $|\bar{\mathcal{N}}| = n - 1$, by the induction hypothesis, the algorithm finds a maximum value flow vector in $\bar{\mathrm{G}}$ after a finite number of iterations and terminates; a contradiction. Hence, the theorem must also hold for the network G with $|\mathcal{N}| = n$, under the induction hypothesis. Since the theorem is true when the network has only two nodes, by induction it is true in general. ∎

Denote the arcs in G by $e_1, \ldots, e_m$. Correspondingly, let $\ell = (\ell_1, \ldots, \ell_m), k = (k_1, \ldots, k_m)$, and $f = (f_1, \ldots, f_m)$ denote the lower bound, capacity, and flow vectors. Given a feasible flow vector $f = (f_t)$ in G define the *partition of arcs corresponding to* f to be $(\mathbf{L}_f, \mathbf{I}_f, \mathbf{U}_f)$, where $\mathbf{L}_f = \{t : f_t = \ell_t\}, \mathbf{I}_f = \{t : \ell_t < f_t < k_t\}, \mathbf{U}_f = \{t : f_t = k_t\}$. Clearly the total number of such distinct partitions is $\stackrel{\leq}{=} 3^{\mathbf{m}}$. Another proof of Theorem 2.5 comes from the result in the following exercise.

Exercise

2.9 Prove that the partitions of arcs corresponding to flow vectors in the sequence of feasible flow vectors generated by the scanning version using a consistent labeling rule applied on G are all distinct. And prove that the number of flow vectors in this sequence is $\stackrel{\leq}{=} n!$, where $n = |\mathcal{N}|$ (Megiddo and Galil [1979]).

The result in Exercise 2.9 shows that the number of flow augmentations carried out in the scanning version with a consistent labeling procedure is $\leqq$ min. $\{n!, 3^m\}$. Even though this number is finite, it grows very rapidly with n, m. In fact, Megiddo and Galil [1979] have constructed an infinite class of maximum value problems with integer capacities and zero lower bounds to demonstrate that the number of flow augmentations in the scanning version under a general consistent labeling procedure may grow exponentially with the size of the problem. Their construction involves the basic network structure of the type in Figure 2.14; however, the arc capacities are different. The network for the rth problem in their class is a combination of r similar structures placed side by side with the leftmost node of each structure coinciding with the rightmost node of the structure to its left, with capacities of all the nonvertical arcs $= 2^r$, and those of vertical arcs $= 2^{r-1}$. This rth problem has $1 + 8r$ nodes and $10r$ arcs. They show that the scanning version with a consistent labeling procedure specified by them and initiated with the zero flow vector requires 2^r flow augmentations before reaching a maximum value flow vector in the rth problem. This is exponential growth with size.

2.3.3   Shortest Augmenting Path Method

Define the length of a simple path to be the number of arcs in it. In this version, due to Edmonds and Karp [1972], the node to be scanned is selected from the list in order to guarantee that the FAP obtained is the shortest among all the FAPs from the source to the sink at that stage. This is the reason for its name. It is also known as the *Edmonds-Karp version of the labeling method.*

In this method, nodes are selected from the list for scanning on a *first labeled first scanned* basis (i.e., nodes are scanned in the order they are labeled). This makes the tree grow in a breadth-first manner; hence the method uses a breadth-first search strategy to find an FAP. It turns out that the FAPs obtained by a breadth-first search strategy are the shortest.

The method is the same as the scanning version of Section 2.3.2 with one minor difference. Here the list is maintained as a queue, from top to bottom. When new nodes are entered into the list in Step 3, they are always entered at the bottom. When a node is to be selected for scanning in Step 2, it is always the topmost node in the list.

As an example, we apply this method on the network in Figure 1.28 with the feasible flow vector entered there. Here the list has to be maintained as an ordered set. We order the nodes in it from left to right (left corresponds to top, and right to bottom). Source node 1 gets labeled with $\emptyset$, list $= \{1\}$. Node 1 is scanned leading to the label of $(1, +)$ for both nodes 2 and 3. Node 1 leaves the list and nodes 2, 3 are added to the list at the right in some order, say in natural order, so the list is $\{2, 3\}$ now. Node 2 is selected for scanning, node 4 gets labeled with $(2, +)$, and node 5 with $(2, -)$. The list is $\{3, 4, 5\}$ now. Scanning of node 3 leads to no new labels. Then node 4 is scanned, leading to the label of $(4, +)$ to the sink node 6, and hence a breakthrough. The FAP with arcs (1, 2), (2, 4), (4, 6), of length 3, can be verified to be a shortest FAP wrt the present flow vector. Flow augmentation can be carried out, and the method can be continued in the same manner.

For the sake of discussion let d_i denote the depth label for labeled node i at any stage of this algorithm. These d_i are not used or maintained in the algorithm. Clearly, d_i is the length of the FAP traced by the current labels, from the source to i at that stage. Also, since a node gets labeled only after its parent, the "first labeled, first scanned" policy used in this algorithm guarantees that the node selected for scanning is always a node with the smallest depth label among those in the list.

Define the *distance* of node i at any stage of this algorithm to be the length of a shortest FAP from source to i wrt the flow vector at that stage, or $+\infty$ if no FAP exists to i.

THEOREM 2.6 *In this version, the FAP from the source to any labeled node* i *traced by the labels is a shortest FAP from the source to* i wrt *the feasible flow vector at that stage.*

Proof   Let $\bar{f}$ be the feasible flow vector at this stage, and let d_i denote the depth label for the labeled node i in the tree being grown. Let d be a positive integer.

Induction Hypothesis   The following statements are true for any $r \leqq d - 1$.

1. During the labeling routine, if the depth label of the node being scanned is r, then the distance of all the unlabeled nodes at this time is $\geqq r + 1$.

2. For all labeled nodes i with $d_i = r$, the FAP from source to i traced by the current labels is a shortest FAP from source to i at that stage.

Statements 1 and 2 are obviously true for $r = 0$. We will now prove that under the induction hypothesis, these statements must also be true for $r = d$.

Let j be a labeled node with $d_j = d$. If the parent of j is p, then $d_p = d - 1$. Applying Statement 1 to the time when p was being scanned, we conclude that the distance of j at this stage is $\geqq d$. But the FAP from source to j traced by the current labels has length d, so it is a shortest FAP from source to j at this stage. This proves that Statement 2 must hold for $r = d$.

Let x be an unlabeled node at the time that j was being scanned; x must also be unlabeled earlier when p was being scanned, which implies by Statement 1 under the induction hypothesis that the distance of x must be $\geqq d$. However, since x remained unlabeled when all the labeled nodes of depth $\leqq d - 1$ were scanned, the distance of x cannot be d. So, the distance of x must be $\geqq d + 1$. This proves that Statement 1 also holds when $r = d$.

Hence, by induction, the statements in the induction hypothesis are true for all r. This proves the theorem. ∎

Now we will study the worst case computational complexity of this version. Let f^0 denote the initial feasible flow vector and f^u the flow vector obtained after u trees are grown and the growth of the $(u+1)$th tree is about to begin, for $u \geqq 1$. So, f^u is the current feasible flow vector during the growth of the $(u+1)$th tree. Let θ_i^{u+1} be the length of a shortest FAP from the source $\check{s}$ to i, and ϑ_i^{u+1} the length of a shortest FAP from i to the sink $\check{t}$ wrt f^u. These quantities are defined to be $+\infty$ if an FAP of the type in its definition does not exist. So, from Theorem 2.6, if i is a labeled node in the $(u+1)$th tree, its depth in this tree is θ_i^{u+1}.

THEOREM 2.7 *For all nodes i and $u \geqq 1$ we have $\theta_i^{u+1} \geqq \theta_i^u$, and $\vartheta_i^{u+1} \geqq \vartheta_i^u$.*

Proof We will first prove $\theta_i^{u+1} \geqq \theta_i^u$. If this is not true, there must exist a p and some nodes i, for which $\theta_i^{p+1} < \theta_i^p$. Obviously $p \geqq 1$. Let r be a node satisfying $\theta_r^p > \theta_r^{p+1} = \min. \{\theta_i^{p+1} : i \text{ such that } \theta_i^{p+1} < \theta_i^p\}$.

The source node $\check{s}$ satisfies $\theta_{\check{s}}^u = 0$ for all $u \geqq 1$. Also, for any $u \geqq 1$, if $\theta_i^u = 0, i$ must be $\check{s}$. Since $\theta_r^{p+1} < \theta_r^p$, we have $r \neq \check{s}$, and hence $\theta_r^{p+1} \geqq 1$. Let the label on node r in the $(p+1)$th tree growth routine be $(j, +)$. (A proof similar to the following holds when the second symbol in the label is $-$.) Then (j, r) is an in-tree arc in the $(p+1)$th tree, so $f_{jr}^p < k_{jr}$, and from Theorem 2.6 we have

$$\theta_r^{p+1} = \theta_j^{p+1} + 1 \tag{2.4}$$

From (2.4) and the choice of r, we have $\theta_j^{p+1} \geqq \theta_j^p$. So

$$\theta_r^{p+1} \geqq \theta_j^p + 1 \tag{2.5}$$

Now, suppose $f_{jr}^{p-1} < k_{jr}$. By the manner in which scanning is done, it is clear that $\theta_r^p \leqq \theta_j^p + 1$. This and (2.5) together imply that $\theta_r^p \leqq \theta_j^p + 1 \leqq \theta_r^{p+1}$; this is a contradiction. Hence $f_{jr}^{p-1} = k_{jr}$. This, and the fact that (j, r) is a forward in-tree arc in the $(p+1)$th tree growth routine, together imply that (j, r) must have been a reverse arc in the FAP identified during the pth tree growth routine. By Theorem 2.6 and the manner in which labeling is done, this implies $\theta_j^p = \theta_r^p + 1$. This together with (2.5) implies that $\theta_r^{p+1} \geqq \theta_r^p + 2$, a contradiction to the choice of r. Hence we must have $\theta_i^{u+1} \geqq \theta_i^u$ for all i and $u \geqq 1$. A similar proof holds for $\vartheta_i^{u+1} \geqq \vartheta_i^u$ for all i and $u \geqq 1$. ∎

THEOREM 2.8 *Let m, n be the number of arcs, nodes respectively in the network $G = (\mathcal{N}, \mathcal{A}, \ell, k, \check{s}, \check{t})$. Beginning with an initial feasible flow vector in G, the shortest augmenting path method terminates with a maximum value flow vector after at most $mn/2$ trees are grown.*

Proof Let (i, j) be an arc on an FAP from $\check{s}$ to $\check{t}$ wrt a flow vector $\tilde{f}$ in G, and $\hat{f}$ the flow vector obtained after augmenting $\tilde{f}$ using this FAP. Then (i, j) is said to be a *critical forward arc* if $\hat{f}_{ij} = k_{ij}$, a *critical reverse arc* if $\hat{f}_{ij} = \ell_{ij}$, and a *critical arc* if it is either a critical forward or reverse arc.

Suppose (i, j) is a critical arc in the FAP obtained during the ath tree growth routine in the shortest augmenting path method applied on G beginning with an initial feasible flow vector, and again for the next time in the bth tree growth routine, $b > a$. Assume that (i, j) was a critical forward arc in the FAP of the ath tree growth routine (a proof similar to the following holds if it is a critical reverse arc on this FAP). So,

$$f_{ij}^a = k_{ij} \tag{2.6}$$

There are two cases to consider. Either $f_{ij}^{b-1} = k_{ij}$, or $< k_{ij}$. Consider the case $f_{ij}^{b-1} = k_{ij}$ first. In this case (i, j) must be a critical reverse arc in the FAP of the bth tree growth routine. So, from the manner in which labeling is done we have $\theta_j^a = \theta_i^a + 1$, and $\theta_i^b = \theta_j^b + 1$. Also, from Theorem 2.7 we have $\theta_j^b \geqq \theta_j^a$. From all these we have $\theta_i^b = \theta_j^b + 1 \geqq \theta_j^a + 1 = \theta_i^a + 2$. Also, from Theorem 2.7 we have $\vartheta_i^b \geqq \vartheta_i^a$. So, in this case we have

$$\theta_i^b + \vartheta_i^b \geqq \theta_i^a + \vartheta_i^a + 2 \tag{2.7}$$

Now consider the case where $f_{ij}^{b-1} < k_{ij}$. From (2.6), we conclude that this can only happen if there exists a w such that $a + 1 \leqq w \leqq b - 1$, and in the FAP obtained during the wth tree growth routine (i, j) is a reverse arc. Using the same arguments as above, in this case we have

$$\theta_i^w + \vartheta_i^w \geqq \theta_i^a + \vartheta_i^a + 2 \tag{2.8}$$

However, since $b > w$, by Theorem 2.7, $\theta_i^b + \vartheta_i^b \geqq \theta_i^w + \vartheta_i^w$. This, together with (2.8) implies that (2.7) holds in this case also. Thus (2.7) holds always.

If node i is on the FAP obtained in the uth tree growth routine, the length of that FAP is $\theta_i^u + \vartheta_i^u$. By (2.7) we therefore conclude that each succeeding FAP obtained during the algorithm in which the arc (i, j) is critical, is longer than the preceding one by at least two arcs. Also, in this algorithm, an FAP can have at most $n - 1$ arcs. So, no arc can appear as a critical arc in an FAP obtained during this version more than $n/2$ times. So, the total number of FAPs obtained during this version cannot exceed $mn/2$. ∎

The result in Theorem 2.8 holds irrespective of whether the data (lower bounds, capacities, and the initial feasible flow vector) is integral, rational, or irrational. Each application of the tree growth routine requires at most $O(m)$ effort in terms of comparisons, additions, or lookups. So, the overall computational effort required by the shortest augmenting path method is at most $O(nm^2)$. Since $m \leqq n(n - 1)$, this is at most $O(n^5)$. Zadeh [1972] has constructed networks with n nodes and m arcs on which this method requires $O(nm)$ flow augmentations to find a maximum value flow.

2.4 Multipath Labeling Methods Beginning with a Feasible Flow Vector

For the maximum value flow problem in the directed single commodity flow network $G = (\mathcal{N}, \mathcal{A}, \ell, k, \breve{s}, \breve{t})$, we discuss here a new class of augmentation methods that use all shortest FAPs simultaneously. In each FAP used in a step, the orientation of the reverse arcs is changed so that the FAP gets transformed into a chain from $\breve{s}$ to $\breve{t}$. When these chains corresponding to all the FAPs used in a step are put together, we get an acyclic network known as an *auxiliary network* wrt the present flow vector in G; it will be a partial subnetwork of the residual network at this stage, containing all the nodes and arcs that lie on at least one shortest chain from $\breve{s}$ to $\breve{t}$ in it.

The auxiliary network wrt the flow vector $\bar{f}$ is denoted by $\mathrm{AN}(\bar{f}) = (\mathrm{N}(\bar{f}), \mathrm{A}(\bar{f}))$. Since it is a partial subnetwork of the residual network $G(\bar{f})$, arcs in it carry $+$ or $-$ labels. So, $\mathrm{A}(\bar{f})$ is partitioned into $\mathrm{A}^+(\bar{f}) \cup \mathrm{A}^-(\bar{f})$. All lower bounds in $\mathrm{AN}(\bar{f})$ are 0. If $(i,j) \in \mathrm{A}^+(\bar{f})$, it has a $+$ label and capacity $\kappa_{ij} = k_{ij} - \bar{f}_{ij}$ and it corresponds to $(i,j) \in \mathcal{A}$ which satisfies $\bar{f}_{ij} < k_{ij}$. If $(i,j) \in \mathrm{A}^-(\bar{f})$, it has a $-$ label and capacity $\kappa_{ij} = \bar{f}_{ji} - \ell_{ji}$ and it corresponds to $(j,i) \in \mathcal{A}$ which satisfies $\bar{f}_{ji} > \ell_{ji}$. Every chain in $\mathrm{AN}(\bar{f})$ from $\breve{s}$ to $\breve{t}$ becomes an FAP wrt $\bar{f}$ in G when the orientations of the $-$ labeled arcs in it are reversed.

All these methods find a maximal or blocking flow vector in $\mathrm{AN}(\bar{f})$ using efficient special procedures. To avoid confusion with flow vectors in G, we will use the symbol $g = (g_{ij})$ to denote flow vectors in auxiliary networks. The maximal flow vector in the auxiliary network is then used to revise the flow vector in G, and the whole process is repeated. Here are the steps in these methods.

THE GENERAL MULTIPATH METHOD

Step 1 Initialization Initiate the algorithm with a feasible flow vector in G.

Step 2 Auxiliary network construction Let $\bar{f}$ be the present flow vector. Construct $\mathrm{AN}(\bar{f})$. This may lead to two possible outcomes: (1) The conclusion that $\bar{f}$ is of maximum value. In this case terminate the algorithm. (2) $\mathrm{AN}(\bar{f})$ itself.

Step 3 Find a maximal feasible flow vector in the auxiliary network Use a procedure to find a maximal feasible flow vector $\hat{g}$ in $\mathrm{AN}(\bar{f})$.

Step 4 Augmentation Compute the new flow vector $\hat{f} = (\hat{f}_{ij})$ in G by

$$
\hat{f}_{ij} = \begin{cases}
\bar{f}_{ij} & \text{if } (i,j) \text{ is not an arc in } \mathrm{AN}(\bar{f}) \\[2mm]
\bar{f}_{ij} + \hat{g}_{ij} & \text{if } (i,j) \text{ is a } + \text{ arc in } \mathrm{AN}(\bar{f}) \\[2mm]
\bar{f}_{ij} - \hat{g}_{ij} & \text{if } (j,i) \text{ is a } - \text{ arc in } \mathrm{AN}(\bar{f})
\end{cases}
$$

$\hat{f}$ is a feasible flow vector in G of value $\bar{v} + \hat{w}$, where $\hat{w}$ is the value of $\hat{g}$ in AN($\hat{f}$). Go to Step 2 with $\hat{f}$.

So, for each method we need only to describe the method to be used for constructing the auxiliary network, and the procedure to be used for finding a maximal feasible flow vector in it.

2.4.1   Dinic's Method

In this method, the auxiliary network is called the *layered network*. Nodes in it are partitioned into nonempty subsets $\mathcal{N}_0, \mathcal{N}_1, \ldots$ called *layers*, and every arc in it joins a node in some layer to a node in the next. The procedure for constructing it builds a layer at a time and terminates when (i) $\check{t}$ lies in a layer (in this case this will be the last layer), or (ii) $\check{t}$ is not in any layer so far, and there are no nodes that can be included in the next layer (this implies that the present flow vector in G has maximum value). So, if the procedure completes the construction of the layered network, the last layer contains $\check{t}$. The *length* of a layered network is the number of layers in it.

PROCEDURE FOR CONSTRUCTING THE LAYERED NETWORK

Let $\bar{f}$ be the present feasible flow vector in G of value $\bar{v}$.

Defining the initial layer   Define $\mathcal{N}_0 = \{\check{s}\}$.

Constructing the next layer   Let $\mathcal{N}_r$ be the last layer constructed so far. Define $\mathbf{X}_r = \bigcup_{h=0}^{r} \mathcal{N}_h$, $\bar{\mathbf{X}}_r = \mathcal{N} \backslash \mathbf{X}_r$, and $A_{r+1}(\bar{f}) = A_{r+1}^+(\bar{f}) \cup A_{r+1}^-(\bar{f})$, where

$$A_{r+1}^+(\bar{f}) = \{(i,j) : i \in \mathcal{N}_r, j \in \bar{\mathbf{X}}_r, (i,j) \in \mathcal{A}, \text{ and } \bar{f}_{ij} < k_{ij}\}$$

$$A_{r+1}^-(\bar{f}) = \{(i,j) : i \in \mathcal{N}_r, j \in \bar{\mathbf{X}}_r, (j,i) \in \mathcal{A}, \text{ and } \bar{f}_{ji} > \ell_{ji}\}$$

If $A_{r+1}(\bar{f}) = \emptyset$, $\bar{f}$ has maximum value in G and $[\mathbf{X}_r, \bar{\mathbf{X}}_r]$ is a minimum capacity cut separating $\check{s}$ and $\check{t}$ in G. Terminate the whole method.

If $A_{r+1}(\bar{f}) \neq \emptyset$, define the next layer to be $\mathcal{N}_{r+1} = \{j : j \text{ is the head of some arc in } A_{r+1}(\bar{f})\}$, and include all the arcs in $A_{r+1}(\bar{f})$ in the layered network. Each arc $(i,j) \in A_{r+1}^+(\bar{f})$ gets a + label, and has capacity $\kappa_{ij} = k_{ij} - \bar{f}_{ij}$. Each arc $(i,j) \in A_{r+1}^-(\bar{f})$ gets a − label, and has capacity $\kappa_{ij} = \bar{f}_{ji} - \ell_{ji}$. The lower bounds for all the arcs in the layered network are always zero.

If $\check{t} \in \mathcal{N}_{r+1}$, define N($\bar{f}$) = $\mathbf{X}_r \cup \mathcal{N}_{r+1}$, and A($\bar{f}$) to be the set of all arcs included so far. (N($\bar{f}$), A($\bar{f}$)) is the layered network; terminate the construction procedure.

If $\check{t} \notin \mathcal{N}_{r+1}$, repeat this step for constructing the next layer.

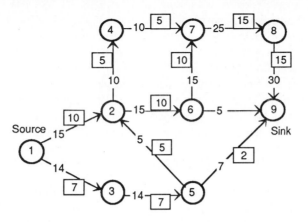

Figure 2.15 All lower bounds are 0, and capacities are entered on the arcs. Nonzero flow amounts are entered in boxes by the side of the arcs.

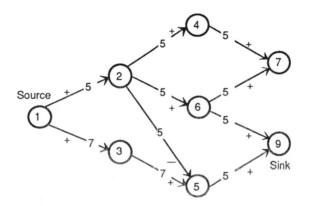

Figure 2.16 The layered network. Capacities are marked on the arcs. Arc labels, $+, -$ are also entered. Length of this layered network is 3.

If $A_{r+1}(\bar{f}) = \emptyset$, we have $\check{s} \in X_r$ and $\check{t} \in \bar{X}_r$, so $[X_r, \bar{X}_r]$ is a cut separating $\check{s}$ and $\check{t}$. Besides, $A_{r+1}(\bar{f}) = \emptyset$ implies that $\bar{f}_{ij} = k_{ij}$ for all $(i,j) \in (X_r, \bar{X}_r)$ and that $\bar{f}_{ij} = \ell_{ij}$ for all $(i,j) \in (\bar{X}_r, X_r)$, i.e., $\bar{f}(X_r, \bar{X}_r) - \bar{f}(\bar{X}_r, X_r) = k(X_r, \bar{X}_r) - \ell(\bar{X}_r, X_r)$. These facts together with the results in Theorems 2.1 and 2.2 imply that in this case $\bar{f}$ is a maximum value flow and that $[X_r, \bar{X}_r]$ is a minimum capacity cut separating $\check{s}$ and $\check{t}$ in G.

Since all the arcs in a layered network are forward arcs joining a node in a layer to the next layer, it is an acyclic network.

L, the length of the layered network wrt $\bar{f}$, is clearly the length (in terms of the number of arcs) of a shortest FAP in G from $\check{s}$ to $\check{t}$ wrt $\bar{f}$. $L \overset{\leq}{=} n - 1$. Every chain from $\breve{s}$ to $\breve{t}$ in the layered network has length L. If $i \in \mathcal{N}_h$, any chain from $\breve{s}$ to i in the layered network has length h arcs, and any chain from i to $\breve{t}$ has L $-h$ arcs. And every chain from $\breve{s}$ to $\breve{t}$ in the layered network corresponds to a shortest FAP in G wrt the present flow vector.

In the procedure for constructing the layered network, we may have to examine each arc in G at most twice, once from each of its nodes. So, the computational effort needed to construct the layered network is at most O(m), where $m = |\mathcal{A}|$.

As an example consider the network in Figure 2.15 with an initial feasible flow vector $\bar{f}$ of value 17 marked on it. The layered network wrt $\bar{f}$ in this example is given in Figure 2.16, with the nodes in each layer aligned vertically.

DINIC'S PROCEDURE FOR FINDING A MAXIMAL FLOW IN THE LAYERED NETWORK

Let $\mathcal{L} = ($ N, A, 0, $\kappa, \breve{s}, \breve{t})$ denote the layered network. The procedure begins with the flow vector $g^0 = 0$ in $\mathcal{L}$. It augments the flow using FACs detected by *depth first search*. It maintains a subset **F** of unsaturated arcs satisfying the property that if there is an FAC from $\breve{s}$ to $\breve{t}$ in $\mathcal{L}$ wrt the present flow vector g, all the arcs on it must lie in **F**. Arcs are deleted from **F** whenever they become saturated or whenever it becomes clear that there exists no FAC containing them. Termination occurs when **F** becomes $\emptyset$. Let $\mathcal{N}_0, \ldots, \mathcal{N}_L$ be the layers in $\mathcal{L}$.

Step 1 Initialization　　Start with the flow vector g^0 in $\mathcal{L}$. Let **F** = A.

Step 2 Begin depth first search　　Label $\breve{s}$ with $\emptyset$ and make it the *current node*.

Step 3 Search for an arc incident out of the current node　　Let $i \in \mathcal{N}_r$ be the current node. Look for an arc incident out of i in **F**. If none, go to Step 6 if $i = \breve{s}$, or to Step 5 otherwise. If there is such an arc, suppose (i, j) is the one selected. Clearly $j \in \mathcal{N}_{r+1}$. Label j with predecessor index i. If $j = \breve{t}$, an FAC has been found; go to Step 4. Otherwise make j the new current node replacing i from this status, and repeat this step.

Step 4 Flow augmentation　　Find the FAC by a backwards trace of node labels beginning with $\breve{t}$ and add its residual capacity to the flow amounts on all its arcs. Delete all the saturated arcs in the new flow vector from **F**. Erase the labels on all the nodes and go back to Step 2.

Step 5 Arc deletion from F　　Since i is the current node, and there exists no arc incident out of i in **F**, there exists no FAC from $\breve{s}$ to $\breve{t}$ through i wrt the present flow vector. Let p be the predecessor index of i. Delete all arcs incident into i from **F**; make p the current node replacing i from this status, and go back to Step 3

Step 6 Termination　　The present flow vector $\bar{g}$ is a maximal flow vector in $\mathcal{L}$. Terminate.

Discussion

When the procedure reaches Step 6, $\check{s}$ is the current node and there exists no arc incident out of it in **F**. This implies that there exists no FAC from $\check{s}$ to $\check{t}$ in $\mathcal{L}$ wrt the present flow vector $\bar{g}$ (i.e., it is a maximal flow vector).

Whenever Steps 4 or 5 occur, at least one arc is deleted from **F**. So, the total number of times that either Step 4 or 5 can occur is $m = |A|$. The amount of work in between two consecutive occurrences of either Step 4 or 5 is at most that of L consecutive node labelings, which is at most O(n), where n is the number of nodes in $\mathcal{L}$. Thus the overall computational effort in this procedure is at most O(mn).

As an example consider the layered network $\mathcal{L}$ in Figure 2.16. To find a maximal flow vector in this by Dinic's procedure, we begin with $g^0 = 0$, and **F** = set of all arcs in $\mathcal{L}$. We label the nodes in the order 1, 2, 4, 7, and then Step 5 occurs. Arcs (4, 7), (6, 7) get deleted from **F** and 4 becomes the current node. Again Step 5 occurs, and the arc (2, 4) gets deleted from **F**. Now 2 becomes the current node, and nodes 5, 9 are labeled in this order next. We have an FAC consisting of arcs (1, 2), (2, 5), (5, 9), with residual capacity 5. The new flow vector is given in Figure 2.17.

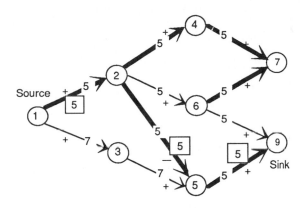

Figure 2.17 Nonzero flow amounts are entered in boxes by the side of the arcs. Thick arcs are those deleted from **F** at this stage.

At this stage the set **F** consists of only the thin arcs in Figure 2.17. When the procedure continues, after some labeling, arcs (3, 5), (1, 3) get deleted from **F** in that order, and the procedure terminates. The flow vector in Figure 2.17 is maximal.

THEOREM 2.9 *The layered network construction step has to be carried out at most* $(n-1)$ *times in this method before a maximum value flow vector is found in* G, *where* $n = |\mathcal{N}|$.

Proof We will prove that each successive layered network is strictly longer than the previous one in this method. Let $\mathcal{L}_u$ denote the uth layered network constructed in this method, and L_u its length, for $u = 1, 2, \ldots$. Let $f^{u-1} = (f_{ij}^{u-1})$ be the feasible flow vector in G at the beginning of construction of $\mathcal{L}_u, u = 1, 2,$ $\ldots$. Let $\mathcal{N}_0^u, \mathcal{N}_1^u, \ldots, \mathcal{N}_{L_u}^u$ be the layers in $\mathcal{L}_u$. Fix u. We will now prove that $L_{u+1} > L_u$.

There exists a chain of length L_{u+1} in $\mathcal{L}_{u+1}$ from $\breve{s}$ to $\breve{t}$. Let $\mathcal{C}$ be such a chain and suppose the sequence of nodes on this chain is $\breve{s} = i_0, i_1, \ldots, i_{-1+L_{u+1}}, i_{L_{u+1}} = \breve{t}$. Then $i_r \in \mathcal{N}_r^{u+1}$ for all $r = 0$ to L_{u+1}. We consider two cases.

Case 1 All the nodes $i_r, r = 0$ to L_{u+1} appear in $\mathcal{L}_u$: In this case suppose $i_r \in \mathcal{N}_p^u$. We will now show that $r \overset{\geq}{=} p$ by induction on r. For $r = 0$, $i_0 = \breve{s} \in \mathcal{N}_0^u$, so the statement is true.

Induction Hypothesis For all $a \overset{\leq}{=} r$, if $i_a \in \mathcal{N}_b^u$, then $a \overset{\geq}{=} b$.

We will now show that under the induction hypothesis, the statement there must also hold for $a = r + 1$. $i_r \in \mathcal{N}_p^u$ and suppose $i_{r+1} \in \mathcal{N}_d^u$. By the induction hypothesis, $r \overset{\geq}{=} p$. We have to show that these facts imply that $r + 1 \overset{\geq}{=} d$. If $d \overset{\leq}{=} p+1$, we are done. Suppose $d > p+1$. This implies that (i_r, i_{r+1}) is not an arc in $\mathcal{L}_u$. So, the flow on the arc in G corresponding to (i_r, i_{r+1}) in $\mathcal{L}_{u+1}$ must have remained unchanged as we move from f^{u-1} to f^u, and since (i_r, i_{r+1}) is an arc in $\mathcal{L}_{u+1}$, but not $\mathcal{L}_u$, we have a contradiction. So d cannot be $> p + 1$. Hence the statement in the induction hypothesis must also be true for $a = r+1$. By induction it is true for all a.

Since $i_{L_{u+1}} = \breve{t}$, and $\breve{t} \in \mathcal{N}_{L_u}^u$, we have $L_{u+1} \overset{\geq}{=} L_u$. If $L_{u+1} = L_u$, by the above statement the entire chain $\mathcal{C}$ must be in $\mathcal{L}_u$. So, $\mathcal{C}$ is in both $\mathcal{L}_u$ and $\mathcal{L}_{u+1}$. If g^u is the maximal flow vector obtained in $\mathcal{L}_u$ in Dinic's method, at least one arc in $\mathcal{C}$ must be saturated in g^u, and by the augmentation step in this method, this arc cannot be in $\mathcal{L}_{u+1}$, which is a contradiction. So, $L_{u+1} > L_u$ in this case.

Case 2 Not all the nodes $i_r, r = 0$ to L_{u+1} appear in $\mathcal{L}_u$: i_0 and $i_{L_{u+1}}$ ($\breve{s}$ and $\breve{t}$ respectively) are in both $\mathcal{L}_u$ and $\mathcal{L}_{u+1}$. Let $r + 1$ be the smallest value of d such that i_d does not appear in $\mathcal{L}_u$. So, $0 < r + 1 < L_{u+1}$; i_r appears in $\mathcal{L}_u$, in the layer $\mathcal{N}_p^u$, say; and (i_r, i_{r+1}) is an arc in $\mathcal{L}_{u+1}$ but not in $\mathcal{L}_u$. Let (i, j) be the arc in G corresponding to (i_r, i_{r+1}) in $\mathcal{L}_{u+1}$. Since (i_r, i_{r+1}) is not an arc in $\mathcal{L}_u$, we must have $f_{ij}^u = f_{ij}^{u-1}$. This and the fact that (i_r, i_{r+1}) is an arc in $\mathcal{L}_{u+1}$ imply that the only possible reason for (i_r, i_{r+1}) not being an arc in $\mathcal{L}_u$ must be that $p + 1 = L_u$. By the inductive argument of Case 1, $r \overset{\geq}{=} p$. Thus $r + 1 \overset{\geq}{=} L_u$, and therefore $L_{u+1} > L_u$ in this case too.

The length of any layered network is at most $n - 1$, where $n = |\mathcal{N}|$. Since each successive layered network obtained in the method is strictly longer than the previous, the maximum number of layered networks constructed in Dinic's method before termination is at most $(n - 1)$. ∎

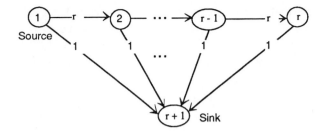

Figure 2.18 Network for the rth problem in the class. All lower bounds are 0. The horizontal arcs have capacity r, and others have capacity 1.

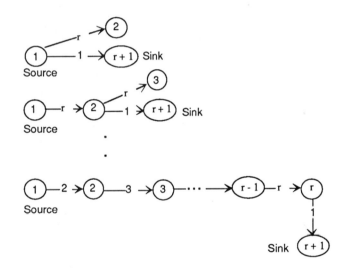

Figure 2.19 The various layered networks generated. All arcs are + arcs. Capacities are entered on the arcs.

We will now provide a class of examples due to Waissi [1985] in which Dinic's method constructs $(n-1)$ layered networks in an n node network. The network for the rth problem in the class has $r+1$ nodes and is given in Figure 2.18. The arc flow capacity is r for all arcs $(i, i+1), i = 1$ to $r-1$, and 1 for all arcs $(i, r+1), i = 1$ to r. To find a maximum value flow in it, beginning with the 0-flow vector, Dinic's method generates r layered networks, shown in Figure 2.19. There is a unique simple chain from source to sink in each layered network, and the unique maximal flow in this layered network consists of a flow of 1 unit on each arc of this chain and zero flow on the other arcs. The value of the flow vector in the original network goes up by 1 after augmentation using the maximal flow vector in each successive layered network, finally reaching the maximum value of r.

The effort required to construct a layered network and to find a maximal flow vector in it have already been shown to be O(mn), where $m = |\mathcal{A}|, n = |\mathcal{N}|$. By Theorem 2.9, at most $n-1$ layered networks are constructed in the method before termination. Thus the overall effort in Dinic's method is at most O(mn^2) $\leqq$ O(n^4).

2.4.2   Dinic-MKM Method

The auxiliary network used in this method is also Dinic's layered network. But to find a maximal flow in each layered network, it uses an algorithm of V. M. Malhotra, M. P. Kumar, and S. N. Maheswari [1978]. Let $\mathcal{L} = (N, A, 0, \kappa, \check{s}, \check{t})$ be the layered network. This algorithm does not use FACs; it changes the flow vector by operations called *flow pushing* and *pulling*. It maintains a set of nodes **Y** and a set of arcs **F** in $\mathcal{L}$. **Y** will always be the set of nodes on arcs in **F**. As the algorithm progresses, saturated arcs are deleted from **F**. Nodes are also deleted from **Y**, and whenever a node is deleted from **Y**, all the arcs incident at it are deleted from **F**. The following property will always hold: If there is an FAC from $\check{s}$ to $\check{t}$ in $\mathcal{L}$ wrt the present flow vector g, all the nodes on it must lie in **Y**, and all the arcs on it must lie in **F**.

If $\bar{g}$ is the present flow vector in $\mathcal{L}$, and **Y**,**F** are the present sets, for each $i \in$ **Y**, define wrt $\bar{g}$, **Y**, **F**

$$\alpha(i) = \text{ in-potential of } i = \sum(\kappa_{ji} - \bar{g}_{ji} : \text{ over } j \text{ s. t. } (j,i) \in \mathbf{F})$$

$$\beta(i) = \text{ out-potential of } i = \sum(\kappa_{ij} - \bar{g}_{ij} : \text{ over } j \text{ s. t. } (i,j) \in \mathbf{F})$$

$$\rho(i) = \text{flow-potential of } i = \begin{cases} \text{min. } \{\alpha(i), \beta(i)\} & \text{for } i \neq \check{s}, \check{t} \\ \beta(i) & \text{for } i = \check{s} \\ \alpha(i) & \text{for } i = \check{t} \end{cases}$$

Whenever the flow vector g or the sets **Y**, **F** change, these flow potentials have to be updated.

THE MKM ALGORITHM FOR FINDING A MAXIMAL FLOW
IN A LAYERED NETWORK

Step 1 Initialization   Start with $g^0 = 0$ in $\mathcal{L}$. Let **Y** = N, **F** = A. Compute $\rho(i)$ for all $i \in$ **Y**. If $\rho(i) > 0$ for all $i \in$ **Y**, go to Step 2. If $\rho(i) = 0$ for at least one $i \in$ **Y**, go to Step 6.

Step 2 Reference node selection   Find $\rho = $ min. $\{\rho(i) : i \in \mathbf{Y}\}$. Let $p \in$ **Y** be a node which attains this minimum, break ties arbitrarily. Node p is the present *reference node*, and ρ is the present *reference potential*.

Step 3 Flow pushing and flow pulling   Let p be the reference node and ρ the reference potential.

FLOW PUSHING Begin at p, and push an excess flow of ρ out of p. This requires increasing the flow on the arcs in **F** which are in the forward star of p, one by one, saturating them one after the other, until the total increase reaches ρ. In this process, at most one outgoing arc has flow increased on it but remains unsaturated.

Now the excess flow has been pushed from node p to its adjacent nodes in **Y** in the next layer. If i is one of these nodes, and the flow increase on the arc (p, i) was γ, push the excess flow of γ out of node i in exactly the same way. Repeat with all adjacent nodes of p which received excess flow. Then repeat this process for nodes in the next layer that received excess flow, and continue this way until all the excess flow of ρ units reaches $\breve{t}$. In this process we can never get stuck with excess supply that cannot be pushed out of a node, because of the definition of ρ, which implies that the potential of every node in **Y** is $\overset{\geq}{=} \rho$.

FLOW PULLING Pull an excess flow of ρ units into p. This requires increasing the flow on arcs in **F** incident into p, one by one, saturating them one after the other, until the total increase reaches ρ, making sure that at most one incoming arc has flow added to it but remains unsaturated. If the flow increase on an arc (j, p) was δ, pull an excess flow of δ into node j in exactly the same way. Repeat with all adjacent nodes of p from which excess flow was pulled into p. Then repeat this process for nodes in the preceding layer from which excess flow was pulled, and continue the same way until all the excess flow of ρ units is pulled out of $\breve{s}$. Again we can never get stuck in this pulling process.

After the flow pushing and pulling is completed, we again have a feasible flow vector in $\mathcal{L}$.

Step 4 Updating the sets Y, F after Step 3 Delete all the saturated arcs from **F**. If all the arcs into or out of a node i are deleted from **F**, delete that node i from **Y** and all arcs incident at node i from **F**.

Step 5 Updating the potentials Update the in and out-potentials and the flow potential of all the nodes in **Y** wrt the present flow vector in $\mathcal{L}$, and the present sets **Y, F**. If $\rho(i) > 0$ for all $i \in \mathbf{Y}$, go to Step 2; otherwise go to Step 6.

Step 6 Updating Y, F after Step 5 If there are $i \subset \mathbf{Y}$ with $\rho(i) = 0$, go to Step 7 if either $\rho(\breve{s})$ or $\rho(\breve{t})$ is zero. If both $\rho(\breve{s}), \rho(\breve{t})$ are > 0, delete all nodes i for which $\rho(i) = 0$ from **Y** and all the arcs incident at such nodes from **F**. Go to Step 5.

Step 7 Termination Now, $\rho(\breve{s})$, or $\rho(\breve{t})$ has become 0. So, the present flow vector is a maximal flow vector in $\mathcal{L}$. Terminate.

Discussion

An example of flow pushing and pulling is given in Figures 2.20 and 2.21. Node 5 is the reference node with reference potential of 17 units. Only nodes adjacent to node 5 in the present set **Y** are shown in Figure 2.20. In pushing the flow out of node 5, we begin saturating the arcs incident out of node 5 in **F** one by one in some order, say, from top to bottom, until a total of 17 units is pushed. In pulling we do the same thing with arcs incident into node 5. The situation after the pushing and pulling at node 5 is indicated in Figure 2.21.

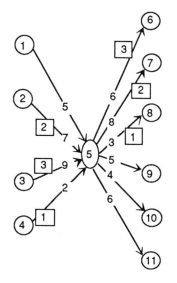

Figure 2.20 Arcs incident at reference node 5 are shown with their capacities and nonzero flow amounts in boxes.

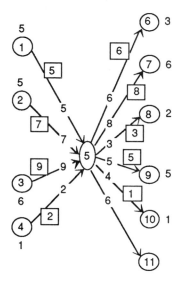

Figure 2.21 Flows on arcs incident at 5 after pushing and pulling. The amount by the side of each node is the amount to be pulled into or pushed out of it.

When we reach Step 7 in this algorithm, the flow potential of either $\breve{s}$ or $\breve{t}$ is 0. Then there exists no FAC in $\mathcal{L}$ from $\breve{s}$ to $\breve{t}$ wrt the present flow vector. Hence the flow vector in $\mathcal{L}$ at that time is a maximal flow vector.

Whenever Step 3 is completed (i.e., after pushing reaches all the way to $\breve{t}$ and pulling reaches all the way to $\breve{s}$) flow conservation holds at all the nodes and the flow vector becomes feasible. The net result of this step is to increase the flow value by the reference potential. Then either all the incoming arcs or all the outgoing arcs at the reference node have become saturated. These get deleted from **F**, and the reference node gets deleted from **Y** when we move over to Step 4. Likewise, each time an $i \in \mathbf{Y}$ with $\rho(i) = 0$ is noticed in Step 6, all the arcs at that node are deleted from **F** and that node is deleted from **Y**. Thus Steps 3 and 4, or Step 6 with an $i \in \mathbf{Y}$ satisfying $\rho(i) = 0$, can occur at most n times before termination, where n is the number of nodes in $\mathcal{L}$.

Let b_r be the number of arcs deleted from **F** during the rth time that a node is deleted from **Y**. If it has occurred in Step 6, the effort involved in executing that step is $O(b_r)$. If it has occurred in Step 4, the effort in flow pushing and pulling in the preceding Step 3 is at most $O(n + b_r)$, because in Step 3, at each node, at most one outgoing or incoming arc has flow added to it, but remains unsaturated in this step. Thus the effort needed during the rth execution of Steps 3 and 4, or Step 6 with an $i \in \mathbf{Y}$ satisfying $\rho(i) = 0$, is at most $O(n + b_r)$. Summing over all these executions, we find that the overall effort is at most $O(\sum_{r=1}^{n}(n + b_r))$. Since an arc deleted from **F** is never again considered for flow change during the algorithm, $\sum_r b_r \leq m$, where m is the number of arcs in $\mathcal{L}$. Thus the overall effort needed by this algorithm is at most $O(n^2 + m) = O(n^2)$.

The practical efficiency of the MKM algorithm for finding a maximal flow in a layered network $\mathcal{L} = (N, A)$ can be improved considerably by applying a routine known as the *backward pass routine* to find a partial subnetwork of $\mathcal{L}$ known as the *referent* before applying the algorithm. Let $\mathcal{N}_0, \ldots, \mathcal{N}_L$ be the layers in $\mathcal{L}$. This routine takes L+2 steps, and it prunes all the nodes and arcs in $\mathcal{L}$ which are not contained on any chain from $\breve{s}$ to $\breve{t}$ in $\mathcal{L}$ very efficiently. The remaining portion of the layered network after this pruning is completed is called the *referent*.

BACKWARD PASS ROUTINE

Step 0   Define $\hat{\mathcal{N}}_L = \{\breve{t}\}$.

General step r, $r = 1$ to L   When we reach this step, the sets of nodes $\hat{\mathcal{N}}_L, \ldots,$ $\hat{\mathcal{N}}_{L-r+1}$ would have been obtained. Define $\hat{\mathcal{N}}_{L-r} = \{i : i \in \mathcal{N}_{L-r}, \text{ and there}$ is a $j \in \hat{\mathcal{N}}_{L-r+1}$ s.t. $(i, j) \in A\}$, $\hat{A}_{L-r+1} = \{(i, j) : i \in \hat{\mathcal{N}}_{L-r}, j \in \hat{\mathcal{N}}_{L-r+1},$ $(i, j) \in A\}$.

$\hat{\mathcal{N}}_{L-r} \neq \emptyset$, as otherwise the referent is empty (i.e., there is no chain in from $\breve{s}$ to $\breve{t}$ in $\mathcal{L}$). This cannot happen from the manner in which the layered network is constructed. Go to the next step.

Step L+1   Let $\hat{N} = \bigcup_{r=0}^{L} \hat{\mathcal{N}}_r$, $\hat{A} = \bigcup_{r=1}^{L} \hat{A}_r$. $\hat{\mathcal{L}} = (\hat{N}, \hat{A})$ is the referent. Terminate.

Discussion

As an example, consider the layered network in Figure 2.16. When the backward pass is applied on it, nodes 7, 4 and arcs (4, 7), (6, 7), (2, 4) get pruned as they do not lie on any chain from node 1 (source) to node 9 (sink) in this network.

In the MKM algorithm for finding a maximal flow in $\mathcal{L}$, the pruning performed in the backward pass gets carried out at the beginning during the sequence of steps consisting of Step 1, followed by consecutive pairs of Steps 6 and 5, and finally another Step 6, before the method goes to Step 2 for the first time. But this pruning is carried out much more efficiently in the backward pass. Thus the practical efficiency of the MKM algorithm for maximal flow in $\mathcal{L}$ improves considerably if the backward pass is carried out immediately after Step 1 in the MKM algorithm if there is at least one $i \in \mathbf{Y}$ with $\rho(i) = 0$ at that time, instead of going to Step 6 from Step 1. Then, when the referent is obtained, begin MKM algorithm again in Step 1 with the referent.

By Theorem 2.9, the layered network construction step has to be carried out at most $(n-1)$ times in the Dinic-MKM method before a maximum value flow vector is found in G beginning with a feasible flow vector. The construction of each layered network and finding a maximal flow vector in it by MKM algorithm needs an effort of at most $O(m) + O(n^2) = O(n^2)$, where n, m are the number of nodes and arcs in G. Thus the overall effort for finding a maximum value flow vector in G beginning with a feasible flow vector by Dinic-MKM method is at most $O(n^3)$.

2.5   The Preflow Push Algorithm

In this section we consider the problem of finding a maximum value flow in the directed single commodity flow network G=$(\mathcal{N}, \mathcal{A}, 0, k, \breve{s}, \breve{t})$ where $k > 0, |\mathcal{N}| = n, |\mathcal{A}| = m$. All the algorithms discussed so far for this problem maintain a feasible flow vector throughout, and augment the flow value along augmenting paths, either one path at a time or all shortest augmenting paths at once using the layered network. An alternative method for this problem based on the concept of preflows has been initiated by Karzanov [1974]. A *preflow* in G is a vector $g = (g_{ij}; (i,j) \in \mathcal{A})$ which is bound feasible (i.e., $0 \leqq g \leqq k$) but in which, for each $i \in \mathcal{N}$, the amount of material flowing into node i is only required to be greater than or equal to the amount flowing out of node i. That is, a preflow g in G may not satisfy the flow conservation equations; instead it satisfies

$$g(\mathcal{N}, i) - g(i, \mathcal{N}) \geqq 0, \quad \text{for all } i \in \mathcal{N} \backslash \{\breve{s}, \breve{t}\}$$

Here we will describe an algorithm for the maximum value flow problem in G by A.V. Goldberg and R.E. Tarjan [1986, 1988] that maintains a preflow, and pushes local flow excess towards the sink. Only when the algorithm terminates does the preflow become a flow, and then it is a maximum value flow. In this section the symbol g denotes a preflow in G.

In Section 2.6, we will describe how the maximum value flow in a network with nonzero lower bounds on arc flows, can be found using the preflow push algorithm twice in two phases.

Given a preflow g in G, for each $i \in \mathcal{N} \backslash \{\breve{s}, \breve{t}\}$, the excess in g at node i is defined to be $e(i) = g(\mathcal{N}, i) - g(i, \mathcal{N})$, it is the net flow into node i in the preflow g. A node $i \in \mathcal{N} \backslash \{\breve{s}, \breve{t}\}$ is said to be an *active node* wrt g if $e(i) > 0$. The algorithm works by pushing the excess from active nodes to $\breve{t}$ or to nodes estimated to be closer to $\breve{t}$.

However, if $\check{t}$ is not reachable from an active node, the algorithm pushes the excess there to nodes estimated to be closer to $\check{s}$. The algorithm terminates when there are no active nodes; at that time the preflow is a feasible flow of maximum value.

In the residual network $G(g) = (\mathcal{N}, \mathcal{A}(g), 0, \kappa)$ wrt g, an estimate of the distance of a node from $\check{t}$ in terms of the number of arcs, is kept by maintaining a node label, called the *distance label*, denoted by $d(i)$, which is always a nonnegative integer or $+\infty$. Since it estimates the distance from the node to $\check{t}$, we define $d(\check{t}) = 0, d(\check{s}) = n$ always. In the algorithm, the preflow, distance label vector pair, g, d, are required to always satisfy

$$d(\check{t}) = 0, d(\check{s}) = n$$

$$d(i) \stackrel{\leq}{=} d(j) + 1, \quad \text{for every arc } (i, j) \text{ in } G(g) \tag{2.9}$$

The purpose of this definition is to make sure that if $d(i) < n$ for any i, then $d(i)$ is a lower bound on the actual distance from i to $\check{t}$ in $G(g)$; and if $d(i) \stackrel{\geq}{=} n$, then $d(i) - n$ is a lower bound on the actual distance from $\check{s}$ to i in $G(g)$. Conditions (2.9) are called the *validity conditions* for the distance label vector d.

Given the preflow g and distance label vector d, an arc (i, j) in $G(g)$ is said to be *admissible* wrt g, d, if i is an active node and $d(j) = d(i) - 1$.

PREFLOW PUSH ALGORITHM

Initialization Define the initial preflow g^0 in G by $g^0_{\check{s}j} = k_{\check{s}j}$ for all $j \in \mathbf{A}(\check{s})$, and $g^0_{pq} = 0$ for all other arcs $(p, q) \in \mathcal{A}$. The simplest choice for the initial distance node labels is $d^0 = (d^0(i))$ where $d^0(\check{s}) = n, d^0(i) = 0$ for all $i \in \mathcal{N} \backslash \{\check{s}\}$. A better choice which improves the practical efficiency of the algorithm is to determine $d^0(i)$ to be the distance labels obtained in a backward breadth-first-search of $G(g^0)$ starting at node $\check{t}$.

General step Let g, d be the present preflow, distance label pair. If there are no active nodes, terminate; g is a maximum value feasible flow vector in G. Otherwise perform one of the two following operations in any order.

PUSH Select an active node i and an admissible arc (i, j) in $G(g)$ incident out of i. If (i, j) has a $+$ label in $G(g)$, increase g_{ij} by $\varepsilon = \min\{e(i), \kappa_{ij}\}$; if (i, j) has a $-$ label in $G(g)$, decrease g_{ji} by ε. This push is said to be *saturating* if $\varepsilon = \kappa_{ij}$; otherwise it is *nonsaturating*. This push has the effect of decreasing $e(i)$ by ε and increasing $e(j)$ by ε if $j \neq \check{t}$.

RELABEL Select an active node i which has no admissible arc incident out of it in $G(g)$. So, we have $d(i) \stackrel{\leq}{=} d(j)$ for every j such that $(i, j) \in \mathcal{A}(g)$. It can be shown that we will have $0 < d(i) < n$. Replace $d(i)$ by $\min . \{d(j)+1 : j$ such that $(i, j) \in \mathcal{A}(g)\}$. This relabel operation resets $d(i)$ to the largest value allowed by the validity conditions (2.9). It creates at least one admissible arc at i on which a push operation can be carried out next.

Discussion

1. At any stage of the algorithm, if node i is an active node, clearly either a push or a relabel operation can be carried out at it.

2. If g, d are the present preflow, distance label pair, and a push operation is carried out leading to the preflow g^1 (a relabel operation is carried out leading to distance label vector d^1), then it can be verified that g^1, d (g, d^1) satisfy (2.9). So, (2.9) will hold throughout the algorithm.

3. If g, d are the present pair at some stage of the algorithm, there exists no chain from $\check{s}$ to $\check{t}$ in the residual network $G(g)$. This can be seen from the following. If there is a chain from $\check{s}$ to $\check{t}$ in $G(g)$, there must be a simple chain. Suppose $1 = \check{s}, 2, \ldots, l + 1 = \check{t}$ is the sequence of nodes on a simple chain in $G(g)$. From (2.9), we have $d(i) \leqq d(i + 1) + 1$ for $i = 1$ to l. Hence $d(\check{s}) = d(1) \leqq d(\check{t}) + l < n$ since $d(\check{t}) = 0$ and $l \leqq n - 1$, which contradicts $d(\check{s}) = n$.

4. If g, d are the present pair and i is an active node, then there is a chain from i to $\check{s}$ in $G(g)$. Suppose this is not the case. Let $\mathbf{X} = \{j : j \in \mathcal{N}$ such that there is a chain from i to j in $G(g)\}$. So, $\check{s} \notin \mathbf{X}$, and hence $\overline{\mathbf{X}} = \mathcal{N}\backslash\mathbf{X} \neq \emptyset$. From the definition of $G(g)$ and $\mathbf{X}$, we must therefore have $g_{pq} = k_{pq}$ for all $(p, q) \in \mathcal{A}$ satisfying $p \in \mathbf{X}, q \in \overline{\mathbf{X}}$; and $g_{pq} = 0$ for all $(p, q) \in \mathcal{A}$ satisfying $p \in \overline{\mathbf{X}}, q \in \mathbf{X}$. Now $e(\mathbf{X}) = g(\mathcal{N}, \mathbf{X}) - g(\mathbf{X}, \mathcal{N}) = g(\overline{\mathbf{X}}, \mathbf{X}) - g(\mathbf{X}, \overline{\mathbf{X}}) = -g(\mathbf{X}, \overline{\mathbf{X}})$ by the above, and since $g \geqq 0$ we have $e(\mathbf{X}) \leqq 0$. But $e(i) > 0$ since $i \in \mathbf{X}$ is an active node and $e(p) \geqq 0$ for all p as g is a preflow, there is a contradiction in $e(\mathbf{X}) \leqq 0$.

5. For all $i \in \mathcal{N}, d(i)$ never decreases during the algorithm, since distance labels change only when relabeling is done, and from the facts mentioned in the relabel operations.

6. Throughout the algorithm $d(i) \leqq 2n - 1$ for all $i \in \mathcal{N}$. Since $d(\check{t}) = 0, d(\check{s}) = n$ always, this statement is true for $\check{s}$ and $\check{t}$. Let $i \neq \check{s}$ or $\check{t}$. Suppose i is active at some stage. Then by 4, there exists a simple chain on which the sequence of nodes is $i_o = i, i_1, i_2, \ldots, i_l = \check{s}$, say, in the residual network $G(g)$ at that stage. So $l \leqq n - 1$, and (i_r, i_{r+1}) is an arc in $G(g)$ for all $r = 0$ to $l - 1$. So by (2.9), $d(i_r) \leqq d(i_{r+1}) + 1$. Hence $d(i) = d(i_0) \leqq d(i_l) + l \leqq d(\check{s}) + n - 1 = 2n - 1$. So for all active nodes $d(i) \leqq 2n - 1$ always. Since the algorithm changes distance labels for only active nodes, the statement holds for all $i \in \mathcal{N}$.

7. The total number of relabeling operations carried out is at most $(2n - 1)$ per node and $(2n - 1)(n - 2)$ during the entire algorithm. A relabeling operation carried out at a node i increases $d(i)$ by at least one. So, these facts follow from 6.

8. The total number of saturating push operations carried out is at most $2nm$. Consider an arc $(i,j) \in \mathcal{A}$ and a saturating push from i to j on it. The next push operation on this arc (which will decrease the flow on it, so it will be from j to i) cannot happen until $d(j)$ increases by at least 2. And similarly for the next time in the other direction again. Since $d(i) + d(j) \geqq 1$, when the first push between i and j occurs, and $d(i) + d(j) \leqq 4n - 3$ when the last such push occurs (by 6), the above fact implies that the total number of saturating pushes on (i,j) is at most $2n - 1$. So, the total number of saturating pushes over all edges is at most $(2n - 1)m < 2nm$.

9. The total number of nonsaturating pushing operations is at most $4n^2m$ in the algorithm. A nonsaturating push at an active node i with the admissible arc (i,j) in $G(g)$, makes node i inactive, and at that time $d(j) = d(i) - 1$, hence $d(i)$ must have been $\geqq 1$. Hence if we define $L = \sum(d(i)$: over active nodes i); L decreases by at least 1 in each nonsaturating push operation.

   Consider a saturating push at an active node i with the admissible arc (i,j) in $G(g)$. This might make node j active, hence by 6, this operation may increase L by at most $2n - 1$. So by 8, the total increase in L due to saturating push operations is at most $(2n - 1)2nm$.

   In a relabeling operation carried out on a node i, its distance label increases by $\gamma \geqq 1$, and it increases L by at most γ. By 5 and 6, the total increase in $d(i)$ for any node i during the entire algorithm by relabeling operations is at most $2n$. Since relabeling operations are carried out only at nodes $\neq \breve{s}$ or $\breve{t}$, the total increase in L by relabeling operations during the entire algorithm is at most $2n(n - 2)$.

   Initially $L \geqq 0$, and at termination, $L = 0$. Hence, the total of decreases in L during the algorithm and the total number of nonsaturating pushes, is $\leqq$ the total of increases in L during the algorithm, which is at most $(2n - 1)2nm + 2n(n - 2) \leqq 4n^2m$.

10. From 7, 8, and 9 we see that the total number of basic operations carried out in this algorithm is at most $O(n^2m)$.

    The running time of this algorithm depends on the order in which the push and relabel operations are applied and on the other details of the implementation. The simple scheme which selects an active node, maintains the set of residual arcs incident at it in some order, carries out the push operation at this node using these arcs in order, one after the other, and then relabels that node, can be shown to have running time of $O(n^2m)$. It has been shown that maintaining the set of active nodes as a queue and selecting the node for push/relabel operations using a first-in first-out rule, the worst case running time of the algorithm improves to $O(n^3)$. Using different rules for selecting nodes for push/relabel operations (for example, selecting the active node with the highest distance label, etc.) and exploiting the other flexible features of this algorithm, many different versions of the algorithm with improved complexity bounds have been obtained. From a theoretical worst case computational complexity aspect, the best version so far, based on the dynamic tree data structures, has a running time bound of $0(nm \log(n^2/m))$.

2.6   Phase 1 For Problems With $\ell \neq 0$

Let $G = (\mathcal{N}, \mathcal{A}, \ell, k, \check{s}, \check{t})$ be a directed single commodity flow network with $0 \leq \ell \leq k$. In this section we will discuss an algorithm for the Phase 1 problem of finding a feasible flow vector in G, if one exists.

Let G^1 be the network G with an additional artificial arc $(\check{t}, \check{s})$ with lower bound 0 and capacity $+\infty$. If f is a feasible flow vector of value v in G, define $f^1_{\check{t}\check{s}} = v$, and if $f^1 = (f, f^1_{\check{t}\check{s}})$, then f^1 is a feasible circulation in G^1. Conversely, given any feasible circulation f^1 in G^1, let $f^1_{\check{t}\check{s}}$ be the flow amount in f^1 on the artificial arc $(\check{t}, \check{s})$, and f the vector obtained by deleting the entry $f^1_{\check{t}\check{s}}$ from f^1. Then f is a feasible flow vector in G of value $f^1_{\check{t}\check{s}}$. Hence, the problem of finding a feasible flow vector in G is equivalent to that of finding a feasible circulation in G^1. For this problem in G^1, the original source and sink nodes $\check{s}, \check{t}$ do not play any special role, they are like any other node since it is a circulation problem. We will again transform this problem into that of finding a maximum value flow on a further augmented network G^* in which the lower bounds for all the arc flows are 0. In G, for $i \in \mathcal{N}$ define $\alpha_i = \ell(\mathcal{N}, i)$, $\beta_i = \ell(i, \mathcal{N})$. Add artificial source and sink nodes s^* and t^* to G^1. Introduce an artificial arc (s^*, i) with capacity α_i for each $i \in \mathcal{N}$ such that $\alpha_i \neq 0$, and an artificial arc (i, t^*) with capacity β_i for each $i \in \mathcal{N}$ such that $\beta_i \neq 0$. Change the capacity on each original arc $(i, j) \in \mathcal{A}$ to $k_{ij} - \ell_{ij}$. Make the lower bounds for flows on all the arcs 0. Let G^* be the resulting network. Notice that the source and sink nodes in G^* are the artificial nodes s^*, t^* respectively (the source and the sink nodes $\check{s}, \check{t}$ in G are transit nodes in G^* like all other nodes in $\mathcal{N}$).

Beginning with the initial feasible flow vector of 0, find a maximum value flow, f^*, say, in G^*. Let the value of f^* be v^*. The following conclusions hold.

1. Find $\ell(\mathcal{N}, \mathcal{N})$ in the original network G. If $v^* < \ell(\mathcal{N}, \mathcal{N})$, there exists no feasible circulation in G^1, and consequently no feasible flow vector in G (see Theorem 2.10 below).

2. If $v^* = \ell(\mathcal{N}, \mathcal{N})$, define $\hat{f}_{ij} = f^*_{ij} + \ell_{ij}$ for each $(i, j) \in \mathcal{A}$. Then $\hat{f} = (\hat{f}_{ij} : (i, j) \in \mathcal{A})$ is a feasible flow vector in G, and $(\hat{f}, f^*_{\check{t}, \check{s}})$ is a feasible circulation in G^1.

THEOREM 2.10 *A feasible circulation in* G^1, *and consequently a feasible flow vector in* G, *exist iff* v^*, *the maximum value in* G^* *is* $\ell(\mathcal{N}, \mathcal{N})$.

Proof Suppose f^* is a maximum value flow vector in G* of value $v^* = \ell(\mathcal{N}, \mathcal{N})$. Then all the arcs of the form $(s^*, i), (i, t^*)$ in G* are saturated in f^*, and the statements in 2 above can be verified to be true.

To show the converse, assume that $f = (f_{ij})$ is a feasible flow vector of value v in G. Define a flow vector f^* in G* by: $f_{ij}^* = f_{ij} - \ell_{ij}$ for all $(i, j) \in \mathcal{A}$, $f_{t\hat{s}}^* = v$, and $f_{s^*i}^* = \alpha_i, f_{it^*}^* = \beta_i$ for all such arcs in G*. Verify that f^* is feasible for G* and that it saturates all the arcs of the form $(s^*, i), (i, t^*)$. This implies that f^* is a maximum value flow in G* and that its value is $\ell(\mathcal{N}, \mathcal{N})$. ∎

As an example consider the network G in Figure 2.22. Data on the arcs is lower bound, capacity in that order. The corresponding networks G^1, G* are drawn in Figures 2.23 and 2.24. A maximum value flow in G* is entered in Figure 2.24, it saturates all the arcs of the form $(s^*, i), (i, t^*)$ and thus satisfies the condition in Theorem 2.10. The feasible flow vector in G constructed from this maximum value flow in G* is $(f_{12}, f_{13}, f_{23}, f_{32}, f_{24}, f_{34}) = (3, 4, 1, 2, 4, 3)$.

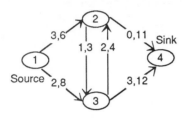

Figure 2.22 Network G

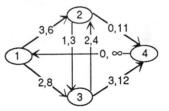

Figure 2.23 Network G^1

THEOREM 2.11 *CONDITIONS FOR THE EXISTENCE OF A FEASIBLE CIRCULATION* Let G $= (\mathcal{N}, \mathcal{A}, \ell, k)$ *be a directed single commodity flow network, where* $0 \leqq \ell \leqq k$. *A feasible circulation exists in G iff*

$$k(\mathbf{X}, \bar{\mathbf{X}}) \geqq \ell(\bar{\mathbf{X}}, \mathbf{X}), \text{ for all } \mathbf{X} \subset \mathcal{N}, \bar{\mathbf{X}} = \mathcal{N} \backslash \mathbf{X} \qquad (2.10)$$

Proof Introduce the artificial source and sink nodes, s^*, t^*, and arcs (s^*, i) with capacity $\ell(\mathcal{N}, i)$, (i, t^*) with capacity $\ell(i, \mathcal{N})$, for each $i \in \mathcal{N}$. Make the lower bounds on all the arcs zero, and change the capacity on $(i, j) \in \mathcal{A}$ to $k_{ij} - \ell_{ij}$. Let the resulting network be G* $= (\mathcal{N}^*, \mathcal{A}^*, 0, k^*, s^*, t^*)$. From Theorem 2.10 we know that a feasible circulation exists in G iff the maximum flow value from s^* to t^* in G* is $\ell(\mathcal{N}, \mathcal{N})$. A necessary and sufficient condition for this is that the capacity

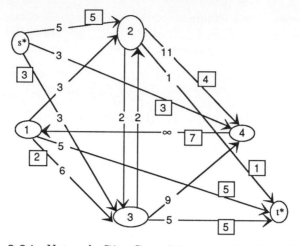

Figure 2.24 Network G*. Capacities are entered on the arcs. Nonzero flows are shown in boxes by the side of the arcs.

of every cut separating s^* and t^* in G* is $\geqq \ell(\mathcal{N},\mathcal{N})$. Let $[\,\mathbf{X}^*,\bar{\mathbf{X}}^*\,]$ be such a cut, and let $\mathbf{X} = \mathbf{X}^*\backslash\{s^*\}$, $\bar{\mathbf{X}} = \mathcal{N}\backslash\,\mathbf{X}$. The capacity of the cut $[\,\mathbf{X}^*,\bar{\mathbf{X}}^*]$ in G* is $\sum(k^*_{ij}$: over $(i,j) \in \mathcal{A}^*$, with $i \in \mathbf{X}^*$, $j \in \bar{\mathbf{X}}^*) = \sum(\,k_{ij} - \ell_{ij}$: over $(i,j) \in \mathcal{A}$ with $i \in\mathbf{X},\ j \in \bar{\mathbf{X}}) + \sum(\,\ell(\mathcal{N},j)$: over $j \in \bar{\mathbf{X}}) + \sum(\ell(i,\mathcal{N})$: over $i \in \mathbf{X}) = k(\mathbf{X}, \bar{\mathbf{X}})\, - \,\ell(\mathbf{X},\ \bar{\mathbf{X}}) + \ell(\mathcal{N},\bar{\mathbf{X}}) + \ell(\mathbf{X},\mathcal{N}) = k(\mathbf{X},\ \bar{\mathbf{X}}) + \ell(\bar{\mathbf{X}},\bar{\mathbf{X}}) + \ell(\mathbf{X},\mathcal{N})$. This cut capacity is $\geqq \ell(\mathcal{N},\mathcal{N})$, iff (2.10) holds. Therefore, a feasible circulation exists in G iff (2.10) holds. ∎

Let $|\mathcal{N}| = n$. There are 2^n conditions in (2.10). Thus the result in Theorem 2.11 is not practically useful for verifying the existence of feasible circulations in G unless n is small. Fortunately, the problem of finding either a feasible circulation in G, or a subset $\mathbf{X}\subset \mathcal{N}$ violating (2.10) can be carried out with at most $O(n^3)$ effort by applying the methods discussed earlier to find a maximum value flow in the augmented network G* discussed in the proof of Theorem 2.11. Suppose f^* is a maximum value flow in G* of value v^*. If $v^* = \ell(\mathcal{N},\mathcal{N})$, define $f = (f_{ij})$ by $f_{ij} = f^*_{ij} + \ell_{ij}$ for each $(i,j) \in \mathcal{A}$. It is a feasible circulation in G. If $v^* < \ell(\mathcal{N},\mathcal{N})$ (this happens if some of the arcs of the form $(s^*,i),(i,t^*)$ remain unsaturated in f^*), let $[\mathbf{Y}^*, \bar{\mathbf{Y}}^*]$ be a minimum capacity cut in G*. Then the subset $\mathbf{X} = \mathbf{Y}^*\backslash\{s^*\}$ can be verified to violate (2.10).

Conditions (2.10) have a practical interpretation. They require a sufficient escape capacity from the set of nodes $\mathbf{X}$ to disperse the flow forced into the set by the lower bound constraints on arc flows. They provide a useful infeasibility analysis procedure. If $\mathbf{X} \subset \mathcal{N}$ violates (2.10), either the capacities on arcs in $(\mathbf{X}, \bar{\mathbf{X}})$ have to be increased or the lower bounds on arcs in $(\bar{\mathbf{X}}, \mathbf{X})$ have to be reduced in order to remedy the situation.

Necessary and sufficient conditions for the existence of feasible flow vectors in any single commodity flow network can be derived by applying Theorem 2.11 to an appropriate modification of the network. As an example consider the directed single commodity flow network $G = (N, A, \ell, k, V)$, where V is the specified vector of exogenous flows at the nodes. Let $S = \{i : V_i > 0\}$, $N = \{i : V_i < 0\}$. S, N are the sets of source and sink nodes respectively in G. A flow vector in G is feasible if it satisfies the constraints in (1.20). Modify G by introducing the artificial nodes s, t, arcs (s, i) with lower bound and capacity equal to V_i for each $i \in S$, arcs (j, t) with lower bound and capacity both equal to $|V_j|$ for each $j \in N$, and the arc (t, s) with lower bound 0 and capacity ∞. Let $\hat{G} = (\hat{N}, \hat{A}, \hat{\ell}, \hat{k})$ denote the modified network. Clearly a feasible flow vector exists in G iff a feasible circulation exists in $\hat{G}$. Condition (2.10) corresponding to $X = \{s, t\}$, N respectively lead to $V(S) \geqq -V(N)$, $-V(N) \geqq V(S)$. Since $V_i = 0$ for all $i \in N \backslash (S \cup N)$, these conditions are equivalent to

$$V(N) = 0 \tag{2.11}$$

which is the same as (1.21). Let $X \subset N$, $\bar{X} = N \backslash X$. Condition (2.10) corresponding to the subset of nodes X, or $X \cup \{s\}$, or $X \cup \{s, t\}$, can be verified to lead to

$$k(X, \bar{X}) - \ell(\bar{X}, X) \geqq V(X), \text{ for all } X \subset N, \bar{X} = N \backslash X \tag{2.12}$$

Thus (2.11) and (2.12) are the necessary and sufficient conditions for the existence of a feasible circulation in $\hat{G}$, and consequently a feasible flow vector in G. If (2.11) holds, the procedure discussed above to find a feasible circulation in $\hat{G}$, either finds a feasible flow vector in G or a subset X violating (2.12).

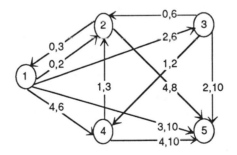

Figure 2.25 Data on the arcs is lower bound, capacity, in that order.

Exercises

2.10 Find a maximum value flow from 1 to 5 on the network in Figure 2.25.

2.11 Find feasible flow vectors in the networks in Figures 2.26 and 2.27. Data on the arcs is lower bound, capacity in that order. The exogenous flow at each node is entered by its side.

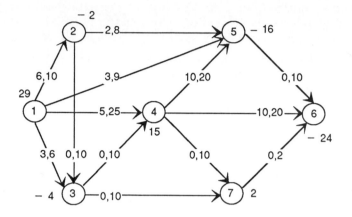

Figure 2.26

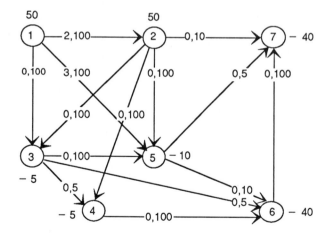

Figure 2.27

2.12 G $= (\mathcal{N}, \mathcal{A}, 0, k)$ is a directed single commodity flow network. **S**, **N** are the subsets of source, sink nodes respectively, and all the nodes in $\mathcal{N}\backslash(\mathbf{S} \cup \mathbf{N})$ are transient nodes. For each source node $i \in \mathbf{S}$, a positive quantity a_i is given; it is the maximum amount of material available at i for shipping out. For each sink node $j \in \mathbf{N}$, a positive quantity b_j is given; it is the minimum amount of material required to be delivered at j. Prove that a feasible flow vector satisfying all these constraints exists in G iff

$$k(\mathbf{X}, \bar{\mathbf{X}}) \geqq b(\mathbf{N} \cap \bar{\mathbf{X}}) - a(\mathbf{S} \cap \bar{\mathbf{X}}), \text{ for all } \mathbf{X} \subset \mathcal{N}, \bar{\mathbf{X}} = \mathcal{N} \backslash \mathbf{X}$$

Also, in this problem, if all a_i, b_j, k_{ij} are integers and a feasible flow vector exists, then prove that an integer feasible flow vector exists.

2.13 $G = (\mathcal{N}, \mathcal{A}, \ell, k)$ is a directed single commodity network with $\ell \leqq k$. For each $i \in \mathcal{N}, a_i$ and b_i are given integers satisfying $a_i \leqq b_i$. It is required to find a flow vector $f = (f_{ij})$ in G satisfying $\ell \leqq f \leqq k$ and $a_i \leqq f(i, \mathcal{N}) - f(\mathcal{N}, i) \leqq b_i$ for each $i \in \mathcal{N}$. Prove that a flow vector satisfying these conditions exists in G iff

$$k(\mathbf{X}, \bar{\mathbf{X}}) \geqq \ell(\bar{\mathbf{X}}, \mathbf{X}) + \text{max.} \{a(\mathbf{X}), -b(\bar{\mathbf{X}})\}, \text{ for all } \mathbf{X} \subset \mathcal{N}, \bar{\mathbf{X}} = \mathcal{N} \backslash \mathbf{X}$$

Discuss an efficient procedure that will either find a feasible flow vector if one exists, or determine a subset of nodes $\mathbf{X}$ violating the above condition.

2.14 Consider a round robin tournament between n baseball teams, with each team playing every other team exactly β times. For $i = 1$ to n, let α_i be the number of wins for the ith team at the conclusion of the tournament. Derive necessary and sufficient conditions on a given set of nonnegative integers $\alpha_1, \ldots, \alpha_n$ in order that they represent a possible win record (D. Gale).

2.15 $\tilde{f}$ is a non-integral circulation in a directed single commodity flow network G. Prove that there exists an integral circulation $\bar{f}$ in G satisfying $|\bar{f}_{ij} - \tilde{f}_{ij}| < 1$ for all arcs (i, j) in G. Discuss an efficient algorithm for finding such an $\bar{f}$.

How to find a maximum value flow in G with $\ell \neq 0$ using the preflow push algorithm

There is one significant difference between networks with zero lower bounds and those with nonzero lower bounds. If the lower bound vector is zero and the capacity vector is nonnegative, there is always a feasible flow vector, since the vector 0 is itself feasible. If the lower bound vector is nonzero, it is possible that there is no feasible flow vector.

The preflow push algorithm discussed in Section 2.5 is based on preflows. It obtains a sequence of preflows and terminates only when the preflow becomes a feasible flow vector, at which time it will be a maximum value flow. That's why it cannot detect infeasibility directly and is thus applied directly only for solving the maximum value flow problem in networks with zero lower bounds. Let *Algorithm 1* refer to any such algorithm. Here we show that Algorithm 1 can be used to find the maximum value flow in the network $G = (\mathcal{N}, \mathcal{A}, \ell, k, \breve{s}, \breve{t})$ with $\ell \neq 0$, by using it in two phases. This two-phase procedure from Yi and Murty [1991] is the following:

PROCEDURE

Phase I   In this phase, try to find a feasible flow vector in G. As discussed earlier, this problem can be transformed into that of finding a maximum value flow in an augmented network G^* in which lower bounds for all arc flows are zero. Because of this, a maximum value flow in G^* can be found by Algorithm 1 directly. From this either conclude that there is no feasible flow vector in G and terminate or obtain a feasible flow vector in it.

Suppose a feasible flow vector $\bar{f} = (\bar{f}_{ij})$ of value $\bar{v}$ has been found in G.

Phase II   Find a maximum value feasible flow vector in the residual network $G(\bar{f}) = (\mathcal{N}, \mathcal{A}(\bar{f}), 0, \kappa = (\kappa_{ij}), \breve{s}, \breve{t})$ with the same nodes $\breve{s}, \breve{t}$ as source, sink nodes. Since the lower bounds on all arc flows are zero in $G(\bar{f})$; this can be carried out by Algorithm 1. Let $\bar{h} = (\bar{h}_{pq})$ be the maximum value flow vector obtained in $G(\bar{f})$, and ω its value. Lower bounds in the residual network $G(\bar{f})$ are 0, and for a pair of nodes p, q, if there are arcs (p, q) and (q, p) both with positive flows in $\bar{h}$, then the flows on them can be canceled (i.e., replace the larger of $\bar{h}_{pq}, \bar{h}_{qp}$ by their difference and the smaller by 0) and at least one of these flows converted to 0, leading to another feasible flow vector in $G(\bar{f})$ of the same maximum value. We assume that this is done. So, without loss of generality, we assume that for any pair of nodes, the vector $\bar{h}$ has positive flows in at most one of the two orientations for arcs joining them.

Define a flow vector $\hat{f} = (\hat{f}_{ij})$ in G, where for $(i, j) \in \mathcal{A}$, $\hat{f}_{ij}$ is given by

$$
\hat{f}_{ij} = \begin{cases} \bar{f}_{ij} + \bar{h}_{ij} & \text{if there is a + labeled arc } (i,j) \text{ in} \\ & G(\bar{f}), \text{ corresponding to } (i,j) \in \mathcal{A}, \\ & \text{with } \bar{h}_{ij} > 0. \\ \bar{f}_{ij} - \bar{h}_{ji} & \text{if there is a } - \text{ labeled arc } (j,i) \text{ in} \\ & G(\bar{f}), \text{ corresponding to } (i,j) \in \mathcal{A}, \\ & \text{with } \bar{h}_{ji} > 0. \\ \bar{f}_{ij} & \text{otherwise} \end{cases}
$$

Then $\hat{f}$ is a maximum value feasible flow vector in G and its value is $\bar{v} + \omega$. Terminate.

THEOREM 2.12 *The flow vector $\hat{f}$ obtained in Phase II of the above procedure is a maximum value feasible flow vector in G.*

    Proof   Since $\bar{f}$ is a feasible flow vector in G of value $\bar{v}$, and $\bar{h}$ is a feasible flow vector of value ω in $G(\bar{f})$, the fact that $\hat{f}$ is a feasible flow vector of value $\bar{v} + \omega$

follows from the flow conservation equations satisfied by $\bar{f}$ and $\bar{h}$ in the respective networks G and $G(\bar{f})$ and by the definition of upper bounds in $G(\bar{f})$.

Now, to show that $\hat{f}$ is a mximum value flow in G, suppose it is not true. Then there must exist an FAP from $\check{s}$ to $\check{t}$ wrt $\hat{f}$ in G. Suppose it is $\mathcal{P}$. We will now show that using $\mathcal{P}$ we can construct an FAP from $\check{s}$ to $\check{t}, \mathcal{P}_1$, in $G(\bar{f})$ wrt $\bar{h}$.

1. If (i,j) is a forward arc on $\mathcal{P}$ with $\hat{f}_{ij} = \bar{f}_{ij}$, we have $\bar{f}_{ij} < k_{ij}$, so arc (i,j) exists in $G(\bar{f})$ with $+$ label and capacity $k_{ij} - \bar{f}_{ij} > 0$, and since $\hat{f}_{ij} = \bar{f}_{ij}$, we must have $\bar{h}_{ij} = 0$. So put (i,j) as a forward arc on $\mathcal{P}_1$.

2. If (i,j) is a forward arc on $\mathcal{P}$ with $\hat{f}_{ij} > \bar{f}_{ij}$, from the definition of $\hat{f}$, (i,j) must be a $+$ labeled arc in $G(\bar{f})$ with $\bar{h}_{ij} = \hat{f}_{ij} - \bar{f}_{ij} > 0$, and since $\hat{f}_{ij} = \bar{f}_{ij} + \bar{h}_{ij} < k_{ij}$, we have $\bar{h}_{ij} < k_{ij} - \bar{f}_{ij} = \kappa_{ij}$. So put (i,j) as a forward arc on $\mathcal{P}_1$.

3. If (i,j) is a forward arc on $\mathcal{P}$ with $\hat{f}_{ij} < \bar{f}_{ij}$, from the definition of $\hat{f}$, (j,i) must be a $-$ labeled arc in $G(\bar{f})$ with $\bar{h}_{ji} > 0$. Put (j,i) as a reverse arc on $\mathcal{P}_1$.

4. If (i,j) is a reverse arc on $\mathcal{P}$ with $\hat{f}_{ij} = \bar{f}_{ij}$, we have $\bar{f}_{ij} > \ell_{ij}$, so arc (j,i) must be a $-$ labeled arc in $G(\bar{f})$ with capacity $\kappa_{ji} = \bar{f}_{ij} - \ell_{ij} > 0$, and from the definition of $\hat{f}$, $\bar{h}_{ji} = 0$. Put (j,i) as a forward arc on $\mathcal{P}_1$.

5. If (i,j) is a reverse arc on $\mathcal{P}$ with $\hat{f}_{ij} > \bar{f}_{ij}$, from the definition of $\hat{f}$, (i,j) must be a $+$ labeled arc in $G(\bar{f})$ with $\bar{h}_{ij} = \hat{f}_{ij} - \bar{f}_{ij} > 0$. Put (i,j) as a reverse arc on $\mathcal{P}_1$.

6. If (i,j) is a reverse arc on $\mathcal{P}$ with $\hat{f}_{ij} < \bar{f}_{ij}$, from the definition of $\hat{f}$, (j,i) must be a $-$ labeled arc in $G(\bar{f})$ with $\bar{h}_{ji} = \bar{f}_{ij} - \hat{f}_{ij} > 0$; and since $\bar{f}_{ij} - \bar{h}_{ji} = \hat{f}_{ij} > \ell_{ij}$, we have $\bar{h}_{ji} < \bar{f}_{ij} - \ell_{ij} = \kappa_{ji}$. Put (j,i) as a forward arc on $\mathcal{P}_1$.

It can be verified that the path $\mathcal{P}_1$ constructed using statements 1 to 6 above is a path from $\check{s}$ to $\check{t}$ in $G(\bar{f})$, with the forward, reverse orientations for arcs on it as specified in these statements, and that it is an FAP from $\check{s}$ to $\check{t}$ in $G(\bar{f})$ with respect to $\bar{h}$. This contradicts the hypothesis that $\bar{h}$ is a maximum value flow from $\check{s}$ to $\check{t}$ in $G(\bar{f})$. So there does not exist any FAP from $\check{s}$ to $\check{t}$ in G with respect to $\hat{f}$. Hence $\hat{f}$ is a maximum value flow in G. ∎

The theorem shows that the two-phase procedure described here always finds a maximum value feasible flow vector in the given network G.

2.7 Sensitivity Analysis

Let $G = (\mathcal{N}, \mathcal{A}, \ell = 0, k, \check{s}, \check{t})$ be a directed single commodity flow network. Consider a particular arc $(i,j) \in \mathcal{A}$. Sensitivity analysis in G deals with the problem of deriving the maximum value flow as a function of k_{ij} as it varies from 0 to ∞, while all the other data remains unchanged.

To avoid confusion, denote k_{ij} by ξ and let $v(\xi)$ be the maximum value of flow in G as a function of ξ. Let $\mathbf{U}_1$ ($\mathbf{U}_2$) denote the set of all cuts separating $\check{s}$ and $\check{t}$ in G that contain (i, j) as a forward arc (do not contain (i, j) as a forward arc).

The capacity of any cut in $\mathbf{U}_2$ is unaffected by changes in the capacity ξ of arc (i, j). Suppose the minimum capacity among cuts in $\mathbf{U}_2$ is c_2; c_2 is defined to be $+\infty$ if $\mathbf{U}_2 = \emptyset$.

Let c_1 be the minimum capacity of cuts in $\mathbf{U}_1$ when $\xi = 0$. Since all cuts in $\mathbf{U}_1$ contain (i, j) as a forward arc, the minimum capacity among cuts in $\mathbf{U}_1$ as a function of ξ is $c_1 + \xi$.

Hence the minimum capacity of cuts separating $\check{s}$ and $\check{t}$ in G, as a function of ξ is min. $\{c_1 + \xi, c_2\}$. Thus $v(\xi) = \min. \{c_1 + \xi, c_2\}$.

Therefore, if $c_1 \overset{>}{=} c_2$, $V(\xi) = c_2$ for every $\xi \overset{>}{=} 0$. In this case there is a minimum capacity cut separating $\check{s}$ and $\check{t}$ in G which does not contain (i, j) as a forward arc, for every $\xi \overset{>}{=} 0$.

Suppose $c_1 < c_2$. In this case $v(\xi) = c_1 + \xi$ for $0 \overset{<}{=} \xi \overset{<}{=} c_2 - c_1$, and in this interval for ξ there is a minimum capacity cut separating $\check{s}$ and $\check{t}$ in G, which contains (i, j) as a forward arc. For $\xi > c_2 - c_1$, $v(\xi) = c_2$. So, $v(\xi)$ increases as ξ increases from 0 to $c_2 - c_1$, and beyond $c_2 - c_1$ it does not change. The number $c_2 - c_1$ is therefore called the *critical capacity* of arc (i, j) and denoted by k_{ij}^*. So, we have $c_1 = v(0), c_2 = v(\infty)$, and $v(\xi) = \min. \{v(0) + \xi, v(\infty)\}$, for all $\xi \overset{>}{=} 0$. See Figure 2.28.

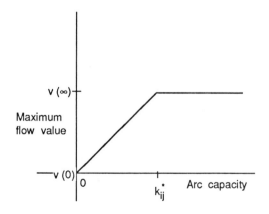

Figure 2.28 The maximum flow value as a function of the capacity of arc $(i, j).k_{ij}^*$ is the critical capacity.

To compute the critical capacity $k_{ij}^* = v(\infty) - v(0)$, we have to solve two maximum value flow problems, one with the capacity of this arc set at ∞, and the other with this capacity set at 0. We have the following facts.

1. Whenever $k_{ij} < k_{ij}^*$, (i, j) is a forward arc in every minimum capacity cut separating $\check{s}$ and $\check{t}$. If $k_{ij} > k_{ij}^*$, (i, j) is not a forward arc in any minimum capacity cut separating $\check{s}$ and $\check{t}$. If $k_{ij} = k_{ij}^*$, there exists a minimum capacity cut separating $\check{s}$ and $\check{t}$ that contains (i, j) as a forward arc, and another that does not.

2. Destroying an arc in a network is equivalent to reducing its capacity to 0. The amount by which the maximum flow value would decrease if arc (i, j) is destroyed is min. $\{k_{ij}, k_{ij}^*\}$.

2.8 Exercises

2.16 In the single commodity flow network in Figure 2.29 all lower bounds are zero, and capacities are entered on the arcs. Find a maximum value flow, a minimum capacity cut, and the critical capacities of arcs $(3, 5)$, $(2, 6)$.

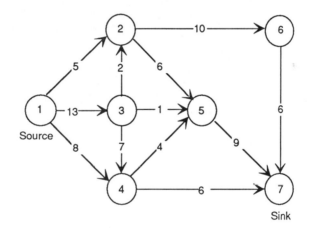

Figure 2.29

2.17 Discuss an efficient scheme for finding an arc, the destruction of which reduces the maximum flow value from the source to the sink the most.

2.18 Let k_{ij}, k_{ij}^* be the present capacity and critical capacities respectively of arc (i, j). Maximum $\{k_{ij}^* - k_{ij}, 0\}$ is the *scope* for increasing the maximum flow value by developing the arc (i, j). Discuss an efficient scheme for finding an arc with maximum scope from those in a specified subset of arcs.

2.19 Let (i, j), (p, q) be two arcs in a directed single commodity flow network $G = (\mathcal{N}, \mathcal{A}, 0, k, \check{s}, \check{t})$. Let $v(\xi, \eta)$ denote the maximum flow value in G as a function of $\xi = k_{ij}$, $\eta = k_{pq}$, in the region $\xi \geq 0$, $\eta \geq 0$, while all the other data remains unchanged.

Prove that $v(\xi, \eta) = \min\ \{v(0,0)+\xi+\eta, v(0,\infty)+\xi, v(\infty,0)+\eta, v(\infty,\infty)\}$. Using this, show that in the nonnegative quadrant of the $\xi, \eta-$ plane, $v(\xi, \eta)$ is a piecewise linear function, dividing this quadrant into at most four convex regions in each of which $v(\xi, \eta)$ is linear. Illustrate with a numerical example.

2.20 Using the notation of Exercise 2.19, prove that for all rectangles with vertices $(\xi, \eta), (\xi + h, \eta), (\xi, \eta + r), (\xi + h, \eta + r)$ in the $\xi, \eta-$ nonnegative quadrant, the difference quotient $(v(\xi + h, \eta + r) - v(\xi + h, \eta) - v(\xi, \eta + r) + v(\xi, \eta))/hr$ always has the same sign.

2.21 Let $[\mathbf{X},\ \bar{\mathbf{X}}]$ be a cut, and f a feasible flow vector in the directed single commodity flow network $G = (\mathcal{N}, \mathcal{A}, \ell, k, V)$. Prove that the net flow across the cut $[\mathbf{X}, \bar{\mathbf{X}}]$ in f is $V(\mathbf{X})$.

2.22 $\bar{f} = (\bar{f}_{ij})$ is a feasible flow vector of value $\bar{v}$ in a directed single commodity flow network $G = (\mathcal{N}, \mathcal{A}, \ell, k, \check{s}, \check{t})$ with $0 < \ell < k < \infty$ and $|\mathcal{A}| = m$. δ is a given small positive number, much smaller than any $k_{ij} - \ell_{ij}$. We plant a rooted tree at $\check{s}$ and grow it by the following labeling rules.

Forward labeling   If i is labeled, j is unlabeled, and $(i, j) \in \mathcal{A}$ and $\bar{f}_{ij} \overset{\leq}{=} k_{ij} - \delta$; label j with $(i, +)$.

Reverse labeling   If i is labeled, j is unlabeled, and $(j, i) \in \mathcal{A}$ and $\bar{f}_{ij} \overset{\geq}{=} \ell_{ij} + \delta$; label j with $(i, -)$.

The tree growth routine has terminated without $\check{t}$ ever getting labeled. $\mathbf{X}$ is the set of labeled nodes, and $\bar{\mathbf{X}}$ its complement at termination. Prove that the maximum flow value in G is $\overset{\leq}{=} \bar{v} + m\delta$, and that the minimum capacity of cuts separating $\check{s}$ and $\check{t}$ in G is $\overset{\geq}{=}$ capacity of $[\mathbf{X}, \bar{\mathbf{X}}]$ - $m\delta$.

2.23 In Figure 2.30, we show the arcs incident at the reference node 14 in a layered network in which a maximal flow is being found by the MKM algorithm. The number on each arc is its capacity in this layered network. Nonzero flow amounts at this stage are entered in little boxes by the side of the arcs. Of the nodes shown, only 10 to 20 are in the set $\mathbf{Y}$ at this stage, 8, 9, 21, 22 are not. Clearly, the arcs incident at nodes 8, 9, 21, 22 are not in the set $\mathbf{F}$ at this stage. Compute the reference potential at this stage. Do flow pushing and pulling at node 14, and indicate the new flow amounts on the arcs shown in Figure 2.30. Also indicate how much flow pushing or pulling has to be carried out at each of the adjacent node of 14 as this step is continued.

2.24 Discuss the main difference in the strategies employed by the following algorithms: (a) Ford-Fulkerson labeling algorithm, (b) Edmonds-Karp labeling algorithm, (c) Dinic's algorithm for finding a maximum value flow in a directed single commodity flow network. Are all three algorithms guaranteed to solve the problem always? If not, mention the conditions under which they can solve the problem. What is the worst case computational complexity of each of these algorithms?

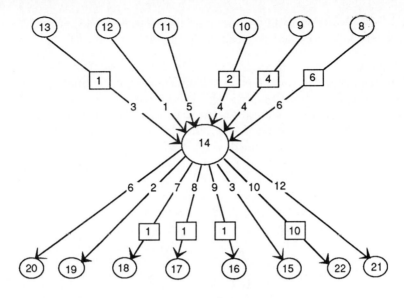

Figure 2.30

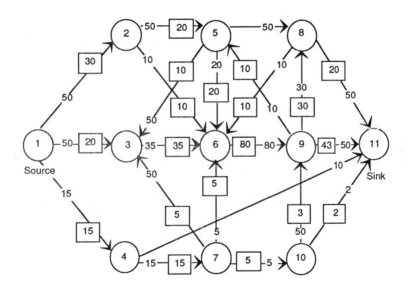

Figure 2.31

Consider the network in Figure 2.31. All lower bounds are zero, and the capacities are entered on the arcs. Nonzero flow amounts in a feasible flow vector are marked in little boxes by the side of the arcs. Draw the layered network wrt this flow vector, and obtain the referent by carrying out any pruning that is necessary.

Use the MKM algorithm to find a maximal flow vector in the referent. Augment the flow vector in the original network using this maximal flow.

2.25 Consider the following variant of the initial version of the labeling method for finding the maximum value flow in the directed single commodity flow network $G = (\mathcal{N}, \mathcal{A}, \ell, k, \check{s}, \check{t})$ beginning with the feasible flow vector $f^o = (f^0_{ij})$, called *"Capacity."* At each stage choose the next node to be labeled by the following procedure: let $f = (f_{ij})$ be the feasible flow vector in G at that stage. Define $\mathbf{E}^+ = \{(i,j) : i$ labeled, j unlabeled, $(i,j) \in \mathcal{A}$, and $f_{ij} < k_{ij}\}$, $\mathbf{E}^- = \{(i,j) : j$ labeled, i unlabeled, $(i,j) \in \mathcal{A}$ and $f_{ij} > \ell_{ij}\}$. Define $\varepsilon_{ij} = k_{ij} - f_{ij}$ for $(i,j) \in \mathbf{E}^+$, or $f_{ij} - \ell_{ij}$ for $(i,j) \in \mathbf{E}^-$. Select the next arc to be made in-tree to be $(p,q) \in \mathbf{E}^+\cup \mathbf{E}^-$ satisfying $\varepsilon_{pq} = \text{maximum } \{\varepsilon_{ij} : (i,j) \in \mathbf{E}^+\cup \mathbf{E}^-\}$. If $(p,q) \in \mathbf{E}^+$, label q with the label $(p,+)$; and if $(p,q) \in \mathbf{E}^-$, label p with $(q,-)$. Prove the following about this algorithm *"Capacity."*

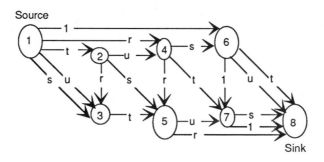

Figure 2.32

(i)   In each iteration, the FAP obtained gives the largest possible flow augmentation among all FAPs from $\check{s}$ to $\check{t}$ wrt f at that stage.

(ii) When ℓ, k, f^o are all integer vectors, it terminates with the maximum value flow vector after at most $0(m(\log m + \log \bar{c}))$ flow augmentations, where $m = |\mathcal{A}|$ and $\bar{c}$ is the average of $(k_{ij} - \ell_{ij})$ over arcs in G if all k_{ij} are finite, or the logarithm of the finite capacity of any cut in G if some k_{ij} is infinite. This establishes that the algorithm is polynomially bounded.

(iii) Consider the network $G_r = (\mathcal{N}, \mathcal{A}, 0, k^r, \check{s}, \check{t})$ which is the same as the network discussed in Example 2.1, except that the capacity of arc (i,j) is now $k^r_{ij} = \lfloor 2^r k_{ij} \rfloor$, where k_{ij} is the real capacity of (i,j) as defined in Example 2.1. Here $\log(\bar{c}) = O(r)$. Beginning with the zero flow vector in prove that the algorithm "Capacity" takes $0(r)$ flow augmentations to find the maximum value flow in G_r, establishing that on networks with integer data and integer initial feasible flow vector, the computational effort needed by this algorithm may continue to grow with the size of the capacity data, indefinitely, on the same network.

(iv) Consider the network in Figure 2.32 with capacity data entered on the arcs, where $\alpha = \frac{1}{2}(-1 + \sqrt{5}), r = \alpha, s = \alpha/2, t = (1 + \alpha)/2$ and $u = 1/2$. All lower bounds are 0. Beginning with the initial flow vector $f^\circ = 0$, prove that the algorithm "Capacity" produces an infinite sequence of feasible flow vectors converging to a maximum value flow vector in an infinite number of iterations. This establishes that on networks with nonrational data "Capacity" may not find a maximum value flow within a finite number of iterations.

(v) Prove that the sequence of flows constructed by "Capacity" always converges to a maximum value flow vector, whether the data is rational or not.

(Queyranne [1980])

2.26 Rectilinear Distance Facility Location Located at $(a_i, b_i), i = 1$ to m in $\mathbb{R}^2$ are m existing factories. We need to determine $(x_j, y_j), j = 1$ to n, optimum locations for n new facilities to minimize the weighted sum of rectilinear distances

$$z(x,y) = \sum_{j=1}^{n}\sum_{i=1}^{m} w_{ji}(|x_j - a_i| + |y_j - b_i|) +$$

$$\frac{1}{2}\sum_{j=1}^{n}\sum_{r=1}^{n} v_{jr}(|x_j - x_r| + |y_j - y_r|)$$

where $x = (x_1 \ldots, x_n)^T, y = (y_1, \ldots, y_n)^T, w_{ji} \geqq 0, v_{jr} = v_{rj} \geqq 0$ for all r, j, i. $z(x,y)$ can be written as $g(x) + h(y)$, so problem equivalent to minimizing $g(x)$ and $h(y)$ separately. We consider the problem of minimizing $g(x)$.

$$g(x) = \sum_{j=1}^{n}\sum_{i=1}^{m} w_{ji}|x_j - a_i| + \frac{1}{2}\sum_{j=1}^{n}\sum_{r=1}^{n} v_{jr}|x_j - x_r|.$$

(i) Prove that there exists an $x^* = (x_j^*)$ minimizing $g(x)$ in which $x_j^* \in \{a_1, \ldots, a_m\}$ for all $j = 1$ to n.

(ii) Consider the special case in which $a_i = i, i = 1$ to m. Let $f(x)$ be the value of $g(x)$ in this special case. So, consider the problem:

$$\text{minimize} \quad f(x)$$

$$\text{subject to} \quad x_j \in \{1, \ldots, m\}, \ j = 1 \text{ to } n \qquad (2.13)$$

For any $p \in \{1, \ldots, m\}$ and $x = (x_j)$ feasible to (2.13), let $S_p = \{j : x_j \leqq p\}, \overline{S}_p = \{j : x_j > p\}$. Prove that $f(x) = \sum_{p=1}^{m-1} c_p(S_p, \overline{S}_p)$ where

$$c_p(S_p, \overline{S}_p) = \sum_{j \in S_p}\sum_{i=p+1}^{m} w_{ji} + \sum_{j \in \overline{S}_p}\sum_{i=1}^{p} w_{ji} + \sum_{j \in S_p}\sum_{r \in \overline{S}_p} v_{jr}.$$

Consider the undirected network $G_p = (\mathcal{N}, \mathcal{A}, 0, k^p)$ with $\mathcal{N} = \{s, 1, \ldots, n, t\}, \mathcal{A} = \{(s; j), (j; t) : \text{for all } j = 1 \text{ to } n\} \cup \{(j_1; j_2); \text{ for all } j_j \neq j_2 \in \{1, \ldots, n\}\}$, called the p locale network, where the capacity vector k^p is given by the following:

$$k^p_{s;j} = \sum_{i=1}^{p} w_{ji}, k^p_{j;t} = \sum_{i=p+1}^{m} w_{ji}, j = 1 \text{ to } n$$

$$k^p_{j_1;j_2} = v_{j_1,j_2} \quad \text{for all} \quad j_1 \neq j_2 \in \{1, \ldots, n\}.$$

If $\mathbf{X} \cup \overline{\mathbf{X}}$ is a partition of $\{1, \ldots, n\}$, show that the capacity of the cut $(\{s\} \cup \mathbf{X}; \{t\} \cup \overline{\mathbf{X}})$ in G_p is $c_p(\mathbf{X}, \overline{\mathbf{X}})$. Prove that $x^o = (x^o_j)$ is an optimum solution of (2.13) iff for each $p = 1$ to $m - 1$, the cut $(\{s\} \cup \mathbf{X}_p; \{t\} \cup \overline{\mathbf{X}}_p)$, where $\mathbf{X}_p = \{j : x^o_j \overset{\leq}{=} p\}, \overline{\mathbf{X}}_p = \{j : x^o_j \overset{\geq}{=} p + 1\}$, is a minimum capacity cut in G_p. Conversely, if $(\{s\} \cup \mathbf{Y}_p; \{t\} \cup \overline{\mathbf{Y}}_p)$ is a minimum capacity cut in G_p, show that there exists an optimum solution of (2.13), $x^* = (x^*_j)$ satisfying $x^*_j \overset{\leq}{=} p$ for all $j \in \mathbf{Y}_p$ and $x^*_j \overset{\geq}{=} p + 1$ for all $j \in \overline{\mathbf{Y}}_p$. From this, show that (2.13) can be solved as a minimum capacity cut problem, if $m = 2$.

(iii) Consider (2.13) when $m \overset{\geq}{=} 3$. Let $N_1, \ldots, N_q$ be a partition of $\{1, \ldots, n\}$ and let $B_1, \ldots, B_q$ be integers satisfying $1 \overset{\leq}{=} B_1 \overset{\leq}{=} B_2 \overset{\leq}{=} \cdots \overset{\leq}{=} B_q \overset{\leq}{=} m$. Consider the problem:

$$\text{minimize } f(x)$$

$$\text{subject to } B_s \overset{\leq}{=} x_j \overset{\leq}{=} B_{s+1}, \text{ for } j \in N_s, s = 1 \text{ to } q \qquad (2.14)$$

$$x_j \text{ integer for all } j.$$

Prove that for any $s = 1$ to q, the optimum values of x_j for $j \in N_s$ are independent of the actual locations of those facilities (new and old) located outside the interval B_s to B_{s+1}, but are dependent on which of these facilities are located at points $\overset{\leq}{=} B_s$ and which are located at points $\overset{\geq}{=} B_{s+1}$.

		w_{ji}		
old facility	$i = 1$	2	3	4
new facility $j = 1$	3	2	3	1
2	1	2	1	2
3	9	1	2	2

		v_{jr}	
$r =$	1	2	3
$j = 1$	·	3	2
2	3	·	1
3	2	1	·

Using this, show that (2.14) can be decomposed into q problems of the same form as (2.13), where for each s all new facilities $j \in \cup(N_p; p = 1$ to $s - 1)$ and $j \in \cup(N_p : p = s+1$ to $q)$ are treated as old facilities located at B_s, B_{s+1}, respectively and old facilities $i \leqq B_s$, and $i \geqq B_{s+1}$ are treated as old facilities located at B_s, B_{s+1} respectively. From these results, develop an algorithm for solving (2.13) by finding a minimum capacity cut in each of the p local networks G_p, for $p = 1$ to $m - 1$. Solve the numerical problem (2.13) with $m = 4, n = 3$, and the data in the tables given above.

(iv) Now consider the original problem of minimizing $g(x)$. Assume that $a_i s$ are ordered so that $a_1 < a_2 < \ldots < a_m$. If $x^o = (x_j^o)$ is an optimum solution of (2.13), then prove that $x^* = (x_j^*)$, where $x_j^* = a_i$ when $x_j^o = i$, is an optimum solution of this problem.

(Picard and Ratliff [1978])

2.27 The Sharing Problem $G = (\mathcal{N}, \mathcal{A}, 0, k)$ is a single commodity directed flow network with $S \subset \mathcal{N}$ as the set of source nodes with a_i units material available at source node $i \in S$; and $D \subset \mathcal{N}$ as the set of sink nodes with w_j being the weight of sink node $j \in D$ for allocation of material in a shortage situation. Under shortage, an equitable distribution of material should attempt to maximize the minimum weighted net flow reaching the sink nodes in D. Discuss an approach for solving this problem. Solve the numerical problem in Figure 2.33 using this approach(Brown [1979]).

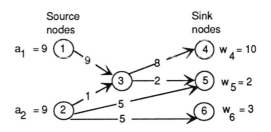

Figure 2.33 All lower bounds are 0. Capacities are entered on the arcs.

 2.28 D is a finite set of points. For each $d \in D$, c_d is the cost of choosing it. $c = (c_d) > 0$ is given. $\triangle$ is a class of subsets of **D**. For each element $\sigma \in \triangle$, p_σ is the profit of choosing that element. $p = (p_\sigma) > 0$ is given. A *selection* is a collection of elements from $\triangle$ together with all points of **D** which belong to this collection. If $\{\sigma_1, \ldots, \sigma_r\} \subset \triangle$ is the set of elements in a selection, its value is defined to be $\sum_{i=1}^r p_{\sigma_i} - \sum(c_d : \text{over } d \in \cup_{i=1}^r \sigma_i)$. It is required to find a selection of maximum value. Consider a directed bipartite network $G = (\mathcal{N}_1, \mathcal{N}_2; \mathcal{A}, 0, k)$ with $\mathcal{N}_1 = \{\breve{s}\} \cup D$, $\mathcal{N}_2 = \{\breve{t}\} \cup \triangle$ where $\breve{s}, \breve{t}$ are source and sink nodes, and arcs

$(\check{s}, \sigma)$ with capacity $k_{s\sigma} = p_\sigma$, for each $\sigma \in \Delta$, $(d, \check{t})$ with capacity $k_{dt} = c_d$, for each $d \in \mathbf{D}$, and (d_i, σ_j) with capacity ∞, for each $d_i \in \sigma_j$, for each $\sigma_j \in \Delta$. Show that there is a one-to-one correspondence between selections, and cuts in G separating $\check{s}$ and $\check{t}$ which contain no forward arcs of the type (d_i, σ_j). Using this, show that a maximum value selection corresponds to a minimum capacity cut separating $\check{s}$ and $\check{t}$ in G, and hence can be found efficiently by finding a maximum value flow from $\check{s}$ to $\check{t}$ in G.

As an application, consider the following problem of a transport undertaking. They are considering the installation of freight handling terminals at locations d_1, d_2, d_3, and d_4. The cost of installing the terminals at any of these locations is \$5 million. The existence of terminals at certain pairs of terminals permits a service to be operated between those terminals, which is associated with a net profit. These pairs are: (d_1, d_2) with a profit of \$2 million, $(d_1; d_3)$ with a profit of \$9 million, $(d_2; d_3)$ with a profit of \$4 million, and $(d_3; d_4)$ with a profit of \$6 million. Find a selection of services and terminals to maximize excess of profit over cost. (Rhys [1970], Balinski [1970] both of Chapter 1, Murchland [1968])

2.29 Let $G = (\mathcal{N}, \mathcal{A}, 0, k, 1, n)$ be a directed single commodity flow network with $|\mathcal{N}| = n$, and nodes $1, n$ as the source, sink nodes respectively, with $k > 0$. Any partition of the nodes in $\mathcal{N}$ into $\mathbf{X}, \overline{\mathbf{X}}$ with $1 \in \mathbf{X}, n \in \overline{\mathbf{X}}$, can be represented by the 0-1 vector $x = (x_1, x_2, \ldots, x_n)$ defined on $\mathcal{N}$ by

$$x_i = \begin{cases} 1, & \text{if } i \in \mathbf{X} \\ 0, & \text{if } i \in \overline{\mathbf{X}} \end{cases}$$

and conversely any $0 - 1$ vector x defined on $\mathcal{N}$ with $x_1 = 1, x_n = 0$ defines such a partition of $\mathcal{N}$.

(i)   If $[\mathbf{X}, \overline{\mathbf{X}}]$ is a cut separating 1 and n in G, associated with the 0-1 vector $x = (x_i)$ on $\mathcal{N}$, prove that its capacity $k(\mathbf{X}, \overline{\mathbf{X}}) = \sum (k_{ij} x_i (1 - x_j))$; over$(i, j) \in \mathcal{A}$.

(ii)   From the above result, show that the quadratic 0-1 integer programming problem, where $q_{jr} \geqq 0$ for all j and r, can be solved very efficiently by solving a maximum value flow problem on a related network. Based on this, develop an efficient algorithm for this quadratic programming problem.

$$\text{minimize} \sum_{j=1}^{n} p_j y_j - \sum_{j=1}^{n} \sum_{r=1}^{n} q_{jr} y_j y_r$$

$$\text{subject to } y_j = 0 \text{ or } 1 \text{ for all } j = 1 \text{ to } n$$

(Picard and Ratliff [1975])

2.30 Let $\Gamma = \{1, \ldots, n\}$, $\{\mathbf{S}_1, \ldots, \mathbf{S}_r\}$ is a family of subsets of Γ each of cardinality $\geqq 2$. Consider the following 0-1 nonlinear programming problem.

$$\text{maximize } z(x) \;=\; \sum_{t=1}^{r} a_t (\Pi_{j \in \mathbf{S}_t} x_j) + \sum_{j=1}^{n} c_j x_j$$

$$\text{subject to } x_j \;\in\; \{0, 1\}, \text{ for all } j \in \Gamma. \qquad (2.15)$$

where $(a_1, \ldots, a_r) > 0$. Now define new variables $y_1, \ldots, y_r$, with y_t corresponding to the set $\mathbf{S}_t$ for $t = 1$ to r. Show that (2.15) is equivalent to the following 0-1 integer program (2.16). Show that (2.16) is a "selection problem" as it is defined in Exercise 2.28. In (2.16) if $c_j \geqq 0$ for some j, prove that $x_j = 1$ in an optimum solution. Hence such variables can be fixed equal to 1 and the problem size reduced. In the sequel we assume that $c_j < 0$ for all j. Create a network G with node set $\mathcal{N} = \{\check{s}, \mathbf{S}_1, \ldots, \mathbf{S}_r, 1, \ldots, n, \check{t}\}$. Arcs in it are $(\check{s}, \mathbf{S}_t)$ with capacity a_t for $t = 1$ to r; $(\mathbf{S}_t, j)$ for each $j \in \mathbf{S}_t$, $t = 1$ to r with capacity ∞; and $(j, \check{t})$ with capacity $-c_j$ for $j = 1$ to n. All lower bounds in G are 0.

$$\text{maximize } \sum_{t=1}^{r} a_t y_t + \sum_{j=1}^{r} c_j x_j$$

$$\text{subject to } y_t \leqq x_j, \text{ for all } j \in \mathbf{S}_t, \text{ for all } t = 1 \text{ to } r \qquad (2.16)$$

$$y_t = 0 \text{ or } 1 \text{ for all } t = 1 \text{ to } r$$
$$x_j = 0 \text{ or } 1 \text{ for all } t = 1 \text{ to } n.$$

Prove that the variables x_j which take on a value of 1 in an optimum solution of (2.15) or (2.16) correspond to labeled vertices $j \in \mathcal{N}$ after a maximum value flow in G from $\check{s}$ to $\check{t}$ has been found by any of the labeling algorithms. Solve the following numerical problem using this approach.

$$\begin{aligned}
\max z(x) - \quad & 2x_1 x_2 + 2x_1 x_2 x_3 + 6x_1 x_2 x_4 \\
& + x_2 x_3 x_4 + x_1 x_2 x_3 x_4 - 2x_1 \\
& - x_2 - 5x_3 - 2x_4
\end{aligned}$$

$$\text{subject to } x_j = 0 \text{ or } 1, \text{ for } j = 1 \text{ to } 4.$$

(Picard and Queyranne [1982a])

2.31 Let $G = (\mathcal{N}, \mathcal{A}, \ell, k, \check{s}, \check{t})$ be a directed connected single commodity flow network with $k > \ell$. Among all minimum capacity cuts separating $\check{s}$ and $\check{t}$ in G, it is required to find one satisfying one of the following additional properties: (i) it has the smallest number of forward arcs, (ii) it has the smallest number of reverse arcs, or (iii) it has the smallest number of arcs. Develop efficient algorithms for these problems (Hamacher [1982]).

2.32 Consider the connected directed single commodity flow network $G = (\mathcal{N}, \mathcal{A}, 0, k, \check{s}, \check{t})$ where $k > 0$. Let f^* be a maximum value feasible flow vector in G of value v^*, and f a feasible flow vector of value $v < v^*$. Define a feasible flow vector g in the residual network wrt f, $G(f) = (\mathcal{N}, \mathcal{A}(f), 0, \kappa, \check{s}, \check{t})$ by the following:

1. For each $(i,j) \in \mathcal{A}$ satisfying $0 < f_{ij} < k_{ij}$, we have $(i,j) \in \mathcal{A}(f)$ and $(j,i) \in \mathcal{A}(f)$, make $g_{ij} = $ maximum $\{0, f_{ij}^* - f_{ij}\}, g_{ji} = $ maximum $\{0, f_{ij} - f_{ij}^*\}$.

2. For each $(i,j) \in \mathcal{A}$ satisfying $f_{ij} = 0$, we have $(i,j) \in \mathcal{A}(f)$, make $g_{ij} = f_{ij}^*$.

3. For each $(i,j) \in \mathcal{A}$ satisfying $f_{ij} = k_{ij}$, we have $(j,i) \in \mathcal{A}(f)$, make $g_{ji} = -(f_{ij}^* - f_{ij})$.

Prove that g is a maximum value feasible flow vector in $G(f)$. Prove that if the node partition $[\mathbf{X}, \overline{\mathbf{X}}]$ defines minimum capacity cut separating $\breve{s}$ and $\breve{t}$ in G, then it defines a minimum capacity cut separating $\breve{s}$ and $\breve{t}$ in $G(f)$ (Ramachandran [1987]).

2.33 The Maximum Flow Problem is Not Easier in an Acyclic Network Than in a General Network. Let $G = (\mathcal{N}, \mathcal{A}, 0, k, \breve{s}, \breve{t})$ be a connected directed single commodity flow network. Without any loss of generality we assume that every node and arc in G lies on at least one chain from $\breve{s}$ to $\breve{t}$ (otherwise such things could be deleted).

Find a depth first search spanning tree $\mathbb{T}$ rooted at $\breve{s}$ in G and let $\mathbf{B}$ be the set of arcs in G that are back arcs wrt $\mathbf{T}$. Get a new network $G^* = (\mathcal{N}\mathcal{A}^*, 0, k^*, \breve{s}, \breve{t})$ from G, and a feasible flow vector g in it by doing the following for each $(i,j) \in \mathbf{B}$: Replace (i,j) by (j,i) but keep its capacity the same, k_{ij}. Define the flow g_{ji} on this new arc to be its capacity k_{ij}. Introduce new arcs $(\breve{s},j)$, $(i,\breve{t})$ also of the same capacity k_{ij}, and make the flow on both of them equal to this capacity. Verify that the resulting network G^* is acyclic, that the flow defined on it, g, has value $\sum(k_{ij} :$ over $(i,j) \in \mathbf{B})$ and that it saturates all the newly introduced arcs.

Let $G^*(g)$ be the residual network of G^* wrt g. Verify that $G^*(g)$ is G together with some additional arcs either incident into $\breve{s}$ or incident out of $\breve{t}$. So, any partition of the nodes in $\mathcal{N}$ that defines a minimum capacity cut separating $\breve{s}$ and $\breve{t}$ for $G^*(g)$ also defines a minimum capacity cut for G. By Exercise 2.32, the node partition that induces a minimum capacity cut separating $\breve{s}$ and $\breve{t}$ in G^* induces also a minimum capacity cut for $G^*(g)$. Using these facts show that the problem of finding a minimum capacity cut separating the source and sink in the general directed network G reduces in linear time to a corresponding problem in an acyclic network. Then show that the maximum value flow problem is a general directed network reduces in linear time to a corresponding problem in an acyclic network (Ramachandran [1987]).

2.34 Consider the single commodity flow network $G = (\mathcal{N}, \mathcal{A}, 0, k, \breve{s}, \breve{t})$. A subset $\mathbf{A} \subset \mathcal{A}$, $|\mathbf{A}| = r$ is said to be the set of r most vital arcs in this network, if the simultaneous removal of the arcs in $\mathbf{A}$ results in the greatest decrease in the maximum flow value in the remaining network, among all subsets of arcs of cardinality r. Prove that the r most vital arcs in G are the r largest capacity arcs in a particular cut separating $\breve{s}$ and $\breve{t}$ in G. Develop an algorithm for finding such a set of arcs (Ratliff, Sicilia and Lubore [1975]).

2.35 Consider the following discrete quadratic programming problem. Define $\mathcal{N} = \{1, \ldots, n\}$, $\mathcal{A} = \{(i,j) : i,j \in \mathcal{N}$ such that $d_{ij} \neq 0\}$, $G = (\mathcal{N}, \mathcal{A}, k)$, where for $(i,j) \in \mathcal{A}$, $k_{i;j} = d_{ij}$. Define the capacity of a cut in G to be the sum of k_{ij} over (i,j) in the cut. Show that this quadratic program is equivalent to the problem of finding a maximum capacity cut in G.

$$\text{minimize} \sum_{i=1}^{n-1} \sum_{j=i+1}^{n} d_{ij} x_i x_j$$

$$\text{subject to } x_i \in \{-1, 1\}, \text{ for each } i = 1 \text{ to } n.$$

(Barahona [1982])

2.36 Let $G = (\mathcal{N}, \mathcal{A}, 0, k)$ be a connected undirected single commodity flow network, with $k > 0$. Let $v(x,y)$ denote the maximum flow value in this network with x as the source node and y as the sink node. Hence $v(x,y)$ is a positive valued function defined for pairs of distinct nodes of G, which will be called a *flow value function* of G. Prove the following:

(i) $v(x,y) = v(y,x)$, for all $x \neq y \in \mathcal{N}$.

(ii) $v(x,y) \geq \min\{v(x,z), v(z,y)\}$, for every three distinct nodes x, y, z in $\mathcal{N}$.

(iii) $v(x_1, x_r) \geq \min\{v(x_1, x_2), v(x_2, x_3), \ldots, v(x_{r-1}, x_r)\}$, for any sequence of distinct nodes $x_1, \ldots, x_r$ in G.

(iv) From (ii) show that among $v(x,y), v(x,z)$ and $v(z,y)$, at least two must be equal, and the third is $\geq$ their common value.

(v) Let $|\mathcal{N}| = n$. Let H be the complete undirected network on $\mathcal{N}$ with $v(x;y)$ as the length of the edge $(x;y)$, for $x \neq y \in \mathcal{N}$. Let T be a maximum length spanning tree in H. Using (iii) show that the length of every out-of-tree edge in H must be equal to the length of some in-tree edge. Thus prove that of the $n(n-1)$ flow values $v(x,y)$ in G, there are at most $(n-1)$ numerically distinct values.

(vi) Given any positive symmetric function $v(\cdot, \cdot)$ defined over pairs of distinct nodes in $\mathcal{N}$, satisfying the "traingle" inequality in (ii), prove that a spanning tree spanning the nodes in $\mathcal{N}$ can be constructed and capacities of edges in this tree defined in such a way that $v(\cdot, \cdot)$ is the flow value function on this tree.

(vii) Devise a scheme to determine the flow value function $v(\cdot, \cdot)$ on G by the successive solution of at most $(n-1)$ maximum value flow problems.

(Gomory and Hu [1961])

2.37 The Allocation of Specialists to Hospitals in a Region   There are four hospitals, h_1, h_2, h_3, h_4, in a region. There are seven specialties for which there are plans to develop additional facilities (i.e., hospital beds dedicated to those specialties) in the region, these are s_1 to s_7 as indicated in the table given below. Lower and upper bounds are imposed on the total number of beds allocated to (i) each speciality in each hospital, and (ii) to each hospital. Determine one feasible allocation of speciality beds to hospitals satisfying all these constraints, using a network formulation. Discuss how one can determine an "optimum allocation" among all feasible allocations of specialty beds to hospitals in this problem (Duncan [1979]).

Specialty	Lower/upper bounds on no. of beds allocated to specialty in hospital				Total beds allocated to specialty
	h_1	h_2	h_3	h_4	
ENT s_1	17/28	7/11	-	-	28
Dental surgery s_2	3/5	1/2	-	-	5
Plastic surgery s_3	24/39	10/15	-	-	39
Rheumatology s_4	4/10	2/5	2/8	4/9	14
T. & O. surgery s_5	80/97	34/51	-	-	131
General surgery s_6	51/136	50/71	34/92		183
General medicine s_7	51/139	40/73	50/111	28/75	187
Lower/upper bounds on number of beds allocated to hospital	0/247	0/176	0/183	0/100	

T. & O. is Traumatic and Orthopedic

2.38   There are n jobs ordered as $1, 2, \ldots, n$, to be processed on either of two available machines. Job i has processing time of p_i, q_i respectively, depending on whether it is processed on machine A or B. In the subset of jobs assigned to each machine, they can only be processed in the order of lowest number job first. The flow time of a job is defined to be the duration of time lapse from the beginning (i.e., time point 0) till its processing is completed. It is required to assign the jobs to the two machines so as to minimize the sum of flow times of all the jobs. Define $x_i = 1$, if job i is assigned to machine A, 0 if it is assigned to B. The 0-1 assignment vector is $x = (x_1, \ldots, x_n)^T$. Show that the total flow time corresponding to the assignment x is $x^T C x + \sum_{j=1}^{n} w_j$, where $C = (c_{ij})$ is a symmetric matrix given by

$$w_j = \sum_{i=1}^{j} q_j$$

$$c_{ij} = \begin{cases} (p_i + q_i)/2, & \text{if } j > i \\ -[w_j + (n - j + 1)q_j - (p_j + q_j)], & \text{if } j = i, i = 1 \text{ to } n \\ (p_j + q_j)/2, & \text{if } j < i \end{cases}$$

So, the problem of finding an optimum assignment of jobs to machines A, B is equivalent to

$$\text{minimize } x^T C x$$
$$\text{subject to } x_i = 0 \text{ or } 1 \text{ for all } i$$

Augment the matrix C into a symmetric matrix $C' = (C'_{ij})$ of order $(n+1) \times (n+1)$ by adding a dummy row (row 0) and dummy column (column 0) so that the sum of all entries in each row and in each column of C' is zero. Thus

$$C'_{oi} = C'_{io} = -[\sum_{j=1}^{i}((p_j - q_j)/2) + (n - i + 1)(p_i - q_i)/2], \; i = 1 \text{ to } n$$

$$C'_{oo} = -\sum_{i=1}^{n} C'_{oi}, \text{ and } C'_{ij} = C_{ij} \text{ for } i, i = 1 \text{ to } n.$$

Let $X = (x_0, x_1, \ldots, x_n)^T$. Show that the above optimum assignment problem is equivalent to

$$\text{minimize} X^T C' X$$
$$\text{subject to } x_i = 0 \text{ or } 1, \; i = 0, 1, \ldots, n.$$

Formulate this as the problem of finding a maximum capacity cut in a network for which the set of nodes is $\mathcal{N} = \{0, 1, \ldots, n\}$. Using the special property that $C'_{ij} = r_i = (p_i + q_i)/2$ for all $j > i$, develop an efficient direct algorithm for solving this maximum capacity cut problem. Discuss how to solve the job assignment problem using this algorithm.

Solve the numerical problem with $n = 5$ and $p = (p_i) = (2, 6, 7, 9, 8)$, $q = (q_i) = (4, 4, 11, 3, 14)$ using this approach (Lakshminarayanan, Lakshmanan, Papineau, and Rochette [1979]).

2.39 Let $G = (\mathcal{N}, \mathcal{A}, 0, k, \breve{s}, \breve{t})$ be a directed single commodity flow network, with the capacity vector $k \geqq 0$. Define $y = (y_{ij} : (i, j) \in \mathcal{A})$ to be a vector of variables associated with the arcs in G. Consider the following problem

$$\text{minimize } \sum(k_{ij} y_{ij} : \text{ over } (i, j) \in \mathcal{A})$$

$$\text{subject to } \sum(y_{ij} : \text{ over } (i, j) \in \mathcal{C}) \geqq 1, \text{ for chains } \mathcal{C} \text{ from } \breve{s} \text{ to } \breve{t} \text{ in G}$$

$$y_{ij} = 0 \text{ or } 1 \text{ for all } (i, j) \in \mathcal{A}.$$

Show that this is the arc-chain formulation of the problem of finding a minimum capacity cut separating $\breve{s}$ and $\breve{t}$ in G. Show that the LP relaxation obtained by replacing the 0-1 constraints on y by $y \geqq 0$ has an optimum solution which satisfies $y_{ij} = 0$ or 1 for all (i, j). Give an interpretation for the dual of the LP relaxation discussed above, as an alternative formulation of the maximum value flow problem in G.

2.40 Let $G = (\mathcal{N}, \mathcal{A}, 0, k, \breve{s}, \breve{t})$ be a directed connected single commodity flow network. Suppose we are given a feasible flow vector of maximum value from $\breve{s}$ to $\breve{t}$ in G. Then the labeling procedure discussed in the proof of Theorem 2.3 can be used to generate a minimum capacity cut separating $\breve{s}$ and $\breve{t}$ in G with a computational effort of at most $0(|\mathcal{A}|)$. Conversely suppose we are given a minimum capacity cut separating $\breve{s}$ and $\breve{t}$ in G. Is there a procedure that can use this information to generate a maximum value flow from in G efficiently? (Picard and Queyranne [1982b])

2.41 Maximum Weighted Closure of a Network   Let $G = (\mathcal{N}, \mathcal{A})$ be a directed network with $w = (w_i : i \in \mathcal{N})$ as the vector of vertex weights that may be of any sign. A *closure* of G is any subset $\mathbf{U} \subset \mathcal{N}$ satisfying the property that $i \in$ $\mathbf{U}$ and $(i, j) \in \mathcal{A}$ imply that $j \in \mathbf{U}$ also. A closure of G is also called a *hereditary subset*, or *initial subset* or a *selection*. The closure has application in the selection of contingent investments. Suppose we are given a set of projects (represented by nodes in a network) and a set of contingency relations among them. Project i is contingent to project j means that if we decide to select project i, then we must also select project j (this is represented by an arc (i, j) in the network). Every project is associated with a net profit (which may be negative for a project presumably useful to the selection of other more profitable projects). The problem of selecting the subset of projects to implement in order to maximize net profit then becomes that of finding a maximum weight closure in the network. Define the variables y_i for $i \in \mathcal{N}$ by $y_i = 1$ if i is included in the closure, or $y_i = 0$ otherwise. Show that the problem of finding a maximum weight closure in G is equivalent to the 0-1 nonlinear program.

$$\begin{aligned} \text{maximize } \sum(w_i y_i &\quad : \quad \text{over } i \in \mathcal{N}) \\ \text{subject to } \quad y_i(1 - y_j) &= \quad 0, \text{ for each } (i, j) \in \mathcal{A} \\ y_i &= \quad 0 \text{ or } 1 \text{ for all } i \in \mathcal{N}. \end{aligned}$$

When λ is sufficiently large ($\lambda > 1 + \sum_{i \in \mathcal{N}} |w_i|$), show that this problem is equivalent to the 0-1 quadratic program (QP) given below. Augment G into a network G' by the following procedure. Make the capacity of all the arcs in G equal to λ. Introduce a new source node $\breve{s}$ and a new sink node $\breve{t}$. For each $i \in \mathcal{N}$ if $w_i \geqq 0$ include on arc $(\breve{s}, i)$ with capacity w_i; if $w_i < 0$ include $(i, \breve{t})$ with capacity $-w_i$. The lower bounds on all the arcs in G' is 0. Let $[\mathbf{X}, \overline{\mathbf{X}}]$ be a minimum capacity cut separating $\breve{s}$ and $\breve{t}$ in G'. Show that the incidence vector of $\mathbf{X}\backslash\{s\}$ is an optimum solution of the 0-1 QP below, and that $\mathbf{X}\backslash\{s\}$ is a maximum weight closure of G.

$$\begin{aligned} \text{maximize } \sum_{i \in \mathcal{N}} w_i y_i &\quad - \quad \lambda \sum (y_i(1 - y_j) : \text{over } (i, j) \in \mathcal{A}) \\ y_i &= \quad 0 \text{ or } 1 \text{ for all } i \in N. \end{aligned}$$

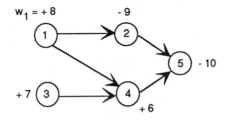

Figure 2.34

Find a maximum weight closure in the network in Figure 2.34 with the minimum cut approach outlined above (Picard and Queyranne [1982b]).

2.42 Activity Selection Game   A is a set of activities. Selection of activity i yields a profit w_i (of arbitrary sign), and selection of activity i without the selection of activity j leads to a penalty of $\lambda_{ij} \geqq 0$. It is required to find a subset of activities $U \subset A$ which yields the maximum net profit. Show that this problem is equivalent to finding a $y = (y_i :\ i \in A)$ that solves

$$\text{maximize } [\sum_{i \in A} w_i y_i - \sum_{i,j \in A, i \neq j} \lambda_{ij} y_i (1 - y_j)]$$

subject to $y_i = 0$ or 1 for all $i \in \mathbf{A}$.

Show how this problem can be transformed into a minimum capacity cut problem in a network, using the approach discussed in Exercises 2.29 and 2.41 (Picard and Queyranne [1982 b], Topkis [1980]).

2.43 Binary Posynomial Maximization   Let $y = (y_1, \ldots, y_n)^T$ be a vector of binary variables and let $\mathbf{S}_1, \ldots, \mathbf{S}_r$ be r distinct nonempty subsets of $\{1, \ldots, n\}$. The function $P(y) = \sum_{t=1}^{r} c_t (\Pi_{i \in \mathbf{S}_t} y_i)$ is said to be a posynomial in y if $c_t > 0$ for all $t = 1$ to r. Maximizing $P(y)$ has the trivial solution $y = \mathbf{e}$, the vector of all 1's. However, the problem

$$\text{maximize } P(y) - by$$

subject to $y_i = 0$ or 1 for all i

where b is an arbitrary real $n-$vector, is nontrivial, it is called the binary posynomial maximization problem.

Generate a directed network $\mathrm{G} = (\mathcal{N}, \mathcal{A})$ with node weight vector w by the following procedure: Associate a node in $\mathcal{N}$ with the set $\mathbf{S}_t$ for each $t = 1$ to r, make its weight $w_t = c_t$ if $|\mathbf{S}_t| \geqq 2$, or $c_t - b_i$ if $\mathbf{S}$ is the singleton set $\{y_i\}$. For each $i = 1$ to n such that the singleton set $\{y_i\}$ is not among the sets $\mathbf{S}_1, \ldots, \mathbf{S}_r$, associate a node in $\mathcal{N}$ with its weight $-b_i$. Let $\mathcal{N}_1$, be the set of nodes introduced so far. For each $\mathbf{Q} \subset \{1, \ldots, n\}$ associated with which there is a node in $\mathcal{N}_1$, introduce a new node associated with the set $\mathbf{Q} \backslash \{j\}$ if it is not there already, and an arc directed from the node associated with $\mathbf{Q}$ to this node, for each $j \in \mathbf{Q}$. Make the weights of all the nodes for which the weight is not defined above equal to zero. $\mathcal{N}$ is the set of all the nodes and $\mathcal{A}$ is the set of all the arcs defined above.

Show that the above problem is equivalent to the maximum weight closure problem (defined in Exercise 2.41) in G (Picard and Queyranne [1982 b]).

2.44 Let $G = (\mathcal{N}, \mathcal{A}, 0, k, \check{s}, \check{t})$ be a directed connected network with $k > 0$. Using the transformation of the minimum capacity cut problem into a binary quadratic programming problem discussed in Exercise 2.29 and the fact that the maximum capacity cut problem is NP-hard, show that cardinality constrained minimum cut problems (i.e., the two problems, finding a minimum capacity cut $[(\mathbf{X}, \bar{\mathbf{X}})]$ separating $\check{s}$ and $\check{t}$ in G with either the constraint that $|(\mathbf{X}, \bar{\mathbf{X}})| = r$ or the constraint $|\mathbf{X}| = r$, for specified r) are NP-hard problems (Picard and Queyranne [1982 b]).

2.45 A nursing staff scheduling problem   There are three departments in a hospital, each of which operates on three shifts daily, with staff members working one shift per day. The daily requirements are spelled out in the table given below. It is required to determine the minimum number of nurses to staff these three departments subject to the constraints and bounds described above. Formulate this as a flow problem and find an optimum solution for it.

Shift	Minimum no. of nurses required in shift	Lower bound-upper bound for nurses in shift in department		
		1	2	3
1	26	6-8	11-12	7-12
2	24	4-6	11-12	7-12
3	19	2-4	10-12	5-7
minimum no. of nurses required in dept. over all these shifts put together		13	32	22

(Khan and Lewis [1987])

2.46 Another Version of a Maximum Cut Problem   Let $G = (\mathcal{N}, \mathcal{A})$ be a directed network with $c = (c_{ij})$ as the vector of arc weights. Define a cut in G to be a partition of $\mathcal{N}$ into $[\mathbf{X}, \bar{\mathbf{X}}]$, and its value to be $c(\mathbf{X}, \bar{\mathbf{X}}) - c(\bar{\mathbf{X}}, \mathbf{X},)$. Define a node $i \in \mathcal{N}$ to be an *overall source* if $c(i, \mathcal{N}) > c(\mathcal{N}, i)$, *overall sink* if $c(\mathcal{N}, i,) > c(i, \mathcal{N})$, and *neutral* if it is neither an overall source nor an overall sink. Under these definitions, prove that a cut $[\mathbf{X}, \bar{\mathbf{X}}]$ is a maximum value cut in G if every overall source is in the set $\mathbf{X}$ and every overall sink is in the set $\bar{\mathbf{X}}$. From this result construct an efficient algorithm to find a maximum value cut in G by the definition given here (Farley and Proskurowski [1982]).

2.47 A Maximum Value Cut Problem in a Tree   Let $\mathbb{T}$ be a tree in which every line is an arc. Let $c = (c_{ij})$ be the vector of arc weights in $\mathbb{T}$. Define a cut in

$\mathbb{T}$ to be a partition of its nodes into $(\mathbf{X}, \overline{\mathbf{X}})$, and its value to be $c(\mathbf{X}, \overline{\mathbf{X}})$. Develop a linear-time algorithm to find a maximum value cut in $\mathbb{T}$ under this definition (Farley and Proskurowski [1982]).

2.48 $G = (\mathcal{N}, \mathcal{A}, 0, k)$ is a connected directed capacitated network. For a pair of distinct nodes i, j in $\mathcal{N}$, let $v_{i,j}$ denote the capacity of minimum capacity cut separating i and j. For $i_1, \ldots, i_r \in \mathcal{N}$, prove that $v_{i_1, i_r} \geq$ min. $\{v_{i_t, i_{t+1}} : t = 1, \ldots, r-1\}$. Prove that the set of distinct values of $v_{ij}, i \neq j \in \mathcal{N}$ is at most $|\mathcal{N}| - 1$.

2.49 $G = (\mathcal{N}, \mathcal{A}, 0, k, \breve{s}, \breve{t})$ is a directed single commodity flow network. $\mathbf{E} = \{(i_1, j_1), \ldots, (i_r, j_r)\} \subset \mathcal{A}$. It is required to check whether $\mathbf{E}$ is a subset of a set of forward arcs of a cut separating $\breve{s}$ and $\breve{t}$ in G, and if so to find such a cut of a minimum capacity. Develop necessary conditions, and an efficient algorithm for this problem based on a maximum value flow formulation.

2.50 Preemptive Scheduling of Jobs with Due Dates $T_1, \ldots, T_n$ are n tasks to be processed on a single machine. Task T_i is decomposed into two subtasks denoted by M_i (mandatory subtask) and O_i (optional subtask). The processing times of M_i, O_i are m_i, σ_i respectively. The time at which M_i becomes ready for execution is r_i, and O_i becomes ready for execution when M_i is completed. The deadline for T_i (the time at which T_i must be completed) is d_i. For each i, M_i must be executed to completion, but preemption is allowed. O_i can be executed any time within the interval between its ready time and the deadline. It is terminated at the deadline d_i even if it is not completed.

A schedule is an assignment of the tasks $M_i, O_i, i = 1$ to n to the machine in disjoint intervals of time. There are precedence constraints specified for processing the tasks. This defines a precedence successor relationship among them. If T_j is a successor of T_i, then execution of M_j cannot begin until M_i is completed.

A schedule is said to be a valid schedule if each M_i is executed to completion in the time interval $[r_i, \infty]$ and the precedence constraints between all tasks are obeyed. A valid schedule is said to be feasible if each M_i is completed in the interval $[r_i, d_i]$. Let p_i denote the machine time allotted to O_i in a feasible schedule (since the processing of O_i is terminated at d_i, p_i may be $\leq \sigma_i$). If $p_i = \sigma_i$, the task T_i is said to be precisely scheduled in this schedule. Otherwise, if $p_i < \sigma_i$, the error of the task T_i in this schedule is defined to be $\epsilon_i = \sigma_i - p_i$ (this portion of O_i is essentially discarded if this schedule is implemented). The total error in the schedule is $\sum_{i=1}^{n} \epsilon_i$. A feasible schedule is said to be a precise schedule if all the ϵ_i are zero in it.

Let A_i denote the set of all successors (i.e., descendents) of T_i in the precedence relationship. Define $\bar{d}_i = \min\{d_i, \min\{d_j : j \in A_i\}\}$. Working with the modified deadlines $\bar{d}_i$ instead of d_i allows the precedence constraints to be ignored temporarily (from an invalid schedule in which portions of T_i are scheduled after some portion of T_j for $T_j \in A_i$, a valid schedule can be constructed by appropriate exchange of the time segments).

Let $a_1, \ldots, a_g$ be the strictly increasing sequence of all the distinct entries in the set $\{r_1, \ldots, r_n, \bar{d}_1, \ldots, \bar{d}_n\}$, so $g \leqq 2n$. This sequence divides time into $g-1$ intervals $[a_h, a_{h+1}]$ for $h = 1$ to $g - 1$. The length of the hth interval is $t_h = a_{h+1} - a_h$.

Define a directed network $G(\delta) = (\mathcal{N}, \mathcal{A})$ by the following. $\mathcal{N}$ contains a source node $\check{s}$, a sink node $\check{t}$, two nodes called T_i and T_i^1 for each task T_i, nodes called M_i, O_i for each i, a node called $[a_h, a_{h+1}]$ representing the hth interval defined above for each h, and another vertex called I. $\mathcal{A}$ contains the following arcs: $(\check{s}, T_i)$ for each $i = 1$ to n with capacity $\tau_i = m_i + \sigma_i$; (T_i, M_i) with capacity m_i and (T_i, O_i) with capacity σ_i for $i = 1$ to n; (M_i, T_i^1) with capacity m_i, and (O_i, T_i^1) with capacity σ_i, for $i = 1$ to n; $(T_i^1, [a_h, a_{h+1}])$ with capacity t_h for each h such that $r_i \leqq a_h$ and $\bar{d}_i \geqq a_{h+1}$ (this implies that the task T_i can be scheduled in this interval); $([a_h, a_{h+1}], \check{t})$ with capacity t_h for each h; (O_i, I) with capacity σ_i for each $i = 1$ to n; and $(I, \check{t})$ with capacity δ. All lower bounds for arch flows in $G(\delta)$ are 0.

Define $G^1(\delta) = (\mathcal{N}, \mathcal{A}^1)$ to be the network obtained from $G(\delta)$ by deleting all the arcs (T_i, O_i) and (O_i, T_i^1) for $i = 1$ to n, from it.

(i) Show that a feasible schedule exists if v = maximum flow value from $\check{s}$ to $\check{t}$ in $G^1(0)$, is $\sum_{i=1}^n m_i$. If $v < \sum_{i=1}^n m_i$, no feasible schedule exists. Given a feasible flow vector in $G^1(0)$ of value $\sum_{i=1}^n m_i$, discuss a procedure for generating a feasible schedule from it.

(ii) Assume that a feasible schedule exists. Show that a precise schedule exists if F = the maximum flow value from $\check{s}$ to $\check{t}$ in $G(0)$ is $\sum_{i=1}^n (m_i + \sigma_i) = u$. Given a feasible flow vector in $G(0)$ of value u, discuss a procedure for generating a precise schedule from it.

(iii) Suppose a feasible schedule exists but not a precise schedule. Let $\delta = u - F$. Show that a feasible schedule with minimum total error can be constructed from a maximum value flow in $G(\delta)$ and that its total error is δ.

(iv) Find a feasible schedule with minimum total error in the problem with the following data, using this approach. $n = 5$. T_1 precedes both T_2 and T_3, T_2 precedes T_4, and T_3 precedes T_5.

	r_i	d_i	m_i	O_i
$i = 1$	0.0	0.6	0.2	0.2
2	0.2	0.7	0.1	0.3
3	0.4	1.0	0.2	0.3
4	1.2	1.5	0.1	0.2
5	0.6	2.0	0.5	0.3

(v) Suppose we are given a weight w_i which measures the relative importance of the task $T_i, i = 1$ to n. Develop a formulation for the problem of finding a feasible schedule that minimizes the total weighted error, $\sum_{i=1}^{n} w_i \varepsilon_i$, as a minimum cost flow problem.

(Shih, Liu, Chung and Gillies [1989])

2.51 A large building, or building complex, occupied by hundreds of people at the same time, may contain several floors, have several corridors in each floor, and have a combination of elevators and stairways connecting the floors. Hotels, hospitals, schools and universities, libraries, large office complexes, malls, entertainment facilities, etc., are examples of such buildings. In an emergency situation, such a building has to be evacuated in a short time period. In evaluating building designs, an important characteristic is the number of people that can be moved out of the building per unit time using all available exits. This depends on the capacities of the corridors, doors, stairways, and elevators of the building; it also depends on the distribution, or expected concentration of people in various areas of the building.

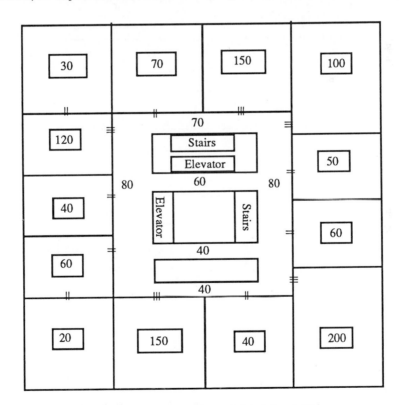

Figure 2.35 Floor plan of third (top) Floor.

In a network model of emergency evacuation, the corridors, a point on a line where two corridors intersect, the points where a corridor leads to a stairway or an elevator, or the rooms, halls, or the labs where people collect can all be represented as nodes. In Figure 2.35, the floor plan of the third floor of a three-floor building is shown, with all the relevant data.

Each room contains a little box with the number of people expected in that room entered inside it. Doors are marked with two or three lines (|| or |||) with capacities of evacuating 50 and 75 persons/minute respectively. Each stairway has a capacity of evacuating 100 persons/minute, while each elevator has a capacity of evacuating 12 person/minute. The capacities of corridors (persons/minute) are entered along the corridor. The floor plans of the second and first (ground) floors are given in Figures 2.36 and 2.37 respectively.

The entrance door marked with four lines (||||) has a capacity of evacuating 500 persons/minute. Formulate the problem of determining the maximum number of persons that can be evacuated from this building per minute in case of an emergency as a network flow problem, clearly showing the network on which the problem is posed.

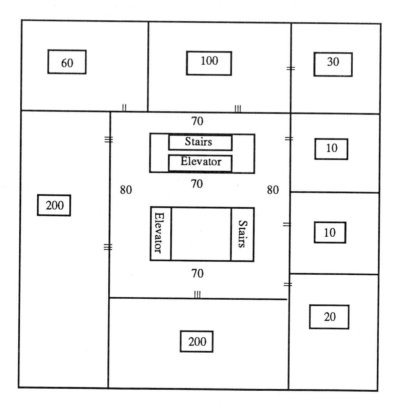

Figure 2.36 Floor plan of second Floor.

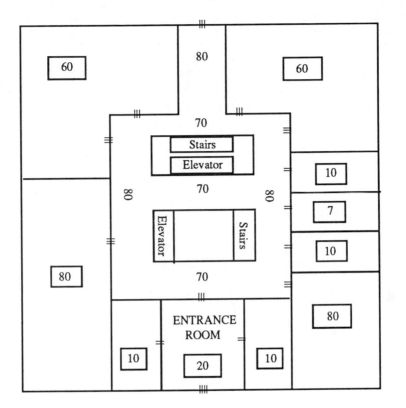

Figure 2.37 Floor plan of first (ground) Floor.

2.52 There are $p = nm$ students to be formed into n groups of m students each. Each group will work on a separate project. For every pair of distinct students i, j we are given $u_{ij} > 0$, the *utility* of assigning both of them to the same group, whichever group it may be. The $p \times p$ matrix $u = (u_{ij})$ is given. It is required to form the groups so that the total utility is as large as possible. Give a formulation for this problem and check whether it is NP-hard.

2.53 Hall's Theorem $\mathbf{C} = \{\mathbf{S}_1, \ldots, \mathbf{S}_r\}$ is a class of nonempty subsets of $\Gamma = \{1, \ldots, n\}$. A subset Δ of r distinct elements of Γ, $\Delta = \{i_1, \ldots, i_r\}$ is said to be an SDR (*system of distinct representatives*) for the class $\mathbf{C}$ if $i_h \in \mathbf{S}_h$ for each $h = 1$ to r, in this case i_h is said to represent $\mathbf{S}_h$ for $h = 1$ to r. As an example, for the class $\{\mathbf{S}_1 = \{1, 2, 3, 4\}, \mathbf{S}_2 = \{3, 4, 5, 6\}\}$, the set $\{1, 4\}$ is an SDR as 1 can represent $\mathbf{S}_1$ and 4 can represent $\mathbf{S}_2$ in it. But the set $\{3\}$ is not an SDR for this class even though 3 is in both $\mathbf{S}_1$ and $\mathbf{S}_2$.

Formulate the problem of finding an SDR for $\mathbf{C}$ as a flow problem. Draw the networks, and find the SDRs for the classes $\{\mathbf{S}_1 = \{2, 4, 5\}, \mathbf{S}_2 = \{1, 5\}, \mathbf{S}_3 = \{3, 4\}, \mathbf{S}_4 = \{3, 4\}\}$, $\{\mathbf{S}_1 = \{1, 2\}, \mathbf{S}_2 = \{2\}, \mathbf{S}_3 = \{2, 3, 4, 5\}, \mathbf{S}_4 = \{1, 2\}\}$

respectively, by solving the corresponding flow problems. Prove that an SDR exists for the class **C** iff every union of u sets of **C** contains at least u distinct elements, for $u = 1, 2, \ldots, r$. (Ford and Fulkerson [1962 of Chapter 1])

2.54 Let $f^1 = (f^1_{ij})$, $f^2 = (f^2_{ij})$ be two feasible flow vectors in $G = (\mathcal{N}, \mathcal{A}, \ell, k, \breve{s}, \breve{t})$ of values v_1, v_2 respectively. Generate a flow vector h in the residual network $G(f^1)$ by the following rules: (a) For each $(i, j) \in \mathcal{A}$ satisfying $f^1_{ij} > f^2_{ij}$, we must have $\ell_{ij} \leqq f^2_{ij} < f^1_{ij}$, so the arc (j, i) exists in $G(f^1)$ with a $-$ label, make $h_{ji} = f^1_{ij} - f^2_{ij}$. (b) For each $(i, j) \in \mathcal{A}$ satisfying $f^1_{ij} < f^2_{ij}$, we must have $f^1_{ij} < f^2_{ij} \leqq k_{ij}$, so the arc (i, j) exists in $G(f^1)$ with a $+$ label, make $h_{ij} = f^2_{ij} - f^1_{ij}$. (c) On all the arcs in $G(f^1)$ on which a flow amount is not defined by (a), (b) above, make the flow in the vector h to be zero. Then show that h is a feasible flow vector in the residual network $G(f^1)$ with flow value from $\breve{s}$ to $\breve{t}$ of $v_2 - v_1$.

Similarly, if $f^1 = (f^1_{ij})$ is a feasible flow vector in G of value v_1, and $h = (h_{pq})$ a feasible flow vector in the residual network $G(f^1)$ of value ω from $\breve{s}$ to $\breve{t}$, define a flow vector $\hat{f} = (\hat{f}_{ij})$ in G, where $\hat{f}_{ij}$ is given by

$$
\hat{f}_{ij} = \begin{cases}
f^1_{ij} + h_{ij} & \text{if there is a } + \text{ labeled arc } (i, j) \text{ in} \\
& G(f^1), \text{ corresponding to } (i, j) \in \\
& \mathcal{A}, \text{ with } h_{ij} > 0. \\
f^1_{ij} - h_{ji} & \text{if is a } - \text{ labeled arc } (j, i) \text{ in } G(f^1), \\
& \text{corresponding to } (i, j) \in \mathcal{A}, \text{ with} \\
& h_{ji} > 0. \\
f^1_{ij} & \text{otherwise}
\end{cases}
$$

Then show that $\hat{f}$ is a feasible flow vector in G and its value is $v_1 + \omega$. Use these results to provide an alternate proof of Theorem 2.12

2.55 The Minimum Value Flow Problem   Consider the single commodity flow network $G = (\mathcal{N}, \mathcal{A}, \ell, k, \breve{s}, \breve{t})$ with $0 \leqq \ell \leqq k$, in which some or all of the capacities k_{ij} may be infinite.

(i)   Assuming that $k = \infty$ and $\ell > 0$, develop an efficient algorithm for finding a feasible flow vector in G, if one exists; by specializing the methods of Section 2.6 as applied to G. In this case show that the maximum value flow from $\breve{s}$ to $\breve{t}$ in G is infinite.

(ii)   When, $\ell \neq 0$, $f = 0$ is not a feasible flow vector in G. So, in this case, the problem of finding a *minimum value flow* (i.e., one which has the least possible value among all feasible flow vectors in G) is of interest. Given a feasible flow vector $\bar{f} = (\bar{f}_{ij})$ of value $\bar{v}$ in G, a path $\mathcal{P}$ from $\breve{s}$ to $\breve{t}$ in G satisfying

$$(i, j) \text{ is a forward arc on} \mathcal{P} \quad \text{implies} \quad \bar{f}_{ij} > \ell_{ij}$$
$$(i, j) \text{ is a reverse arc on} \mathcal{P} \quad \text{implies} \quad \bar{f}_{ij} < k_{ij}$$

is known as a *flow reduction path* (FRdP) from $\check{s}$ to $\check{t}$ wrt $\bar{f}$. Given a FRdP $\mathcal{P}$ from $\check{s}$ to $\check{t}$ wrt $\bar{f}$, the flow reduction step using it generates the new flow vector $\hat{f} = (\hat{f}_{ij})$ where $\epsilon = $ min. $\{\bar{f}_{ij} - \ell_{ij}:$ over (i,j) forward on $\mathcal{P}\} \cup \{k_{ij} - \bar{f}_{ij}:$ over (i,j) reverse on $\mathcal{P}\}$, and

$$\hat{f}_{ij} = \begin{cases} \bar{f}_{ij} - \epsilon & \text{if } (i,j) \text{ is forward on } \mathcal{P} \\ \bar{f}_{ij} + \epsilon & \text{if } (i,j) \text{ is reverse on } \mathcal{P} \\ \bar{f}_{ij} & \text{otherwise} \end{cases}$$

Show that the new flow vector $\hat{f}$ is a feasible flow vector in G of value $\hat{v} = \bar{v} - \epsilon$.

Prove that a feasible flow vector f in G is a minimum value feasible flow vector iff there exists no FRdP from $\check{s}$ to $\check{t}$ wrt it. Using this result, develop algorithms analogous to those discussed in this chapter, to find a minimum value feasible flow vector in G. Apply your algorithm to find minimum value feasible flow vectors in the networks in Figures 2.22, 2.25, 2.26, and 2.27.

(iii)   Prove that if a feasible flow vector exists in G, then the minimum value in G is equal to the maximum of $\ell(\mathbf{X}, \overline{\mathbf{X}}) - k(\overline{\mathbf{X}}, \mathbf{X})$ over all $\mathbf{X} \subset \mathcal{N}$, $\overline{\mathbf{X}} = \mathcal{N} \backslash \mathbf{X}$, such that $\check{s} \in \mathbf{X}$, $\check{t} \in \overline{\mathbf{X}}$. This result is the analogue of Theorem 2.4 to the minimum value flow problem.

2.56 G $ = (\mathcal{N}, \mathcal{A}, \ell, k)$ is a directed connected single commodity flow network and (p, q) is a selected arc in G. It is required to find a feasible circulation $f = (f_{ij})$ in G which (i) maximizes f_{pq}, (ii) minimizes f_{pq}. Discuss methods for solving these problems. Apply your methods to solve these problems in the network in Figure 2.23 with $(p, q) = (4, 1)$.

Comment 2.1   Ford and Fulkerson [1962 of Chapter 1] state that the problem of maximizing flow from one point to another in a capacity-constrained network was posed to them in the spring of 1955 by T. E. Harris and General F.S. Ross. Shortly afterwards the maximum flow minimum cut theorem was established by Ford and Fulkerson in their 1956 paper listed in this chapter's references. The relationship of this theorem to the duality theorem of linear programming was recognized almost at the same time. The scanning version of the labeling algorithm was devised by Ford and Fulkerson in 1957. Johnson [1966 of Chapter 5] suggested a supplemental rule to guarantee finite termination of this algorithm for arbitrary data. Edmonds and Karp [1972] demonstrated that if the node to scan from the list is selected by the FIFO rule, then the worst case computational complexity of this algorithm is of order $O(nm^2)$ where n, m are the number of nodes, arcs in the network. This is the shortest augmenting path implementation. Zadeh [1972] constructed examples to show that this bound is tight in dense networks.

Multipath methods using all available shortest augmenting paths in each iteration were introduced by Dinic [1970] and refined by Malhotra, Kumar and Maheswari [1978], and others. Methods based on preflows were introduced by Karzanov

[1974]. Goldberg and Tarjan [1986, 1988] developed simple and elegant preflow-push algorithms based on push/relabel steps. These methods have several advantages.

These are the algorithms that have given good computational performance. However, the literature on algorithms for the maximum value flow problem is vast. Cherkasky [1977] and Galil [1980] made improvements on Karzanov algorithm. Galil and Naamad [1980] and Sleater and Tarjan [1983] discuss improvements in Dinic's algorithm using new data structures. Gabow [1985] discusses an approach based on scaling. Cheriyan and Maheswari [1987] and Ahuja, Orlin, and Tarjan [1987] discuss variants of the Goldberg-Tarjan algorithm. Goldfarb and Hao [1990] developed an $O(mn^2)$ complexity primal simplex variant for this problem. Ramachandran [1987] has shown that in a theoretical worst-case sense, the minimum capacity cut and the maximum value flow problems are not easier in an acyclic network than in a network that is not acyclic.

So far, no one has succeeded in developing an algorithm with the worst case computational complexity of $O(nm)$ for the maximum value flow problem in sparse networks. So, the search goes on.

For results of computational studies comparing the various algorithms, see Cheung [1980], Glover, Klingman, Mote, and Whitman [1979], Imai [1983], and Hamacher [1979]. But these studies were carried out before the development of preflow-push algorithms. Currently the computational performance of preflow-push algorithms is being evaluated by several groups.

Accounts of the contributions of D. R. Fulkerson can be found in Billera and Lucas [1978] and Hoffman [1978].

A variety of applications of the minimum capacity cut problem are discussed in Picard and Ratliff [1975] and Picard and Queyranne [1982a, 1982b]. Several other applications of the maximum value flow, minimum capacity cut problems are discussed in many papers in the following references, notable among those are included among the exercises given above.

Necessary and sufficient conditions for the feasibility of flow and circulation problems are due to Gale [1957] and Hoffman [1960].

2.9   References

R. K. AHUJA and J. B. ORLIN, Sept.-Oct. 1989, "A Fast and Simple Algorithm for the Maximum Flow Problem," *OR*, 37, no. 5 (748-759).

R. K. AHUJA, J. B. ORLIN, and R. E. TARJAN, 1987, "Improved Time Bounds for the Maximum Flow Problem," Sloan School of Management, MIT, Cambridge, MA.

F. BARAHONA, 1982, "On the Computational Complexity of the Ising Spin Models," *Journal of Physics A: Mathematics and General*, 15(3241-3250).

L. J. BILLERA and W. F. LUCAS, 1978, "Delbert Ray Fulkerson: August 14, 1924 - January 10, 1976," *MPS*, 8(1-16).

J. R. BROWN, Mar.-Apr. 1979, "The Sharing Problem," *OR*, 27, no. 2(324-340).

V. CABOT, R. L. FRANCIS, and M. A. STARY, 1970, "A Network Flow Solution to a Rectilinear

Distance Facility Location Problem," *AIIE Transactions*, 2(132-141).

J. CHERIYAN and S. N. MAHESWARI, 1989, "Analysis of Preflow Push Algorithms for Maximum Network Flow," *SIAM J. on Computing*, 18(1057-1086).

R. V. CHERKASKY, 1977, "Algorithms of Construction of Maximal Flow in Networks with Complexity $O(n^2\sqrt{m})$ Operations," *Mathematical Methods of Solution of Economical Problems*, 7(112-125).

T. CHEUNG, 1980, "Computational Comparison of Eight Methods for the Maximum Network Flow Problem," *ACM Transactions on Mathematical Software*, 6(1-16).

E. A. DINIC, 1970, "Algorithms for Solution of a Problem of Maximum Flow in Networks with Power Estimation," *Soviet Mathematics Doklady*, 11(1277-1280).

I. B. DUNCAN, Nov. 1979, "The Allocation of Specialities to Hospitals in a Health District," *JORS*, 30, no. 11(953-961).

J. EDMONDS and R. M. KARP, 1972, "Theoretical Improvements in Algorithmic Efficiency for Network Flow Problems," *JACM*, 19(248-264).

P. ELIAS, A. FEINSTEIN, and C. E. SHANNON, 1956, "Note on Maximum Flow Through a Network," *IRE Transactions on Information Theory*, IT-2(117-119).

A. M. FARLEY and A. PROSKUROWSKI, Dec. 1982, "Directed Maximal-Cut Problems," *IPL*, 15, no. 5(238-241).

L. R. FORD Jr. and D. R. FULKERSON, 1956, "Maximum Flow Through a Network," *Canadian Journal of Mathematics*, 8(399-404).

H. N. GABOW, 1985, "Scaling Algorithms for Network Problems," *Journal of Computer and System Sciences*, 31(148-168).

D. GALE, 1957, "A Theorem on Flows in Networks," *Pacific Journal of Mathematics*, 7(1073-1082).

Z. GALIL, 1980, "An $O(n^{5/3}m^{2/3})$ Algorithm for the Maximal Flow Problem," *Acta Informatica*, 14(221-242).

Z. GALIL and A. NAAMAD, 1980, "An $O(nmlog^2 n)$ Algorithm for the Maximal Flow Problem," *Journal of Computer and System Sciences*, 21(203-217).

G. GALLO, M. D. GRIGORIADIS, and R. E. TARJAN, Feb. 1989, "A Fast Parametric Maximum Flow Algorithm," *SIAM J. on Computing*, 18, no. 1(30-55).

F. GLOVER, D. KLINGMAN, J. MOTE and D. WHITMAN, 1979, "Comprehensive Computer Evaluation and Enhancement of Maximum Flow Algorithms," Graduate School of Business Administration, University of Colorado, Boulder, CO, extended abstract in *DAM*, 2 (1980)(251-254).

A. V. GOLDBERG and R. E. TARJAN, 1986, "A New Approach to the Maximum Flow Problem," *Proceedings of the 18th Symposium on the Theory of Computing*,(136-146).

A. V. GOLDBERG and R. E. TARJAN, Oct. 1988, "A New Approach to the Maximum Flow Problem," *JACM*, 35, no. 4(921-940).

D. GOLDFARB and J. HAO, Aug. 1990, "A Primal Simplex Algorithm that Solves the Maximum Flow Problem in at Most nm Pivots and $O(n^2 m)$ Time," *MP*, 47, no. 3(353-365).

R. E. GOMORY and T. C. HU, Dec. 1961, "Multi-terminal Network Flows," *Journal of SIAM*, 9, no. 4(551-570).

D. GUSFIELD, C. MARTEL and D. FERNANDEZ-BACA, 1987, "Fast Algorithms for Bipartite Network Flow," *SIAM J. on Computing*, 16(237-251).

H. HAMACHER, 1979, "Numerical Investigations on the Maximal Flow Algorithm of Karzanov,"

Computing, 22(17-29).

H. HAMACHER, 1982, "Determining Minimal Cuts with a Minimal Number of Arcs," *Networks*, 12, no.4(493-504).

A. J. HOFFMAN, 1960,"Some recent applications of the Theory of Linear Inequalities to Extremal Combinatorial Analysis," *Proceedings of the Symposia on Applied Math.*, 10(113-128).

A. J. HOFFMAN, 1978, "Ray Fulkerson's Contributions to Polyhedral Combinatorics," *MPS*, 8(17-23).

H. IMAI, 1983, "On the Practical Efficiency of Various Flow Algorithms," *Journal of the OR Society of Japan*, 26(61-82).

A. V. KARZANOV, 1974, "Determining the Maximal Flow in a Network by the Method of Preflows," *Soviet Mathematics Doklady*, 15(434-437).

M. R. KHAN and D. A. LEWIS, 1987, "A Network Model forNursing Staff Scheduling," *Zeittschrift fur Operations Research*, 31, no. 6(B161- B171).

S. LAKSHMINARAYANAN, R. LAKSHMANAN, R. L. PAPINEAU and R. ROCHETTE, Aug. 1979,"Order Preserving Allocation of Jobs to Two Non-identical Parallel Machines: A Solvable Case of the Maximum Cut Problem," *INFOR*, 17, no. 3(230-241).

V. M. MALHOTRA, M. P. KUMAR and S. N. MAHESWARI, 1978, "An O(n^3) Algorithm for Finding Maximum Flows in Networks," *IPL*, 7(277-278).

N. MEGIDDO and Z. GALIL, 1979, "On Fulkerson's Conjecture About Consistent Labeling Processes," *MOR*, 4(265-267).

J. D. MURCHLAND, 1968, "Rhy's Combinatorial Station Selection Problem," London Graduate School of Business.

J. C. PICARD and H. D. RATLIFF, 1975, "Minimum Cuts and Related Problems," *Networks*, 5(357-370).

J. C. PICARD and H. D. RATLIFF, May-June 1978, "A Cut Approach to the Rectilinear Distance Facility Location Problem," *OR*, 26, no. 3(422-433).

J. C. PICARD and M. QUEYRANNE, 1982a, "A Network Flow Solution to Some Nonlinear 0-1 Programming Problems with Applications to Graph Theory," *Networks*, 12(141-159).

J. C. PICARD and M. QUEYRANNE, Nov. 1982b, "Selected Applications of Maximum Flows and Minimum Cuts in Networks," *INFOR*, 20, no. 4(394-422).

M. QUEYRANNE, 1980, "Theoretical Efficiency of the Algorithm ' Capacity' for the Maximum Flow Problem," *MOR*, 5(258-266).

V. RAMACHANDRAN, 1987, "The Complexity of Minimum Cut and Maximum Flow Problems in an Acyclic Network," *Networks*, 17, no. 4(387-392).

H. D. RATLIFF, G. T. SICILIA and S. H. LUBORE, Jan. 1975, "Finding the n Most Vital Links in Flow Networks," *MS*, 21, no. 5(531-539).

W. K. SHIH, J. W. S. LIU, J. Y. CHUNG and D. W. GILLIES, July 1989, "Scheduling Tasks with Ready Times and Deadlines to Minimize Average Error," *Operating Systems Review*, 23, no. 3(14-28).

D. SLEATOR and R. E. TARJAN, 1983, "A Data Structure for Dynamic Trees," *Journal of Computer and System Sciences*, 24(362-391).

D. SLEATOR and R. E. TARJAN, 1985, "Self-adjusting Binary Search Trees," *JACM*, 32(652-686).

R. E. TARJAN, 1984, "A Simple Version of Karzanov's Blocking Flow Algorithm," *OR Letters*,

2(265-268).

D. M. TOPKIS, 1980, "Activity Selection Games and the Minimum Cut Problem," Technical report, Bell Labs. Holmdel, NJ.

A. TUCKER, May 1977, "A Note on Convergence of the Ford-Fulkerson Flow Algorithm," *MOR*, 2(143-144).

G. R. R. WAISSI, 1985, "Acyclic Network Generation and Maximal Flow Algorithms for Single Commodity Flow," Ph. D. dissertation, Dept. of Civil Engineering, University of Michigan, Ann Arbor, Mich.

R. D. WOLLMER, "Removing Arcs From A Network," *OR*, 12, no. 6(934-940).

T. YI and K G MURTY, 1991, "Finding Maximum Flows in Networks with Nonzero Lower Bounds Using Preflow Methods," Tech. report, IOE Dept., University of Michigan, Ann Arbor, Mich.

N. ZADEII, 1972, "Theoretical Efficiency of Edmonds-Karp Algorithm For Computing Maximal Flows," *JACM*, 19(184-192).

Chapter 3

Primal-Dual and Dual Algorithms for the Assignment and Transportation Problems

The assignment and transportation problems are single commodity minimum cost flow problems on pure bipartite networks. Primal-dual algorithms are a class of methods for LPs with the following characteristic features:

1. They maintain a dual feasible solution, and a primal vector (this vector is primal infeasible until termination) that together satisfy all the complementary slackness optimality conditions for the original problem throughout the algorithm. This primal vector is usually feasible to a relaxation of the primal problem (typically this is obtained by changing the equality constraints in the problem to "$\overset{\scriptscriptstyle\leq}{=}$" inequalities).

2. In each step, the algorithm either performs (a) below, or (b) if this is not possible.

   (a) Keeps the dual solution fixed, and tries to alter the primal vector to bring it closer to primal feasibility while continuing to satisfy the complementary slackness conditions together with the present dual solution.

   (b) Keeps the primal vector fixed, and changes the dual feasible solution. The aim of this is to get a new dual feasible solution satisfying two conditions. The first is that the new dual feasible solution satisfies the complementary slackness conditions together with the present primal vector. The second is that it makes it possible to get a new primal vector closer to primal feasibility when the algorithm continues.

3. As the algorithm progresses, the primal vector moves closer and closer to primal feasibility. In other words, there is a measure of primal infeasibility which improves monotonically during the algorithm.

There are two possible conclusions at termination. One occurs if the primal vector being maintained becomes primal feasible at some stage; then it is an optimum solution. The second occurs if a primal infeasibility criterion is satisfied at some stage.

For the assignment and transportation problems it is easy to obtain an initial dual feasible solution. And the task in (a) above is a maximum value flow problem on a subnetwork known as the *admissible* or *equality subnetwork* wrt the present dual feasible solution. These facts make the primal-dual methods particularly attractive to solve them. The blossom algorithms discussed in Chapter 10 for matching and edge covering problems are primal-dual algorithms that are generalizations of the Hungarian method of the next section to those problems. The primal-dual approach can be used to solve a general LP, however, for these general problems it seems to offer no particular advantage over the primal simplex algorithm.

3.1 The Hungarian Method for the Assignment Problem

The data in an assignment problem of order n is the cost matrix $c = (c_{ij})$ of order $n \times n$. Given c the problem is to find $x = (x_{ij})$ of order $n \times n$ to

$$\text{Minimize } z(x) = \sum_{i=1}^{n} \sum_{j=1}^{n} c_{ij} x_{ij}$$

$$\text{Subject to } \sum_{j=1}^{n} x_{ij} = 1 \quad \text{for } i = 1 \text{ to } n \tag{3.1}$$

$$\sum_{i=1}^{n} x_{ij} = 1 \quad \text{for } j = 1 \text{ to } n$$

$$x_{ij} \geqq 0 \quad \text{for all } i, j$$

$$\text{and } x_{ij} = 0 \text{ or } 1 \quad \text{for all } i, j \tag{3.2}$$

Every feasible solution of (3.1) and (3.2) is an assignment of order n and vice versa. See (1.17) for an assignment of order 4. Every BFS of (3.1) satisfies (3.2). So, if (3.1) is solved by the simplex method ignoring (3.2), the optimum solution obtained will satisfy (3.2) automatically. The Hungarian method does not use basic vectors for (3.1), but it maintains (3.2) throughout.

In an assignment x, if a particular $x_{ij} = 1$, the cell (i,j) is said to have an *allocation* (in this case row i is said to be *allocated to*, or *matched with*, column j in x). A *partial assignment* of order n is a 0-1 square matrix of order n which contains at most one nonzero entry of 1 in each row and column. Here is a partial assignment of order 3.

$$\begin{pmatrix} 0 & 1 & 0 \\ 0 & 0 & 0 \\ 0 & 0 & 0 \end{pmatrix}$$

Clearly, a partial assignment is a feasible solution for a relaxed version of (3.1) and (3.2) in which the equality constraints in (3.1) are replaced by the corresponding "$\leq$" inequalities. The Hungarian method moves among partial assignments in which the number of allocations keeps on increasing as the algorithm progresses.

Represent each row and each column of an $n \times n$ array by a node. Join each row node to each column node by an edge, leading to a complete bipartite network (see Figure 3.1). Make the cost of the edge joining row node i with column node j, c_{ij}. The set of allocated cells in any partial assignment (assignment) corresponds to a matching (perfect matching) in this bipartite network, and vice versa. For example, the wavy subnetwork in Figure 3.1 is the perfect matching corresponding to the assignment in (1.17). Hence, the assignment problem (3.1) and (3.2) is equivalent to that of finding a minimum cost perfect matching in this bipartite network. That's why it is also known as the *bipartite minimum cost perfect matching problem*.

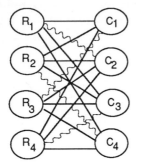

Figure 3.1 R_i is the node corresponding to row i, C_j is the node corresponding to column j.

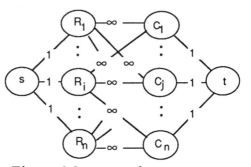

Figure 3.2 s, t are the super source and super sink respectively. All lower bounds are 0, and capacities are entered on the arcs. Cost coefficients of arcs incident at s, t are 0. All arcs are directed from left to right. Cost coefficient of arc (R_i, C_j) is c_{ij}.

Direct all the lines in this bipartite network from the row node to the column node. Treat all the row nodes as sources each with a supply of one unit, and the column nodes as sinks with a demand of one unit. Then (3.1) is the problem of finding a minimum cost feasible flow in this network. An allocation in cell (i,j) corresponds to a flow of one unit on the arc connecting row node i with column node

j and vice versa. We can introduce a supersource s and supersink t and transform this into a minimum cost flow problem on the network in Figure 3.2, in which all the arcs incident at s, t are required to be saturated.

In large scale applications, for each i, a subset of $\{1, \ldots, n\}$ is usually specified, and x_{ij} can be equal to 1 only when j is in that subset; otherwise it has to be 0. This can be handled by defining c_{ij} to be $+\infty$ whenever x_{ij} is required to be 0. In the corresponding networks in Figures 3.1 and 3.2, the arc (i, j) (i.e., (R_i, C_j)) is included only if x_{ij} can be equal to 1 in the problem. Thus the network is no longer the complete bipartite network. Let m denote the number of arcs in the network (i.e., the number of variables x_{ij} which can assume the value 1). The assignment problem is said to be *sparse* if m is small compared to n^2, and *dense* if m is close to n^2.

In this section we present an implementation of the primal-dual approach for the assignment problem known as the *Hungarian method*. It is described using arrays for ease of understanding, but computer implementations are usually based on the corresponding network. The arc joining row node i to column node j is omitted if x_{ij} is required to be 0 in the problem.

The dual of (3.1) is

$$\text{Maximize} \quad \sum_{i=1}^{n} u_i \ + \ \sum_{j=1}^{n} v_j$$

$$\text{Subject to} \quad u_i + v_j \ \lessgtr \ c_{ij}, \ i, j = 1 \text{ to } n \tag{3.3}$$

Denote the objective value of assignment x with c as the cost matrix by $z_c(x)$. Let c' be the matrix obtained by subtracting a real number α from every element in a row or a column of c. Since each assignment contains a single nonzero entry of 1 in each row and column, we have $z_c(x) = \alpha + z_{c'}(x)$. So, the set of optimum assignments that minimize $z_c(x)$ is the same as the set of optimum assignments that minimize $z_{c'}(x)$. Hence, for solving the assignment problem, we can replace c by c'. We use this idea repeatedly. Let $u = (u_1, \ldots, u_n)$, $v = (v_1, \ldots, v_n)$ be such that $c_{ij}^0 = c_{ij} - u_i - v_j \geqq 0, i, j = 1$ to n. This is the condition for u, v to be dual feasible, and in this case c_{ij}^0 are the dual slacks, $c^0 = (c_{ij}^0)$ is known as the *reduced cost matrix*, and $r_0 = \sum_{i=1}^{n} u_i + \sum_{j=1}^{n} v_j$ as the *total reduction* wrt u, v. The matrix c^0 is obtained by subtracting u_i from each entry in row i of c for $i = 1$ to n, and then subtracting v_j from each entry in column j of the resulting matrix, for $j = 1$ to n. By the above argument, for each assignment x we have $z_c(x) = r_0 + z_{c^0}(x)$. Since $c^0 \geqq 0$, $z_{c^0}(x) \geqq 0$, and so r_0 is a lower bound for the minimum objective value in (3.1) whenever u, v is dual feasible. If we can find an assignment x which has allocations only in those cells in which the entries in c^0 are zero, then $z_{c^0}(x) = 0$, and hence x is an optimum assignment. To find such an assignment, we define the cell (i, j) or the corresponding arc (i, j) (we will use this notation to denote the arc joining row node i to column node j) in the network in Figure 3.2 to be *admissible* (or an *equality cell* or *equality arc* respectively) if $c_{ij}^0 = c_{ij} - u_i - v_j = 0$. When all inadmissible arcs are removed from the network in Figure 3.2, we get the *admissible subnetwork* or the *equality subnetwork* wrt the dual feasible solution u, v. A maximum value flow from s to t in the equality subnetwork corresponds to a partial assignment having the maximum number of allocations among admissible cells. If this is an assignment, it is clearly optimal to (3.1) and (3.2); otherwise we get a partial assignment x satisfying

$$x_{ij}(c_{ij} - u_i - v_j) = 0, \text{ for all } i, j \qquad (3.4)$$

The complementary slackness optimality conditions for (3.1) and its dual (3.3) are (3.4). The Hungarian method maintains $x, (u, v)$ always satisfying (3.2), dual feasibility constraints in (3.3), and (3.4). When x satisfies (3.1), it is an optimum assignment and the method terminates. If the maximum value flow in the equality subnetwork does not saturate all the arcs incident at t, there exists no assignment which has allocations among admissible cells only. In this case the Hungarian method goes to a dual solution change routine. After this change the new reduced cost matrix will contain some new cells with zero entries in columns which have no allocations at present, and the procedure is repeated. During the method, each row and column of the array (i.e., each node in the network implementation) may be in three possible states: *unlabeled, labeled and unscanned, labeled and scanned.* The *list* is always the set of current labeled and unscanned rows and columns.

THE HUNGARIAN METHOD

Step 0 The initial dual feasible solution   If some dual feasible solution is available, use it as the initial one; otherwise define it to be $(u^1 = (u_i^1), v^1 = (v_j^1))$ where $u_i^1 = \min\{c_{ij} : j = 1 \text{ to } n\}$, $v_j^1 = \min\{c_{ij} - u_i^1 : i = 1 \text{ to } n\}$, for $i, j = 1$ to n. List $= \emptyset$. Go to Step 1 with any partial assignment containing allocations only among admissible cells in the reduced cost matrix wrt the initial dual solution (this could be 0, containing no allocations).

Step 1 Tree growth routine

Substep 1   Label each row without an allocation with $(s, +)$, and include it in the list.

Substep 2   If list $= \emptyset$, tree growth has terminated and there is a nonbreakthrough. The present set of allocations contains the maximum number possible among admissible cells; go to Step 3. Otherwise, select a row or column from the list for scanning and delete it from the list.

Forward labeling   Scanning row i consists of labeling each unlabeled column j for which (i, j) is an admissible cell, with the label (row i, +).

Reverse labeling   To scan column j, check whether it has an allocation. If the row in which that allocation occurs is unlabeled so far, label it with (column j, −).

If any column without an allocation has been labeled, there is a break-through; go to Step 2. Otherwise, include all newly labeled rows and columns in the list, and repeat this Substep 2.

Step 2 Allocation change routine Suppose column j, which does not have an allocation, has been labeled. Trace its predecessor path using the labels. Delete present allocations in cells corresponding to reverse arcs, and add allocations in cells corresponding to forward arcs of this path. If all the columns have allocations now, these allocations define an optimum assignment; terminate. Otherwise, chop down the present trees (i.e., erase the labels on all the rows and columns) and go back to Step 1.

Step 3 Dual solution change routine Compute δ, minimum value of reduced cost coefficient among cells in labeled rows and unlabeled columns. δ will be > 0. If $\delta = +\infty$; this can only happen if some x_{ij} are constrained to be 0 in the problem; there is no feasible assignment, terminate. If δ is finite, add it to the value of u_i in all labeled rows and subtract it from the value of v_j in all labeled columns. Compute the new reduced cost coefficient in each cell. Retain the present labels on all the labeled rows and columns, but include all the labeled rows in the list, and resume tree growth by going to Substep 2 in Step 1.

Discussion

When solving small problems by hand, a good initial partial assignment in Step 0 can be obtained by making an allocation in an admissible cell in the initial reduced cost matrix that is not yet struck off, in a row or column containing only one such cell if possible, or in any admissible cell not yet struck off otherwise; striking off all other admissible cells in the row and column of the allocated cell; and repeating this process with the remaining admissible cells.

Also, to find δ in Step 3, draw a straight line in the present reduced cost matrix through each unlabeled row and each labeled column, then these straight lines cover all the admissible cells (if there is an admissible cell without a line through it, its column would have been labeled when its row was scanned, a contradiction, see Array 3.1). Hence every reduced cost coefficient not covered by a straight line is > 0, and δ is the minimum of these entries. So, δ will always be > 0. The number of allocations at this stage can be verified to be equal to the number of straight lines drawn. To get the new reduced cost coefficients, subtract δ from the entry in the present reduced cost matrix in each cell in a labeled row and unlabeled column (i.e., those cells without a straight line through them), and add δ to the entry in each cell in an unlabeled row and labeled column (i.e., those cells at the intersection of two straight lines). From the definition of δ this implies that all the new reduced cost coefficients are ≥ 0, i.e., the new dual solution is dual feasible. Let α be (the number of labeled rows $-$ the number of labeled columns) at present. Add $\delta\alpha$ to the total reduction; this updates it.

Array 3.1 Summary of the Position When the Hungarian Method Reaches the Dual Solution Change Routine. $(\tilde{u}, \tilde{v}, \overset{\approx}{c} = (\overset{\approx}{c}_{ij} = c_{ij} - \tilde{u}_i - \tilde{v}_j))$, are the dual solution and reduced cost matrix before the change. $(\hat{u}, \hat{v}, \hat{c} = (\hat{c}_{ij} = c_{ij} - \hat{u}_i - \hat{v}_j))$ are the corresponding things after the change.

	Block of labeled cols.	Block of unlabeled cols.	Alloca-tions	St. lines	Dual change
Block of la-beled rows	Each col. here has an alloca-tion among labeled rows (there is breakthrough otherwise). $\hat{c}_{ij} = \overset{\approx}{c}_{ij}$ here, so, admissibil-ity pattern re-mains unchanged.	No admissible cells here (one col. would be labeled otherwise). $\overset{\approx}{c}_{ij} > 0$, here. $\delta = $ Min.$\{\overset{\approx}{c}_{ij}:$ (i,j) here$\} > 0$. $\hat{c}_{ij} = \overset{\approx}{c}_{ij} - \delta$ here. New ad-missible cells created here, this allows tree growth.	Some rows have no alloca-tion.		$\hat{u}_i = \tilde{u}_i +$ δ for i here.
Block of unla-beled rows	No allocation here (otherwise a row here could be labeled). $\hat{c}_{ij} = \overset{\approx}{c}_{ij} + \delta$ here. All cells here become inadmissible next.	Each row here contains an allocation in these cols. $\hat{c}_{ij} = \overset{\approx}{c}_{ij}$ here, so admissibi-lity pattern re-mains un-changed here.	Each row has alloca-tion.	Draw through each row.	$\hat{u}_i = \tilde{u}_i$ for i here.
Alloca-tions	Each col. has allocation.	Some cols. have no allocation.			
St. lines	Draw through each col.				
Dual change	$\hat{v}_j = \tilde{v}_j - \delta$ for j here	$\hat{v}_j = \tilde{v}_j$ for j here.			

Cells that have allocations at present remain admissible in the new reduced cost matrix, and hence (3.4) continues to hold. New admissible cells are created among labeled rows and unlabeled columns, so, when the list is made equal to the set of all labeled rows, and tree growth resumed, at least one new column will be labeled.

See Array 3.1

Suppose δ came out to be $+\infty$ in Step 3 at some stage. Let $\mathbf{J}$ be the set of all cells (i, j) with row i labeled, and column j unlabeled at this stage. So, the present reduced cost coefficient $\bar{c}_{ij} = +\infty$ for all cells $(i, j) \in \mathbf{J}$ (i.e., x_{ij} is required to be 0 for every $(i, j) \in \mathbf{J}$). Even if all cells not in $\mathbf{J}$ are made admissible, no more labeling can be carried out, and the current nonbreakthrough continues to hold. This implies that the present partial assignment contains the maximum number of allocations possible under the constraint that x_{ij} must be 0 for all $(i, j) \in \mathbf{J}$, hence there is no feasible assignment in the problem.

Consider the Hungarian method applied to solve an assignment problem of order n. Whenever Step 2 is carried out, the number of allocations increases by 1. Thus Step 2 is carried out at most n times in the algorithm. From Array 3.1 we see that at least one new column gets labeled when tree growth is resumed after a dual solution change step. Thus Step 3 can occur at most n times between two consecutive occurrences of Step 2. The effort needed to carry out Step 3 (updating the reduced cost coefficients) and the following tree growth, before going to Step 2 or 3 again is at most $O(n^2)$. Thus the effort between two consecutive occurrences of Step 2 is $O(n^3)$, and therefore the entire method takes at most $O(n^4)$. Later on we show that the method can be implemented so that its worst case computational complexity is at most $O(n^3)$. If the infeasibility criterion is never satisfied, at termination we will have an assignment x and a dual feasible solution (u, v) which together satisfy the complementary slackness conditions (3.4), so x is an optimum assignment, and (u, v) is an optimum dual solution.

EXAMPLE 3.1 Illustration of the allocation change routine

In this example $n = 6$ and Array 3.2 contains all the relevant information. Admissible cells are those with a zero in the upper left corner. Allocations are marked with a $\square$ in the cell. All other information is omitted.

When column 6 without an allocation is labeled (Row 6, $+$) we had a breakthrough, so we put a new allocation in the cell $(6, 6)$. Now look at the label on row 6, which is (Col 1, $-$). Thus we delete the allocation in cell $(6, 1)$. Continuing this way using the labels, we put a new allocation in $(3, 1)$, delete the one in $(3, 4)$, add on allocation in $(4, 4)$, and reach row 4 labeled $(s, +)$, implying that the allocation change routine is complete. The allocation change path indicated by the labels on Array 3.2 is shown in Figure 3.3. The wavy edges in Figure 3.3 correspond to allocated cells in Array 3.2, on this allocation change path. This path is clearly an alternating path (nodes in it correspond alternatively to unallocated cells, allocated cells in Array 3.2). It is called the alternating predecessor path of column 6 traced by the labels.

Array 3.2

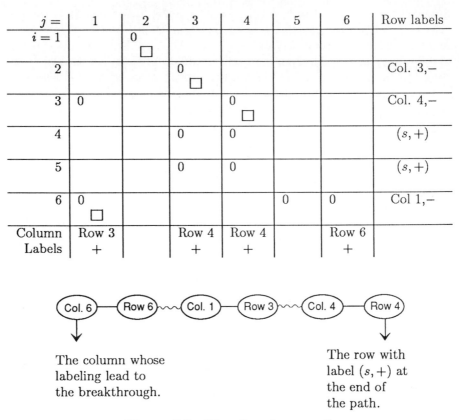

$j =$	1	2	3	4	5	6	Row labels
$i = 1$		0 ☐					
2			0 ☐				Col. 3,−
3	0			0 ☐			Col. 4,−
4			0	0			$(s,+)$
5			0	0			$(s,+)$
6	0 ☐				0	0	Col 1,−
Column Labels	Row 3 +		Row 4 +	Row 4 +		Row 6 +	

The column whose labeling lead to the breakthrough.

The row with label $(s,+)$ at the end of the path.

Figure 3.3 Allocation change path.

The allocation change routine reverses the roles of unallocated and allocated cells along this path. It has the effect of increasing the number of allocations by 1. Hence a path like this is called an *augmenting path* in the admissible network. An augmenting path in the admissible network wrt the present allocations is an alternating path of unallocated and allocated arcs, joining a column node and a row node both of which have no allocated arcs incident at them. The tree growth routine discussed above is an efficient scheme to look for such an augmenting path. When a breakthrough occurs, it is an indication that an augmenting path has been identified. In this case the tree is said to have become an *augmenting tree*. The augmenting path is the predecessor path of the column node without an allocated arc incident at it, whose labeling lead to the breakthrough. If a nonbreakthrough occurs, it is an indication that no augmenting path exists in the admissible network wrt the present allocations.

The new allocations after the allocation change are shown in Array 3.3. The new partial assignment has five allocations, one more than the previous. Now the labels on all the rows and columns are erased, and the algorithm begins another labeling cycle to see whether yet another allocation can be squeezed in among the present admissible cells.

Array 3.3

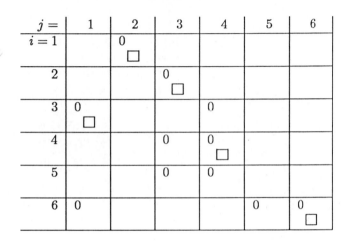

$j =$	1	2	3	4	5	6
$i = 1$		0 □				
2			0 □			
3	0 □			0		
4			0	0 □		
5			0	0		
6	0				0	0 □

EXAMPLE 3.2

Here we solve the assignment problem of order 4 with the original cost matrix $c = (c_{ij})$ given below.

	c_{ij}				Initial	$c_{ij} - u_i$			
$j =$	1	2	3	4	u_i	$j = 1$	2	3	4
$i = 1$	15	22	13	4	4	11	18	9	0
2	12	21	15	7	7	5	14	8	0
3	16	20	22	6	6	10	14	16	0
4	6	11	8	5	5	1	6	3	0
					Initial				
					$v_j \rightarrow$	1	6	3	0

Total Reduction $= 32$

First reduced cost matrix with initial allocations

$j =$	1	2	3	4	u_i	Row labels
$i = 1$	$10 = \bar{c}_{11}$	12	6	0	4	$s, +$
2	4	8	5	0 □	7	Col. 4, $-$
3	9	8	13	0	6	$s, +$
4	0	0	0 □	0	5	
v_j	1	6	3	0		
Col. labels				Row 3, $+$		

Nonbreakthrough
$\delta = 4$
Labeling resumed below

Second reduced cost matrix

$j =$	1	2	3	4	u_i	Row labels
$i = 1$	6	8	2	0	8	$s, +$
2	0	4	1	0 □	11	Col. 4, $-$
3	5	4	9	0	10	$s, +$
4	0	0	0 □	4	5	
v_j	1	6	3	-4		
Col. labels	Row 2 $+$			Row 3, $+$		

Breakthrough
since col. 1 is labeled
Total reduction $= 40$

Second reduced cost matrix with new allocations and new labels

$j =$	1	2	3	4	u_i	Row labels
$i = 1$	6	8	2	0		$s, +$
					8	
2	0 □	4	1	0	11	
3	5	4	9	0 □	10	Col. 4, $-$
4	0	0	0 □	4	5	
v_j	1	6	3	-4		
Col. labels				Row 1, +		

Nonbreakthrough
$\delta = 2$
Labeling resumed below

Third reduced cost matrix

$j =$	1	2	3	4	u_i	Row labels
$i = 1$	4	6	0	0	10	$s, +$
2	0 □	4	1	2	11	
3	3	2	7	0 □	12	Col. 4, $-$
4	0	0	0 □	6	5	Col. 3, $-$
v_j	1	6	3	-6		
Col. labels		Row 4 +	Row 1 +	Row 1, +		

Breakthrough
since col. 2 labeled.
Total reduction $= 42$

**Third reduced cost matrix
with new allocations**

$j =$	1	2	3	4	u_i
$i = 1$	4	6	0 □	0	10
2	0 □	4	1	2	11
3	3	2	7	0 □	12
4	0	0 □	0	6	5
v_j	1	6	3	-6	

Now we have a full assignment, $\{(1, 3), (2, 1), (3, 4), (4, 2)\}$ among the admissible cells, which is an optimum assignment for the problem. Its cost (wrt the original cost matrix) can be verified to be 42, which is also the total reduction at this stage. An optimum dual solution is the (u, v) from the final reduced cost matrix.

An O(n^3) Implementation of the Hungarian Method

An O(n^3) implementation of the Hungarian method is obtained by not updating the entire reduced cost matrix but only portions of it that are needed, after each dual solution change step. In this implementation an *index* is maintained on each unlabeled column. It is an ordered pair of the form $[t_j, p_j]$ for column j, where p_j is the minimum current reduced cost coefficient in this column among labeled rows, and t_j is the number of a labeled row in which this minimum occurs. The index is maintained in each column as long as it remains unlabeled. The moment the column gets labeled in the method its index is erased. So, at each visit to Step 3 in the method, $\delta = \min \{p_j$: over columns j unlabeled at that time$\}$.

Define a *stage* in the Hungarian method to begin either after getting the initial partial assignment or after an allocation change has been carried out, and to end when the next allocation change has been completed. So there are at most n stages in the method, and Step 3 occurs at most n times in each stage. All reduced cost coefficients are computed once at the beginning of each stage using the dual solution at that time. The reduced cost coefficient is not computed again during this stage in any cell in this implementation.

We will now describe how the work in a stage is organized in this implementation. At the beginning of the stage, compute the entire reduced cost matrix $\bar{c} = (\bar{c}_{ij} = c_{ij} - u_i - v_j)$ wrt the present (u, v). Then carry out Step 1 as usual, and if this leads to Step 2, this stage is completed after it. On the other hand, if this leads to Step 3, before going there, compute for each unlabeled column j, $p_j = \min \{\bar{c}_{ij}$: row i labeled$\}$, and let t_j be an i that ties for this minimum (break ties arbitrarily), and index this column with $[t_j, p_j]$. Then enter Step 3. Each time you have to carry Step 3 in this algorithm, compute $\delta = \min\{p_j:$ over columns j unlabeled at that time$\}$. Subtract δ from the p_j index of each unlabeled column j. Let $\mathbf{L}$ be the set of all unlabeled columns j for which p_j became 0 as a result of this operation. For each $j \in \mathbf{L}$, (t_j, j) is a new admissible cell. Put all labeled rows in the list, and resume tree growth (scanning the present labeled rows first) treating $\{(t_j, j) : j \in \mathbf{L}\}$ as the set of new admissible cells among the unlabeled columns. In this tree growth step, each column $j \in \mathbf{L}$ will get labeled. Whenever a column gets labeled, erase its index. Whenever a new row, say row i, is labeled do the following. Because of the way the dual variables are changed in Steps 3 of the algorithm, for each unlabeled column j the present v_j is the same as at the beginning of this stage, and the same thing is true for u_i since row i remained unlabeled so far and got labeled just now. So, the present reduced cost coefficient in cell (i, j) in unlabeled columns j in this row i is the same as at the beginning of this stage, $\bar{c}_{ij}$. For each unlabeled column j at this time let $[t_j, p_j]$ be the present index on it. If $\bar{c}_{ij} = 0$ label column j with the label (row i, +). If $p_j > \bar{c}_{ij} > 0$ change the index on column j to $[i, \bar{c}_{ij}]$; if $p_j \overset{\leq}{=} \bar{c}_{ij} > 0$ leave the index on column j unchanged. Repeat this same process each time Step 3 has to be carried out in this stage.

The efficiency of this implementation stems from the fact that it updates the indices on unlabeled columns during a stage without recomputing the reduced cost coefficients after each dual solution change. Since the reduced cost coefficient in any cell is computed once only in each stage, the computational effort per stage can be verified to be at most $O(n^2)$. Since there are at most n stages, the overall computational complexity of the Hungarian method with this implementation is at most $O(n^3)$.

Algorithm to Obtain a Maximum Cardinality Independent Set of Admissible Cells

Suppose we are given a subset of cells of the $p \times n$ transportation array called *admissible cells*. A subset of admissible cells is said to be *independent* if no two cells in it lie in the same row or column. Let a line refer to either a row or column of the array. A subset of lines is said to *cover* all the admissible cells (or to *form a covering set of lines*) if each admissible cell is contained in a row or column (or both) from the subset. Every subset of an independent set is obviously independent, but an independent set may loose its independence when a new cell is introduced into it. So, the problem of finding a maximum cardinality independent set is mathematically interesting. Similarly, a covering set of lines continues to possess this property when new lines are introduced into it, but may no longer remain a covering set if lines are deleted from it. Hence the problem of finding a minimum cardinality covering set of lines is a mathematically interesting problem. The famous König-Egerváry Theorem in bipartite network theory (see Exercise 3.13) states that the cardinalities of maximum cardinality independent sets of cells and minimum cardinality covering sets of lines are always equal.

When an allocation is made in each cell of an independent set, we get a partial assignment. So, a maximum cardinality independent set can be obtained by finding a partial assignment with the maximum number of allocations among admissible cells. For this we can start with 0 allocations in the $p \times n$ array, and carry out Steps 1 and 2 of the Hungarian method until the tree growth routine ends in a nonbreakthrough. The set of cells with allocations at termination is a maximum cardinality independent set, and the set consisting of unlabeled rows and labeled columns is a minimum cardinality covering set of lines.

Exercises

3.1 Solve the assignment problems with the following cost matrices, to minimize cost.

$$
\begin{pmatrix}
10 & 11 & 10 & 5 & 6 & 4 & 3 \\
5 & 26 & 14 & 18 & 15 & 10 & 10 \\
6 & 22 & 18 & 17 & 15 & 8 & 8 \\
2 & 14 & 16 & 16 & 24 & 25 & 12 \\
4 & 15 & 19 & 10 & 8 & 14 & 11 \\
10 & 22 & 22 & 15 & 28 & 24 & 12 \\
8 & 18 & 21 & 18 & 18 & 18 & 14
\end{pmatrix}
,
\begin{pmatrix}
121 & 6 & 17 & 9 & 8 & 10 \\
8 & 33 & 45 & 15 & 20 & 31 \\
5 & 19 & 30 & 16 & 14 & 22 \\
7 & 22 & 35 & 25 & 27 & 26 \\
2 & 10 & 24 & 18 & 31 & 14 \\
4 & 12 & 31 & 17 & 18 & 18
\end{pmatrix}
$$

3.2 The coach of a swim team needs to assign swimmers to a 200-yard medley relay team. The "best times" (in seconds for 50 yards) achieved by his five swimmers in each of the strokes are given below. Which swimmer should the coach assign to each of the four strokes?

Stroke	Carl	Chris	Ram	Tony	Ken
Backstroke	37.7	32.9	33.8	37.0	35.4
Breast Stroke	43.4	33.1	42.2	34.7	41.8
Butterfly	33.3	28.5	38.9	30.4	33.6
Freestyle	29.2	26.4	29.6	28.5	31.1

3.3 There are n boys and n girls. Friendship between boys and girls is a mutual relationship (i.e., a boy and a girl are either friends of each other or not). b_j is the number of boys among the n who are friends of girl j, and a_i is the number of girls among the n who are friends of boy $i, i, j = 1$ to n. If all the a_i and b_j are equal to a positive number γ, prove that it is possible to form n boy-girl friendly couples.

3.4 A colonel has five positions to fill and five eligible candidates to fill them. The number of years of experience of each candidate in each field is given in the following table. How should the candidates be assigned to positions to give the greatest total years of experience for all jobs?

Candidate	Position				
	Adjutant	Intelli.	Operations	Supply	Training
1	3	5	6	2	2
2	2	3	5	3	2
3	3	0	4	2	2
4	3	0	3	2	2
5	0	3	0	1	0

3.5 Find a minimum cost assignment wrt the cost matrix given below. No allocations are allowed in cells with a dot in them.

$$
\begin{pmatrix}
. & 12 & . & 6 & 9 & 6 & . & . \\
. & 8 & . & 4 & 3 & 8 & . & . \\
. & 3 & . & 18 & 3 & 19 & . & . \\
. & 1 & . & 6 & 5 & 11 & . & . \\
5 & 1 & 13 & 4 & 5 & 6 & 1 & 2 \\
. & 13 & . & 12 & 3 & 1 & . & . \\
3 & 12 & 3 & 7 & 13 & 6 & 8 & 3 \\
13 & 4 & 1 & 5 & 5 & 5 & 4 & 9
\end{pmatrix}
$$

3.6 Let $c = (c_{ij})$ where $c_{ij} = (i-1)(j-1)$, for $i, j = 1$ to n; $\bar{c} = (\bar{c}_{ij})$, where $\bar{c}_{ij} = (i+j-n)(i+j-n-1)/2$, for $i, j = 1$ to n; $\bar{x} = (\bar{x}_{ij})$ where $\bar{x}_{ij} = 1$ if $i + j = n + 1$, 0 otherwise, for $i, j = 1$ to n.

For $n = 5$, solve the assignment problem with c as the cost matrix by the Hungarian method. Show that $\bar{c}$ is the final reduced cost matrix at termination, and that $\bar{x}$ is the optimum assignment. Show that these results are true for any $n \overset{\geq}{=} 2$, and that the Hungarian method goes through $(n-1)(n-2)/2$ breakthroughs and nonbreakthroughs put together before solving this problem. (Silver [1960], Machol and Wien [1977]).

3.7 Prove that every feasible solution x of (3.1) can be expressed as a convex combination of assignments of order n.

3.8 The HQ of a large company has n gates, each of which is manned by a night watchman every night. There are n night watchmen on the payroll, and they are rotated among the gates for security reasons. A planning period consists of r nights, and $x = (x_{ij})$ is a nonnegative integral matrix satisfying $\sum_{j=1}^{n} x_{ij} = r, \sum_{i=1}^{n} x_{ij} = r$, for all $i, j = 1$ to n. It is required to assign night watchman to gates over the period (one watchman per gate per night) so that the ith watchman is assigned to watch the jth gate for exactly x_{ij} nights. Prove that an assignment like that exists, and develop an efficient algorithm to find it. Apply your algorithm on the problem in which $n = 5, r = 30$ and x is the following matrix, and generate an assignment of the 5 night watchman to the 5 gates over the 30 nights of the planning period, satisfying the conditions mentioned above.

$$x = \begin{pmatrix} 3 & 10 & 9 & 5 & 3 \\ 8 & 2 & 5 & 10 & 5 \\ 11 & 4 & 10 & 5 & 0 \\ 0 & 7 & 3 & 4 & 16 \\ 8 & 7 & 3 & 6 & 6 \end{pmatrix}$$

(D. Gale)

3.9 An assignment problem of order 9 is being solved by the $O(n^3)$ implementation of the Hungarian method discussed above. Relevant information on the array when a nonbreakthrough has just occurred is given in the following array. □ indicates an allocation in that cell in the present partial assignment. In some of the cells the original cost coefficient is given at the bottom right corner. The present labels on all the labeled rows and labeled columns, and the indices on the unlabeled columns are given. The present dual feasible solution is also given. Continue the application of the method until the next breakthrough occurs.

$j =$	1	2	3	4	5	6	7	8	9	label	u_i
$i = 1$										s, +	200
2										s, +	200
3	□									1, −	200
4			□							3, −	200
5		□								2, −	200
6				□	400	400	400	400	400		200
7				400	□	400	400	400	350		200
8				400	400	□	400	400	400		200
9				400	400	400	□	400	400		200
label	1, +	1, +	2, +								
index				[2,40]	[3,40]	[2,80]	[4,100]	[5,110]	[5,110]		
v_j	100	100	100	100	100	100	100	100	100		

3.10 Prove that the number of labeled columns plus the number of unlabeled rows is equal to the number of allocated cells, at the occurrence of a nonbreakthrough in the Hungarian method. At the same point in the method, prove that the number of labeled rows minus the number of labeled columns is $\geqq$ the number of rows without allocations. Using this prove that the total reduction strictly increases whenever the dual solution changes in the method.

Give a proof of the primal infeasibility criterion in the Hungarian method (that there is no feasible assignment if $\delta = +\infty$ in some dual solution change step) using the duality theorem of LP.

3.11 Consider an assignment problem of order n in which x_{ij} is required to be 0 whenever $j \notin S_i$ for each $i = 1$ to n, where $S_1, \ldots, S_n$ are all subsets of $\{1, \ldots, n\}$ which are given. Determine the necessary and sufficient conditions that these sets have to satisfy, for the existence of a feasible assignment to the problem.

3.12 Evaluate the worst case computational complexity of the Hungarian method to solve a sparse assignment problem of order n as a function of order n and m ($n^2 - m$ is the number of variables required to be 0 in the problem).

3.13 Let $\mathbf{Q}$ be a specified subset of admissible cells in the $p \times n$ array. Set up a directed network $G = (\mathcal{N}, \mathcal{A})$ where $\mathcal{N} = \{\check{s}, R_1, \ldots, R_p, C_1, \ldots, C_n, \check{t}\}$ ($\check{s}, \check{t}$, are the supersource, supersink respectively, and R_i, C_j correspond to row i, col j of the array, for $i = 1$ to p, $j = 1$ to n), $\mathcal{A} = \{(\check{s}, R_i) : i = 1 \text{ to } p\} \cup \{(R_i, C_j) : i, j \text{ s. t. } \text{cell } (i,j) \in \mathbf{Q} \} \cup \{(C_j, \check{t}) : j = 1 \text{ to } n\}$, as in Figure 3.2. Make the lower bound for flow on all the arcs in $\mathcal{A}$ zero, and the capacities for flow on all the arcs $(\check{s}, R_i)$ and $(C_j, \check{t})$ equal to 1, and ∞ for all the arcs of the form $(R_i, C_j) \in \mathcal{A}$.

(i) Prove that the maximum cardinality among independent sets of admissible cells in the array is equal to the maximum value of flow from $\check{s}$ to $\check{t}$ in G. Given an integral maximum value flow vector in G, discuss how to construct a maximum cardinality independent set of admissible cells in the array from it, and vice versa.

(ii) Let $[\mathbf{X}, \bar{\mathbf{X}}]$ be a cut separating $\check{s}$ and $\check{t}$ in G. Prove that the capacity of this cut is finite iff there are no arcs of the form $(R_i, C_j) \in \mathcal{A}$ with $R_i \in \mathbf{X}$, $C_j \in \bar{\mathbf{X}}$ and $(i,j) \in \mathbf{Q}$. Using this prove that in this case, the set of lines $\{\text{Row } i : i \in \bar{\mathbf{X}}\} \cup \{\text{Column } j : j \in \mathbf{X}\}$ is a covering set of lines, and that the cardinality of this covering set is equal to the capacity of this cut $[\mathbf{X}, \bar{\mathbf{X}}]$ in G. Conversely, given a covering set of lines in the array, prove that $[\mathbf{X}, \bar{\mathbf{X}}]$, where $\mathbf{X} = \{\check{s}\} \cup \{\text{Row } i: \text{row } i \text{ is not in the covering set}\} \cup \{\text{Column } j: \text{column } j \text{ is in the covering set}\}$, and $\bar{\mathbf{X}} = \mathcal{N} \setminus \mathbf{X}$, is a cut separating $\check{s}$ and $\check{t}$ in G whose capacity is equal to the cardinality of this covering set of lines. Using these prove that the problem of finding a minimum cardinality cover for $\mathbf{Q}$ in the array is equivalent to that of finding a minimum capacity cut separating $\check{s}$ and $\check{t}$ in G.

(iii) Using these prove the König-Egerváry Theorem which states that the maximum cardinality among independent sets is equal to the minimum cardinality among covering sets of lines, by applying the maximum flow minimum cut theorem on G.

(iv) Prove that the covering set of lines obtained at termination of the algorithm discussed above is a minimum cardinality covering set of lines.

3.14 $c = (c_{ij})$ is the original cost matrix for an assignment problem of order n. For any cell (i, j) in the $n \times n$ array, let $a(i, j)$ denote a minimum cost assignment among those containing an allocation in cell (i, j). For a given r, the following method finds $a(r, 1), \ldots, a(r, n)$.

Step 1   Find a minimum cost assignment with c as the cost matrix by the Hungarian method. Let $a_1 = \{(1, j_1), \ldots, (n, j_n)\}$ be the optimum assignment obtained, and $\bar{c}$ the final reduced cost matrix.

Step 2   Let α be a positive number $>$ every entry in $\bar{c}$. Add α to all the elements in row r of $\bar{c}$. In the resulting matrix, subtract α from each entry in column j_r.

Step 3   Find a most negative entry in the present matrix. Suppose it appears in row i and has absolute value β. Add β to every entry in row i of the present matrix. In the resulting matrix subtract β from every entry in column j_i.

Step 4   Repeat Step 3 as often as necessary, until the present matrix becomes $\geqq 0$. Then go to Step 5.

Step 5   Let $\tilde{c} = (\tilde{c}_{ij})$ be the matrix obtained at the end. For $1 \leqq q \leqq n$, there exists an assignment with an allocation in cell (r, q) and all the other allocations in cells (i, j) with $\tilde{c}_{ij} = 0$. Any such assignment is $a(r, q)$.

(i)   Prove that Step 3 has to be repeated exactly $n - 1$ times before going to Step 5 in this method.

(ii)   Prove the statement in Step 5, that for each $q = 1$ to n, there exists an assignment with an allocation in cell (r, q), and all other allocations in cells with 0 entries in $\tilde{c}$, and any such assignment is indeed a minimum cost assignment among those with an allocation in (r, q).

(iii)   Derive the worst case computational complexity of this method. Apply the method on the problem with $n = 5$, $r = 1$ and

$$c = \begin{pmatrix} 0 & 3 & 9 & 4 & 0 \\ 2 & 0 & 7 & 0 & 11 \\ 5 & 15 & 0 & 16 & 12 \\ 4 & 0 & 18 & 0 & 17 \\ 0 & 20 & 21 & 13 & 0 \end{pmatrix}$$

(Kreuzberger and Weiterstadt, [1971], "Eine Methode zur Bestimmung mehrerer Losungen Furdas Zuordnungsproblem," *Angewandte Informatik*, 13, no. 9 (407-414), in German)

Comment 3.1   The Hungarian method for the assignment problem is due to Kuhn [1955]. The name for the method recognizes the work of the Hungarian mathematicians J. Egerváry and D. König (the König-Egervráy Theorem, see Exercise 3.13) which is the basis for the method. Each maximum value flow problem encountered in the method is of the König-Egerváry type (i.e., that of finding a maximum cardinality set of independent admissible cells). The $O(n^3)$ implementation of the Hungarian method is due to Lawler [1976 of Chapter 1].

3.1.1 Minimal Chain Decompositions in Partially Ordered Sets

Suppose we are given a finite set of n elements, **P**. We will number the elements serially and represent each element by its number, thus $\mathbf{P} = \{1, \ldots, n\}$. A strict partial order on **P** is an order relation between some pairs of elements of **P**, denoted by $\succ$, satisfying : $i \not\succ i$ for all i; $i \succ j$ implies $j \not\succ i$; and the following property called *transitivity* : for i, j, h, $i \succ j$, $j \succ h$ implies $i \succ h$. This relationship makes **P** a *partially ordered set* or *poset*, which we also denote by the same symbol **P**. The partial order can be represented by a directed network with the elements in **P** as its nodes and arcs (i, j) if $i \succ j$. Normally if $i \succ j$ and $j \succ h$, we have $i \succ h$ by transitivity, but we do not include the arc (i, h) in this network, even though it is quite harmless to do so. Thus whenever $i \succ j$, either there is an arc (i, j), or there exists a chain from node i to node j in the network; and conversely. Thus $i \succ j$ iff there exists a chain from i to j in this network representation. The network thus constructed has no directed circuits by the properties of the partial order, see Figure 3.4 for an example.

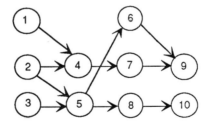

Figure 3.4

We define a *chain* in this poset to be a set of one or more elements $i_1, i_2, \ldots, i_t$ of **P** satisfying $i_1 \succ i_2 \succ \ldots \succ i_t$; thus it corresponds to a chain in the network when there are two or more elements in it. However, a single element by itself (that is, a single node in the network) is also considered a chain of the poset.

A *decomposition* of this poset is a partition of the set **P** into chains which are mutually disjoint. The trivial decomposition of **P** into n one-element chains is an example of a decomposition. A decomposition with the smallest number of chains in it is said to be a *minimal decomposition*. We will use the symbol $\triangle$ to denote decompositions and $|\triangle|$ to denote the number of disjoint chains in $\triangle$.

A subset $\mathbf{N} \subset \mathbf{P}$ is said to be *a set of unrelated elements* in the poset if for every pair $i, j \in \mathbf{N}$, neither $i \succ j$, nor $j \succ i$.

As an example, consider the poset on $\mathbf{P} = \{1 \text{ to } 10\}$ represented by the network in Figure 3.4. In this poset, the set of all j satisfying $2 \succ j$ is $\{4, 5, 6, 7, 8, 9, 10\}$. The sets $\{1, 2, 3\}$, $\{6, 7, 8\}$ are both sets of unrelated elements, but the set $\{5, 6, 7, 8\}$ is not since $5 \succ 6$. $\{\mathcal{C}_1, \mathcal{C}_2, \mathcal{C}_3, \mathcal{C}_4\}$ is a chain decomposition of this poset where

$$
\begin{aligned}
\mathcal{C}_1 &= 1, (1,4), 4, (4,7), 7, (7,9), 9 \\
\mathcal{C}_2 &= 2, (2,5), 5, (5,8), 8, (8,10), 10 \\
\mathcal{C}_3 &= 3; \quad \mathcal{C}_4 = 6
\end{aligned}
\tag{3.5}
$$

In any decomposition, each node in a set of unrelated elements $\mathbf{N}$ has to appear on a different chain (as otherwise there will be two nodes i, j in $\mathbf{N}$ such that either $i \succ j$ or $j \succ i$, a contradiction). Hence the number of disjoint chains in any decomposition of $\mathbf{P}$ is $\geqq$ the cardinality of any set of unrelated elements in $\mathbf{P}$. Hence we have the result.

$$
\left.
\begin{array}{c}
\text{the number of chains in a} \\
\text{minimal decomposition of } \mathbf{P}
\end{array}
\right\} \geqq \left\{
\begin{array}{c}
\text{the maximum number of} \\
\text{mutually unrelated elements of } \mathbf{P}
\end{array}
\right.
\tag{3.6}
$$

The finite case of the well-known Dilworth's chain decomposition theorem for posets asserts that equality holds in (3.6). Here we show how to prove this result as a corollary of the König-Egerváry theorem, through a bipartite network formulation. Also, we show that a minimal decomposition of $\mathbf{P}$ and a maximum cardinality set of unrelated elements in $\mathbf{P}$ can both be obtained using the algorithms discussed earlier. These results are due to Fulkerson [1956].

Construct the bipartite network $\mathbf{G} = (\mathcal{N}_1, \mathcal{N}_2; \mathcal{A})$ where $\mathcal{N}_1 = \{R_1, \ldots, R_n\}$, $\mathcal{N}_2, = \{C_1, \ldots, C_n\}$, and $\mathcal{A} = \{(R_i, C_j) : i \succ j\}$. Notice that $\mathbf{G}$ contains arcs corresponding to all order relations implied by transitivity, that is, $(R_j, C_j) \in \mathcal{A}$ whenever $i, j \in \mathbf{P}$ are such that $i \succ j$. Corresponding to $\mathbf{G}$, set up an $n \times n$ array with R_i, C_j associated with row i, column j of the array, and the arc (R_i, C_j) in $\mathbf{G}$ associated with the cell (i, j) in the array. Make the cell (i, j) in the array admissible iff (R_i, C_j) is an arc in $\mathbf{G}$. Then an independent set of admissible cells in the array corresponds to a matching in $\mathbf{G}$ and vice versa. If $\mathbf{M} = \{(R_{i_1}, C_{i_2}), (R_{i_3}, C_{i_4}), \ldots, (R_{i_{2l-1}}, C_{i_{2l}})\}$ is a matching in $\mathbf{G}$ with $|\mathbf{M}| = l$, $\{(i_1, i_2), (i_3, i_4), \ldots, (i_{2l-1}, i_{2l})\}$ is the corresponding independent set of admissible cells in the associated array, and vice versa. We can group the distinct elements in the set $\{i_1, i_2, \ldots, i_{2l}\}$ into chains, each one passing through two or more elements. Since $\mathbf{M}$ is a matching in $\mathbf{G}$, these chains will be disjoint. Let r be the number of these chains, and let $p_1, p_2, \ldots, p_r$ be the numbers of elements on them. So, the numbers of arcs on these chains are

$p_1 - 1, \ldots, p_r - 1$ respectively, and hence $l = |\mathbf{M}| = (p_1 - 1) + \ldots + (p_r - 1)$. The total numbers of elements of $\mathbf{P}$ which do not appear on any of these chains is $n - (p_1 + \ldots + p_r) = n_1$. Add a one-element chain at each of these elements to the set of r chains obtained above. This leads to a decomposition $\triangle$ of $\mathbf{P}$, with $|\triangle| = n_1 + r$. Now, $n = (n - (p_1 + \ldots + p_r)) + p_1 + \ldots + p_r = n_1 + r + ((p_1 - 1) + \ldots + (p_r - 1)) = |\triangle| + |\mathbf{M}|$. So, corresponding to any matching $\mathbf{M}$ in G, we can construct a decomposition $\triangle$ of $\mathbf{P}$ such that $|\triangle| = n - |\mathbf{M}|$.

As an example, consider the poset represented by the acyclic network in Figure 3.4. Here $n = 10$, and the array corresponding to this partially ordered set is given below, with admissible cells marked by a x.

Array 3.4

	C_1	C_2	C_3	C_4	C_5	C_6	C_7	C_8	C_9	C_{10}
R_1				x			x		x	
R_2			x	x	x	x	x	x	x	x
R_3				x	x		x	x	x	
R_4						x			x	
R_5					x		x	x	x	
R_6									x	
R_7									x	
R_8										x
R_9										
R_{10}										

An independent set of admissible cells in Array 3.4 is $\mathbf{M} = \{(1, 4), (2, 5), (4, 7), (5, 8), (7, 9), (8, 10)\}$. This independent set of admissible cells corresponds to the chain decomposition $\triangle = \{C_1, C_2, C_3, C_4\}$ of the partially ordered set, where C_1, C_2, C_3, C_4 are given in (3.5). It can be verified that $|\mathbf{M}| + |\triangle| = n = 10$ here.

Let $\mathbf{X} = \{R_{t_1}, \ldots, R_{t_w}; C_{j_1}, \ldots, C_{j_q}\}$ be a set of lines (rows and columns) in the array corresponding to the poset $\mathbf{P}$, which covers all the admissible cells. Suppose $\mathbf{X}$ is a proper cover, that is, no subset of $\mathbf{X}$ covers all the admissible cells in the array. Then we claim that $t_1, \ldots, t_w, j_1, \ldots, j_q$ are all distinct. To see this, suppose $t_1 = j_1$. Since $\mathbf{X}$ is a proper cover, there must exist an r such that $R_r \notin \mathbf{X}$ and (R_r, C_{j_1}) is an admissible cell. Similarly there must exist an s such that $C_s \notin \mathbf{X}$ and (R_{t_1}, C_s) is admissible. By transitivity and the assumption that $t_1 = j_1$, it follows that (R_r, C_s) is an admissible cell, and since neither R_r, nor C_s is in $\mathbf{X}$, this contradicts the assumption that $\mathbf{X}$ covers all admissible cells. So, elements in $\{t_1, \ldots, t_w; j_1, \ldots, j_q\}$ are all distinct. Let $\mathbf{U} = \{1, \ldots, n\} \setminus \{t_1, \ldots, t_w; j_1, \ldots, j_q\}$. Since $\mathbf{X}$ covers all admissible cells, the elements in $\mathbf{U}$ are mutually unrelated, and by the definition of $\mathbf{U}$ we have $\{t_1, \ldots, t_w, j_1, \ldots, j_q\} \cup \mathbf{U} = \mathbf{P}$, and hence $|\mathbf{X}| + |\mathbf{U}| = n$. So, corresponding to any covering set $\mathbf{X}$ of lines in the array, we can construct a mutually unrelated set of elements $\mathbf{U}$ of $\mathbf{P}$ such that $|\mathbf{U}| = n - |\mathbf{X}|$.

As an example, consider the poset represented by the acyclic network in Figure 3.4. Array 3.4 corresponds to it. In Array 3.4, all admissible cells are covered by the set of lines $\{R_1, R_2, R_3, R_4, R_5, R_8, C_9\}$. This covering set of lines leads to the set of unrelated elements $\mathbf{U} = \{6, 7, 10\}$ in $\mathbf{P}$.

THEOREM 3.1 *(Dilworth's Theorem) The number of chains in a minimal decomposition of a finite poset* $\mathbf{P}$ *is equal to the maximum number of mutually unrelated elements in* $\mathbf{P}$.

Proof Construct the array corresponding to $\mathbf{P}$ as described above. In this array find a maximum cardinality set of independent admissible cells $\hat{\mathbf{M}}$, and a minimum cardinality set of covering lines $\hat{\mathbf{L}}$ covering all the admissible cells. Obtain the chain decomposition $\hat{\triangle}$ of $\mathbf{P}$ corresponding to $\hat{\mathbf{M}}$, and a set $\hat{\mathbf{U}}$ of mutually unrelated elements of $\mathbf{P}$ corresponding to $\hat{\mathbf{L}}$ by the procedures described above. We have $|\hat{\triangle}| = |\mathbf{P}| - |\hat{\mathbf{M}}|$ and $|\hat{\mathbf{U}}| = |\mathbf{P}| - |\hat{\mathbf{L}}|$, and by König-Egerváry theorem $|\hat{\mathbf{M}}| = |\hat{\mathbf{L}}|$. So $|\hat{\triangle}| = |\hat{\mathbf{U}}|$. But by (3.6) we have $|\triangle| \geqq |\mathbf{U}|$ for every chain decomposition $\triangle$ and set of unrelated elements $\hat{\mathbf{U}}$ in $\mathbf{P}$ and $|\hat{\triangle}| = |\hat{\mathbf{U}}|$. This implies that $\hat{\triangle}$ is a minimal decomposition of $\mathbf{P}$ and $\hat{\mathbf{U}}$ is a maximum cardinaltiy set of mutually unrelated elements in $\mathbf{P}$, and since $|\hat{\triangle}| = |\hat{\mathbf{U}}|$, the theorem is proved. ∎

Hence to find a minimal chain decomposition of a finite poset $\mathbf{P}$, we construct the array corresponding to it as described above, and then find a maximum cardinality set of independent admissible cells, $\mathbf{M}$, in it, using the algorithm discussed earlier. A minimal chain decomposition of $\mathbf{P}$ is obtained directly using $\mathbf{M}$ as described above.

3.1.2   The Bottleneck Assignment Problem

In the assignment problem (3.1), the objective function is the sum of the costs of all the allocations. This objective function may not be appropriate in some practical applications. As an example, consider the application discussed in Exercise 3.15 at the end of this section. Minimizing the sum of the inconvenience of all the salesmen may not please a particular salesman if his own individual inconvenience turns out to be large in the optimum assignment. A better objective in this case, is to minimize the maximum inconvenience experienced by any salesman. This leads to a problem known as the *bottleneck assignment problem* or the *min-max assignment problem*. If $c = (c_{ij})$ is the square cost matrix of order n, in this problem our aim is to find an assignment $x = (x_{ij})$ that minimizes the function $f(x) = $ maximum $\{c_{ij} : i, j$ such that $x_{ij} = 1\}$ over the set of all assignments of order n. We discuss an algorithm for this problem now. It uses the tree growth steps in the Hungarian method. To avoid confusion we refer to steps in this method as items. In this method the value of z^r is like a threshold. At each stage, all cells with cost coefficient $\leqq$ threshold are admissible. If there is no assignment among the set of admissible cells, the threshold is increased to the next higher level. The method terminates as soon as a full assignment is located among the set of admissible cells. Hence, this method is known as the *threshold method* for the bottleneck assignment problem. It is due to Gross [1959]. The assignment at termination is an optimum assignment, and the terminal value of z^r is the optimum objective value in the bottleneck problem.

THE THRESHOLD METHOD

Item 0   Define $u_i = \min. \{c_{i1}, \dots, c_{in}\}$, $i = 1$ to n, $v_j = \min. \{c_{1j}, \dots, c_{nj}\}$, $j = 1$ to n, $z^1 = \max. \{u_1, \dots, u_n; v_1, \dots, v_n\}$. Define the set of *admissible cells* to be $\{(i,j) : i, j \text{ s. t. } c_{ij} \leqq z^1\}$. With this set of admissible cells and any partial assignment with allocations only among admissible cells (for example, the 0 partial assignment), enter the tree growth routine (Step 1) of the Hungarian method and make allocation changes as breakthroughs occur, until either a full assignment is obtained or a nonbreakthrough occurs. If a full assignment is obtained, it is an optimum assignment; terminate. If a nonbreakthrough occurs, go to Item 1.

Item 1   Let r be the number of the present visit to this iteration. Define $z^{r+1} = \min.\{c_{ij} : (i,j) \text{ is inadmissible at this stage}\}$. Let the new set of admissible cells be $\{(i,j) : c_{ij} \leqq z^{r+1}\}$. Make the list = set of all labeled rows, and resume tree growth by going to Substep 2 in Step 1 of the Hungarian method, and continuing as in Item 0.

Exercises

3.15 There are n salesmen to be assigned to n markets on a one to one basis. c_{ij} measures the inconvenience experienced by the ith salesman if he is assigned to market j (for example, c_{ij} may be the daily commuting distance for him under this assignment). It is required to find an assignment that minimizes the maximum inconvenience experienced by any salesman. Solve this problem when $c = (c_{ij})$ is the matrix given below.

$$c = \begin{pmatrix} 3 & 9 & 15 & 20 & 5 \\ 13 & 16 & 7 & 8 & 9 \\ 19 & 12 & 13 & 14 & 15 \\ 7 & 17 & 8 & 9 & 13 \\ 9 & 15 & 16 & 12 & 11 \end{pmatrix}$$

3.16 Each visit to Item 1 in the the threshold method increases the number of admissible cells in the array by at least one. Hence Item 1 occurs at most n^2 times in the method. Using this, derive the worst case computational complexity of this method.

3.17 An assembly line consists of workstations $1, \ldots, n$, each staffed by an operator. Each unit is processed on each workstation as it passes along, it cannot move past a workstation until its processing there is completed. There are n laborers, and c_{ij} is the number of seconds that the jth laborer takes to process a unit at workstation i. It is required to determine how the laborers should be assigned to the workstations on a one to one basis, so as to maximize the productivity of the line. Formulate this problem. Obtain an optimum solution when $c = (c_{ij})$ is the following matrix.

$$c = \begin{pmatrix} 9 & 7 & 18 & 13 & 14 & 16 \\ 16 & 12 & 11 & 23 & 5 & 19 \\ 13 & 9 & 25 & 19 & 17 & 18 \\ 18 & 16 & 10 & 9 & 13 & 11 \\ 12 & 12 & 8 & 18 & 8 & 9 \\ 6 & 11 & 7 & 19 & 9 & 12 \end{pmatrix}$$

3.2   The Primal-Dual Method for the Uncapacitated Balanced Transportation Problem

A commodity has to be shipped from p sources (ith source has a_i units available to ship, $i = 1$ to p) to n markets (jth market needs b_j units, $j = 1$ to n). c_{ij} is the unit transportation cost ($/unit) on the route from source i to market j, $i = 1$ to p, $j = 1$ to n. The data satisfies

$$a_i, b_j > 0, \text{ for all } i, j; \text{ and } \sum a_i = \sum b_j$$

x_{ij} denotes the amount of material shipped from source i to market j. These x_{ij} are the decision variables in the problem. It is

$$
\begin{aligned}
\text{Minimize } z(x) \quad &= \quad \sum_{i=1}^{p} \sum_{j=1}^{n} c_{ij} x_{ij} \\
\text{Subject to } \sum_{j=1}^{n} x_{ij} \quad &= \quad a_i, \ i = 1 \text{ to } p \qquad\qquad (3.7) \\
\sum_{i=1}^{p} x_{ij} \quad &= \quad b_j, \ j = 1 \text{ to } n \\
x_{ij} \quad &\gtreqless \quad 0, \text{ for all } i, j
\end{aligned}
$$

It is the problem of finding a minimum cost flow vector saturating all the arcs leading to the super sink in the bipartite network in Figure 3.5. All lower bounds are 0. Data on each arc is capacity; unit cost coefficient in that order. x_{ij} is the flow on the arc joining source i and market j. By identifying the cell (i, j) in the $p \times n$ transportation array with the arc joining source i to market j in the network, all the computations can actually be carried out on the array itself.

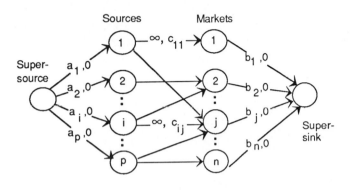

Figure 3.5

In practical applications, each source may not be able to ship to all the sinks. If sink j is too far away from source i, it is realistic to specify that source i cannot ship to sink j (i.e., that $x_{ij} = 0$). There may also be other practical reasons why a source cannot ship to some sinks. Thus, for each source, a subset of sinks to which it can ship is specified, and all flows from that source to sinks outside that set are required to be zero. This can be handled by defining c_{ij} to be $+\infty$ whenever x_{ij} is required to be zero in the formulation (3.7). In the corresponding bipartite network in Figure 3.5, the arc (i, j) is not included if x_{ij} is required to be 0; thus it is no longer the complete bipartite network. We denote by m the number of source to market arcs in this network. Clearly $m \leqq pn$, the transportation problem is said to be *sparse* if m is small compared to pn, and *dense* if m is close to pn.

The primal-dual algorithm is described using arrays for ease of understanding, but computer implementations are usually based on the network formulation as described above, this saves in memory requirements and running time for solving practical problems which are almost always sparse. The dual problem is

$$\text{Maximize} \sum a_i u_i + \sum b_j v_j$$

$$\text{Subject to} \quad u_i + v_j \leqq c_{ij} \text{ for all } i, j \quad (3.8)$$

Let $\bar{c}_{ij} = c_{ij} - u_i - v_j$, for $i = 1$ to p, $j = 1$ to n. These are the *reduced cost coefficients* wrt (u, v), and (u, v) is dual feasible iff they are all $\geqq 0$. The complementary slackness conditions for optimality in these problems are

$$x_{ij}\bar{c}_{ij} = 0 \text{ for all } i, j. \tag{3.9}$$

The cell (i, j) in the array (the corresponding arc (i, j) in the bipartite network in Figure 3.5) is said to be an *admissible* or *equality cell* (*admissible* or *equality arc*) wrt (u, v) if $\bar{c}_{ij} = 0$; otherwise it is *inadmissible*. The network obtained by deleting all the inadmissible arcs from Figure 3.5 is known as the *admissible* or *equality subnetwork* wrt (u, v). The complementary slackness conditions require that the flow amounts should be 0 on all the inadmissible arcs. The flow problem, known as the *restricted primal* at this stage, is to find a maximum value flow from the super source to the super sink in the equality subnetwork. It is equivalent to

$$
\begin{aligned}
\text{Maximize } &\sum (x_{ij} \quad : \quad \text{over } (i, j) \text{ admissible }) \\
\text{Subject to } &\sum_{j=1}^{n} x_{ij} \;\;\leqq\;\; a_i, \; i = 1 \text{ to } p \\
&\sum_{i=1}^{p} x_{ij} \;\;\leqq\;\; b_j, \; j = 1 \text{ to } n \\
&x_{ij} \;\;\geqq\;\; 0, \;\; \text{if } (i, j) \text{ admissible, } 0 \text{ otherwise}
\end{aligned}
\tag{3.10}
$$

The primal-dual algorithm maintains vectors $x, (u, v)$ which always satisfy the constraints in (3.10), (3.8), (3.9). When the x vector satisfies the equality constraints in (3.7), it is an optimum solution and the method terminates.

THE CLASSICAL PRIMAL-DUAL METHOD FOR THE UNCAPACITATED BALANCED TRANSPORTATION PROBLEM

Step 0 Initialization   If some dual feasible solution is available, use it as the initial one, otherwise define it to be $(u^1 = (u_i^1), v^1 = (v_j^1))$ where $u_i^1 = \min \{ c_{ij} : j = 1 \text{ to } n \}$, $v_j^1 = \min \{ c_{ij} - u_i^1 : i = 1 \text{ to } p \}$, for $i = 1$ to p, $j = 1$ to n. List $= \emptyset$. Define $x^1 = 0$. Go to Step 1.

Step 1 Tree growth routine   Let $\tilde{x} = (\tilde{x}_{ij})$ be the present flow.

Substep 1   Label each row i satisfying $\sum_j \tilde{x}_{ij} < a_i$ with $(s, +)$, and include it in the list.

Substep 2   If list $= \emptyset$, tree growth has terminated and there is a nonbreakthrough. The present flow is a maximum value flow in the admissible subnetwork, go to Step 3. Otherwise, select a row or column from the list for scanning and delete it from the list.

Forward labeling   Scanning row i consists of labeling each unlabeled column j such that (i, j) is an admissible cell, with the label (row i, +).

Reverse labeling To scan column j, label all unlabeled rows i satisfying $\tilde{x}_{ij} > 0$ with (column j, $-$).

If any column without an allocation has been labeled, there is a break-through, go to Step 2. Otherwise, include all newly labeled rows and columns in the list, and repeat this Substep 2.

Step 2 Flow change routine Suppose column j satisfying $\alpha = b_j - \sum_i \tilde{x}_{ij} > 0$ has been labeled. Trace its predecessor path using the labels, suppose it ends with row i with the label of $(s, +)$. This path is an FAP from row i to column j in the equality subnetwork. Call it $\mathcal{P}$. Let $\beta = a_i - \sum_j \tilde{x}_{ij}$, and ϵ the residual capacity of $\mathcal{P}$. Define $\gamma = \min\{\alpha, \beta, \epsilon\}$. Carry out flow augmentation by the amount γ on the FAP $\mathcal{P}$, and get the new flow $\hat{x}$. If $\hat{x}$ is feasible to (3.7), it is an optimum solution; terminate. Otherwise, chop down the present trees (i.e., erase the labels on all the rows and columns) and go back to Step 1.

Step 3 Dual solution change routine Same as Step 3 of the Hungarian Method.

EXAMPLE 3.3

Consider the balanced transportation problem with data given in the following array. An initial dual feasible solution $(u = (u_i), v = (v_j))$ is also given in the array.

$j =$	1	2	3	4	5	6	a_i	u_i
$i = 1$							4	
	$c_{11} = 5$	3	7	3	8	5		3
2							3	
	5	6	12	5	7	11		5
3							3	
	2	8	3	4	8	2		2
4							7	
	9	6	10	5	10	9		5
b_j	3	3	6	2	1	2		
v_j	0	0	1	0	2	0		

The reduced cost coefficients, $\bar{c}_{ij} = c_{ij} - u_i - v_j$ are entered in the upper left corners of the cells in the following array. All cells in which $\bar{c}_{ij} = 0$ are admissible cells, and they are marked with a little box in the middle. An initial flow among the admissible cells is obtained by inspection, and the flow amounts, when nonzero, are entered in the little boxes.

$j =$	1	2	3	4	5	6	a_i	u_i	Row label
$i = 1$	$2 = \bar{c}_{11}$	0 ⊡3	3	0 ⊡1	3	2	4	3	Col. 4, -
2	0 □+θ	1	6	0 □	0 ⊡1	6	3	5	s, +
3	0 ⊡3 −θ	6	0 □+θ	2	4	0 □	3	2	Col. 1, -
4	4	1	4	0 ⊡1	3	4	7	5	s, +
b_j	3	3	6	2	1	2			
v_j	0	0	1	0	2	0			
Col. label	Row 2 +		Row 3 +	Row 2 +	Row 2 +				

$j =$	1	2	3	4	5	6	a_i	u_i	Row label
$i = 1$	$2 = \bar{c}_{11}$	0 ⊡3	3	0 ⊡1	3	2	4	3	Col. 4, -
2	0 ⊡2	1	6	0 □	0 ⊡1	6	3	5	
3	0 ⊡1	6	0 ⊡2	2	4	0 □	3	2	
4	4	1	4	0 ⊡1	3	4	7	5	s, +
b_j	3	3	6	2	1	2			
v_j	0	0	1	0	2	0			
Col. label		Row 1 +		Row 4 +					

In the first array at the top, rows 2 and 4 have additional material to be shipped, so they are labeled with $(s, +)$. Columns 3 and 6 have unfulfilled requirements. The labeling routine is applied, and it ends in a breakthrough with column 3 labeled.

We move to the flow change routine. Entries of $+\theta$ and $-\theta$ are made in admissible cells as indicated by the labels. The value of θ should be min.$\{6 =$ unfulfilled requirement in column 3; $2 =$ additional material available at row 2; $3 =$ flow amount in cell $(3, 1)$ with a $-\theta$ entry$\} = 2$. The new flow vector is recorded in the next array together with the present reduced cost coefficients. The old labels are erased and the labeling routine is applied again. The row and column labels obtained are recorded on the array.

We have a nonbreakthrough. So, we move to the dual solution change routine. $\delta = 2$. The new dual feasible solution is $\hat{u}, \hat{v}$ marked in the following array. In this array we show the new reduced cost coefficients in each cell. The new admissible cells are marked with a little box in the middle. Notice that all the cells with positive flow amounts in the present flow vector remain admissible. The present flow amounts in cells with positive flow are entered inside the box in them.

$j =$	1	2	3	4	5	6	a_i	$\hat{u}_i$
$i = 1$	0	0 [3]	1	0 [1]	1	0 [ ]	4	5
2	0 [2]	3	6	2	0 [1]	6	3	5
3	0 [1]	8	0 [2]	4	4	0 [ ]	3	2
4	2	1	2	0 [1]	1	2	7	7
b_j	3	3	6	2	1	2		
$\hat{v}_j$	0	-2	1	-2	2	0		

The algorithm now resumes labeling using the new admissible cells. It can be continued in the same manner until an optimum solution is obtained.

Discussion

If δ turned out to be $+\infty$ in Step 3 at some stage, the flow at that stage is a maximum value flow, not only in the equality subnetwork at that stage but in the entire original bipartite network. This case can only occur if some of the x_{ij} were required to be 0 in the original problem. Since this flow leaves the demand at some markets unfulfilled, it implies that there is no feasible solution to the problem.

Array 3.5 Summary of Position at Occurrence of Nonbreakthrough. $(\tilde{u}, \tilde{v}, \overset{\approx}{c} = (\overset{\approx}{c}_{ij} = c_{ij} - \tilde{u}_i - \tilde{v}_j)$, are the dual solution and reduced cost matrix before the change. $(\hat{u}, \hat{v}, \hat{c} = (\hat{c}_{ij} = c_{ij} - \hat{u}_i - \hat{v}_j)$ are the corresponding things after the change. $\tilde{x} = (\tilde{x}_{ij})$ is the present flow.

	Block of labeled cols.	Block of unlabeled cols.	Supply position	Dual change
Block of labeled rows	$\hat{c}_{ij} = \overset{\approx}{c}_{ij}$ here, so admissibility pattern remains unchanged	No admissible cells here (a col. would be labeled otherwise). So $\overset{\approx}{c}_{ij} > 0$ here. $\delta = \text{Min}\{\overset{\approx}{c}_{ij}: (i,j)$ here$\} > 0$. $\hat{c}_{ij} = \overset{\approx}{c}_{ij} - \delta$ here. New admissible cells created here, and their cols. get labeled next.	Material available at some rows here.	$\hat{u}_i = \tilde{u}_i + \delta$ here.
Block of unlabeled rows	$\tilde{x}_{ij} = 0$ here (otherwise a row could be labeled). $\hat{c}_{ij} = \overset{\approx}{c}_{ij} + \delta$ here. All cells here become inadmissible next.	$\hat{c}_{ij} = \overset{\approx}{c}_{ij}$ here, so admissibility pattern remains unchanged.	$\sum_j \tilde{x}_{ij} = a_i$ here; otherwise a row could be labeled.	$\hat{u}_i = \tilde{u}_i$ here.
Requirement position	$\sum_i \tilde{x}_{ij} = b_j$ here; otherwise it would be a breakthrough.	Some cols. with unfulfilled requirements.		
Dual change	$\hat{v}_j = \tilde{v}_j - \delta$ here	$\hat{v}_j = \tilde{v}_j$ here.		

For arbitrary c and arbitrary (i.e., not necessarily rational) positive a_i, b_j satisfying $\sum a_i = \sum b_j$, the method terminates in a finite number of steps if the list is maintained as a queue and the node for scanning in Substep 2 of Step 1 is selected by the FIFO (first in first out) rule. To see this, we notice from the facts in Array 3.5 that when labeling is resumed after each occurrence of Step 3, at least one new column gets labeled. So, in this algorithm, after at most n consecutive occurrences of Step 3, Step 2 must occur, and then the total flow $\sum\sum x_{ij}$ strictly increases.

Each equality subnetwork appearing in the algorithm consists of all the nodes (source nodes with their availabilities, sink nodes with their requirements) and equality arcs corresponding to the subset of equality cells wrt the dual feasible solution at that stage. We can associate any equality subnetwork with the subset of cells corresponding to the equality arcs in it. Denote by $\vartheta(\mathbf{E})$ the value of the maximum value flow in the equality subnetwork associated with the subset of cells $\mathbf{E}$.

The equality subnetwork changes in the algorithm after each occurrence of Step 3. The first time that Step 3 occurs after an equality subnetwork associated with a subset $\mathbf{E}$ of cells appears for the first time in the algorithm, the total flow $\sum\sum x_{ij}$ becomes equal to $\vartheta(\mathbf{E})$, and this happens after at most $O(m(p+n))$ consecutive occurrences of Steps 1 and 2, by the results in Section 2.3.3. If the algorithm did not terminate then, Step 3 will occur, and it could repeat at most n times consecutively, followed by another occurrence of Steps 1 and 2, at which time the total flow strictly exceeds $\vartheta(\mathbf{E})$.

So, after the equality subnetwork defined by the subset of cells $\mathbf{E}$ first appears in the algorithm, there is a consecutive run of at most $O(m(p+n))$ Steps 1 and 2 ; followed by a consecutive run of at most n Steps 3, and then another occurrence of Steps 1 and 2. After these are over, the equality subnetwork defined by the subset of cells $\mathbf{E}$ can never reappear in the algorithm. Since there are only a finite number of subsets of cells, this implies that the algorithm is a finite algorithm.

This primal-dual method for the transportation problem is practically efficient and is useful for doing sensitivity analysis when the a_i and b_j change. Experience indicates that in solving large problems, computer implementations of the primal algorithm using tree labels, discussed in Chapter 5, are superior to those of the primal-dual method discussed here. Newer variants of the primal-dual method are discussed in Chapter 5, they are competitive with other algorithms for solving large scale minimum cost flow problems.

To study the worst case computational complexity of this algorithm, we will assume that all the a_i and b_j are positive integers and let $\sum_i a_i = \sum_j b_j = \gamma$. In this case, every breakthrough leads to an increase in the total flow value, $\sum_i \sum_j x_{ij}$, by at least 1. When tree growth is resumed after each dual solution change, at least one new column gets labeled. Hence between two consecutive breakthroughs, there are at most n nonbreakthroughs. It can be verified that the overall computational effort in the primal-dual method in this case is bounded above by $O(\gamma(n+p)^2)$. This grows exponentially with the size of the problem, since γ grows exponentially with the number of digits needed to store the a_i, b_j. However there is a *scaling* or *digit-by-digit implementation* of the primal-dual method which is polynomially bounded, which we discuss next.

Polynomially Bounded Scaling Implementation

Consider the case where the a_i, b_j are all rational numbers. By selecting the unit for measuring the commodity appropriately, the problem can then be modified into one in which all the a_i, b_j are positive integers. We assume that this has been done. Let $\sum_i a_i = \sum_j b_j = \gamma$. No assumptions are made about c_{ij}.

Let q be the smallest positive integer such that a_i and $b_j \leqq 2^q$ for all i, j (i.e., each a_i, b_j have at most q digits in its binary expansion, and q is the smallest integer with this property). The scaling implementation deals with a sequence of $q + 1$ problems called subproblems all of which are transportation problems with the same cost matrix (and hence the same set of nodes and arcs) as (3.7), but with scaled down availabilities and requirements which approximate those of (3.7) to successively more digits of precision. Initially availabilities and requirements, and hence flow augmentations are on a coarser scale than in the original problem, but by the end of the sequence all the data converts to the original data, and the terminal solution, if one is obtained, will be an optimum solution of the original problem. The final solution of each approximate problem leads to a good initial flow for the next approximate problem in the sequence. The worst case computational complexity of the scaling implementation is proportional to the number of digits in the binary encoding of the a_i, b_j.

We consider the general problem in which some of the x_{ij} may be required to be 0. For $i = 1$ to p, let $\Gamma_i = \{j: x_{ij}$ can have a positive value in the solution$\}$; and for $j = 1$ to n, let $\Omega_j = \{i: x_{ij}$ can have a positive value in the solution$\}$. So, source i is allowed to ship only to sinks in Γ_i for all i, and sink j can receive shipments only from sources in Ω_j for all j. Let $\mathbf{F} = \{(i, j) : x_{ij}$ can have a positive value in the solution$\} = \cup_{i=1}^p \{(i, j) : j \in \Gamma_i\} = \cup_{j=1}^n \{(i, j) : i \in \Omega_j\}$. So, the cost coefficients c_{ij} are only defined for $(i, j) \in \mathbf{F}$ in the problem, for $(i, j) \notin \mathbf{F}$ we treat c_{ij} to be $+\infty$. Hence the original problem to be solved is

$$\text{Minimize} \sum_{(i,j)\in\mathbf{F}} c_{ij} \; x_{ij}$$

$$\text{Subject to} \sum_{j\in\Gamma_i} x_{ij} = a_i, i = 1 \text{ to } p$$

$$\sum_{i\in\Omega_j} x_{ij} = b_j, j = 1 \text{ to } n \qquad (3.11)$$

$$x_{ij} \overset{\geq}{=} 0, \text{ for } (i, j) \in \mathbf{F}$$

Let $d_i, g_j, i = 1$ to p, $j = 1$ to n, be non-negative integers. In the scaling implementation, the subproblems are all transportation models in the form (3.12), where Δ is a specified positive integer. In the algorithm, d_i will be of the form $\lfloor \frac{a_i}{2^s} \rfloor$ and g_j will be of the form $\lfloor \frac{b_j}{2^s} \rfloor$, where s ranges from q to 0 in stages. When some of the x_{ij} are required to be zero even if the original problem (3.11) is feasible, the subproblems in some stages may be infeasible for some Δ. In any feasible solution of (3.12), Δ is known as the *total flow* in that solution. It is the total amount of material reaching the sinks from the sources in that solution.

$$\text{Minimize} \sum_{(i,j)\in \mathbf{F}} c_{ij}\ x_{ij}$$

$$\text{Subject to} \sum_{j\in \Gamma_i} x_{ij} \lesseqgtr d_i, i = 1 \text{ to } p$$

$$\sum_{i\in \Omega_j} x_{ij} \lesseqgtr g_j, j = 1 \text{ to } n \qquad (3.12)$$

$$\sum_i \sum_j x_{ij} = \Delta$$

$$x_{ij} \gtreqless 0, \text{ for } (i,j) \in \mathbf{F}$$

The dual of the original problem (3.11) is

$$\text{Maximize} \sum a_i u_i + \sum b_j v_j$$

$$\text{Subject to } u_i + v_j \lesseqgtr c_{ij} \text{ for all } (i,j) \in \mathbf{F} \qquad (3.13)$$

Denoting the dual variables corresponding to the constraints in (3.12) by π_i, μ_j, δ in that order, the dual of (3.12) is

$$\text{Maximize } \delta\Delta - \sum d_i \pi_i - \sum g_j \mu_j$$

$$\text{Subject to } -\pi_i - \mu_j + \delta \lesseqgtr c_{ij}, \text{ for all } (i,j) \in \mathbf{F} \qquad (3.14)$$

$$\pi_i, \mu_j \gtreqless 0, \text{ for all } i, j.$$

The complementary slackness optimality conditions in the primal dual pair (3.12), (3.14) are

$$(c_{ij} + \pi_i + \mu_j - \delta)x_{ij} = 0, \text{ for all } (i,j) \in \mathbf{F} \qquad (3.15)$$

$$\pi_i(d_i - \sum_{j=1}^{n} x_{ij}) = 0, i = 1 \text{ to } p \qquad (3.16)$$

$$\mu_j(g_j - \sum_{i=1}^{p} x_{ij}) = 0, j = 1 \text{ to } n \qquad (3.17)$$

We call the pair $(x, (\pi, \mu, \delta))$ feasible to (3.12), (3.14) an *extreme pair* if they together satisfy (3.15), whether they satisfy (3.16), (3.17) or not. An extreme pair for (3.12),(3.14) is said to be a *maximum extreme pair* if the total flow $\sum(x_{ij} : \text{over} (i,j) \in \mathbf{F})$ is the maximum value that can be attained with the supply at source (row) node i limited to at most d_i, and the amount that can be shipped to sink (column) node j limited to at most g_j, $i = 1$ to p, $j = 1$ to n.

If $(\overline{x}, (\overline{\pi}, \overline{\mu}, \overline{\delta}))$ is a maximum extreme pair when $d_i = a_i, g_j = b_j$ for all i, j, and the total flow in $\overline{x}$ is $\gamma = \sum a_i$, then (3.16), (3.17) hold automatically in this pair. In this case, let $\overline{u}_i = \overline{\delta} - \overline{\pi}_i, \overline{v}_j = -\overline{\mu}_j$ for all i, j, then $(\overline{x}, (\overline{u}, \overline{v}))$ is an optimum pair for the original problem. This fact is used in this implementation.

For $r = 0$ to q, define

$$d_i^r = \lfloor \frac{a_i}{2^{q-r}} \rfloor, g_j^r = \lfloor \frac{b_j}{2^{q-r}} \rfloor, i = 1 \text{ to } p, j = 1 \text{ to } n.$$

The scaling implementation solves a sequence of $(q+1)$ subproblems. For $r = 0$ to q, the rth subproblem is (3.12) with $d_i = d_i^r, g_j = g_j^r$ for all i, j. Each subproblem is solved by the primal dual algorithm, walking along extreme pairs only, until a maximum extreme pair is obtained for it. This will be recognized in the algorithm when, either the capacity at all the source (row) nodes is used up, or all the capacity at the sink (column) nodes is used up, or if $\delta = \infty$ in a dual solution change step. It then moves to the next subproblem, beginning with an initial extreme pair for it constructed from the last pair for the present subproblem. In the qth (i.e., final) subproblem, we have $d_i = a_i, g_j = b_j$ for all i, j, and since $\sum a_i = \sum b_j$, a maximum extreme pair for it leads to a solution for the original problem (3.11) if it is feasible, as shown later.

For $r < q$ we may not have $\sum_i d_i^r = \sum_j g_j^r$, but we apply the primal dual algorithm to find a maximum extreme pair for the rth subproblem. While solving this problem the algorithm maintains the pair $(x, (u, v))$ where x satisfies (3.12) with $d_i = d_i^r, g_j = g_j^r$ for all i, j; u, v satisfy (3.13) always, and these vectors together satisfy (3.9). There is a vector (π, μ, δ) feasible to (3.14) and satisfying $-\pi_i - \mu_j + \delta = u_i + v_j$ for all i, j; that can be obtained from (u, v) by the formulas $\pi_i = \nu - u_i, \mu_j = \nu - v_j, \delta = 2\nu$ for all i, j, where $\nu = \max.\{u_1, \ldots, u_p; v_1, \ldots, v_n\}$, and the pair $(x, (\pi, \mu, \delta))$ will then be an extreme pair for (3.12), (3.14).

Let G denote the bipartite network for the original problem (3.11) as in Figure 3.5, in which the arc joining source node i to market node j exists iff $(i, j) \in$ F. For $r = 0$ to q, let G_r denote the same network as G, but with the capacities a_i replaced by d_i^r, and capacities b_j replaced by g_j^r, for all i, j.

Subproblem 0 begins with $x^0 = 0, u^0 = (u_i^0 = \min.\{c_{ij} : j = 1 \text{ to } n\}), v^0 = (v_j^0 = \min.\{c_{ij} - u_i^0 : i = 1 \text{ to } p\})$. The rth subproblem is terminated when flow value in the network G_r from the supersource to the supersink (which is the total flow in the solution x at that stage) reaches the maximum value possible in G_r recognized as described above. If the original problem is on the complete bipartite network (i.e., all variables x_{ij} are allowed to take positive values), this will happen when either the supply d_i^r at each source i is used up or the demand g_j^r at each market j is met. If some variables are required to be zero in the original problem, another signal for the termination of the rth subproblem is δ becoming ∞ in a dual solution change step. Let $\tilde{x}^r, (\tilde{u}^r, \tilde{v}^r)$ be the pair at the termination of subproblem r. The total flow in $\tilde{x}^r$ is

$$\sum_i \sum_j \tilde{x}^r = \gamma_r \qquad (3.18)$$

Define, for $r = 0$ to q

$$\Delta_r = \text{min.} \left\{ \sum_i d_i^r, \sum_j g_j^r \right\} \qquad (3.19)$$

So, $\Delta_q = \gamma$, and $\gamma_r \overset{\leq}{=} \Delta_r$ for all r. For $r = 0$ to $q - 1$, if $\gamma_r < \Delta_r - (n + p)$, terminate the algorithm. In this case the original problem (3.11) has no feasible solution (see Lemma 3.2 given below). Otherwise, initiate subproblem $r + 1$ with the pair $x^{r+1} = 2\tilde{x}^r$, $(u^{r+1} = \tilde{u}^r, v^{r+1} = \tilde{v}^r)$, this pair satisfies (3.13), (3.9) since $\tilde{x}^r, (\tilde{u}^r, \tilde{v}^r)$ do. Thus all the pairs obtained during the implementation will be extreme pairs. At the end of subproblem q, we have the maximum extreme pair $\tilde{x}^q, (\tilde{u}^q, \tilde{v}^q)$ in G, with total flow $= \gamma_q$. If $\gamma_q = \gamma$, $\tilde{x}^q, (\tilde{u}^q, \tilde{v}^q)$ together satisfy (3.11), (3.13), (3.9), so they form an optimum primal dual pair for the original problem (3.11). If $\gamma_q < \gamma$, the original problem (3.11) is infeasible.

Flow changes occur at most Δ_0 times in subproblem 0. If the algorithm is continued after subproblem r, $2\tilde{x}^r$ is the initial primal vector for solving the $(r+1)$th subproblem. By Lemma 3.2 given below, there will be at most $\Delta_{r+1} - 2\Delta_r + 2(n + p)$ flow changes while solving the $(r + 1)$th subproblem. So, the total number of flow augmentations during the entire scaling implementation is at most $\Delta_0 + \sum_{r=0}^{q-1}(\Delta_{r+1} - 2\Delta_r) + 2q(n + p)$.

i	a	d^0	d^1	d^2	d^3	d^4	d^5	d^6
1	38	0	1	2	4	9	19	38
2	41	0	1	2	5	10	20	41
3	23	0	0	1	2	5	11	23
4	15	0	0	0	1	3	7	15
Total	117	0	2	5	12	27	57	117

j	1	2	3	4	Total
b	25	29	30	33	117
g^0	0	0	0	0	0
g^1	0	0	0	1	1
g^2	1	1	1	2	5
g^3	3	3	3	4	13
g^4	6	7	7	8	28
g^5	12	14	15	16	57
g^6	25	29	30	33	117

$\Delta_r = 0, 1, 5, 12, 27, 57, 117$
respectively for $r = 0$ to 6.

As an example we consider a problem with $p = n = 4$. The cost matrix is not shown, but the vectors a, b, and d^r, g^r are given for all r in the table above. $\gamma = \sum a_i = \sum b_j = 117$. Max.$\{a_i, b_j : \text{all } i, j\} = 41$, and the smallest integer q satisfying the property that all $a_i, b_j \overset{\leq}{=} 2^q$ is 6. So, we have to solve seven subproblems, these correspond to $r = 0$ to 6.

In this example $\Delta^0 + \sum_{r=0}^{q-1}(\Delta_{r+1} - 2\Delta_r) = 15$. So, to solve this problem, the scaling implementation needs at most $15 + 96 = 111$ flow augmentations, this compares with the maximum of 117 that may be needed in the direct implementation of the primal dual method.

LEMMA 3.1 $0 \leqq \Delta_0 \leqq max.\{p, n\}; 0 \leqq \Delta_{r+1} - 2\Delta_r \leqq max.\{ p, n \}$, for all $r \geqq 0$.

Proof From the definition of q, all d_i^0, g_j^0 are 0 or 1. So, $0 \leqq \Delta_0 \leqq max.\{p, n\}$. Let ξ, η be any positive integers. We have

$$2\lfloor \frac{\eta}{2^\xi} \rfloor \leqq \lfloor \frac{\eta}{2^{\xi-1}} \rfloor \leqq 2\lfloor \frac{\eta}{2^\xi} \rfloor + 1$$

Applying this we conclude that

$$0 \leqq \sum_{i=1}^{p} \lfloor \frac{a_i}{2^{q-r-1}} \rfloor - 2\sum_{i=1}^{p} \lfloor \frac{a_i}{2^{q-r}} \rfloor \leqq p$$

$$0 \leqq \sum_{j=1}^{n} \lfloor \frac{b_j}{2^{q-r-1}} \rfloor - 2\sum_{j=1}^{n} \lfloor \frac{b_j}{2^{q-r}} \rfloor \leqq n$$

These inequalities, and the definitions of Δ_r, Δ_{r+1}, imply this lemma. ∎

LEMMA 3.2 *If the original problem (3.11) has a feasible solution, then* $\gamma_r \geqq \Delta_r - (p+n)$, *for all* r.

Proof Let $\hat{x} = (\hat{x}_{ij})$ be a feasible solution for (3.11). Then, for all i, j

$$\sum_{j \in \Gamma_i} (\hat{x}_{ij}/2^r) = a_i/2^{q-r}$$

$$\sum_{i \in \Omega_j} (\hat{x}_{ij}/2^r) = b_j/2^{q-r}$$

So, if the amount of material that can leave source (row) node i is $a_i/2^{q-r}$, and that reaching market (column) node j is $b_j/2^{q-r}$, for all i, j; then all the arcs incident at the supersource and the supersink will be exactly saturated in a maximum value flow ($\hat{x}/2^r$ is such a flow). However, the maximum amount of material that can leave source (row) node i is limited to at most $d_i^r = \lfloor a_i/2^{q-r} \rfloor \geqq (a_i/2^{q-r}) - 1$, and that reaching market (column) node j is limited to at most $g_j^r = \lfloor b_j/2^{q-r} \rfloor \geqq (b_j/2^{q-r}) - 1$ in G_r. So the maximum flow value in G_r is $\geqq \Delta_r - (p+n)$, i.e., $\gamma_r \geqq \Delta_r - (p+n)$. ∎

THEOREM 3.2 *Assuming that all the a_i, b_j are positive integers, the overall computational effort for solving (3.11) by the scaling implementation of the primal dual algorithm is at most $O(L(n+p)^2 (max.\{p,n\}))$, where L is the sum of the binary digits in all the a_i and b_j.*

Proof q defined earlier is $< L$. Lemmas 3.1 and 3.2 imply that the total number of flow changes in any subproblem is at most $4 \max.\{p,n\}$. So, the total number of flow augmentations in the scaling implementation is at most $4(q+1)(\max.\{p,n\})$ $\leq 4L(\max.\{p,n\})$. We have already seen that the computational effort between two consecutive occurrences of flow augmentation in any subproblem is at most $O(p+n)^2$. Hence the result follows. ∎

So, the scaling implementation of the primal dual algorithm is a polynomially bounded algorithm for solving the transportation problem (3.7) with some specified subset of variables x_{ij} set equal to 0 if desired, in which the overall computational effort is bounded above by a low degree polynomial in the size of the problem.

Comment 3.2   This scaling technique was introduced by Edmonds and Karp [1972 of Chapter 2]. It led to the first polynomial time algorithm for minimum cost flow problems.

Exercises

3.18 Complete the solution of the problem in Example 3.3 by the primal-dual algorithm

3.19 Develop a primal-dual method to solve the following capacitated transportation problem.

$$\text{Minimize } z(x) \ = \ \sum_{i=1}^{p} \sum_{j=1}^{n} c_{ij} x_{ij}$$

$$\text{Subject to } \sum_{j=1}^{n} x_{ij} \ \leqq \ a_i, i = 1 \text{ to } p$$

$$\sum_{i=1}^{p} x_{ij} \ = \ b_j, j = 1 \text{ to } n$$

$$0 \leqq x_{ij} \ \leqq \ k_{ij}, \text{ for all } i, j$$

3.20 Whenever a nonbreakthrough occurs in the primal-dual algorithm for the balanced transportation problem (3.7), prove that the quantity $\sum (a_i$: over labeled rows $i) - \sum(b_j$: over labeled cols. $j) > 0$. Using this prove that the dual objective function $\sum_i a_i u_i + \sum_j b_j v_j$ strictly increases whenever the dual solution changes in this algorithm. Also prove the following using the duality theorem of LP: While solving (3.7) with some variables x_{ij} constrained to be 0, by the primal-dual algorithm, if δ turns out to be $+\infty$ in some dual solution change step, the problem is infeasible.

3.3   Transformation of the Single Commodity Minimum Cost Flow Problem into a Sparse Balanced Transportation Problem

The balanced transportation problem is a special case of the single commodity minimum cost flow problem on a bipartite network. We will now show that every single commodity minimum cost flow problem, even on a nonbipartite network, can be transformed into an uncapacitated balanced transportation problem which is sparse.

Consider the single commodity minimum cost flow problem on the directed network $G = (\mathcal{N}, \mathcal{A}, 0, k, c, \check{s}, \check{t}, \bar{v})$ for shipping $\bar{v}$ units from $\check{s}$ to $\check{t}$ at minimum cost. Let $|\mathcal{N}| = n, |\mathcal{A}| = m$. Construct a bipartite network $H = (\mathcal{N}_1, \mathcal{N}_2; \mathbf{A})$ as follows. For each $(i, j) \in \mathcal{A}$ put a node corresponding to it in $\mathcal{N}_1$ which we conveniently denote by the symbol "ij." So, $|\mathcal{N}_1| = m$. Make $\mathcal{N}_2 = \mathcal{N}$. $\mathcal{N}_1, \mathcal{N}_2$ are respectively the sets of source and sink nodes in H. Define $\mathbf{A} = \{(ij, i), (ij, j)$: for each $(i, j) \in \mathcal{A}\}$, so $|\mathbf{A}| = 2m$. The lower bounds and capacities for arcs in $\mathbf{A}$ are 0, $+\infty$ respectively. For each $(i, j) \in \mathcal{A}$, in H the unit cost coefficient on the arc (ij, i) is 0, and that on the arc (ij, j) is c_{ij}; and the availability of material at the source node ij in $\mathcal{N}_1$ is k_{ij}. For each $i \in \mathcal{N}$, the requirement at the sink node $i \in \mathcal{N}_2$ in H is b_i where

$$b_i = \begin{cases} k(i, \mathcal{N}) - \bar{v} & \text{if } i = \check{s} \\ k(i, \mathcal{N}) + \bar{v} & \text{if } i = \check{t} \\ k(i, \mathcal{N}) & \text{if } i \neq \check{s} \text{ or } \check{t} \end{cases}$$

Unless $k(\check{s}, \mathcal{N}) \geqq \bar{v}$, the flow problem in G is infeasible. So, we assume that this condition holds. This implies that all the availabilities at the sources, and the requirements at the sinks, in H, are $\geqq 0$. In H, each source node is joined to exactly two sink nodes. See Figure 3.6. The sum of all the availabilities at the source nodes in H, as well as the sum of all the requirements at the sink nodes, are both equal to $k(\mathcal{N}, \mathcal{N})$. Thus the minimum cost flow problem in H satisfying all the availability, requirement constraints at the nodes, is a sparse uncapacitated balanced transportation problem.

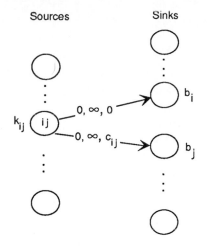

Figure 3.6 Bipartite network H. Data on arcs is lower bound, capacity, unit cost coefficient, in that order. Data by the side of the nodes is availability (for sources), requirement (for sinks).

We will now prove the equivalence of the minimum cost flow problems in G and H. Let $f = (f_{ij})$ be a feasible flow vector in G. Define the corresponding flow vector in H to be x where for each $(i, j) \in \mathcal{A}, x_{ij,j} = f_{ij}$, and $x_{ij,i} = k_{ij} - f_{ij}$. Then it can be verified that x is a feasible flow vector in H, having the same cost as f in G. Conversely, if x is a feasible flow vector in H, define the flow vector $f = (f_{ij})$ in G by: for each $(i, j) \in \mathcal{A}, f_{ij} = x_{ij,j}$. It can be verified that f is a feasible flow vector in G which has the same cost as x in H. So, the minimum cost flow problem in G is equivalent to the minimum cost flow problem in H, which is a sparse uncapacitated balanced transportation problem. Using this transformation, and the scaling implementation to solve the resulting transportation problem, we get a polynomially bounded algorithm for the minimum cost flow problem in G. Primal-dual algorithms that solve the minimum cost flow problem in G directly will be discussed later in Chapter 5.

3.4 Dual Simplex Signature Methods for the Assignment Problem

Consider the LP

$$
\begin{array}{lll}
\text{Minimize} & c_B x_B & + & c_D x_D \\
\text{Subject to} & B x_B & + & D x_D = b \\
& x_B & , & x_D \geqq 0
\end{array}
$$

in which the coefficient matrix $A = (B \vdots D)$ is of order $m \times n$ and rank m. The LP is written with the variables partitioned into basic, nonbasic parts, x_B, x_D. The primal basic solution associated with the basic vector x_B is $\bar{x} = (\bar{x}_B = B^{-1}b, \bar{x}_D = 0)$. It is *primal feasible* if $\bar{x} \geqq 0$. The dual basic solution associated with this basic vector is $\bar{\pi} = c_B B^{-1}$. This basic vector is *dual feasible* if the associated reduced cost vector $\bar{c} = (\bar{c}_B, \bar{c}_D) = (0, c_D - \bar{\pi}D) \geqq 0$. The dual simplex algorithm for solving this LP is always initiated with a dual feasible basic vector x_B. If x_B is also primal feasible, it is an optimum basic vector, and the algorithm terminates. Otherwise, the algorithm checks to see if a condition for primal infeasibility is satisfied, and if so, it again terminates. If neither of these two events has occurred, it selects a primal variable x_j whose value $\bar{x}_j$ in the primal basic solution $\bar{x}$ associated with x_B satisfies $\bar{x}_j < 0$, as the *dropping variable* from the present basic vector x_B, for a dual simplex pivot step. The nonbasic variable to replace it, called the *entering variable* is selected by the dual simplex minimum ratio test, whose function is to guarantee that the next basic vector will also be dual feasible. Then the whole process is repeated with the new basic vector. In the dual simplex algorithm, the dual objective function πb is monotone nondecreasing. See Murty [1983 of Chapter 1] for a detailed description of the dual simplex algorithm. In this section we will discuss a new class of algorithms for the assignment problem called *signature methods*, which are related to the dual simplex algorithm in spirit.

We consider the assignment problem (3.1) of order n associated with the cost matrix $c = (c_{ij})$, a minimum cost flow problem on the bipartite network G = $(\mathcal{N}_R, \mathcal{N}_C, \mathcal{A})$, where $\mathcal{N}_R = \{R_1, \ldots, R_n\}$ = set of row nodes, $\mathcal{N}_C = \{C_1, \ldots, C_n\}$ = set of column nodes, and $\mathcal{A} = \{(R_i, C_j): i, j$ s.t. an allocation is allowed in cell $(i, j)\}$. The arc (R_i, C_j) in G is associated with the variable x_{ij} in the problem. The dual variables are u_i, v_j associated with the nodes R_i, C_j respectively, $i, j = 1$ to n. Each basic vector for (3.1) corresponds to a spanning tree in G and vice versa, in-tree arcs are associated with basic variables, and out-of-tree arcs with nonbasic variables. A spanning tree $\mathbb{T}$ in G has exactly $2n - 1$ arcs, and the dual basic solution corresponding to it, denoted by $u(\mathbb{T}) = (u_i(\mathbb{T})), v(\mathbb{T}) = (v_j(\mathbb{T}))$, can be obtained by arbitrarily fixing the value of one dual variable (e.g. $u_1(\mathbb{T}) = 0$) and then solving the equations

$$u_p(\mathbb{T}) + v_q(\mathbb{T}) = c_{pq}, \text{ for each in-tree arc } (R_p, C_q)$$

by back substitution. The matrix $\bar{c}(\mathbb{T}) = (\bar{c}_{ij}(\mathbb{T}))$, where $\bar{c}_{ij}(\mathbb{T}) = c_{ij} - u_i(\mathbb{T}) - v_j(\mathbb{T})$ is the matrix of reduced or relative cost coefficients wrt $\mathbb{T}$. $\mathbb{T}$ is dual feasible if $\bar{c} \geqq 0$, dual infeasible otherwise.

The primal basic solution of (3.1) corresponding to $\mathbb{T}$, $x(\mathbb{T}) = (x_{ij}(\mathbb{T}))$ is obtained by setting $x_{ij}(\mathbf{T}) = 0$ whenever (R_i, C_j) is an out-of-tree arc, in the system of equality constraints in (3.1), and then solving the remaining system by back substitution. $x(\mathbf{T})$ is integer, but it may not be $\geqq 0$. $\mathbb{T}$ is primal feasible if $x(\mathbb{T}) \geqq 0$ (in this case $x(\mathbb{T})$ will be an assignment), primal infeasible otherwise.

We denote by $\mathcal{P}(p, q, \mathbb{T})$, the unique path between the nodes p, q in $\mathbb{T}$.

The *row signature vector (column signature vector)* of a spanning tree in G is the vector of its row node degrees (column node degrees). If $(d_1, \ldots, d_n)$ is the row or column signature vector of a spanning tree in G, clearly, $d_i \overset{\geq}{=} 1$ for all $i = 1$ to n, and $\sum_{i=1}^{n} d_i = 2n - 1$. Hence, in every signature vector there is at least one entry equal to 1. It corresponds to a terminal node. For example, the row, column signature vectors of the spanning tree in Figure 3.10 are (2, 1, 2, 2, 2, 2), (1, 3, 4, 1, 1, 1) respectively. Both these vectors have at least one entry of 1.

THEOREM 3.3 *A spanning tree* $\mathbb{T}$ *in* G *which contains exactly one entry of 1 in either its row or column signature vectors is a primal feasible spanning tree.*

Proof   Let $d = (d_1, \ldots, d_n)$ be the row signature vector of $\mathbb{T}$, and suppose it contains a unique 1 entry. A similar proof holds if the column signature vector contains a single 1 entry.

Let i^* be the unique number such that $d_{i^*} = 1$. Since $d_i \overset{\geq}{=} 1$ for all i and $\sum_{i=1}^{n} d_i = 2n - 1$, the hypothesis implies that $d_i = 2$ for all $i \neq i^*$. For each $i \neq i^*$, the path $\mathcal{P}(R_{i^*}, R_i, \mathbb{T})$ contains exactly one edge incident at R_i, and since $d_i = 2$, there must be exactly one edge in $\mathbb{T}$ incident at R_i which is not on the path $\mathcal{P}(R_{i^*}, R_i, \mathbb{T})$. Let $\sigma_{i^*} = j^*$, where (R_{i^*}, C_{j^*}) is the unique in-tree arc incident at R_{i^*}, and for $i \neq i^*$, let σ_i be j, where j is such that (R_i, C_j) is the unique in-tree arc incident at R_i not on $\mathcal{P}(R_{i^*}, R_i, \mathbb{T})$. Clearly, $\sigma_i \neq \sigma_h$ for any $i \neq h$, as otherwise there will be a cycle in $\mathbb{T}$, a contradiction. So, $\{(i, \sigma_i) : i = 1 \text{ to } n\}$ is an assignment, and since all these cells correspond to in-tree edges, this assignment is the basic solution of (3.1) corresponding to $\mathbb{T}$, hence $\mathbb{T}$ is a primal feasible tree. ∎

As an example consider the spanning tree in Figure 3.10 with its row signature vector (2, 1, 2, 2, 2, 2). R_2 is the unique terminal row node. The primal basic solution corresponding to this tree is the assignment {(1, 1), (2, 3), (3, 2), (4, 5), (5, 4), (6, 6)}, which corresponds to the set of solid lines in Figure 3.10.

Signature methods are always initiated with a dual feasible spanning tree in G and dual feasibility is maintained throughout. They go through a sequence of pivot steps; each pivot step moves from a tree to an adjacent tree that differs from it in a single arc. The dropping arc from the present tree is always selected by a special dropping arc selection rule described in the algorithm. The entering arc to replace the dropping arc is determined so as to maintain dual feasibility. Termination occurs when either primal infeasibility is established, or when a spanning tree with exactly one terminal row node, or one terminal column node is obtained. In the latter case the primal basic solution corresponding to the final tree is an optimum assignment. The method tries to reduce the number of terminal row nodes (or terminal column nodes) to one. This number is monotone nonincreasing during the method.

A *stage* in a signature method begins with a dual feasible spanning tree in G with more than one terminal row node (or column node, if the method works with column signature vectors) and is completed when this number decreases by 1. Each stage may consist of several pivot steps. Since G is bipartite and $\mathbb{T}$ has $2n-1$ arcs, the total number of terminal row nodes, or column nodes in any spanning tree in G, is at most $n-1$. So, the algorithm has at most $n-2$ stages.

The dropping arc choice is defined by the method, but once the dropping arc is selected, the choice of the entering arc to replace it is carried out by the same procedure in all the signature methods. We describe this next.

The Entering Arc Choice Rule

Let $\mathbb{T}$ be the present dual feasible spanning tree in G, and let (w, l) denote the in-tree arc that has been selected as the dropping arc, w, l are the row and column nodes on it. Deletion of the arc (w, l) from $\mathbb{T}$ leaves two distinct subtrees, a $\mathbb{T}^w$ containing node w, and a $\mathbb{T}^l$ containing node l. If $\mathbb{T}$ is drawn as a rooted tree with w as the root node, the family of node l is exactly the set of nodes in $\mathbb{T}^l$, and the set of nodes in $\mathbb{T}^w$ is the complement of this set. Let $\mathbf{X} = \mathbf{R}^w$, $\bar{\mathbf{X}} = \mathbf{R}^l$ ($\mathbf{Y} = \mathbf{C}^w$, $\bar{\mathbf{Y}} = \mathbf{C}^l$) be the set of row (column) nodes in $\mathbb{T}^w, \mathbb{T}^l$ respectively. $(\mathbf{X}, \bar{\mathbf{X}})$, $(\mathbf{Y}, \bar{\mathbf{Y}})$, are the partitions of row, column nodes in this pivot step. They are uniquely determined by $\mathbb{T}$ and (w, l).

Since (w, l) is an in-tree arc, $\tilde{\bar{c}}_{wl}$, its present relative cost coefficient, is 0, but it becomes nonnegative in the next tree. Suppose its new value is δ. Let $\tilde{u}_i, \tilde{v}_j, \tilde{\bar{c}}_{ij}$ denote the present quantities in the tree $\mathbb{T}$. It can be verified that the unique solution of the system: $u_1 = 0, c_{pq} - u_p - v_q = 0$ for each arc $(R_p, C_q) \in \mathbb{T}$ except the arc (w, l), and $= \delta$ for arc (w, l), is $(\hat{u}, \hat{v})$ marked in the following Array 3.6, and $\hat{\bar{c}}_{ij}$ are the reduced cost coefficients wrt it.

Array 3.6

Block of rows ↓	Block of cols.		New dual solution $\hat{u}$
	$\mathbf{Y} = \mathbf{C}^w$	$\bar{\mathbf{Y}} = \mathbf{C}^l$	
$\mathbf{X} = \mathbf{R}^w$	$\hat{\bar{c}}_{ij} = \tilde{\bar{c}}_{ij}$ in this block	$\hat{\bar{c}}_{ij} = \tilde{\bar{c}}_{ij} + \delta$ in this block	$\hat{u}_i = \tilde{u}_i$ in this block
$\bar{\mathbf{X}} = \mathbf{R}^l$	$\hat{\bar{c}}_{ij} = \tilde{\bar{c}}_{ij} - \delta$ in this block	$\hat{\bar{c}}_{ij} = \tilde{\bar{c}}_{ij}$ in this block	$\hat{u}_i = \tilde{u}_i + \delta$ in this block
New dual solution $\hat{v}$	$\hat{v}_j = \tilde{v}_j$ in this block	$\hat{v}_j = \tilde{v}_j - \delta$ in this block	

Dual feasibility is maintained in the new tree if $\hat{\bar{c}}_{ij} \geqq 0$ for all i, j. For this we must choose

$$\delta = \text{ min.} \{\tilde{\bar{c}}_{ij} = c_{ij} - \tilde{u}_i - \tilde{v}_j : R_i \in \bar{\mathbf{X}}, C_j \in \mathbf{Y}\}$$

Since the present tree $\mathbb{T}$ is dual feasible, this δ will be $\geqq 0$. Consequently, the entering arc in this pivot step is selected to be an arc (g, h) with $g \in \bar{\mathbf{X}}$ and $h \in \mathbf{Y}$, which attains the minimum for δ in the above equation. The new tree is obtained by replacing (w, l) in $\mathbb{T}$ with (g, h).

In the above equation, δ will be ∞ only if all the cells (i, j) with $i \in \bar{\mathbf{X}}$, and $j \in \mathbf{Y}$ are constrained to have no allocation in them. As before, this is an indication that there exists no feasible assignment, and the method terminates if this happens.

Selection of the Initial Dual Feasible Spanning Tree

If the assignment problem being solved is on the complete bipartite network (i.e., every variable x_{ij} is eligible to have value 1), an initial dual feasible spanning tree can be taken to be the tree $\mathbb{T}^0$ consisting of the following arcs: all arcs $(R_p, C_j), j = 1$ to n, for some p, and arcs (R_i, C_{q^i}) for $i \neq p$, where q^i is an index attaining the minimum for $u_i^0 = \text{min.} \{ c_{ij} - c_{pj} : j = 1 \text{ to } n \}$. Let $u_p^0 = 0$, and $v_j^0 = c_{pj}$, for $j = 1$ to n. It can be verified that (u^0, v^0) is the dual basic solution associated with $\mathbb{T}^0$, and that it is dual feasible. The row signature vector of $\mathbb{T}^0$ is $(n, 1, \dots, 1)$. Hence it has $n - 1$ terminal row nodes.

In dense or sparse problems, other methods for obtaining an initial dual feasible spanning tree are discussed in Murty and Witzgall [1977], and Section 13.6, Chapter 13 in Murty [1983 of Chapter 1]. In highly sparse problems, it may be necessary to include some artificial arcs (among those which correspond to cells that are constrained not to have any allocations) in the network, associated with very large positive cost coefficients (equal to $\beta = 1 + n(\text{max.} \{|c_{ij}| : (i, j) \text{ can have an allocation}\}))$, in order to get an initial dual feasible spanning tree. If any of these artificial cells have an allocation in the final optimum assignment, it is an indication that the original problem has no feasible assignment.

The only remaining thing needed to describe a signature method is the dropping arc selection strategy to be used. We discuss several signature methods next.

3.4.1   Signature Method 1

This method is initiated with a dual feasible spanning tree. It is entirely guided by the row signature vector, the column signature vector and the primal basic solution are never explicitly used in carrying out the algorithm.

The method seeks a tree whose row signature vector contains a single 1 and otherwise 2s. Given a dual feasible spanning tree $\mathbb{T}$ with more than one terminal row node, it selects one of these nodes, say t, and designates it as the *target row node*. There must be at least one row node in $\mathbb{T}$ of degree $\geqq 3$. It selects one of these, say s, as the *source row node*. Once the target row node t and the source row node s in the tree $\mathbb{T}$ are selected, the dropping arc in this method is always the arc incident at s on the path $\mathcal{P}(s, t, \mathbb{T})$. If it is the arc (s, l), l is a column node with degree $\geqq 2$ in $\mathbb{T}$, as the path $\mathcal{P}(s, t, \mathbb{T})$ itself contains two arcs incident at l. Let (g, h) be the entering arc to replace (s, l), determined as described above, and $\mathbb{T}^1$ the new tree obtained after this pivot step. Let $d(s), d(g), d^1(s), d^1(g)$ be the degrees of s, g in $\mathbb{T}, \mathbb{T}^1$ respectively. Then $d^1(s) = d(s) - 1, d^1(g) = d(g) + 1$. So, g which is the row node on the entering arc, is not a terminal row node in $\mathbb{T}^1$. If g was a terminal row node in $\mathbb{T}$, the number of terminal row nodes has decreased by one in this pivot step, this completes a stage in the method. If the number of terminal row nodes in $\mathbb{T}^1$ is one, it is primal feasible, and hence optimal, and the method terminates. Otherwise, it goes to the next stage with $\mathbb{T}^1$.

If g was not a terminal row node in $\mathbb{T}$, continue the stage with $\mathbb{T}^1$. Keep the same target row node t, but make g the next source row node and carry out the next pivot step. In the next pivot step, the set $\bar{\mathbf{X}}$ in the row partition becomes smaller (it loses node g) and the set $\mathbf{Y}$ in the column partition becomes larger.

So, the target row node remains the same in a stage, but the source row node keeps changing. The set $\bar{\mathbf{X}}$ gets smaller, and the set $\mathbf{Y}$ gets larger, until in the last pivot step of the stage the row node in the entering arc is a terminal row node before that arc enters. The row nodes that drop off from $\bar{\mathbf{X}}$ during this stage are all nonterminal row nodes, so, if the initial tree in this stage had r terminal row nodes, the number of pivot steps in it will be $\leqq n - r$. If carried out directly, the computational effort in each pivot step (to compute δ and update all the reduced cost coefficients) is at most $O(n^2)$, and hence the total effort in this stage may be $O((n - r)n^2)$, or $O(n^3)$. However, using the technique described under the $O(n^3)$ implementation of the Hungarian method, this stage can be implemented in such a way that the computational effort in it is at most $O(n^2)$. In this implementation, all the reduced cost coefficients are computed in the first pivot step of the stage, and for each row node R_i in the set $\bar{\mathbf{X}}$ at that time an index of the form $[w_i, p_i]$ is defined, where $p_i = $ minimum current reduced cost coefficient in R_i among columns in the present set $\mathbf{Y}$, and w_i is a column in $\mathbf{Y}$ where this minimum occurs. In each pivot step of the stage, we do the following work.

(i)　$\delta = $ minimum of p_i in the indices of rows in the set $\bar{\mathbf{X}}$ in that step. Computing this δ therefore takes only $O(n)$ effort using these indices and if the row $R_q \in \bar{\mathbf{X}}$ attains this minimum (break ties arbitrarily) and the present index on R_q is $[w_q, p_q]$, then the arc (R_q, w_q) is the entering arc in this pivot step.

(ii)　Update the dual solution as indicated in Array 3.6

(iii)　Get the new sets $\bar{\mathbf{X}}, \mathbf{Y}$. Eliminate the indices on all the rows no longer in $\bar{\mathbf{X}}$.

(iv)　Subtract δ from the p_i entry in the index of each row node in the new $\bar{\mathbf{X}}$.

(v)　For each column C_j that just joined the set $\mathbf{Y}$, compute the correct reduced cost coefficient $\bar{c}'_{ij} = c_{ij} - u'_i - v'_j$, where (u', v') is the new dual solution, for rows R_i in the new set $\bar{\mathbf{X}}$ only, and if the present p_i on this row satisfies $p_i > \bar{c}'_{ij}$ change the index for R_i to $[C_j, \bar{c}'_{ij}]$; otherwise leave this index unchanged.

Under this implementation, in each cell, the reduced cost coefficient is computed once at the beginning of the stage, and at most once more during the entire stage. It can be verified that the overall computational effort during a stage is at most $O(n^2)$ under this implementation. Each stage reduces the number of terminal row nodes by one, hence there can be at most n stages in the method. So, the overall computational effort in this method is at most $O(n^3)$, which is of the same order as that of the Hungarian method.

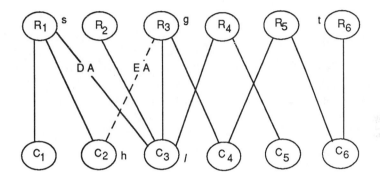

Figure 3.7 DA, EA denote the dropping arc, entering arc respectively.

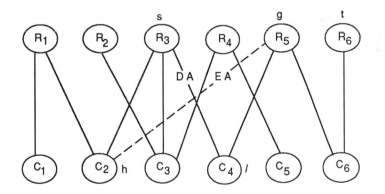

Figure 3.8 DA, EA denote the dropping arc, entering arc respectively.

As an example consider the problem in which $n = 6$, and the dual feasible spanning tree $\mathbf{T}$ at the beginning of a stage is the one in Figure 3.7. Orientations of the arcs are not shown in the figure, each line is directed from the row node to the column node on it. To keep the presentation simple, we do not include the arc cost coefficients in this illustration, but we give the entering arc in each step. $\mathbb{T}$ has two terminal row nodes, R_2 and R_6, of which R_6 has been selected as the target row node t. R_1 of degree 3 in $\mathbb{T}$ has been chosen as the source row node s. (R_1, C_3), the first arc on the path $\mathcal{P}(R_1, R_6, \mathbb{T})$ is the dropping arc in this pivot step. We have $\mathbf{R}^s = \mathbf{X} = \{R_1\}$, $\mathbf{C}^s = \mathbf{Y} = \{\,C_1, C_2\}$, $\mathbf{R}^l = \bar{\mathbf{X}} = \{R_2, R_3, R_4, R_5, R_6\,\}$, $\mathbf{C}^l = \bar{\mathbf{Y}}$ $= \{\,C_3, C_4, C_5, C_6\}$. Suppose the arc (R_3, C_2) marked with a dashed line in Figure 3.7 is the entering arc. So, in the notation used above for describing the pivot step, $g = R_3, h = C_2$. Since $g = R_3$ is not a terminal row node in Figure 3.7, the stage continues. The next tree $\mathbb{T}^1$ is drawn in Figure 3.8.

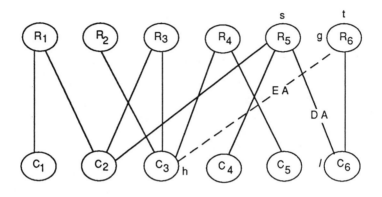

Figure 3.9 DA, EA are dropping arc, entering arc respectively.

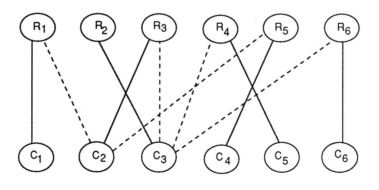

Figure 3.10

R_6 continues to be the target row node t. The new source row node is R_3. The dropping arc is (R_3, C_4). $\bar{\mathbf{X}} = \{R_5, R_6\}$. $\mathbf{Y} = \{C_1, C_2, C_3, C_5\}$. It can be verified that the set $\bar{\mathbf{X}}$ became smaller, and $\mathbf{Y}$ became larger. In $\mathbb{T}^1$ the entering arc is (R_5, C_2), so $g = R_5$. This is not a terminal row node in $\mathbb{T}^1$, so the stage continues. The next tree $\mathbb{T}^2$ is drawn in Figure 3.9. In $\mathbb{T}^2$, $\bar{\mathbf{X}} = \{R_6\}$, $\mathbf{Y} = \{C_1, C_2, C_3, C_4, C_5\}$, and (R_6, C_3) is the entering edge. Since $g = R_6$ is a terminal row node in $\mathbf{T}^2$ (it is actually the target row node), with the pivot step in it, the stage is completed. The next tree $\mathbf{T}^3$ is drawn in Figure 3.10. The row signature vector in this tree is (2, 1, 2, 2, 2, 2), so this is an optimum tree. The solid arcs in Figure 3.10 define an optimum assignment.

The method is executed without ever computing the primal basic solution corresponding to any tree, except for the very last tree to get the optimum assignment. Let $\tilde{\mathbb{T}}, \hat{\mathbb{T}}$ be two consecutive trees obtained in the method in that order, with $x(\tilde{\mathbb{T}}) = (x_{ij}(\tilde{\mathbb{T}}))$, $x(\hat{\mathbb{T}}) = (x_{ij}(\hat{\mathbb{T}}))$, the associated primal basic solutions. Let (s, l) be the dropping arc from $\tilde{\mathbb{T}}$, which is replaced by the entering arc (g, h) in the pivot step that led to $\hat{\mathbb{T}}$. Let $\mathbb{C}$ be the fundamental cycle of (g, h) wrt $\tilde{\mathbb{T}}$. It contains the dropping arc (s, l). Orient $\mathbb{C}$ so that the dropping arc (s, l) is a reverse arc. Let θ be the flow amount on the arc (s, l) in $x(\tilde{\mathbb{T}})$. In $x(\hat{\mathbb{T}})$, add θ to the flow amounts on all the forward arcs of $\mathbb{C}$, and subtract θ from the flow amounts on all the reverse arcs of $\mathbb{C}$, this leads to $x(\hat{\mathbb{T}})$. In the same way, the primal basic solution corresponding to each tree obtained in the method can be computed by updating the previous basic solution along the fundamental cycle of the entering arc.

We will now show that if this method is initiated with the dual feasible spanning tree $\mathbb{T}^0$ discussed above, then it is very similar to the dual simplex algorithm in its choice of the dropping arc in each pivot step. This follows from the following results. We assume that the initial spanning tree is $\mathbb{T}^0$ constructed with $p = 1$.

1. Let $\mathbb{T}$ be a tree obtained during the method, and $(\mathbf{X}, \bar{\mathbf{X}})$ the partition of row nodes in the pivot step carried in it. Then every row node in $\bar{\mathbf{X}}$ has degree 1 or 2 in $\mathbb{T}$. The reason for this is the following. From the initial spanning tree $\mathbb{T}^0$ and the manner in which the method is executed, it is clear that R_1 is the only row node whose degree can be > 2 in any tree obtained during the method. If $\mathbb{T}$ is the first tree in any stage, $\mathbf{X}$ contains the source row node s which is R_1, as this is the only row node with degree > 2 in $\mathbb{T}$, the result holds. If $\mathbb{T}$ is not the first tree in a stage, the result follows from this and the fact that the set $\bar{\mathbf{X}}$ gets smaller as we move from one pivot step to the next in a stage.

2. Let (s, l) be the dropping arc in a pivot step carried out on a spanning tree $\mathbb{T}$ in this method. Then $x_{sl}(\mathbb{T}) \stackrel{<}{=} 0$. To see this, let $(\mathbf{X}, \bar{\mathbf{X}})$, $(\mathbf{Y}, \bar{\mathbf{Y}})$ be the row node and column node partition in the pivot step carried out on $\mathbb{T}$. $\bar{\mathbf{X}}, \bar{\mathbf{Y}}$ are the sets of row, column nodes in the component $\mathbb{T}^l$ containing column node l when (s, l) is deleted from $\mathbb{T}$. By 1, each node in $\bar{\mathbf{X}}$ has degree 1 or 2 in $\mathbb{T}$, let n_1, n_2 be the numbers of these nodes with degrees 1, 2 respectively. This implies that $\mathbb{T}^l$ has $n_1 + n_2$ row nodes, and $n_1 + 2n_2$ arcs. Since $\mathbb{T}^l$ is a tree, these facts imply that the number of column nodes in it, $|\bar{\mathbf{Y}}| = n_1 + 2n_2 + 1 - (n_1 + n_2) = n_2 + 1$.

$x(\mathbb{T})$ is obtained by substituting $x_{ij} = 0$ for all (i,j) corresponding to arcs not in $\mathbb{T}$ in the system of equality constraints in (3.1) and then solving the remaining system for the values of x_{ij} for (i,j) corresponding to arcs in $\mathbb{T}$. From this, and the fact that $|\bar{\mathbf{X}}| = n_1 + n_2$, $|\bar{\mathbf{Y}}| = 1 + n_2$, and that node l is the column node on the dropping arc (s, l), we get the following relations by summing the equality constraints in (3.1) over rows in $\mathbb{T}^l$, and columns in $\mathbb{T}^l$ separately.

$$\sum(x_{ij}(\mathbb{T}): \text{ over } (i,j) \text{ corresponding to arcs in } \mathbb{T}^l) = n_1 + n_2$$

$$\sum(x_{ij}(\mathbb{T}): \text{ over } (i,j) \text{ corresponding to arcs in } \mathbb{T}^l) = n_2 + 1 - x_{sl}(\mathbb{T})$$

So, $n_2 + 1 - x_{sl}(\mathbb{T}) = n_1 + n_2$, which implies that $x_{sl}(\mathbb{T}) = 1 - n_1 \overset{\leq}{=} 0$, since $\mathbb{T}^l$ contains the target node which has degree 1 in $\mathbb{T}$.

3. The dual objective value is nondecreasing during the method. This follows from the result in 2.

These results imply that if Signature method 1 is initiated with the spanning tree $\mathbb{T}^0$, then it can be viewed as a dual simplex method, even though it does not always choose pivots in the usual dual simplex way, since the primal basic value associated with the dropping arc may sometimes be 0 instead of being < 0.

Signature method 1 can of course be initiated with any dual feasible spanning tree in G. However, in this case the primal basic value associated with the dropping arc may not be $\overset{\leq}{=} 0$ in some pivot steps (the proof of this property given above is based on the fact that all the trees obtained under the method have at most one row node of degree > 2 which may not hold in this general case). So, the dual objective value may increase, decrease, or remain unchanged in a pivot step in this general version, and hence this general version cannot always be interpreted as a dual simplex method, even though it moves from a dual feasible tree to an adjacent dual feasible tree in each pivot step. This general version also takes no more than n stages, and the computational effort in it is bounded above by $O(n^3)$.

Exercises

3.21 Let $\mathbb{T}$ be a spanning tree in G whose row signature vector $d = (d_i)$ contains a unique entry equal to 1, which is d_p. For each $j = 1$ to n, define μ_j to be the i where i is such that (R_i, C_j) is the arc incident at C_j on the path $\mathcal{P}(R_p, C_j, \mathbb{T})$. Prove that the primal basic solution associated with $\mathbb{T}$ is the assignment $\{(\mu_j, j): j = 1 \text{ to } n\}$.

3.22 Prove that if two spanning trees of G correspond to different assignments, they cannot have the same signature vectors.

3.23 Let $\mathbb{T}$ be a spanning tree in G whose row signature vector contains a single 1 entry, and $\bar{\mathbb{T}}$ the corresponding cotree. Prove that $(\mathbb{T}, \bar{\mathbb{T}}, \emptyset)$ is a strongly feasible partition (see Section 5.5.1 for definition) for (3.1) when the root node is the unique terminal row node in $\mathbb{T}$.

3.24 Solve the assignment problem of order 5 with the following cost matrix, to find a minimum cost assignment.

$$\begin{pmatrix} 14 & 18 & 15 & 10 & 10 \\ 18 & 17 & 15 & 8 & 8 \\ 16 & 16 & 24 & 25 & 12 \\ 19 & 10 & 8 & 14 & 11 \\ 22 & 15 & 28 & 24 & 12 \end{pmatrix}$$

3.25 Consider a spanning tree $\mathbb{T}$ in G, in which all row nodes R_2 to R_n have degree $\leqq 2$, and only R_1 has degree > 2. Prove that the associated primal basic solution $(x_{ij}(\mathbb{T}))$ is primal feasible iff $x_{1j}(\mathbb{T}) \geqq 0$ for all j. (Balinski [1985])

3.26 A Worst Case Example Consider the assignment problem of order n with the cost matrix $c = (c_{ij})$ where $c_{ij} = (n-i)(j-1)$ for each i, j. When initiated with the dual feasible spanning tree $\mathbb{T}^0$ discussed above, corresponding to $p = 1$, show that Signature method 1 requires $(n-1)(n-2)/2$, or $O(n^2)$ pivot steps to solve this problem. (Balinski [1985, 1986])

3.4.2   An Inductive Signature Method

Consider an assignment problem of order n with $c = (c_{ij} : i, j = 1 \text{ to } n)$ as the cost matrix. For $r = 1$ to n, define $c^r = (c_{ij} : i, j = 1 \text{ to } r)$, and the assignment problem of order r to be the one with row nodes $R_1, \ldots, R_r$, column nodes $C_1, \ldots, C_r$, and cost matrix c^r.

Suppose we have an optimum tree $\mathbb{T}(r)$ for problem r. Let $u^r = (u_1^r, \ldots, u_r^r), v^r = (v_1^r, \ldots, v_r^r)$ be the dual basic solution associated with $\mathbb{T}(r)$. Define $\hat{u}_{r+1} = \min \{ c_{r+1,j} - v_j^r : j = 1 \text{ to } r \}$, and let q be a j which attains this minimum, break ties arbitrarily. Define $\hat{u}^{r+1} = (u_1^r, \ldots, u_r^r, \hat{u}_{r+1})$. Define $\hat{v}_{r+1} = \min \{ c_{i,r+1} - \hat{u}_i^{r+1} : i = 1 \text{ to } r+1 \}$, and let w be $r+1$ if it attains this minimum, otherwise w is any i which attains this minimum. Define $\hat{v}^{r+1} = (v_1^r, \ldots, v_r^r, \hat{v}_{r+1})$. In the bipartite network for problem $r+1$ define the spanning tree $\mathbb{T}(r+1)$ to be the one consisting of all the arcs in $\mathbb{T}(r)$ and the two arcs (R_{r+1}, C_q) and (R_w, C_{r+1}). The dual basic solution corresponding to $\mathbb{T}(r+1)$ in problem $r+1$ is $(\hat{u}^{r+1}, \hat{v}^{r+1})$. This can be verified to be dual feasible. If $w = r+1$, $\mathbb{T}(r+1)$ has exactly one terminal row node (the one which is a terminal row node in $\mathbb{T}(r)$, since R_{r+1} has degree 2 in $\mathbb{T}(r+1)$ in this case), and hence an optimum tree for problem $r+1$. If $w \neq r+1$, $\mathbb{T}(r+1)$ has exactly two terminal row nodes (these are R_{r+1} and the terminal row node in $\mathsf{T}(r)$). So, problem $r + 1$ can be solved in this case by Signature method 1 initiated with $\mathsf{T}(r+1)$, in just one stage, this takes a computational effort of at most $O(r^2)$.

This method proceeds inductively on r. It is initiated with the problem of order 2 for which the solution is trivially obtained. For $r = 2$ to $n - 1$, it solves problem $r + 1$ beginning with an optimum tree obtained for problem r. The final problem is the original problem. The overall computational effort is $\sum O(r^2) = O(n^3)$.

Suppose we are solving a sparse assignment problem using this method. After solving problem r, it may be necessary to rearrange the remaining rows (and columns in the same order) and select row $r + 1$ among them appropriately, in order to guarantee that $\mathsf{T}(r)$ can be extended to an initial dual feasible spanning tree for problem $r + 1$, using only cells in which allocations are permitted; or artificial cells can be introduced as mentioned earlier.

The worst case upper bound on the computational effort of this inductive version can be shown to be achieved on the problem with cost matrix $c = (c_{ij})$ where $c_{ij} = ij$ for all i, j. The Hungarian method and Signature method 1 initiated with T^0 also require the worst case upper bound computational effort on this problem.

3.4.3   Signature Method 2 : A Dual Simplex Method

We now discuss signature methods which are strictly dual simplex methods on all counts. In these methods, the initial dual feasible spanning tree in G is T^0 discussed above constructed with $p = 1$. Here all the trees will be treated as rooted trees with R_1 permanently designated as the root node (because in the initial tree, R_1 has degree n). These methods use both the row and column signature vectors.

An arc e in a spanning tree in G is said to be an *odd arc* if the son(e) is the column node on it, *even arc* otherwise. Also given two nodes p, q we say that p *is higher (lower) than* q if p is an ancestor (descendent) of q. Likewise, given two in-tree arcs e, e', we say that e is *higher (lower) than* e' if e is on the predecessor path of parent(e') (e' is on the predecessor path of parent(e)).

We discuss two versions of this method based on the following dropping arc choice rules.

DROPPING ARC CHOICE RULE 1   If a stage has just been completed and the present spanning tree is T, terminate if it has a unique terminal column node, since T is optimal then. Otherwise, select any odd arc (R_i, C_j) in T where C_j has degree $\stackrel{>}{=} 3$ in T, as the dropping arc in the initial pivot step for the next stage. If you are not at the beginning of a stage, the preceding pivot step must have caused the degree of the column node, say l, on the entering arc in that step to increase to 3 or more. Select the dropping arc to be the unique odd arc incident at l in T.

DROPPING ARC CHOICE RULE 2   Let T be the present dual feasible but primal infeasible spanning tree in G. Define PC(T) to be the set of all in-tree arcs that tie for the most negative primal basic value among all in-tree arcs on which the column node has degree $\geqq 3$. If you are beginning a new stage, select the dropping arc to be the highest arc in $\text{PC}(\mathbb{T})$. If you are not at the beginning of a stage, the preceding pivot step must have caused the degree of the column node, say h, of the entering arc in that step to increase to 3 or more. Choose the dropping arc to be the unique odd arc on which the column node has degree $\geqq 3$ that is highest above h in $\mathbb{T}$.

Dropping arc choice rule 2 helps to increase the sets $\mathbf{X}$, $\mathbf{Y}$ very rapidly, and thus gain efficiency. See Balinski [1986] for a proof that in both these versions, the primal basic value associated with the dropping arc is always $\leqq -1$. Hence these versions are strictly dual simplex methods. Balinski [1985, 1986] initiated the signature methods, and the Signature methods 1,2 are both due to him. The inductive signature method is due to Goldfarb [1985].

3.5 Other Methods for the Assignment Problem

There are several other methods for the assignment problem. The algorithms of Chapter 5 can be specialized to the assignment problem. For example the specialization of the algorithms of Section 5.6 leads to relaxation methods for the assignment problem (Bertsekas [1981]). The method of Section 5.8.1 specialized to the assignment problem leads to the shortest augmenting path algorithm for it.

3.6 Algorithm for Ranking Assignments in Nondecreasing Order of Cost

Consider the assignment problem of order n with $c = (c_{ij})$ as the cost matrix. Here we discuss an efficient method that actually ranks the assignments in nondecreasing order of cost starting with a minimum cost one. In each step it obtains one new assignment in the ranked sequence with a computational effort of at most $O(n^3)$, and can be continued as long as necessary, and terminated whenever a sufficient number of assignments in the ranked sequence have been obtained.

We will denote the assignment $x = (x_{ij})$ by the set $\{(i,j) : x_{ij} = 1\}$. Correspondingly, we write $(i,j) \in x, or \notin x$ to indicate that $x_{ij} = 1$ or 0 respectively. Let $a(1)$ denote a minimum cost assignment, and $a(1), a(2), \ldots, a(r), \ldots$ the ranked sequence of assignments satisfying for all $r \geqq 2$

$$a(r) = \text{ a min. cost assignment excluding } a(1), \ldots, a(r-1)$$

In the algorithm we will use subsets of assignments called *nodes*. A node is a nonempty subset of assignments a of the form

$$\begin{aligned}
\mathbf{N} &= \{(i_1, j_1), \ldots, (i_r, j_r); \overline{(m_1, p_1)}, \ldots, \overline{(m_u, p_u)}\} \\
&= \{a : (i_1, j_1) \in a, \ldots, (i_r, j_r) \in a; \qquad (3.20)\\
&\qquad (m_1, p_1) \notin a, \ldots, (m_u, p_u) \notin a\}
\end{aligned}$$

The cells $(i_1, j_1), \ldots, (i_r, j_r)$ are *specified to be contained in*, and the cells (m_1, p_1), $\ldots, (m_u, p_u)$ are *specified to be excluded from* each assignment in $\mathbf{N}$. In the definition of the node $\mathbf{N}$, $i_1, \ldots, i_r$ will all be distinct, and the same property will hold for $j_1, \ldots, j_r$. Also, in all nodes generated in the algorithm, all the specified to be excluded cells will belong to the same row of the array, i.e., $m_1 = m_2 = \ldots = m_u$ in $\mathbf{N}$. The matrix obtained by striking off rows $i_1, \ldots, i_r$ and columns $j_1, \ldots, j_r$ from c and replacing the entries in positions $(m_1, p_1), \ldots, (m_u, p_u)$ by infinity or a very large positive number, is known as *the remaining cost matrix corresponding to node* $\mathbf{N}$ and is denoted by $c_{\mathbf{N}}$. A minimum cost assignment in $\mathbf{N}$ can be found by solving the assignment problem of order $n - r$ with $c_{\mathbf{N}}$ as the cost matrix. Let $x_{\mathbf{N}}, z_{\mathbf{N}}$ denote a minimum cost assignment in $\mathbf{N}$ and its objective value.

One of the operations performed in the algorithm is that of *partitioning a node using a minimum cost assignment in it*. Let $\mathbf{N}$ be the node in (3.20) and $x_{\mathbf{N}} = \{(i_1, j_1), \ldots, (i_r, j_r), (s_1, t_1), \ldots, (s_{n-r}, t_{n-r})\}$ be an optimum assignment in it. Each of $(s_1, t_1), \ldots, (s_{n-r}, t_{n-r})$ should be distinct from $(m_1, p_1), \ldots, (m_u, p_u)$ from the definition of $\mathbf{N}$. Let

$$\begin{aligned}
\mathbf{N}_1 &= \{(i_1, j_1), \ldots, (i_r, j_r); \overline{(m_1, p_1)}, \ldots, \overline{(m_u, p_u)}, \overline{(s_1, t_1)}\} \\
\mathbf{N}_2 &= \{(i_1, j_1), \ldots, (i_r, j_r); (s_1, t_1); \overline{(m_1, p_1)}, \ldots, \overline{(m_u, p_u)}, \overline{(s_2, t_2)}\}
\end{aligned}$$

$$\vdots$$

$$\mathbf{N}_{n-r-1} = \{(i_1, j_1), \ldots, (i_r, j_r); (s_1, t_1), \ldots, (s_{n-r-2}, t_{n-r-2});$$

$$\overline{(m_1, p_1)}, \ldots, \overline{(m_u, p_u)}, \overline{(s_{n-r-1}, t_{n-r-1})}\}$$

The partitioning of $\mathbf{N}$ using $x_{\mathbf{N}}$ generates the mutually disjoint subnodes $\mathbf{N}_1, \ldots, \mathbf{N}_{n-r-1}$, and the partition itself is

$$\mathbf{N} = \{x_{\mathbf{N}}\} \cup \cup_{v=1}^{n-r-1} \mathbf{N}_v \qquad (3.21)$$

The algorithm maintains a *list* which is a set of nodes. Each node in the list is stored together with a minimum cost assignment in it and its objective value.

THE ASSIGNMENT RANKING ALGORITHM

Initial step   Find a minimum cost assignment $a(1) = \{(1, j_1), \ldots, (n, j_n)\}$, say. Let the list at this stage be $\{\overline{(1, j_1)}\}$, $\{(1, j_1); \overline{(2, j_2)}\}, \ldots, \{(1, j_1), \ldots,$

$(n-2, j_{n-2}); \overline{(n-1, j_{n-1})}\}$. A minimum cost assignment in each of these nodes is found, and it is stored together with the node in the list. Go to the next step.

General step   Suppose $a(1), \ldots, a(r)$ in the ranked sequence have already been obtained, and the list of nodes at this stage is $M_1, \ldots, M_\ell$. From the manner in which these are generated, $M_1, \ldots, M_\ell$ will be mutually disjoint, and their union will be the set of all assignments excluding $a(1), \ldots, a(r)$. Let x_{M_d} be an optimum assignment in the node M_d and z_{M_d} its objective value, for $d = 1$ to ℓ. So, the next assignment in the ranked sequence is $a(r+1) = x_{M_d}$ for a d satisfying $z_{M_d} = \min. \{z_{M_1}, \ldots, z_{M_\ell}\}$. If $a(r+1)$ is the last assignment in the ranked sequence that is required, the algorithm terminates here. Otherwise, delete M_d from the list, and partition it using x_{M_d}. Let $M_{d,1}, \ldots, M_{d,f}$ be the subnodes generated. Find a minimum cost assignment in each of them, add each of these subnodes to the list and go to the next step.

Discussion

If a predetermined number, h, of assignments in the ranked sequence are required, only the h nodes that are associated with the least objective values are stored in the list and the rest are pruned. If it is desired to obtain all the assignments whose cost is $\leqq$ some predetermined number, α, only those nodes in which the minimum objective value is $\leqq \alpha$ are stored in the list, and the rest are pruned.

As an example consider the assignment problem of order 10 with the following cost matrix.

$$
c = \begin{pmatrix}
7 & 51 & 52 & 87 & 38 & 60 & 74 & 66 & 0 & 20 \\
50 & 12 & 0 & 64 & 8 & 53 & 0 & 46 & 76 & 42 \\
27 & 77 & 0 & 18 & 22 & 48 & 44 & 13 & 0 & 57 \\
62 & 0 & 3 & 8 & 5 & 6 & 14 & 0 & 26 & 39 \\
0 & 97 & 0 & 5 & 13 & 0 & 41 & 31 & 62 & 48 \\
79 & 68 & 0 & 0 & 15 & 12 & 17 & 47 & 35 & 43 \\
76 & 99 & 48 & 27 & 34 & 0 & 0 & 0 & 28 & 0 \\
0 & 20 & 9 & 27 & 46 & 15 & 84 & 19 & 3 & 24 \\
56 & 10 & 45 & 39 & 0 & 93 & 67 & 79 & 19 & 38 \\
27 & 0 & 39 & 53 & 46 & 24 & 69 & 46 & 23 & 1
\end{pmatrix}
\qquad (3.22)
$$

The minimum cost assignment in this example is $a(1) = \{(1, 9), (2, 7), (3, 3), (4, 8), (5, 6), (6, 4), (7, 10), (8, 1), (9, 5), (10, 2)\}$, with an objective value of 0. The list at the end of the initial step consists of the following nodes. The minimum objective value in each node is also recorded.

$$
\begin{aligned}
M_1 &= \{\overline{(1,9)}\}, z_{M_1} = 10 \\
M_2 &= \{(1,9), \overline{(2,7)}\}, z_{M_2} = 14
\end{aligned}
$$

$$\mathbf{M_3} = \{(1,9),(2,7),\overline{(3,3)}\}, z_{\mathbf{M_3}} = 14$$
$$\mathbf{M_4} = \{(1,9),(2,7),(3,3),\overline{(4,8)}\}, z_{\mathbf{M_4}} = 1$$
$$\mathbf{M_5} = \{(1,9),(2,7),(3,3),(4,8),\overline{(5,6)}\}, z_{\mathbf{M_5}} = 15$$
$$\mathbf{M_6} = \{(1,9),(2,7),(3,3),(4,8),(5,6),\overline{(6,4)}\}, z_{\mathbf{M_6}} = 53$$
$$\mathbf{M_7} = \{(1,9),(2,7),(3,3),(4,8),(5,6),(6,4),\overline{(7,10)}\}, z_{\mathbf{M_7}} = 45$$
$$\mathbf{M_8} = \{(1,9),(2,7),(3,3),(4,8),(5,6),(6,4),(7,10),\overline{(8,1)}\}, z_{\mathbf{M_8}} = 47$$
$$\mathbf{M_9} = \{(1,9),(2,7),(3,3),(4,8),(5,6),(6,4),(7,10),(8,1),\overline{(9,5)}\}, z_{\mathbf{M_9}} = 56$$

Comparing the values of $z_{\mathbf{M_1}}$ to $z_{\mathbf{M_9}}$, we find that $a(2)$ is a minimum cost assignment in $\mathbf{M_4}$. It is $a(2) = \{(1, 9), (2, 7), (3, 3), (4, 2), (5, 6), (6, 4), (7, 8), (8, 1), (9, 5), (10, 10)\}$ with an objective value of 1. If it is required to find $a(3)$, then $\mathbf{M_4}$ should be partitioned using $a(2)$ and the algorithm continued.

Comment 3.3   This ranking algorithm has been taken from Murty [1968]. The same approach has been extended to rank the chains between a pair of nodes in a directed network (Section 4.7); to rank the spanning trees in an undirected network (Section 9.3); to rank the cuts in a capacitated network (Hamacher [1982], Hamacher, Picard and Queyranne [1984]); and in general to rank the solutions of any discrete optimization problem (Lawler [1972]).

3.7   Exercises

3.27 Consider the assignment ranking example given above for the assignment problem of order 10 with the cost matrix c given in (3.22). Find $a(3)$ to $a(6)$ in this example.

3.28 There are n jobs with given processing durations $t_i, i = 1$ to n; and job starting times, $s_i, i = 1$ to n. Each job must be processed without interruption on any one of the unlimited set of identical machines. Each machine can process any job, but no more than one job at a time. Formulate the problem of determining the smallest number of machines to process all the jobs, as one of finding a minimal chain decomposition of a poset. Solve the problem with the following data. (Gertsbakh and Stern [1978]).

i	1	2	3	4	5	6	7	8	9	10
t_i	30	25	10	18	65	7	9	10	3	18
s_i	4	30	50	68	7	19	8	110	150	88

3.29 An Application in Job Scheduling   There are n jobs. For $i = 1$ to n, the processing of the ith job has to start at specified time a_i and must be finished at time $b_i(> a_i)$, $t_i = b_i - a_i$ being the processing time for this job. All the jobs can be processed by a type of machine, of which several copies are available. The set-up time required for a machine to process job j after processing job i is $r_{ij} \overset{\geq}{=} 0$, where the (r_{ij}) satisfy the triangle inequality : $r_{ij} \overset{\leq}{=} r_{ip} + r_{pj}$, for all i, j, p. Define a partial order $\not\prec$ on the set of jobs by $i \not\prec j$ iff $b_i + r_{ij} \overset{\leq}{=} a_j$ (i.e., $i \not\prec j$ iff j can be processed by the same machine after it processes job i). Verify that this satisfies all the conditions for being a partial order. It is required to find the minimum number of machines needed to meet the given job schedule. Formulate this as the problem of finding a minimal chain decomposition of a poset. Solve the numerical problem with the following data: $n = 7, r_{ij} = 4$ for all $i \neq j$,

i	1	2	3	4	5	6	7
a_i	0	2	19	12	11	29	37
b_i	9	8	23	25	22	33	47

(Ford and Fulkerson [1962 of Chapter 1]).

3.30 Consider n men and n women such that each man-woman pair is either 'compatible' or 'incompatible.' If there is no way to match the men and women into n compatible couples, then prove that for some $p > 0$, there is a subset of p men who together are compatible with only r women where $r \overset{\leq}{=} p - 1$.

3.31 Consider the transportation problem with an additional constraint on the left-hand side, where n, M and (c_{ij}) are data. Prove that this problem is equivalent to that of minimizing z subject to the system of constraints on the right-hand side.

$$\text{minimize } z = \sum_{i=1}^{n} \sum_{1=j}^{n} c_{ij} x_{ij} \qquad \sum_{j=1}^{n+1} x_{ij} = 1, i = 1 \text{ to } n$$

$$\text{subject to } \sum_{j=1}^{n} x_{ij} \overset{\leq}{=} 1, i = 1 \text{ to } n \qquad \sum_{i=1}^{n+1} x_{ij} = 1, j = 1 \text{ to } n$$

$$\sum_{i=1}^{n} x_{ij} \overset{\leq}{=} 1, j = 1 \text{ to } n \qquad \sum_{j=1}^{n+1} x_{n+1,j} = M + n$$

$$\sum_{i=1}^{n} \sum_{j=1}^{n} x_{ij} \overset{\leq}{=} n - M \qquad \sum_{i=1}^{n+1} x_{i,n+1} = M + n$$

$$x_{ij} \overset{\geq}{=} 0, \text{ for all } i, j \qquad x_{ij} \overset{\geq}{=} 0$$

$$0 \overset{\leq}{=} x_{n+1,n+1} \overset{\leq}{=} n.$$

(Glover, Klingman, and Phillips [1984])

3.32 Let c be the cost matrix for an assignment problem of order n, for which $\bar{x}$ is an optimum assignment, and $(\bar{u}, \bar{v})$ an optimum dual solution. Let c' be a matrix obtained by changing the values in a single row, or a single column of c. Beginning with $\bar{x}, (\bar{u}, \bar{v})$, show that an optimum assignment with c' as the cost matrix can be obtained with a computational effort of at most $0(n^2)$. (Weintraub [1973])

3.33 Let a, b be two given vectors in $\mathbb{R}^n$. It is required to find a permutation of the vector b which brings it as close to a as possible. This is equivalent to finding a permutation P of order n which minimizes $\| a - Pb \|$. For any $p-$ norm $(1 \leqq p \leqq \infty) \| \cdot \|$, show that this problem can be transformed into an assignment problem.

3.34 Consider the cost minimizing assignment problem with the cost matrix (c_{ij}) where $c_{ij} = u_i v_j$, with $u_1 \geqq u_2 \geqq \ldots \geqq u_n > 0$ and $0 < v_1 \leqq v_2 \leqq \ldots \leqq v_n$. Prove that the unit matrix is an optimum assignment for this problem.

3.35 Consider an $m \times n$ transportation problem. Let a *block* of cells in this problem refer to a subset of cells in the array of the following forms: a subset of cells within a single row or a single column of the array, or the set of all cells in a subset of rows or a subset of columns of the array. Suppose there are additional constraints in the problem, where each constraint is either a lower bound or an upper bound on the sum of flows in all the cells in some block. Show that the overall problem can still be posed as a minimum cost network flow problem.

3.36 Consider the following school timetable problem. There are n classes, m teachers, and p time periods in which lectures could be scheduled every week. Each period is one hour long. Following data is available. Discuss a method for constructing a timetable for class-teacher meetings over the available periods each week, subject to the constraints given, so that as many of the meetings as possible are scheduled. (DeWerra [1971])

$$\alpha_i \ = \ \text{number of periods that class } i \text{ should meet per week,}$$
$$i = 1 \text{ to } n$$

$$\beta_j \ = \ \text{number of periods that teacher } j \text{ should teach per week,}$$
$$j = 1 \text{ to } n$$

$$e_{it} \ = \ \begin{cases} 1, & \text{if class } i \text{ is unavailable for lecture during period } t \\ & \text{every week (they may have other non-lecture} \\ & \text{activities scheduled for that period)}, \ i = 1 \text{ to,} n \\ 0, & \text{otherwise} \end{cases}$$

$$d_{jt} \ = \ \begin{cases} 1, & \text{if teacher } j \text{ is unavailable to lecture in period } t \\ & \text{every week, } j = 1 \text{ to } m, t = 1 \text{ to } p \\ 0, & \text{otherwise.} \end{cases}$$

3.37 Consider the assignment problem (3.1), (3.2). There may be many assignments which are optimal to this problem. Define a *second best valued assignment* in this problem to be an assignment whose objective value is strictly greater than that of an optimum assignment but has minimum cost among all such assignments. Develop a suitable modification of the partitioning routine using an optimum assignment to the problem, discussed in Section 3.6, to find a second best valued assignment with a computational effort of at most $O(n^3)$. (Matsui, Tamura, and Ikebe [1991]).

3.38 Assignment Using Choice Lists Giving numerical measures for preferences is hard, it is more natural for preferences to be expressed by choice lists without actual numerical measures. Consider a situation involving people $P_1, \ldots, P_n$ and items $x_1, \ldots, x_n$. Each person gives his/her *choice list* which is the list of the items in decreasing order of preference.

A choice list is a *linear ordering* if each item in the list is strictly preferred over those appearing later (example: x_3, x_2, x_1, x_4; here x_3 is strictly preferred over x_2, etc.) It is a *weak preference ordering* if some set of consecutive items in the list are considered equal in the individuals choice. We will enclose these within brackets (example: $x_4, (x_2, x_3), x_1$; here x_4 is strictly preferred over x_2 or x_3, x_2 and x_3 are equally preferred and either of these is strictly preferred over x_1).

(i)   If each person gives his/her choice list, but items have no choice lists for people, develop an algorithm for assigning each person a different item, so that each gets his/her highest favored item as far as possible. Apply this algorithm to the problem in which $n = 4$ and the choice lists are

$$
\begin{aligned}
P_1 &: \quad x_4, (x_2, x_3), x_1 \\
P_2 &: \quad (x_1, x_3), (x_2, x_4) \\
P_3 &: \quad x_3, x_2, x_1, x_4 \\
P_4 &: \quad (x_1, x_3, x_4), x_2.
\end{aligned}
$$

(ii) Suppose each person may not list all items in his/her choice list, or the number of people and items may not be equal. Modify the algorithm developed above to assign at most one item per person, so that as many people as possible are assigned their highest favored items as far as possible.

(iii) Consider a case with n people, n items again. Each person gives his/her choice list for items. Also, each item gives its choice list of persons. With respect to these choice lists, an assignment a of items to people is said to be a *stable assignment* if there exists no pair (P_i, x_j) without an allocation in a such that both P_i and x_j prefer each other to their partners in a. Develop an algorithm for finding a stable assignment of items to people.

(Wilson [1977], Gale and Shapley [1962])

3.39 Stable Assignment Problem A group consists of boys, $b_1, \ldots, b_n$; and girls, $g_1 \ldots, g_n$. Each person lists the persons of the other sex in the order in which he/she prefers them, this is called that person's choice list. An assignment of boys to girls, a, is said to be a stable assignment with respect to these choice lists, if there is no pair (b_i, g_j) without an allocation in a, such that both b_i and g_j prefer each other to their partners in a. Gale and Shapley [1962] proposed the following algorithm for finding a stable assignment.

"To start, let each girl propose to her favorite boy. Each boy who receives more than one proposal rejects all but his favorite from among those proposed to him. However, he does not accept her yet, but keeps her on a string to allow for the possibility that someone better may come along later.

We are now ready for the second stage. Those girls who were rejected now propose to their second choices. Each boy receiving proposals chooses his favorite from the group consisting of the new proposers and the girl on his string, it any. He rejects all the rest and again keeps the favorite in suspense.

We proceed in the same manner. Those who are rejected at the second stage propose to their next choices, and the boys again reject all but the best proposals they have had so far. As soon as the last boy gets his proposal the "courtship" is declared over, and each boy is now required to accept the girl on his string." Remember that this algorithm terminates as soon as every boy receives at least one proposal.

(i)   Prove that this algorithm terminates with a stable assignment.

(ii)   In the assignment obtained under this method, prove that at most one girl ends up with her last choice as a partner.

(iii)   Prove that this algorithm terminates after at most $n^2 - 2n + 2$ stages.

(iv)   Apply this algorithm when $n = 5$, and the choice lists are:

$$
\begin{array}{l|l}
b_1 : g_4, g_3, g_2, g_1, g_5, & g_1 : b_1, b_2, b_3, b_4, b_5 \\
b_2 : g_3, g_2, g_1, g_5, g_4, & g_2 : b_4, b_1, b_2, b_3, b_5 \\
b_3 : g_2, g_1, g_5, g_4, g_3, & g_3 : b_3, b_4, b_1, b_2, b_5 \\
b_4 : g_1, g_5, g_4, g_3, g_2, & g_4 : b_2, b_3, b_4, b_1, b_5 \\
b_5 : \text{arbitrary} & g_5 : b_1, b_2, b_3, b_4, b_5
\end{array}
$$

(v)   Verify that the algorithm described above goes through exactly $n^2 - 2n + 2$ stages before termination when the choice lists are as specified below.

$$
\begin{array}{l|l}
b_1 : g_{n-1}, g_{n-2}, \ldots, g_1, g_n, & g_1 : b_1, b_2, \ldots, b_{n-1}; b_n \\
b_2 : g_{n-2}, g_{n-3}, \ldots, g_1, g_n, g_{n-1} & g_2 : b_{n-1}, b_1, b_2, \ldots, b_{n-2}, b_n \\
\ \vdots & \\
& g_3 : b_{n-2}, b_{n-1}, b_1, \ldots, b_{n-3}; b_n \\
b_{n-1} : g_1, g_n, g_{n-1}, \ldots, g_2 & \ \vdots \\
b_n : \quad \text{arbitrary} & g_{n-1} : b_2, b_3, \ldots, b_{n-1}, b_1; b_n \\
& g_n : b_1, b_2, \ldots, b_{n-1}, b_n
\end{array}
$$

(vi) If the method described above goes through the upper bound of $n^2 - 2n + 2$ stages before termination, prove that one girl must get her last choice and the other girls must get their second to last choices in the assignment obtained. Also, in this case prove that the boy who received the last proposal must be the last choice of all the girls. Also in this case prove that each boy, except possibly the last one to receive a proposal, must get his first choice.

(vii) Show that the upper bound on the number of proposals made in the above algorithm before termination is $n + (n-1)^2 = n^2 - n + 1$, and that this is attained in the problem with the following data, even though the method does not go through the upper bound on the number of stages in this problem.

$$n = 4 \text{ and the choice lists are given below}$$

$$
\begin{array}{l|l}
b_1 : g_3, g_2, g_1, g_4 & g_1 : b_1, b_3, b_2, b_4 \\
b_2 : g_1, g_4, g_2, g_3 & g_2 : b_2, b_1, b_3, b_4 \\
b_3 : g_2, g_1, g_3, g_4, & g_3 : b_2, b_3, b_1, b_4 \\
b_4 : \text{arbitrary} & g_4 : b_1, b_3, b_2, b_4
\end{array}
$$

(viii) In the same manner consider the following choice lists.

$$
\begin{array}{l|l}
b_1 : g_3, g_4, g_5, \ldots, g_1, g_2, & g_1 : b_{n-1}, b_1, b_2, b_3, \ldots, b_{n-2}, b_n \\
b_2 : g_4, g_5, g_6, \ldots, g_2, g_3, & g_2 : b_1, b_2, b_3, b_4, \ldots, b_{n-1}, b_n \\
b_3 : g_5, g_6, g_7, \ldots, g_3, g_4, & g_3 : b_2, b_3, b_4, b_5 \ldots, b_1, b_n \\
\vdots & \vdots \\
b_{n-1} : g_2, g_3, g_4, \ldots, g_n, g_1, & g_{n-1} : b_{n-2}, b_{n-1}, b_1, b_2, \ldots, b_{n-3}, b_n \\
b_n : g_1, g_2, g_3, \ldots, g_{n-1}, g_n, & g_n : b_{n-1}, b_1, b_2, b_3, \ldots, b_{n-2}, b_n
\end{array}
$$

For these lists prove the following: (1) in stage 1, all girls propose and the only girl whose proposal is rejected is g_1. (2) At every stage at most one proposal is made. (3) At stage $1 < i < n^2 - 2n + 2$, girl g_r will make her tth proposal to the boy b_s where r, s, t can be expressed in terms of i and n as

$$
r = \begin{cases} (i-1) \bmod n & \text{if } (i-1) \bmod n \neq 0 \\ n & \text{otherwise} \end{cases}
$$
$$
t = 1 + \lceil (i-1)/n \rceil, \quad \text{and}
$$

$$
s = \begin{cases} t + r - 2 & \text{for } t + r - 2 < n \\ t + r - 1 - n & \text{otherwise.} \end{cases}
$$

This proposal is accepted and a girl $g_{r'}$ is jilted by boy b_s where $r' = r + 1$, if $r \neq n$; 1 otherwise. (4) At stage $n^2 - 2n + 2$, g_1 makes her nth proposal to b_n and that proposal is accepted. (5) This takes $n^2 - 2n + 2$ stages.

(Itoga [1978], Kapur and Krishnamoorty [1985])

3.40 Consider the problem discussed in Exercise 3.39. Each person lists all the persons of the other sex in decreasing order of his/her preference. Show that every stable assignment $x = (x_{ij})$ of boys to girls is an extreme point of the set of feasible solutions of the following system and vice versa.

$$\sum_{j=1}^{n} x_{ij} = 1, i = 1 \text{ to } n$$

$$\sum_{i=1}^{n} x_{ij} = 1, j = 1 \text{ to } n$$

$$x_{pq} + \sum (x_{pj} \quad : \quad \text{over girls } j \text{ preferred over girl } q \text{ by boy } p) +$$

$$\sum (x_{iq} \quad : \quad \text{over boys } i \text{ preferred over boy } p \text{ by girl } q) \geqq 1, p, q = 1 \text{ to } n$$

$$x_{ij} \geqq 0, \text{ for all } i, j.$$

(Vande Vate [1989])

3.41 The following is a minimum cost network flow model for a 3-period production planning problem with inventory and backorder bounds. k_1, k_2, k_3 are production capacities; and d_1, d_2, d_3 are the demands in the three periods. s_1, s_2 are respectively the inventory limits from periods 1 to 2, and 2 to 3; b_1, b_2 are the backorder bounds in periods 1 and 2. These data provide the capacities for arcs in the following network, all lower bounds are 0. The cost data is not shown. Transform this problem into an equivalent uncapacitated transportation problem. (Evans [1985]).

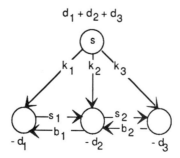

Figure 3.11

3.42 Allocation of Candidates to Jobs   There are m jobs, and n candidates. For each candidate, we are given a nonempty subset of jobs (called *candidate's jobset*) to which that candidate could be allocated. Each candidate must be allocated to exactly one job in his/her job-set, but each job can be allotted any number of candidates. Let $r_1 \ldots, r_m$ be the number of candidates allocated to various jobs in an allocation. For the ith job, we are given a monotonic strictly decreasing cost function $f_i(r_i), i = 1$ to m. Arrange the costs $f_1(r_1), \ldots, f_m(r_m)$ in non-increasing order. The resulting vector is called the *ranked cost vector* associated with the allocation. It is required to find an allocation which corresponds to a lexico minimum of the ranked cost vector subject to the conditions discussed above.

As a numerical example consider $m = 3, n = 5$, and the job-sets of candidates 1 to 5 to be $\{1\}, \{1,2\}, \{1,3\}, \{2,3\}, \{3\}$ respectively. Let the cost functions $f_i(r_i)$ for jobs $i = 1$ to 3 be $\frac{1}{r_1}, \frac{2}{r_2}, \frac{5}{r_3}$ respectively. Consider the two allocations listed below

Alloc.	Job allocated to cand. j					r_1	r_2	r_3	$f_1(r_1)$	$f_2(r_2)$	$f_3(r_3)$
	$j = 1$	2	3	4	5						
a_1	1	2	3	3	3	1	1	3	1	2	$\frac{5}{3}$
a_2	1	1	1	2	3	3	1	1	$\frac{1}{3}$	2	5

The ranked cost vectors associated with the allocations a_1, a_2 are $(2, \frac{5}{3}, 1), (5, 2, \frac{1}{3})$ respectively, and hence by the lexico minimum criterion a_1 is better than a_2. In fact in this numerical example it can be shown that a_1 is an optimum allocation for the problem.

(i) Model this problem using a bipartite network. Show that each allocation of candidates to jobs corresponds to a subnetwork of this bipartite network.

(ii) Given an allocation a, define an alternating path wrt it, to be a path beginning and terminating with job nodes, with successive edges having a common node and alternately belonging/not belonging to a.

Given a feasible allocation a and an alternating path $\mathcal{P}$ wrt it, define the operation of rematching a using $\mathcal{P}$ to be that of obtaining an allocation a^1 by (A) including in a allocations corresponding to edges in $\mathcal{P}$ which are not already in a, and (B) deleting all allocations common to a and $\mathcal{P}$. Show that the resulting allocation a^1 will be feasible.

(iii) Define an alternating path $\mathcal{P}$ wrt a feasible allocation a, to be an improving alternating path wrt a, if rematching a using $\mathcal{P}$ leads to an allocation a^1 whose ranked cost vector is lexico smaller than that of a.

Prove that a feasible allocation corresponds to a lexico minimum ranked cost vector iff there exists no improving alternating path wrt it.

(iv) Develop an algorithm which finds an optimum allocation by stepwise improvement of a feasible allocation.

(v)   Consider the following instance that arises in the army. $m = 2$, the jobs are driving and cooking. $n = 3600$ candidates who are reservists. Of the 3600; 900 can only drive, 2100 can only cook, and the remaining 600 can do both.

The army needs 1200 man months of driving time and 1800 man months of cooking time annually. Let r_1, r_2 be the number of candidates allocated to driving, cooking respectively. Then $f_1(r_1) = 1200/r_1$, this is the months of army service per annum for drivers. Likewise $f_2(r_2) = 1800/r_2$ is the months of army service for cooks. Beginning with the feasible allocation with $r_1 = 900, r_2 = 2700$, obtain an optimum allocation.

(Cramer and Pollatschek [1979])

3.43 A problem of interest in core management of pressurized water nuclear reactors is to find an optimal allocation of the given fuel assemblies which may be differing on their burn-up states, to particular locations in the reactor, such that each full assembly is assigned to a location and vice versa, under a constraint on the power distribution form factor. This gives rise to an assignment problem subject to one additional constraint, of the following form

$$\text{minimize } z = \sum_{i=1}^{n} \sum_{j=1}^{n} c_{ij} \; x_{ij}$$

$$\text{subject to } \sum_{j=1}^{n} x_{ij} \; = 1, i = 1 \text{ to } n$$

$$\sum_{i=1}^{n} x_{ij} \; = 1, j = 1 \text{ to } n$$

$$x_{ij} \; \geqq 0, \text{ for all } i, j \qquad (3.23)$$

$$\sum_{i=1}^{n} \sum_{j=1}^{n} d_{ij} \; x_{ij} \; \leqq b$$

$$\text{and } \quad x_{ij} \quad \text{integer for all } i, j$$

where, without any loss of generality, we can assume that $c_{ij}, d_{ij} \geqq 0$ for all i, j and $b > 0$. (3.23) is an integer program. Show that it is an NP-hard problem. Develop a practically efficient method for solving it, based on Lagrangian relaxation and assignment ranking. Apply this method to solve the numerical problem with the following data: $n = 4, b = 26$.

$$(c_{ij}) = \begin{pmatrix} 1 & 5 & 7 & 4 \\ 9 & 7 & 9 & 9 \\ 5 & 5 & 11 & 5 \\ 8 & 7 & 8 & 5 \end{pmatrix}, (d_{ij}) = \begin{pmatrix} 9 & 6 & 4 & 8 \\ 7 & 5 & 9 & 6 \\ 5 & 8 & 7 & 11 \\ 6 & 3 & 2 & 10 \end{pmatrix}$$

(Aggarwal [1985], Gupta and Sharma [1981])

3.44 In some transportation models, the costs of different shipments are borne by different individuals, and the sum of all the costs is not a good objective to minimize. In these models, a better objective is to minimize the maximum cost incurred by any single individual, called the bottleneck objective function. This leads to:

$$\text{minimize } (\max \{c_{ij} \quad : \quad (i,j) \text{ such that } x_{ij} > 0\})$$

$$\text{subject to } \sum_{j=1}^{n} x_{ij} \leqq a_i, i = 1 \text{ to } m \tag{3.24}$$

$$\sum_{i=1}^{n} x_{ij} \geqq b_j, j = 1 \text{ to } n$$

$$x_{ij} \geqq 0, \text{ for all } i, j$$

(i) Develop an efficient algorithm for this problem. Apply this algorithm on the numerical problem with the following data.

$$c_{ij}$$

$j =$	1	2	3	4	5	6	7	a_i
$i = 1$	10	6	4	8	8	10	0	27
2	4	2	2	2	8	1	0	26
3	14	12	9	4	9	3	0	26
4	4	7	6	9	1	4	0	27
b_j	19	17	17	15	10	8	20	

(ii) A commonly encountered constraint in distribution problems is the requirement that the entire demand of each customer must be supplied from a single source or supplier. Consider the problem (3.24) with such an additional constraint. Develop a practical approach for solving this combined problem. Apply this approach on the numerical problem with the data given above.

(Nagalhout and Thompson [1984])

3.45 Classroom Allocation to Courses   A big university has a total of L classrooms of varying capacities (the capacity of a room is the number of students it can accommodate) spread over various buildings. In a term the university is planning to offer a total of N courses. Each course meets for a total of 2, 3, or 4 hours per week (this is known as the *number of credit hours for the course*) and this may all be in a single session on one day of the week, or split into several sessions (each of the sessions are either one, or one and a half, or three hours in length) over several days of the week. For example a 3 credit hour course may meet in one session of three hours say from 7-10 PM on one day; or in three hourly sessions say from 2 to 3 PM on three different days of the week; etc. For each course the number of credit hours, and the sessions in which they will meet (i.e. on which days the course meets and the beginning and ending time of the session on each day) has already been determined and all this information is given. The expected enrollment in each course is available, and using this and other information, the university has compiled a subset $\mathbf{X}_i$ of classrooms in which they would prefer to hold course i, $i = 1$ to N.

The constraints in allotting classrooms to courses are the following: At any point of time there can be only one course allotted to a classroom; also, if a course consists of several sessions during a week, all the sessions must meet in the same classroom. It is required to allot classrooms to courses from their preferred subset, in such a way that the number of courses for which allocations are made is maximized (i.e., the number of courses for which you are unable to allot a classroom from its preferred subset, should be as low as possible). Model this problem.

3.46 Bottleneck Assignment Problem with Node Weights   Let $a_1, \ldots,$ a_n ; $b_1, \ldots, b_n$ be weights associated with the rows; and columns of the $n \times n$ assignment array corresponding to a bipartite network G. For each $i = 1$ to n, let $\mathbf{S}_i = \{j : (i; j) \text{ is an edge in G}\}$, and correspondingly for each $j = 1$ to n let $\mathbf{P}_j = \{i : j \in \mathbf{S}_i\}$. Consider the following bottleneck assignment problem: find $x = (x_{ij})$ to

$$
\text{minimize (maximum } \{(a_i + b_j) \quad x_{ij} \quad : i, j = 1 \text{ to } n\})
$$

$$
\text{subject to } \sum_{j \in \mathbf{S}_i} x_{ij} = 1, i = 1 \text{ to } m
$$

$$
\sum_{i \in \mathbf{P}_j} x_{ij} = 1, j = 1 \text{ to } n \qquad (3.25)
$$

$$
x_{ij} = 0 \text{ or } 1, \text{ for all } i, j
$$

$$
x_{ij} = 0, \text{ for } j \notin \mathbf{S}_i
$$

(i)   Show that this is a special case of the bottleneck assignment problem discussed in Section 3.1.5 in which $c_{ij} = a_i + b_j$ for $j \in \mathbf{S}_i, \infty$ for $j \notin \mathbf{S}_i$.

(ii) Consider the special case of (3.25) in which all $\mathbf{S}_i$ and $\mathbf{P}_j$ are $\{1, \ldots, n\}$. In this case order the $a_i s$ in nonincreasing order, and $b_j s$ in nondecreasing order. Suppose these orders are $a_{i_1} \geq a_{i_2} \geq \ldots \geq a_{i_n}$ and $b_{j_1} \leq b_{j_2} \leq \ldots \leq b_{j_n}$. In this case, show that the assignment $\{(i_1, j_1), (i_2, j_2), \ldots, (i_n, j_n)\}$ is an optimum solution of (3.25).

(iii) Consider cell (i, j) of the $n \times n$ assignment array *admissible* if $j \in \mathbf{S}_i$, *inadmissible* otherwise. Develop a dual algorithm for (3.25) which starts with an infeasible assignment (i.e., one containing some allocations in inadmissible cells), while maintaining optimality conditions, and tries to reduce the infeasibility in each iteration.

(iv) Let $\bar{x} = (\bar{x}_{ij})$ be a feasible assignment to (3.25), and $\mathbf{M} = \{(i; j); \bar{x}_{ij} = 1\}$ be the corresponding matching in G. The cell (r, s) is said to be a bottleneck cell (corresponding to a bottleneck edge in G) with respect to $\bar{x}$ if $\bar{x}_{rs} = 1$ and $a_r + b_s = \max\{a_i + b_j; (i,j) \text{ such that } \bar{x}_{ij} = 1\}$.

A decreasing alternating path from r to s is G with respect to M or $\bar{x}$ is a path from r to s satisfying the following properties: (1) it does not contain $(r; s)$, (2) edges in it are alternately in M and not in M, (3) for every edge $(i; j)$ on it which is not from M, we have $a_i + b_j < a_r + b_s$. When $(r; s)$ is a bottleneck cell with respect to $\bar{x}$ and there is no decreasing alternating path from r to s, prove that $\bar{x}$ is an optimum solution for (3.25)

If a decreasing alternating path exists from r to s, define $\hat{x} = (\hat{x}_{ij})$ by

$$
\hat{x}_{ij} = \begin{cases} 1 - \bar{x}_{ij} & \text{for all } (i,j) \text{ on the path.} \\ 0, & \text{for } (i,j) = (r,s). \\ \bar{x}_{ij} & \text{for all } (i,j) \neq (r,s) \text{ not on the path.} \end{cases}
$$

Then show that $\hat{x}$ is a better assignment for (3.25) than $\bar{x}$. Using these results develop a primal algorithm for (3.25) that starts with and maintains feasible assignments and strictly decreases the objective value in each iteration by finding a decreasing alternating path and using it as above. Discuss an initialization phase for this algorithm if an initial feasible assignment is not available.

(Lawler [1976 of Chapter 1], Carraresi and Gallo [1984])

3.47 The Bus Driver Rostering Problem Consider an m day time horizon, with n shifts in each day. Each shift on each day is to be manned by a single bus driver. w_{pj} denotes the weight of the jth shift on day $p, j = 1$ to $n, p = 1$ to m, this may be the time duration or some other measure of the workload for that shift. There are n drivers. This problem is concerned with the assignment of drivers to shifts over the days of the horizon so that each driver receives an even balance of each type of shift. It can be formulated as a bottleneck problem, to minimize the maximum total weight of the shifts assigned to a driver.

Represent the jth shift of pth day by a node numbered (p, j). This node has weight w_{pj}. Let $\mathbf{S}_p(j)$ be the set of shifts in day $p+1$ that can be assigned to a driver who has been assigned shift j in day p by union rules or other work constraints. Draw an arc from node (p, j) to each node $(p + 1, i)$ for $i \in \mathbf{S}_p(j)$. Let $\mathbf{P}_p(i)$ be the set of shifts in day p which could have been assigned in day p to a driver to whom shift i has been assigned in day $p + 1$. The results is a layered acyclic network of the form shown in Figure 3.12.

A feasible work assignment to a single worker corresponds to a chain from a node in the first layer to a node in the mth layer in this network. Its total workload being given by the sum of the weights of the nodes on the chain. The problem of finding work assignments to all the n workers, such that the maximum workload is minimized, can be formulated as the problem of finding, in the network in Figure 3.12, n node disjoint chains from layer 1 to layer m so as to minimize the longest among these chains. Give a mathematical formulation of this problem using 0-1 variables.

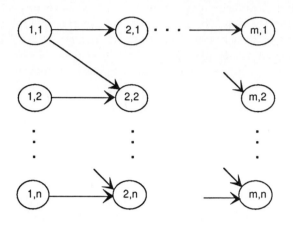

Figure 3.12

Prove that a feasible solution to this problem exists iff for each $p = 1$ to $m - 1$, the constraints

$$\sum_{j \in \mathbf{S}_p(i)} x_{ij}^p = 1, i = 1 \text{ to } n$$
$$\sum_{i \in \mathbf{P}_p(j)} x_{ij}^p = 1, j = 1 \text{ to } n$$
$$x_{ij}^p = 0 \text{ or } 1, \text{ for all } i, j$$

are feasible. If $m = 2$, show that this problem reduces to a bottleneck assignment problem of the form discussed in Exercise 3.46.

When $m > 2$, develop a heuristic algorithm for this problem based on the bottleneck assignment problem discussed in Exercise 3.46. (Carraresi and Gallo [1984])

3.48 A Vehicle Scheduling Application   This application is concerned with the optimal assignment of vehicles to time-tabled trips so that each trip is carried out by one vehicle subject to some constraints. A trip is defined by a quadruple (τ_i, l_i, o_i, d_i) where

τ_i = scheduled start time of the ith trip.

l_i = duration or length of the ith trip.

o_i = the origin or the start terminal for the ith trip.

d_i = the destination terminal for the ith trip.

Suppose the time-table consists of n trips, denoted by $\zeta_1, \ldots, \zeta_n$. In addition to these regular trips, deadheading trips are allowed between terminals. δ_{ij} denotes the duration of a deadheading trip from d_i to $o_j, i, j = 1$ to $n, i \neq j$. The ordered pair of trips (ζ_i, ζ_j) is said to be a *compatible pair* if $\tau_i + l_i + \delta_{ij} + \varepsilon \overset{\leq}{=} \tau_j$, where $\varepsilon \overset{\geq}{=} 0$ is a tolerance parameter to absorb possible delays. If (ζ_i, ζ_j) is compatible, it is clearly feasible to have trips ζ_i, ζ_j operated in sequence by the same vehicle. A *vehicle duty* is a sequence $\vartheta = (\zeta_{i_1}, \zeta_{i_2}, \ldots, \zeta_{i_r})$ of trips satisfying the property that every consecutive pair of trips in this sequence is compatible, all these trips can be operated by the same vehicle. A feasible vehicle schedule is a family $\{\vartheta_1, \ldots, \vartheta_g\}$ of vehicle duties such that each trip $\zeta_1, \ldots, \zeta_n$ belongs to exactly one $\vartheta_h, h = 1$ to g.

Construct a bipartite network $G = (\mathbf{S}, \mathbf{T}, \mathcal{A})$ where $\mathbf{S} = \{s_1, \ldots, s_n\}$, $\mathbf{T} = \{t_1, \ldots, t_n\}$ and $\mathcal{A} = \{(s_i, t_j);$ for all i, j such that (ζ_i, ζ_j) is a compatible pair$\}$.

(i)   A vehicle duty $\vartheta = (\zeta_{i_1}, \zeta_{i_2}, \ldots, \zeta_{i_r})$ can be represented in G by the set of arcs $\{(s_{i_1}, t_{i_2}), (s_{i_2}, t_{i_3}), \ldots, (s_{i_{r-1}}, t_{i_r})\}$ which can be verified to be a matching in G. A vehicle duty containing only one trip $\{\zeta_j\}$ can be represented by leaving both nodes s_j, t_j as exposed nodes in G. Using this, show that there is a one-to-one correspondence between vehicle schedules and matchings in G.

Consider the example in which there are 4 terminals a,b,c,d, and 5 trips with the following data: $\varepsilon = 0$

Trip	τ_i	l_i (minutes)	o_i	d_i
ζ_1	7:10 a.m.	20	a	b
ζ_2	7:20 a.m.	20	c	d
ζ_3	7:40 a.m.	25	b	a
ζ_4	8:00 a.m.	30	d	c
ζ_5	8:35 a.m.	30	c	d

$$(\delta_{ij}) = \begin{array}{c|cccc} \text{\textit{from}} \quad \text{\textit{to}} & a & b & c & d \\ \hline a & 0 & 15 & 20 & 20 \\ b & 15 & 0 & 25 & 10 \\ c & 20 & 25 & 0 & 15 \\ d & 20 & 10 & 15 & 0 \end{array}$$

Construct the network G for this example and obtain the vehicle schedule corresponding to the matching $\{(s_1, t_3), (s_2, t_4), (s_4, t_5)\}$ in it.

(ii)   Define the set of arcs $\mathcal{A}^* = \{(s_i, t_j); (s_i, t_j) \notin \mathcal{A}$ defined above, and either $i = j$ or $\tau_i \overset{\geq}{=} \tau_j + l_j + \delta_{ji}\}$. Let G′ denote the network obtained by augmenting G with the additional arcs $\mathcal{A}^*$, and introducing a unit exogenous supply at each $s_i, i = 1$ to n (these are now source nodes), and a unit demand at each $t_j, j = 1$ to n, and lower bound of 0 and capacity of 1 on each arc in $\mathcal{A} \cup \mathcal{A}^*$.

If $f = (f_{s_i, t_j})$ is an integer feasible flow vector in G′, it defines an assignment or perfect matching in G′. If $f_{s_i, t_j} = 1$ then the arc (s_i, t_j) is in the perfect matching; and in addition if $(s_i, t_j) \in \mathcal{A}$ make the trips ζ_i and ζ_j belong in that order to the same vehicle duty, and if $(s_i, t_j) \in \mathcal{A}^*$ then make ζ_i and ζ_j to be the last and first trips respectively of a vehicle duty (in this case they may or may not belong to the same vehicle duty). Also, if $f_{s_i, t_i} = 1$ then make ζ_i as the only trip of a vehicle duty. Under this convention, show that every feasible integer flow vector in G' (or a perfect matching in G') corresponds to a feasible vehicle schedule where the number of vehicles used is equal to the number of units of flow (or arcs in the perfect matching) on arcs from the set $\mathcal{A}^*$ and vice versa. Draw the network G' for the example given in (i) and illustrate this point.

(iii) Now consider the case in which all the vehicles are housed in one depot (common in small size transit companies, or those in which the service area is partitioned into zones with each zone assigned to one single depot). Assume that there are no constraints other than compatibility and that all vehicles are of the same type.

Formulate the problem of finding a vehicle schedule to minimize the fleet size as an assignment problem.

If it is required to find a vehicle schedule that minimizes the operational costs (these include deadheading travel costs, and cost of any idle time between the end of a trip and the starting of the next), formulate the problem of finding it as an assignment problem.

Also discuss how one can find a vehicle schedule that minimizes a combination of the above two costs, or one that minimizes the operational cost subject to the constraint that the fleet size is the minimum.

(iv) Now consider the case in which there are multiple depots, each with a given capacity for the number of vehicles it can house. Formulate this multiple depot problem as a 0-1 integer program and discuss heuristic approaches to obtain reasonable solutions for it.

(Carraresi and Gallo [1984])

3.49 Let G=(S,T; $\mathcal{A}$) be a bipartite network with $|S| = |T|$. E $\subset \mathcal{A}$ is a specified subset of edges in G. It is required to determine whether there exists a perfect matching in G containing at most r edges from **E**. Formulate this as a minimum cost network flow problem.

3.50 Let G=($\mathcal{N}, \mathcal{A}$) be a directed acyclic network. Define a *chain cover* for G to be a node disjoint union of chains in G (degenerate chains consisting of a single node by itself being admissible) which contains all the nodes in $\mathcal{N}$. The size of a chain cover for G is defined to be the number of chains in it. The *chain covering number* for G is the size of a minimum size chain cover.

(i)   Show that a chain cover for G is of minimum size iff it contains the maximum number of arcs among all chain covers.

(ii) Based on the result in (i), a greedy approach for finding a minimum size chain cover in G, consists of successively determining longest chains. Show that this greedy approach fails to give a minimum size chain cover in the following network (minimum size is 4, greedy approach gives a cover of size 5).

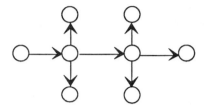

Figure 3.13

(iii) Obtain a new network $G' = (\mathcal{N}', \mathcal{A}')$ from G by the following procedure. Replace each node i in $\mathcal{N}$ by a pair of nodes i' and i''. Each are incident into i in G becomes an arc incident into i', and each arc incident out of i becomes an arc incident out of i'', in G'. So, every node in G' has either zero in-degree or zero out-degree. Show that G' is bipartite.

(iv) Show that the transformation in (iii) converts the arcs in any collection of node-disjoint chains in G into a matching in G'. Using the fact that G is acyclic, show that every matchings in G' corresponds to a unique set of arcs of some node-disjoint collection of chains in G.

From these results show that a minimum size chain cover in G can be obtained by finding a maximum cardinality matching in G'.

(Boesch and Gimpel [1977])

3.51 Let $G=(\mathcal{N}, \mathcal{A}, \breve{s}, \breve{t})$ be an acyclic network with $\breve{s}$ as the source node with in-degree zero, and $\breve{t}$ as the sink node with out-degree zero. An $\breve{s} - \breve{t}$ chain cover for the nodes of G is a collection of chains from $\breve{s}$ to $\breve{t}$ in G such that each node $i \in \mathcal{N}$ is contained on at least one chain in the collection, its size is defined to be the number of chains in it. A minimum $\breve{s} - \breve{t}$ chain cover for nodes in G is one of the smallest possible size, let α denote its size.

A pair of distinct nodes $i, j \in \mathcal{N}$ are said to be incomparable in G, if there exists no chain from i to j or from j to i in G. A subset $\mathbf{X} \subset \mathbf{N}$ is an incomparable node set in G if every pair of distinct nodes in $\mathbf{X}$ is incomparable. Let β be the cardinality of a maximum cardinality incomparable node set in G.

Transform G into G' by splitting each node $i \in \mathcal{N} \backslash \{\breve{s}, \breve{t}\}$ into two nodes i', i'' and adding the arc (i', i''), all arcs incident into (out of) i in $\mathcal{A}$ will be incident into i' (out of i'') in G'. Define lower bounds for flows on arcs in G' to be 1 for all arcs of the form (i', i'') and 0 for all other arcs, and capacities to be $+\infty$ on all the arcs.

Find a minimum value feasible flow vector $\bar{f}$ in G', let its value be $\bar{v}$. Then prove that $\bar{v} = \alpha = \beta$. Use this to develop an algorithm for finding a minimum $\check{s} - \check{t}$ chain cover for nodes in G.

This provides a method based on the minimum value flow problem for finding α and β. An alternate method for the same based on maximum cardinality matching is discussed in Section 3.1.4. (Ntafos and Hakimi [1979])

3.52 Let G $=(\mathcal{N}, \mathcal{A}, \check{s}, \check{t})$ be an acyclic network with $\check{s} =$ the source node with in-degree zero, and $\check{t} =$ the sink node with out-degree zero. An $\check{s} - \check{t}$ chain cover for arcs in G is a collection of chains from $\check{s}$ to $\check{t}$ in G that contains all the arcs in $\mathcal{A}$, its size is the number of chains in it. Let ϑ be the size of a minimum size $\check{s} - \check{t}$ chain cover for arcs in G.

Two arcs e_1, e_2 in G are said to be incomparable if there is no chain from $\check{s}$ to $\check{t}$ containing both of them. An incomparable arc set in G is a set of arcs every pair of which are incomparable. Let δ be the cardinality of a maximum cardinality incomparable arc set in G.

Prove that $\vartheta = \delta$. As in Exercise 3.51, show that a minimum size $\check{s} - \check{t}$ chain cover for arcs in G can be found by the minimum value flow method by applying it directly on G by placing a lower bound of one unit on each arc in G.

Let $\{e_1, \ldots, e_{|\mathcal{A}|}\}$ be the arcs in $\mathcal{A}$, and $\mathcal{N}'' = \{1, \ldots, |\mathcal{A}|\}$ with $j \in \mathcal{N}''$ corresponding to the arc e_j in $\mathcal{A}$. Define $\mathcal{A}'' = \{(i, j) : i, j \in \mathcal{N}'', $ and there is a chain in G from the head node of e_i to the tail node of $e_j\}$. Let G'' $= (\mathcal{N}'', \mathcal{A}'')$.

Show that G'' is also acyclic. Show that a minimum size $\check{s} - \check{t}$ chain cover for arcs in G, and a maximum cardinality incomparable arc set in G, can both be found by applying the maximum matching method discussed in Section 3.1.1 to the network G''. (Ntafos and Hakimi [1979])

3.53 Let G $= (\mathcal{N}, \mathcal{A})$ be a directed connected network. Define a chain cover for the nodes in G to be a set of chains (not necessarily simple or even elementary) such that each node in $\mathcal{N}$ is contained on at least one chain in the set. Let $P_N(\text{G})$ denote the cardinality of a minimum cardinality chain cover for nodes in G.

A node $i \in \mathcal{N}$ reaches (is reached from) a node j if there is a chain from i to j (from j to i) in G. A set $\mathbf{X} \subset \mathcal{N}$ of nodes in an incomparable node set if there is no chain in G between any pair of distinct nodes in $\mathbf{X}$. Let $I_N(\text{G})$ denote the cardinality of a maximum cardinality incomparable node set in G.

Since any two vertices in a strongly connected component are mutually reachable, it follows that there always exists a chain containing all the nodes in a strongly connected component. Hence, for the purpose of finding a minimum cardinality chain cover for nodes, each strongly connected component in G can be replaced by a single new vertex.

Find all the strongly connected components (SCCs) in G; suppose there are r of them. Construct a new network G' with node set $\{1, \ldots, r\}$, in which node i corresponds to the ith SCC in G. Introduce an arc (i, j) in G' if there exists a chain from some node in the ith SCC in G, to some node in the jth SCC. Show that

G' is acyclic. Prove that $P_N(G) = P_N(G')$ and that $I_N(G) = I_N(G')$. Using this prove that $P_N(G) = I_N(G)$. Discuss an efficient algorithm for finding a minimum cardinality chain cover for nodes, and a maximum cardinality incomparable node set in G, using the results in Exercise 3.52. (Ntafos and Hakimi [1979])

3.54 Let $G = (\mathcal{N}, \mathcal{A})$ be a connected directed network. Define chain covers for arcs in G, reachability among arcs, incomparable arc sets, the same way it was done for nodes in Exercise 3.53. Let $P_A(G)$ be the cardinality of a minimum cardinality chain cover for arcs, and let $I_A(G)$ be the cardinality of a maximum cardinality incomparable arc set in G. Prove that $P_A(G) = I_A(G)$, and discuss an efficient method for finding a minimum cardinality chain cover for arcs, and a maximum cardinality incomparable arc set in G, using the results in Exercises 3.52, 3.53. (Ntafos and Hakimi [1979])

Comment 3.4 The Hungarian method for the assignment problem, a combinatorial procedure, was developed by H. Kuhn [1955]. Since the main ideas underlying the method came from Egerváry's proof of the König-Egerváry theorem concerning bipartite graphs, he called it the "Hungarian method." The primal-dual setting has made it possible to solve this problem by combinatorial methods through a sequence of maximum value flow problems. The method generalizes directly to the transportation problem, and to minimum cost flow problems in capacitated networks which are not necessarily bipartite (see Chapter 5). The extension of the primal-dual approach to solve minimum cost matching problems in nonbipartite networks required a new technique, namely the characterization of the convex hull of matching incidence vectors by a system of linear inequalities. This was done by J. Edmonds [1965 of Chapter 10], the resulting combinatorial algorithms for matching problems are discussed in Chapter 10.

The $O(n^3)$ implementation of the Hungarian method is due to Lawler [1976 of Chapter 1].

When the method of Section 5.8.1 for minimum cost flows is specialized to the assignment problem, it leads to a shortest augmenting path method for it. This has been discussed by Derigs and Metz [1986], Jonker and Volgenant [1987], and Tomizawa [1972]. Related methods based on successive shortest chains using a relaxation approach have been discussed by Dinic and Kronrod [1969], Engquist [1982], and Hung and Rom [1980]. The primal algorithm of Balinski and Gomory [1964] maintains a feasible assignment and reaches an optimum assignment by augmenting flows along negative cost cycles, this method is again based on shortest chain computations.

Several variants of the primal simplex method for solving the assignment problem have been discussed in the literature. Barr, Glover, and Klingman [1977] discuss the finite version based on using strongly feasible partitions (see Chapter 5), and report good computational performance of it. Akgul [1987], Hung [1983], and Roohy-Laleh [1981] discuss polynomial time variants of the primal simplex algorithm for the assignment problem.

Signature methods for the assignment problem have been proposed by Balinski[1985, 1986]. Goldfarb [1985] developed the inductive signature method. Recently signature methods have been generalized to solve transportation problems by Paparrizos [1990].

Computational studies on these different algorithms for the assignment problem by different groups at different times have rated the Hungarian method, relaxation methods, shortest augmenting path methods, as having produced excellent performance. Carpento, Martello, and Toth [1988], and Burkard and Derigs [1980 of Chapter 1] present FORTRAN implementations of assignment algorithms for dense and sparse cases.

Metric and Maybee [1973] present a FORTRAN implementation of the assignment ranking algorithm.

The scaling technique for modifying the primal-dual method into a polynomially bounded algorithm, by applying it on a sequence of better and better approximations of the original problem, was introduced by Edmonds and Karp [1972 of Chapter 2]. However, this technique is mainly of theoretical interest, as the performance of the resulting algorithm in computational tests does not compare favorably with that of other minimum cost flow algorithms discussed in Chapter 5.

Algorithms for stable assignments are discussed in Gale and Shapley [1962], Irving, Leather, and Gusfield [1987], Kapur and Krishnamoorty [1981], and Wilson [1977]. These algorithms are not discussed in the text, but are presented among various exercises given above.

3.8   References

V. AGGARWAL, 1985, "A Lagrangian Relaxation Method for the Constrained Assignment Problem," *COR*, 12, no. 1(97-106).

M. AKGUL, 1987, "A Genuinely Polynomial Primal Simplex Algorithm for the Assignment Problem," Working paper IEOR 87-07, Bilikent University, Ankara, Turkey.

M. AKGUL, 1988, "A Sequential Dual Simplex Algorithm for the Linear Assignment Problem," *OR Letters*, 7(155-158).

M. AKGUL and O. EKIN, June 1990, "A Dual Feasible Forest Algorithm for the Linear Assignment Problem," Research Report IEOR-9011, Bilkent University, Ankara, Turkey.

M. L. BALINSKI, 1985, "Signature Methods for the Assignment Problem," *OR*, 33(527-537).

M. L. BALINSKI, March 1986, "A Competitive (Dual)Simplex Method for the Assignment Problem," *MP*, 34, no. 2(125-141).

M. L. BALINSKI and R. E. GOMORY, 1964, "A Primal Method for the Assignment and Transportation Problems," *MS*, 10(578-593).

R. S. BARR, F. GLOVER and D. KLINGMAN, 1977, "The Alternating Path Basis Algorithm for Assignment Problems" *MP*, 13(1-13).

D. BERTSEKAS, 1981, "A New Algorithm for the Assignment Problem," *MP*, 21(152-171).

F. T. BOESCH and J. F. GIMPEL, April 1977, "Covering the Points of a Digraph With Point-Disjoint Paths and its Application to Code Optimization," *JACM*, 24, no. 2(192-198).

G. CARPANETO, S. MARTELLO and P. TOTH, 1988, "Algorithms and Codes for the Assignment Problem," (193-224) in B. Simeone, et. al. (Eds.), *FORTRAN Codes for Network Optimization, AOR, 13.*

G. CARPANETO and P. TOTH, 1987, "Primal-Dual Algorithms for the Assignment Problem," *DAM*, 18(137-153).

P. CARRARESI and G. GALLO, 1984, "Network Models for Vehicle and Crew Scheduling," *EJOR*, 16(139-151).

P. CARRARESI and G. GALLO, 1984, "A Multi-Level Bottleneck Assignment Approach to the Bus Drivers' Rostering Problem," *EJOR*, 16(163-173).

J. CRAMER and M. POLLATSCHEK, May 1979, "Candidate Job Allocation Problem With a Lexicographic Objective," *MS*, 25, no. 5(466-473).

U. DERIGS and A. METZ, 1986, "An In-Core/Out-of-Core Method for Solving Large Scale Assignment Problems," *Zeitschrift für Operations Research*, 30, no. 5(A181-A195).

U. DERIGS and U. ZIMMERMANN, 1978, "An Augmenting Path Method for Solving Linear Bottleneck Assignment Problems," *Computing*, 19(285-295).

D. DEWARRA, March 1971, "Construction of School Timetables by Flow Methods," *INFOR*, 9, no. 1(12-22).

E. A. DINIC and M. A. KRONROD, 1969, "An Algorithm for Solution of the Assignment Problem," *Soviet Math. Doklady*, 10(1324-1326).

J. EGERVÁRY, 1931, "Matrixok Kombinatorikus Tulajdonságairól," *Mat. és Fiz. Lapok*, 38(16-28).

M. ENGQUIST, Nov. 1982, "A Successive Shortest Path Algorithm for the Assignment Problem," *INFOR*, 20, no, 4(370-384).

J. R. EVANS, March 1985, "On Equivalent Transportation Models for Production Planning Problems," *IIE Transactions*, 17, no, 1(102-104).

L. R. FORD and D. R. FULKERSON, 1957, "A Primal-Dual Algorithm for the Capacitated Hitchcock Problem," *NRLQ*, 4(47-54).

D. R. FULKERSON, 1956, "Note On Dilworth's Decomposition Theorem for Partially Ordered Sets," *Proc. Amer. Math. Soc.*, 7(701-702).

D. GALE and L. S. SHAPLEY, 1962, "College Admissions and the Stability of Marriage," *American Math. Monthly*, 69(9-15).

I. GERTSBAKH and H. I. STERN, Jan. - Feb. 1978, "Minimal Resources for Fixed and Variable Job Schedules," *OR*, 26, no. 1(68-85).

F. GLOVER, D. KLINGMAN and N. PHILLIPS, 1984, "An Equivalent Subproblem Relaxation for Improving the Solution of a Class of Transportation Scheduling Problems," *EJOR*, 17(123-124).

D. GOLDFARB, 1985, "Efficient Dual Simplex Methods for the Assignment Problem," *MP*, 33(187-203).

O. GROSS, March 1959, "The Bottleneck Assignment Problem," P-1630, The RAND Corp.

A. GUPTA and J. SHARMA, 1981, "Tree Search Method for Optimal Core Management of Pressurized Water Reactors," *COR*, 8(263-266).

H. HAMACHER, Nov. 1982, "An $O(kn^4)$ Algorithm for finding the k Best Cuts in a Network," *OR Letters*, 1, no. 5(186-189).

H. HAMACHER, J. C. PICARD and M. QUEYRANNE, March 1984, "On Finding the k Best

Cuts in a Network," *OR Letters*, 2, no. 6(303-305).

H. HAMACHER, J. C. PICARD and M. QUEYRANNE, 1984, "Ranking the Cuts and Cut-Sets of a Network," *Annals of Discrete Applied Mathematics*, 19(183-200).

M. S. HUNG, 1983, "A Polynomial Simplex Method for the Assignment Problem," *OR*, 31(595-600).

M. S. HUNG and W. D. ROM, 1980, "Solving the Assignment Problem by Relaxation," *OR*, 28(969-982).

R. W. IRVING, P. LEATHER and D. GUSFIELD, July 1987, "An Efficient Algorithm for the 'Optimal' Stable Marriage," *JACM*, 34, n0. 3(532-543).

S. Y. ITOGA, Aug. 1978, "The Upper bound for the Stable Marriage Problem," *JORS*, 29, no. 8(811-814).

R. JONKER and T. VOLGENANT, Oct. 1986, "Improving the Hungarian Assignment Algorithm," *OR Letters*, 5, no. 4(171-175).

R. JONKER and A. VOLGENANT, 1987, "A Shortest Augmenting Path Algorithm for Dense and Sparse Linear Assignment Problems," *Computing*, 38(325-340).

D. KAPUR and M. S. KRISHNAMOORTHY, July 1981, "Worst Case Choice for the Stable Marriage Problem," *IPL*, 21, no. 1(27-30).

P. KLEINSCHMIDT, C. W. LEE, and H. SCHANNATH, Mar. 1987, "Transportation Problems Which Can be Solved by the Use of Hirsch Paths for the Dual Problems," *MP*, 37, no. 2(153-168).

D. KÖNIG, 1950, "*Theorie der Endlichen und Unendlichen Graphen*," Chelsea Publishing Co. NY.

H. W. KUHN, 1955, "The Hungarian Method for the Assignment Problem," *NRLQ*, 2(83-97).

E. L. LAWLER, Mar. 1972, "A Procedure for Computing the k Best Solutions to Discrete Optimization Problems and its Application to the Shortest Path Problem," *MS*, 18(401-405).

R. E. MACHOL and M. WIEN, April 1977, "Errata," *OR*, 25, no. 2(364).

T. MATSUI, A. TAMURA, and Y. IKEBE, March 1991, "Algorithms for finding a kth best valued assignment," Tech. report, Dept. of Industrial Administration, Science University of Tokyo, Tokyo, Japan.

L. B. METRICK and J. D. MAYBEE, 1973, "Assignment Ranking," Tech. Report 73-7, Dept. IOE, University of Michigan, Ann Arbor, MI.

K. G. MURTY, May-June 1968, "An Algorithm for Ranking All the Assignments in Order of Increasing Cost," *OR*, 16, no. 3(682-687).

K. G. MURTY and C. WITZGALL, 1977, "Dual Simplex Method for the Uncapacitated Transportation Problem Using Tree Labelings" Tech. Report 77-9, Dept. IOE, University of Michigan, Ann Arbor, MI.

R. V. NAGELHOUT and G. L. THOMPSON, 1984, "A Study of the Bottleneck Single Source Transportation Problem," *COR*, 11, no. 1(25-36).

S. C. NTAFOS and S. L. HAKIMI, Sept. 1979, "On Path Cover Problems in Digraphs and Applications to Program Testing," *IEEE Transactions on Software Engineering*, SE-5, no. 5(520-529).

K. PAPARRIZOS, 1990, "Generalization of a Signature Method to Transportation Problems," Math. Democritus University of Thrace, Xanthi, Greece.

E. ROOHY-LALEH, 1981, "Improvements to the Theoretical Efficiency of the Network Simplex Method," Ph. D thesis, Carleton University.

R. SILVER, 1960, "An Algorithm for the Assignment Problem," *CACM*, 3(603-606).

N. TOMIZAWA, 1972, "On Some Techniques Useful for Solution of Transportation Network Problems," *Networks*, 1(179-194).

J. H. VANDE VATE, 1989, "Linear Programming Brings Marital Bliss," *OR Letters*, 8(147-153).

A. WEINTRAUB, 1973, "The Shortest and the k-Shortest Routes as Assignment Problems," *Networks*, 3(61-73).

L. B. WILSON, 1977, "Assignment Using Choice Lists," *ORQ*, 28, no. 3(569-578).

Chapter 4

Shortest Chain Algorithms

The problem of finding a shortest chain from a specified origin (or source) node to a specified destination (or sink) node is a fundamental problem that appears in many applications. It generates essential information in transportation, routing, and communications applications; here the lengths of the arcs are the geographical distances (or the travel times) associated with them. This problem appears as a subproblem in each step in solving multicommodity flow problems using the revised simplex method on an arc chain formulation (Section 5.11); in this application the lengths of the arcs are numbers derived by the algorithm, and they depend on the dual solution in that step. In critical path methods (CPM, see Chapter 7) arcs in the acyclic project network correspond to jobs, the length of each is the negative of the time duration required to complete the corresponding job, and the shortest chain problem appears in the task of scheduling the various jobs over time. There are also applications in equipment replacement (Garcia-Diaz and Liebman [1980]), preparation of travel time and distance charts (Golden and Magnanti [1978]), vehicle routing and scheduling (Christofides, Mingozzia and Toth [1981], Golden and Magnanti [1978]), capacity planning (Doulbiez and Ras [1975]), design and/or expansion of transportation and communication networks (Golden and Magnanti [1978], Schwartz and Stern [1980]), and in several other areas. The shortest chain problem also appears as a subproblem in algorithms for the solution of other network flow and discrete optimization problems. In this chapter we discuss various algorithms for finding shortest chains in a network.

What we call a "chain" is called a "path" in some books. So, in the literature, the shortest chain problem is often referred to as the "shortest path problem" or the "shortest route problem."

Suppose there is an edge $(i; j)$ in the original network of length $c_{i;j}$. If $c_{i;j} \geq 0$ we replace $(i; j)$ by the pair of arcs $(i, j), (j, i)$ both of length $c_{i;j}$. On the other hand, if $c_{i;j} < 0$, this operation creates a negative length circuit $(i, j), (j, i)$ in the transformed network (notice that this is not a circuit in the original network; see Section 1.2.2). As explained later, negative length circuits in the network create

244

difficulties for solving shortest chain problems in it; hence we cannot replace a negative length edge $(i;j)$ by the pair of arcs $(i,j), (j,i)$ of the same length. When the original network contains negative length edges, but otherwise no negative length circuits, the problem of finding shortest simple chains in it requires special techniques based on matching algorithms (Tobin [1975]) but we will not discuss them because of their limited applicability.

Hence, we assume that if there are any edges in the original network, their lengths are $\geqq 0$, and we will replace each of them by a pair of arcs of the same length as described above. So in the sequel we assume that the network on which our shortest chain problem is defined is a directed connected network $G = (\mathcal{N}, \mathcal{A}, c)$ with $c = (c_{ij})$ as the vector of arc lengths, and $|\mathcal{N}| = n$.

The length of any chain is the sum of the lengths of arcs in it. If there are parallel arcs with tail node i and head node j in G, only the shortest among them will be used on the shortest chains. Call this the arc (i, j) and eliminate all other arcs parallel to it from further consideration. We continue to denote the set of arcs by $\mathcal{A}$. Let $m = |\mathcal{A}|$.

We will discuss the problem of finding the shortest chains from a fixed origin to all the other nodes in G. Given a chain containing some circuits, all the arcs on these circuits can be eliminated, leaving a simple chain; this elimination will not increase the length of the chain if the lengths of these circuits are $\geqq 0$. Hence, if a chain exists from the origin to j, and G contains no negative length circuits, there exists a shortest chain from the origin to j which is simple. All shortest chains found by algorithms discussed in this chapter will be simple chains.

If shortest chains from the origin to all the other nodes in G exist, by the result in Exercise 4.1 later on, there exists an outtree rooted at the origin such that the unique path in it between the origin and any other node is a shortest chain to that node. Such an outtree is called a *shortest chain tree* (also called a *shortest path tree* in the literature). The shortest chain tree stored using the predecessor labels, provides a convenient data structure for storing the shortest chains out of the origin.

Another problem that we will consider is that of finding shortest chains between every pair of nodes in G.

4.1 LP Formulation of the Unconstrained Shortest Chain Problem as a Minimum Cost Flow Problem

To find a shortest chain from 1 to n in $G = (\mathcal{N}, \mathcal{A}, c)$ is equivalent to the following problem of sending one unit of flow from 1 to n at minimum cost with $0, \infty, c$ as the lower bound, capacity and unit cost coefficient vectors for arc flows.

Minimize $\sum (c_{ij} f_{ij} : \text{over } (i, j) \in \mathcal{A})$

Subject to $\quad - f(i, \mathcal{N}) + f(\mathcal{N}, i) \; = \; \begin{cases} -1 & \text{if } i = 1 \\ 0 & \text{if } i \neq 1, n \\ 1 & \text{if } i = n \end{cases}$ $\qquad$ (4.1)

$f_{ij} \geqq 0$, for all $(i, j) \in \mathcal{A}$

The chain may traverse through an arc any number of times since this is an unconstrained shortest chain problem. The flow on each arc represents the number of times the chain traverses it (if the capacity of an arc is made $= 1$, then we get the flow formulation of the constrained shortest chain problem in which you cannot pass through that arc more than once). (4.1) has a redundant constraint that can be eliminated; we choose it to be the one corresponding to the origin, node 1. The dual of the resulting problem involves dual variables which are node prices π_i. It is

Maximize $\pi_n \; - \; \pi_1$

Subject to $\quad \pi_j \; - \; \pi_i \leqq c_{ij}, \quad \text{for each } (i, j) \in \mathcal{A}$ $\qquad$ (4.2)

$\pi_1 \; = \; 0.$

Every basic vector for (4.1) consists of flow variables associated with arcs in a spanning tree in G. A feasible basic vector for it corresponds to a spanning tree $\mathbb{T}$ such that the unique path in $\mathbb{T}$ from 1 to n is a chain and vice versa. In the corresponding BFS the flow amounts are equal to 1 on all the arcs of that chain, and 0 on all the other arcs. Hence, in every BFS $f = (f_{ij})$ of (4.1), all f_{ij} are 0 or 1, and $\{(i, j) : f_{ij} = 1\}$ is the set of arcs on a simple chain from 1 to n. Thus each extreme point of the set of feasible solutions of (4.1), i.e., each BFS of (4.1), is the incidence vector of the set of arcs on a simple chain from 1 to n in G and vice versa.

Suppose there is a negative length circuit $\overrightarrow{\mathbb{C}}$ in G. Summing the dual constraints in (4.2) corresponding to arcs (i, j) on $\overrightarrow{\mathbb{C}}$ leads to the inconsistent inequality $0 \leqq$ a negative number. So, the dual problem (4.2) is infeasible, and by the duality theorem of LP, if (4.1) is feasible, the objective value in it is unbounded below. The solution of (4.1) which makes the objective function unbounded below corresponds to a flow that traverses around $\overrightarrow{\mathbb{C}}$ an infinite number of times. Hence, when the network contains a negative length circuit, unconstrained shortest chain algorithms will detect one such circuit and terminate with the unboundedness conclusion.

As an example consider the network in Figure 4.1. Arc lengths are entered on the arcs. A feasible flow vector $f(\mu)$, for $\mu \geqq 0$, is marked with the flow on each arc entered inside a box if it is $\neq 0$. As $\mu \to \infty$ the objective value of $f(\mu) \to -\infty$.

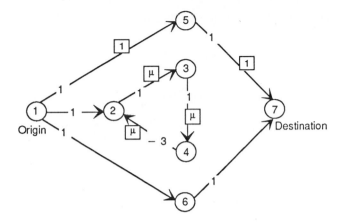

Figure 4.1 Network with a negative length circuit containing nodes 2, 3, 4.

Even when G contains a negative length circuit, the constrained shortest chain problem of finding a minimum length simple chain from 1 to n is a well-posed problem. (4.1) has a finite number of BFSs, each associated with a simple chain from 1 to n in G. Hence this problem is equivalent to that of finding a minimum cost BFS among the finite number of BFSs of the unbounded LP (4.1). In Section 2.2 we have seen that the maximum capacity cut problem is also a problem of this type. In general these are NP-hard combinatorial optimization problems. At the moment the only known algorithms for tackling them are enumerative algorithms whose computational requirements grow exponentially with the size of the problem in the worst case.

The network in Figure 4.1 has the special property that any node which lies on a negative length circuit does not lie on any chain from the origin to the destination. Under this property it is possible to solve the shortest simple chain problem efficiently; see Exercise 4.3.

In the shortest simple chain problem, unboundedness in the presence of negative length circuits occurs due to traversal around such a circuit an infinite number of times. From this, one may be tempted to think that the simple technique of imposing an upper bound of 1 on all the variables in (4.1) will take care of this difficulty. This capacitated problem has an integer optimum solution as long as there is a chain from 1 to n in G; suppose it is $\bar{f} = (\bar{f}_{ij})$. The set of arcs (i, j) on which $\bar{f}_{ij} = 1$ defines a chain from 1 to n which we denote by C_0. C_0 may not be simple; it may include some negative length circuits. Eliminate all circuits from C_0; suppose this leaves the simple chain C_1 from 1 to n. Unfortunately, C_1 may not be a shortest simple chain from 1 to n in G as illustrated by the network in Figure 4.2. Arc lengths are entered on the arcs. The shortest simple chain from 1 to 7 in this network is 1, (1, 6), 6, (6, 7), 7, of length 20. Problem (4.1) with capacities of 1 on all the arcs in this network has the optimum flow with flow amount of 1 on all the thick arcs, and 0 on others. The set of thick arcs includes the negative length circuit containing nodes 3,5,4. Elimination of this circuit from the set of thick arcs leaves the simple chain 1, (1, 2), 2, (2, 3), 3, (3, 7), 7, which is not a shortest simple chain from 1 to 7 in this network.

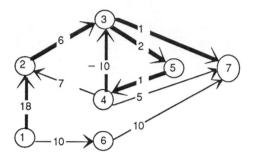

Figure 4.2

Let $\hat{f} = (\hat{f}_{ij})$ be a basic feasible flow vector for (4.1) and $\hat{\pi} = (\hat{\pi}_i)$, a node price vector in G. Let $\hat{\mathcal{C}}$ be the simple chain from 1 to n in G corresponding to $\hat{f}$; i.e., $\hat{f}$ is the incidence vector of $\hat{\mathcal{C}}$. By the duality theory of LP $\hat{f}, \hat{\pi}$ are respectively optimal to (4.1), (4.2), iff

$$\hat{\pi}_n = \text{opt. obj. value in (4.2)} = \text{opt. obj. value in (4.1)}$$
$$= \text{length of a shortest chain from } 1 \text{ to } n \text{ in G} \tag{4.3}$$

$$\hat{\pi}_j - \hat{\pi}_i = c_{ij} \text{ whenever } \hat{f}_{ij} = 1, \text{ i.e., for all } (i,j) \text{ on } \hat{\mathcal{C}} \tag{4.4}$$

$$\hat{\pi}_j - \hat{\pi}_i \leq c_{ij} \text{ for all } (i,j) \in \mathcal{A} \tag{4.5}$$

(4.5) are just the conditions for the node price vector $\hat{\pi}$ to be dual feasible. Conversely, given a dual feasible node price vector $\hat{\pi}$, if there exists a chain from 1 to n consisting only of arcs (i,j) for which $\hat{\pi}_j - \hat{\pi}_i = c_{ij}$, then that chain is a shortest chain.

Assume that G contains no negative length circuits. One can derive the above facts directly without appealing to the duality theory of LP. For this, define $\hat{\pi}_1 = 0$, and for $i \in \mathcal{N}, i \neq 1$, let $\hat{\pi}_i$ be ∞ if there exists no chain from 1 to i in G, or the length of a shortest chain from 1 to i otherwise. For $(i,j) \in \mathcal{A}$, if $\hat{\pi}_i < \infty$, $\hat{\pi}_i + c_{ij}$ is the length of a chain from 1 to j (it consists of a shortest chain from 1 to i and then the arc (i,j)), and hence $\hat{\pi}_i + c_{ij} \geq \hat{\pi}_j$, since $\hat{\pi}_j$ is the length of a shortest chain from 1 to j. So, $\hat{\pi} = (\hat{\pi}_i)$ must satisfy (4.5). Also the shortest chains themselves consist of arcs satisfying these conditions as equations. Also, given π feasible to (4.2) or (4.5), any chain from 1 to q consisting of arcs for which (4.5) holds as an equation must have length π_q, and hence is a shortest chain from 1 to q.

From this discussion it is clear that the value of the dual variable π_i in an optimum dual solution can be interpreted as being the length of a shortest chain from 1 to i in G. And, the vector $\hat{\pi} = (\hat{\pi}_i)$ of shortest chain lengths out of node 1 satisfy

$$\hat{\pi}_1 = 0$$

$$\hat{\pi}_j = \min.\{\hat{\pi}_i + c_{ij} : i \in B_j\}, \text{ for all } j \in \mathcal{N}, j \neq 1$$

(4.6)

(4.6) are known as the *Bellman-Ford equations* for the shortest chain problem with node 1 as the origin node. They represent necessary conditions for the vector of shortest chain lengths out of node 1. Conversely, if the vector $\pi = (\pi_i)$ satisfies the Bellman-Ford equations, and there is a chain from 1 to i of length π_i, then that chain is a shortest chain from 1 to i.

(4.1) is the LP formulation for finding a shortest chain from node 1 to the specified destination, node n. But its dual (4.2) involves variables which represent the lengths of the shortest chains from 1 to all the other nodes in G. Therefore it usually happens that to find a shortest chain from 1 to a specified destination, one may have to find the shortest chains from 1 to all the other nodes in the network.

Exercises

4.1 G $= (\mathcal{N}, \mathcal{A}, c)$ is a directed network with the origin at node 1. The simple chain $\mathcal{C}_i$ is a shortest chain from 1 to i in G. For every node p on $\mathcal{C}_i$ prove that $\mathcal{C}_i(p)$, the portion of $\mathcal{C}_i$ between 1 and p, is a shortest chain from 1 to p in G.

Let the simple chain $\mathcal{C}_j$ be a shortest chain from 1 to $j \neq i$, and $p \neq 1$ a common node on $\mathcal{C}_i$ and $\mathcal{C}_j$. Prove that $\mathcal{C}_i(p)$ and $\mathcal{C}_j(p)$ must have the same length. See Figure 4.3. Using this prove that there must exist simple chains from 1 to i, j respectively which are shortest chains to i, j satisfying: either these chains have no common node other than the origin 1; or if they have any common node $p \neq 1$, the portions of both these chains between 1 and p are identical.

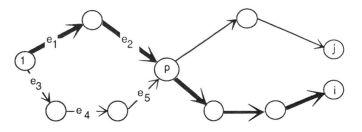

Figure 4.3 $\mathcal{C}_i$ is thick, and $\mathcal{C}_j$ is normal. $\mathcal{C}_i(p)$ consists of arcs e_1, e_2; $\mathcal{C}_j(p)$ of e_3, e_4, e_5.

Using this prove that if shortest chains exist in G from 1 to all the other nodes, then there exists a spanning outtree, T, rooted at 1, on which all the paths from the root are shortest chains. Hence show that in this case there exists an optimum feasible basic vector for (4.1), corresponding to an outtree rooted at 1 which is a shortest chain tree.

4.2 $G = (\mathcal{N}, \mathcal{A}, c)$ is a connected directed network with c as the vector of arc lengths, and $|\mathcal{N}| = n$. Show that the following minimum cost flow problem is a formulation for the problem of finding shortest chains from the origin, 1, to all the other nodes in G.

$$\text{Minimize} \quad \sum (c_{ij} f_{ij} : \text{ over } (i,j) \ \in \ \mathcal{A})$$

$$\text{Subject to} \quad f(i, \mathcal{N}) - f(\mathcal{N}, i) \ = \ \begin{cases} n-1 & , \text{ if } i = 1 \\ -1 & , \text{ if } i \neq 1 \end{cases} \qquad (4.7)$$

$$f_{ij} \overset{\geq}{=} 0, \quad \text{for all } (i,j) \ \in \ \mathcal{A}$$

Prove that every BFS for (4.7) is nondegenerate. Show that the dual of (4.7) is

$$\text{Maximize} \quad \sum \ (\pi_i - \pi_1)$$

$$\text{Subject to} \quad c_{ij} - \ (\pi_j - \pi_i) \ \overset{\geq}{=} 0, \quad \text{for all } (i,j) \in \mathcal{A}$$

Show that we can set $\pi_1 = 0$ in the dual problem, and in that case, π_i in an optimum dual solution gives the shortest chain distance from the origin to node i.

Suppose a shortest chain tree T_1 for this problem is given. Now consider the same problem with a different origin node, say node 2. For the flow version of the new problem, T_1 is dual feasible but primal infeasible. Develop a special adaptation of the dual simplex algorithm of LP to compute a shortest chain tree rooted at node 2, starting with T_1 (Florian, Nguyen, and Pallatino [1981]).

4.3 $G = (\mathcal{N}, \mathcal{A}, c)$ is a directed network with c as the arc length vector. Suppose it is required to find a shortest simple chain from 1 to n in G that has no common nodes with any negative length circuit. Develop a procedure for this problem using any of the shortest chain algorithms discussed in later sections. If there are no common nodes between any chain from 1 to n and any negative length circuit in G, show that this procedure finds a shortest chain from 1 to n in G.

4.4 Suppose there are some negative length circuits in $G = (\mathcal{N}, \mathcal{A}, c)$. Solve (4.1) with the capacity constraint of 1 on all the arcs, let $\bar{f} = (\bar{f}_{ij})$ be an optimum integer flow vector for this problem. If $\bar{f}$ is the incidence vector of a simple chain from 1 to n, prove that it is a shortest simple chain from 1 to n in G.

4.2   Label Setting Methods for Shortest Chains from a Specified Origin

Let $G = (\mathcal{N}, \mathcal{A}, c)$ be the connected directed network with node 1 as the origin for this problem. Here we discuss an algorithm due to E. W. Dijkstra [1959]. The arc length vector c must be $\geqq 0$ for using this algorithm. It builds the shortest chain tree in a connected fashion one arc per step. Once an arc is included in this tree, it is not removed later on. The arc that is included in each step is the best (i.e., shortest) among all the arcs joining an in-tree node to an out-of-tree node at that stage. So, the arc selection in each step is a *greedy selection* and the algorithm itself is known as a *greedy algorithm*. We now provide the main result on which the algorithm is based.

THEOREM 4.1 *Let* $G = (\mathcal{N}, \mathcal{A}, c)$ *be a directed connected network with arc length vector* $c \geqq 0$. *Let* $\mathbb{T}$ *be an outtree rooted at the origin, node 1, spanning the nodes in the set* $\mathbf{X} \subset \mathcal{N}$, *which is a shortest chain tree. For each* $i \in \mathbf{X}$, *let* π_i *be the length of a shortest chain from 1 to i in* G. *Let* $p \in \mathbf{X}$, $q \in \bar{\mathbf{X}} = \mathcal{N} \setminus \mathbf{X}$ *satisfy:* $(p, q) \in \mathcal{A}$, *and*

$$\pi_p + c_{pq} = \text{ min. } \{\pi_i + c_{ij} : \text{ over } i \in \mathbf{X}, j \in \bar{\mathbf{X}}, (i, j) \in \mathcal{A}\} \qquad (4.8)$$

Then adding the arc (p, q) *to* $\mathbb{T}$ *and defining* $\pi_q = \pi_p + c_{pq}$, *extends* $\mathbb{T}$ *into* $\mathbb{T}'$ *which is a shortest chain tree spanning the nodes in* $\mathbf{X} \cup \{q\}$.

Proof Let $\mathcal{C}_0$ denote the chain obtained by adding the arc (p, q) at the end of the chain from 1 to p in $\mathbb{T}$. Let $\mathcal{C}$ be any chain from 1 to q. Let (r, s) be the first arc in the cut $(\mathbf{X}, \bar{\mathbf{X}})$ as you travel along $\mathcal{C}$. Let $\mathcal{C}_1$ be the portion of $\mathcal{C}$ from 1 to r, and $\mathcal{C}_2$ the portion from s to q. Then $\mathcal{C}$ consists of the chains $\mathcal{C}_1$ and $\mathcal{C}_2$ and the arc (r, s). So, the length of $\mathcal{C} = c_{rs} +$ length of $\mathcal{C}_1 +$ length of $\mathcal{C}_2 \geqq c_{rs} +$ length of $\mathcal{C}_1$ (since length of $\mathcal{C}_2 \geqq 0$ as $c \geqq 0$) $\geqq c_{rs} + \pi_r$ (since π_r is the shortest chain length from 1 to r) $\geqq \pi_p + c_{pq}$ (by (4.8)) = length of $\mathcal{C}_0$. Hence $\mathcal{C}_0$ is a shortest chain from 1 to q. ∎

Beginning with the trivial tree consisting of the single node 1, the result in Theorem 4. 1 can be used repeatedly to obtain in $(n - 1)$ steps the shortest chain tree in G, adding one arc per step. At the stage when the tree spans r nodes, selecting the next arc to add to the tree can take up to $O(r(n - r))$ additions and comparisons, since there may be that many arcs in the cut at that stage. If carried out directly, the computational effort in the entire algorithm may therefore be $\sum_r O(r(n - r)) = O(n^3)$. Dijkstra observed that repeated minimization over the cut leads to repeated examination of arcs, and pointed out that cut examination can be replaced by setting and updating node labels called *temporary labels* on out-of-tree nodes. By this, each arc is examined precisely once in the algorithm, and the overall computational effort will be at most $O(n^2)$.

Nodes may be in three possible states: *permanently labeled, temporary labeled,* or *unlabeled* in this algorithm. The label in node i is always of the form $(P(i), d_i)$, where $P(i)$ is the predecessor index of i, and d_i is the length of the present chain to i, which is the predecessor path of i in reverse order. A node becomes permanently labeled when it is included in the shortest chain tree, and then its label will not change subsequently. Hence, this method and all variants of it are called *label setting methods*. For each permanently labeled node, the present chain from 1 to it will be a shortest chain. X, Y, N denote the sets of permanently labeled, temporary labeled, unlabeled nodes respectively. In each step, one node is transferred from Y to X. The labels on nodes in Y are updated in each step.

THE LABEL SETTING METHOD

Step 1   Label the origin, 1, with the permanent label $(\emptyset, 0)$. Label each $j \in A_1$ with the temporary label $(1, c_{1j})$. $X = \{1\}$, $Y = A_1$, $N = \mathcal{N} \backslash (X \cup Y)$. Go to Step 2.

Step 2   If $Y = \emptyset$ go to Step 3. If $Y \neq \emptyset$, find an $i \in Y$ for which the distance index d_i in the label is the least among all the nodes in Y at this stage. Break ties for this i arbitrarily. Make the current label on i permanent and move it from Y to X. If the label on i is $(P(i), \pi_i)$, $(P(i), i)$ is the arc included in the shortest chain tree in this step.

For each $j \in Y$, let d_j be its present distance index. If $(i, j) \in A$ and $d_j > \pi_i + c_{ij}$, change the label on j to $(i, \pi_i + c_{ij})$, otherwise leave the temporary label on j unchanged.

Temporary label each $j \in N \cap A_i$ with $(i, \pi_i + c_{ij})$ and move it from N to Y.

If $X \neq \mathcal{N}$ repeat this Step 2.

Step 3   The present labels define a shortest chain tree spanning the nodes in X. Since $Y = \emptyset$, if $X \neq \mathcal{N}$, there exists no chain from 1 to any node in the set N at this stage. Terminate.

Discussion

As an example consider the network G in Figure 4.4 with arc lengths entered on the arcs. Figures 4.4 to 4.9 illustrate the various steps in the label setting method to find a shortest chain tree rooted at node 1 in this network. Permanent labels are marked in bold face, temporary labels in regular style. In-tree arcs are marked with thick lines. Figure 4.9 contains the shortest chain tree rooted at node 1.

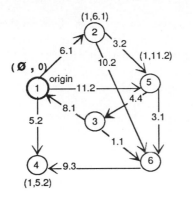
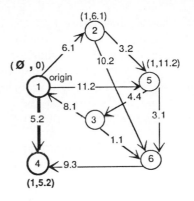

Figures 4.4, 4.5

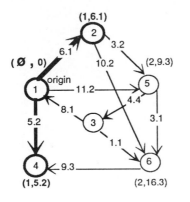
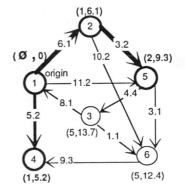

Figures 4.6, 4.7

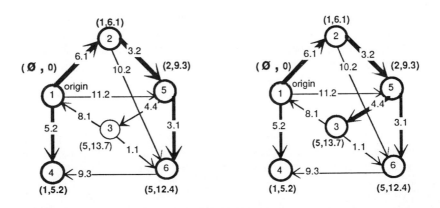

Figures 4.8, 4.9

Suppose all the c_{ij} are equal to 1. In this case the shortest chain problem is that of finding a chain from the origin to the destination consisting of the smallest number of arcs. In the shortest augmenting path method of Section 2.3.3 for the maximum value flow problem, in each step the problem of finding a shortest augmenting path is exactly a problem of this type in the residual network at that stage. It can be verified that the breadth first search method for it discussed in Section 2.3.3 is a specialization of this label setting method to it.

THEOREM 4.2 *Assuming that $c \geqq 0$, the in-tree chain from 1 to any permanently labeled node is a shortest chain.*

Proof   Assume that the following statements hold at some stage.

(i)   For each $j \in \mathbf{X}$, the present in-tree chain to j is a shortest chain.

(ii) For each $j \in \mathbf{Y}$, the present chain from 1 to j traced by the current labels (permanent or temporary) traverses only through nodes in $\mathbf{X}$ before reaching j, and has minimum length among all chains satisfying this property.

Let $i \in \mathbf{Y}$ be the temporarily labeled node whose label is made permanent next. For $j \in \mathbf{Y}$ let d_j be the present distance label on j. Then by the choice of i

$$d_i = \text{min. } \{d_j : j \in \mathbf{Y}\} \tag{4.9}$$

Suppose the present chain to i is not a shortest. Let $\mathcal{C}$ be a shortest chain from 1 to i. Then length of $\mathcal{C}$ must be $< d_i$, and by (i) it must pass through some nodes $j \in \mathbf{Y}$, $j \neq i$ before reaching i, suppose $q \in \mathbf{Y}$ is the first such node on $\mathcal{C}$. Let δ be the length of the portion of $\mathcal{C}$ from q to i, $\delta \geqq 0$ since $c \geqq 0$. The portion of $\mathcal{C}$ from 1 to q is a shortest chain to q, and all nodes on it other than q are from $\mathbf{X}$, so, by (ii) its length is d_q. So, length of $\mathcal{C} = d_q + \delta \geqq d_q \geqq d_i$ by (4.9). Since the length of $\mathcal{C} < d_i$ by the hypothesis, this is a contradiction. So, the present chain to i must be a shortest chain, thus (i) continues to hold after transferring i from $\mathbf{Y}$ to $\mathbf{X}$. By the updating process of temporary labels, (ii) also continues to hold after this transfer.

Clearly (i), (ii) hold initially. By the above arguments and induction, (i), (ii) hold after each step, and the assertion in the theorem remains valid. ∎

THEOREM 4.3 *The computational effort in this method is bounded above by $O(n^2)$ where $n = |\mathcal{N}|$.*

Proof   Each time Step 2 is executed, the choice of node i requires at most $(n-2)$ comparisons, and the updating of temporary labels requires an additional $(n-2)$ additions and comparisons at most. Step 2 is executed at most $(n-1)$ times during the algorithm. Hence the result follows. ∎

The label setting method does not work unless $c \geqq 0$. One can verify this by applying it on the network in Figure 4.10.

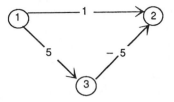

Figure 4.10

A major computational burden in this method is the repeated search among temporary labels to find the minimum. Special data structures and sorting techniques have been developed to make this search efficient; see Denardo and Fox [1979], Dial [1965], Dial, Glover, Karney, and Klingman [1979], Fredman and Tarjan [1987], Imai and Iri [1984], and Johnson [1973, 1977].

Once a point is permanently labeled in this method, the shortest chain from the origin to that point is known. Thus, if shortest chains from the origin to only a specified subset of points are desired, the method can be terminated as soon as all those points are permanently labeled. This is an important advantage of label setting methods. The label correcting methods discussed in the next section do not have this advantage; there the full shortest chain tree has to be found before any shortest chain is known with certainty.

4.3 Label Correcting Methods for Shortest Chains from a Specified Origin

Let $G = (\mathcal{N}, \mathcal{A}, c)$ be the connected directed network with node 1 as the origin for this problem. Here we discuss variants of the primal simplex method applied to the LP formulation (4.1). Every basis for (4.1) corresponds to a spanning tree in G, and a pivot step in it corresponds to exchanging an out-of-tree arc with an in-tree arc in its fundamental cycle. These methods always maintain a spanning outtree rooted at the origin, and change it in each step by changing the labels on one or more nodes. Hence these methods are known as *label correcting methods*.

For each $j \in \mathcal{N}$ such that $(1, j) \notin \mathcal{A}$, these methods usually introduce an artificial arc $(1, j)$ associated with a large length α. Taking $\alpha = 1 + n(\max.\{|c_{pq}| : (p, q) \in \mathcal{A}\})$ is sufficient. After this change, let $\mathbb{T}_0$ be the spanning outtree rooted at 1 consisting of the arcs $(1, j)$ for each $j \neq 1$. $\mathbb{T}_0$ is normally used as the initial outtree in most of these methods. If an artificial arc $(1, j)$ is contained in the shortest chain tree obtained at the termination of the method, it implies that there exists no chain from 1 to that particular node j in the original network. Thus, eliminating all the artificial arcs from the final outtree obtained in the method leaves a shortest chain tree spanning all the nodes that can be reached from 1 by a chain in the original network.

These methods work for general c, so we do not require $c \geqq 0$ in this section. Each of them will terminate after a finite number of steps with either a negative length circuit, or a shortest chain tree. These methods maintain node labels of the form $(P(i), d_i)$ on node i, for each $i \in \mathcal{N}$, where $P(i)$ is the predecessor index of i. At any stage, we denote the set of arcs $\{(P(i), i): i \neq 1\}$ by the symbol **E**. As long as the node labels represent a tree, **E** is the set of arcs in the outtree at that stage; in this case for each $i \in \mathcal{N}$, d_i in the label on i is either the length of the present chain from 1 to i, or a number $\geqq$ that length. When a negative length circuit is obtained in the method, the node labels may no longer represent a spanning outtree; at that time **E** will form two or more connected components, an outtree (not spanning) rooted at 1, and one or more disjoint negative length circuits. We will first discuss the classical primal method for the shortest chain problem to provide the basic ideas behind this class of algorithms.

THE CLASSICAL PRIMAL METHOD FOR THE SHORTEST CHAIN PROBLEM

In this method node labels always define a spanning outtree and they are of the form $(P(i), \pi_i)$ where π_i is always the length of the present chain to i. Artificial arcs are eliminated from the network once they leave the outtree.

Initialization   Begin with the spanning outtree T_0 discussed above, and the node labels that go with it.

General step   For each $i \in \mathcal{N}$ let $(P(i), \pi_i)$ be the current label on it. Look for an arc $(i, j) \in \mathcal{A}$ violating the dual constraint corresponding to it in (4.2), i.e., satisfying $\pi_j > \pi_i + c_{ij}$. If there is no such arc, the present outtree is a shortest chain tree; terminate. Otherwise, with this arc (i, j) do the following. Let $\delta = \pi_j - \pi_i - c_{ij}$. So, $\delta > 0$.

> **Ancestor checking operation**   Check whether j is an ancestor of i in the present tree. If it is, the circuit consisting of the arc (i, j) and the portion of the predecessor path of i between j and i is a negative length circuit of length $-\delta$; terminate. Otherwise continue.

> **Label correction**   Replace the in-tree arc incident into node j by (i, j); i.e., change the label on j to $(i, \pi + c_{ij})$. This has the effect of reducing the length of the chain to j by δ.

> **Correcting the distance index of descendents**   For each descendent t of j leave its predecessor index unchanged, but change its distance from the present π_t to $\pi_t - \delta$.

Go to the next step.

Discussion

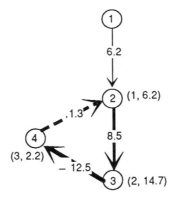

Figure 4.11

As an example consider the network in Figure 4.11. Numbers on the arcs are their lengths, and node labels are given by the side of nodes. Arc (4, 2) violates dual feasibility as $\pi_2 = 6.2 > \pi_4 + c_{42} = 2.2 + 1.3 = 3.5$. Here $\delta = 2.7$. Node 2 is on the predecessor path of node 4; this identifies the thick negative length circuit of length -2.7, and the method terminates.

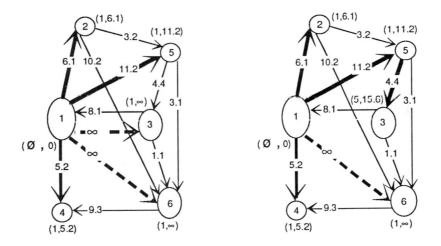

Figures 4.12, 4.13 Node 1 is the origin.

As another example consider the network in Figure 4.4 with arc lengths entered on the arcs. In Figure 4.12, we introduce the dashed artificial arcs (1, 3), (1, 6) with length ∞. The various outtrees obtained in applying this method on this network are given in Figures 4.12 to 4.15. Outtrees are marked by thick lines and in each step the node labels are entered by the side of the nodes.

Each outtree corresponds to a unique π-vector, and in each step the π_i for at least one node i strictly decreases. So, no outtree can reappear, and the method

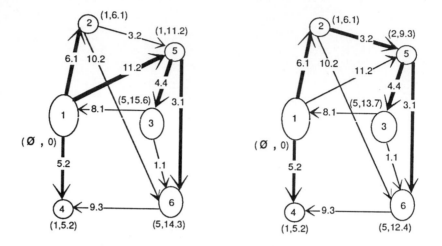

Figures 4.14, 4.15

must terminate after at most a finite number of steps. If a negative length circuit is not discovered, at termination we have an outtree in which the optimality conditions discussed earlier hold; hence it is a shortest chain tree.

The computational complexity of this method depends critically on the procedure used to look for an arc (i, j) violating dual feasibility. Under some rules for this search, the computational requirements in the method grow exponentially with the size of the problem in the worst case. There are some search procedures under which the method turns out to be nicely polynomially bounded. We will discuss some of these efficient implementations. Some of these implementations select a node i first by some rule, and then they examine successively all arcs in the forward star (or reverse star) of this node for dual feasibility. This operation is called *branching out* of node i. Candidate nodes for branching out are maintained in a list.

The ancestor checking operation, and correcting the distance index of descendents, each adds an $O(n)$ effort per step to the worst case computational complexity. For correcting the distance indices of descendents efficiently, the predecessor indices alone are not adequate; we need to use some other list structures discussed in Chapter 1 to represent the tree and we need to update them in every step. This makes each step expensive. That's why in the implementations discussed below, these operations are not carried out.

When arc (i, j) violates dual feasibility, and we do not carry out the ancestor checking operation, j may be an ancestor of i and we may not detect it. If j is an ancestor of i, and we carry out the label correction on node j, the node labels will no longer represent a spanning tree; from then on the set **E** will consist of a negative length circuit containing node j. These implementations have special rules to detect the occurrence of this possibility.

If the operation of correcting the distance index of descendents is not carried out in each step, in the label $(P(i), d_i)$ on node i, d_i may not be equal to the length of the present chain to node i; it will be a number $\geqq$ that length. However this will be detected when all the arcs on this chain are examined, and the distance indices will get corrected eventually, and if a negative length circuit is not identified, the distance indices on the nodes become the actual lengths of the chains at termination.

Elimination of the ancestor checking and correcting the distance labels of descendents operations in each step makes it possible for the following implementations to have a better worst case computational complexity bound. Clearly, the ancestor checking operation should be eliminated if it is known that no negative length circuits exist in G. Otherwise, computational tests reveal that the extra work in carrying out these operations is worthwhile as it avoids scanning nodes just for correcting their labels, and definitely pays off in better average performance, particularly as the density of the network increases.

The Bellman-Ford-Moore Label Correcting Algorithm

We will now discuss an implementation known as the *Bellman-Ford-Moore label correcting algorithm* or the *BFM method* which can be interpreted as an iterative approach for solving the Bellman-Ford equations (4.6). It goes through several iterations; the distance indices in the rth iteration are denoted by $\pi_i^r, i \in \mathcal{N}, r = 1, 2, \ldots$. In each iteration it corrects the distance labels on all the nodes using recursive equations of the form (4.10) given below. So, the method is one of successive approximations for (4.6), obtaining the $(r+1)$th order approximation from the rth.

THE BFM METHOD

Initialization   Begin with the spanning outtree T_0 discussed above, and the node labels that go with it. Set iteration count equal to 1.

General iteration $r + 1$ for $r \geqq 1$ Let $(P(i), \pi_i^r)$ be the label on node $i \in \mathcal{N}$ at the end of the previous step. For each $j \in \mathcal{N}$ compute

$$\pi_j^{r+1} = \text{min.} \{\pi_j^r; \pi_i^r + c_{ij} : \text{over } i \in \mathsf{B}_j\} \tag{4.10}$$

If $\pi_j^{r+1} = \pi_j^r$ for all $j \in \mathcal{N}$; terminate, the present labels define a shortest chain tree rooted at node 1. If this condition does not hold, for each j such that $\pi_j^{r+1} < \pi_j^r$ do the following: Let $i = u$ be an index attaining the minimum in (4.10). Change the label on j to (u, π_j^{r+1}). If $r + 1 \leqq n - 1$ go to the next iteration. Otherwise (i.e., if $r + 1 = n$) a negative length circuit has been detected; identify it in the present set E of arcs, and terminate.

Discussion

For each $j \in \mathcal{N}$ and $r \geqq 0$, π_j^r here represents the length of a shortest chain from 1 to j among those that contain no more than r arcs. This interpretation is clearly valid for $r = 1$. Set up an induction hypothesis that this statement is true for some r. Now we will show that this implies that it must be true for $r + 1$ too. A shortest chain from 1 to j of no more than $r + 1$ arcs either contains $\leqq r$ arcs (in which case its length is π_j^r by the induction hypothesis) or contains $r + 1$ arcs and has some final arc incident into j, say (u, j). The portion of this chain from 1 to u must be a shortest chain to u containing r arcs, so the length of this chain to j is $\pi_u^r + c_{uj}$. Hence minimizing $\pi_u^r + c_{uj}$ over $u \in B_j$, which gives π_j^{r+1}, yields the length of the shortest chain to j with $r + 1$ or less arcs in the latter case. So, the statement must hold for $r + 1$ too, and by induction it holds for all r.

If there is no negative length circuit in G, with the introduction of the artificial arcs in Step 1, there exists an unconstrained shortest chain to j with $\leqq (n - 1)$ arcs for each $j \in \mathcal{N}$, and hence π_j^r will stabilize at the length of the unconstrained shortest chain to j in $\leqq (n - 1)$ iterations. Such stability (i.e., $\pi_j^{r+1} = \pi_j^r$ for each $j \in \mathcal{N}$) will not be reached after n iterations only if there is a negative length circuit in G. If stability is attained, we then have an outtree in which the optimality conditions discussed earlier hold, and hence it is a shortest chain tree.

In each iteration all the nodes are examined. Examining node j in an iteration requires $O(|B_j|)$ additions and comparisons. So, the computational effort in an iteration is $O(m)$, and since there are at most n iterations in the method, the overall computational effort is at most $O(nm)$.

As mentioned above, if the BFM method goes through n iterations without stabilizing, it is an indication that a negative length circuit has been detected. There are other conditions signaling the detection of a negative length circuit in this method. One is that $\pi_1^r < 0$ for any r (1 is the origin node). This signifies the detection of a negative length circuit containing node 1. Another condition is that $\pi_j^{r+1} < \pi_j^r$ holds for at least $(n - r)$ distinct nodes j, for some $r = 1$ to $n - 1$; see Exercise 4.6. When any of these conditions hold, the negative length circuit itself can be located among the set of arcs E at that stage, and the method terminated.

The Dynamic Breadth-First Search Label Correcting Algorithm

Let the *present label depth of node j*, denoted by a_j, be the number of arcs on the present chain to j. Node labels on node j in this algorithm are of the form $(P(j), d_j, a_j')$, where $P(j)$, d_j have the same meaning as before, and a_j' is the *label depth index of node j*; it may not be equal to the label depth of j in intermediate stages since the correction operation on descendents is not carried out, but it will be corrected and will become equal to the true label depth on all the nodes, at termination. $a_j' \leqq a_j$ for all j always, as the label depth index of a node can only increase in this algorithm. Like breadth-first search, this variant performs a sequence of iterations in which in iteration r, only those nodes with label depth index r are branched out. Since the label depth index of a node can change during the algorithm, this variant can be viewed as a dynamic breadth-first search algorithm.

At any stage, for each h, $n(h)$ denotes the number of nodes j for which $a'_j = h$. In each iteration of this algorithm, nodes from a specified subset called *list* are selected according to a criterion, for branching out. The iteration is completed when all the eligible nodes in the list are branched out. While this is going on, a new subset of nodes called *next list* is being built. The next list at the end of an iteration becomes the list for the next iteration.

THE DBFS LABEL CORRECTING ALGORITHM

Initialization   Begin with the spanning outtree T_0 discussed above, and the node labels that go with it. $a'_1 = 0$ and $a'_j = 1$ for all $j \neq 1$. List $= \mathcal{N} \backslash \{1\}$. Next list $= \emptyset$. $n(0) = 1, n(1) = n - 1$ and $n(h) = 0$ for all $h > 1$. Set iteration count equal to 1. Go to iteration 1.

General iteration r: **1. Select a node to branch out**   If list $= \emptyset$ go to Step 3 given below. Otherwise select a node i from the list to branch out, and delete it from the list. Let the present label on i be $(P(i), d_i, a'_i)$. If $a'_i = r$ go to Step 2. Otherwise repeat this step.

   2. Branching out of selected node i   Do the following for each $j \in A_i$. Let the present label on j be $(P(j), d_j, a'_j)$. If $d_j \stackrel{\leq}{=} d_i + c_{ij}$ continue. If $d_j > d_i + c_{ij}$, change the predecessor index of j to i, its distance label to $d_i + c_{ij}$, and if $a'_j \neq r + 1$ subtract 1 from $n(a'_j)$.

   If $n(a'_j) = 0$, the network contains a negative length circuit. One such circuit can be identified by beginning at j and tracing a path using the predecessor labels until a node repeats.

   Otherwise change a'_j to $r + 1$, add 1 to $n(r + 1)$, and include j in next list.

   After all this work is completed return to Step 1.

   3. Set up candidate set for next iteration   If next list $= \emptyset$, the present labels define a shortest chain tree rooted at the origin, node 1; terminate. Otherwise make next list into list, next list $= \emptyset$, and go to the next iteration.

Discussion

First assume that there is no negative length circuit in G. For each $j \in \mathcal{N}$ let π_j^* denote the unknown length of a shortest chain from 1 to j, and b_j the smallest number of arcs in a shortest chain from 1 to j. Define $\mathbf{L}(r) = \{ j : b_j = r \}$.

Observe that at the start of iteration r, the list consists entirely of those nodes i for which $a'_i = r$, since in the previous step these and only these nodes are put

4.9 Consider the following implementations of the primal method for finding a shortest chain tree in $G = (\mathcal{N}, \mathcal{A}, c)$ rooted at node 1. Ancestor checking and correcting the distance index on descendents operations are not carried out in each step. These implementations go through several iterations. In each iteration they select nodes for branching out always from the top of a sequence S_1 of nodes, in which nodes are arranged from top to bottom. As the iteration is progressing, all the nodes whose labels have changed are arranged in another sequence S_2 according to one of the following disciplines. These implementations differ in the way nodes j whose labels have been revised during the iteration are entered in the sequence list S_2.

FIFO/NO MOVE   If j is not in S_2, it is inserted at the bottom of S_2. If j is already in S_2, it is left in its current position.

FIFO/MOVE   If j is not in S_2, it is inserted at the bottom of S_2. If j is already in S_2, it is moved from its present position to the bottom of S_2.

Both these implementations begin iteration 1 with $S_1 = \{1\}$ after initialization. If there are no negative length circuits in G, prove that these implementations terminate with a shortest chain tree after at most n iterations, with the sequence S_2 becoming $\emptyset$ at the end of the last iteration. Using this result show that these implementations can be operated so that their worst case computational complexity to find either a negative length circuit or a shortest chain tree is $O(nm)$ (Shier and Witzgall [1981]).

4.10 Consider the following implementations of the primal method for finding a shortest chain tree in $G = (\mathcal{N}, \mathcal{A}, c)$ rooted at node 1. Correcting the distance index on descendents operation is not carried out in each step. The implementations select nodes for branching out always from the top of a sequence of nodes in which nodes are arranged from top to bottom. As the algorithm is progressing, each node j whose label has changed is added to the sequence according to one of the following disciplines.

LIFO/NO MOVE   If j is not in the sequence at this time, it is added at the top. If it is already in the sequence it is left in its current position.

LIFO/MOVE   If j is not in the sequence at this time, it is added at the top. If it is already in the sequence, it is moved from its position to the top.

Both these implementations begin iteration 1 with $S_1 = \{1\}$ after initialization. Even when there is no negative length circuit in G, show that the number of nodes branched out before termination in these implementations can grow exponentially with n in the worst case (Shier and Witzgall [1981]).

4.4 Shortest Chains From a Specified Origin in an Acyclic Network

Let $G = (\mathcal{N}, \mathcal{A}, c = (c_{ij}))$ be an acyclic network with c as the vector of arc lengths, $|\mathcal{N}| = n$, and $|\mathcal{A}| = m$. We assume that the nodes in G are numbered serially using an acyclic numbering, i.e., $i < j$ for each $(i,j) \in \mathcal{A}$. If it is required to find the shortest chains from an origin, node p, since there are no arcs (j, p) for $j < p$, there exist no chains from such nodes j to p. Hence all nodes $j < p$ and all arcs incident to them can be eliminated. The resulting network is again acyclic, and the nodes can be renumbered in it by subtracting $p - 1$ from the present number. This transforms the problem into one in an acyclic network with node 1 as the origin. Hence in the sequel we assume that node 1 is the origin.

Since G contains no circuits, there are no negative length circuits even if $c < 0$, and the following special algorithm always finds the shortest chains in it. The algorithm is a recursive procedure that comes from the application of the principle of optimality of dynamic programming to the problem. Nodes are labeled in it in the order 1 to n at the rate of one per step, and all node labels assigned are permanent labels.

ACYCLIC SHORTEST CHAIN ALGORITHM

Step 1 Label the origin, node 1, with $(\emptyset, 0)$. Go to Step 2.

General step r, for $r \geqq 2$ At this stage all nodes $1, \dots, r - 1$ would have been labeled already. Let the distance index on node i be $\pi_i, i = 1$ to $r - 1$. Find

$$\pi_r = \min. \{\pi_i + c_{ir} : i \in \mathrm{B}_r\} \tag{4.11}$$

Use the convention that the minimum in the empty set is ∞. Let $P(r)$ be equal to one of the i that attain the minimum in (4.11) if π_r is finite. Label node r with $(P(r), \pi_r)$. If $r = n$, terminate. Otherwise go to the next step.

Discussion

Since $i < j$ for all $(i, j) \in \mathcal{A}$, (4.11) guarantees that the following conditions, which are the optimality conditions established earlier for the present outtree to be a shortest chain tree, will hold.

$$\pi_j - \pi_i \begin{cases} \leqq c_{ij} & \text{for all } (i, j) \in \mathcal{A} \\ = c_{ij} & \text{if } i = P(j) \end{cases}$$

So, the final outtree is a shortest chain tree rooted at 1. For $i \in \mathcal{N}$ if $\pi_i = \infty$ there exists no chain from 1 to i in G; otherwise π is the length of a shortest chain to i. The computational effort in Step r of the algorithm is $|\mathrm{B}_r|$ additions and comparisons, so the overall computational complexity of this algorithm is $O(m)$.

As an example, consider the acyclic network in Figure 4.17, with arc lengths entered on the arcs. Nodes are given an acyclic numbering. Node labels obtained in the algorithm are entered by the side of the nodes in Figure 4.17.

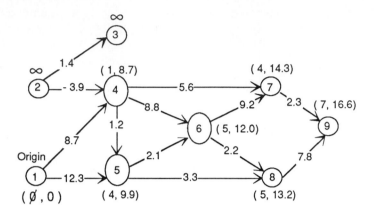

Figure 4.17

This algorithm finds an important application in critical path methods on project networks discussed in Chapter 7.

4.5   Matrix Methods for Shortest Chains Between All Pairs of Nodes

Here we consider the problem of computing shortest chains between every pair of points in the directed network $G = (\mathcal{N}, \mathcal{A}, c)$. This can be solved by applying the algorithms of Section 4.3 with each node as the origin separately, but the algorithms discussed in this section will find all these shortest chains simultaneously.

All the shortest chains and their lengths can be stored very conveniently by maintaining two square matrices, a *label matrix* and a *distance matrix* . Both the rows and columns of each matrix are associated with nodes in $\mathcal{N}$. The entry (called the *label*) in the label matrix in the row of node i and the column of node j, is the predecessor of j in the chain from i to j. So, if this entry is p, (p, j) is the last arc in the present chain from i to j. The remaining arcs on this chain can be traced by looking up the the label in the row of i and the column of p, and continuing until node i is reached. The entry in the distance matrix in the row of i and the column of j is the length of the present chain from i to j.

For $i, j \in \mathcal{N}$, if $(i, j) \notin \mathcal{A}$, introduce an artificial arc (i, j) with length equal to ∞, or a large positive number as discussed before. The methods discussed in this section work with the distance matrix using matrix theoretic operations. Hence, these methods are classified as *matrix methods*, and they terminate either by finding a negative length circuit, or by finding all shortest chains. In the latter case if the shortest chain obtained from a node p to a node q contains an artificial arc, it implies that there is no chain from p to q in the original network.

An Inductive Algorithm for Finding All Shortest Chains

This algorithm proceeds inductively on the number of nodes in the network. It takes exactly n steps. In the rth step, it obtains all the shortest chains in the partial network induced by the subset of nodes $\{1, \ldots, r\}$. In the $(r+1)$th step it brings node $r+1$ into the set of included nodes. This algorithm is due to Dantzig [1967].

INDUCTIVE ALGORITHM

Step 1 Begin with the partial network of node 1 alone. The initial matrices are

<table>
<tr><td colspan="2">Label Matrix</td><td></td><td colspan="2">Distance Matrix</td></tr>
<tr><td></td><td>to 1</td><td></td><td></td><td>to 1</td></tr>
<tr><td>from 1</td><td>1</td><td></td><td>from 1</td><td>0</td></tr>
</table>

General step $r+1$ for $r \geqq 1$ Let $(L_{ij}^r : i, j = 1 \text{ to } r)$, (d_{ij}^r) be the label and distance matrices at the end of Step r. L_{ii}^r will be i and d_{ii}^r will be 0 for all $i = 1$ to r. Compute

$$d_{i,r+1}^{r+1} = \text{min.} \ \{c_{i,r+1}; d_{ij}^r + c_{j,r+1}, j = 1 \text{ to } r, j \neq i\} \quad (4.12)$$

$$d_{r+1,i}^{r+1} = \text{min.} \ \{c_{r+1,i}; c_{r+1,j} + d_{ji}^r, j = 1 \text{ to } r, j \neq i\} \quad (4.13)$$

$$d_{r+1,r+1}^{r+1} = \text{min.} \ \{0; d_{r+1,j}^{r+1} + d_{j,r+1}^{r+1}, j = 1 \text{ to } r\} \quad (4.14)$$

For $i = 1$ to r, define

$$L_{i,r+1}^{r+1} = \begin{cases} i, \text{ if } d_{i,r+1}^{r+1} = c_{i,r+1} \\ \\ \text{or a } j \text{ attaining min. in (4.12) otherwise} \end{cases} \quad (4.15)$$

$$L_{r+1,i}^{r+1} = \begin{cases} r+1, \text{ if } d_{r+1,i}^{r+1} = c_{r+1,i} \\ \\ \text{or a } j \text{ attaining min. in (4.13) otherwise} \end{cases} \quad (4.16)$$

Counting the self-loop at node $r+1$ of length 0 as a simple circuit, $d_{r+1,r+1}^{r+1}$ is the length of the shortest simple circuit containing node $r+1$ in the partial network induced by the subset of nodes $\{\,1,\ldots,r+1\,\}$. If $d_{r+1,r+1}^{r+1} < 0$, let p be a j attaining the min.in (4.14). Let $\mathcal{C}_1$ be the chain consisting of only the arc $(r+1,p)$ if $d_{r+1,p}^{r+1} = c_{r+1,p}$; otherwise it consists of the arc $(r+1,q)$ and the chain from q to p determined by the label matrix at the end of Step r, where $q = L_{r+1,p}^{r+1}$. Similarly, let $\mathcal{C}_2$ be the chain consisting of only the arc $(p,r+1)$ if $d_{p,r+1}^{r+1} = c_{p,r+1}$; otherwise it consists of the chain from p to u determined by the label matrix at the end of Step r, and the arc $(u,r+1)$, where $u = L_{p,r+1}^{r+1}$. Then the simple circuit $\overrightarrow{\mathbb{C}}_1$ obtained by combining the chains $\mathcal{C}_1$ and $\mathcal{C}_2$ is a negative length circuit of length $d_{r+1,r+1}^{r+1}$; terminate.

If $d_{r+1,r+1}^{r+1} = 0$, define $L_{r+1,r+1}^{r+1} = r+1$. For $i,j = 1$ to r, define

$$d_{i,j}^{r+1} \;=\; \text{min.} \; \{d_{ij}^{r}; d_{i,r+1}^{r+1} + d_{r+1,j}^{r+1}\} \qquad (4.17)$$

$$L_{ij}^{r+1} \;=\; \begin{cases} L_{ij}^{r}, & \text{if } d_{i,j}^{r+1} = d_{i,j}^{r+1} \\[2mm] L_{r+1,j}^{r+1}, & \text{otherwise} \end{cases}$$

$(L_{ij}^{r+1} : i,j = 1$ to $r+1)$, $(d_{ij}^{r+1} : i,j = 1$ to $r+1)$ define the new label and distance matrices respectively. If $r+1 < n$ go to the next step; otherwise terminate.

Discussion

Consider the label and distance matrices at the end of Step r. If no negative length circuits have been detected up to this stage, the following facts (i), (ii) hold by the manner in which the label and distance matrices are updated during the algorithm. (iii) follows by using induction, it establishes the validity of the algorithm.

(i) For all i,j,h between 1 to r,

$$d_{i,h}^{r} \overset{\leq}{=} d_{ij}^{r} + d_{jh}^{r} \qquad (4.18)$$

i.e., the entries in the distance matrix satisfy the *triangle inequality.*

(ii) If node j appears on the chain from i to h traced by the labels in the label matrix, (4.18) holds as a strict equation.

(iii) For all i,j between 1 to r the chain from i to j traced by the labels in the label matrix is a shortest chain in the partial network induced by the subset of nodes $\{1,\ldots,r\}$, and d_{ij}^{r} is its length.

The computational effort in Step $r+1$ can be verified to be $O((r+1)^2)$. Hence, the overall computational effort in this algorithm is $O(\sum (r+1)^2) = O(n^3)$.

The Floyd-Warshall Algorithm for All Shortest Chains

Let $G = (\mathcal{N}, \mathcal{A}, c)$ be the directed network in which the problem is being solved. Number the nodes in G serially 1 to $n = |\mathcal{N}|$. On any simple chain, nodes different from the initial and terminal nodes are called *intermediate nodes*. A simple chain has no intermediate nodes iff it consists of a single arc.

This algorithm maintains label and distance matrices of order $n \times n$ throughout, they store the current chains and their lengths respectively. The algorithm terminates after at most n steps. The distance matrix obtained at the end of Step r is denoted by (d_{ij}^r), where d_{ij}^r will be equal to the length of a shortest chain from node i to node j among those chains satisfying the condition that every intermediate node on them is from the set $\{1, \ldots, r\}$ (i, j themselves may or may not be from this set). The label matrix at that stage storing these constrained shortest chains is denoted by (L_{ij}^r). The main computational tool used repeatedly in this algorithm is called *the triangle (or triple) operation* given in (4.19) for nodes i, j and fixed node $r + 1$.

FLOYD-WARSHALL ALGORITHM

Initialization   Define the initial label and distance matrices to be $(L_{ij}^0), (d_{ij}^0)$ respectively, where, for all i, j, $L_{ij}^0 = i$, and $d_{ij}^0 = c_{ij}$ for $i \neq j$, 0 if $i = j$. Go to Step 1.

General Step $r+1$ for $r \geq 0$ At this stage we have the matrices $(L_{ij}^r), (d_{ij}^r)$. For each $i, j \in \mathcal{N}$ compute

$$d_{ij}^{r+1} = \text{min.} \; \{d_{ij}^r, d_{i,r+1}^r + d_{r+1,j}^r\} \tag{4.19}$$

$$L_{ij}^{r+1} = L_{ij}^r \; \text{if} \; d_{ij}^{r+1} = d_{ij}^r; \; L_{r+1,j}^r \; \text{otherwise}$$

$(L_{ij}^{r+1}), (d_{ij}^{r+1})$ are the new label and distance matrices. If $d_{ii}^{r+1} < 0$ for any $i \in \mathcal{N}$, the circuit obtained by putting the present chain from i to $r + 1$ together with that from $r + 1$ to i, is a negative length circuit, terminate. If $d_{ii}^{r+1} = 0$ for all i, and $r + 1 = n$, the present chains are the shortest chains; terminate. Otherwise, go to the next step.

Discussion

We will now show that if d_{ii}^r stayed $= 0$ for all $i \in \mathcal{N}$ and $r = 1$ to n, then (d_{ij}^n) is the matrix of shortest chain lengths in G. For this we set up an induction hypothesis that (d_{ij}^r) is the matrix of shortest chain lengths subject to the constraint that every intermediate node on every chain is numbered $\overset{\leq}{=} r$. This hypothesis clearly holds for $r = 0$, by initialization.

We will now prove that under the induction hypothesis, the statement in it also remains valid when r is replaced by $r + 1$. For any $i, j \in \mathcal{N}$, a shortest chain from i to j with intermediate nodes numbered $\overset{\leq}{=} r + 1$, either does not contain node $r + 1$ (in this case we must have $d_{ij}^{r+1} = d_{ij}^r$ by the triple operations carried out in Step $r + 1$), or it contains node $r + 1$. In the latter case, since $d_{i,r+1}^r, d_{r+1,j}^r$ are both shortest chain distances with intermediate nodes numbered $\overset{\leq}{=} r$ by the induction hypothesis, $d_{i,r+1}^r + d_{r+1,j}^r$ is the shortest chain distance from i to j with intermediate nodes numbered $\overset{\leq}{=} r + 1$. Hence in either case, the statement in the induction hypothesis is also valid for $r + 1$.

Hence, by induction, (d_{ij}^n) is the matrix of shortest chain lengths in G in this case.

In each step there are n^2 additions and n^2 comparisons to be performed. Hence the overall computational complexity of this method is $O(n^3)$.

4.6   Sensitivity Analysis in the Shortest Chain Problem

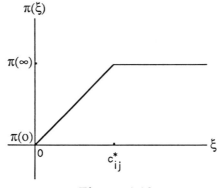

Figure 4.18

Consider the problem of finding a shortest chain from node 1 to node n in the directed network G $= (\mathcal{N}, \mathcal{A}, c)$. Let (i, j) be a particular arc in $\mathcal{A}$ whose length c_{ij} is a parameter ξ which can vary, while all the other data remains unchanged. Let $\pi(\xi)$ denote the length of a shortest chain from 1 to n in G as a function of ξ.

If (i, j) is not contained on a shortest chain from 1 to n when $\zeta = 0$, then $\pi(\xi) = \pi(0)$ for every $\xi \overset{\geq}{=} 0$, and the same chain not containing (i, j) remains a shortest for all $\xi \overset{\geq}{=} 0$.

If (i, j) is contained on a shortest chain from 1 to n when $\xi = 0$, this chain remains a shortest one as ξ increases from 0 to $\pi(\infty) - \pi(0)$. If $\xi > \pi(\infty) - \pi(0)$, the shortest chain changes, and (i, j) is not contained on any shortest chain from 1 to n in this range.

So, $\pi(\xi) = \text{min. } \{ \pi(0) + \xi, \pi(\infty) \}$ for all $\xi \geqq 0$. See Figure 4.18. The quantity $\pi(\infty) - \pi(0)$ where $\pi(\xi)$ changes slope, is called the *critical length of arc* (i, j), and denoted by c_{ij}^*. Destroying an arc in a network is equivalent to making its length ∞. If arc (i, j) with present length c_{ij} and critical length c_{ij}^* is destroyed, the length of the shortest chain from 1 to n goes up by max. $\{0, c_{ij}^* - c_{ij}\}$.

Exercises

4.11 Find the shortest chains between all pairs of nodes in the mixed network in Figure 4.19. The number on each line is its length.

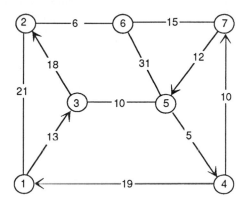

Figure 4.19

4.12 Develop an efficient scheme for finding an arc in G, the destruction of which increases the length of the shortest chain from the origin to the destination by as much as possible.

4.13 Suppose G is the street network in a city, with nodes representing street intersections, and arcs representing street segments joining pairs of adjacent nodes. The length of each arc is the driving time through the corresponding segment under normal conditions. By making improvements on an arc, its length can be reduced. Discuss how to pick a single arc for making improvements, to reduce the length of the shortest chain from the origin to the destination by as much as possible.

4.7   Algorithm for Ranking Chains in Nondecreasing Order of Length

Consider the directed network $G = (\mathcal{N}, \mathcal{A}, c)$ with $|\mathcal{N}| = n, |\mathcal{A}| = m$, and nodes 1, n as the origin and destination respectively. We assume that there are no negative length circuits in G. Here we discuss an adaptation of the ranking approach discussed in Section 3.6 to rank the chains from 1 to n in G in nondecreasing order of length, due to Lawler [1972]. Let $\mathbf{S}$ = set of all chains from 1 to n in G, simple or not; and $\mathbf{P}$ = set of all simple chains from 1 to n in G. The algorithm can rank either the chains in $\mathbf{S}$, or only those in $\mathbf{P}$. It obtains one new chain in the ranked sequence per step with a computational effort of at most $O(n^3)$.

We define a *special subset* to be the set of chains from 1 to n in G (from $\mathbf{S}$ or $\mathbf{P}$ as desired) containing all the arcs in a specified *initial segment* which is a chain from 1 to some node, and not containing any arc from a specified subset of *excluded arcs*. If the specified initial segment is $\mathcal{C}_1 = 1, (1, i_1), i_1, \ldots, (i_p, r), r$; and the specified subset of excluded arcs is $\mathbf{E}$, we denote the special subset by the symbol $\mathbf{F} = [\mathcal{C}_1; \bar{\mathbf{E}}]$. It contains all chains (from $\mathbf{S}$ or $\mathbf{P}$ as desired) obtained by combining $\mathcal{C}_1$ with a chain from r to n in G not containing any arc in $\mathbf{E}$.

A shortest chain in a special subset $\mathbf{F} = [\mathcal{C}_1; \bar{\mathbf{E}}]$, where $\mathcal{C}_1 = 1, (1, i_1), i_1, \ldots, (i_p, r), r$; from $\mathbf{S}$ or $\mathbf{P}$,can be found efficiently. Let $\tilde{G}$ be the network obtained by deleting each of the arcs in the set $\mathbf{E}$ from $\mathcal{A}$. Let $\hat{G}$ be the network obtained by deleting each of the arcs incident at the nodes $1, i_1, \ldots, i_p$ and all the arcs incident into node r from $\tilde{G}$. For ranking the chains in $\mathbf{S}$, find a shortest chain from r to n in $\tilde{G}$ by any method discussed earlier. If there is no chain from r to n in $\tilde{G}$, $\mathbf{F} = \emptyset$. If a shortest chain from r to n in $\tilde{G}$ is found, add it at the end of the specified initial segment $\mathcal{C}_1$ to get a shortest chain from 1 to n in $\mathbf{F}$ from $\mathbf{S}$. For ranking the chains in $\mathbf{P}$, carry out the same procedure replacing $\tilde{G}$ by $\hat{G}$, to get a shortest chain in $\mathbf{F}$ from $\mathbf{P}$. The important thing is that the shortest chain from r to n, found in either of these cases, is a simple chain (since the algorithms discussed earlier deal only with simple chains) and hence consists of at most $n - 1$ arcs.

The algorithm uses an operation called *partitioning a special subset using a shortest chain in it.* Let $\mathbf{F} = [1, (1, i_1), i_1, \ldots, (i_p, r), r ; \bar{\mathbf{E}}]$ be the special subset, and $\mathcal{C}_2 = 1, (1, i_1), i_1, \ldots, (i_p, r), r, (r, i_{p+1}), i_{p+1}, \ldots, (i_{p+h}, n), n$, be a shortest chain in $\mathbf{F}$ obtained as above. As mentioned above, $h + 1$, the number of arcs on $\mathcal{C}_2$ between the nodes r to n is at most $n - 1$. Partitioning $\mathbf{F}$ using $\mathcal{C}_2$ expresses $\mathbf{F} \setminus \{\mathcal{C}_2\}$ as a disjoint union of $h + 1$ special subsets. These special subsets are

$$
\begin{aligned}
\mathbf{F}_1 &= [1, (1, i_1), \ldots, (i_p, r), r; \overline{\mathbf{E} \cup \{(r, i_{p+1})\}}] \\
\mathbf{F}_2 &= [1, (1, i_1), \ldots, (i_p, r), r, (r, i_{p+1}), i_{p+1}; \overline{\mathbf{E} \cup \{(i_{p+1}, i_{p+2})\}}] \\
&\ \ \vdots \\
\mathbf{F}_{h+1} &= [1, (1, i_1), \ldots, (i_p, r), r, (r, i_{p+1}), \ldots, (i_{p+h-1}, i_{p+h}), i_{p+h}; \overline{\mathbf{E} \cup \{(i_{p+h}, n)\}}]
\end{aligned}
$$

THE RANKING ALGORITHM

Step 1   Find a shortest chain in G from 1 to n using any of the algorithms discussed earlier. If there is no chain from 1 to n, terminate. Otherwise, by our assumption that there is no negative length circuit in G, a shortest simple chain will be obtained; let it be $\mathcal{C}^1 = 1, (1, j_1), j_1, \ldots, (j_u, n), n$. We would have obtained a shortest chain tree rooted at node 1. Nodes not on this tree cannot be reached from 1 by a chain, so eliminate them and all the arcs incident at them from further consideration. Let π_i be the length of the chain from 1 to i in the shortest chain tree for remaining nodes i. For each remaining arc (i, j), we replace c_{ij} by $c_{ij} - (\pi_j - \pi_i) \geqq 0$, this has the effect of subtracting the same constant $(\pi_n - \pi_1)$ from the length of every chain from 1 to n and hence does not affect the ranking of these chains. With this change the arc length vector becomes nonnegative, even though the original vector may not be, and the efficient Dijkstra's method can be used to compute shortest chains in subsequent steps of the algorithm. Generate the following $u + 1$ ($\leqq n - 1$) special subsets.

$$[\overline{\{(1, j_1)\}}]$$
$$[1, (1, j_1), j_1; \overline{\{(j_1, j_2)\}}]$$
$$\vdots$$
$$[1, (1, j_1), \ldots, (j_{u-1}, j_u), j_u; \overline{\{(j_u, n)\}}]$$

Find a shortest chain in each, and discard any special subsets among these which are empty. Arrange the remaining special subsets together with the shortest chain in each, in a list, in increasing order of the length of the shortest chain from 1 to n in it, top to bottom. Go to the next step.

General step $p+1$, for $p \geqq 1$ If the list is empty, there are no more chains from 1 to n in G; terminate. Otherwise, delete the topmost special subset, say **F**; from the list. The shortest chain in **F**, $\mathcal{C}^{p+1}$, say, is the next, i.e., $(p + 1)$th chain in the ranked sequence. If you have enough chains in the ranked sequence already, terminate. Otherwise, partition **F** using $\mathcal{C}^{p+1}$. Find a shortest chain from 1 to n in each of the special subsets generated in this partitioning. Discard any empty ones among these, but include the nonempty ones together with the shortest chain in each, in the list in their proper position according to the length of the shortest chain from 1 to n in it. Go to the next step.

Discussion

At the end of Step 1, the list has at most $(n - 1)$ special subsets. In each step, the top special subset is taken out, and at most $(n - 1)$ newly generated special subsets are added to the list. Hence there will be at most $p(n-1)$ special subsets in the list at the end of Step p. Hence, the size of the list grows linearly with the number of chains ranked.

In each step the shortest chains in at most $(n-1)$ new special subsets need to be computed, each with a computational effort of at most $O(n^2)$, as pointed out above. Hence the computational effort in each step is at most $O(n^3)$, plus the effort needed to insert the new special subsets in the list. Thus the overall effort needed to find p chains in the ranked sequence is at most $O(pn^3)$.

4.8   Exercises

4.14 0-1 Knapsack Problem   It is required to determine a subset of n available objects to load into a knapsack to maximize the value loaded subject to the knapsack's weight capacity constraint. The jth object has weight w_j kg. and value $\$v_j$, $j = 1$ to n, and the knapsack's capacity by weight is w_0 kg. All data are positive integers and $w_j \leqq w_0$ for all $j = 1$ to n. Objects cannot be broken; they have to be either loaded whole or left out.

Let G be a network in which the nodes are integer points $[p,q]$ in $\mathbb{R}^2$ for $0 \leqq p \leqq w_0$ and $1 \leqq q \leqq n$, and a sink node $\check{t}$. The arcs in G are: $([w_0,1], [w_0,2])$ with length 0, $([w_0,1], [w_0 - w_1, 2])$ with length v_1; for $2 \leqq q \leqq n-1$, $([p,q], [p,q+1])$ with length 0 for $0 \leqq p \leqq w_0$, $([p,q], [p - w_q, q+1])$ with length v_q for $w_q \leqq p \leqq w_0$; and $([p,n], \check{t})$ with length 0 for $0 \leqq p \leqq w_0$.

Show that G is acyclic and that the knapsack problem is equivalent to that of finding a longest chain from $[w_0, 1]$ to $\check{t}$ in G.

4.15 We have already seen that the problem of finding a longest simple chain from node 1 to node n in a directed network G = $(\mathcal{N}, \mathcal{A}, c)$ with $c > 0$ is a hard problem in general. Using this show that if $a \geqq 0, b > 0$ are two edge cost vectors in G, the problem of finding a simple chain $\mathcal{C}$ from 1 to n that minimizes the ratio objective function $(\sum(a_{ij} : \text{over } (i,j) \in \mathcal{C}))/ (\sum(b_{ij} : \text{over } (i,j) \in \mathcal{C}))$ is a hard problem (Ahuja, Batra, and Gupta [1983]).

4.16 Relationship between Assignment and Shortest Chain Problems
Consider the problem of finding a shortest chain from 1 to n in a directed connected network G = $(\mathcal{N}, \mathcal{A}, c)$. Define the $n \times n$ matrix $D = (d_{ij})$, where $d_{ij} = c_{ij}$ if $(i,j) \in \mathcal{A}$, $= 0$ if $i = j$, and $= \infty$ if $i \neq j$ and $(i,j) \notin \mathcal{A}$.

Find a minimum cost assignment with D as the cost matrix. Prove that the unit matrix is a minimum cost assignment in this problem iff there exist no negative length circuits in G. Conversely show that if the unit matrix is not optimal to this problem, then a negative length circuit in G can be identified from an optimum assignment.

Suppose there are no negative length circuits in G. Change d_{n1} to $-M$ where M is a very large positive number, and find a minimum cost assignment wrt the modified matrix D. If there exists no chain from 1 to n in G prove that the unit matrix is a minimum cost assignment for this problem and conversely. Otherwise, the allocations in a minimum cost assignment for this problem correspond to a circuit in G containing the arc $(n, 1)$ and a shortest chain from 1 to n, and self-loops at the remaining nodes (Weintraub [1973]).

4.17 Suppose a shortest chain from node 1 to node n in the directed connected network $G = (\mathcal{N}, \mathcal{A}, c)$ without any self-loops or negative length circuits, has been found using an algorithm for the assignment problem with the approach discussed in Exercise 4.16. From this information, if any of the following types of changes occur in G, show that a shortest chain from 1 to n in the modified network can be obtained with a computational effort of at most $O(n^2)$: (a) new arcs are added, all incident at an existing node in G, (b) a subset of arcs incident at a node in G are deleted, (c) a node and all the arcs incident at it in G are deleted, (d) a new node is added to G together with some new arcs incident at it, (e) the lengths of arcs incident at a node have been modified (Weintraub [1973]).

4.18 Negative Length Simple Cycles in Undirected Networks   Let $\bar{\mathrm{E}}$ be the set of negative length edges in an undirected network $G = (\mathcal{N}, \mathcal{A}, c = (c_{i;j}))$ with c as the vector of edge lengths. Assume that $\bar{\mathrm{E}} \neq \emptyset$, and let $\bar{\mathcal{N}}$ be the set of nodes on edges in $\bar{\mathrm{E}}$. If there is a simple cycle in $(\bar{\mathcal{N}}, \bar{\mathrm{E}})$, it is a negative length simple cycle in G. So, assume that there is no simple cycle in $(\bar{\mathcal{N}}, \bar{\mathrm{E}})$; i.e., it is a forest.

Let $\bar{\mathcal{N}}_o$ be the set of odd degree nodes in $(\bar{\mathcal{N}}, \bar{\mathrm{E}})$. Construct the complete undirected network $H = (\bar{\mathcal{N}}_o, A, d = (d_{ij}))$, where $A = \{ (i;j) : i \neq j, i, j \in \bar{\mathcal{N}}_o \}$, and d is the vector of edge lengths in H with d_{ij} being the length of a shortest path between i and j, $\mathcal{P}_{ij}$, in $(\mathcal{N}, \mathcal{A})$ with $(|c_{ij}|)$ as the vector of edge lengths. Let $\bar{\mathrm{M}}$ be a minimum cost perfect matching in H, and let $P(\bar{\mathrm{M}})$ be the union of the sets of edges on the paths $\mathcal{P}_{ij}$ for $(i;j) \in \bar{\mathrm{M}}$.

Prove that G contains a nonpositive length simple cycle if $\bar{\mathrm{E}} \not\subset P(\bar{\mathrm{M}})$. Prove that G contains no negative length simple cycles if $\bar{\mathrm{E}} \subset P(\bar{\mathrm{M}})$. Using these results develop an efficient algorithm for detecting a negative length simple cycle in G if one exists. Also, develop an efficient algorithm for finding shortest paths between every pair of nodes in G assuming that there are no negative length simple cycles in G (Tobin [1975]).

4.19 Find the shortest chain tree rooted at node 1 in the network in Figure 4.20 using the LIFO sequence-list driven labeling methods discussed in Exercise 4.10. Data on each arc is its length.

In general consider the following sequence of networks of which the one in Figure 4.20 is the fifth. The first network consists of two nodes 1, 2, and two arcs (1, 2) and (2, 1) each with length 1. To get the $(r + 1)$th network from the rth do the following: Add node $r + 2$. Add arcs $(r + 2, i)$ for $i = 2$ to $r + 1$ with length $c_{r+2,i}$ same as $c_{1,i}$ in the rth network. Change $c_{1,i}$ to $c_{1,i} + 2^{r-1} + 1$ for $i = 2$ to $r + 1$.

Add an arc $(1, r+2)$ with length 1. Show that node 2 will be branched out 2^{r-2} times in these methods on the rth network (Kershenbaum [1981]).

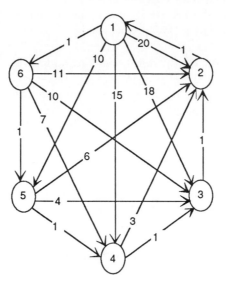

Figure 4.20

4.20 $G = (\mathcal{N}, \mathcal{A})$ is a given directed network. Given three nodes s, t, p in G, show that the problem of checking whether there exists a simple chain from s to t via p is an NP-complete problem.

Given nodes s, t in G, show that the problem of checking whether there exists a simple chain from s to t consisting of an even number of arcs is an NP-complete problem (Lapaugh and Papadimitriou [1984]).

4.21 $G = (\mathcal{N}, \mathcal{A}, c)$ is a directed connected network with $c \overset{\geq}{=} 0$. **X, Y** are respectively the sets of permanently labeled and temporarily labeled nodes at some stage in the process of finding the shortest chain tree rooted at node 1 in G by Dijkstra's algorithm. Let μ_i be the present distance index on node i. For every $j \in \mathbf{X}$, prove that min. $\{\mu_i : i \in \mathbf{Y}\} \overset{\geq}{=} \mu_j$. Hence show that the distance index of a node at the time that it gets permanently labeled is $\overset{\geq}{=}$ the distance indices of nodes already in **X** at that time.

4.22 In the network in Figure 4.21 with $2r+1$ nodes and $3r$ arcs, show that every one of the 2^r simple chains from node 1 to node $2r+1$ is a shortest chain.

4.23 A Constrained Shortest Chain Problem   An air carrier is trying to cover r flights with m planes which is known to be inadequate. a_i, b_i are the scheduled start time (at origin), and finish time (at destination) of ith flight, $i = 1$ to r. Construct a directed network G with nodes $L_1, \ldots, L_r$ for flights, and

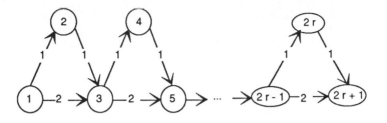

Figure 4.21

$P_1, \ldots, P_m$ for planes. Include arcs (L_i, P_j) and (P_j, L_i) of length 0 for each i, j. For each i, j such that the destination of L_i and the origin of L_j are the same, include an arc (L_i, L_j) of length equal to the delay that occurs for flight L_j if the same aircraft performs L_j after L_i.

Show that any chain in G passing through every node once corresponds to an assignment of planes to flights; and that the length of such a chain is exactly the total delay in all the flights corresponding to that assignment. Hence show that the problem of covering all the r flights with m planes, to minimize total delay, is the problem of finding a shortest chain in G subject to the constraint that it must pass through each node exactly once (Teodorovic and Guberinic [1984]).

4.24 Application to Set Partitioning Problem   The following are three important 0-1 integer programming models, in which the data consists of a 0-1 integer matrix $A = (a_{ij})$ of order $m \times n$ and a column vector e of all 1's.

Set partitioning problem		Set covering problem		Set packing problem	
min.	cx	min.	cx	min.	cx
s. to	$Ax = e$	s. to	$Ax \geq e$	s. to	$Ax \leq e$
all	$x_j = 0$ or 1	all	$x_j = 0$ or 1	all	$x_j = 0$ or 1

Consider the special case of the set partitioning problem in which each column of A consists of a single consecutive segment of ones, i.e., for each $j = 1$ to n there exists $g_j \leq h_j$ such that $a_{ij} = 1$ for $g_j \leq i \leq h_j$ and 0 for all other i. Such problems arise in crew scheduling applications where each crew must do a single stretch of duty comprising a consecutive set of trips, known as one-part duty crew scheduling problems. Construct the network $G = (\mathcal{N}, \mathcal{A})$ where $\mathcal{N} = \{1, \ldots, m+1\}$ with node i corresponding to row i of A for $1 \leq i \leq m$, and $\mathcal{A}$ consists of arcs $(g_j, h_j + 1)$ of length c_j corresponding to $A_{.j}$ for $j = 1$ to n.

Show that G is acyclic, and that this special case of the set partitioning problem is equivalent to the problem of finding a shortest chain in G from 1 to $m + 1$.

A set partitioning problem is infeasible when an exact cover does not exist, but in practical applications relaxations that *overcover* or *undercover* with a possible penalty expense are typically permitted. So, consider the following extended set partitioning problem, where $(p_i, q_i) \geqq 0$ for all i. Show that when A satisfies the special property mentioned above, this extended set partitioning problem can also be formulated as a shortest chain problem in a network. Develop this formulation.

$$\text{min.} \quad \sum_j c_j \ x_j \ + \sum_i (p_i u_i + q_i \sigma_i)$$

$$\text{s. to} \quad \sum_j a_{ij} \ x_j \ + u_i - \sigma_i = 1, i = 1 \text{ to } m$$

$$x_j \ , u_i = 0 \text{ or } 1 \text{ for all } i, j$$

$$\sigma_i \ \geqq 0 \text{ and integer for all } i$$

When A satisfies the special property mentioned above, both the set covering and the set packing problems can also be formulated as short chain problems. Derive these formulations (Darby-Dowman and Mitra [1985], and Shepardson and Marsten [1980]).

4.25 Manpower Planning   A construction company requires the following number of steel erectors over a 6-month horizon.

Month	March	April	May	June	July	August
Erectors needed	4	6	7	4	6	2

In February there were 3 steel erectors on site, and in September this number has to be the same. It costs $1000 to hire an erector, and $1600 to terminate the services of one. Each erector's contract is decided on a month-to-month basis at the start of each month. No more than 3 erectors can be hired anew at the start of any month. Under a union agreement no more than a third of the current manpower can be terminated at the end of a month. The cost of keeping a surplus erector is $3000 per month. When there is a shortage, it has to be made up in overtime; this costs $6000 per erector per month. Overtime cannot exceed 25% of normal time. It is required to determine how many erectors to hire or terminate each month in order to minimize the total cost of hiring, terminating, surplus and shortage costs over the horizon, subject to the given constraints. Formulate this as a shortest chain problem in an acyclic network and find an optimum solution (Clark and Hastings [1977]).

4.26 Replacement Problem   This problem arises in planning over a 5-year horizon at a plant. They have a machine which will be 3 years old at the beginning of the first year of the horizon. The machine is inspected at the beginning of each year and is either overhauled or replaced with a new machine. The cost of a new machine is $40,000. The cost of overhaul and the scrap value of the machine depend on its age as given below.

Age (years)	1	2	3	4
Overhaul cost ($1000 units)	15	6	18	
Scrap value ($1000 units)	20	10	5	2

Company's policy is that the machine must be replaced at age 4 years. Also, assume that at the end of the 5-year horizon, the then current machine will be scrapped.

It is required to find the policy that minimizes the present value of the cost of overhauling or replacing the machine, when the costs are discounted at the rate of 15% per year. Formulate this as a shortest chain problem in an acyclic network and find an optimum solution (Clark and Hastings [1977]).

4.27 Detecting Negative Cost Circuits by Node Elimination   Consider the problem of finding negative cost circuits in the directed connected network G $= (\mathcal{N}, \mathcal{A}, c)$. $\mathcal{A}$ may contain some self-loops. Look for negative cost self-loops. If

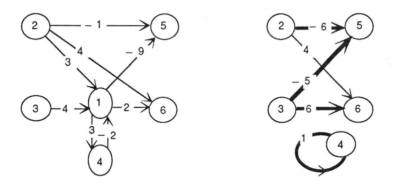

Figure 4.22 Elimination of node 1.

there are none, or if self-loops are not to be considered as circuits, eliminate all self-loops in G. Define $(i, j) \in \mathcal{A}$ as an *original arc* and associate it with the chain from i to j consisting of that arc only. During the algorithm, new arcs are created by coalescing chains in the original network into arcs. These will be called *created arcs* ; each of them is associated with a chain in G. Some of the created arcs may be self-loops; each of them is associated with a circuit in G.

Select a node, say 1. If $\mathbf{A}_1$ or $\mathbf{B}_1$ is empty, 1 is not on any circuit; eliminate it and all arcs incident at it. If both $\mathbf{A}_1, \mathbf{B}_1$ are nonempty, for each $i \in \mathbf{A}_1, j \in \mathbf{B}_1$ do the following: If (i, j) does not exist now, introduce the created arc (i, j) with cost coefficient $c_{i1} + c_{1i}$, associated with the chain $i, (i, 1), 1, (1, j), j$ in G (if $i = j$ this created arc will be a self-loop at i and the chain in G associated with it is a circuit). If (i, j) already exists, and $c_{ij} \overset{\le}{=} c_{i1} + c_{1j}$ leave it as it is. If $c_{ij} > c_{i1} + c_{1j}$, change its cost coefficient to $c_{i1} + c_{1j}$ and change the chain associated with it to that from i to j obtained by combining the chains associated with $(i, 1)$ and $(1, i)$ in that order.

Now eliminate node 1 and all the arcs incident at it, and let the resulting network be G^1. See Figure 4.22 where the cost coefficients are entered on the arcs, and created arcs are thick.

The same process is repeated on the network G^1, and the process continued yielding networks $G^2, \ldots$, with decreasing number of nodes. Prove that G has a negative cost circuit iff G^r has either a negative cost self-loop or a negative cost circuit. So, if G has a negative cost circuit, it will be detected by one of the networks in the sequence having a negative cost self-loop. The circuit associated with such a self-loop is a negative cost circuit in G. Apply this approach to detect any negative cost circuits in the following networks with cost coefficients entered on the arcs. Analyze the computational complexity of this algorithm (Chen [1975]).

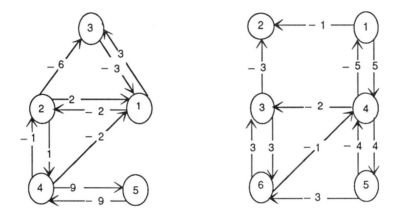

Figure 4.23

4.28 Let $G = (\mathcal{N}, \mathcal{A})$ be an undirected network. For $i, j \in \mathcal{N}$ define $d_{ij} = 0$ if $i = j$, $= \infty$ if there is no path from i to j in G, and $=$ the number of edges on a path with the smallest number of edges between i and j in G otherwise. Develop efficient methods for computing the matrix (d_{ij}), and for updating this matrix when an edge or a node is added or deleted (Cheston and Corneil [1982]).

4.29 We consider a vehicle routing problem involving also fleet size and mix decisions. T is the number of vehicle types. a_r, f_r are respectively the capacity and the fixed cost of acquiring vehicle type r, $r = 1$ to T. This data satisfies $a_1 < a_2 < \ldots < a_T$, $f_1 < \ldots < f_T$. 0 denotes the depot, and 1 to n are the customers. All routes originate and terminate at the depot. c_{ij} is the cost of travel from i (depot or customer) to j, assumed to be symmetric and independent of vehicle type used. The customers are ordered so that $c_{01} > c_{02} > \ldots > c_{0n}$. d_i is the demand at customer i, a positive integer, for $i = 1$ to n, and $D = \sum d_i$. Splitting of a customer's demand between two vehicles is allowed. Develop a method for obtaining a lower bound for the minimum total cost (sum of fixed costs of acquiring the vehicles and the variable costs of travel) for making deliveries to meet the customer's demands, by solving a shortest chain problem on a directed network with $D + 1$ nodes (Golden, Assad, Levy, and Gheysens [1984]).

4.30 Minimum Bottleneck Cost Chains   Let $G = (\mathcal{N}, \mathcal{A}, c, \check{s}, \check{t})$ be a directed network. Define the bottleneck cost of a chain $\mathcal{C}$ from $\check{s}$ to $\check{t}$ to be max. $\{c_{ij} : (i, j) \in \mathcal{C}\}$. Develop an efficient algorithm for finding a minimum bottleneck cost chain from $\check{s}$ to $\check{t}$ in G. What is its computational complexity? Develop a special procedure for solving this problem when G is an acyclic network, that exploits the acyclic structure.

4.31 Optimal Sum-Bottleneck (SB) Chains   Let $G = (\mathcal{N}, \mathcal{A}, \check{s}, \check{t})$ be a directed network which is doubly weighted, i.e., each arc (i, j) has a *sum weight*

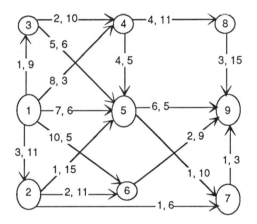

Figure 4.24 The entries on arc (i, j) are σ_{ij}, β_{ij}.

σ_{ij}, and a *bottleneck weight* β_{ij} associated with it. The sum weight of a chain $\mathcal{C}$, $\sigma(\mathcal{C})$ is the sum of σ_{ij} over arcs (i, j) on it. The bottleneck weight of $\mathcal{C}$, $\beta(\mathcal{C})$, is the maximum of β_{ij} over arcs (i, j) on it. The SB weight of $\mathcal{C}$ is defined to be equal to max. $\{ \sigma(\mathcal{C}), \beta(\mathcal{C}) \}$. Develop an efficient algorithm for finding a minimum SB weight chain from $\check{s}$ to $\check{t}$ in G and derive its computational complexity. Apply this algorithm on the network in Figure 4.24 with $\check{s} = 1$, and $\check{t} = 9$ (Bokhari [1988]).

4.32 The Jogger's Problem   Let $G = (\mathcal{N}, \mathcal{A})$ be a connected road network at the disposal of a jogger beginning his jog at node $1 \in \mathcal{N}$. Each line in the network possesses some positive measure of undesirability. The jogger's problem is to identify a simple circuit containing node 1 for his jog, of minimum total undesirability. The route is required to be a simple circuit, so it cannot consist of the back-and-forth traversal of a single edge. In the directed case it cannot just consist of a pair of arcs of the form $(1, i)$, $(i, 1)$; the requirement is that it must consist of 3 or more nodes.

If $\mathcal{C}_{1p}$ denotes the chain from 1 to p obtained when the last arc, $(p, 1)$ say, from an optimum simple circuit is deleted, show that $\mathcal{C}_{1p}$ must be a minimum undesirable chain from 1 to p not containing the arc $(1, p)$. Use this to develop an algorithm for the jogger's problem, and determine its computational complexity.

Now consider several different versions of the problem.

VERSION 1:   Here G is undirected, and the undesirability of an edge is independent of the direction of travel. Let $\mathbb{T}$ be a shortest (i.e., min. cost) path tree rooted at 1 using the undesirability ratings as the edge cost coefficients. Prove that there exists an optimum jogger's route in which all but one of the edges are from $\mathbb{T}$, and the other, say (x, y), is such that x and y have no common ancestor in $\mathbb{T}$ other than 1. Use this to develop a more efficient algorithm than that discussed above for this version.

VERSION 2:   Even for a two-way street, the undesirability measure may depend on the direction of travel (e.g., if there is a gradient); then it becomes necessary to represent it by a pair of arcs, one in each direction. In this case G becomes directed, and arcs $(i, j), (j, i)$ may exist with different undesirability measures. From the requirements of the route, at most one of these two arcs $(i, j), (j, i)$ can appear on the jog for any i, j. In this version prove that there is an optimum jogger's route which is a concatenation of the form $\mathcal{C}_{1x}, (x, y), \mathcal{C}_{y1}$ where $\mathcal{C}_{1x}, \mathcal{C}_{y1}$ are minimum undesirability chains from 1 to x, and y to 1 respectively. Using this develop an efficient algorithm in this case.

VERSION 3 :   G is undirected as in Version 1. We are given the length of each edge in G in addition to its undesirability measure. There is a constraint that the total length on the circuit must be $\geqq$ some specified quantity. Develop an efficient algorithm for the jogger's problem under this constraint.
(Bird [1981]).

4.33 G $= (\mathcal{N}, \mathcal{A})$ is an acyclic network in which a subset of nodes $\mathbf{X} \subset \mathcal{N}$ have weights w_i associated with them. It is required to find a chain from 1 to n in G such that the sum of the weights of the $\mathbf{X}$-nodes belonging to the chain is maximum. Develop an efficient $O(|\mathcal{N}|)$ complexity algorithm for this problem (Kundu [1978]).

Comment 4.1   As the shortest chain problem (called shortest path problem in some books) is so fundamental for modeling distribution and routing applications, it has been a major focus of research in network optimization, and the literature on it is vast. Deo and Pang [1984] contains an extensive bibliography.

The first label setting algorithm is due to Dijkstra [1959], and independently by Dantzig [1960] and Whiting and Hillier [1960]. Its worst case computational complexity of $O(n^2)$ is the best possible running time in dense networks, since any algorithm for this problem must examine every arc. However, in sparse networks, it is possible to improve the performance of this algorithm through the use of appropriate data structures. Dial's implementation [1965] is very popular and gives excellent computational performance. For other improvements in Dijkstra's algorithm see Denardo and Fox [1979], Dial, Glover, Karney, and Klingman [1979], Fredman and Tarjan [1987], and Johnson [1973, 1977].

Label correcting methods were introduced by Ford [1956], Moore [1957] and Bellman [1958]. Many improvements in the basic algorithm are discussed in Gilsinn and Witzgall [1973], Glover, Glover, and Klingman [1984], Glover, Klingman, and Phillips [1985], Glover, Klingman, Phillips, and Schneider [1985], Goldfarb, Hao, and Kai [1989a, 1989b], Pape [1974], Shier and Witzgall [1981], and Yen [1970]. With these improvements, label correcting methods are very efficient in practice, particularly in sparse networks.

Computational studies comparing various shortest chain algorithms may be found in Denardo and Fox [1979], Dial, Glover, Karney, and Klingman [1979], Gallo and Pallatino [1988], Gilsinn and Witzgall [1973], Glover, Klingman, Phillips, and Schneider [1985], Imai and Iri [1984], Kelton and Law [1978], Pape [1974], and Van Vliet [1978].

4.9   References

R. K. AHUJA, J. L. BATRA, and S. K. GUPTA, May 1983, "Combinatorial Optimization With Rational Objective Functions: A Communication," *MOR*, 8, no. 2(314).

R. BELLMAN, 1958, "On a Routing Problem," *QAM*, 16(87-90).

R. S. BIRD, Dec. 1981, "The Jogger's Problem," *IPL*, 13, no.(114-117).

S. H. BOKHARI, Jan. 1988, "Partitioning Problems in Parallel, Pipelined and Distributed Computing," *IEEE Transactions on Computers*, 37, no. 1(48-57).

I-NGO CHEN, June 1975, "A node Elimination Method for Finding Negative Cycles in a Directed Graph," *INFOR*, 13, no. 2(147-158).

G. A. CHESTON and D. G. CORNEIL, Aug. 1982, "Graph Property Update Algorithms and Their Application to Distance Matrices," *INFOR*, 20, no. 3(178-201).

N. CHRISTOFIDES, A. MINGOZZIA, and P. TOTH, 1981, "Exact Algorithms for the Vehicle Routing Problem, Based on Spanning Trees and Shortest Path Relaxations," *MP*, 20(255-282).

J. A. CLARK and N. A. J. HASTINGS, 1977, "Decision Networks," *ORQ*, 28, no. 1i(51-68).

G. B. DANTZIG, 1960, "On the Shortest Route Through a Network," *MS*, 6(187-190).

G. B. DANTZIG, 1967, "All Shortest Routes in a Graph," in P. Rosentiehl (Ed.), *Theory of Graphs*, Dunod, Paris(91-92).

K. DARBY-DOWMAN and G. MITRA, Aug. 1985, "An Extension of Set Partitioning With Applications to Scheduling Problems," *EJOR*, 21, no. 2(200-205).

E. DENARDO and B. FOX, 1979, "Shortest-route Methods; 1. Reaching, Pruning and Buckets," *OR*, 27(161-186).

N. DEO and C. - Y. PANG, 1984, "Shortest Path Algorithms: Taxonomy and Annotation," *Networks*, 14(275-323).

R. DIAL, 1965, "Algorithm 360: Shortest Path Forest With Topological Ordering," *CACM*, 12(632-633).

R. DIAL, F. GLOVER, D. KARNEY, and D. KLINGMAN, 1979, "A Computational Analysis of Alternative Algorithms and Labeling Techniques for Finding Shortest Path Trees," *Networks*, 9(215-248).

E. DIJKSTRA, 1959, "A Note on Two Problems in Connection With Graphs," *Numerische Mathematik*, 1(269-271).

P. DOULBIEZ and M. RAS, 1975, "Optimal Network Capacity Planning: A Shortest Path Scheme," *OR*, 23(810-818).

S. DREYFUS, 1969, "An Appraisal of Some Shortest-path Algorithms," *OR*, 17(395-412).

J. EDMONDS, 1970, "Exponential Growth of the Simplex Method for Shortest Path Problems," unpublished report, University of Waterloo.

M. FLORIAN, S. NGUYEN, and S. PALLOTINO, 1981, "A Dual Simplex Algorithm for Finding All Shortest Paths," *Networks*, 11(367-378).

R. W. FLOYD, 1962, "Algorithm 97, Shortest Path," *CACM*, 5(345).

L. R. FORD, Jr. 1956, "Network Flow Theory," Rand report P-923.

M. L. FREDMAN and R. E. TARJAN, 1987, "Fibonacci Heaps and Their Uses in Improved Network Optimization Algorithms," *JACM*, 34(596-615).

G. GALLO and S. PALLOTINO, 1988, "Shortest Path Algorithms," in B. Simeone, P. Toth, G. Gallo, F. Maffioli, and S. Pallotino (Eds.), *Fortran Codes for Network Optimization, AOR*, 13(3-79).

A. GARCIA-DIAZ and J. LIEBMAN, 1980, "An Investment Staging Model for a Bridge Replacement Problem," *OR*, 28(736-753).

J. GILSINN and C. WITZGALL, 1973, "A Performance Comparison of Labeling Algorithms for Calculating Shortest Path Trees," Technical note 772, NBS, Washington, D.C.

F. GLOVER, R. GLOVER, and D. KLINGMAN, 1984, "Computational Study of an Improved Shortest Path Algorithm," *Networks*, 14, no. 1(25-36).

F. GLOVER, D. KLINGMAN, and N. PHILLIPS, 1985, "A New Polynomially Bounded Shortest Path Algorithm," *OR*, 33(65-73).

F. GLOVER, D. KLINGMAN, N. PHILLIPS, and R. F. SCHNEIDER, 1985, "New Polynomial Shortest Path Algorithms and Their Computational Attributes," *MS*, 31(1106-1128).

B. GOLDEN, A. ASSAD, L. LEVY, and F. GHEYSENS, 1984, "The Fleet Size and Mix Vehicle Routing Problems," *COR*, 11, no. 1(49-66).

B. GOLDEN and T. MAGNANTI, 1978, "Transportation Planning: Network Models and Their Implementation," *Studies in Operations Management*, North Holland, Amsterdam(365-518).

D. GOLDFARB, J. HAO, and S. R. KAI, July-Aug. 1990, "Efficient Shortest Path Simplex Algorithms," *OR*, 38, no. 4(79 - 91).

D. GOLDFARB, J. HAO, and S. R. KAI, 1991, "Shortest Path Algorithms Using Dynamic Breadth-First Search," *Networks*, 21(29-50).

H. IMAI and M. IRI, 1984, "Practical Efficiencies of Existing Shortest Path Algorithms and a New Bucket Algorithm," *Journal of the OR Society of Japan*, 27(43-58).

D. B. JOHNSON, July 1973, "A Note on Dijkstra's Shortest Path Algorithm," *JACM*, 20(385-388).

D. B. JOHNSON, Jan. 1977, "Efficient Algorithms for Shortest Paths in Sparse Networks," *JACM*, 24, no. 1(1-13).

W. D. KELTON and A. M. LAW, 1978, "A Mean-time Comparison of Algorithms for all Pairs Shortest Path Problem With Arbitrary Arc Lengths," *Networks*, 8(97-106).

A. KERSHENBAUM, 1981, "A Note on Finding Shortest Path Trees," *Networks*, 11, no. 4,(399-400).

M. KLEIN and R. K. TIBREWALA, Feb. 1973, "Finding Negative Cycles," *INFOR*, 11, no. 1(59-65).

S. KUNDU, Jan. 1978, "Note on a Constrained-path Problem in Program Testing," *IEEE Transactions on Software Engineering*, SE-4, no. 1(75-76).

A. S. LAPAUGH and C. H. PAPADIMITRIOU, 1984, "The Even Path Problem for Graphs and Digraphs," *Networks*, 14, no. 4(507-513).

E. L. LAWLER, Mar. 1972, "A Procedure for Computing the *k* Best Solutions to Discrete Optimization Problems and its Application to the Shortest Path," *MS*, 18(401-405).

K. MALIK, A. K. MITTAL, and S. K. GUPTA, Aug. 1989, "The *k*-most Vital Arcs in the Shortest Path Problem," *OR Letters*, 8, no. 4(223-227).

G. J. MINTY, 1958, "A Variant on the Shortest Route Problem," *OR*, 6(882-883).

E. F. MOORE, 1957, "The Shortest Path Through a Maze," in *Proc. of the International Symposium on the Theory of Switching, Part 2*, The Annals of the Computation Laboratory of Harvard Univ. 30(285-292).

U. PAPE, 1974, "Implementation and Efficiency of Moore-algorithms for the Shortest Route Problem," *MP*, 7(212-222).

A. R. PIERCE, 1975, "Bibliography on Algorithms for Shortest Path, Shortest Spanning Tree and Related Circuit Routing Problems," *Networks*, 5(129-143).

M. SCHWARTZ and T. STERN, 1980, "Routing Techniques Used in Computer Communication Networks," *IEEE Transactions on Comm.*, COM-28(539-552).

F. SHEPARDSON and R. E. MARSTEN, 1980, "A Lagrangian Relaxation Algorithm for the Two Duty Period Scheduling Problem," *MS*, 3(274-281).

D. SHIER and C. WITZGALL, 1981, "Properties of Labeling Methods for Determining Shortest Path Trees," *Journal of Research of the NBS*, 86(317-330).

D. TEODOROVIC and S. GUBERINIC, Feb. 1984, "Optimal Dispatching Strategy on an Airline Network After a Schedule Perturbation," *EJOR*, 15, no. 2(178-182).

R. L. TOBIN, 1975, "Minimal Complete Matchings and Negative Cycles," *Networks*, 5, no. 4(371-387).

D. VAN VLIET, 1978, "Improved Shortest Path Algorithms for Transport Networks," *Transportation Research*, 12(7-20).

S. WARSHALL, 1962, "A Theorem on Boolean Matrices," *JACM*, 9(11-12).

A. WEINTRAUB, 1973, "The Shortest and the k-shortest Routes as Assignment Problems," *Networks*, 3(61-73).

P. D. WHITTING and J. A. HILLIER, 1960, "A Method for Finding the Shortest Route Through a Road Network," *ORQ*, 11(37-40).

J. YEN, 1970, "An Algorithm for Finding Shortest Routes From All Source Nodes to a Given Destination in General Networks," *QAM*, 27(526-530).

Chapter 5

Algorithms for Minimum Cost Flow Problems in Pure Networks

This chapter will consider algorithms for minimum cost flow problems in pure networks. We begin by considering pure single commodity linear static minimum cost flow problems in Sections 5.1 to 5.8. The assignment, transportation, and shortest chain problems discussed in Chapters 3, 4 are special cases of these problems. Section 5.9 treats the case of a piecewise linear convex cost function. In Section 5.10 we consider dynamic flow problems in pure networks briefly. Finally, in Section 5.11 we present the arc-chain approach for solving multicommodity flow problems in pure networks.

As discussed in Chapter 1, we assume without any loss of generality that our problems are defined on the directed network $G = (\mathcal{N}, \mathcal{A}, \ell = (\ell_{ij}), k = (k_{ij}), c = (c_{ij}))$, with ℓ, k, c as the lower bound, capacity, and unit cost coefficient vectors for flows on the arcs in $\mathcal{A}$. We assume that $k \geqq \ell \geqq 0$. Some, or all, of the entries in k may be ∞. If a feasible flow vector exists in G, it will be shown later that the cost function is unbounded below on the set of feasible solutions iff there exists an uncapacitated negative cost circuit (i.e., a negative cost circuit consisting only of arcs (i, j) with $k_{ij} = \infty$). By forcing a flow amount of ∞ around an uncapacitated negative cost circuit, the cost can be driven to $-\infty$. We saw this phenomenon also in the shortest chain problem in Chapter 4.

5.1   Different Types of Single Commodity Minimum Cost Flow Models

A common problem, occurring on a directed network $G = (\mathcal{N}, \mathcal{A}, \ell, k, c, \breve{s}, \breve{t}, \overline{v})$, is to ship a specified quantity, $\overline{v}$ units of the commodity from the specified sources $\breve{s}$, to the specified sink $\breve{t}$ in G at minimum cost. It is to find $f = (f_{ij} : (i,j) \in \mathcal{A})$ to

$$\text{Minimize} \sum (c_{ij} f_{ij} : \text{over } (i,j) \in \mathcal{A})$$

$$\text{Subject to} \quad -f(i, \mathcal{N}) + f(\mathcal{N}, i) = \begin{cases} -\overline{v} & \text{if } i = \breve{s} \\ 0 & \text{if } i \neq \breve{s}, \breve{t} \\ \overline{v} & \text{if } i = \breve{t} \end{cases} \tag{5.1}$$

$$\ell_{ij} \leq f_{ij} \leq k_{ij}, \text{ for all } (i,j) \in \mathcal{A}$$

A special case of (5.1) has $\ell = 0$, and $k_{ij} = \infty$ for all $(i,j) \in \mathcal{A}$. If it is feasible, and there are no negative cost circuits in G, an optimum flow for this special problem is obtained by sending all the $\overline{v}$ units along a shortest chain from $\breve{s}$ to $\breve{t}$ with c as the arc length vector. So, this special case is equivalent to a shortest chain problem.

Minimum Cost Circulation Problem

Another problem, known as the *minimum cost circulation problem* , is that of finding a minimum cost circulation in a directed network $G = (\mathcal{N}, \mathcal{A}, \ell, k, c)$. It is to find $f = (f_{ij})$ to

$$\text{Minimize} \sum (c_{ij} f_{ij} : \text{over } (i,j) \in \mathcal{A})$$

$$\text{Subject to} \quad -f(i, \mathcal{N}) + f(\mathcal{N}, i) = 0, \text{ for each } i \in \mathcal{N} \tag{5.2}$$

$$\ell_{ij} \leq f_{ij} \leq k_{ij}, \text{ for all } (i,j) \in \mathcal{A}$$

(5.1) is equivalent to a minimum cost circulation problem on an augmented network G′ obtained by including in G a new arc $(\breve{t}, \breve{s})$ associated with lower bound and capacity both equal to $\overline{v}$, and unit cost coefficient of 0.

Minimum Cost Flow Model with Exogenous Flow Values

Another model, (5.3), is to find a feasible flow vector $f = (f_{ij})$ in the directed network $G = (\mathcal{N}, \mathcal{A}, \ell, k, c, V = (V_i))$, which is a minimum cost flow satisfying specified exogenous flows at the nodes, given by the vector V. In this model, node i is a *source node* if $V_i > 0$, a *sink node* if $V_i < 0$, and an *intermediate* or *transit node* if $V_i = 0$. (5.3) can easily be transformed into a problem of type (5.1) on an augmented network obtained by including a super source and a super sink in G as in Section 2.1.

Minimize $\sum (c_{ij} f_{ij} : \text{over } (i,j) \in \mathcal{A})$

Subject to $-f(i, \mathcal{N}) + f(\mathcal{N}, i) = -V_i$, for each $i \in \mathcal{N}$ (5.3)

$\ell_{ij} \leqq f_{ij} \leqq k_{ij}$, for all $(i,j) \in \mathcal{A}$

The transportation problem is a special case of (5.3) in which the network is bipartite, every node is either a source or a sink node, and all the arcs are directed from a source to a sink node. In a directed network, a node i is called a *shipping node* if $\mathbf{B}_i = \emptyset$, a *receiving node* if $\mathbf{A}_i = \emptyset$, and a *transshipment node* if both $\mathbf{B}_i, \mathbf{A}_i \neq \emptyset$. The transportation problem is a minimum cost flow problem on a network containing no transshipment nodes and vice versa. A minimum cost flow problem on a directed network containing some transshipment nodes is called a *transshipment problem* in the literature.

A necessary condition for (5.3) to be feasible, obtained by summing all the equality constraints in it, is

$$\sum (V_i : \text{over } i \in \mathcal{N}) = 0 \qquad (5.4)$$

In this model, finite nonzero lower bounds on the variables can easily be transformed into zeros. If $\ell_{ij} \neq 0$ and finite for some $(i,j) \in \mathcal{A}$, transform f_{ij} by substituting $f_{ij} = f'_{ij} + \ell_{ij}$, where f'_{ij} is the new variable replacing f_{ij}. This leads to the transformation of the data on arc (i,j) as shown in Figure 5.1. Verify that in the transformed problem, (5.4) continues to hold if it does so in the original problem. A similar transformation can be carried out for every arc with finite nonzero lower bound.

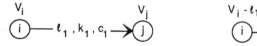
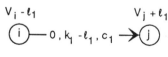

Original arc with data. Lower   Transformed arc with 0
bound is finite.   as lower bound.

Figure 5.1

Suppose (5.4) does not hold and $U = \sum (V_i : \text{over } i \in \mathcal{N}) \neq 0$. In this case we have *excess supply* if $U > 0$, *shortage* otherwise, and the equality constraints in (5.3) cannot be satisfied exactly. If there is excess supply, introduce an artificial sink node $\check{t}$ with exogenous flow value, $V_{\check{t}} = -U$; and artificial arcs $(i, \check{t})$ with lower bound, capacity, unit cost coefficient equal to 0, ∞, 0 respectively, for each source node i. Flows on each of these artificial arcs represent the amount of material unutilized in G at the source nodes on those arcs. Data on the augmented network now satisfy (5.4). We can find an optimum flow vector in G which meets all the requirements at the sink nodes exactly at minimum cost, while leaving U units of material unutilized at the source nodes, by solving the problem of type (5.3) in the augmented network.

If there is shortage, there will be an unfulfilled demand of $|U|$ units. Introduce an artificial source node $\breve{s}$ with an exogenous flow of $|U|$ units; and artificial arcs $(\breve{s}, j)$ with lower bound, capacity, unit cost coefficient equal to 0, ∞, $c_{\breve{s}j}$ respectively for each sink node j. Flow on the artificial arc $(\breve{s}, j)$ represents the unfulfilled requirement at node j. All $c_{\breve{s}j}$ are made 0 if it is just required to find how the existing supply can be used to meet as much of the demand as possible at minimum shipping cost, without giving any special preference to any of the sink nodes. Otherwise $c_{\breve{s}j}$ can be taken to be the per-unit shortage at sink j if such data is available and it is desired to minimize the sum of the shipping costs and the costs of shortage. Or, we can make $c_{\breve{s}j}$ to be suitable positive weights to reflect the priorities for fulfilling the demands at sinks j. With these modifications, data on the augmented network satisfy (5.4), and we can solve the problem of type (5.3) on it.

A General Minimum Cost Flow Model

In this model **S**, **T**, the sets of source, sink nodes in $G = (\mathcal{N}, \mathcal{A}, \ell, k, c)$ are specified. For each $i \in$ **S**, we are given numbers $a'_i \geqq a_i > 0$, and the net amount of material shipped out of i is required to be between a_i and a'_i. For each $j \in$ **T**, we are given numbers $b'_j \geqq b_j > 0$, and the net amount of material reaching j is required to be between b_j and b'_j. So, this model is a generalization of the model (5.3), with the exogenous flow amounts V_is being themselves variables subject to lower and upper bound constraints. Introduce two new nodes, $\breve{s}, \breve{t}$, a supersource and supersink. For each $i \in$ **S**, introduce the arc $(\breve{s}, i)$ with lower bound, capacity, cost coefficient equal to $a_i, a'_i, 0$ respectively. For each $j \in$ **T**, introduce the arc $(j, \breve{t})$ with lower bound, capacity, cost coefficient equal to $b_j, b'_j, 0$ respectively. Introduce the arc $(\breve{t}, \breve{s})$ with lower bound, capacity, cost coefficient equal to 0, ∞, 0 respectively. Clearly, this general model is equivalent to the minimum cost circulation problem of the form (5.2) in the augmented network.

The Maximum Profit Flow Problem

This problem, (5.5), is the same as (5.1), with the exception that each unit shipped to the sink can be sold there at a *premium* (this is the difference between the selling prices per unit of the material at the sink and the source) of λ. In this problem the data typically satisfies $\ell = 0, k > 0, c \geqq 0$. The objective is to maximize the net profit which is the total premium minus the shipping cost. The flow value v is also a variable in this problem, and typically it is required to be solved as a parametric problem with λ as a nonnegative parameter. For any λ, (5.5) can be posed as a minimum cost circulation problem of the form (5.2) by introducing the arc $(\breve{t}, \breve{s})$ into the network and making v the flow variable associated with it.

$$\text{Maximize } P(f, \lambda) \;=\; \lambda v - \sum (c_{ij} f_{ij} : \text{ over } (i,j) \in \mathcal{A})$$

$$\text{Subject to} \quad - f(i, \mathcal{N}) + f(\mathcal{N}, i) = \begin{cases} -v & \text{if } i = \check{s} \\ 0 & \text{if } i \neq \check{s}, \check{t} \\ v & \text{if } i = \check{t} \end{cases} \qquad (5.5)$$

$$0 \stackrel{\le}{=} f_{ij} \stackrel{\le}{=} k_{ij}, \text{ for all } (i,j) \in \mathcal{A}$$

In each of the flow models discussed above on the directed network G with $c = (c_{ij})$ as the cost vector, c can be replaced by the reduced cost vector $\bar{c} = (\bar{c}_{ij} = c_{ij} - (\pi_j - \pi_i))$, where $\pi = (\pi_i)$ is any node price vector in G. For any feasible flow vector f, $cf = \bar{c}f +$ a constant, where the constant is independent of f, but depends only on π and the data in the problem such as V etc. Hence replacing c by $\bar{c}$ does not change the set of optimum solutions, and thus leads to an equivalent problem. Some algorithms make use of this idea.

We have seen that the various flow models discussed above are equivalent. In presenting algorithms, this gives us the freedom to select any of these models, and describe the algorithm as it applies to that model.

5.2   Optimality Conditions

Complementary Slackness Optimality Conditions

Consider the problem (5.3) in the network G. The equality constraints in it are $-Ef = -V$, where E is the node-arc incidence matrix of G. Each of these constraints corresponds to a node, and so the dual variable associated with that constraint can be interpreted as the price of that node in the dual problem. So, the variables in the dual problem are node prices. Given the node price (row) vector π, the complementary slackness optimality conditions (see (1.10)) for (5.3) and its dual are stated in terms of f and the vector $c - (-\pi E)$. From the definition of E, it can be seen that $-\pi E$ is the row vector of tensions $(\pi_j - \pi_i : (i,j) \in \mathcal{A})$ on the arcs in $\mathcal{A}$ wrt π. From this, it can be verified that these conditions (1.10) simplify to (5.6) given below for this primal dual pair. In the same manner, in each of the problems (5.1), (5.2), or (5.3) on G, a feasible flow vector $f = (f_{ij})$ is optimal iff there exists a vector π of dual variables (or node prices, or node potentials) such that f, π together satisfy: for each $(i,j) \in \mathcal{A}$

$$\begin{aligned} &\text{if } \pi_j - \pi_i \;>\; c_{ij} \text{ then } k_{ij} \text{ is finite and } f_{ij} = k_{ij} \\ &\text{if } \pi_j - \pi_i \;<\; c_{ij} \text{ then } f_{ij} = \ell_{ij} \end{aligned} \qquad (5.6)$$

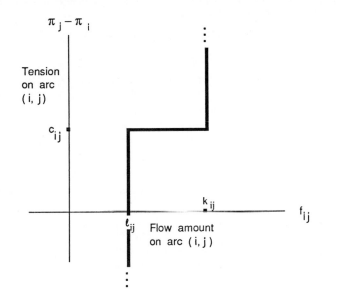

Figure 5.2 C.S. diagram for an arc (i, j) with finite capacity.

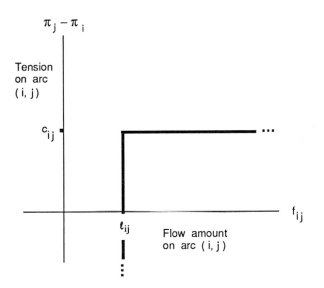

Figure 5.3 C.S. diagram for an arc (i, j) with infinite capacity.

These conditions, known as the *complementary slackness optimality conditions*, or *c.s. conditions* in short, can be illustrated in a diagram known as the *complementary slackness diagram* or *c.s. diagram* for arc (i, j). See Figures 5.2, 5.3, in which f_{ij} is plotted on the horizontal axis, and the tension $\pi_j - \pi_i$ is plotted on the vertical axis. The feasible flow vector-dual vector pair (f, π) satisfies the c.s. conditions for arc (i, j) iff the point $(f_{ij}, \pi_j - \pi_i)$ lies on the chair-shaped curve in Figures 5.2, 5.3.

Optimality Conditions in Terms of Negative Cost Residual Cycles

Consider any of the single commodity minimum cost flow problems (5.1), (5.2), or (5.3) in the directed network $G = (\mathcal{N}, \mathcal{A}, \ell, k, c)$. Given any path, or an oriented cycle in G, define its cost to be = (the sum of the costs of forward arcs in it) − (the sum of the costs of the reverse arcs in it). Theorem 5.1 given below establishes that a feasible flow vector for this problem is optimal iff there exists no negative cost residual cycle wrt it. The proof of this theorem requires a couple of lemmas which we state and prove first.

LEMMA 5.1 G *is a directed network with c as the vector of cost coefficients on the arcs, and* $g^0 = (g^0_{ij}) \geqq 0$ *is a circulation in it satisfying* $cg^0 < 0$. $\mathbf{W}_0 = \{(i, j) : g^0_{ij} > 0\}$. *Then there exists a negative cost simple circuit among the set of arcs* $\mathbf{W}_0$.

Proof Since g^0 is a circulation, if node i lies on an arc in $\mathbf{W}_0$, there must exist at least one arc in $\mathbf{W}_0$ in both the forward and reverse stars of i. Start with such a node i and arcs of the form $(j_1, i), (i, p_1)$ in $\mathbf{W}_0$. Using the same statement again and again trace a chain of the form $i, (i, p_1), p_1, (p_1, p_2), p_2, (p_2, p_3), \ldots$ and a path of the form $i, (j_1, i), j_1, (j_2, j_1), j_2, (j_3, j_2), \ldots$, both beginning at i and consisting of arcs from $\mathbf{W}_0$ only; until either the chain and the path have a common node, say s; or a node is repeated in the chain or the path. The path traced from i to s is actually a chain from s to i in reverse order. When either of these events occur, either the chain, or the path, or both of them put together, have a simple circuit, say $\overrightarrow{\mathbb{C}}_0$ from the set of arcs $\mathbf{W}_0$. If the cost of $\overrightarrow{\mathbb{C}}_0$ is < 0, we are done. Otherwise let $\alpha = \min. \ \{ \ g^0_{ij} : (i, j) \ \text{on} \ \overrightarrow{\mathbb{C}}_0 \ \}$, subtract α from g^0_{ij} for each arc (i, j) on $\overrightarrow{\mathbb{C}}_0$ and let $g^1 = (g^1_{ij})$ be the resulting flow vector in G. Verify that g^1 is again a circulation in G and $g^1 \geqq 0$. The cost of g^1 is $cg^1 = cg^0 - \alpha$ (cost of $\overrightarrow{\mathbb{C}}_0$) < 0 since $cg^0 < 0, \alpha > 0$ and the cost of $\overrightarrow{\mathbb{C}}_0$ is $\geqq 0$. Also, $\mathbf{W}_1 = \{(i, j) : g^1_{ij} > 0\} \subset \mathbf{W}_0$ and $|\mathbf{W}_1| \leqq |\mathbf{W}_0| - 1$. Now apply the same procedure on $\mathbf{W}_1$, and repeat in the same way. This leads to a sequence of nonnegative circulations $g^1, g^2, \ldots$ in G, all of negative cost with $\mathbf{W}_r = \{(i, j) : g^r_{ij} > 0\}$ satisfying $|\mathbf{W}_r| \leqq |\mathbf{W}_{r-1}| - 1$ for all r, and $\mathbf{W}_0 \supset \mathbf{W}_1 \supset \mathbf{W}_2 \supset \ldots$. So, for some $r \leqq |\mathbf{W}_0| - 2$, the simple circuit found when the procedure is applied on $\mathbf{W}_r$ must have negative cost. This circuit is a negative cost simple circuit in $\mathbf{W}_0$, proving the lemma. ∎

LEMMA 5.2 *Let G be a directed network with c as the vector of arc cost coefficients, and* $\overline{g} = (\overline{g}_{ij})$ *a negative cost circulation in G (i.e.,* $\overline{g}$ *satisfies flow conser-*

vation at all the nodes), but $\overline{g}$ may not be $\geqq 0$. $\mathbf{W} = \{ (i,j) : \overline{g}_{ij} \neq 0 \}$. *Then there exists a negative cost oriented cycle $\mathbb{C}$ among the set of arcs $\mathbf{W}$ such that $\overline{g}_{ij} > 0$ on all forward arcs (i,j) on $\mathbb{C}$, and $\overline{g}_{ij} < 0$ on all reverse arcs (i,j) on $\mathbb{C}$.*

Proof If arc (i,j) is such that $\overline{g}_{ij} > 0$, label it with $+$ and leave it as it is. If (i,j) is such that $\overline{g}_{ij} < 0$ reverse its orientation (i.e., replace it with (j,i)), change the cost coefficient on it to $-c_{ij}$, and the flow on it to $-\overline{g}_{ij}$, and label it with $-$. Let the resulting network and the flow vector on it be $\hat{G}, \hat{g}$. Then $\hat{g} \geqq 0$, it is a circulation in $\hat{G}$, and its cost in $\hat{G} =$ the cost of $\overline{g}$ in G, which is < 0. So, by Lemma 5.1 there exists a negative cost simple circuit $\overrightarrow{\mathbb{C}}$ in $\hat{G}$ among the set of arcs on which $\hat{g}$ has positive flow. Verify that changing the orientations of the $-$ labeled arcs on $\overrightarrow{\mathbb{C}}$ converts it into a simple cycle $\mathbb{C}$ in G. Orient $\mathbb{C}$ so that the $+$ labeled arcs on it are forward arcs, and verify that it satisfies all the properties stated in the lemma. $\blacksquare$

THEOREM 5.1 *Let $\overline{f} = (\overline{f})$ be a feasible flow vector in $G = (\mathcal{N}, \mathcal{A}, \ell, k, c)$ for a minimum cost flow problem of the form (5.1), (5.2), or (5.3). $\overline{f}$ is a minimum cost feasible flow vector for this problem iff there exists no negative cost residual cycle wrt $\overline{f}$ in G.*

Proof Suppose $\mathbb{C}_0$ is a negative cost residual cycle wrt $\overline{f}$ in G. Let ϵ_0 be the residual capacity of $\mathbb{C}_0$. Let $\hat{f}$ be the flow vector obtained by increasing (decreasing) the flow amount in $\overline{f}$ on forward (reverse) arcs of $\mathbb{C}_0$ by ϵ_0. Since these flow changes are made along the arcs on a simple cycle, we have $\hat{f}(i,\mathcal{N}) - \hat{f}(\mathcal{N},i) = \overline{f}(i,\mathcal{N}) - \overline{f}(\mathcal{N},i)$ for all $i \in \mathcal{N}$. From this and the definition of ϵ_0, it follows that $\hat{f}$ is a feasible flow vector and $c\hat{f} = c\overline{f} + \epsilon_0$ (cost of $\mathbb{C}_0$) $< c\overline{f}$ since $\epsilon_0 > 0$ and the cost of $\mathbb{C}_0$ is < 0, so $\overline{f}$ is not a minimum cost flow for the problem.

To prove the "only if" part, suppose $\overline{f}$ is a feasible but not a minimum cost flow vector for the problem. Then there must exist a feasible flow vector f^0 whose cost is strictly less than that of $\overline{f}$. Define $g = (g_{ij}) = f^0 - \overline{f}$. g is a negative cost circulation in G, but it may not be nonnegative or satisfy the lower or upper bound conditions on the arcs. Let $\mathbf{W}^+ = \{(i,j) : g_{ij} > 0\}$, $\mathbf{W}^- = \{(i,j) : g_{ij} < 0\}$, $\mathbf{W} = \mathbf{W}^+ \cup \mathbf{W}^-$. Since both f^0 and $\overline{f}$ are feasible flow vectors in G, we have

$$k_{ij} \geqq f_{ij}^0 > \overline{f}_{ij}, \text{ for all } (i,j) \in \mathbf{W}^+$$

$$\ell_{ij} \leqq f_{ij}^0 < \overline{f}_{ij}, \text{ for all } (i,j) \in \mathbf{W}^-$$

By applying Lemma 5.2 to g in G, we conclude that there must exist a negative cost oriented cycle $\mathbb{C}_0$ consisting of arcs from $\mathbf{W}$ satisfying: $g_{ij} > 0$ for forward arcs on $\mathbb{C}_0$ and $g_{ij} < 0$ for reverse arcs on $\mathbb{C}_0$. Verify that $\mathbb{C}_0$ is a negative cost residual cycle wrt $\overline{f}$. $\blacksquare$

An equivalent statement to Theorem 5.1 is that a feasible flow vector f in G for problems (5.1), (5.2), or (5.3) is optimal iff the residual networks $G(f)$ or $G(f, \pi)$ for any node price vector π contain no negative cost circuits.

Canceling a Residual Cycle

Let f be a feasible flow vector in the directed network $G = (\mathcal{N}, \mathcal{A}, \ell, k, c)$ for (5.1), (5.2), or (5.3). Let $\mathbb{C}$ be a residual cycle wrt f of residual capacity α. Let $\hat{f}$ be the flow vector in G obtained by increasing (decreasing) the flow amount in f on forward (reverse) arcs of $\mathbb{C}$ by α. Clearly $\hat{f}$ is also a feasible flow vector for the problem. The operation of obtaining $\hat{f}$ from f is called *canceling the residual cycle* $\mathbb{C}$ *in* f. $c\hat{f} = cf + \alpha$ (cost of $\mathbb{C}$). Since $\alpha > 0$, canceling a negative cost residual cycle strictly reduces the cost. As an example consider the network in Figure 5.7 later on. The data on the arcs is the lower bound, capacity, cost coefficient, in that order. A feasible flow vector of value 12 and cost 126 is marked in Figure 5.7 with the flow on each arc entered inside a box by the side of the arc if it is nonzero. The cycle 1, (1, 3), 3, (3, 5), 5, (2, 5), 2, (1, 2), 1 with (1, 2), (2, 5) as reverse arcs and (1, 3), (3, 5) as forward arcs, is a negative cost residual cycle wrt this flow vector. Its residual capacity is min. $\{10 - 7, 4 - 3, 6, 5\} = 1$, and its cost is -7. Canceling this residual cycle leads to the feasible flow vector marked in Figure 5.8 with cost 119.

Many of the algorithms for solving minimum cost flow problems, such as the out-of-kilter algorithm (Section 5.3), the primal simplex algorithm (Section 5.4), the Goldberg-Tarjan algorithm (Section 5.8), are all based on the operation of finding and canceling negative cost cycles repeatedly.

Some Results on Optimum Solutions

LEMMA 5.3 *In (5.1) suppose $\ell = 0, k > 0$ and there exists no negative cost circuit in G. Let $\delta = $ min. $\{k_{ij} : (i, j) \in \mathcal{A}\}$. If there is a chain from $\check{s}$ to $\check{t}$ in G, any flow vector which sends a flow amount of $\overline{v}$ along all the arcs of a shortest chain from $\check{s}$ to $\check{t}$ with c as the vector of arc lengths, is an optimum flow for (5.1) for all $0 \leqq \overline{v} \leqq \delta$.*

Proof Let $\mathcal{C}$ be any such shortest chain from $\check{s}$ to $\check{t}$. By the results in Chapter 4, there exists a node price vector $\tilde{\pi} = (\tilde{\pi}_i)$ such that

$$\tilde{\pi}_j - \tilde{\pi}_i \begin{cases} = c_{ij} \text{ for all arcs } (i, j) \text{ on } \mathcal{C} \\ \geqq c_{ij} \text{ for all other arcs.} \end{cases} \tag{5.7}$$

Define $\tilde{f}_{ij} = \overline{v}$ for (i, j) on $\mathcal{C}$, $= 0$ otherwise, and let $\tilde{f} = (\tilde{f}_{ij})$. Then $\tilde{f}$ is feasible to (5.1) and by (5.7), $(\tilde{f}, \tilde{\pi})$ satisfy (5.6). So, $\tilde{f}$ is optimal to (5.1) for this value $\overline{v}$.

∎

Under the hypothesis in Lemma 5.3, one is tempted to think of the following scheme for solving (5.1) in G for any $\bar{v} \gtreqless 0$. The scheme begins with an initial optimum flow vector of small value obtained as in the proof of Lemma 5.3. Then it tries to augment flow successively on the cheapest available chain from $\check{s}$ to $\check{t}$ in each step, dropping arcs from further consideration once they become saturated, until the value reaches $\bar{v}$. Since flow augmentation is carried out only along chains, from Chapter 2 we know that when (5.1) is feasible, this scheme may not even find a feasible flow vector at termination. However, it seems highly intuitive that if a feasible flow vector is obtained at termination of this scheme, it will be a minimum cost flow. Unfortunately, this may not be true, as in the network in Figure 5.4 constructed by Mike Plantholt. All lower bounds are 0, and the data on the arcs is the capacity, unit cost coefficient, in that order. We require a minimum cost flow of value 2. With the cost coefficients as the lengths, the shortest chain from 1 to 6 is 1, (1, 2), 2, (2, 5), 5, (5, 6), 6. The capacity of this chain is 1, so, applying the above scheme, we get the initial flow vector of $f^1 = (f^1_{12}, f^1_{13}, f^1_{24}, f^1_{25}, f^1_{34}, f^1_{35}, f^1_{46}, f^1_{56}) = (1, 0, 0, 1, 0, 0, 0, 1)$ of value $v^1 = 1$. In all the flow vectors, we will order the arcs in the same order as in f^1. In f^1 arcs (1, 2), (2, 5), (5, 6) are saturated, which we eliminate from further consideration. The shortest chain from 1 to 6 in the remaining network is 1, (1, 3), 3, (3, 4), 4, (4, 6), 6 with a cost of 1003. Augmenting the flow on each of the arcs of this chain by 1 leads to the flow vector $f^2 = (1, 1, 0, 1, 1, 0, 1, 1)$, of value 2 and cost 1006. f^2 is not a minimum cost flow vector for this problem since the flow vector $f = (1, 1, 1, 0, 0, 1, 1, 1)$ is feasible and has a cost of only 8.

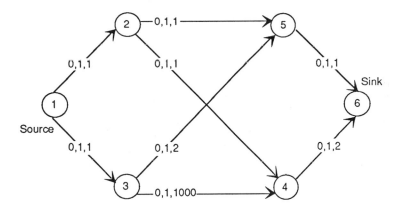

Figure 5.4

Given a minimum cost feasible flow vector f of value v, to get minimum cost flow vectors of value $> v$ one may have to reduce the flow amounts in f on some of the arcs, and reroute those amounts. The above scheme never reduces the flow amount on any arc and that's why it didn't work. However, see Exercise 5.21 for minimum cost flow problems on a very special class of networks, for which this scheme works.

We have seen that flow augmentation along a minimum-cost FAC may not preserve optimality. However, we now show that flow augmentation along any minimum cost FAP does always preserve optimality.

THEOREM 5.2 *Let* $\overline{f}$ *be a minimum cost flow vector in* $G = (\mathcal{N}, \mathcal{A}, \ell, k, c, \check{s}, \check{t})$ *of value* $\overline{v}$. $G(\overline{f}) = (\mathcal{N}, \mathcal{A}(\overline{f}), 0, \kappa, c')$ *is the residual network wrt* $\overline{f}$.

(i)   *The cost of any chain from* $\check{s}$ *to* $\check{t}$ *in* $G(\overline{f})$ *is the same as the cost of the corresponding FAP in* G. *So, every shortest chain from* $\check{s}$ *to* $\check{t}$ *in* $G(\overline{f})$ *with* c' *as the arc length vector, corresponds to a minimum cost FAP from* $\check{s}$ *to* $\check{t}$ *wrt* $\overline{f}$ *in* G.

(ii)   *If there exists no chain from* $\check{s}$ *to* $\check{t}$ *in* $G(\overline{f})$, $\overline{f}$ *is a maximum value flow in* G.

(iii)   *Let* C_0 *be a shortest chain from* $\check{s}$ *to* $\check{t}$ *in* $G(\overline{f})$. *Its capacity is* $\delta = min.$ $\{\kappa_{pq} : (p,q) \text{ on } C_0\}$. *Define a flow vector* $f(\lambda) = (f_{ij}(\lambda) : (i,j) \in \mathcal{A})$ *in* G, *where*

$$f_{ij}(\lambda) = \begin{cases} \overline{f}_{ij} & \text{if } (i,j) \text{ does not correspond to an arc on } C_0 \\ \overline{f}_{ij} + \lambda & \text{if } (i,j) \text{ corresponds to a + arc on } C_0 \\ \overline{f}_{ij} - \lambda & \text{if } (i,j) \text{ corresponds to a - arc on } C_0 \end{cases}$$

Then $\delta > 0$, *and* $f(\lambda)$ *is a minimum cost feasible flow vector in* G *of value* $\overline{v} + \lambda$, *for all* $0 \leqq \lambda \leqq \delta$.

Proof   (i) follows directly from the definitions. If there is no chain from $\check{s}$ to $\check{t}$ in G($\overline{f}$), there exists no FAP from $\check{s}$ to $\check{t}$ wrt $\overline{f}$ in G, hence $\overline{f}$ is a maximum value feasible flow vector in G by Theorem 2.3, establishing (ii).

Clearly $f(\lambda)$ is a feasible flow vector in G of value $\overline{v} + \lambda$, for all $0 \leqq \lambda \leqq \delta$. Since C_0 is a shortest chain in G($\overline{f}$), by the results in Chapter 4 there must exist a node price vector $\mu = (\mu_i : i \in \mathcal{N})$ satisfying

$$\mu_q - \mu_p \leqq c'_{pq}, \text{ for all } (p,q) \in \mathcal{A}(\overline{f}) \tag{5.8}$$

and (5.8) holds as an equation for each arc on C_0.

If $(i,j) \in \mathcal{A}$ is such that $\ell_{ij} < \overline{f}_{ij} < k_{ij}$, both (i,j) and (j,i) are in $\mathcal{A}(\overline{f})$; by applying (5.8) to both these arcs we have $\mu_j - \mu_i = c_{ij}$.

If (i,j) is such that $\overline{f}_{ij} = \ell_{ij}$, and $\ell_{ij} < k_{ij}$, then (i,j) is in $\mathcal{A}(\overline{f})$ but not (j,i), and from (5.8) we have $\mu_j - \mu_i \leqq c_{ij}$. Similarly, if (i,j) is such that $\overline{f}_{ij} = k_{ij}$, and $\ell_{ij} < k_{ij}$, then (j,i) is in $\mathcal{A}(\overline{f})$ but not (i,j), and from (5.8) we have $\mu_j - \mu_i \geqq c_{ij}$.

For all arcs $(i, j) \in \mathcal{A}$ which correspond to an arc in $\mathcal{C}_0, \mu_j - \mu_i = c_{ij}$, since (5.8) holds as an equation for those arcs.

These facts together imply that $f(\lambda), \mu$ together satisfy the complementary slackness optimality conditions (5.6). So, $f(\lambda)$ is an optimum feasible flow vector for all $0 \leqq \lambda \leqq \delta$. ∎

Since shortest chains in $G(\overline{f})$ and $G(\overline{f}, \pi)$ for any node price vector π are the same, the results in Theorem 5.2 continue to hold if we replace $G(\overline{f})$ by $G(\overline{f}, \pi)$. Given a minimum cost flow vector for (5.1) for some $\overline{v}$, these results can be used to find minimum cost flow vectors of higher values by augmenting successively along cheapest FAPs until the desired value is reached. This is the *incremental*, or *build-up approach* for solving this problem. It is the basis for the shortest augmenting path method for minimum cost flows (Sections 5.3, 5.5). This approach is also useful when it is required to solve (5.1) parametrically, treating $\overline{v}$ as a parameter. Also, since the maximum profit flow problem (5.5) is basically a parametric problem, the algorithm for it discussed in Section 5.3 is derived by applying this approach to that problem.

There are three approaches on which most of the practical minimum cost flow algorithms are based. They are: (1) the shortest augmenting path, or the incremental approach, (2) the negative cost residual cycle approach, and (3) the primal-dual approach. We discuss several of these algorithms next.

Exercises

5.1 Let $\overline{f}$ be a minimum cost feasible flow vector for (5.1) in G, and $\mathcal{P}$ a shortest (i.e., minimum cost) augmenting path wrt $\overline{f}$ from $\check{s}$ to $\check{t}$ of residual capacity ϵ. Let $f(\lambda)$ be the flow vector obtained by augmenting the flow by λ along $\mathcal{P}$, for $0 \leqq \lambda \leqq \epsilon$. Let $\mathcal{C}$ be the shortest chain in $G(\overline{f})$ corresponding to $\mathcal{P}$. If the residual network $G(f(\lambda))$ has a negative cost circuit $\overrightarrow{\mathbb{C}}$, show that $\overrightarrow{\mathbb{C}}$ must have some common arcs with $\mathcal{C}$ (use Theorem 5.1) and that $\overrightarrow{\mathbb{C}} \cup \mathcal{C}$ contains a chain in $G(\overline{f})$ shorter than $\mathcal{C}$, a contradiction. Hence provide an alternate proof of (iii) of Theorem 5.2 using Theorem 5.1.

5.2 Show that the minimum cost flow problem (5.3) can be transformed into a problem of the same type on an augmented network in which the lower bounds associated with all the arcs are 0, and all the capacities are all ∞.

5.3 $G = (\mathcal{N}, \mathcal{A}, 0, k, \check{s}, \check{t})$ is a directed connected single commodity flow network, with the following additional features. On each arc $(i, j) \in \mathcal{A}$, any flow amount $\leqq$ the capacity k_{ij} goes through absolutely free of cost. It is possible however to send on this arc any amount of flow $>$ the capacity, at a cost of $\$d_{ij}$ per additional unit. $d = (d_{ij})$ is given. The second feature is that we get a reward of $\$\mu$ for each unit of material reaching $\check{t}$ from $\check{s}$. It is required to find a conservative flow vector (i.e., one satisfying flow conservation equations at every node) in G that maximizes the net return. Formulate this as a minimum cost circulation problem.

5.4 Consider the minimum cost flow problem (5.1). $[\mathbf{X}, \overline{\mathbf{X}}]$ is an arbitrary cut separating $\check{s}$ and $\check{t}$ in G. Determine the new cost vector $c' = (c'_{ij})$, where $c'_{ij} = c_{ij} + \alpha$ if $(i, j) \in (\mathbf{X}, \overline{\mathbf{X}})$, $c_{ij} - \alpha$ if $(i, j) \in (\overline{\mathbf{X}}, \mathbf{X})$, or c_{ij} otherwise, for some α. Consider the same problem with c changed to c'. Is there any relationship between the sets of optimum solutions for the two problems? Explain why.

5.3   The Out-of-Kilter Algorithm

Consider any of the minimum cost flow problems discussed in Section 5.1 on the directed network $G = (\mathcal{N}, \mathcal{A}, \ell, k, c)$ with $|\mathcal{N}| = n, |\mathcal{A}| = m$. To solve it, the OK method can be initiated with an arbitrary flow vector, node price vector pair (f, π). The practical efficiency of the algorithm improves considerably if the initial f is a feasible flow vector; this can be obtained using the methods discussed in Chapter 2. If initiated with a feasible flow vector, all flow vectors in the algorithm will be feasible. The method alternates between a flow change subroutine (during which the node price vector remains unchanged) and a node price change subroutine (during which the flow vector remains unchanged). So, in each step, on each arc (i, j), the point $(f_{ij}, \pi_j - \pi_i)$ either moves horizontally (flow change), or vertically (node price change), and it always moves closer to the chair-shaped c.s. diagram for that arc. We first discuss the version of the algorithm that begins with an initial feasible flow vector.

The *kilter status* of an arc (i, j) wrt a feasible flow vector, node price vector pair $(f = (f_{ij}), \pi = (\pi_i))$ is determined by the position of the point $(f_{ij}, \pi_j - \pi_i)$ on the c.s. diagram. There are five possible states which are determined by the following conditions. (i, j) is an:

$$\alpha - \text{arc} \quad if \qquad \pi_j - \pi_i < c_{ij} \text{ and } f_{ij} = \ell_{ij}$$

$$\beta - \text{arc} \quad if \qquad \pi_j - \pi_i = c_{ij} \text{ and } \ell_{ij} \leqq f_{ij} \leqq k_{ij}$$

$$\gamma - \text{arc} \quad if \quad \pi_j - \pi_i > c_{ij} \text{ and } k_{ij} \text{ is finite and } f_{ij} = k_{ij}$$

$$a - \text{arc} \quad if \qquad \pi_j - \pi_i < c_{ij} \text{ and } f_{ij} > \ell_{ij}$$

$$b - \text{arc} \quad if \qquad \pi_j - \pi_i > c_{ij} \text{ and } f_{ij} < k_{ij}$$

See Figure 5.5. The pair (f, π) satisfies the c.s. conditions (5.6) corresponding to $\alpha-, \beta-, \gamma-$arcs, hence these arcs are said to be *in kilter*. It violates the c.s.

conditions (5.6) corresponding to a-, b-arcs; hence these arcs are said to be *out-of-kilter*.

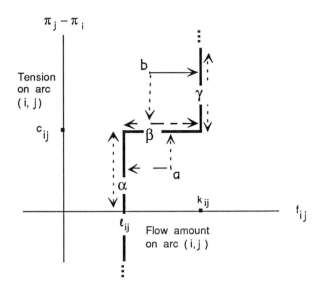

Figure 5.5 Kilter classes and possible changes in f_{ij}, $\pi_j - \pi_i$ in each class for a capacitated arc (i,j).

The β-arcs can be further classified into 3 distinct classes by the flow amount on them. In the pair (f, π) a β-arc (i,j) is: an *upper boundary* or *saturated β-arc* if k_{ij} is finite and $f_{ij} = k_{ij}$; an *interior β-arc* if $\ell_{ij} < f_{ij} < k_{ij}$; a *lower boundary β-arc* if $f_{ij} = \ell_{ij}$.

For each arc (i,j) we define a number called its *killer number*, denoted by $KN(i,j)$, wrt a pair (f, π), to measure how far away the point $(f_{ij}, \pi_j - \pi_i)$ is from satisfying the c.s. condition. $KN(i,j)$ is always > 0 if (i,j) is out-of-kilter, 0 if it is in-kilter. The kilter numbers are not used in the execution of the algorithm, but only in proving its finite termination property. One possible definition is:

$$KN(i,j) = \begin{cases} 0 & \text{if } (i,j) \text{ is an } \alpha\text{-}, \beta\text{-, or } \gamma\text{-arc} \\ f_{ij} - \ell_{ij} & \text{if } (i,j) \text{ is an } a\text{-arc} \\ k_{ij} - f_{ij} & \text{if } (i,j) \text{ is a capacitated } b\text{-arc} \end{cases} \tag{5.9}$$

With this definition, at any stage of the algorithm, the sum of the kilter numbers on all the arcs is a measure (in terms of flow units) of the extent of nonoptimality of the current pair (f, π). Other definitions of kilter numbers are sometimes used; these are given later. In this algorithm, the kilter number of every arc will be monotone nonincreasing, and the algorithm terminates whenever all of them become 0.

Only certain types of flow and tension changes are permitted in this algorithm in order to make sure that in-kilter arcs always stay in-kilter, and out-of-kilter arcs always move closer to the in-kilter status. These permissible changes are summarized below (also see Figure 5.5).

$a-$arc f_{ij} can only decrease, up to ℓ_{ij},
               $\pi_j - \pi_i$ can increase only, up to c_{ij}

$b-$arc f_{ij} can only increase, up to k_{ij},
               $\pi_j - \pi_i$ can decrease only, up to c_{ij}

$\beta-$arc f_{ij} can change freely within ℓ_{ij} to k_{ij} ,
               $\pi_j - \pi_i$ cannot change for interior β-arcs
                        can decrease arbitrarily for lower boundary β-arcs
                        can increase arbitrarily for upper boundary β-arcs

$\alpha-$arc f_{ij} can't change, $\pi_j - \pi_i$ can decrease arbitrarily, or increase up to c_{ij}

$\gamma-$arc f_{ij} can't change, $\pi_j - \pi_i$ can increase arbitrarily, or decrease up to c_{ij}

Notice that an uncapacitated b-arc can never be brought any closer to in-kilter status by flow changes only; it can come closer to kilter only by reducing its tension up to its cost coefficient. For this reason, when there are uncapacitated arcs in G, we select the initial node price vector by a special procedure to guarantee that there will never be any uncapacitated b-arcs.

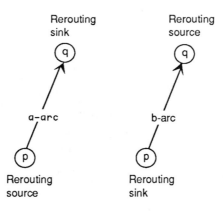

Figure 5.6   Definition of rerouting source and sink on distinguished arc (p, q).

In each stage the algorithm selects an out-of-kilter arc and tries to bring it closer to kilter; this arc is called the *distinguished arc* in that stage. Suppose it is (p, q). The algorithm first tries to bring it into kilter through flow change. If (p, q) is an *a*-arc, the flow on it, f_{pq}, has to be decreased. In order to maintain feasibility, any decrease in f_{pq} has to be sent from p to q by a different route. Hence, this flow change operation is called *flow rerouting*, p is the *rerouting source* and q is the *rerouting sink* for it.

If the distinguished arc (p, q) is a *b*-arc, f_{pq} has to be increased, and so the rerouting source, sink are q, p respectively. See Figure 5.6.

Let $(f = (f_{ij}), \pi)$ be the current pair and (p, q) the distinguished arc. Denote the rerouting source, sink respectively by $\bar{s}, \bar{t}$. Define $\nu = f_{pq} - \ell_{pq}$ $(k_{pq} - f_{pq})$ if (p, q) is an *a*-arc (*b*-arc). To bring (p, q) into kilter by flow change alone we need to reroute ν units from $\bar{s}$ to $\bar{t}$. For this, we need to find an FRP (*flow rerouting path*) from $\bar{s}$ to $\bar{t}$ in G which is an admissible path in the sense that flow changes on it are permissible as described above; i.e., all forward (reverse) arcs on it, in which the flow amount increases (decreases) are *b*- or β-arcs (*a*- or β-arcs). The algorithm tries to find a shortest (in terms of the number of arcs on it) FRP from $\bar{s}$ to $\bar{t}$ using the tree growth routine described in Section 2.3.3, paying attention to allowable flow changes on the various arcs while growing the tree.

So, any FRP that is obtained in the algorithm will be a simple path, and it can be verified that the distinguished arc (p, q) cannot be contained on it. If an FRP, $\mathcal{P}$ say, is obtained, its *capacity* is $\epsilon_1 = $ min. $\{k_{ij} - f_{ij} : (i, j)$ a forward arc on $\mathcal{P}$ $\} \cup \{ f_{ij} - \ell_{ij} : (i, j)$ a reverse arc on $\mathcal{P}\}$, and let $\epsilon_0 = $ min. $\{ \nu, \epsilon_1\}$. Flow rerouting using $\mathcal{P}$ consists of adding ϵ_0 to the flow amounts on all the forward arcs on $\mathcal{P}$ and (p, q) if it is a *b*-arc, and subtracting ϵ_0 from the flow amounts on all the reverse arcs on $\mathcal{P}$ and (p, q) if it is an *a*-arc.

Since the distinguished arc joins the rerouting source and sink, when we combine $\mathcal{P}$ with the distinguished arc we get a simple cycle, $\mathbb{C}$ say. Orient $\mathbb{C}$ so that all the forward, reverse arcs on $\mathcal{P}$ remain forward, reverse arcs respectively. Then verify that $\mathbb{C}$ is a residual cycle wrt f, and that its residual capacity is ϵ_0 computed above. Also, all forward arcs on $\mathbb{C}$ are *b*- or β-arcs, and all reverse arcs on it are *a*- or β-arcs. So, we have

$$\pi_j - \pi_i \begin{cases} \overset{\geq}{=} c_{ij} & \text{for forward arcs } (i, j) \text{ on } \mathbb{C} \\ \overset{\leq}{=} c_{ij} & \text{for reverse arcs } (i, j) \text{ on } \mathbb{C} \end{cases} \tag{5.10}$$

Multiplying (5.10) by $+1$ over forward arcs (i, j) on $\mathbb{C}$, and by -1 over reverse arcs (i, j) on $\mathbb{C}$, and summing, we get $0 >$ cost of $\mathbb{C}$. Hence, $\mathbb{C}$ is a negative cost residual cycle wrt f. It can also be verified that the operation of flow rerouting using $\mathcal{P}$ is exactly the same as canceling the negative cost residual cycle $\mathbb{C}$ in f.

If the tree growth routine is unable to find an FRP it will terminate with a set of labeled nodes **X** containing the rerouting source $\bar{s}$ and not $\bar{t}$. This implies that at this stage, it is impossible to bring the distinguished arc any closer to kilter by flow changes. So, we try to do it by node price changes. Let $\overline{\mathbf{X}}$ be the complement of **X**. Forward arcs of the cut $[\mathbf{X}, \overline{\mathbf{X}}]$ at this stage can be *a*-, α-, γ-, or saturated β-arcs, but cannot be *b*-arcs (otherwise the node on that *b*-arc in $\overline{\mathbf{X}}$ would have been labeled in the tree growth process). Similarly reverse arcs in the cut $[\mathbf{X}, \overline{\mathbf{X}}]$

can be b-, γ-, α-, or lower boundary β-arcs, but cannot be a-arcs for similar reasons. Define

$$\mathbf{A}^1 \;=\; \{(i,j) : (i,j) \text{ is a forward } a\text{-, or } \alpha\text{-arc in } (\mathbf{X}, \overline{\mathbf{X}}) \} \qquad (5.11)$$

$$\mathbf{A}^2 \;=\; \{(i,j) : (i,j) \text{ is a reverse } b\text{-, or } \gamma\text{-arc in } (\overline{\mathbf{X}}, \mathbf{X}) \} \qquad (5.12)$$

$$\delta \;=\; \text{min. } \{|\pi_j - \pi_i - c_{ij}| : (i,j) \in \mathbf{A}^1 \cup \mathbf{A}^2\} \qquad (5.13)$$

It can be verified that the distinguished arc (p,q) is contained in $\mathbf{A}^1 \cup \mathbf{A}^2$, so $\mathbf{A}^1 \cup \mathbf{A}^2 \neq \emptyset$. Also, on each arc in $\mathbf{A}^1$ ($\mathbf{A}^2$) it is permissible to increase (decrease) the tension. These facts imply that δ is a finite positive quantity. It can also be verified that δ is the largest amount by which the tension on all the arcs in $\mathbf{A}^1$ can be increased, and that on all the arcs in $\mathbf{A}^2$ decreased simultaneously, while keeping all these changes permissible as defined above.

The new node price vector will be a vector of the form $\hat{\pi} = (\hat{\pi}_i)$ where $\hat{\pi}_i = \pi_i$ if $i \in \mathbf{X}$, or $= \pi_i + \theta$ if $i \in \overline{\mathbf{X}}$; for a positive quantity θ to be selected suitably. As a result of this change, the tension on an arc does not change if both nodes on it are either in $\mathbf{X}$ or in $\overline{\mathbf{X}}$, increases by θ if the arc is in $(\mathbf{X}, \overline{\mathbf{X}})$, and decreases by θ if that arc is in $(\overline{\mathbf{X}}, \mathbf{X})$. These changes are permissible as defined earlier, but we must make sure that the tension on α-, a-arcs does not increase beyond the cost coefficient, and that on γ-, b-arcs does not decrease below the cost coefficient. For this we need to have $\theta \leqq \delta$. So, δ is the largest value that θ can have, and giving it this value brings the distinguished arc closest to the in-kilter status possible by these changes; hence we choose $\theta = \delta$ for defining the new node price vector. Verify that in the new pair $(f, \hat{\pi})$, the kilter number of every arc in G is either the same as before, or smaller, and all in-kilter arcs remain in kilter. The kilter status of the arcs in the cut $[\mathbf{X}, \overline{\mathbf{X}}]$ may change; this is determined for all arcs in the cut. If (p, q) still remains out-of-kilter, it continues to be the distinguished arc, and the algorithm now resumes tree growth from where it was left earlier. When this is done, if there are any α-arcs or unsaturated a-arcs in $(\mathbf{X}, \overline{\mathbf{X}})$, or γ-arcs or b-arcs with flow $>$ lower bound in $(\overline{\mathbf{X}}, \mathbf{X})$, which tied for the minimum in (5.13) in the definition of δ, all the unlabeled nodes on them will get labeled. And the method continues.

We are now ready to describe the OK algorithm. We will begin with the routine for selecting the initial node price vector.

SELECTION OF THE INITIAL NODE PRICE VECTOR IN THE OK ALGORITHM

If there are no uncapacitated arcs, i.e., if k is finite, the initial node price vector π^0 can be selected arbitrarily. So, assume that $\mathbf{U} = \{(i,j) : k_{ij} = \infty\} \neq \emptyset$. In this case we will select π^0 so that

$$\pi_j^0 - \pi_i^0 \overset{\le}{=} c_{ij} \text{ for all } (i,j) \in \mathbf{U} \tag{5.14}$$

(5.14) guarantees that there will be no uncapacitated b-arcs in the algorithm. Let $\mathbf{V}$ be the set of nodes on arcs in $\mathbf{U}$, and $c^1 = (c_{ij} : (i,j) \in \mathbf{U})$. $\mathbf{P} = (\mathbf{V}, \mathbf{U}, c^1)$ is the partial subnetwork consisting of the uncapacitated arcs in G. If $c^1 \overset{\ge}{=} 0$, select $\pi^0 = (\pi_i^0)$ where $\pi_i^0 = 0$ if $i \in \mathbf{V}$, and arbitrary if $i \notin \mathbf{V}$.

If $c^1 \not\geq 0$, select any node in $\mathbf{V}$, say i_1, and find a shortest chain tree rooted at i_1 in P by any of the efficient algorithms of Section 4.3. If this algorithm terminates by discovering a negative cost circuit $\overset{\rightarrow}{\mathbb{C}}$ in P, our minimum cost flow problem in G is unbounded below (start with any feasible flow, and add an amount η to the flow amount on each arc of $\overset{\rightarrow}{\mathbb{C}}$, as $\eta \to \infty$ the cost of this feasible flow $\to -\infty$), so terminate.

If the unboundedness termination did not occur, the shortest chain algorithm will obtain a shortest chain tree rooted at i_1 in P spanning all the nodes that can be reached from i_1 by a chain in it. Let this tree be $\mathbb{T}_1$. For each node i in $\mathbb{T}_1$ select π_i^0 to be the shortest chain length (with c^1 as the arc length vector) from i_1 to i in $\mathbb{T}_1$. If $\mathbb{T}_1$ does not span all the nodes in $\mathbf{V}$, any arc in $\mathbf{U}$ joining a node in $\mathbf{V}$ outside of $\mathbb{T}_1$ and a node in $\mathbb{T}_1$ must be directed towards the node in $\mathbb{T}_1$ (since there are no chains from i_1 to any node outside of $\mathbb{T}_1$ in P). Delete all the nodes in $\mathbb{T}_1$ and all the arcs incident at them from P and suppose this leads to $\mathbf{P}^1 = (\mathbf{V}^1, \mathbf{U}^1, c'^1)$. Select any node i_2 in $\mathbf{V}^1$ and find a shortest chain tree rooted at i_2 in $\mathbf{P}^1$; let it be $\mathbb{T}_2$. Let μ be the vector of shortest chain lengths from i_2 in $\mathbb{T}_2$. For each node i in $\mathbb{T}_2$ choose π_i^0 to be $\mu_i + \vartheta$, where ϑ is a positive quantity, same for all nodes in $\mathbb{T}_2$, selected so that $\pi_j^0 - \pi_i^0 \overset{\le}{=} c_{ij}$ holds for all arcs $(i,j) \in \mathbf{U}$ with i in $\mathbb{T}_2$ and j in $\mathbb{T}_1$. If $\mathbb{T}_2$ does not span the nodes in $\mathbf{V}^1$, repeat the same process. If $c^1 \not\geq 0$, and there is no negative cost circuit in P, by a suitable choice of ϑ in each step of this procedure, it is possible to select a spanning tree in each connected component of P, so that the node price vector (π_j^0) satisfying (5.14) in P obtained in the procedure is the node price vector corresponding to these trees.

If $c^1 \not\geq 0$, another method for obtaining an initial node price vector π^0 satisfying (5.14) is the following. Add an artificial origin node s to the set $\mathbf{V}$, and arcs (s, i) of cost 0 for each $i \in \mathbf{V}$. Let $\mathbf{V}^2 = \mathbf{V} \cup \{s\}$, $\mathbf{U}^2 = \mathbf{U} \cup \{(s, i) : i \in \mathbf{V}\}$, and c^2 the cost vector on $(\mathbf{V}^2, \mathbf{U}^2)$ with cost coefficients of arcs in $\mathbf{U}$ given by those in c^1 and for those of the form (s, i) given by 0. In $(\mathbf{V}^2, \mathbf{U}^2)$ with arc length vector c^2, find a shortest chain tree rooted at s. Since there is an arc of the form (s, i) for each $i \in \mathbf{V}$, every node in $(\mathbf{V}^2, \mathbf{U}^2)$ can be reached from s by a chain. So, the shortest chain algorithm on $(\mathbf{V}^2, \mathbf{U}^2)$ will terminate either by finding a negative cost circuit or a shortest chain tree rooted at s. In the former case, let the circuit be $\overset{\rightarrow}{\mathbb{C}}$. Since there are no arcs incident into s in $(\mathbf{V}^2, \mathbf{U}^2)$, $\overset{\rightarrow}{\mathbb{C}}$ cannot contain the node s, so it is a negative cost circuit in the subnetwork P. In the latter case, let $\nu_s = 0, \nu_i$ for $i \in \mathbf{V}$ be the shortest chain distances from s to i in the shortest chain tree obtained.

From the optimality conditions for the shortest chain tree, we verify that selecting $\pi_i^0 = \nu_i$ for all $i \in \mathbf{V}$ satisfies (5.14).

Applying either of these methods, we will find a node price vector π^0 in G satisfying (5. 14), if there is no negative cost circuit consisting only of uncapacitated arcs in it. The version of the out-of-kilter algorithm discussed next is initiated with such a node price vector and a feasible flow vector.

THE OUT-OF-KILTER ALGORITHM INITIATED
WITH A FEASIBLE FLOW VECTOR

Step 1 Initialization   Let f^0 be a feasible flow vector for the problem. Select a node price vector by the routine described above. If a negative cost circuit consisting only of uncapacitated arcs is discovered in this process, the cost function is unbounded below in the problem; terminate. Otherwise, let π^0 be the node price vector obtained. (f^0, π^0) is the initial pair. If this pair satisfies the c.s. conditions, it is an optimum pair; terminate. Otherwise go to Step 2.

Step 2 Selecting a distinguished arc   Select an out-of-kilter arc, say (p, q) as the distinguished arc. Go to Step 3.

Step 3 Labeling routine   Identify the rerouting source $\bar{s}$ and label it with $\emptyset$. List $= \{\bar{s}\}$. Go to Substep 1.

> **Substep 1 Select the node to scan from list**   If list $= \emptyset$ go to Step 5. Otherwise select the node from the top of the list to scan, delete it from the list, and go to Substep 2.

> **Substep 2 Scanning**   Let i be the node to be scanned. Let $(f = (f_{ij}), \pi)$ be the present pair.

>> **(i) Forward labeling**   Label each node j unlabeled at this stage, such that (i, j) is either a b- or β-arc and $f_{ij} < k_{ij}$ with $(i, +)$ and include it at the bottom of the list.

>> **(ii) Reverse labeling**   Label each node j unlabeled at this stage, such that (j, i) is either an a- or β-arc and $f_{ji} > \ell_{ji}$ with $(i, -)$ and include it at the bottom of the list.
>> If the rerouting sink $\bar{t}$ is now labeled, there is a breakthrough, and an FRP from $\bar{s}$ to $\bar{t}$ has been identified; go to Step 4. Otherwise, go back to Substep 1.

Step 4 Flow rerouting   Find the FRP, $\mathcal{P}$, from $\bar{s}$ to $\bar{t}$, by a backward trace of the labels beginning at $\bar{t}$. When $\mathcal{P}$ is combined with the distiguished arc we get a negative cost residual cycle, $\mathbb{C}$. Orient $\mathbb{C}$ so that all forward, reverse arcs on $\mathcal{P}$ remain forward, reverse arcs respectively on $\mathbb{C}$. Cancel $\mathbb{C}$ in f, leading to the new flow vector $\hat{f}$. Find the new kilter status of each arc on $\mathbb{C}$ wrt the new pair $(\hat{f}, \pi)$. If all the arcs in the network are now in-kilter, the present pair is an optimum pair; terminate. If there are still some out-of-kilter arcs, but the distinguished arc is now in-kilter, go back to Step 2. If the distinguished arc continues to be out-of-kilter, erase the labels on all the nodes and go back to Step 3 with the same distinguished arc.

Step 5 Node price vector change   Let $\mathbf{X}$ be the set of labeled nodes at this stage, and $\overline{\mathbf{X}}$ its complement. Find the quantity δ as in (5.11), (5.12), (5.13). Define the new node price vector to be $\hat{\pi} = (\hat{\pi}_i)$ where $\hat{\pi}_i = \pi_i$ for $i \in \mathbf{X}$, and $= \pi_i + \delta$ for $i \in \overline{\mathbf{X}}$. Find the new kilter status of each arc in the cut $[\mathbf{X}, \overline{\mathbf{X}}]$ wrt the new pair $(f, \hat{\pi})$. If all the arcs in the network are now in-kilter, the present pair is an optimum pair; terminate. If there are still some out-of-kilter arcs, but the distinguished arc is now in-kilter, go back to Step 2. If the distinguished arc continues to be out-of-kilter, make the list $= \mathbf{X}$, and resume tree growth by going to Substep 1 in Step 3.

EXAMPLE 5.1

The network $G = (\mathcal{N}, \mathcal{A}, \ell = 0, k, c, \check{s} = 1, \check{t} = 6, \overline{v} = 12)$ for a minimum cost flow problem is given in Figure 5.7. Data on each arc are lower bound, capacity, cost coefficient, in that order. A feasible flow vector of specified value 12 is entered inside little boxes by the side of the arcs with nonzero flow amounts. Node prices in an initial node price vector are entered by the side of the nodes. The kilter status of each arc is also marked on the arcs.

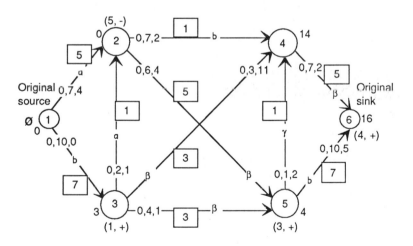

Figure 5.7 Arc (1, 2) is the distinguished arc. Nodes 1, 2 are the rerouting source, sink.

We select the a-arc (1, 2) as the distinguished arc. The rerouting source, sink are 1, 2 respectively. The labeling routine is applied and the labels are entered by the side of the nodes. Breakthrough has occurred, an FRP has been identified, and the negative cost residual cycle is 1, (1, 3), 3, (3, 5), 5, (2, 5), 2, (1, 2), 1, with residual capacity 1 and cost −7. The new flow vector obtained when this cycle is canceled is shown in Figure 5.8, together with the new kilter status of all the arcs. Notice that this step has resulted in a change in the objective value of −7.

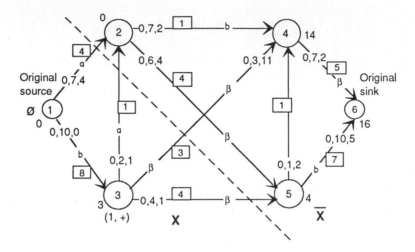

Figure 5.8   Arc (1, 2) is the distinguished arc. 1, 2 are the rerouting source, sink. Labeling ends in a nonbreakthrough with the cut [**X** = {1, 3 }, $\overline{\mathbf{X}}$ = { 2, 4, 5, 6 }].

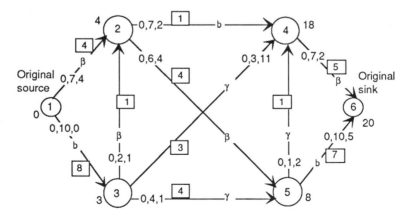

Figure 5.9 Arc (1, 2) is now in kilter.

Arc (1, 2) continues to be an a-arc in Figure 5.8, so it remains the distinguished arc. Labeling routine is entered afresh resulting in the node labels shown in Figure 5.8. It ends in a nonbreakthrough, with the cut [**X** = { 1, 3 }, $\overline{\mathbf{X}}$ = { 2, 4, 5, 6 }]. **A**^{1} = { (1, 2), (3, 2) }, **A**^{2} = ∅. So, δ = min. { 4, 4} = 4. The new node price vector and the new kilter status of each arc are shown in Figure 5.9. Arcs (2, 4), (5, 6) are still out-of-kilter. One of these can be selected as the distinguished arc and the algorithm continued.

An optimum flow vector, node price vector pair for this problem is $(f = (f_{12}, f_{13}, f_{32}, f_{34}, f_{35}, f_{25}, f_{24}, f_{54}, f_{46}, f_{56}), \pi = (\pi_1, \text{ to } \pi_6)) = ((6, 6, 2, 0, 4, 1, 7, 0, 7, 5), (11, 15, 11, 21, 19, 24)))$.

Discussion

When Step 4 occurs in the algorithm, if the cycle canceled is $\mathbb{C}$ with residual capacity ϵ and cost γ, the cost function changes by $\epsilon\gamma$ which is < 0 as $\epsilon > 0$ and $\gamma < 0$. Hence each time Step 4 occurs, the objective value strictly decreases.

Whenever Step 4 occurs, the kilter status of the arcs on the cycle $\mathbb{C}$ canceled may change (there won't be any changes for arcs not on $\mathbb{C}$ as the flow remains unaltered on them). $KN(i, j)$ decreases by the residual capacity of $\mathbb{C}$ for all a-, b-arcs on $\mathbb{C}$. An a-arc on $\mathbb{C}$ may remain an a-arc, or become an α-arc. A b-arc may remain a b-arc or become a γ-arc. β-arcs on $\mathbb{C}$ remain β-arcs, but their classification into lower boundary, interior, saturated β-arcs may change.

Whenever Step 5 occurs, the kilter status for arcs in the cut $[\mathbf{X}, \overline{\mathbf{X}}]$ may change, but there won't be any change for arcs not in this cut. Tension increases by δ for each arc in $(\mathbf{X}, \overline{\mathbf{X}})$, and decreases by δ for each arc in $(\overline{\mathbf{X}}, \mathbf{X})$. An a-arc in $(\mathbf{X}, \overline{\mathbf{X}})$ may remain an a-arc or become a β-arc. An α-arc in $(\mathbf{X}, \overline{\mathbf{X}})$ may remain an α-arc or become a lower boundary β-arc. A γ-arc in $(\mathbf{X}, \overline{\mathbf{X}})$ remains a γ-arc, and a saturated β-arc in $(\mathbf{X}, \overline{\mathbf{X}})$ becomes a γ-arc. A b-arc in $(\overline{\mathbf{X}}, \mathbf{X})$ may remain a b-arc or become a β-arc. A γ-arc in $(\overline{\mathbf{X}}, \mathbf{X})$ may remain a γ-arc or become a saturated β-arc. A lower boundary β-arc in $(\overline{\mathbf{X}}, \mathbf{X})$ becomes an α-arc.

Finiteness of the OK Algorithm

1. First consider the case where ℓ, k and the initial feasible flow vector f^0 are all integer vectors (c may be arbitrary). Let L denote the sum of the kilter nunbers $KN(i, j)$ over all arcs $(i, j) \in \mathcal{A}$. Let Δ_0 be the value of L in the initial pair (f^0, π^0). By our assumption, if (f^0, π^0) is not an optimum pair, Δ_0 is a positive integer. Also, all flow vectors obtained will be integer vectors, $KN(i, j)$ will be a positive integer for all out-of-kilter arcs (i, j), and whenever Step 4 is carried out in the algorithm, L decereases by at least 1.

   At an occurrence of Step 5 in this algorithm, there are two possibilities. One, denoted by P1, is that either an a-arc in $(\mathbf{X}, \overline{\mathbf{X}})$, or a b-arc in $(\overline{\mathbf{X}}, \mathbf{X})$ ties for the minimum in (5.13) for the definition of δ in that step. Under this possibility these arcs become β-arcs and come into kilter as a result of the node price change in this step, and L undergoes a decrease of at least 1 right away. The second possibility, denoted by P2, is that the only arcs which tie for the minimum in (5.13) in this step are some α-arcs in $(\mathbf{X}, \overline{\mathbf{X}})$, or γ-arcs in

$(\overline{\mathbf{X}}, \mathbf{X})$. Under this possibility, all unlabeled nodes on these arcs get labeled once tree growth is resumed after completing this step, and the number of labeled nodes increases by at least 1.

Thus if Step 5 occurs with possibility P1, L undergoes a decrease of at least 1 right in this step. Step 5 with possibility P2 can occur consecutively at most n times; then it must be followed by Step 4, which results in a decrease of at least 1 in L.

From these facts we verify that the algorithm is finite in this case, and that its overall complexity is at most $O(n^2\Delta_0)$, but this grows exponentially with the size of data in the worst case. See Zadeh [1973a, 1973b] for examples in which this exponential growth of computational effort occurs.

2. Now consider the case where c and the initial node price vector π^0 are integer vectors, but ℓ, k, f^0 may be real. In this case the quantity δ in Step 5 of the algorithm will always be a positive integer, and all the node price vectors obtained will be integral. Here we will define the kilter number of an arc (i, j) to be $KN_1(i, j)$ where

$$KN_1(i,j) = \begin{cases} 0, \text{ if } (i,j) \text{ is an } \alpha\text{-},\beta\text{-, or } \gamma\text{-arc} \\ |c_{ij} - (\pi_j - \pi_i)|, \text{ if } (i,j) \text{ is an } a\text{-, or } b\text{-arc} \end{cases} \qquad (5.15)$$

$KN_1(i, j)$ is 0 for all in-kilter arcs, and a strict positive integer for all out-of-kilter arcs. Let L_1 denote the sum of $KN_1(i, j)$ over all $(i, j) \in \mathcal{A}$, and Δ_1 the value of L_1 in the initial pair (f^0, π^0). Whenever Step 5 occurs in this algorithm, the distinguished arc at that stage is in the set $\mathbf{A}^1 \cup \mathbf{A}^2$ and its kilter number decreases by δ, so L_1 decreases by at least 1. Consider the interval between two consecutive occurrences of Step 5. Only flow changes take place in this interval. Two types of events can occur in this interval. One is the transformation of a b-arc into a γ-arc by flow increase until saturation. The other is the transformation of an a-arc into an α-arc by flow decrease to its lower bound. Both these events result in an out-of-kilter arc coming into kilter, and hence a strict decrease of L_1 by a positive integer. In between two consecutive events, or between the beginning of the interval and the first event, or between the last event and the end of the interval, the kilter status of every arc remains unchanged, and hence the subnetwork on which flow changes are permitted remains the same. The work during this time can be posed as a maximum value flow problem in that permissible subnetwork, which can be solved with at most $O(n^3)$ effort. So, between two consecutive instances of the value of L_1 decreasing (by either Step 5, or one of the two events mentioned above between two consecutive occurrences of Step 5) in this algorithm, the computational effort expended is at most $O(n^3)$. And each time L_1 decreases, it decreases by a positive integer. Hence the overall complexity of the algorithm in this case is at most $O(n^3\Delta_1)$.

3. We now consider the general case in which all the data may be real. For any feasible flow vector f, define $V_i = f(i, \mathcal{N}) - f(\mathcal{N}, i)$ for each $i \in \mathcal{N}$. In the problems (5.1), (5.2), (5.3) that we are considering, all these V_i are specified data and do not depend on the particular feasible f.

Consider a stage during which a particular arc, (p, q) say, is the distinguished arc. Let f be the feasible flow vector, and $[\mathbf{X}, \overline{\mathbf{X}}]$ the cut when the algorithm has just reached Step 5 some time during this stage. So, at this time there are no b- or unsaturated β-arcs in $(\mathbf{X}, \overline{\mathbf{X}})$; or a- or non-lower boundary β-arcs in $(\overline{\mathbf{X}}, \mathbf{X})$ (otherwise labeling can continue and we would not have arrived at Step 5). Define $\mathbf{H}_1, \mathbf{H}_2, \mathbf{H}_3$ to be the set of α-, saturated β-, γ-arcs respectively in $(\mathbf{X}, \overline{\mathbf{X}})$; $\mathbf{H}_4, \mathbf{H}_5, \mathbf{H}_6$ to be the set of α-, lower boundary β-, γ-arcs respectively in $(\overline{\mathbf{X}}, \mathbf{X})$; $\mathbf{H}_7$ to be the set of a-arcs in $(\mathbf{X}, \overline{\mathbf{X}})$ other than (p, q); $\mathbf{H}_8$ to be the set of b-arcs in $(\overline{\mathbf{X}}, \mathbf{X})$ other than (p, q); and $\mathbf{H}_0 = \{(p, q)\}$. So, $\mathbf{H}_0$ to $\mathbf{H}_8$ is a partition of the cut $[\mathbf{X}, \overline{\mathbf{X}}]$.

Let $g_0 = \sum (f_{ij} : \text{over } (i, j) \in \mathbf{H}_7) - \sum (f_{ij} : \text{over } (i, j) \in \mathbf{H}_8)$. Let $\omega = -1$ if (p, q) is an a-arc, $+1$ if (p, q) is a b-arc. We have

$$\sum_{i \in \mathbf{X}} V_i = V(\mathbf{X}) \quad = \quad f(\mathbf{X}, \overline{\mathbf{X}}) - f(\overline{\mathbf{X}}, \mathbf{X})$$

$$= \quad g_0 + \sum (\ell_{ij} : \text{ over } (i, j) \in \mathbf{H}_1)$$

$$+ \sum (k_{ij} : \text{ over } (i, j) \in \mathbf{H}_2 \cup \mathbf{H}_3)$$

$$- \sum (\ell_{ij} : \text{ over } (i, j) \in \mathbf{H}_4 \cup \mathbf{H}_5)$$

$$- \sum (k_{ij} : \text{ over } (i, j) \in \mathbf{H}_6) - \omega f_{pq}$$

Define $g_1 = V(\mathbf{X}) - \sum (\ell_{ij} : \text{over } (i, j) \in \mathbf{H}_1) - \sum (k_{ij} : \text{over } (i, j) \in \mathbf{H}_2 \cup \mathbf{H}_3) + \sum (\ell_{ij} : \text{over } (i, j) \in \mathbf{H}_4 \cup \mathbf{H}_5) + \sum (k_{ij} : \text{over } (i, j) \in \mathbf{H}_6)$. Then we have

$$g_0 = g_1 + \omega f_{pq} \tag{5.16}$$

g_1 depends only on the data in the problem, the set $\mathbf{X}$ of nodes, and the partition of the arcs in the set $[\mathbf{X}, \overline{\mathbf{X}}] \setminus (\mathbf{H}_0 \cup \mathbf{H}_7 \cup \mathbf{H}_8)$ into $\mathbf{H}_1$ to $\mathbf{H}_6$, and not on the flow vector. The quantity ωf_{pq} on the right-hand side of (5.16) strictly increases during flow change steps in this stage for bringing (p, q) into kilter. By the permissible flow changes in the algorithm, the quantity g_0 on the left hand side of (5.16) can only decrease during flow change steps in this stage. So, (5.16) implies that any given partition of the set of nondistinguished arcs in the cut $[\mathbf{X}, \overline{\mathbf{X}}]$ into $\mathbf{H}_1$ to $\mathbf{H}_8$ can appear at most once when the algorithm arrives at Step 5 during this stage. Since there are only a finite number of cuts separating the rerouting source and sink, and finite number of partitions of the nondistinguished arcs in each into the sets $\mathbf{H}_1$ to $\mathbf{H}_8$, this implies that the overall computational effort involved in this stage for bringing the distinguished arc (p, q) into kilter is finite. There are at most m out-of-kilter arcs to be brought into kilter, and once an arc comes into kilter, it remains in kilter subsequently. This shows that the overall computational effort in the algorithm is finite.

The Parametric Value Minimum Cost Flow Problem

Suppose it is required to solve the minimum cost flow problem (5.1) in $G = (\mathcal{N}, \mathcal{A}, \ell, k, c, \check{s}, \check{t}, v)$ treating the flow value v as a parameter for all possible values of v. Let $g(v)$ denote the minimum objective value in this problem as a function of v.

THEOREM 5.3 *Let $(\overline{f}, \overline{\pi})$ be an optimum pair of value $\overline{v}$ in G. So, all arcs in G are in-kilter (i.e., $\alpha-, \beta-,$ or γ-arcs) in the pair $(\overline{f}, \overline{\pi})$. Any FAP, $\mathcal{P}_0$ say, from $\check{s}$ to $\check{t}$ wrt $\overline{f}$ consisting of β-arcs only has cost $= \overline{\pi}_{\check{t}} - \overline{\pi}_{\check{s}}$, and is a least cost FAP from $\check{s}$ to $\check{t}$ wrt $\overline{f}$.*

Proof   Since all the arcs (i, j) on $\mathcal{P}_0$ are β-arcs, they satisfy $\overline{\pi}_j - \overline{\pi}_i = c_{ij}$. So, the cost of $\mathcal{P}_0 = \sum (c_{ij}: (i, j)$ forward on $\mathcal{P}_0) - \sum(c_{ij}: (i, j)$ reverse on $\mathcal{P}_0) = \sum((\overline{\pi}_j - \overline{\pi}_i): (i, j)$ forward on $\mathcal{P}_0) - \sum ((\overline{\pi}_j - \overline{\pi}_i): (i, j)$ reverse on $\mathcal{P}_0) = \overline{\pi}_{\check{t}} - \overline{\pi}_{\check{s}}$.

If $\mathcal{P}$ is any other FAP from $\check{s}$ to $\check{t}$ wrt $\overline{f}$, all forward arcs (i, j) on it must be α- or β-arcs and hence satisfy $\overline{\pi}_j - \overline{\pi}_i \overset{\le}{=} c_{ij}$; and all reverse arcs (i, j) on it must be β- or γ-arcs and hence satisfy $\overline{\pi}_j - \overline{\pi}_i \overset{\ge}{=} c_{ij}$. Substituting these in the expression for the cost of $\mathcal{P}$, we see that the cost of $\mathcal{P} \overset{\ge}{=} \overline{\pi}_{\check{t}} - \overline{\pi}_{\check{s}} =$ the cost of $\mathcal{P}_0$. ∎

THE PARAMETRIC VALUE MINIMUM COST FLOW ALGORITHM

Step 1 Initialization   Start with an optimum pair of some value, say (f^1, π^1) of value v^1. To find optimum pairs for $v > v^1$, define the flow changing source $\overline{s} = \check{s}$, and flow changing sink $\overline{t} = \check{t}$. Go to Step 2.

Step 2 Labeling routine   Label $\overline{s}$ with $\emptyset$. Make list $= \{\, \overline{s} \,\}$. Go to Step 3.

Step 3 Select a node for scanning   If list $= \emptyset$ go to Step 6. Otherwise, select the topmost node from the list for scanning. Delete this node from the list and go to Step 4.

Step 4 Scanning   Let $(f = (f_{ij}), \pi = (\pi_i))$ be the present optimum pair of value v. Let i be the node being scanned.

  Substep 1 Forward labeling   For each unlabeled node j such that (i, j) is a β-arc satisfying $f_{ij} < k_{ij}$, label it with $(i, +)$ and include it at the bottom of the list.

Substep 2 Reverse labeling   For each unlabeled node j such that (j, i) is a β-arc satisfying $f_{ji} > \ell_{ji}$, label it with $(i, -)$ and include it at the bottom of the list.

If $\bar{t}$ is labeled there is a breakthrough; go to Step 5. Otherwise, go back to Step 3.

Step 5 Flow augmentation   Trace the FAP from $\bar{s}$ to $\bar{t}$ using the labels. Suppose it is $\mathcal{P}$ with residual capacity ϵ. For $0 \overset{\leq}{=} \theta \overset{\leq}{=} \epsilon$ define the flow vector $f(\theta) = (f_{ij}(\theta))$ obtained by augmenting the flow along $\mathcal{P}$ by amount θ. Then $(f(\theta), \pi)$ is an optimum pair of value $v + \theta$ for all $0 \overset{\leq}{=} \theta \overset{\leq}{=} \epsilon$. In this range the slope of $g(v)$ is $\pi_{\bar{t}} - \pi_{\bar{s}}$. To find optimum pairs for values beyond $v + \epsilon$, erase all the node labels and go back to Step 2 with the new pair $(f(\epsilon), \pi)$.

Step 6 Node price change   Let (f, π) be the present pair, $\mathbf{X}$ the set of labeled nodes, and $\overline{\mathbf{X}}$ its complement. Define $\mathbf{A}^1 = \{(i, j) : (i, j)$ a forward α-arc in $(\mathbf{X}, \overline{\mathbf{X}})\}$, $\mathbf{A}^2 = \{(i, j) : (i, j)$ a reverse γ-arc in $(\overline{\mathbf{X}}, \mathbf{X})\}$. If $\mathbf{A}^1 \cup \mathbf{A}^2 = \emptyset$ f is a maximum value flow in G, terminate. If $\mathbf{A}^1 \cup \mathbf{A}^2 \neq \emptyset$, let δ = min. $\{|\pi_j - \pi_i - c_{ij}| : (i, j) \in \mathbf{A}^1 \cup \mathbf{A}^2\}$. Find the new node price vector $\hat{\pi} = (\hat{\pi}_i)$ where $\hat{\pi}_i = \pi_i$ if $i \in \mathbf{X}$, $= \pi_i + \delta$ if $i \in \overline{\mathbf{X}}$. $(f, \hat{\pi})$ is the new optimum pair, find the new kilter status of each arc in the cut $[\mathbf{X}, \overline{\mathbf{X}}]$ wrt it. Make the list $= \mathbf{X}$ and return to Step 3.

Discussion

Since all the FAPs used in this algorithm are least cost FAPs at the time they are used, this is a shortest augmenting path method for finding optimum flows of increasing value, implemented using the OK method. Once initiated with an optimum pair, the method preserves the in-kilter status of every arc; hence all the pairs obtained in the method are optimal pairs for their value. To find optimum pairs of value $<$ the initial value v^1, begin with the original pair (f^1, π^1), and choose $\check{t}, \check{s}$ as the flow changing source, sink respectively, and repeat the same process. Here the FAPs identified will actually be flow reducing paths in the original network, and the flow value in G will decrease with each flow change. This process terminates when the minimum possible flow value in G is reached.

For all v between two consecutive node price change steps in this algorithm $g(v)$ is linear with slope $\pi_{\bar{t}} - \pi_{\bar{s}}$. By the node price updating formula we see that at each node price change step, this slope changes by δ in that step. Hence the slope of $g(v)$ is nondecreasing with v, establishing its convexity. It is piecewise linear convex. So, slope of $g(v)$ changes whenever node prices change in the algorithm, i.e., in Step 6. From LP theory we know that the number of distinct values of the slope of $g(v)$ is finite, so, Step 6 occurs only a finite number of times. In between two consecutive occurrences of Step 6, the kilter status of each arc remains unchanged, and hence the work carried out in this interval is the solution of a maximum value flow problem on the β-arc subnetwork at that stage; this takes only finite effort by the results in Chapter 2. So, this algorithm finds optimum pairs for all possible values after a finite amount of computation.

OK Algorithm Initiated with an Arbitrary Flow Vector

We consider the minimum cost flow problems (5.1), (5.2), (5.3) on the directed network $G = (\mathcal{N}, \mathcal{A}, \ell, k, c)$. Here we discuss the version of the OK algorithm that can be used to solve this problem beginning with a flow vector f that satisfies the conservation conditions at all the nodes (i.e., the equality conditions in the problem), but may violate the bounds on the nodes. Let (f, π) be such a pair where f is a flow vector of this type. We have the following new kilter statuses for arcs wrt (f, π), in addition to the ones discussed earlier. All these new statuses are out-of-kilter statuses. For each new status we also specify the permissible flow changes on them, and the appropriate definition of the kilter number $KN(i,j)$. There are no constraints on tension changes on arcs with these new statuses. For arcs with the kilter statuses defined earlier, permissible flow and tension changes remain exactly the same as before.

New Kilter Status for (i, j) wrt (f, π)

Status	Condition	Permissible Flow Change	$KN(i,j)$
a_1-arc	$f_{ij} < \ell_{ij}, \pi_j - \pi_i < c_{ij}$	Increase up to ℓ_{ij}	$\ell_{ij} - f_{ij}$
a_2-arc	$f_{ij} > k_{ij}, \pi_j - \pi_i < c_{ij}$	Decrease up to ℓ_{ij}	$f_{ij} - \ell_{ij}$
β_1-arc	$f_{ij} < \ell_{ij}, \pi_j - \pi_i = c_{ij}$	Increase up to k_{ij}	$\ell_{ij} - f_{ij}$
β_2-arc	$f_{ij} > k_{ij}, \pi_j - \pi_i = c_{ij}$	Decrease up to ℓ_{ij}	$f_{ij} - k_{ij}$
b_1-arc	$f_{ij} < \ell_{ij}, \pi_j - \pi_i > c_{ij}$	Increase up to k_{ij}	$k_{ij} - f_{ij}$
b_2-arc	$f_{ij} > k_{ij}, \pi_j - \pi_i > c_{ij}$	Decrease up to k_{ij}	$f_{ij} - k_{ij}$

Let f^0 be the initial flow vector satisfying node conservation at all the nodes. Obtain an initial node price vector π^0 by the special procedure described earlier to satisfy $\pi_j^0 - \pi_i^0 \leqq c_{ij}$ for all uncapacitated arcs (i, j). (f^0, π^0) is the initial pair.

As before the algorithm selects an out-of-kilter arc as the distinguished arc, and tries to bring it into kilter by flow rerouting and node price changes alternately. In flow rerouting steps, the flow rerouting source, sink on the distinguished arc (p, q) are p, q respectively if (p, q) is a b_2-, β_2-, a_2-, or a-arc; or q, p respectively if it is an a_1-, b_1-, β_1-, or b-arc.

The labeling step is the same as before with the following exceptions. Node j can be forward labeled $(i, +)$ while scanning node i if $f_{ij} < k_{ij}$ and (i, j) is a b-, β-, a_1-, β_1-, or b_1-arc. Node j can be reverse labeled $(i, -)$ while scanning node i if $f_{ji} > \ell_{ji}$ and (j, i) is an a-, β-, a_2-, β_2-, or b_2-arc.

In the node price change step it may happen that $\mathbf{A}^1 \cup \mathbf{A}^2 = \emptyset$. If this occurs, it can be verified that the cut defined by the labeled and unlabeled nodes at that stage provides an instance where the conditions for the existence of a feasible flow vector (Chapter 2) are violated; hence terminate with the conclusion that no feasible flow vector exists in G.

It can be verified that all the finiteness proofs hold good for this version of the OK algorithm too.

Other Sensitivity Analysis

If changes take place in c_{ij}, ℓ_{ij}, or k_{ij} after an optimum pair is obtained, redefine the kilter status of arc (i, j) using the new values, and apply the OK algorithm again until this arc comes into kilter. Thus the OK algorithm is a very convenient tool for doing sensitivity analysis on minimum cost flow problems.

A Polynomially Bounded Implementation of the OK Algorithm Based on Scaling

We consider the minimum cost circulation problem (5.2) in the directed network $G = (\mathcal{N}, \mathcal{A}, \ell, k, c)$, with $|\mathcal{N}| = n$, $|\mathcal{A}| = m$. Since all the other minimum cost flow problems discussed in Section 5.1 can be transformed into this, it covers all those models. We assume that ℓ, k are integer vectors, and that $0 \overset{\leq}{=} \ell \overset{\leq}{=} k$. Let p be the smallest positive integer such that all the ℓ_{ij} and the finite capacities k_{ij} are all $\overset{\leq}{=} 2^{\mathbf{P}}$. Here we present an implementation of the OK algorithm based on the scaling technique discussed earlier in Section 3. 2 in the context of the transportation problem, that can solve this problem with a computational complexity of $O((p + 1)mn^2)$.

The implementation applies the OK algorithm on a series of $(p+1)$ subproblems which provide successively closer approximations to the original problem. The subproblems differ in only the lower bound and capacity vectors. $\ell^r = (\ell^r_{ij})$, $k^r = (k^r_{ij})$, where $(\ell^r_{ij}) = \lfloor \ell_{ij}/2^{\mathbf{P}-r} \rfloor$, $(k^r_{ij}) = \lceil k_{ij}/2^{\mathbf{P}-r} \rceil$, are the lower bound and capacity vectors for the rth subproblem for $r = 0$ to p. Since $\ell^r_{ij} \overset{\leq}{=} \ell_{ij}/2^{\mathbf{P}-r}$, and $k^r_{ij} \overset{\geq}{=} k_{ij}/2^{\mathbf{P}-r}$, for all $r = 0$ to p, and for each $(i, j) \in \mathcal{A}$, if f is a feasible circulation for the original problem, $f/2^{\mathbf{P}-r}$ is a feasible circulation for the rth subproblem. Thus if the original problem is feasible, so is every subproblem.

If there are some uncapacitated arcs in G, determine the initial node price vector $\pi^0 = (\pi^0_i)$ so that $\pi^0_j - \pi^0_i \overset{\leq}{=} c_{ij}$ for all (i, j) with $k_{ij} = \infty$. This takes $O(nm)$ effort using the efficient methods discussed in Chapter 4. This either finds an uncapacitated negative cost circuit, or finds the vector π^0. In the former case, we terminate, since the objective function is unbounded below in (5.2) if it is feasible.

The 0th subproblem is to find a minimum cost circulation problem in $G^0 = (\mathcal{N}, \mathcal{A}, \ell^0, k^0, c)$. $\ell^0 = 0$ and $k^0_{ij} = 0$ or 1 for all (i, j) with k_{ij} finite. So, $f^0 = 0$ is a feasible circulation in G^0. Use (f^0, π^0) as the initial pair to solve the 0th subproblem by the OK algorithm. The kilter number $KN(i,j)$ wrt (f^0, π^0) is 0 or 1 for all arcs (i,j). So, solving the 0th problem by the OK algorithm takes at most $O(mn^2)$ effort.

In general, suppose we have an optimum pair (f^{r-1}, π^{r-1}) for the $(r-1)$th subproblem. The rth subproblem is to find a minimum cost circulation in $G^r = (\mathcal{N}, \mathcal{A}, \ell^r, k^r, c)$. Solve it by the OK algorithm using $(2f^{r-1}, \pi^{r-1})$ as the initial pair. Since (f^{r-1}, π^{r-1}) is an optimum pair for the $(r-1)$th subproblem, we have for all $(i,j) \in \mathcal{A}$

$$\pi_j^{r-1} - \pi_i^{r-1} > c_{ij} \quad \text{implies} \quad f_{ij}^{r-1} = k_{ij}^{r-1} \tag{5.17}$$

$$\pi_j^{r-1} - \pi_i^{r-1} < c_{ij} \quad \text{implies} \quad f_{ij}^{r-1} = \ell_{ij}^{r-1} \tag{5.18}$$

Arcs (i,j) satisfying (5.17), (5.18) are the γ-, α-arcs respectively at the end of the $(r-1)$th subproblem. Since k_{ij}^r is either $2k_{ij}^{r-1} - 1$ or $2k_{ij}^{r-1}$, if (i,j) satisfies (5.17), it will either be a γ-arc or a b-arc with $KN(i,j)$ of 1 in the pair $(2f^{r-1}, \pi^{r-1})$ for the rth subproblem. By similar arguments it can be verified that every arc has kilter number of 0 or 1 wrt the initial pair in the rth subproblem. So, solving the rth subproblem by the OK algorithm takes a computational effort of at most $O(mn^2)$.

We continue this way, using $(2f^{r-1}, \pi^{r-1})$ as the initial pair for the rth subproblem, where (f^{r-1}, π^{r-1}) is the optimal pair obtained for the $(r-1)$th subproblem, for $r = 1$ to p. Since the initial flow vector may not always be feasible, we may have to use the general version of the OK algorithm that can begin with an infeasible flow satisfying node conservation but not the bounds. If any of the subproblems in the sequence turns out to be infeasible, the original problem must be infeasible too; terminate. Otherwise, the pth subproblem is the original problem itself, and when we come to it and solve it, we will have an optimum pair for it.

The overall computational effort in this implementation is bounded above by $O(pmn^2)$, so it is polynomially bounded. This implementation is basically of theoretical interest; it shows a way of solving minimum cost circulation problems in polynomial time using scaling and the OK algorithm.

The Parametric Maximum Profit Flow Problem

Consider the maximum profit flow problem (5.5) in the directed network $G = (\mathcal{N}, \mathcal{A}, \ell = 0, k, c, \check{s}, \check{t}, \lambda)$, where $k > 0, c \geqq 0$ and λ is the premium per unit material reaching $\check{t}$ from $\check{s}$. The problem is to be solved parametrically in λ in the range $\lambda \geqq 0$. The c.s. optimality conditions for the feasible flow vector, node price vector pair (f, π) are: for each $(i,j) \in \mathcal{A}$

$$\pi_j - \pi_i > c_{ij} \quad \text{implies} \quad k_{ij} \text{ is finite and } f_{ij} = k_{ij}$$

$$\tag{5.19}$$

$$\pi_j - \pi_i < c_{ij} \quad \text{implies} \quad f_{ij} = 0$$

$$\pi_{\check{t}} - \pi_{\check{s}} = \lambda \tag{5.20}$$

Therefore, a feasible flow vector, node price vector pair (f, π) is an optimum pair for a λ iff (5.20) holds, and every arc in the network has kilter status α-, β-, or γ- wrt (f, π) as defined earlier. This is also a consequence of the fact that if (f, π) is an optimum pair for (5.5) for λ, and the value of f is v, then f must minimize $\sum c_{ij} f_{ij}$ among all feasible flow vectors of value v. So, (5.5) can be solved by the OK algorithm for the parametric value minimum cost flow problem discussed earlier. That algorithm generates a series of optimal pairs (f, π) wrt which all the arcs are α-, β-, or γ-arcs, and hence f will be an optimum feasible flow vector for $\lambda = \pi_{\check{t}} - \pi_{\check{s}}$, the current value of λ. $\lambda = \pi_{\check{t}} - \pi_{\check{s}}$ changes only whenever a node price change step occurs in this algorithm. In between two consecutive node price change steps, flow changes may occur several times with the consequent increase in flow value. Each FAP obtained in that interval will be a least-cost FAP at that stage, of cost = the current λ, and all the flow vectors generated are therefore alternate optimum feasible flow vectors for the current value of λ.

The algorithm is initiated with $(f^0 = 0, \pi^0 = 0)$ which is an optimum pair for $\lambda = \lambda_0 = 0$.

In a general step let $(\overline{f}, \overline{\pi} = (\overline{\pi}_i))$ be the current optimum pair for $\lambda = \overline{\lambda} = \overline{\pi}_{\check{t}} - \overline{\pi}_{\check{s}}$. The parametric value minimum cost flow algorithm is continued. If an FAP, $\mathcal{P}$ say, from $\check{s}$ to $\check{t}$ wrt $\overline{f}$ is identified at this stage, its cost will be $\overline{\lambda}$, and let ϵ be its residual capacity. We get the new feasible flow vector $f(\theta)$ by carrying out flow augmentation by amount θ on this FAP. Each unit of this quantity arrives at $\check{t}$ at an incremental cost of $\overline{\lambda}$ which is canceled by the premium earned by this unit when it reaches there; hence the net profit does not change. So, for $0 \overset{\leq}{=} \theta \overset{\leq}{=} \epsilon$ $f(\theta)$ is an alternate optimum feasible flow vector for $\lambda = \overline{\lambda}$.

If this $\epsilon = \infty$ (this can only happen if $\mathcal{P}$ has no reverse arcs, i.e., $\mathcal{P}$ is an FAC, and all arcs on it are uncapacitated), $f(\theta)$ is feasible for all $\theta \overset{\geq}{=} 0$, and the profit associated with $f(\theta)$ diverges to ∞ as $\theta \to \infty$, for any $\lambda > \overline{\lambda}$, terminate.

If ϵ is finite, continue with the new optimum pair $(f(\epsilon), \overline{\pi})$.

Suppose a nonbreakthrough occurs when the current optimum pair is $(\hat{f}, \hat{\pi} = (\hat{\pi}_i)), \lambda = \hat{\lambda} = \hat{\pi}_{\check{t}} - \hat{\pi}_{\check{s}}$. Let $\mathbf{X}$ be the set of labeled nodes at this stage and $\overline{\mathbf{X}}$ its complement. Determine the sets $\mathbf{A}^1, \mathbf{A}^2$ as in the algorithm. If $\mathbf{A}^1 \cup \mathbf{A}^2 = \emptyset$, then we have $\hat{f}_{ij} = k_{ij}$ for all $(i, j) \in (\mathbf{X}, \overline{\mathbf{X}})$, and $\hat{f}_{ij} = 0$ for all $(i, j) \in (\overline{\mathbf{X}}, \mathbf{X})$, and so $[\mathbf{X}, \overline{\mathbf{X}}]$ is a minimum capacity cut separating $\check{s}$ and $\check{t}$ in G, and $\hat{f}$ is a maximum value flow. Since all the arcs are in-kilter in the present pair, we know that $\hat{f}$ is a maximum value flow that minimizes $\sum c_{ij} f_{ij}$ among all maximum value flows. This implies that $\hat{f}$ remains a maximum profit flow for all $\lambda \overset{\geq}{=}$ its present value $\hat{\lambda}$. Terminate.

If $\mathbf{A}^1 \cup \mathbf{A}^2 \neq \emptyset$, compute δ as in the algorithm and define the new node price vector $\pi(\Delta) = (\pi_i(\Delta))$ where $\pi_i(\Delta) = \hat{\pi}_i$ for $i \in \mathbf{X}$, or $= \hat{\pi}_i + \Delta$ for $i \in \overline{\mathbf{X}}$. It can be verified that $(\hat{f}, \pi(\Delta))$ satisfies (5.19) for all $0 \leqq \Delta \leqq \delta$. So, $(\hat{f}, \pi(\Delta))$ is an optimum pair, i.e., $\hat{f}$ is an optimum flow for $\lambda = \hat{\lambda} + \Delta$, for all $0 \leqq \Delta \leqq \delta$. Continue the algorithm with the new pair $(\hat{f}, \pi(\delta))$ and $\lambda = \hat{\lambda} + \delta$. Verify that at least one arc which is an α-arc in $(\mathbf{X}, \overline{\mathbf{X}})$, or a γ-arc in $(\overline{\mathbf{X}}, \mathbf{X})$ in the pair $(\hat{f}, \hat{\pi})$, will become a β-arc in the new pair $(\hat{f}, \pi(\delta))$, and when tree growth is resumed, the node in $\overline{\mathbf{X}}$ on that arc will get labeled. Hence the total number of consecutive occurrences of a nonbreakthrough before a breakthrough occurs can never exceed n.

Let $Q(\lambda)$ denote the maximum profit, i.e., the optimum objective value in the maximum profit flow problem as a function of the premium λ. In this algorithm, there may be consecutive node price changes without any flow change in between. (f^0, π^0) is the initial pair. Let (f^r, π^r) denote the pair after the rth node price change step is completed and tree growth is about to be resumed again. Let v^r be the value of f^r. v^r is monotonic nondecreasing with r. Let $g_1 =$ smallest r such that $f^r \neq f^0$, and for $d \geqq 1$ let $g_{d+1} =$ smallest $r > g_d$ such that $f^r \neq f^{g_d}$. So, there is no change in the flow vector f^r for $g_d \leqq r \leqq g_{d+1}$ for all d. $\lambda_0 = 0$, and let $\lambda_r = \pi_{\hat{t}}^r - \pi_{\hat{s}}^r$. For $\lambda_0 = 0 \leqq \lambda \leqq \lambda_{g_1-1}$, $Q(\lambda) = 0$. For $\lambda_{g_d-1} \leqq \lambda \leqq \lambda_{g_{d+1}-1}$, $Q(\lambda)$ is linear with slope v_{g_d}. It is piecewise linear, and convex since its slope is nondecreasing with λ. See Figure 5.10.

$f^0 = 0$ is optimal for $\lambda_0 = 0 \leqq \lambda \leqq \lambda_{g_1-1}$. For $d \geqq 1$, f^{g_d} is optimal for $\lambda_{g_d-1} \leqq \lambda \leqq \lambda_{g_{d+1}-1}$.

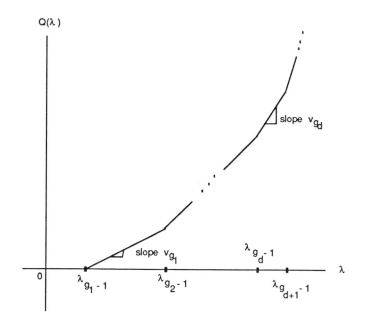

Figure 5.10 $Q(\lambda)$ is piecewise linear and convex

5.4 The Primal Network Simplex Method

We consider the minimum cost flow problem (5.3) in the directed network $G = (\mathcal{N}, \mathcal{A}, \ell, k, c, V)$. If there is an arc (p, q) for which $\ell_{pq} = k_{pq} = \alpha$, say, fix $f_{pq} = \alpha$, delete (p, q) from further consideration, and change V_p to $V_p - \alpha$ and V_q to $V_q + \alpha$. Repeat with other arcs of this type. After this we will have $\ell_{ij} < k_{ij}$ for each remaining arc (i, j). If the network is not connected, each connected component can be solved separately. So, in the sequel we assume that G is connected, that $\ell < k$, and that the necessary condition for feasibility, (5.4), holds. Let $|\mathcal{N}| = n, |\mathcal{A}| = m$.

A basic solution for this problem corresponds to a *partition* of the set of arcs $\mathcal{A}$ into $(\mathbb{T}, \mathbf{L}, \mathbf{U})$, where $\mathbb{T}$ is a spanning tree in G, and $(\mathbf{L}, \mathbf{U})$ is a partition of the out-of-tree arcs subject to the condition that all the arcs in $\mathbf{U}$ are capacitated. The *primal basic solution* corresponding to this partition is $f = (f_{ij})$ where $f_{ij} = \ell_{ij}$ for $(i, j) \in \mathbf{L}$, $= k_{ij}$ for $(i, j) \in \mathbf{U}$, and $(f_{ij} : (i, j) \in \mathbb{T})$ are determined so as to satisfy the equality constraints in (5.3) after fixing the flow amounts on all the arcs in $\mathbf{L} \cup \mathbf{U}$ as defined above. In the partition $(\mathbb{T}, \mathbf{L}, \mathbf{U})$, arcs in $\mathbb{T}$ are called *basic arcs*, and those in $\mathbf{L} \cup \mathbf{U}$ are called *nonbasic arcs*. The partition $(\mathbb{T}, \mathbf{L}, \mathbf{U})$ is said to be a *primal feasible partition* if the flow amounts in the corresponding primal basic solution on all the in-tree arcs satisfy the bound constraints in them, *primal infeasible partition* otherwise.

Because of (5.4), the system of equality constraints in (5.3) is redundant; any one of the constraints from them can be eliminated as a redundant constraint. We select the constraint corresponding to node n, say, as the one to eliminate, and fix n as the root node. This has the effect of setting the node price of n, π_n to 0 in the dual problem.

The *dual basic solution* corresponding to a partition $(\mathbb{T}, \mathbf{L}, \mathbf{U})$, is the node price vector $\pi = (\pi_i)$ where

$$
\begin{aligned}
\pi_n &= 0 \quad (n \text{ is the root node in } \mathbb{T}) \\
\pi_j - \pi_i &= c_{ij} \text{ for all } (i, j) \in \mathbb{T}
\end{aligned}
\tag{5.21}
$$

The node price π_i in this dual basic solution can be verified to be equal to the cost of the predecessor path of node i, treated as a path from root node to i (see Exercise 5.7).

The optimality conditions for a primal feasible partition $(\mathbb{T}, \mathbf{L}, \mathbf{U})$ to be optimal to the problem are

$$
\bar{c}_{ij} = c_{ij} - (\pi_j - \pi_i) \begin{cases} \geqq 0 \text{ for all } (i, j) \in \mathbf{L} \\ \leqq 0 \text{ for all } (i, j) \in \mathbf{U} \end{cases}
\tag{5.22}
$$

$\bar{c}_{ij}$ is called the *reduced (or relative) cost coefficient* of arc (i, j) wrt the partition $(\mathbb{T}, \mathbf{L}, \mathbf{U})$. From the fact that $c_{pq} = \pi_q - \pi_p$ for all in-tree arcs (p, q) in $\mathbb{T}$, it can be verified that for any out-of-tree arc $(i, j), \bar{c}_{ij}$ is the cost of its fundamental cycle wrt $\mathbb{T}$, oriented such that (i, j) is a forward arc in it.

The *primal network simplex algorithm* is the specialization of the bounded variable primal simplex algorithm of LP to our problem. It needs an initial primal feasible partition, $(\mathbb{T}_0, \mathbf{L}_0, \mathbf{U}_0)$, say. So, to implement this algorithm we need an efficient scheme to do the following.

1. Find the primal and dual basic solutions associated with the initial partition $(\mathbb{T}_0, \mathbf{L}_0, \mathbf{U}_0)$.

The general step in this algorithm, called a *pivot step*, begins with the primal feasible partition, $(\mathbb{T}, \mathbf{L}, \mathbf{U})$, say, the associated feasible flow vector, node price vector pair (f, π), obtained at the end of the previous step, and carries out the following work.

2. Identify the set of arcs, $\mathbf{E}$ say, violating the optimality conditions (5.22) for the current partition $(\mathbb{T}, \mathbf{L}, \mathbf{U})$. $\mathbf{E}$ is called the *set of arcs which are eligible to enter the basic set* in this step. If $\mathbf{E} = \emptyset$, the present partition and the pair (f, π) are optimal; terminate. Otherwise, one of the arcs from $\mathbf{E}$ is selected as the *entering arc* by an *entering arc selection (or pivot choice) rule*.

   There are several entering arc selection rules, and extensive computational studies have been carried out to determine which work best. One which performed very well is the *outward-node most negative evaluator rule*. This rule examines arcs leaving out of the nodes until it encounters the first node containing an arc violating the optimality criterion. It then selects the outward arc at this node which violates the optimality criterion by the largest amount, as the entering arc.

3. If the entering arc selected is (i, j), an effort is made to increase the flow amount on it if it is from $\mathbf{L}$, or decrease the flow amount if it is from $\mathbf{U}$, while leaving the flow amounts on all the other arcs in $\mathbf{L} \cup \mathbf{U}$ unchanged. This involves making flow changes on the fundamental cycle, $\mathbb{C}$, of (i, j) wrt $\mathbb{T}$. $\mathbb{C}$ is called the *pivot cycle* in this pivot step. Orient $\mathbb{C}$ such that the entering arc (i, j) is a forward (reverse) arc if (i, j) is in $\mathbf{L}$ $(\mathbf{U})$. With this orientation, it can be verified that the cost of $\mathbb{C}$ is $-|\bar{c}_{ij}|$, hence $\mathbb{C}$ is a negative cost cycle. Find θ, the residual capacity of $\mathbb{C}$; it is known as the *primal simplex minimum ratio* in this pivot step. The set of arcs, $\mathbf{D}$ on $\mathbb{C}$ which tie in the inequality for defining its residual capacity, is called the *set of arcs eligible to be dropping arcs* in this pivot step.

   If $\theta = \infty$ (this can only happen if there are no reverse arcs on $\mathbb{C}$, i.e., it is a negative cost circuit, and all arcs on it are uncapacitated), the objective function is unbounded below in our problem; terminate.

   If $\theta = 0$, $\mathbb{C}$ is not a residual cycle, and the pivot step is said to be a *degenerate pivot step*. In this case define $\hat{f} = f$.

   If $\theta > 0$ and finite, $\mathbb{C}$ is a negative cost residual cycle, and the pivot step is said to be a *nondegenerate pivot step*. Cancel $\mathbb{C}$ in f and let the new flow vector be $\hat{f}$.

Select a dropping arc (p, q) say, from $\mathbf{D}$. If $(p, q) = (i, j)$ (this can only happen if $\theta > 0$; this happens if the flow amount on the entering arc moves from its lower bound to its capacity or vice versa in the flow change just carried out) move (i, j) to the appropriate set in $\mathbf{L}$, $\mathbf{U}$ based on the new flow amount on it. Let $(\mathbb{T}_1 = \mathbb{T}, \mathbf{L}_1, \mathbf{U}_1)$ be the new partition, and $(\hat{f}, \hat{\pi} = \pi)$ the new pair.

If $(p, q) \neq (i, j)$, let $\mathbb{T}_1$ be the tree obtained by replacing (p, q) in $\mathbb{T}$ by (i, j), delete (i, j) from $\mathbf{L}$ or $\mathbf{U}$ where it was before, and insert (p, q) in the appropriate set among these depending on the value of $\hat{f}_{pq}$, and let $(\mathbb{T}_1 = \mathbb{T}, \mathbf{L}_1, \mathbf{U}_1)$ be the new partition.

If $\mathbb{T}_1 \neq \mathbb{T}$, update the node labels and the node price vector. Then go to the next step.

If a primal feasible partition is not known initially, the primal simplex algorithm described above cannot be applied directly. This is where the *primal simplex method* comes in; one should clearly distinguish it from the primal simplex algorithm. It selects an arbitrary spanning tree $\mathbb{T}$ in G (the algorithms discussed in Section 1.2.2 can be used for this), and by partitioning the out-of-tree arcs into $\mathbf{L}$, $\mathbf{U}$ arbitrarily, generates an initial partition $(\mathbb{T}, \mathbf{L}, \mathbf{U})$. If $(\mathbb{T}, \mathbf{L}, \mathbf{U})$ is primal feasible, the original problem is solved by initiating the primal simplex algorithm with it. Otherwise, the primal simplex method begins a Phase I which modifies the lower bounds and capacities on arcs in $\mathbb{T}$ so that $(\mathbb{T}, \mathbf{L}, \mathbf{U})$ becomes primal feasible on the modified network. A *Phase* I *objective function*, which measures the extent of infeasibility of the current partition to the original problem, is constructed. The *Phase* I *problem* is that of minimizing the Phase I objective function on the modified network. Since the current partition is primal feasible on the modified network, the Phase I problem can be solved by the primal simplex algorithm initiated with it. When the Phase I problem is solved, we will either conclude that the original problem has no feasible solution, or obtain a primal feasible partition for it. In the later case, we go into a Phase II in which the original problem is solved by the primal simplex algorithm initiated with the primal feasible partition for it obtained at the end of Phase I.

We will now describe how the various operations in the primal simplex method can be carried out efficiently.

To Compute the Primal and Dual Basic Solutions Associated with a Given Partition

Let $(\mathbb{T}, \mathbf{L}, \mathbf{U})$ be a partition in G for (5.3), and let $\hat{f} = (\hat{f}_{ij})$ be the primal basic solution associated with it. We know that $\hat{f}_{ij} = \ell_{ij}$ for $(i, j) \in \mathbf{L}$, and $= k_{ij}$ for $(i, j) \in \mathbf{U}$. We substitute these values in the system of equality constraints in (5.3) and then obtain $\hat{f}_{ij}$ for $(i, j) \in \mathbb{T}$ by applying a procedure to solve the remaining system by back substitution. This procedure processes each non-root node exactly once. Once a node is processed, the flow amounts in $\hat{f}$ on all the arcs incident at it are known. At any stage, $\mathbf{Y}$ denotes the set of processed nodes at that stage.

The procedure also maintains another set of nodes $\mathbf{X}$ satisfying the property that for each $i \in \mathbf{X}$, flow amounts on all the arcs incident at i are already known at this time except on the arc joining it to its immediate predecessor.

The procedure is initiated with $\mathbf{X}$ = set of non-root terminal nodes in $\mathbb{T}$, $\mathbf{Y} = \emptyset$. In a general step, select a node from $\mathbf{X}$, say i, and delete it from $\mathbf{X}$. Find the flow amount on the arc joining i and its immediate predecessor from the equation $f(i, \mathcal{N}) - f(\mathcal{N}, i) = V_i$ and the known flow amounts on all the other arcs incident at i. Include i in $\mathbf{Y}$. If all the brothers of i are already in $\mathbf{Y}$, and the immediate predecessor of i is not the root, include it in $\mathbf{X}$. If $\mathbf{X} = \emptyset$, the flow vector has been completely determined; terminate. Otherwise, go to the next step.

The node price vector $\hat{\pi}$ associated with the spanning tree $\mathbb{T}$ is determined from (5.21). These equations are solved by back substitution beginning at the root node, and going down in increasing order of level.

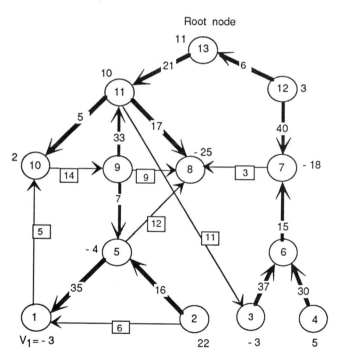

Figure 5.11

EXAMPLE 5.2

Consider the network in Figure 5.11. The exogenous flow at each node is entered by its side if it is nonzero. $\mathbb{T}$ is the spanning tree of thick arcs. c_{ij} is entered on arc (i, j). The flow amount on each out-of-tree arc, which is either the lower bound or the capacity depending on whether the arc is in **L** or **U**, is entered in a little box by the side of the arc. All the other data is omitted.

To find the primal basic solution corresponding to this partition, we begin with $X = \{1,2,3,4,8,10\}$, the set of non-root leaf nodes in $\mathbb{T}$, and $Y = \emptyset$. We select 1 from X for processing. There is a demand of 3 units at node 1, 6 units arriving along arc $(2, 1)$, and 5 units leaving through arc $(1, 10)$. For conservation to hold at node 1, the flow amount on in-tree arc $(5, 1)$ must be 2 units. Transfer 1 from X to Y and proceed. Continuing, we are lead to the following flow amounts on in-tree arcs: $(f_{5,1}, f_{2,5}, f_{3,6}, f_{4,6}, f_{6,7}, f_{12,7}, f_{12,13}, f_{9,5}, f_{9,11}, f_{11,10}, f_{11,8}, f_{13,11})$ $= (2, 16, 8, 5, 13, 8, -5, 2, 3, 7, 1, 6)$.

The node price vector associated with $\mathbb{T}$ is $\pi = (\pi_{13}, \text{ to } \pi_1) = (0, -6, 21, 26, -12, 38, 34, 19, -5, -11, -18, -21, 30)$.

Updating the Tree Labels and the Node Price Vector

Let $\mathbb{T}$ denote the current spanning tree in G associated with the tree labels $P(i)$, $S(i)$, $YB(i)$, $EB(i)$ for $i \in \mathcal{N}$. Let $\mathbb{T}'$ be the new spanning tree obtained after the entering arc replaces the dropping arc. The root node is never changed in this implementation. For $i \in \mathcal{N}$ let $P'(i)$, $S'(i)$, $YB'(i)$, $EB'(i)$ denote the tree labels corresponding to $\mathbb{T}'$. We now define some symbols used in the updating process. See Figure 5.12.

i_1, j_1 j_1 specifically denotes the node on the entering arc whose predecessor path contains the dropping arc, and i_1 is the other node on the entering arc.

i_0 apex node, it is the first common node on the predecessor paths of i_1 and j_1.

i_2, j_* being an in-tree arc, the dropping arc consists of a parent and a son node; these are called i_2, j_* respectively. So, $P(j_*) = \pm i_2$.

$j_1, j_2, \ldots, j_{t+1} = j_*$ the sequence of nodes on the predecessor path of j_1 up to j_*.

The portion of the predecessor path of j_1 in $\mathbb{T}$ up to the node j_* is known as the *pivot stem* in this pivot step. The updating formulas can be viewed as those arising from a gravity model of the rooted tree. Think of the rooted tree $\mathbb{T}$ standing with the root at the top and successive levels going down. $\mathbb{T}'$ is obtained by introducing the entering arc into $\mathbb{T}$, and then snipping off the dropping arc. When this is done, the points along the path containing the entering arc and the pivot stem fall down by gravity, revolving around each node as the path falls down, giving the new tree $\mathbb{T}'$. In this process if some nodes acquire a new son, we assume that this son joins at the left of the existing sons of that point (i.e., as an elder brother of all the existing sons of that point). It is convenient to do the updating in the order indicated below.

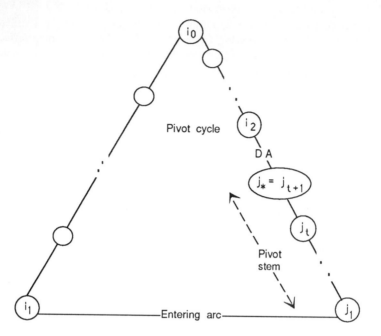

Figures 5.12 Fundamental cycle of the entering arc wrt $\mathbb{T}$. Arc orientations are not shown in the figure. DA is the dropping arc.

Updating the predecessor indices The predecessor indices change only for nodes on the pivot stem. Set $P'(j_1) = +i_1$ or $-i_1$ depending on whether the entering arc is (i_1, j_1) or (j_1, i_1). For $u = 2$ to $t+1$ set $P'(j_u) = $ (minus the sign of $P(j_{u-1}))j_{u-1}$.

Updating the successor indices The successor indices change only for nodes on the entering arc, pivot stem and dropping arc. Set $S'(i_1) = +j_1$ or $-j_1$ depending on whether the entering arc is (j_1, i_1) or (i_1, j_1). For $u = 1$ to t, set $S'(j_u) = $ (sign of $P(j_u))j_{u+1}$. If $S(i_2) \neq \pm j_*$, make $S'(i_2) = S(i_2)$. If $S(i_2) = \pm j_*$, make $S'(i_2) = $ (minus the sign of $P(YB(j_*)))YB(j_*)$. If $S(j_*) \neq \pm j_t$, make $S'(j_*) = S(j_*)$. If $S(j_*) = \pm j_t$, make $S'(j_*) = $ (minus the sign of $P(YB(j_t)))YB(j_t)$.

Updating the brother indices Let $|S(i)|$ denote the successor index of i without its sign. We assume that if a point is removed from the set of immediate successors of a node, then the elder-younger brotherly relationships among the remaining points in this set remain unchanged. The brother indices may change only for nodes in the set $H(\mathbb{T}, j_*)$, the family of j_* in $\mathbb{T}$, and for the nodes $|S(i_1)|$, $YB(j_*)$, and $EB(j_*)$ if these are not empty.

Set $YB'(j_1) = |S(i_1)|$. For each $u = 1$ to $t + 1$, if $EB(j_u) \neq \emptyset$, set $YB'(EB(j_u))$ $= YB(j_u)$; and if $YB(j_u) \neq \emptyset$, set $EB'(YB(j_u)) = EB(j_u)$. If $S(i_1) \neq \emptyset$, set $EB'(|S(i_1)|) = j_1$. For each $u = 1$ to $t+1$, set $EB'(j_u) = \emptyset$, because these points join as the eldest to their new set of brothers. If $S(j_1) \neq \emptyset$, set $EB'(|S(j_1)|) = j_2$. For each $u = 2$ to t if $|S(j_u)| \neq j_{u-1}$, set $EB'(|S(j_u)|) = j_{u+1}$; otherwise, if $YB(j_{u-1})$ $\neq \emptyset$, set $EB'(YB(j_{u-1})) = j_{u+1}$. $YB'(j_2) = S(j_1)$. For each $u = 3$ to $t + 1$, if $|S(j_{u-1})| \neq j_{u-2}$, set $YB'(j_u) = |S(j_{u-1})|$, and if $|S(j_{u-1})| = j_{u-2}$, set $YB'(j_u) = YB(j_{u-2})$. If $YB(j_*) \neq \emptyset$, $EB'(YB(j_*)) = EB(j_*)$. if $EB(j_*) \neq \emptyset$, $YB'(EB(j_*)) = YB(j_*))$. Leave all other brother indices unchanged.

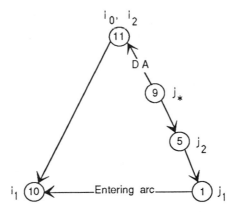

Figure 5.13 DA is the dropping arc.

As an example, consider the spanning tree $\mathbb{T}$ in Figure 1.18, Section 1.2.2. Let (1, 10) be the entering arc into $\mathbb{T}$, and let the in-tree arc (9, 11) be the dropping arc. The pivot cycle for this pivot step is in Figure 5.13, the updated spanning tree $\mathbb{T}'$ is in Figure 5.14, and the updated node labels are shown below.

Updated Tree Labels Corresponding to $\mathbb{T}'$

Node i	1	2	3	4	5	6	7	8	9	10	11	12	13 Root
$P(i)$	-10	-5	-6	-6	-1	-7	+12	+11	-5	+11	+13	-13	$\emptyset$
$S(i)$	+5	$\emptyset$	$\emptyset$	$\emptyset$	+9	+3	+6	$\emptyset$	$\emptyset$	+1	-10	-7	-11
$YB(i)$	$\emptyset$	$\emptyset$	4	$\emptyset$	$\emptyset$	$\emptyset$	$\emptyset$	$\emptyset$	2	8	12	$\emptyset$	$\emptyset$
$EB(i)$	$\emptyset$	9	$\emptyset$	3	$\emptyset$	$\emptyset$	$\emptyset$	10	$\emptyset$	$\emptyset$	$\emptyset$	11	$\emptyset$

If other node labels such as the distance label, thread label, number of successors label, preorder distance label, last successor label, etc. are used in the implementation, they are updated in a manner similar to the above. The updating of the thread label is easy, as it changes only for the nodes on the pivot stem and their eldest and youngest children. See Glover and Klingman [1982], and Glover, Klingman, and Stutz [1974].

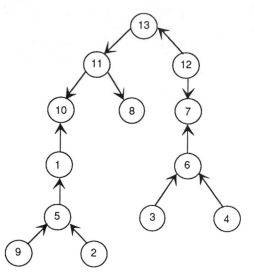

Figure 5.14 The updated spanning tree $\mathbb{T}'$.

Updating the node price vector   Let $\pi = (\pi_i), \pi' = (\pi_i')$ be the node price vectors corresponding to the spanning tree $\mathbb{T}$, the updated spanning tree $\mathbb{T}'$ respectively. $\bar{c}_{ij} = c_{ij} - (\pi_j - \pi_i)$ are the relative cost coefficients wrt π. $\mathbf{H}(\mathbb{T}, j_*)$ is the family of j_* in $\mathbb{T}$. Then

$$\pi_i' = \begin{cases} \pi_i & \text{for } i \notin \mathbf{H}(\mathbb{T}, j_*) \\ \pi_i + \alpha\bar{c}_e & \text{for all } i \in \mathbf{H}(\mathbb{T}, j_*) \end{cases} \tag{5.23}$$

where $\bar{c}_e$ is the relative cost coefficient of the entering arc wrt π, and $\alpha = +1$ if the entering arc is (i_1, j_1), or -1 if it is (j_1, i_1). It can be verified that π' defined by (5.23) satisfies (5.21) for the tree $\mathbb{T}'$; hence it is the node price vector corresponding to $\mathbb{T}'$

As an example, consider the spanning tree $\mathbb{T}$ consisting of the thick arcs in Figure 5.11. The node price vector π corresponding to $\mathbb{T}$ in this network has been obtained in Example 5.2. Let $\mathbb{T}'$ obtained by replacing the dropping arc $(9, 11)$ by the entering arc $(1, 10)$. Suppose the cost coefficient $c_{1,10} = 5$. wrt π, the relative cost coefficient $\bar{c}_{1,10} = 5 - (26 - 30) = 9$. $j_* = 9$ in this example, and the family $\mathbf{H}(\mathbb{T}, 9) = \{\, 9, 5, 1, 2 \,\}$. The new node price vector π' wrt $\mathbb{T}'$ is $(\pi_{13}'$ to $\pi_1') = (0, -6, 21, 26, -21, 38, 34, 19, -14, -11, -18, -30, -21)$.

Updating the node price vector when one cost coefficient changes   Let π be the node price vector corresponding to the spanning tree $\mathbb{T}$ in G. Suppose the cost coefficient c_{pq} of one in-tree arc (p, q) is changed to $c_{pq} + \delta$ while the tree itself and other data remains unchanged. This kind of change occurs during Phase I of the simplex method. Let $\mathbf{H}$ denote the family of son(p, q) in $\mathbb{T}$. $\hat{\pi} = (\hat{\pi}_i)$, the new node price vector is given by: $\hat{\pi}_i = \pi_i$ for all $i \notin \mathbf{H}$, and $= \pi_i + \alpha\delta$ for $i \in \mathbf{H}$; where $\alpha = +1$ if son$(p, q) = q$, and $= -1$ if son$(p, q) = p$.

Setting up the Phase I problem   Let $(\mathbb{T}_0, \mathbf{L}_0, \mathbf{U}_0)$ be an initial partition selected in G, associated with the primal basic solution $f^0 = (f^0_{ij})$. If f^0 is primal infeasible, we set up a Phase I problem whose aim is to find a primal feasible partition in G. Define

$$
\begin{aligned}
\mathbf{K}_1 &= \{(i, j) : (i, j) \in \mathbb{T}_0, f^0_{ij} < \ell_{ij}\} \\
\mathbf{K}_2 &= \{(i, j) : (i, j) \in \mathbb{T}_0, f^0_{ij} > k_{ij}\}
\end{aligned}
$$

Arcs in $\mathbf{K}_1, \mathbf{K}_2$ are called *type 1, 2 arcs* respectively. The Phase I problem is defined on the same network $(\mathcal{N}, \mathcal{A})$ but has modified data: lower bound vector $\ell' = (\ell'_{ij})$, capacity vector $k' = (k'_{ij})$, cost vector $c^* = (c^*_{ij})$, where

$$
\text{for } (i, j) \notin \mathbf{K}_1 \cup \mathbf{K}_2, \quad \ell'_{ij} = \ell_{ij}, \quad k'_{ij} = k_{ij}, \quad c^*_{ij} = 0
$$

$$
\text{for } (i, j) \in \mathbf{K}_1, \quad \ell'_{ij} = f^0_{ij}, \quad k'_{ij} = \ell_{ij}, \quad c^*_{ij} = -1
$$

$$
\text{for } (i, j) \in \mathbf{K}_2, \quad \ell'_{ij} = k_{ij}, \quad k'_{ij} = f^0_{ij}, \quad c^*_{ij} = +1
$$

For the Phase I problem $(\mathcal{N}, \mathcal{A}, \ell', k', c^*, V)$, $(\mathbb{T}_0, \mathbf{L}_0, \mathbf{U}_0)$ is a feasible partition associated with the primal feasible basic solution f^0. Starting with this partition we minimize the Phase I objective function $\sum(c^*_{ij} f_{ij} : \text{over } (i, j) \in \mathcal{A})$ in Phase I, by the primal network simplex algorithm. From the definition of c^*, this has the effect of increasing the flow amounts on type 1 arcs, and decreasing the flow amounts on type 2 arcs, thereby bringing the flow vector closer to feasibility for the original problem. The type 1, 2 arcs play the role of artificial variables in Phase I problems for general LPs.

Whenever the flow vector changes in Phase I, the data and the sets $\mathbf{K}_1, \mathbf{K}_2$ are revised. Suppose $\hat{f} = (\hat{f}_{ij})$ is a new flow vector obtained during Phase I. For each arc (i, j) which was a type 1 arc before $\hat{f}$ was obtained, cancel its type 1 status and make it a regular arc and change $c^*_{ij}, \ell'_{ij}, k'_{ij}$ to $0, \ell_{ij}, k_{ij}$ respectively if $\hat{f}_{ij} = \ell_{ij}$; otherwise just change ℓ'_{ij} to $\hat{f}_{ij}$. Make a corresponding alteration for each arc which was a type 2 arc before $\hat{f}$ was obtained. Hence, the data changes in every nondegenerate step in Phase I.

At some stage during Phase I let f be the present flow vector, and π^* the node price vector computed using the current c^* as the cost vector. If there are no type 1, 2 arcs, f is feasible to the original problem; move to Phase II with the present partition. Even if there are type 1, 2 arcs, Phase I terminates if for each $(i, j) \in \mathcal{A}$

$$f_{ij} = \text{current } \ell'_{ij} \qquad \text{if} \quad \bar{c}^*_{ij} > 0$$
$$f_{ij} = \text{current } k'_{ij} \qquad \text{if} \quad \bar{c}^*_{ij} < 0$$

where $\bar{c}^*_{ij} = c^*_{ij} - (\pi^*_j - \pi^*_i)$. These are the optimality conditions for the Phase I problem. If these conditions hold, and $\mathbf{K}_1 \cup \mathbf{K}_2 \neq \emptyset$, the original problem is infeasible; terminate.

Resolution of Cycling in the Primal Network Simplex Algorithm

With respect to a feasible flow vector $\bar{f} = (\bar{f}_{ij})$ in G for (5.3), an arc $(i, j) \in \mathcal{A}$ is said to be an

$$\begin{array}{rcl}
\text{interior arc} & \text{if} & \ell_{ij} < \bar{f}_{ij} < k_{ij} \\
\text{lower boundary arc} & \text{if} & \ell_{ij} = \bar{f}_{ij} \\
\text{upper boundary or saturated arc} & \text{if} & \bar{f}_{ij} = k_{ij}
\end{array}$$

A primal feasible partition $(\mathbb{T}, \mathbf{L}, \mathbf{U})$ and the associated BFS f are said to be *primal nondegenerate* if all the arcs in $\mathbb{T}$ are interior arcs wrt f, *primal degenerate* if at least one arc in $\mathbb{T}$ is a lower boundary or saturated arc wrt f.

It can be verified that a feasible flow vector $\bar{f}$ for (5.3) is a BFS iff the set of interior arcs wrt $\bar{f}$ forms a forest, and a primal nondegenerate BFS iff this set forms a spanning tree in G.

The residual capacity, θ, of the pivot cycle in a pivot step of the primal simplex algorithm can be verified to be strictly > 0 if the flow vector, f at that stage is primal nondegenerate. If f is degenerate, θ could be 0, and in this case the pivot step becomes a degenerate pivot step. During a nondegenerate pivot step the BFS changes and the objective value strictly decreases. During a degenerate pivot step, the objective value and the BFS do not change, but the spanning tree and the node price vector change. Since the objective value is monotone nonincreasing in this algorithm, if a nondegenerate pivot step occurs in a partition, that partition can never reappear in the sequel. However, the algorithm may go through a sequence of consecutive degenerate pivot steps without any change in the objective value or the flow vector, and return again to the partition at the beginning of this sequence, thus creating a *cycle of degenerate pivot steps*. The algorithm can repeat this cycle indefinitely, and never terminate. This phenomenon is called *cycling under degeneracy in the primal network simplex algorithm*. It is an indefinite repetition of the same finite cycle of degenerate pivot steps again and again, without ever satisfying a termination condition. An example of cycling in an assignment problem of order 4 has been constructed by L. Johnson and is reported in the paper Gassner [1964]. The cost matrix for the problem is

$$c = \begin{pmatrix} 3 & 5 & 5 & 11 \\ 9 & 7 & 9 & 15 \\ 7 & 7 & 11 & 13 \\ 13 & 13 & 13 & 17 \end{pmatrix}$$

It is a minimum cost flow problem on a complete bipartite network of order 4×4, in which all the arcs are uncapacitated. So, a partition for this problem is of the form $(\mathbb{T}, \mathbf{L})$, basic and nonbasic arcs, with the flow amounts on all nonbasic arcs being 0 in the associated basic solution. We denote by the ordered pair (i, j) the arc joining row node i to column node j. A cycle of degenerate pivot steps for the primal network simplex algorithm in this problem is given below. It consists of 12 pivot steps at the end of which we get the initial partition back. In the following table, we give the basic set of arcs in each pivot step.

Pivot step	Basic set of arcs	Entering arc	Dropping arc
1	$\{(1,1), (2,2), (3,3), (4,4), (1,2), (2,3), (3,4)\}$	$(1,3)$	$(2,3)$
2	$\{(1,1), (2,2), (3,3), (4,4), (1,2), (1,3), (3,4)\}$	$(4,2)$	$(1,2)$
3	$\{(1,1), (2,2), (3,3), (4,4), (4,2), (1,3), (3,4)\}$	$(3,2)$	$(3,4)$
4	$\{(1,1), (2,2), (3,3), (4,4), (4,2), (1,3), (3,2)\}$	$(4,1)$	$(4,2)$
5	$\{(1,1), (2,2), (3,3), (4,4), (4,1), (1,3), (3,2)\}$	$(4,3)$	$(1,3)$
6	$\{(1,1), (2,2), (3,3), (4,4), (4,1), (4,3), (3,2)\}$	$(2,1)$	$(4,1)$
7	$\{(1,1), (2,2), (3,3), (4,4), (2,1), (4,3), (3,2)\}$	$(3,1)$	$(3,2)$
8	$\{(1,1), (2,2), (3,3), (4,4), (2,1), (4,3), (3,1)\}$	$(2,4)$	$(2,1)$
9	$\{(1,1), (2,2), (3,3), (4,4), (2,4), (4,3), (3,1)\}$	$(2,3)$	$(4,3)$
10	$\{(1,1), (2,2), (3,3), (4,4), (2,4), (2,3), (3,1)\}$	$(1,4)$	$(2,4)$
11	$\{(1,1), (2,2), (3,3), (4,4), (1,4), (2,3), (3,1)\}$	$(3,4)$	$(3,1)$
12	$\{(1,1), (2,2), (3,3), (4,4), (1,4), (2,3), (3,4)\}$	$(1,2)$	$(1,4)$
13	$\{(1,1), (2,2), (3,3), (4,4), (1,2), (2,3), (3,4)\}$		

It can be verified that all these pivot steps are degenerate pivot steps, and that every one of these basic sets is associated with the same BFS in which f_{ii} is 1 for all $i = 1$ to 4, and flows on all the other arcs are 0. The relative cost coefficient of the entering cell in each pivot step is - 2.

The occurrence of cycling is extremely rare in real-world applications. But, since the possibility of cycling exists, we cannot mathematically guarantee that the primal network simplex algorithm is finite unless methods for resolving it are used. One such method which is purely combinatorial, has been developed originally by Cunningham [1976, 1979]; we will show later that it is exactly the specialization of the lexicographic bounded variable primal simplex algorithm for general LP discussed in Section 11.4 of Murty [1983 of Chapter 1], to the minimum cost flow problem.

Let $(\mathbb{T}, \mathbf{L}, \mathbf{U})$ be a primal feasible partition in G associated with the BFS $f = (f_{ij})$. This partition is said to be a *strongly feasible partition* if for each $(i, j) \in \mathbb{T}$

$$f_{ij} = \ell_{ij} \quad \text{implies} \quad (i, j) \text{ is directed away from root, i.e., } P(j) = +i$$

$$(5.24)$$

$$f_{ij} = k_{ij} \quad \text{implies} \quad (i, j) \text{ is directed towards the root, i.e., } P(i) = -j$$

i.e., a strongly feasible partition is one in which all lower boundary in-tree arcs are directed away from the root node, and all upper boundary in-tree arcs are directed towards the root node. The method for resolving cycling initiates the primal algorithm with a strongly feasible partition and maintains strong feasibility throughout by a special dropping arc selection rule in each pivot step. We will summarize the main features of this method. The details and proofs are mainly of theoretical interest; the interested reader should consult Cunningham [1976, 1979].

Cunningham [1976] has given an efficient procedure for obtaining a strongly feasible partition from an arbitrary feasible partition. Let β be the number of bad arcs, i.e.,boundary in-tree arcs which are wrongly oriented for strong feasibility, in the initial partition. Then the procedure goes through β stages. Each stage tries to replace a bad arc with a nonbasic arc. Each stage may require up to $(n-1)$ such pivot steps and leads to a reduction in the number of bad arcs by 1.

Once the primal algorithm is initiated with a strongly feasible partition, strong feasibility can be preserved by adopting the following dropping arc choice rule in each pivot step (there is no restriction on the entering arc choice).

Dropping Arc Choice Rule to Preserve Strong Feasibility

Let (p, q) be the entering arc in the pivot step in the strongly feasible partition $(\mathbb{T}, \mathbf{L}, \mathbf{U})$ associated with the pair (f, π), and $\mathbb{C}$ the pivot cycle in this pivot step with g as the apex node. Let $\mathbf{D}$ denote the set of eligible dropping arcs in this pivot step.

1. If $\bar{c}_{pq} = c_{pq} - (\pi_q - \pi_p) > 0$ and $f_{pq} = k_{pq}$, select the dropping arc in this pivot step to be the first arc encountered in $\mathbf{D}$ while traveling along $\mathbb{C}$ from g back to g in the direction opposite to the orientation of (p, q).

2. If $\bar{c}_{pq} < 0$ and $f_{pq} = \ell_{pq}$, select the dropping arc in this pivot step to be the first arc encountered in $\mathbf{D}$ while traveling along $\mathbb{C}$ from g back to g in the same direction as (p, q).

This rule identifies the dropping arc uniquely by unambiguous and simple combinatorial rules. If $(\mathbb{T}', \mathbf{L}', \mathbf{U}')$ is the partition obtained after this pivot step operated using this dropping arc choice rule, it will also be strongly feasible. Also, let (f', π') be the basic pair associated with this partition. If this pivot step is a degenerate pivot step (i.e., $f' = f$), then $\pi' \stackrel{\geq}{=} \pi$ and $\sum_{i \in \mathcal{N}} \pi_i' < \sum_{i \in \mathcal{N}} \pi_i$ (these inequalities relating π and π' may not hold if the pivot step is nondegenerate). See Cunningham [1976] for proofs of these results.

The primal network simplex algorithm initiated with a strongly feasible partition, and operated using the above dropping arc choice rule, is called *the method of strongly feasible partitions*. In this method, the entering arc can be chosen arbitrarily among those eligible to enter in each pivot step. In each nondegenerate pivot step of this method, the primal objective value decreases strictly. In each degenerate pivot step, the primal objective value remains unchanged, but $\sum_{i \in \mathcal{N}} \pi_i$ strictly decreases. So, a partition can never reappear in this method, cycling cannot occur, and the method terminates finitely.

Like some of the other methods for resolving cycling in the primal simplex algorithm for general LP, the method of strongly feasible partitions also can be given a perturbation interpretation. Let ϵ be a small positive number, and consider the perturbed problem obtained by changing V_i in (5.3) into V_i^1 for all $i \in \mathcal{N}$, where $V_i^1 = V_i + \epsilon$ for $i \neq n$, and $= V_n - (n-1)\epsilon$ for $i = n$. Then it can be shown that when ϵ is positive but sufficiently small, a partition in G is feasible to the perturbed problem iff it is strongly feasible to the original problem. So, the method of strongly feasible partitions can be viewed as the usual primal network simplex algorithm applied to solve the perturbed problem, treating ϵ as a sufficiently small positive number without giving it a specific value.

The method of strongly feasible partitions can also be shown to be a specialization of the lexicographic bounded variable primal simplex algorithm for general LP (see Section 11.4 in Murty [1983 of Chapter 1]). Consider the bounded variable LP (1.7) discussed in Section 1.2.1. A feasible partition (x_B, x_L, x_U), where $x_B = (x_{i_1}, \ldots, x_{i_m})$ say, associated with the basis B and the BFS $\bar{x} = (\bar{x}_j)$ for (1.7), is said to be a *strongly feasible partition* if for each r

$$\bar{x}_{i_r} = \ell_{i_r} \quad \text{implies} \quad \text{the } r\text{th row of } B^{-1} \text{ is lexico positive}$$

$$(5.25)$$

$$\bar{x}_{i_r} = k_{i_r} \quad \text{implies} \quad \text{the } r\text{th row of } B^{-1} \text{ is lexico negative}$$

Our minimum cost flow problem (5.3) will be in the form (1.7) if we eliminate the flow conservation equation corresponding to the root node n. Let B be the basis associated with a spanning tree $\mathbb{T}$ for the resulting problem. From Theorem 1.8 we know that all nonzero entries in any row of B^{-1} always have the same sign. From this we conclude that the row corresponding to $(p, q) \in \mathbb{T}$ in B^{-1} is lexico positive iff (p, q) is directed away from the root node in $\mathbb{T}$, and lexico negative iff (p, q) is directed towards the root node. Hence for a feasible partition for (5.3), the definition of strong feasibility using (5.24) is identical to that using (5.25).

Also, the dropping arc choice rule in the method of strongly feasible partitions turns out to be exactly the same as the dropping variable choice rule in the lexicographic bounded variable primal simplex algorithm. Hence, the method of strongly feasible partitions is exactly the specialization of the lexicographic bounded variable primal simplex algorithm to our minimum cost flow problem. This specialization is made possible by the result in Theorem 1.8.

Two disadvantages of the method of strongly feasible partitions are that it needs an initial strongly feasible partition, and that the special dropping arc choice rule has to be used in every pivot step, even during sequences of nondegenerate pivot steps, since otherwise strong feasibility will be lost. Another method for resolving cycling in the primal network simplex algorithm developed by Gamble, Conn, and

Pulleyblank [1988] is much simpler computationally. It can be initiated with any primal feasible partition, without the need to convert it into a strongly feasible one. The core of this method is also a special dropping arc choice rule. But the special rule needs to be invoked only when a degenerate pivot step occurs, and can be turned off when the next nondegenerate pivot step occurs. Because of this, finiteness is guaranteed in this method even though the anti-cycling mechanism is only maintained during degenerate pivot steps. See their paper for details.

Simultaneous Resolution of Cycling and Stalling in the Primal Network Simplex Algorithm

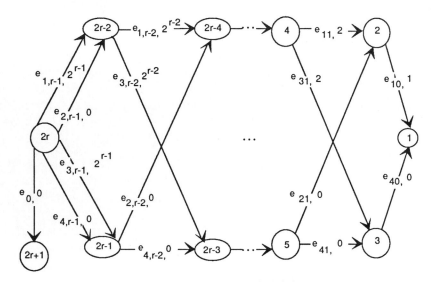

Figures 5.15 Network for the rth problem in the class for $r \geqq 2$. All lower bounds are 0, and capacities are $+\infty$. Source = node $2r$, with 1 unit available, sink = node $2r+1$ with requirement 1 unit. Cost coefficients are entered on the arcs.

Cycling is the infinite repetition of the same cycle of degenerate pivot steps in the primal simplex algorithm. The method of strongly feasible partitions resolves cycling and converts the primal simplex algorithm into a finite algorithm. But the method of strongly feasible partitions could encounter another computationally expensive phenomenon called *stalling*, which is a finite but very long sequence of consecutive degenerate pivot steps, in which the number of steps is not bounded above by any polynomial in $|\mathcal{N}|$ or the size of the problem. An example of stalling has been constructed by Cunningham [1979] using a class of problems developed by J. Edmonds to show that the primal simplex algorithm for the shortest chain problem is an exponential growth algorithm in the worst case. This class has a minimum cost flow problem for each $r \geqq 2$, the rth problem being on a directed network with $2r + 1$ nodes and $4r - 1$ arcs. See Figure 5.15. Arcs are grouped into 5 groups, the bth arc in the ath group being denoted by $e_{a,b}$. For example, $a = 1$ corresponds to the group of arcs at the top of the network in Figure 5.15, etc. All lower bounds are 0, and all capacities are ∞. It is required to ship 1 unit of material from the source node, $2r$, to the sink node, $2r + 1$, at minimum cost. Since it is an uncapacitated minimum cost flow problem with a specified source and sink node, it is a shortest chain problem. Arc cost coefficients are entered on the arcs.

The only feasible flow vector for the rth problem is $\tilde{f}$ with a flow of 1 on the arc e_0, and 0 flow on all the other arcs. Hence $\tilde{f}$ is the optimum solution to the problem.

Fix the root at the source node $2r$. Since all the capacities are ∞, partitions for this problem are of the form $(\mathbb{T}, \mathbb{L})$, where $\mathbb{T}$ is a spanning tree, and every such partition corresponds to the flow vector $\tilde{f}$ and is therefore feasible. e_0 is contained in every spanning tree, and it is an interior arc in every partition since the flow on it is 1, and all the other arcs are lower boundary arcs.

Define the partition $(\mathbb{T}_0, \mathbb{L}_0)$ where $\mathbb{T}_0 = \{e_0, e_{1,0}; \text{ and } e_{1,b}, e_{3,b} \text{ for } b = 1 \text{ to } r - 1 \}$, $\mathbb{L}_0 = \{e_{4,0}; \text{ and } e_{2,b}, e_{4,b} \text{ for } b = 1 \text{ to } r - 1\}$. All arcs in $\mathbb{T}_0$ other than e_0 are lower boundary arcs directed away from the root node; hence $(\mathbb{T}_0, \mathbb{L}_0)$ is a strongly feasible partition. Consider the following order for the arcs in the network in the rth problem.

$$e_{1,0}, e_{4,0}, e_{1,b}, e_{2,b}, e_{3,b}, e_{4,b} \text{ for } b = 1 \text{ to } r - 1, e_0 \tag{5.26}$$

If the rth problem in the class is solved by the method of strongly feasible partitions initiated with the partition $(\mathbb{T}_0, \mathbb{L}_0)$ and executed using the entering arc choice rule: always choose as the entering arc the first arc in the order listed in (5.26) which is eligible to enter; then it goes through a sequence of $3(2^r - 1) - 2r$ consecutive degenerate pivot steps before terminating. See Cunningham [1979] for a proof. This is stalling.

Cunningham [1979] has shown that a simple entering arc selection strategy will resolve stalling in the method of strongly feasible partitions for the class of minimum cost flow problems in pure networks. The requirement is that the entering arc selection strategy examine every arc periodically (say once in every γm pivot steps for some γ) and select it as the entering arc if it is eligible at that time. One such rule is LRC (*least recently considered*) entering arc choice rule. This rule arranges all the arcs in the network in a fixed order, say $e_1, \dots, e_m$, at the beginning of the algorithm. The entering arc in a pivot step is the first eligible arc in the list $e_{t+1}, e_{t+2}, \dots, e_m, e_1, \dots, e_t$, where e_t was the entering arc in the previous step. So, this rule circles through the list of arcs beginning with the entering arc in the previous step, looking for the entering arc. Hence each arc is examined once in every m pivot steps and given a chance to become the entering arc. Under this rule in the method of strongly feasible partitions, it has been shown that the total number of consecutive degenerate pivot steps can never exceed nm, and hence stalling cannot occur. Several other entering arc choice rules to prevent stalling in the method of strongly feasible partitions are discussed in Cunningham [1979].

The related problem in general LP: are there any entering and dropping variable choice rules for the primal simplex algorithm to solve an LP which resolve both cycling and stalling, still remains an open question at this time. It is the most important open question in LP theory. It can be shown that any such scheme consisting of entering and dropping variable choice rules which resolve both cycling and stalling in the primal simplex algorithm for LP at a degenerate BFS, would provide a polynomially bounded pivotal algorithm for checking the feasibility of a system of linear inequalities, and conversely. Hence the question of resolving both cycling and stalling in general LP is intimately related to the more fundamental question: is there a polynomially bounded version of the primal simplex algorithm for solving an LP with rational data?

Exercises

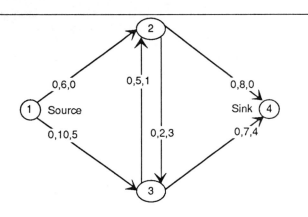

Figure 5.16

5.5 Let $\hat{\pi} = (\hat{\pi}_i)$ be the node price vector associated with the rooted spanning tree $\mathbb{T}$ rooted at node n in G, obtained by solving (5.21). For each $i \in \mathcal{N}$ prove that $-\hat{\pi}_i$ is the cost of the predecessor path of i in $\mathbb{T}$, as a path from i to n.

5.6 Find a minimum cost flow vector of value 5 in the network in Figure 5.16 by the OK algorithm. Draw the curve depicting the optimum objective value as a function of the value.

Find a minimum cost maximum value feasible flow vector in the network in Figure 5.17 by the OK algorithm. In both networks, the data on the arcs is lower bound, capacity, cost coefficient, in that order.

5.7 Let $\hat{f} = (\hat{f}_{ij})$ be the basic solution of (5.3) corresponding to the partition ($\mathbb{T}$, **L**, **U**). Let $(p, q) \in \mathbb{T}$ and $t = \text{son}(p, q)$. Define $\omega = +1$ if $t = q$, -1 if $t = p$. For any $i \in \mathcal{N}$ let $\overline{\mathbf{H}}(\mathbb{T}, i)$ denote the complement of $\mathbf{H}(\mathbb{T}, i)$. Prove that

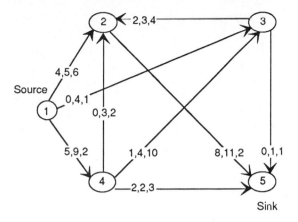

Figure 5.17

$$
\hat{f}_{pq} \;=\; \omega\Big(\sum_{j\in\overline{\mathbf{H}}(\mathbb{T},t)} V_j - \sum(\hat{f}_{ij}:\; over\;(i,j)\in(\mathcal{A}\backslash\mathbb{T})\cap(\overline{\mathbf{H}}(\mathbb{T},t),\mathbf{H}(\mathbb{T},t)))
$$

$$
+\sum(\hat{f}_{ij}:\; over\;(i,j)\in(\mathcal{A}\backslash\mathbb{T})\cap(\mathbf{H}(\mathbb{T},t),\overline{\mathbf{H}}(\mathbb{T},t))))\Big)
$$

5.8 Suppose (p,q) is the entering arc in a pivot step during Phase II of the primal network simplex method for solving (5.3). Prove that the relative cost coefficient $\bar{c}_{pq}$ at that time is the cost of the fundamental cycle of (p,q) wrt the spanning tree at that time.

5.9 In the network in Figure 5.18, $\ell_{ij}, k_{ij}, c_{ij}$ are marked on the arcs in that order. V_i is entered by the side of node i if it is nonzero. A spanning tree $\mathbb{T}$ is marked in thick arcs, and flow amounts chosen on out-of-tree arcs are entered in little squares on them if they are nonzero. Solve this problem by the primal network simplex method beginning with the partition provided by this information.

5.10 Consider the bipartite network representation of the assignment problem of order n on the bipartite network $G = (\mathcal{N}_1 = \{R_1,\ldots,R_n\}, \mathcal{N}_2 = \{C_1,\ldots,C_n\}, \mathcal{A} = \{(R_i,C_j): i,j = 1 \text{ to } n\})$. The arc (R_i,C_j) in $\mathcal{A}$ is directed from R_i to C_j for all i,j, and has lower bound 0 and capacity 1. Each node in $\mathcal{N}_1$ is a source with 1 unit of material available for shipment, and each node in $\mathcal{N}_2$ is a sink with 1 unit of material required. Let $\mathbb{T}$ be a spanning tree in G with column signature vector $(2,\ldots, 2,1)$ with only node C_n having degree 1. Treat $\mathbb{T}$ as a rooted tree with C_n as the root node. Define $\mathbf{L} = \mathcal{A}\backslash\mathbb{T}$ and $\mathbf{U} = \emptyset$.

In the basic solution associated with the partition $(\mathbb{T}, \mathbf{L}, \mathbf{U})$, prove that the flow on an in-tree arc is 1 if the arc is directed towards the root node, and 0 if it is directed away from the root node. Using this prove that $(\mathbb{T}, \mathbf{L}, \mathbf{U})$ is a highly degenerate strongly feasible partition for this problem. (Akgul and Ekin [1990 of Chapter 3]).

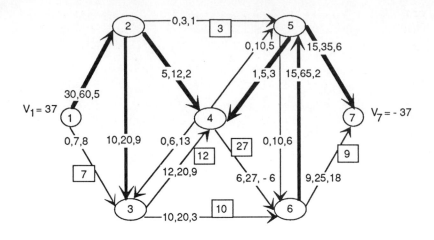

Figure 5.18

5.5   The Shortest Augmenting Path Method

This method is based on successive augmentations along cheapest FAPs. It solves the minimum cost flow problem by repeated application of a shortest chain algorithm in the residual network, each time saturating the FAP in the original network corresponding to the shortest chain obtained in the residual network at that stage. Thus the basic approach behind the algorithm is an *incremental approach*, since the flow value is incremented in each iteration by the least costly increment available at that stage. Hence, the algorithm has also been called a *buildup algorithm*, as it builds up the flow value at minimum cost thereby obtaining a sequence of flows of increasing value, each of minimum cost for its value.

This algorithm is essentially the same as the parametric value minimum cost flow algorithm discussed earlier. There it was based on the OK method; here we will discuss the version based on Dijkstra's shortest chain algorithm.

We consider the minimum cost flow problem in the form (5.1) in the directed network $G = (\mathcal{N}, \mathcal{A}, \ell = 0, k, c, \check{s}, \check{t}.\overline{v})$, where $|\mathcal{N}| = n$, $k > 0$, and $\overline{v}$ is the value required to be shipped from $\check{s}$ to $\check{t}$ at minimum cost. The method obtains a sequence of flow vector, node price vector pairs (f^r, π^r), $r = 0, 1, \ldots$, where π^r will always be dual feasible, and f^r always satisfies the lower bound and capacity conditions on all the arcs, and flow conservation at all the nodes, and its value v^r strictly increases in each step. When v^r reaches $\overline{v}$, primal feasibility is attained and the method terminates. Hence, the method is a dual method.

In each step the method solves a shortest chain problem in the residual network wrt the pair at that time. Since all the cost coefficients in it are $\overset{\geq}{=} 0$, the shortest chain problem can be solved by Dijkstra's method with a computational effort of at most $O(n^2)$. To initiate the method, we need an optimum pair of zero value. For this reason we consider the case where either $c \overset{\geq}{=} 0$, or a node price vector $\pi = (\pi_i)$ is known satisfying $\overline{c}_{ij} = c_{ij} - (\pi_j - \pi_i) \overset{\geq}{=} 0$ for all $(i,j) \in \mathcal{A}$. In the latter case we replace c by $\overline{c} = (\overline{c}_{ij})$; this leads to an equivalent problem. Hence without any loss of generality we consider the case $c \overset{\geq}{=} 0$.

Given an optimum pair (f, π) of value v, let $(c_{ij} = c_{ij} - (\pi_j - \pi_i))$ be the reduced cost vector wrt π. By optimality, we know that the following c.s. conditions hold for each $(i,j) \in \mathcal{A}$

$$
\begin{aligned}
0 < f_{ij} < k_{ij} \quad &\text{implies} \quad \overline{c}_{ij} = 0 \\
\overline{c}_{ij} < 0 \quad &\text{implies} \quad f_{ij} = k_{ij} < \infty \\
\overline{c}_{ij} > 0 \quad &\text{implies} \quad f_{ij} = 0
\end{aligned}
\tag{5.27}
$$

By these optimality conditions, the cost vector $\overline{c}'$ in the residual network $G(f, \pi)$ $= (\mathcal{N}, \mathcal{A}(f), 0, \kappa, \overline{c}')$ is $\overset{\geq}{=} 0$.

THE SHORTEST AUGMENTING PATH ALGORITHM

Initialization   Begin with the optimum pair $(f^0 = 0, \pi^0 = 0)$ of value 0. This is optimal because $c \overset{\geq}{=} 0$.

General step   Let (f^r, π^r) be the present pair of value v^r. Find a shortest chain tree rooted at $\breve{s}$ in $G(f^r, \pi^r)$ using Dijkstra's algorithm. Terminate Dijkstra's algorithm as soon as $\breve{t}$ is permanently labeled.

If there exists no chain from $\breve{s}$ to $\breve{t}$ in $G(f^r, \pi^r)$, there exists no FAP from $\breve{s}$ to $\breve{t}$ in G wrt f^r, and since $v^r < \overline{v}$, there is no feasible flow of value $\overline{v}$ in G; terminate.

If a shortest chain, C_r, from $\breve{s}$ to $\breve{t}$ in $G(f^r, \pi^r)$ has been found, define for $i \in \mathcal{N}$, $\mu_i^r =$ cost of shortest chain from $\breve{s}$ to i in $G(f^r, \pi^r)$ if i is permanently labeled before Dijkstra's algorithm is terminated, or $= \mu_{\breve{t}}^r$ otherwise. Let $\mathcal{P}_r$ be the FAP from $\breve{s}$ to $\breve{t}$ wrt f^r in G corresponding to C_r, and ϵ_r its residual capacity. Let $\gamma_r = \min. \{\epsilon_r, \overline{v} - v^r\}$. Carry out flow augmentation along $\mathcal{P}_r$ by γ_r units, and let f^{r+1} be the new flow vector obtained. Its value is $v^{r+1} = v^r + \gamma_r$. Define $\pi_i^{r+1} = \pi_i^r + \mu_i^r$ for all $i \in \mathcal{N}$ and $\pi^{r+1} = (\pi_i^{r+1})$. If $v^{r+1} = \overline{v}$, (f^{r+1}, π^{r+1}) is an optimum pair in G, terminate. Otherwise go to the next step with this new pair.

Discussion

1. For simplicity of notation, let (f, π) denote the pair satisfying (5.27) at the beginning of a step in this algorithm, and by $(\hat{f}, \hat{\pi})$ the pair obtained at the end of this step. Then $(\hat{f}, \hat{\pi})$ also satisfies (5.27).

To prove this, let $\mathbf{X}$ be the set of permanently labeled nodes at the stage that Dijkstra's algorithm was terminated in this step. Let $\overline{c}_{ij} = c_{ij} - (\pi_j - \pi_i)$, $\hat{\overline{c}}_{ij} = c_{ij} - (\hat{\pi}_j - \hat{\pi}_i)$ be the reduced costs of arc (i, j) wrt the node price vectors $\pi, \hat{\pi}$ respectively. $\overline{c}', \hat{\overline{c}}'$ denote the cost vectors in the residual networks $G(f, \pi)$, $G(\hat{f}, \hat{\pi})$ respectively.

For each $j \in \mathbf{X}$, let μ_j denote the cost of a shortest chain from $\breve{s}$ to j in $G(f, \pi)$. For $j \notin \mathbf{X}$, let $\mu_j = \mu_{\breve{t}}$ as defined in the algorithm. By Theorem 4.2 and the result in Exercise 4.21, we have

$$\mu_j - \mu_i \overset{\leq}{=} \overline{c}'_{ij} \text{ for all arcs } (i, j) \text{ in } G(f, \pi) \tag{5.28}$$

Let $\mathcal{C}$ denote the shortest chain from $\breve{s}$ to $\breve{t}$ in $G(f, \pi)$ identified in Dijkstra's algorithm in this step, and $\mathcal{P}$ the FAP in G wrt f corresponding to $\mathcal{C}$. (5.28) holds as a strict equation for arcs (i, j) on $\mathcal{C}$.

So, for arcs (i, j) on $\mathcal{C}$ we have $\mu_j - \mu_i = \overline{c}'_{ij}$. Therefore if (i, j) on $\mathcal{C}$ has a + label in $G(f, \pi)$, then $(i, j) \in \mathcal{A}$ and $\mu_j - \mu_i = \overline{c}'_{ij} = c_{ij} - (\pi_j - \pi_i)$, and hence $c_{ij} = (\pi_j + \mu_j) - (\pi_i + \mu_i) = (\hat{\pi}_j - \hat{\pi}_i)$. If (i, j) has a $-$ label in $G(f, \pi)$, then $(j, i) \in \mathcal{A}$ and $\mu_j - \mu_i = \overline{c}'_{ij} = -(c_{ji} - (\pi_i - \pi_j))$ and hence $c_{ji} = \hat{\pi}_i - \hat{\pi}_j$. Hence all arcs in G corresponding to arcs on $\mathcal{C}$ continue to satisfy (5.27) in the new pair $(\hat{f}, \hat{\pi})$. In fact, for all arcs (p, q) on $\mathcal{P}$ we have $c_{pq} = \hat{\pi}_q - \hat{\pi}_p$.

For any arc (i, j) in G not corresponding to an arc on $\mathcal{C}$, we have $\hat{f}_{ij} = f_{ij}$. From this and (5.28) it can be verified that all these arcs continue to satisfy (5.27) in the new pair $(\hat{f}, \hat{\pi})$.

2. Using the notation in 1., all the arc cost coefficients in $G(\hat{f}, \hat{\pi})$ are $\overset{\geq}{=} 0$.

   To see this, consider an arc (i, j) in $G(\hat{f}, \hat{\pi})$. If it has a + label, then $(i, j) \in \mathcal{A}$ and $\hat{f}_{ij} < k_{ij}$ and $\hat{\overline{c}}'_{ij} = \hat{\overline{c}}_{ij} = c_{ij} - (\hat{\pi}_j - \hat{\pi}_i) \overset{\geq}{=} 0$, by (5.27). If it has a - label, then $(j, i) \in \mathcal{A}$ and $\hat{f}_{ji} > \ell_{ji}$ and $\hat{\overline{c}}'_{ij} = -\hat{\overline{c}}_{ji} \overset{\geq}{=} 0$ by (5.27). Hence all the arc cost coefficients in $G(\hat{f}, \hat{\pi})$ are $\overset{\geq}{=} 0$.

   From this, we see that in every step of this algorithm, the shortest chain problem to be solved is on a residual network in which the arc cost coefficients are $\overset{\geq}{=} 0$. So, Dijkstra's algorithm can be applied to solve the shortest chain problem in every step.

3. Each FAP used to augment the flow in this algorithm is a least cost FAP among all the FAPs at that stage.

   To see this, let us use the notation in 1. From 1. we know that $c_{pq} = \hat{\pi}_q - \hat{\pi}_p$ for all arcs (p, q) on the FAP $\mathcal{P}$ used to generate $\hat{f}$ from f. Hence the cost of this FAP $\mathcal{P}$ is $\hat{\pi}_{\breve{t}} - \hat{\pi}_{\breve{s}}$.

   Suppose $\mathcal{P}'$ is an FAP from $\breve{s}$ to $\breve{t}$ in G wrt f. If (i, j) is a common arc on $\mathcal{P}$ and $\mathcal{P}'$ we have $c_{ij} = \hat{\pi}_j - \hat{\pi}_i$. If (i, j) is a forward arc on $\mathcal{P}'$ which is not on

$\mathcal{P}$, then $\hat{f}_{ij} = f_{ij} < k_{ij}$ and by the result in (a), $\hat{\pi}_j - \hat{\pi}_i \leqq c_{ij}$. If (i,j) is a reverse arc on $\mathcal{P}'$ which is not on $\mathcal{P}$, then $\hat{f}_{ij} = f_{ij} > \ell_{ij}$ and by the result in 1., $\hat{\pi}_j - \hat{\pi}_i \geqq c_{ij}$. These facts imply that the cost of $\mathcal{P}'$ is $\geqq \hat{\pi}_{\bar{t}} - \hat{\pi}_{\bar{s}} = $ cost of $\mathcal{P}$.

So, the FAP $\mathcal{P}$ is the cheapest among all the FAPs available at the time it is used in the algorithm. Hence the name *shortest augmenting path method* for this algorithm.

4. The finiteness of this algorithm follows from the finiteness of the parametric value minimum cost flow algorithm discussed in Section 5.3.

5. When this method is specialized to solve the assignment problem (see Jonker and Volgenant [1987]), it leads to the shortest augmenting path algorithm for the assignment problem. Each augmenting path used in this method will be an alternating path. Since the assignment problem of order n needs at most n augmentations, the method terminates after at most n steps. The work in each step involves solving a shortest chain problem, which takes at most $O(n^2)$ effort by Dijkstra's method. So, the overall computational complexity of this method for an assignment problem of order n is also $O(n^3)$.

5.6   A Class of Primal-Dual Methods

We consider the minimum cost circulation problem (5.2) in the directed network $G = (\mathcal{N}, \mathcal{A}, \ell, k, c)$ in which ℓ, k, c are all finite integer vectors satisfying $\ell \leqq k$, and $|\mathcal{N}| = n, |\mathcal{A}| = m$. In this section we discuss a class of primal-dual algorithms for this problem. These methods maintain in every step; a flow vector, node price vector pair (f, π) called an *admissible pair* which always satisfies the following conditions

1. f is always an integer, bound feasible flow vector (i.e., $\ell \leqq f \leqq k$). π is always an integer vector.

2. all arcs in $\mathcal{A}$ are always α-, β-, or γ-arcs as defined in Section 5.3, in the pair (f, π).

By 2., we know that an admissible pair satisfies the c.s. optimality conditions. All but the final admissible pair obtained in these methods violate flow conservation at some nodes.

Given a flow vector f in G, define *the deficit at node i in f* to be $d_i = f(\mathcal{N}, i) - f(i, \mathcal{N})$.

Given a node price vector $\pi = (\pi_i)$ in G, for each $(i,j) \in \mathcal{A}$ define $Q_{ij}(\pi_j - \pi_i) = \min.\{g_{ij}(c_{ij} - (\pi_j - \pi_i)) : \text{over } \ell_{ij} \leqq g_{ij} \leqq k_{ij} \}$. Define $Q(\pi) = \sum(Q_{ij}(\pi_j - \pi_i) : \text{over } (i,j) \in \mathcal{A})$. $Q_{ij}(\pi_j - \pi_i)$ is a piecewise linear concave function. A plot of it is shown in Figure 5.19.

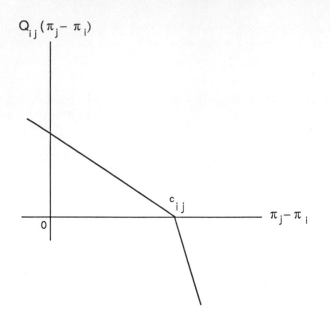

Figure 5.19 The slope of this piecewise linear function is
$-\ell_{ij}$ when tension $< c_{ij}$, and $-k_{ij}$ when tension $> c_{ij}$.

It can be verified that $\sum(d_i : \text{over } i \in \mathcal{N}) = 0$. So, in a flow vector f if some node deficits are $\neq 0$, there must exist a node with a positive deficit, and another with a negative deficit. In an admissible pair (f, π) if deficits are 0 at all the nodes, f is a feasible flow vector, and the pair is an optimal pair. The methods discussed in this section always try to move to a pair in which all node deficits are 0, while maintaining admissibility, by various types of moves described below.

THEOREM 5.4 *Under the assumption that ℓ, k are finite vectors, the dual of the minimum cost circulation problem (5.2) in G is equivalent to the following unconstrained maximization problem*

$$\textit{Maximize } \{Q(\pi) : \textit{ over } \pi \in \mathbb{R}^n\} \tag{5.29}$$

Proof   Let π be the node price vector. Associate the dual variable vectors $\mu = (\mu_{ij}), \nu = (\nu_{ij})$ to the lower, upper bound constraints on f respectively. Then the dual of (5.2) is

$$
\begin{aligned}
\text{Maximize} \quad & \mu\ell - \nu k \\
\text{Subject to} \quad & (\pi_j - \pi_i) + \mu_{ij} - \nu_{ij} = c_{ij}, \text{ for } (i,j) \in \mathcal{A} \\
& \mu, \nu \qquad\qquad\quad \geqq 0
\end{aligned}
\tag{5.30}
$$

Since $\ell \overset{\leq}{=} k$, in an optimum solution of (5.30), we will have $\mu_{ij} = \max.\{0, c_{ij} - (\pi_j - \pi_i)\}$ and $\nu_{ij} = - \min.\{0, c_{ij} - (\pi_j - \pi_i)\}$, for each $(i, j) \in \mathcal{A}$. Using these we can eliminate the variables μ_{ij}, ν_{ij} from the dual problem and express it purely in terms of the node prices π_is. Since at least one of the quantities among max. $\{0, c_{ij} - (\pi_j - \pi_i)\}$, min. $\{0, c_{ij} - (\pi_j - \pi_i)\}$ is always 0, it can be verified that the problem (5.30) expressed in terms of π is exactly equivalent to (5.29). ∎

In the algorithms discussed in this section, we use two criteria for measuring progress. One criterion is the total absolute deficit, $\sum(|d_i| :$ over $i \in \mathcal{N})$, which should be reduced to 0. When it is 0, the current pair is optimal. The second criterion is the dual functional $Q(\pi)$; it should be maximized. When it is maximum, the node price vector π is optimum.

The methods use three types of steps called *flow adjustment steps (FAS)*, *price adjustment steps (PAS)*, and *ascent steps* to solve the problem. All adjustment steps, FAS or PAS, move from one admissible pair (f, π) with a node deficit vector $d = (d_i)$, to another admissible pair $(\hat{f}, \hat{\pi})$ with node deficit vector $\hat{d} = (\hat{d}_i)$ subject to the following conditions

$$\text{If } d_i > 0 \quad \text{then} \quad \hat{d}_i \overset{\leq}{=} d_i \text{ and } \hat{\pi}_i \overset{\leq}{=} \pi_i$$

$$(5.31)$$

$$\text{If } d_i \overset{\leq}{=} 0 \quad \text{then} \quad \hat{d}_i \overset{\leq}{=} 0 \text{ and } \hat{\pi}_i \overset{\geq}{=} \pi_i$$

The adjustment step is called an FAS if $\hat{d}_i < d_i$ for at least one node i with $d_i > 0$, and a PAS if $\hat{\pi}_i > \pi_i$ for at least one node i with $d_i \overset{\leq}{=} 0$.

An ascent step is a move from an admissible pair (f, π) to another admissible pair $(\hat{f}, \hat{\pi})$ satisfying $Q(\hat{\pi}) > Q(\pi)$.

An adjustment step, FAS or PAS, will never increase the total absolute deficit, but an FAS strictly decreases it. To see this let an adjustment step move from the admissible pair (f, π) associated with deficit vector $d = (d_i)$ to $(\hat{f}, \hat{\pi})$ associated with $\hat{d} = (\hat{d}_i)$. We have $\sum d_i = \sum \hat{d}_i = 0$. So,

$$
\begin{aligned}
\frac{1}{2} \sum |d_i| &= \sum (d_i : \text{ over } i \text{ satisfying } d_i > 0) \\
&\overset{\geq}{=} \sum (d_i : \text{ over } i \text{ satisfying } \hat{d}_i > 0) \text{ by (5.31)} \\
&\overset{\geq}{=} \sum (\hat{d}_i : \text{ over } i \text{ satisfying } \hat{d}_i > 0) \\
&= \frac{1}{2} \sum |\hat{d}_i|
\end{aligned}
$$

Also, we see that $\sum |d_i| = \sum |\hat{d}_i|$ iff $\hat{d}_i = d_i$ for all i satisfying $d_i > 0$, and hence this cannot happen in an FAS. Hence an FAS strictly decreases the total absolute deficit, and a PAS never increases it.

An FAS or PAS may decrease the value of $Q(\pi)$. We will call an adjustment step (FAS or PAS) *harmless* if it does not decrease the value of $Q(\pi)$.

An ascent step may not always qualify as an adjustment step. Indeed in an ascent step, the total absolute deficit may increase.

There are several means by which we can look for ways of making adjustment steps. One is to search for the so-called *flow altering paths* or FAℓPs. Given an admissible pair (f, π), an FAℓP wrt it is a path from a node with a positive deficit to one with a negative deficit, all of whose arcs are β-arcs, and all forward arcs have residual capacity, and all reverse arcs have flow $>$ its lower bound. Given an FAℓP wrt the pair (f, π), define its capacity to be $\epsilon = $ min. { deficit of start node, $-$deficit of terminal node, residual capacities of forward arcs, $f_{ij} - \ell_{ij}$ for reverse arcs (i, j) }.

Given an FAℓP $\mathcal{P}$ of capacity ϵ wrt the admissible pair (f, π), the *flow change operation using it* leads to the admissible pair $(\overline{f}, \pi)$, in which $\overline{f}$ is obtained from f by increasing (decreasing) the flow amount on forward (reverse) arcs of $\mathcal{P}$ by ϵ. As a result of this flow change, the deficits of both the start and terminal nodes remain of the same sign (could be 0), but each of their absolute values decrease by ϵ, and the deficits of all other nodes remain unaltered. Since an FAℓP always consists of β-arcs only, the c.s. conditions continue to hold in the new pair. Also, if ℓ, k, f are all integer vectors, ϵ is an integer, and hence the new flow vector $\overline{f}$ will also be an integer vector.

We will now describe a procedure, Procedure 1, which carries out adjustment steps by searching for FAℓPs using a tree growth scheme. To avoid confusion with steps in the main methods, we will call the steps in the procedure *items*.

PROCEDURE 1 :   TO CARRY OUT AN ADJUSTMENT STEP
BY SEARCHING FOR AN FAℓP

Initialization Let (f, π) be the current admissible pair with $d = (d_i)$ as the deficit vector. If $d = 0$, (f, π) is an optimum pair, terminate the method. Otherwise select a negative deficit node i_0 and label it with $\emptyset$. List $= \{i_0\}$. Go to Item 1.

Item 1 Select a node for scanning   If list $= \emptyset$, go to Item 3. Otherwise select a node from the list, say p, for scanning. Delete p from the list, and go to Item 2.

Item 2 Scanning   Let p be the node to be scanned. Find all unlabeled nodes j such that (p, j) is a β-arc with $f_{pj} > \ell_{pj}$, label them with $(p, -)$ and include them in the list. Find all unlabeled nodes i such that (i, p) is a β-arc with $f_{ip} < k_{ip}$, label them with $(p, +)$ and include them in the list.

Terminate this procedure if a node with positive deficit is labeled. If that node is q, the predecessor path of q is an FAℓP wrt (f, π). Carrying out the flow change operation with that FAℓP is an FAS. Since there is no change in the node price vector this is a harmless FAS.

If no node with a positive deficit is labeled go back to Step 1.

Item 3 Node price change   We come to this item if no node with positive
deficit is labeled, and list $= \emptyset$. Let $\mathbf{X} =$ set of labeled nodes, and $\overline{\mathbf{X}}$ its
complement. Let $\mathbf{A}^1$ be the set of γ-arcs in $(\mathbf{X}, \overline{\mathbf{X}})$, and $\mathbf{A}^2$ the set of α-arcs
in $(\overline{\mathbf{X}}, \mathbf{X})$.

If $\mathbf{A}^1 \cup \mathbf{A}^2 = \emptyset$, then $k(\overline{\mathbf{X}}, \mathbf{X}) - \ell(\mathbf{X}, \overline{\mathbf{X}}) = f(\overline{\mathbf{X}}, \mathbf{X}) - f(\mathbf{X}, \overline{\mathbf{X}}) = f(\mathcal{N}, \mathbf{X}) -$
$f(\mathbf{X}, \mathcal{N}) = d(\mathbf{X}) < 0$, by Theorem 2.11 this implies that there is no feasible
circulation in G; terminate the method.

If $\mathbf{A}^1 \cup \mathbf{A}^2 \neq \emptyset$, let $\delta = \min. \ \{|c_{ij} - (\pi_j - \pi_i)| : (i,j) \in \mathbf{A}^1 \cup \mathbf{A}^2\}$. δ is finite
and > 0. Define $\hat{\pi} = (\hat{\pi}_i)$ where $\hat{\pi}_i = \pi_i + \delta$ for $i \in \mathbf{X}$, or $= \pi_i$ for $i \in \overline{\mathbf{X}}$.
Since $d_i \leqq 0$ for all $i \in \mathbf{X}$, changing (f, π) to $(f, \hat{\pi})$ satisfies all the conditions
for being a PAS. Terminate the procedure.

Now we will describe another procedure that tries to make adjustment steps
that involve a single node together with its immediate neighbors, For each $i \in \mathcal{N}$
define the following wrt an admissible pair (f, π) in G.

$$\mathbf{B}_i^+(f, \pi) \ = \ \{j : (j,i) \text{ is a } \beta\text{-arc and } f_{ji} < k_{ji}\}$$
$$\mathbf{A}_i^-(f, \pi) \ = \ \{j : (i,j) \text{ is a } \beta\text{-arc and } f_{ij} > \ell_{ij}\}$$

PROCEDURE 2: SEQUENCE OF SINGLE NODE FAS AND PASs

Single node FASs   Let (f, π) be the present admissible pair with $d = (d_i)$ as
the node deficit vector. Select a node q with $d_q < 0$.

Look for a node $j \in \mathbf{B}_i^+(f, \pi)$ for which $d_j > 0$. If such a node j exists, let ϵ
$= \min. \ \{-d_q, d_j, k_{jq} - f_{jq}\}$. Add ϵ to f_{jq} and d_q; subtract ϵ from d_j. This is
a single node FAS.

If there is no node j like the above, look for a node $p \in \mathbf{A}_i^-(f, \pi)$ for which
$d_p > 0$. If such a node p exists, let $\epsilon = \min. \ \{-d_q, d_p, f_{qp} - \ell_{qp}\}$. Subtract ϵ
from f_{qp}, d_p and add ϵ to d_q. This is a single node FAS too.

Repeat with the new pair and node q if its deficit is still < 0. If its deficit
became 0, select another node with negative deficit and repeat.

Single node PASs   Let (f, π) be the present admissible pair with $d = (d_i)$ as
the node deficit vector. Select a node q with $d_q < 0$. If there are no β-arcs of
form (j, q) with $f_{jq} < k_{jq}$, or (q, j) with $f_{qj} > \ell_{qj}$ go to Item 1; otherwise go
to Item 2.

Item 1   Define $\mathbf{E} = \{(j, q) : f_{jq} < k_{jq}\} \cup \{(q, j) : f_{qj} > \ell_{qj}\}$. If $\mathbf{E} = \emptyset$, all arcs
incident into q are saturated, and all those incident out of q are lower boundary

arcs, and yet $d_q < 0$, so there is no feasible circulation in G; terminate the method.

If $\mathbf{E} \neq \emptyset$, let $\delta = $ min. $\{|c_{uw} - (\pi_w - \pi_u)| : (u, w) \in \mathbf{E}\}$. $\delta > 0$. Add δ to π_q; this operation is a single node PAS.

Item 2   If

$$\epsilon = \sum_{j \in \mathbf{B}_q^+(f,\pi)} (k_{jq} - f_{jq}) + \sum_{j \in \mathbf{A}_q^-(f,\pi)} (f_{qj} - \ell_{qj}) \overset{\leq}{=} -d_q \qquad (5.32)$$

change f_{jq} to k_{jq} for each $j \in \mathbf{B}_q^+(f, \pi)$, and f_{qj} to ℓ_{qj} for each $j \in \mathbf{A}_q^-(f, \pi)$. Both these sets become $\emptyset$ wrt the new flow vector. With the new pair go back to Item 1.

Discussion

The flow change in Item 2 decreases the total absolute deficit iff some node in the original $\mathbf{B}_q^+(f, \pi) \cup \mathbf{A}_q^-(f, \pi)$ had positive deficit; otherwise, the total absolute deficit remains unchanged by this flow change.

It can be shown that the directional derivative of $Q(\pi)$ along the direction of increasing π_q and leaving other node prices unchanged is $\sum(k_{qj}$: over (q, j) is a γ-arc$) + \sum(\ell_{qj}$: over (q, j) an α- or β-arc$) - \sum(k_{jq}$: over (j, q) a γ-, or β-arc$) - \sum(\ell_{jq}$: over (j, q) an α-arc$) = -d_q - \sum(k_{jq} - f_{jq}$: over $j \in \mathbf{B}_q^+(f, \pi))$ $- \sum(f_{qj} - \ell_{qj}$: over $j \in \mathbf{A}_q^-(f, \pi))$. From this it follows that the single node PAS above increases $Q(\pi)$ iff strict inequality holds in (5.32). If (5.32) holds as an equation, the single node PAS at q leaves $Q(\pi)$ unchanged, but the flow change carried out reduces the deficit at node q to 0.

When we are in Item 2 for making a single node PAS at node q, if (5.32) is violated, i.e., if (5.33) holds

$$\sum_{j \in \mathbf{B}_q^+(f,\pi)} (k_{jq} - f_{jq} + \sum_{j \in \mathbf{A}_q^-(f,\pi)} (f_{qj} - \ell_{qj}) > -d_q \qquad (5.33)$$

then no further progress is possible at node q by single node PASs. We can now attempt a single node FAS, but if

$$d_q < 0 \text{ and } d_j \overset{\leq}{=} 0 \text{ for all } j \in \mathbf{B}_q^+(f, \pi) \cup \mathbf{A}_q^-(f, \pi) \qquad (5.34)$$

no further progress can be made at node q with either a single node PAS or FAS. When (5.33), (5.34) both hold, we need to perform FAS or PAS involving multiple nodes. We will now describe a procedure that leads to a multiple node adjustment step that helps also to achieve large price increases consistent with the philosophy of a single node PAS.

PROCEDURE 3: MULTIPLE NODE ADJUSTMENT STEPS

Item 1 Let (f, π) be the current admissible pair with $d = (d_i)$ as the deficit vector. Select a node q with $d_q < 0$ satisfying (5.33), (5.34).

Item 2 Let $\mathbf{D}$ be either $\emptyset$ or a subset of $\mathbf{B}_q^+(f, \pi) \cup \mathbf{A}_q^-(f, \pi)$ satisfying

$$\sum_{j \in \mathbf{B}_q^+(f,\pi) \cap \mathbf{D}} (k_{jq} - f_{jq}) + \sum_{j \in \mathbf{A}_q^-(f,\pi) \cap \mathbf{D}} (f_{qj} - \ell_{qj}) \leqq -d_q \tag{5.35}$$

Let $\overline{\mathbf{D}} = (\mathbf{B}_q^+(f, \pi) \cup \mathbf{A}_Q^-(f, \pi)) \setminus \mathbf{D}$. Because of (5.33) and (5.35) $\overline{\mathbf{D}} \neq \emptyset$. Proceed similar to Procedure 1, with the only difference being that the initial labels will be given only to nodes in the subset $\overline{\mathbf{D}}$ rather than the entire set $\mathbf{B}_q^+(f, \pi) \cup \mathbf{A}_Q^-(f, \pi)$. Label q with $\emptyset$. Label each of the nodes in $\mathbf{B}_q^+(f, \pi) \cap \overline{\mathbf{D}}$ with $(q, +)$, and each of the nodes in $\mathbf{A}_q^-(f, \pi) \cap \overline{\mathbf{D}}$ with $(q, -)$. Put all nodes in $\overline{\mathbf{D}}$ in the list of labeled and unscanned nodes (in this procedure, the root node q is not put in the list unlike Procedure 1)

Continuation now consists of deleting a node p from the list and scanning it as in Procedure 1. If a node with positive deficit is labeled, an FAℓP has been identified; stop tree growth and carry out an FAS using it as under Procedure 1. This procedure terminates after this FAS.

If tree growth stops without a node having positive deficit getting labeled, carry out a PAS using the cut of labeled and unlabeled nodes as under Procedure 1. Then change f_{jq} to k_{jq} for all $j \in \mathbf{B}_q^+(f, \pi) \cap \mathbf{D}$, and f_{qj} to ℓ_{qj} for all $j \in \mathbf{A}_q^-(f, \pi) \cap \mathbf{D}$.

Discussion

If no FAℓP is identified in Procedure 3, it can be verified that the work carried out at the end is a PAS. The flow changes in this PAS increase the deficit of each node in $\mathbf{D}$, while the deficit of q still remains nonpositive because of (5.35). The adjustment step in Procedure 3 is harmless. This procedure coincides with Procedure 1 if $\mathbf{D} = \emptyset$. However, taking $\mathbf{D} \neq \emptyset$ is advantageous if strict inequality holds in (5.35) as this makes the multiple node PAS in Procedure 3 an ascent step.

We will now discuss a procedure for carrying out an ascent step. Let (f, π) be the present admissible pair. Let $\mathbf{S}$ be a proper subset of nodes, and let its incidence vector be $x(\mathbf{S}) = (x_i(\mathbf{S}))$ where $x_i(\mathbf{S}) = 1$ if $i \in \mathbf{S}$, 0 otherwise. Let $\overline{\mathbf{S}}$ be the complement of $\mathbf{S}$. Define $a(\mathbf{S}, \pi) = \sum_{(i,j) \in \mathcal{A}} a_{ij}(\mathbf{S}, \pi)$, where

$$
a_{ij}(\mathbf{S}, \pi) = \begin{cases} \ell_{ij} & \text{if} & (i,j) \in (\mathbf{S}, \overline{\mathbf{S}}) \text{ is an } \alpha - \text{ or } \beta - \text{arc} \\[2ex] k_{ij} & \text{if} & (i,j) \in (\mathbf{S}, \overline{\mathbf{S}}) \text{ is a } \gamma - \text{arc} \\[2ex] -\ell_{ij} & \text{if} & (i,j) \in (\overline{\mathbf{S}}, \mathbf{S}) \text{ is an } \alpha - \text{arc} \\[2ex] -k_{ij} & \text{if} & (i,j) \in (\overline{\mathbf{S}}, \mathbf{S}) \text{ is a } \beta - \text{ or } \gamma - \text{arc} \\[2ex] 0 & & \text{otherwise} \end{cases} \tag{5.36}
$$

It can be verified that $a(\mathbf{S}, \pi)$ is the directional derivative of the dual functional $Q(\pi)$ along the direction $x(\mathbf{S})$. So, if $a(\mathbf{S}, \pi) > 0$, to increase $Q(\pi)$ we can move in the direction $x(\mathbf{S})$, i.e., increase node prices of nodes in $\mathbf{S}$ by equal amounts while keeping node prices outside of $\mathbf{S}$ unchanged. The ascent step uses this fact.

The main feature of this procedure is that the choice of ascent directions is very simple. A node i_0 with nonzero deficit is chosen, and an ascent is attempted along the coordinate direction of π_{i_0}. If such an ascent is not possible and a reduction in the absolute deficit cannot be effected through flow change, an adjacent node of i_0, say i_1, is chosen and an ascent attempted along the sum of the coordinate vectors corresponding to π_{i_0} and π_{i_1}. If such an ascent is not possible, and flow change is not possible either, an adjacent node of either i_0 or i_1 is chosen and the process is continued.

PROCEDURE 4 : ASCENT STEP

Item 1   Let (f, π) be the present admissible pair with nonzero deficit vector $d = (d_i)$. Select a node, say i_0 with $d_{i_0} < 0$. Root a tree at i_0 by labeling it with $\emptyset$. List $= \{i_0\}$, $\mathbf{S} = \emptyset$.

General item   Delete a node, say p, from the list, insert it into $\mathbf{S}$, and scan it. Scanning p involves labeling all unlabeled nodes in $\mathbf{B}_p^+(f, \pi)$ with the label $(p, +)$, and unlabeled nodes in $\mathbf{A}_p^-(f, \pi)$ with the label $(p, -)$ and inserting all these newly labeled nodes in the list.

If a node with positive deficit is labeled, but $a(\mathbf{S}, \pi) \lesseqgtr 0$, an FA$\ell$P has been identified. Carry out an FAS using this FAℓP as under Procedure 1, and terminate this procedure.

On the other hand, if we have $\overline{\mathbf{S}} = \mathcal{N} \setminus \mathbf{S} \neq \emptyset$ and $a(\mathbf{S}, \pi) > 0$, carry out the following ascent step. Let $\mathbf{E} = \{(i,j) : (i,j) \text{ is a } \gamma\text{-arc in } (\mathbf{S}, \overline{\mathbf{S}}), \text{ or an } \alpha\text{-arc in} (\overline{\mathbf{S}}, \mathbf{S})\}$. If $\mathbf{E} = \emptyset$, there is no feasible circulation in G; terminate the method. Otherwise, define

$$
\delta = \text{min.} \{|c_{ij} - (\pi_j - \pi_i)| : (i,j) \in \mathbf{E}\} \tag{5.37}
$$

and the new flow vector, node price vector pair $(\hat{f} = (\hat{f}_{ij}), \hat{\pi} = (\hat{\pi}_i))$, where

$$\hat{f}_{ij} = \begin{cases} k_{ij} \text{ if } (i,j) \text{ is a } \beta\text{-arc in } (\overline{\mathbf{S}}, \mathbf{S}) \text{ and } i \text{ labeled} \\ \\ \ell_{ij} \text{ if } (i,j) \text{ is a } \beta\text{-arc in } (\mathbf{S}, \overline{\mathbf{S}}) \text{ and } j \text{ labeled} \\ \\ f_{ij} \text{ otherwise} \end{cases}$$

$$\hat{\pi}_i = \begin{cases} \pi_i + \delta & \text{if } i \in \mathbf{S} \\ \\ \pi_i & \text{if } i \in \overline{\mathbf{S}} \end{cases}$$

The procedure terminates with the new pair $(\hat{f}, \hat{\pi})$.

If the conditions for neither FAS nor ascent step are satisfied, go to the next item in tree growth.

Discussion

If $\mathbf{E} = \emptyset$ at the occurrence of an ascent step, from (5.36) we have $0 < a(\mathbf{S}, \pi) = -k(\overline{\mathbf{S}}, \mathbf{S}) + \ell(\mathbf{S}, \overline{\mathbf{S}})$, i.e., $k(\overline{\mathbf{S}}, \mathbf{S}) - \ell(\mathbf{S}, \overline{\mathbf{S}}) < 0$, which implies that there is no feasible circulation in G by Theorem 2.11.

If tree growth stops without the conditions for an FAS being satisfied, let $\mathbf{X}$ be the set of labeled nodes at that time. Since all nodes in $\mathbf{X}$ are scanned, $\mathbf{S} = \mathbf{X}$ at this stage. Since the conditions for an FAS were not satisfied so far, $d_i \overset{\leq}{=} 0$ for all $i \in \mathbf{X}$. So, $\overline{\mathbf{X}} = \mathcal{N} \backslash \mathbf{X} = \overline{\mathbf{S}} = \mathcal{N} \backslash \mathbf{S} \neq \emptyset$, since all nodes with positive deficit are in it. We also have $f_{ij} = k_{ij}$ for all β-arcs $(i,j) \in (\overline{\mathbf{S}}, \mathbf{S})$, and $- \ell_{ij}$ for all β arcs $(i,j) \in (\mathbf{S}, \overline{\mathbf{S}})$. By (5.36) this implies that $a(\mathbf{S}, \pi) > 0$ at this stage and an ascent step can therefore be carried out now.

Hence, Procedure 4 will either terminate by carrying out an FAS (which is harmless since there is no change in the node price vector) or by carrying out an ascent step. However, in the ascent step, the total absolute deficit may strictly increase. In this procedure, often the condition for an ascent step is likely to be satisfied well before tree growth stops; hence this procedure tends to terminate earlier than Procedure 3.

ALGORITHMS FOR MINIMUM COST CIRCULATION PROBLEMS

These algorithms can be initiated with any node price vector π, and a bound feasible flow vector selected so that (f, π) is an admissible pair.

The *classical Primal-Dual algorithm* is based solely on Procedure 1. It carries out FAS or PAS by applying Procedure 1 repeatedly until all the node deficits are converted to 0, at which stage we have an optimum pair (f, π). This algorithm is guaranteed to terminate after a finite number of iterations, either with an optimum pair, or with the conclusion that there is no feasible circulation in G.

Other Primal-Dual algorithms consist of a sequence in any order, of FAS, PAS, and ascent steps through Procedures 2, 3, 4. These procedures can be combined in different ways to yield a variety of algorithms. This flexibility allows the construction of algorithms that can be tailored to the problem at hand for maximum effectiveness. In these algorithms, most of the ascent directions tend to be single coordinate directions, leading to the interpretation of these algorithms as coordinate ascent or relaxation methods. This is an important characteristic and a key factor in the practical efficiency of these algorithms. We will now show that any algorithm in this class terminates with an optimum solution in a finite number of steps if the problem is feasible.

THEOREM 5.5 *Suppose the minimum cost circulation problem in G is solved by an algorithm consisting of repeated applications of Procedures 2, 3, 4 in any order beginning with an initial admissible pair (f^0, π^0). The algorithm terminates with an optimum pair after a finite number of steps, if the problem is feasible, under the following assumptions.*

    1. ℓ, k, c, f^0, π^0 are all finite integer vectors

    2. either all steps in the algorithm are adjustment steps (FAS or PAS) or all adjustment steps used are harmless.

Proof   By assumption (a), all the quantities ϵ and δ in every step of the algorithm will always be positive integers, and all the pairs (f, π) obtained in the algorithm are always integer vector pairs.

$Q(\pi)$ is piecewise linear in arc tensions, $\pi_j - \pi_i$, and $a(\mathbf{S}, \pi)$ is the rate of change of $Q(\pi)$ along $x(\mathbf{S})$. The actual change in $Q(\pi + \lambda x(\mathbf{S}))$ as a function of the step size λ is linear up to the point where its slope changes. The value of $\lambda > 0$ closest to 0 where the slope of $Q(\pi + \lambda x(\mathbf{S}))$ changes, is the one for which a new arc incident to $\mathbf{S}$ becomes a β-arc; i.e., it is δ given by (5.37), and hence $\geqq 1$. For all $0 \leqq \lambda \leqq \delta$, we therefore have $Q(\pi + \lambda x(\mathbf{S})) = Q(\pi) + \lambda a(\mathbf{S}, \pi)$. Hence whenever an ascent step is carried out, $Q(\pi)$ increases by a positive integer amount. Therefore, if the problem is feasible, under assumption (b), it is not possible to carry out an infinite number of ascent steps in the algorithm, so after some iteration all steps will be adjustment steps, or else the algorithm will terminate finitely. Hence it is sufficient to prove this theorem under the assumption that all steps are adjustment steps. We will do this now.

Each time an FAS is carried out the total absolute deficit strictly decreases, while each time a PAS is carried out the total absolute deficit does not increase. So, if the algorithm does not terminate finitely, after a finite number of iterations it must execute PAS exclusively. Let (f^r, π^r), with deficit vector $d^r = (d_i^r), r = 1, 2, \ldots$

refer to the sequence of admissible pairs in the algorithm after it began to execute PAS exclusively. In this sequence, the deficits of nodes with positive deficits will be constant, and the deficits of nodes with nonpositive deficits will remain nonpositive. The node price of each node with nonpositive deficit will not decrease, and for at least one of these nodes it strictly increases. Hence $\mathbf{N} = \{\, i : \pi_i^r \to \infty \text{ as } r \to \infty \,\}$ $\neq \emptyset$. Let $\overline{\mathbf{N}}$ be the complement of $\mathbf{N}$.

$d_i^r \stackrel{\leq}{=} 0$ for all $i \in \mathbf{N}$, and this inequality is strict for at least one i, hence $\sum(d_i^r : \text{over } i \in \mathbf{N}) < 0$. By the definition of $\mathbf{N}$ all arcs in $(\overline{\mathbf{N}}, \mathbf{N})$ are γ-arcs, and those in $(\mathbf{N}, \overline{\mathbf{N}})$ are α-arcs. Using this in $\sum(d_i^r : \text{over } i \in \mathbf{N}) < 0$, we have $k(\overline{\mathbf{N}}, \mathbf{N}) - \ell(\mathbf{N}, \overline{\mathbf{N}}) < 0$, implying that there is no feasible circulation in G. ∎

In our description of the procedures, scanning operations started at nodes with negative deficit only. It is possible to start scanning at nodes with positive deficit as well. In practice, it was found that this modification improves the running time by at least a factor of two. The finite termination property for the method, in feasible networks, is preserved, provided that all PASs originating from positive deficit nodes *strictly* improve the dual functional value.

If the primal problem is infeasible, then it is possible for the methods based solely on Procedures 2, 3, 4, as described, to cycle forever. An easy way to circumvent this difficulty is to switch to Procedure 1 after some fixed number of iterations. This guarantees that the method will either conclude that the problem is infeasible, or find an optimum pair, after a finite number of iterations.

Comment 5.1   The new Primal-Dual methods in this section are from Bertsekas [1985], Bertsekas, Hosein, and Tseng [1987], and Bertsekas and Tseng [1988]. These papers provide results from computational experiments comparing the performance of these methods with other Primal-Dual methods, OK methods, primal network simplex and dual network simplex methods, on networks with number of nodes up to 8000, and number of arcs up to 35,000, and randomly generated data. The typical time to solve a minimum cost circulation problem on a network with 100 nodes and 5000 arcs by these methods is 12 seconds on a VAX 11/750 mainframe computer; this increases to 171 seconds when the number of nodes and arcs increases to 3000 and 35,000 respectively.

5.7   The Dual Network Simplex Method for Minimum Cost Flow Problems

We consider the minimum cost flow problem (5.3) in the connected directed network $\mathbf{G} = (\mathcal{N}, \mathcal{A}, \ell, k, c, V)$ where ℓ is finite and $\ell \stackrel{\leq}{=} k$. Let $\mathcal{A}^\infty = \{(i,j) : k_{ij} = \infty\}$. Let $(\mathbb{T}, \mathbf{L}, \mathbf{U})$ be a partition in G associated with the primal and dual basic solution pair $(f = (f_{ij}), \pi = (\pi_i))$. Let $\overline{c} = (\overline{c}_{ij} = c_{ij} - (\pi_j - \pi_i))$ be the reduced cost vector wrt π. $(\mathbb{T}, \mathbf{L}, \mathbf{U})$ is said to be a *dual feasible partition* if

$$\bar{c}_{ij} \begin{cases} \geqq 0 \text{ for all } (i,j) \in \mathcal{A}^{\infty} \\ \geqq 0 \text{ for all } (i,j) \in \mathbf{L} \\ \leqq 0 \text{ for all } (i,j) \in \mathbf{U} \end{cases} \qquad (5.38)$$

If $\mathcal{A}^{\infty} = \emptyset$, we can generate a dual feasible partition in G by selecting an arbitrary spanning tree $\mathbb{T}$ in G, computing the associated $\pi, \bar{c}$ vectors, and then classifying the out-of-tree arcs into $\mathbf{L}$, $\mathbf{U}$ based on the sign of their reduced cost coefficients.

If $\mathcal{A}^{\infty} \neq \emptyset$, we can apply the initialization procedure discussed in Section 5.3 to generate shortest chain trees in each connected component of $(\mathcal{N}, \mathcal{A}^{\infty})$, so that the node price vector corresponding to each of these trees satisfies $\pi_j - \pi_i \leqq c_{ij}$ for all arcs (i,j) in that component. The procedure may terminate either by finding a negative cost circuit in the subnetwork $(\mathcal{N}, \mathcal{A}^{\infty})$, or by finding such trees in each connected component of this subnetwork. In the former case, the dual of (5.3) is infeasible, and the objective value is unbounded below in (5.3) if it is feasible, so, we terminate. In the latter case we augment the shortest chain trees in the connected components of $(\mathcal{N}, \mathcal{A}^{\infty})$ with arcs from $\mathcal{A} \backslash \mathcal{A}^{\infty}$ selected arbitrarily, to get a spanning tree $\mathbb{T}$ in G, and then by partitioning the out-of-tree arcs as mentioned above, generate a dual feasible partition.

The dual feasible partition $(\mathbb{T}, \mathbf{L}, \mathbf{U})$ associated with the reduced cost vector $\bar{c}$ is said to be *dual nondegenerate* if $\bar{c}_{ij} > 0$ for all $(i,j) \in \mathbf{L}$, and $\bar{c}_{ij} < 0$ for all $(i,j) \in \mathbf{U}$; *dual degenerate* otherwise.

The procedure outlined above for finding an initial dual feasible partition is the Phase I of the dual network simplex method. The dual network simplex algorithm needs an initial dual feasible partition to solve (5.3).

THE DUAL NETWORK SIMPLEX ALGORITHM

Initialization   Begin with a dual feasible partition.

General step   Let $(\mathbf{T}, \mathbf{L}, \mathbf{U})$ be the present dual feasible partition associated with the basic solution pair $(f = (f_{ij}), \pi = (\pi_i))$ and reduced cost vector $\bar{c} = (\bar{c}_{ij})$.

    Check primal feasibility of the partition   If $\ell_{ij} \leqq f_{ij} \leqq k_{ij}$ for all $(i,j) \in \mathbb{T}$, $(\mathbf{T}, \mathbf{L}, \mathbf{U})$ is primal feasible and hence optimal, terminate.

    Select dropping arc   Let $\mathbf{D} = $ set of all in-tree arcs (i,j) satisfying either $f_{ij} < \ell_{ij}$ or $f_{ij} > k_{ij}$. $\mathbf{D}$ is the set of arcs *eligible to be dropping arcs* in this step. Select one arc from $\mathbf{D}$, (r,s) say, to be the *dropping* or *leaving* arc.

    Select entering arc   Let $[\mathbf{X}, \overline{\mathbf{X}}]$ be the fundamental cutset of (r,s) wrt $\mathbb{T}$, with $r \in \mathbf{X}$. Define

$$\mathbf{S}^1 = \mathbf{L} \cap (\overline{\mathbf{X}}, \mathbf{X}) \quad , \quad \mathbf{S}^2 = \mathbf{U} \cap (\mathbf{X}, \overline{\mathbf{X}}), \text{ if } f_{rs} < \ell_{rs}$$
$$\mathbf{S}^1 = \mathbf{L} \cap (\mathbf{X}, \overline{\mathbf{X}}) \quad , \quad \mathbf{S}^2 = \mathbf{U} \cap (\overline{\mathbf{X}}, \mathbf{X}), \text{ if } f_{rs} > k_{rs}$$

If $\mathbf{S}^1 \cup \mathbf{S}^2 = \emptyset$, (5.3) is infeasible, terminate.

If $\mathbf{S}^1 \cup \mathbf{S}^2 \neq \emptyset$, define $\delta =$ min. $\{ |\bar{c}_{ij}|:$ over $(i,j) \in \mathbf{S}^1 \cup \mathbf{S}^2\}$. δ is called the *dual simplex minimum ratio* in this pivot step. Let $\mathbf{E}$ be the set of arcs in $\mathbf{S}^1 \cup \mathbf{S}^2$ which tie for the minimum in the definition of δ. $\mathbf{E}$ is the set of arcs *eligible to enter* in this pivot step. Select one arc from $\mathbf{E}$, (p, q) say, as the *entering arc*. Let $\mathbb{C}$ be the fundamental cycle of (p, q) wrt $\mathbb{T}$. $\mathbb{C}$ will contain (r, s). Orient $\mathbb{C}$ so that (r, s) is a forward arc on it.

Delete (p, q) from $\mathbf{L}$ or $\mathbf{U}$ where it was. Replace (r, s) in $\mathbf{T}$ by (p, q) and let $\mathbb{T}'$ be the resulting spanning tree. Include (r, s) in $\mathbf{L}$ if $f_{rs} < \ell_{rs}$, and in $\mathbf{U}$ if $f_{rs} > k_{rs}$. This gives the new partition. Let $\epsilon = \ell_{rs} - f_{rs}$ if $f_{rs} < \ell_{rs}$, or $k_{rs} - f_{rs}$ if $f_{rs} > k_{rs}$. The basic flow vector corresponding to the new partition is obtained from f by adding ϵ to the flow amounts on the forward arcs of $\mathbb{C}$, and subtracting ϵ from the flow amounts on reverse arcs on $\mathbb{C}$. Go to the next step.

The entering arc choice guarantees that dual feasibility is maintained throughout the algorithm. It can be verified that if the entering arc (p, q) replaces the dropping arc (r, s) in a pivot step, then in the flow vector obtained immediately after this step, the flows on both the arcs $(p, q), (r, s)$ satisfy the respective bounds on them.

The flow vector changes in every step of this algorithm. The node price vector changes in a pivot step if the quantity $\delta > 0$, remains unchanged if $\delta = 0$. Hence, a pivot step in this algorithm is said to be *degenerate* if $\delta = 0$ in that step, *nondegenerate* if $\delta > 0$.

In large networks, in particular if $|\mathcal{A}|$ is much larger than $|\mathcal{N}|$, the selection of the entering arc by the dual simplex minimum ratio criterion is a time consuming process, and is inherently wasteful. In all LPs in which the number of variables is much greater than the number of constraints, this problem exists for applying the dual simplex algorithm.

Exercises

5.11 Prove that the dual objective function strictly increases in every nondegenerate pivot step in the dual network simplex method.

5.12 Research Problem It is not known whether cycling can occur under dual degeneracy in the dual network simplex algorithm. Either construct an example of its occurrence, or prove that it cannot occur.

5.13 Develop combinatorial methods for resolving cycling under degeneracy in the dual network simplex algorithm (Partovi[1984]).

5.14 Develop dropping and entering arc choice rules in the dual network simplex algorithm which simultaneously resolve both cycling and stalling in it.

5.8   A Strongly Polynomial Algorithm for Minimum Cost Flow Problems

Consider a single commodity minimum cost flow problem on a directed network $G = (\mathcal{N}, \mathcal{A}, \ell, k, c)$ with $|\mathcal{N}| = n, |\mathcal{A}| = m$. If all the data is rational, the *size* of this problem is defined to be the total number of binary bits of storage needed to store all the data in the problem. The *dimension* of this problem is defined to be the total number of data elements in the statement of the problem, irrespective of whether the data is rational or not. So, the dimension of our minimum cost flow problem is proportional to n and m; it is $5m + n + 2$ (number of nodes and arcs; lower bound, capacity, cost coefficient, head node, and tail node on each arc; and the exogenous flow amount at each node).

The concept of *polynomial boundedness* of an algorithm is defined in the context of problem instances with rational data of finite size. An algorithm for this problem is said to be a *polynomially bounded algorithm* if the total number of arithmetic operations (additions, multiplications and divisions, comparisons, etc.) in the algorithm, when applied on an instance with rational data, is bounded above by a polynomial in the size of the instance; and the size of each of the numbers occurring during the algorithm is always bounded above by a polynomial in the size of the original instance.

The first polynomially bounded algorithms for minimum cost flow problems were developed by Edmonds and Karp [1972 of Chapter 2]; they are based on the scaling technique (see Sections 3.2 and 5.3).

An algorithm for this problem is said to be a *strongly polynomial algorithm* if the total number of arithmetic operations in the algorithm is bounded above by a polynomial in the dimension of the problem even when the algorithm is applied on instances with real (i.e., non-rational) data, as long as the required operations are carried out exactly; and when the algorithm is applied on instances with rational data, the size of each of the numbers occurring during the algorithm is always bounded above by a polynomial in the size of the original instance.

The scaling based polynomially bounded algorithms for minimum cost flow problems discussed in Sections 3.2 and 5.3 are not strongly polynomial since the total number of arithmetic operations in those algorithms grows with the size of the instance, even when the dimension remains fixed. For the maximum value flow problem, strongly polynomial algorithms have been known since the early 1970's. The shortest augmenting path method, Dinic's method, Dinic-MKM method, and the preflow-push method discussed in Chapter 2 are all strongly polynomial algorithms for the maximum value flow problem. However, the problem of developing a strongly polynomial algorithm for the minimum cost flow problem remained open until 1985. In fact in their 1972 paper Edmonds and Karp stated the following challenge: "A challenging open problem is to give a method for the minimum cost flow problem having a bound of computation which is a polynomial in the number of nodes, and is independent of both costs and capacities." The first strongly polynomial algorithm for minimum cost flow problems is due to Tardos [1985]. Since then several strongly polynomial algorithms have been developed for minimum cost flow problems. In this section we discuss a strongly polynomial algorithm for minimum cost flow problems, perhaps the simplest conceptually among all those known, based on canceling minimum mean residual cycles.

We consider the minimum cost circulation problem (5.2) in G. From Theorem 5.1 we know that a feasible circulation f is a minimum cost circulation iff there exist no negative cost residual cycles wrt it. One of the earliest algorithms for finding minimum cost circulations, due to Klein [1967], called the *cycle canceling algorithm*, begins with a feasible circulation and repeats the step of finding a negative cost residual cycle and canceling it. If the cycles to cancel are selected arbitrarily, the computational requirements of this algorithm may grow exponentially with the size, and the algorithm may not even terminate finitely under irrational data (an example of this can be constructed from Ford and Fulkerson's maximum value flow problem discussed in Example 2.1 posed as a minimum cost circulation problem). A natural question to ask is whether the cycle canceling algorithm can be made efficient, in fact strongly polynomial, by a judicious choice of the cycles to cancel. This question has been answered in the affirmative by Goldberg and Tarjan [1989], and we present their results here.

Define the *average* or *mean cost* of a simple cycle to be its cost divided by the number of arcs in it. A simple cycle whose average cost is as small as possible is called a *minimum mean cycle*. A minimum mean cycle can be found by an algorithm of Karp [1978] in at most $O(nm)$ time. Goldberg and Tarjan showed that if the cycle to cancel in the cycle canceling algorithm is selected to be a minimum mean residual cycle in each step, then it terminates after at most $O(nm^2 \log n)$ cycles have been canceled. Hence, with this selection rule, the cycle canceling algorithm solves the minimum cost circulation problem in at most $O(n^2 m^3 \log n)$ time, which makes it an elegant strongly polynomial algorithm.

An Algorithm to Find a Minimum Mean Simple Circuit in a Directed Network

Let H = $(\mathbf{V}, \mathbf{E}, d)$ be a directed network with d as the vector of arc weights (or lengths, or cost coefficients). The vector d is not required to be $\geqq 0$; it can be arbitrary. Let $e_1, \ldots, e_p$ be the arcs in a simple circuit $\overrightarrow{\mathbb{C}}$ in H, with weights $d_1, \ldots, d_p$ respectively. Then the *mean cost (or weight)* of $\overrightarrow{\mathbb{C}}$ is defined to be $a(\overrightarrow{\mathbb{C}}) = \frac{1}{p} \sum (d_t : \text{over } t = 1 \text{ to } p)$. A *minimum mean simple circuit* in H is one which has the least mean cost among all the simple circuits in H. Here we discuss an algorithm for finding a minimum mean simple circuit due to Karp [1978].

If H is not strongly connected, its strong components can be found in $O(|E|)$ effort. The minimum mean simple circuit can be computed separately in each strongly connected component of H, and the best of these is the minimum mean simple circuit in H. So, henceforth we assume that H is strongly connected.

The cost of a minimum mean simple circuit is called the *minimum circuit mean*. Let λ^* denote the minimum circuit mean in H. Let $|V| = n, |E| = m$. Select any node from V, say s, as the origin node. For each node $i \in V$, and $r \geqq 0$ integer define $F_r(i)$ to be the minimum cost of a chain (not necessarily simple) from s to i in H containing exactly r arcs, ∞ if no such chain exists. These quantities $F_r(i)$, for all $i \in V$ and $r = 0$ to n can be computed from the following recursive equations.

$$
\begin{aligned}
F_0(i) &= \quad \infty \text{ for all } i \neq s, \text{ and } 0 \text{ for } i = s \\
F_r(i) &= \quad \min. \{F_{r-1}(j) + d_{ji} : \text{over } j \text{ such that } (j, i) \in E\} \quad (5.39)
\end{aligned}
$$

where d_{ji} is the weight or cost of arc (j, i) in E. Computation of $F_r(i)$ for all $i \in V$, $r = 0$ to n by this recursive method takes $O(nm)$ effort.

ALGORITHM FOR MINIMUM MEAN SIMPLE CIRCUIT IN A STRONGLY CONNECTED NETWORK

Step 1   Compute $F_r(i)$ for all $i \in V$, $r = 0$ to n.

Step 2   Compute λ^* from

$$
\lambda^* = \min._{i \in V} \left\{ \max._{0 \leqq r \leqq n-1} \left\{ \frac{F_n(i) - F_r(i)}{n - r} \right\} \right\} \quad (5.40)
$$

To get a simple circuit yielding the minimum circuit mean, find a minimizing i and r in (5.40), find a minimum cost chain consisting of n arcs from s to i, and extract a simple circuit of $n - r$ arcs within that chain.

Validity of the Minimum Mean Simple Circuit Algorithm

1. First we will prove that (5.40) holds when $\lambda^* = 0$. Since $\lambda^* = 0$, there are no negative cost circuits in H, and there is a 0-cost circuit. Also, H is strongly connected. So, there is a shortest simple chain from s to i consisting of $< n$ arcs for every $i \in V$; let its cost be π_i. Then $F_n(i) \geqq \pi_i$, and $\pi_i = \min. \{ F_r(i) : r = 0 \text{ to } n - 1 \}$. So, $F_n(i) - \pi_i = \max. \{ F_n(i) - F_r(i) : r = 0 \text{ to } n - 1 \} \geqq 0$. Hence

$$\text{max.} \quad \left\{ \frac{F_n(i) - F_r(i)}{n - r} : r = 0, \dots, n - 1 \right\} \geqq 0 \qquad (5.41)$$

Equality holds in (5.41) iff i is such that $F_n(i) = \pi_i$. We will now show that there exists a node i satisfying this condition. Let $\overrightarrow{\mathbb{C}}$ be a simple circuit in H of cost 0, and w a node on it. Let $\mathcal{C}(w)$ be a shortest simple chain from s to w. Then any chain consisting of $\mathcal{C}(w)$ followed by any nonnegative integer number of repetitions of $\overrightarrow{\mathbb{C}}$ is also a shortest chain from s to w in H. But the initial part of any shortest chain must also be a shortest chain from the origin to that point. After some repetitions of $\overrightarrow{\mathbb{C}}$ at the end of $\mathcal{C}(w)$ an initial part consisting of exactly n arcs will appear; suppose its end point is w'. This implies that this part has cost $F_n(w') = \pi_{w'}$. So, (5.41) holds as an equation when $i = w'$. This establishes (5.40) in this case.

2. We now prove that (5.40) holds in general. Let us study the effect of subtracting a constant α from the weight of every arc in H. This subtracts α from λ^*, and subtracts $r\alpha$ from $F_r(i)$ for all i and r. Hence it subtracts α from the right hand side of (5.40). Hence both sides of (5.40) are affected by the same quantity when the cost vector d is translated by a constant. Choosing α to be the unknown λ^* and using the result in 1., establishes (5.40) in general.

3. The computation of $F_r(i)$ using the recursive equations (5.39) can be accomplished with O(nm) effort. Using the $F_r(i)$, λ^* can be computed from (5.40) and a minimum mean simple circuit identified as discussed earlier, with an additional effort of O(n^2). Thus the overall effort in this algorithm for the minimum mean simple circuit is O(nm).

The Minimum Mean Cycle Canceling Algorithm

Let f be a feasible circulation in G. Every residual cycle wrt f in G becomes a simple circuit in the residual network G(f) with the same cost, and vice versa. So, by finding a minimum mean simple circuit in G(f) by the algorithm described above, we can find a minimum mean residual cycle wrt f in G.

THE MINIMUM MEAN CYCLE CANCELING ALGORITHM FOR MINIMUM COST CIRCULATION PROBLEMS

Initialization   Find a feasible circulation in G using the algorithms discussed in Chapter 2.

General step   Let f be the present feasible circulation in G. Find a minimum mean residual cycle wrt f in G, $\mathbb{C}$ say. If the cost of $\mathbb{C}$ is $\geqq 0$, f is a minimum cost circulation in G; terminate. Otherwise cancel $\mathbb{C}$ in f and go to the next step with the new feasible circulation obtained.

Proof of Strong Polynomiality of the
Minimum Mean Cycle Canceling Algorithm

(i) ϵ-optimality For $\epsilon \geqq 0$, a feasible circulation $f = (f_{ij})$ in G is said to be an ϵ-*optimal circulation* if there exists a node price vector $\pi = (\pi_i)$ such that for all $(i, j) \in \mathcal{A}$

$$\text{if} \quad \bar{c}_{ij} = c_{ij} - (\pi_j - \pi_i) \begin{cases} > \epsilon & \text{then } f_{ij} = \ell_{ij} \\ \\ < -\epsilon & \text{then } f_{ij} = k_{ij} \end{cases} \qquad (5.42)$$

By the c.s. optimality conditions, a 0-optimal feasible circulation is a minimum cost circulation in G.

(ii) Let $\alpha > 0$, and (f', π') a feasible circulation, node price vector pair in G satisfying

$$\text{if} \quad \bar{c}'_{ij} = c_{ij} - (\pi'_j - \pi'_i) \begin{cases} \geqq \alpha & \text{then } f'_{ij} = \ell_{ij} \\ \\ \leqq -\alpha & \text{then } f'_{ij} = k_{ij} \end{cases} \qquad (5.43)$$

for every $(i, j) \in \mathcal{A}$. Then for any arc $(p, q) \in \mathcal{A}$ if $|\bar{c}'_{pq}| \geqq n\alpha$, then $\overline{f}_{pq} = f'_{pq}$ in every minimum cost feasible circulation $\overline{f} = (\overline{f}_{ij})$ in G.

To prove this, let $\overline{f}$ be a minimum cost feasible circulation in G such that $\overline{f}_{pq} \neq f'_{pq}$ for an arc $(p, q) \in \mathcal{A}$ satisfying $|\bar{c}'_{pq}| \geqq n\alpha$. Suppose $\overline{f}_{pq} > f'_{pq}$ (a proof similar to the following can be constructed if $\overline{f}_{pq} < f'_{pq}$). If $\bar{c}'_{pq} \leqq -n\alpha$, then $f'_{pq} = k_{pq}$ and $\overline{f}_{pq} > k_{pq}$ cannot hold, so we must have $\bar{c}'_{pq} \geqq n\alpha$. $\overline{f} - f'$ is a nonzero circulation in G; by Lemma 5.2 there exists an oriented simple cycle $\mathbb{C}$ containing the arc (p, q) such that on all forward arcs (i, j) on $\mathbb{C}$ we have $\overline{f}_{ij} - f'_{ij} > 0$, and on all the reverse arcs (i, j) on $\mathbb{C}$ we have $\overline{f}_{ij} - f'_{ij} < 0$. Let $\epsilon = \min. \{ |\overline{f}_{ij} - f'_{ij}| : (i, j)$ on $\mathbb{C} \}$. So, $\epsilon > 0$. Obtain the circulation f'' from $\overline{f}$ by decreasing the flow amount on all the forward arcs of $\mathbb{C}$ by ϵ and by increasing the flow amount on all the reverse arcs of $\mathbb{C}$ by ϵ. f'' is easily verified to be a feasible circulation in G.

On the forward arc (p, q) on $\mathbb{C}$, $\bar{c}'_{pq} \geqq n\alpha$. On all the other forward arcs (i, j) on $\mathbb{C}$ other than (p, q) we have $k_{ij} \geqq \overline{f}_{ij} > f'_{ij}$, i.e., $f'_{ij} < k_{ij}$ which by (5.43) implies that $\bar{c}'_{pq} \geqq -\alpha$. On all the reverse arcs (i, j) on $\mathbb{C}$ we have $\ell_{ij} \leqq \overline{f}_{ij} < f'_{ij}$, i.e., $f'_{ij} > \ell_{ij}$, which by (5.43) implies $\bar{c}'_{pq} \leqq \alpha$. Let r be the number of arcs on $\mathbb{C}$, since $\mathbb{C}$ is simple $r \leqq n$. Cost of $\mathbb{C} = \bar{c}'_{pq} + \sum(\bar{c}'_{ij} :$ over forward arcs $(i, j) \neq (p, q)$ on $\mathbb{C})$ $- \sum(\bar{c}'_{ij} :$ over reverse arcs (i, j) on $\mathbb{C}) \geqq n\alpha - (r-1)\alpha = (n+1-r)\alpha > \alpha > 0$. So cost of $f'' = cf'' = cf' - \epsilon(\text{cost of } \mathbb{C}) < cf'$ since the cost of $\mathbb{C} > 0$ by the above. Since f'' is a feasible circulation, we have a contradiction to the hypothesis that f'

is a minimum cost feasible circulation in G. Hence the statement given above must be true.

(iii) Let $|\mathcal{N}| = n$. If c is an integer vector, for every $0 \overset{\leq}{=} \epsilon < 1/n$, every ϵ-optimal feasible circulation is a minimum cost feasible circulation.

To prove this, let f be an ϵ-optimal feasible circulation in G for some $0 \overset{\leq}{=} \epsilon < 1/n$. Let $\mathbb{C}$ be a residual cycle wrt f. So, by definition, there exists a node price vector $\pi = (\pi_i)$ such that for any arc (i, j) on $\mathbb{C}$

$$\bar{c}_{ij} = c_{ij} - (\pi_j - \pi_i) \begin{cases} \overset{\geq}{=} -\epsilon & \text{if } (i,j) \text{ is a forward arc on } \mathbb{C} \\ \overset{\leq}{=} \epsilon & \text{if } (i,j) \text{ is a reverse arc on } \mathbb{C} \end{cases} \tag{5.44}$$

Since $\mathbb{C}$ is a simple cycle, it has at most n arcs. Hence by (5.44) the cost of $\mathbb{C}$ $= \sum(\bar{c}_{ij} : \text{over forward arcs } (i,j) \text{ on } \mathbb{C}) - \sum(\bar{c}_{ij} : \text{over reverse arcs } (i,j) \text{ on } \mathbb{C})$ $\overset{\geq}{=} -n\epsilon > -1$. However, as c is an integer vector, the cost of $\mathbb{C}$ is an integer, and since it is > -1, it must be $\overset{\geq}{=} 0$. Hence every residual cycle wrt f has a nonnegative cost; by Theorem 5.1, f is a minimum cost feasible circulation.

(iv) Definitions Let f be a feasible circulation in G. Define $\epsilon(f)$ to be the minimum $\epsilon \overset{\geq}{=} 0$ for which f is ϵ-optimal. Define $\mu(f)$ to be the mean cost of a minimum mean residual cycle wrt f.

(v) If f is an ϵ-optimal feasible circulation wrt a node price vector π (i.e., f, π together satisfy (5.42)) then on every arc (i,j) in the residual network $G(f, \pi) = (\mathcal{N}, \mathcal{A}(f), \bar{c}')$, the cost coefficient $\bar{c}'_{ij} \overset{\geq}{=} -\epsilon$. This follows because if (i,j) is an arc in $G(f, \pi)$ with a $+ (-)$ label, then it corresponds to: (i,j) in G satisfying $f_{ij} < k_{ij}$ $((j,i) \text{ in G satisfying } f_{ji} > \ell_{ji})$, and by ϵ-optimality of f wrt π we must have $\bar{c}'_{ij} \overset{\geq}{=} -\epsilon$.

(vi) Relationship Between $\epsilon(f), \mu(f)$ Let f be a feasible circulation in G. Then $\epsilon(f) = \max. \{0, -\mu(f)\}$.

To prove this, let π be a node price vector in G satisfying (5.42) with f for $\epsilon = \epsilon(f)$. Let $\mathbb{C}$ be any residual cycle wrt f, $c(\mathbb{C})$ its cost, and l the number of arcs in it. By the definition of $\epsilon(f)$, (5.42) holds for f, π and $\epsilon = \epsilon(f)$, and the cost of $\mathbb{C}$ wrt $c, \bar{c}$ as cost vectors is the same. So, $c(\mathbb{C}) \overset{\geq}{=} -l\epsilon(f)$, or $c(\mathbb{C})/l \overset{\geq}{=} -\epsilon(f)$, i.e., the mean cost of the residual cycle $\mathbb{C}$ is $\overset{\geq}{=} -\epsilon(f)$. Since this holds for all residual cycles wrt f, we have $\mu(f) \overset{\geq}{=} -\epsilon(f)$, or $\epsilon(f) \overset{\geq}{=} -\mu(f)$.

If $\mu(f) \overset{\geq}{=} 0$, by Theorem 5.1 f is a minimum cost circulation, hence $\epsilon(f) = 0 = \max. \{0, -\mu(f)\}$ in this case.

Now assume $\mu(f) < 0$. Every residual cycle in G wrt f corresponds to a simple circuit in $G(f)$ with the same cost, and vice versa, and $\mu(f)$ is the minimum circuit mean in $G(f) = (\mathcal{N}, \mathcal{A}(f), c')$. Let $\delta = -\mu(f) > 0$. Adding δ to the cost of every arc in $G(f)$ has the effect of adding δ to the mean cost of every circuit in it; hence after this change there will be no negative cost circuits in $G(f)$. Obtain a directed network $H = (\mathcal{N}^1, \mathcal{A}^1, c^1)$ where $\mathcal{N}^1 = \mathcal{N} \cup \{s\}$, s being an artificial origin node. $\mathcal{A}^1 = \mathcal{A}(f) \cup \{(s,i) : i \in \mathcal{N}\}$. c^1, the cost vector on $\mathcal{A}^1$ is defined by making the cost of all arcs of the form (s,i) to be 0, and the cost of all arcs $(i,j) \in \mathcal{A}(f)$ to be $c'_{ij} + \delta$. So, H is a directed network with no negative cost circuits, and all nodes in H can be reached from s by a chain, since (s,i) is an arc in H for all $i \in \mathcal{N}$. Find a shortest chain tree rooted at s in H with c^1 as the arc cost vector, and let ν_i be the cost of the shortest chain from s to i in H for each $i \in \mathcal{N}$. From the optimality properties for shortest chains, we know that $\nu_j - \nu_i \overset{\leq}{=} c'_{ij} + \delta$ for all $(i,j) \in \mathcal{A}(f)$.

If (i,j) is a + labeled arc in $\mathcal{A}(f)$ but there is no arc (j,i) in $\mathcal{A}(f)$, we must have $(i,j) \in \mathcal{A}$ and $f_{ij} = \ell_{ij} < k_{ij}$, and $\nu_j - \nu_i \overset{\leq}{=} c'_{ij} + \delta = c_{ij} + \delta$, or $c_{ij} - (\nu_j - \nu_i) \overset{\geq}{=} -\delta$.

If (i,j) is a - labeled arc in $\mathcal{A}(f)$ but there is no arc (j,i) in $\mathcal{A}(f)$, we must have $(j,i) \in \mathcal{A}$ and $f_{ji} = k_{ji} > \ell_{ji}$, and $\nu_j - \nu_i \overset{\leq}{=} c'_{ij} + \delta = -c_{ji} + \delta$, so $c_{ji} - (\nu_i - \nu_j) \overset{\leq}{=} \delta$.

If (i,j) with a + label, and (j,i) with a - label are both in $\mathcal{A}(f)$, we must have $(i,j) \in \mathcal{A}$, $\ell_{ij} < f_{ij} < k_{ij}$, and $\nu_j - \nu_i \overset{\leq}{=} c'_{ij} + \delta$, and $\nu_j - \nu_i \overset{\leq}{=} c'_{ji} + \delta$, which together imply $-\delta \overset{\leq}{=} c_{ij} - (\nu_j - \nu_i) \overset{\leq}{=} \delta$.

Hence on arcs $(i,j) \in \mathcal{A}$ satisfying $c_{ij} - (\nu_j - \nu_i) \overset{\geq}{=} \delta$, we must have $f_{ij} = \ell_{ij}$; and on arcs $(i,j) \in \mathcal{A}$ satisfying $c_{ij} - (\nu_j - \nu_i) \overset{\leq}{=} -\delta$, we must have $f_{ij} = k_{ij}$. This means that with ν as the node price vector f satisfies (5.42) for $\epsilon = \delta$, i.e., f is δ-optimal, or $(-\mu(f))$-optimal. So, $\epsilon(f) \overset{\leq}{=} -\mu(f)$. We have already proved above that $\epsilon(f) \overset{\geq}{=} -\mu(f)$. So, in this case $\epsilon(f) = -\mu(f) = \max. \{0, -\mu(f)\}$.

Hence, in general, $\epsilon(f) = \max. \{ 0, -\mu(f) \}$.

(vii) Definition Admissible Network   Let (f, π) be a feasible circulation, node price vector pair, and $G(f,\pi) = (\mathcal{N}, \mathcal{A}(f), \overline{c}')$ the residual network wrt it. The *admissible network* wrt (f,π) denoted by $G'(f,\pi)$ is the subnetwork of $G(f,\pi)$ consisting of all the arcs (i,j) in it for which $\overline{c}'_{ij} < 0$.

(viii) Let (f,π) be an ϵ-optimal feasible circulation, node price vector pair in G satisfying (5.42). If the admissible network $G'(f,\pi)$ is acyclic, then f is actually $(1 - 1/n)\epsilon$-optimal.

To prove this, let $\overrightarrow{\mathbb{C}}$ be a simple circuit in $G(f,\pi)$ consisting of l arcs, say. By (v), the cost coefficients of all the arcs in $G(f,\pi)$ are $\overset{\geq}{=} -\epsilon$. However, since $G'(f,\pi)$ is acyclic by hypothesis, $\overrightarrow{\mathbb{C}}$ cannot completely lie in $G'(f,\pi)$, and hence at least one arc on $\overrightarrow{\mathbb{C}}$ has cost in $G(f,\pi) \overset{\geq}{=} 0$. So the mean cost of $\overrightarrow{\mathbb{C}}$ is $\overset{\geq}{=} (l-1)(-\epsilon)/l = -\epsilon + (\epsilon/n) \overset{\geq}{=} -(1 - 1/n)\epsilon$. Thus the mean cost of every simple circuit in $G(f,\pi)$ is $\overset{\geq}{=} -(1 - 1/n)\epsilon$. Thus, $\mu(f) \overset{\geq}{=} -(1 - 1/n)\epsilon$. By (vi), $\epsilon(f) \overset{\leq}{=} (1 - 1/n)\epsilon$, so f is $(1 - 1/n)\epsilon$-optimal.

(ix) Let f be a nonoptimal feasible circulation in G. Canceling a minimum mean cycle in f leads to a feasible circulation $\tilde{f}$ satisfying $\epsilon(\tilde{f}) \leqq \epsilon(f)$.

To prove this, let π be a node price vector wrt which f is ϵ-optimal for $\epsilon = \epsilon(f)$. Let $\mathbb{C}$ be the minimum mean cycle canceled, and $\overrightarrow{\mathbb{C}}$ the simple circuit in $G(f, \pi)$ corresponding to it. By (v), the cost coefficient of arc (i, j) on $G(f, \pi)$, $\overline{c}'_{ij} \geqq -\epsilon(f)$ for all arcs (i, j) on $\overrightarrow{\mathbb{C}}$. But $\overrightarrow{\mathbb{C}}$ is a minimum mean circuit in $G(f, \pi)$, whose cost is $-\epsilon(f)$ by (vi), which implies that $\overline{c}'_{ij} = -\epsilon(f)$ for all (i, j) on $\overrightarrow{\mathbb{C}}$. So, every new residual arc created after the cancellation (such an arc must be the reversal of an arc on $\overrightarrow{\mathbb{C}}$) has cost coefficient of $\epsilon(f)$ in $G(\tilde{f}, \pi)$. So, every arc (i, j) in $G(\tilde{f}, \pi)$ has cost coefficient $\geqq -\epsilon(f)$ too. Hence the minimum circuit mean in $G(\tilde{f}, \pi)$ is $\geqq -\epsilon(f)$. This implies that $\mu(\tilde{f}) \geqq -\epsilon(f)$. Hence $\epsilon(\tilde{f}) \leqq \epsilon(f)$ by (vi).

(**x**) Let f be a feasible circulation in G, and $\tilde{f}$ the feasible circulation obtained after m minimum mean cycle cancellations beginning with f. Then $\epsilon(\tilde{f}) \leqq (1 - 1/n)\epsilon(f)$.

The proof of this goes as follows. Let π be a node price vector wrt which f is ϵ-optimal for $\epsilon = \epsilon(f)$. The admissible network $G'(f, \pi)$ changes with every change in the flow vector. By (v) the cost coefficient of every arc in $G(f, \pi)$ is $\geqq -\epsilon(f)$. Canceling a cycle with all its arcs in the admissible network only adds arcs of positive reduced cost to the residual network and deletes at least one arc from the admissible network, as established in (ix). Consider two cases.

Case 1 Suppose none of the cycles canceled contains an arc (i, j) of nonnegative reduced cost. Then each cancellation deletes at least one arc from the admissible network. So, after m cancellations, the admissible network has no arcs, which implies that the flow vector at that time is optimal; hence $\epsilon(\tilde{f}) = 0$, establishing the result in this case.

Case 2 Suppose some cycle canceled contains an arc of nonnegative reduced cost. Let the first such cycle correspond to the circuit $\overrightarrow{\mathbb{C}}$ in the residual network, and let the flow vector be f^1 just before that cancellation. Using the same arguments as in the proof of (viii) we verify that the mean cost of $\overrightarrow{\mathbb{C}}$ is $\geqq -(1 - 1/n)\epsilon(f)$. Hence $\epsilon(f^1) \leqq (1 - 1/n)\epsilon(f)$ by (vi). By the argument in (ix) applied repeatedly, we have $\epsilon(\tilde{f}) \leqq \epsilon(f^1)$, and hence $\epsilon(\tilde{f}) \leqq (1 - 1/n)\epsilon(f)$ in this case too.

(**xi**) If the cost vector $c = (c_{ij})$ is an integer vector, let $\gamma = \max. \{ |c_{ij}| : (i, j) \in A \}$. Then this algorithm terminates after at most $O(nm\log(n\gamma))$ cancellations.

To prove this, let f^0 denote the initial feasible circulation. Taking $\pi^0 = 0$ we verify that $\epsilon(f^0) \leqq \gamma$. Let f^r denote the feasible circulation obtained in the algorithm after r cancellations. By (iii) we know that if r is such that $\epsilon(f^r) < 1/n$, then f^r is a minimum cost circulation, and the algorithm can terminate. From (x) we know that

$$\epsilon(f^r) \overset{\leq}{=} (1 - 1/n)^{\lfloor \frac{r-1}{m} \rfloor} \epsilon(f^0) \overset{\leq}{=} (1 - 1/n)^{\lfloor \frac{r-1}{m} \rfloor} \gamma$$

and from the above argument it is sufficient to make this $< 1/n$. Hence the maximum number of cancellations required in the algorithm is r where r satisfies

$$(1 - 1/n)^{\lfloor \frac{r-1}{m} \rfloor} < 1/n\gamma \text{ i.e., } \lfloor \frac{r-1}{m} \rfloor \overset{\geq}{=} \frac{-\log(n\gamma)}{\log(1 - 1/n)} \overset{\geq}{=} n \log(n\gamma)$$

since $\log(1 - 1/n) \overset{\leq}{=} -1/n$ for $n > 1$. Hence it is sufficient to take $r = O(nm \log(n\gamma))$. This shows that the algorithm is polynomially bounded when the cost vector is integral.

(xii) **Definition**   For given $\epsilon > 0$, define an arc $(p, q) \in \mathcal{A}$ to be ϵ-fixed iff the flow f_{pq} is the same for all ϵ-optimal feasible circulations $f = (f_{ij})$.

(xiii) Let $\epsilon > 0$ and (f, π) an ϵ-optimal feasible circulation, node price vector pair satisfying (5.42) together. $\bar{c}_{ij} = c_{ij} - (\pi_j - \pi_i)$ is the reduced cost coefficient of $(i, j) \in \mathcal{A}$ wrt π. If $(p, q) \in \mathcal{A}$ satisfies $|\bar{c}_{pq}| \overset{\geq}{=} 2n\epsilon$, then (p, q) is ϵ-fixed.
   This follows from (ii).

(xiv) From (ix) we know that $\epsilon(f)$ is nonincreasing as the algorithm progresses. So, at some stage of the algorithm, if an arc (p, q) becomes fixed, then the flow f_{pq} remains the same in the sequel. Therefore, when all arcs are fixed, the current circulation must be optimal (since an optimal circulation is ϵ-optimal for any $\epsilon \overset{\geq}{=} 0$, and therefore must agree with the current circulation on all the fixed arcs).

(xv) For arbitrary real valued cost vector c, the algorithm terminates after at most $O(nm^2 \log n)$ cancellations.
   To prove this, let $a = mn\lceil (1 + log n) \rceil$. We divide the cancellations in the algorithm into groups of a consecutive cancellations each. We will now show that each group of cancellations fixes the flow on a distinct arc, i.e., the flow on that arc will not change in subsequent iterations.
   Consider a group of cancellations. Let f^0 denote the circulation at the beginning of the first iteration in this group, and f^a the circulation at the end of the last iteration in this group. Let $\epsilon_0 = \epsilon(f^0), \epsilon_a = \epsilon(f^a)$, and let π^a be a node price vector wrt which f^a satisfies the ϵ_a-optimality criterion. Let $\mathbb{C}_0$ be the cycle canceled in the first iteration of the group. By the result in (x), and the choice of a, we have

$$\epsilon_a \overset{\leq}{=} \epsilon_0 (1 - 1/n)^{n\lceil (1 + \log n) \rceil} \overset{\leq}{=} \frac{\epsilon_0}{2n}$$

By (vi) the mean cost of $\mathbb{C}_0$ is $-\epsilon_0$. The mean cost of $\mathbb{C}_0$ remains the same even when the arc cost coefficients are the reduced cost coefficients $c_{ij} - (\pi_j^a - \pi_i^a)$. Since the mean cost of $\mathbb{C}_0$ is $-\epsilon_0$, at least one arc (i, j) on $\mathbb{C}_0$ must have reduced cost $c_{ij} - (\pi_j^a - \pi_i^a) \stackrel{\leq}{=} -\epsilon_0 \stackrel{\leq}{=} -2n\epsilon_a$, and by the arguments in (ix), (xiii), (xiv), the flow on this arc will not change after this group of iterations is completed.

Thus each group of cancellations fixes at least one distinct arc. Hence the algorithm terminates after at most m groups, or $nm^2\lceil (1+\log n) \rceil$ cancellations in all. Hence the total number of cancellations needed in the algorithm is at most $O(nm^2 \log n)$.

Each cancellation involves $O(nm)$ effort to find the cycle to cancel by Karp's algorithm discussed earlier, and $O(m)$ for the cancellation itself, leading to a total effort of $O(nm)$ per iteration. So, the overall computational effort in this algorithm is at most $O(n^2 m^3 \log n)$, establishing the strong polynomiality of the algorithm.

Based on a more flexible selection of cycles to cancel, versions of this algorithm in which the effort per iteration is $O(\log n)$ instead of $O(nm)$ have been developed in Goldberg and Tarjan [1989].

There are several other strongly polynomial algorithms developed for minimum cost flow problems; see Fujishige [1986], Galil and Tardos [1986], Orlin [1984, 1988], and Tardos [1985]. Some of these algorithms are based on making modifications to a non-strongly-polynomial algorithm to convert it into a strongly polynomial algorithm. However, the kind of tricks needed to make an algorithm amenable to a strongly polynomial proof do not normally work well in practical computation. Hence the major impact of these algorithms is mathematical and theoretical.

The question of whether strongly polynomial algorithms exist for the general LP remains an open question. Again, it is one of the most mathematically challenging open questions in LP theory. Outside of the special class of pure minimum cost flow problems, results on this question are practically unknown. The class of generalized network flow problems discussed in Chapter 8 is slightly more general than the class of pure network flow problems. Even for that class no strongly polynomial algorithms are known at the moment.

5.9 Minimum Separable Piecewise Linear Convex Cost Flow Problems

Interval	Slope of $c_{ij}(f_{ij})$ in the interval
ℓ_{ij} to x_{ij}^1	c_{ij}^1
x_{ij}^1 to x_{ij}^2	c_{ij}^2
$\vdots$	$\vdots$
x_{ij}^{r-1} to k_{ij}	c_{ij}^r

Consider a directed single commodity flow network $G = (\mathcal{N}, \mathcal{A}, \ell, k, \check{s}, \check{t}, \overline{v})$ in which the cost of flow on each arc $(i,j) \in \mathcal{A}$ is a piecewise linear convex function

$c_{ij}(f_{ij})$, i.e., the interval ℓ_{ij} to k_{ij} is divided into a finite number, say r, of intervals, and in each interval $c_{ij}(f_{ij})$ is linear with slopes given above.

The break points and slopes satisfy the conditions $\ell_{ij} < x^1_{ij} < x^2_{ij} < \ldots < x^{r-1}_{ij} < k_{ij}$ and $c^1_{ij} < c^2_{ij} < \ldots < c^r_{ij}$; these are the conditions for the continuous piecewise linear function $c_{ij}(f_{ij})$ to be convex. The problem is to find a feasible flow vector which minimizes $\sum(c_{ij}(f_{ij}) :$ over $(i,j) \in \mathcal{A})$, given the break points and slopes associated with each arc.

One way of interpreting the objective function $c_{ij}(f_{ij})$ is the following. Each unit of flow on arc (i,j) between ℓ_{ij} and x^1_{ij} flows at a cost of c^1_{ij} per unit. Once the flow amount on arc (i,j) reaches x^1_{ij}, every additional unit flow on it pays a cost of c^2_{ij} until the flow amount reaches the value x^2_{ij}. This can be handled by treating this flow as if it were flowing through another parallel directed arc from i to j on which the cost coefficient is c^2_{ij} per unit flow. And so on. So, for the cost function given in the above tableau, we replace the original arc (i,j) with r parallel arcs with data as in Figure 5.20. These r parallel arcs correspond to the arc (i,j) in the original network. The flow f_{ij} in the original problem corresponds to the sum of the flow amounts on the parallel arcs corresponding to it in the transformed problem.

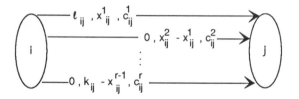

Figure 5.20 Data on the arcs is lower bound, capacity, and unit cost coefficient, in that order.

Each arc in the original network is transformed separately in the same way. The linear minimum cost flow problem in the transformed network is solved by any of the methods discussed earlier. If that problem is infeasible, or unbounded, the same holds for the original piecewise-linear minimum cost flow problem. Otherwise, the flow vector in the original network, corresponding to an optimum solution in the transformed network, is optimal for the original problem.

Suppose the parallel arcs in the transformed network corresponding to the arc (i,j) in the original network, are $(i,j)_1, \ldots, (i,j)_r$ in increasing order of the interval of the original flow variable f_{ij} corresponding to them. Then, because of the monotone increasing property of the slopes, the following property will hold in any optimum flow for the transformed problem

flow on $(i,j)_t$ is 0 unless $(i,j)_1, \ldots, (i,j)_{t-1}$ are all saturated   (5.45)

It is this property which makes the original piecewise linear cost minimization problem equivalent to the linear cost minimization problem in the transformed network.

For any arc (i, j) in the original problem, if $c_{ij}(f_{ij})$ is piecewise linear and continuous, but not convex, then the slopes will not be monotone increasing. In this case, in a minimum cost flow in the transformed network (5.45) may not hold. The original problem is equivalent to the linear minimum cost flow problem in the transformed network, subject to conditions of the form (5.45) for each arc (i, j) ((5.45) holds automatically if the original objective function is convex because of the monotonicity of the slopes). The constraints (5.45) are not linear constraints; finding a minimum cost flow problem subject to them in the transformed network is not an LP. This can of course be transformed into an integer programming problem, or solved directly by enumerative methods. Thus the transformation of the original piecewise linear minimum cost flow problem into the linear minimum cost flow problem on the transformed network is only valid if the piecewise linear function is convex.

5.10 Dynamic Network Flow Problems

In all the flow problems discussed so far, we ignored any consideration of traversal time, and there was at most one flow amount entering each arc; that's why flow vectors in those problems are referred to as *static flows*. In this section we consider flow models called *dynamic flow models*; here the traversal time for each arc in the network is provided as input, and a separate flow amount can enter each arc at its tail node at every integer point of time, even while there may be other units of material that entered this arc earlier still in transit on it. One may either impose a requirement that all the units of the commodity must make the trip from the source to the sink within a given amount of time, or want to determine the maximum amount of the commodity that can reach the sink from the source in a specified amount of time. In dynamic flow models, the transit of each unit of flow has to be completely organized along the time axis. We have to specify when it would enter each arc along the chain that it travels, whether it will be temporarily held over or stored at certain nodes for one or more time periods before it can resume its journey across the next arc, and so forth. Thus, time delays are associated with both arcs (because of transit time across the arc) and nodes (because of any holdover or storage) in this model.

Static problems are those in which a single optimum solution is desired. In dynamic problems we need the optimum solution for each point of time over a time horizon.

Dynamic flow problems occur in many areas. Examples are communication systems and traffic and railway systems. In these systems, the networks used are called *store and forward networks*. Other areas in which these problems arise are production systems, material handling systems, military logistics systems etc. In modern automatic production systems, material flows from one process to the next through conveyor belts or is transshipped automatically by robots. In these systems there may be buffer zones where material may be held over or stored before it is transshipped to the next process, and different processes may have different cycle times. To analyze the flow of material through such a dynamic system one discretizes time by dividing the time horizon into equal intervals called *time slices*; and approximates the processing time of each process, and the transit time between processes by integer multiples of the common discrete time unit (the length of a time slice). The material flow can then be modeled using a dynamic network flow model.

Let $G = (\mathcal{N}, \mathcal{A}, 0, k, \check{s}, \check{t})$ be the network on which the problem is defined, with $|\mathcal{N}| = n$, $|\mathcal{A}| = m$. Here $k = (k_{ij})$ where k_{ij} is the maximum number of units of the commodity that can enter arc (i, j) at its tail node i at the beginning of each time period. We are also given the vector $a = (a_{ij} : (i, j) \in \mathcal{A})$, where a_{ij} is the traversal time across arc (i, j). We assume $k > 0, a \geqq 0$ and that they are both integer vectors. We assume that all the action begins at time point 0, and that material can enter any arc at its tail only at integer points of time in amounts limited by the arc's capacity, and since all transit times are assumed integer, the material will always finish its travel across an arc reach the head node of that arc at an integer point of time.

For each $i \in \mathcal{N}$, we define $a_{ii} = 1$; this represents the fact that material stored or held over at a node is available for change of status (either continue to be stored again, or begin journey across an arc incident out of that node) at the beginning of the next time period.

For each $i \in \mathcal{N}$, k_{ii} denotes the maximum amount of material that can be held over or stored at node i. Let T, a positive integer, denote the time horizon in the problem. It is the maximum number of time periods for which the program has to be worked out. The decision variables in the problem are clearly: for each $(i, j) \in \mathcal{A}$ and each $\tau = 0, 1, \ldots, T$

$$
\begin{aligned}
f(i, j, \tau) &= \quad \text{amount of material entering } (i, j) \text{ at time point } \tau \\
f(i, i, \tau) &= \quad \text{amount of material held over at } i \text{ from time point } \tau \text{ to } \tau + 1
\end{aligned}
$$

The vector $(f(i, j, \tau))$ is known as a *dynamic flow* on the network G. Flow units in it travel across G over time, obeying node and arc capacity constraints and conservation at each node and at each point of time. The node-arc flow vectors discussed earlier, which do not have any concept of time, will be referred to as *static flows* for the sake of distinction.

The constraints that the dynamic flows have to satisfy for feasibility appear more complex than those for static flows. However, in actuality, the T-period dynamic flow on G can be viewed as a static flow on a *time-expanded* version $G(T)$ of G. The procedure for constructing $G(T)$ from G is the following:

1. For each node i in G, there are $T+1$ nodes denoted by $i(\tau)$, $\tau = 0, 1, \dots, T$ in G(T), and arcs $(i(\tau), i(\tau+1))$, $0 \overset{\leq}{=} \tau \overset{\leq}{=} T-1$ with capacity k_{ii} to represent holdover at node i.

2. For each arc (i, j) in G, there are arcs $(i(\tau), j(\tau + a_{ij}))$, $\tau = 0, \dots, T - a_{ij}$ with capacity k_{ij} in G(T).

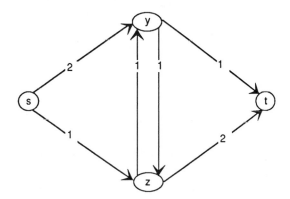

Figure 5.21 The network G, with arc traversal time data.

Associate the flow variable $f(i, j, \tau)$ in the dynamic flow, with the arc $(i(\tau), j(\tau + a_{ij}))$, and the flow variable $f(i(\tau), i(\tau+1), \tau)$ with the arc $(i(\tau), i(\tau+1))$, in G(T). Treat the nodes $\breve{s}(0), \dots, \breve{s}(T)$ as the source nodes, and the nodes $\breve{t}(0), \dots, \breve{t}(T)$ as the sink nodes in G(T) (in fact, since there are hold-over arcs at $\breve{s}$ and $\breve{t}$, we could equally well treat $\breve{s}(0)$ as the only source node, and $\breve{t}(T)$ as the only sink node in G(T).

With these associations, it can be verified that the constraints for the feasibility of the dynamic flow vector in G become the usual flow conservation and capacity constraints for a static node-arc flow vector in the time expanded network G(T). So, feasible static node-arc flow vectors in G(T) become feasible dynamic flows in G with a complete specification of how much material enters each arc or passes through each node at each point of time, and vice versa.

As an example in Figure 5.21 we show a network G consisting of nodes s, t, y, z with traversal time data entered on the arcs. Other data such as capacities are not shown. In Figure 5.22 we show G(5), the 5-period time-expanded version of G. In drawing the time-expanded version in Figure 5.22, we represented time intervals horizontally, and the nodes of the original network vertically at each time point.

Since an item released from a node at a specific time does not return to that location at the same or an earlier time, G(T) cannot contain any circuits, and is therefore acyclic always. The number of nodes in G(T) is $n(T+1)$, and the number of arcs is $(n + m)T + m - \sum(a_{ij} : \text{over } (i, j) \in \mathcal{A})$. In the time-expanded network there are no parallel arcs and it is usually sparse, with its density decreasing as T increases.

A dynamic flow problem is said to be *stationary* if the network parameters such as capacities, arc traversal times, and so on, are constant over time.

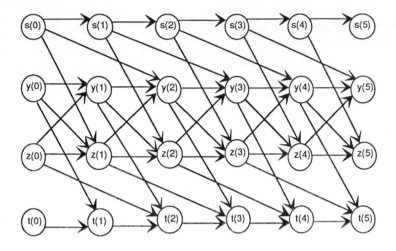

Figure 5.22   The time-expanded network G(5).

Ford and Fulkerson [1962 of Chapter 1] introduced a remarkably simple and effective algorithm for the stationary maximum value dynamic flow problem, i.e., one in which the objective is to maximize the amount of material reaching the sink node. It does not require the construction of the time-expanded network $G(T)$ for solving this problem for any T!. They show that a maximum value dynamic flow in the stationary case can be generated from a static flow f in the smaller network G that maximizes the linear function $w(f, v) = (T+1)v - \sum(a_{ij} f_{ij} : \text{over } (i, j) \in \mathcal{A})$, where v is the value of f in G. Decompose f into a set of arc-chain flows from $\check{s}$ to $\check{t}$ in G. Start each chain flow in this set at time 0 and repeat it after each time period so long as there is enough time left in the horizon for the flow along the chain to arrive at the sink. This leads to a maximum value dynamic flow in this case. We do not discuss the proof of validity of this algorithm here since it only handles a special problem; the interested reader should consult Ford and Fulkerson's book.

In the most general dynamic model, the arc capacities k_{ij} and traversal time a_{ij} may not be constant over time, but may vary. Surprisingly, the technique of time expansion provides a general way of reducing the complicated dynamic situation into the familiar static one in the space-time framework by replicating the physical network over the time domain. But we do pay a price for this transformation, as the size of the time-expanded network blows up linearly with the length of the time horizon studied. If T is large, even a relatively modest network G expanded through T periods can lead to an enormous time-expanded network $G(T)$ which may be computationally very expensive to deal with. Thus, even though the technique of time-expansion is conceptually very nice and simple, the blowup in size makes it unappealing, particularly when the time horizon is large. Unfortunately, efficient techniques such as the one mentioned earlier for the stationary dynamic maximum value flow problem are not known at the moment for other dynamic network flow problems. But research continues with the aim of devising efficient techniques for finding at least approximate solutions.

5.11 Multicommodity Flow Problems

In this section we discuss static multicommodity flow problems, i.e., those involving $p \ (\geqq 2)$ commodities on a directed connected pure network $G = (\mathcal{N}, \mathcal{A}, 0, k)$ with $0 < k < \infty$. To model these problems, we should measure quantities of all commodities in common units, for example, truck loads. The capacities of the arcs are also stated in the same units. Under this, it is possible to add the flows of different commodities on an arc to yield the total flow of all commodities put together on that arc, and we assume that the capacity of the arc applies to this total flow.

By introducing a supersource and supersink for each commodity if necessary, as discussed in Chapter 2, it is possible to formulate the multicommodity flow problem in such a way that each commodity originates at some specified source, and is required to be transported to a specified sink node. Let $\breve{s}^r, \breve{t}^r$ be the source and sink respectively for the rth commodity, $r = 1$ to p. In the node-arc flow model flow conservation must hold at each node for each commodity separately.

We denote the node-arc flow amount of the rth commodity on arc (i, j) by f_{ij}^r, and $f^r = (f_{ij}^r : (i, j) \in \mathcal{A})$ is the flow vector of this commodity, $r = 1$ to p. Let $c^r = (c_{ij}^r : (i, j) \in \mathcal{A})$ be the original cost vector for the rth commodity, where c_{ij}^r is the cost of shipping one unit of this commodity across $(i, j) \in \mathcal{A}$, $r = 1$ to p. Let E be the node-arc incidence matrix of G. Suppose a specified flow value of $\overline{v}^r$ units of the rth commodity is required to be transported from $\breve{s}^r$ to $\breve{t}^r$, $r = 1$ to p. This leads to the following *multicommodity minimum cost flow problem*.

f^1	f^2	$\ldots$	f^p	
I	I	$\ldots$	I	$\leqq k$
. . . .				. . . .
E	0	$\ldots$	0	$= q^1 \overline{v}_1$
0	E	$\ldots$	0	$= q^2 \overline{v}_2$
$\vdots$	$\vdots$	$\ddots$	$\vdots$	$\vdots$
0	0	$\ldots$	E	$= q^p \overline{v}_p$

$$f^r \geqq 0 \text{ for all } r$$

where q^r is a column vector of the hypothetical arc $(\breve{s}_r, \breve{t}_r)$. The constraint at the top representing the arc capacities for the sum of the flows of all the commodities is called the *bundle constraint*. Without the bundle constraint the problem breaks down into several single commodity flow problems, one for each commodity, and can be solved efficiently by the algorithms discussed earlier. The bundle constraint makes the problem difficult, the main difficulty being that of optimally splitting the capacity of each arc among the various commodities. If $|\mathcal{N}| = n, |\mathcal{A}| = m$, this problem is an LP in pm variables, subject to $m + p(n - 1)$ constraints. In applications, the networks tend to be large, $n = 1000$, $m = 10,000$ is very common even in small scale applications. Even if there are only 3 commodities to be considered on such a network, the problem is an LP in 30,000 variables subject to 12,997 constraints, a large scale LP.

Applications for multicommodity flow models are many; most of these tend to be large-scale problems. Its wide applicability has made the multicommodity flow problem a focus of intense research activity by groups working in large scale LP.

The coefficient matrix in this problem is in general not totally unimodular, and hence it may not have an optimum integer solution even if all the data are integral.

One of the earliest approaches proposed for solving multicommodity flow problems based on network methodology is by Ford and Fulkerson [1958] using an arc-chain formulation of the problem. Inspired by it, Dantzig and Wolfe [1960] later extended it into the decomposition principle for general LP. The arc-chain formulation leads to an LP involving only $m + p$ constraints but an enormously large number (growing exponentially with n) of variables. The remarkable thing about this approach is that it is able to process this problem without having the data for all the variables on hand at any point of time. Using the revised simplex format, the method solves the problem by maintaining the data corresponding to only $m + p$ variables in the problem at any time, and in each step it generates the data corresponding to exactly one additional variable as needed. Hence the approach is called a *column generation approach*. The column generation approach is of fundamental importance in large scale optimization and combinatorial optimization.

$$\text{Minimize } \sum_{r=1}^{p} \sum_{h=1}^{d_r} g_h^r x_h^r$$

$$\text{subject to } \sum_{r=1}^{p} (\sum (x_h^r \quad : \text{ over } h \text{ s. t. } C_h^r \text{ contains } e_u)) +$$

$$y_u = k_u, u = 1 \text{ to } m \qquad (5.46)$$

$$\sum_{h=1}^{d_r} x_h^r = \bar{v}_r, r = 1 \text{ to } p$$

$$\text{all } x_h^r \quad , y_u \gneqq 0$$

In the parlance of the decomposition principle of LP, we decompose along the dotted line in the original problem. This leads to the arc-chain formulation of the problem. Denote the arcs in $\mathcal{A}$ by $e_1, \ldots, e_m$; and their capacities by $k_1, \ldots, k_m$. For $r = 1$ to p, let d_r be the total number of distinct simple chains from $\breve{s}_r$ to $\breve{t}_r$ in G (typically this is likely to be a very large number); let these chains themselves be $C_1^r, \ldots, C_{d_r}^r$. Let x_h^r be the amount of the rth commodity shipped along C_h^r, $h = 1$ to

d_r, $r = 1$ to p; these are the variables in the arc-chain formulation. Let $g_h^r = \sum(c_u^r$: over u s.t. e_u is on C_h^r) = cost of the chain C_h^r for the rth commodity. Then the arc-chain formulation of this minimum cost flow problem is (5.46) given above.

The number of constraints in (5.46) is $m + p$, and hence basic vectors for it consist of $m + p$ variables. To get in initial feasible basic vector, we introduce the artificial variables $y_{m+1}, \ldots, y_{m+p}$ and construct the following Phase I problem.

$$\text{Minimize } w = \sum_{r=1}^{p} y_{m+r}$$

$$\text{subject to } \sum_{r=1}^{p}(\sum(x_h^r \quad : \text{ over } h \text{ s. t. } C_h^r \text{ contains } e_u)) +$$

$$y_u = k_u, u = 1 \text{ to } m \tag{5.47}$$

$$\sum_{h=1}^{d_r} x_h^r \quad +y_{m+r} = \overline{v}_r, r = 1 \text{ to } p$$

$$\text{all } x_h^r \quad , y_u \geqq 0$$

$y = (y_1, \ldots, y_{m+p})$ is an initial feasible basic vector for (5.47) with which Phase I is initiated. The original column vector of y_t in (5.47) is the tth column vector of I, the unit matrix of order $m + p$, for $t = 1$ to $m + p$. The original column vector of the arc-chain flow variable x_h^r associated with the simple chain C_h^r in (5.47) is $(a_{1h}, \ldots, a_{m+p,h})^T$, where $a_{ih} = 1$ if the arc e_i is on C_h^r, 0 otherwise, for $i = 1$ to m; and $a_{m+t,h} = 1$ if $t = r$, 0 otherwise, for $t = 1$ to p.

If $(\sigma_1, \ldots, \sigma_{m+p})$ is the Phase I dual basic solution corresponding to a feasible basic vector for (5.47), the Phase I relative cost coefficient wrt this basic vector, of y_u is $-\sigma_u$, and that of x_h^r is $-\sigma_{m+r} - \sum(\sigma_u$: over u s. t. e_u is on C_h^r).

Thus to check whether the present basic vector satisfies the Phase I termination criterion in the simplex method, we need to check whether for each $r = 1$ to p

$$-\sigma_{m+r} - \sum(\sigma_u : \text{ over } u \text{ s. t. } e_u \text{ is on } C_h^r) \geqq 0 \tag{5.48}$$

i.e., for all $h = 1$ to d_r, the length of every simple chain from $\breve{s}_r$ to $\breve{t}_r$ in G with $(-\sigma_1, \ldots, -\sigma_m)$ as the vector of arc lengths, is $\geqq \sigma_{m+r}$. Even though d_r is very large, this can be carried out efficiently by finding the shortest chain from $\breve{s}_r$ to $\breve{t}_r$. Hence, even though there are a lot of variables in (5.47), we can check whether any feasible basic vector for it satisfies the primal simplex termination criterion, or otherwise select an entering variable to carry out a primal simplex pivot step, using the following procedure. First check whether $-\sigma_u < 0$ for some u between 1 to m; if so, select y_u as the entering variable into the present basic vector. If $-\sigma_u \geqq 0$ for all $u = 1$ to m, find the shortest chains from $\breve{s}_r$ to $\breve{t}_r$ in G with $(-\sigma_1, \ldots, -\sigma_m)$ as the vector of arc lengths, for $r = 1$ to p in some order. Since $(-\sigma_1, \ldots, -\sigma_m) \geqq 0$ at this stage, Dijkstra's algorithm can be used to solve these shortest chain problems. If the length of the shortest chain from $\breve{s}_r$ to $\breve{t}_r$ is $\geqq \sigma_{m+r}$ for all $r = 1$ to p, then Phase I termination criterion is satisfied by the present basic vector; terminate Phase I. If the present Phase I objective value $w > 0$, the original problem is infeasible. If $w = 0$, go to Phase II with the present basic vector. If, for some r, the length of the shortest chain from $\breve{s}_r$ to $\breve{t}_r$, $C^r_{h_0}$ say, is $< \sigma_{m+r}$, associate the variable x_{h_0} with $C^r_{h_0}$ and choose $x^r_{h_0}$ as the entering variable into the present basic vector for a Phase I primal simplex pivot step, and continue.

In Phase II the procedure is very similar. Let $(\gamma_1, \ldots, \gamma_{m+p})$ be the Phase II dual basic solution corresponding to a feasible basic vector for (5.47). The Phase II relative cost coefficient wrt the present basic vector, of y_u is $-\gamma_u$, for $u = 1$ to m, and of x^r_h is $g^r_h - \gamma_{m+r} - \sum(\gamma_u$: over u s. t. e_u is on $C^r_h)$. Hence, if $-\gamma_u < 0$ for some u between 1 to m, select y_u as the entering variable into the present basic vector for a primal simplex pivot step. If $-\gamma_u \geqq 0$ for all $u = 1$ to m; for $r = 1$ to p compute a shortest chain from $\breve{s}_r$ to $\breve{t}_r$ in G with $(-\gamma_1, \ldots, -\gamma_m)$ as the vector of arc lengths using Dijkstra's algorithm. If for some r, $C^r_{h_1}$ is the shortest chain obtained in this process and its length is $< \gamma_{m+r} - g^r_{h_1}$, select the arc-chain flow variable $x^r_{h_1}$ associated with it as the entering variable into the present basic vector for a primal simplex pivot step. Otherwise, if the length of the shortest chain $C^r_{h_1}$ from $\breve{s}_r$ to $\breve{t}_r$ obtained in this process is $\geqq \gamma_{m+r} - g^r_{h_1}$ for all $r = 1$ to p, then the Phase II optimality criterion is satisfied and the present BFS is an optimum solution to the problem.

In each step we need to solve at most p shortest chain problems, which takes $O(pn^2)$ effort by Dijkstra's method. Once an entering variable is selected, if its original column is d, we need to find its updated column $\bar{d}$, by solving the system of equations $B\bar{d} = d$, where B is the present basis consisting of the original columns of the present basic variables, for carrying out the primal simplex pivot step. B is of order $(m + p) \times (m + p)$. If m is large (say, $\geqq 10{,}000$) unless techniques that can take advantage of its special structure (it is a 0-1 matrix, and is usually very sparse) are used, finding the updated column of the entering variable itself could be a big computational burden.

There are several other approaches proposed for solving multicommodity flow problems. At the moment, the most practically useful algorithms for these problems seem to be interior point methods for LP implemented on parallel processing supercomputers taking advantage of the special nature of the coefficient matrix (all entries being 0, ± 1) and its sparsity. See Adler, Resende, Veiga and Karmarkar [1989], and Kapoor and Vaidya [1985, 1988].

5.12   Exercises

5.15 G $= (\mathcal{N}, \mathcal{A})$ is a connected network. If $\mathbb{T}_1$ and $\mathbb{T}_2$ are any two spanning trees in G, and e_1 is a line in $\mathbb{T}_1$ not in $\mathbb{T}_2$, then show that there exists a line e_2 in $\mathbb{T}_2$ not in $\mathbb{T}_1$ such that replacing e_1 by e_2 in $\mathbb{T}_1$ leads to another spanning tree.

5.16 Consider the assignment problem, (3.1), of order n. Let $\bar{x}$ be the assignment $\{(1, a_1), \ldots, (n, a_n)\}$ where $(a_1, \ldots, a_n)$ is a permutation of $(1, \ldots, n)$. Let $G = (\mathcal{N}_1, \mathcal{N}_2; \mathcal{A})$ be the bipartite network where $\mathcal{N}_1 = \{1, \ldots, n\}, \mathcal{N}_2 = \{1, \ldots, n\}, \mathcal{A} = \mathcal{N}_1 \times \mathcal{N}_2$. Let $(\mathbb{T}, L)$ be a partition in G corresponding to the assignment $\bar{x}$. $\mathbb{T}$ has $2n - 1$ arcs of G, and L has $n^2 - 2n + 1 = (n-1)^2$ nonbasic arcs. $\bar{x}$ is a degenerate BFS of (3.1), $\mathbb{T}$ consists of the arcs $(i, a_i), i = 1$ to n, and $n - 1$ other arcs corresponding to basic arcs with zero flow. Given the partition $(\mathbb{T}, L)$, we are interested in counting the number of possible pivots in this partition which are nondegenerate, while attempting to enter each of the $(n-1)^2$ nonbasic arcs in L one at a time. In particular, we are interested in determining the maximum and minimum number of potential nondegenerate pivots.

(i) Of all the alternate partitions representing $\bar{x}$, prove that the maximum number of nondegenerate pivots admitted by any such partition is $\frac{1}{2}n(n-1)$. Show that this maximum number is achieved by the partition $(\mathbb{T}, L)$ only if $\mathbb{T}$ is a path. One such partition is obtained when $\mathbb{T}$ consists of the arcs $\{(i, a_i) : i = 1 \text{ to } n\} \cup \{(i, a_{i+1}) : i = 1 \text{ to } n - 1\}$.

(ii) Of all the alternate partitions representing $\bar{x}$, prove that the minimum number of nondegenerate pivots admitted by any such partition is $(n-1)$. Show that this minimum number is achieved by the partition $(\mathbb{T}, L)$ where $\mathbb{T}$ consists of the arcs $\{(i, a_i) : i = 1 \text{ to } n\} \cup \{(i, a_1) : i = 2 \text{ to } n\}$.

(Bazaraa and Sherali [1982])

5.17 Determining single commodity equilibrium trade flow. This problem deals with a simple equilibrium model of interregional trade in a single commodity. We have the following data: N = number of regions, $a_i > 0$ is the equilibrium price in the ith region in the absence of imports and exports, $b_i > 0$ is the elasticity of supply and demand in the ith region, c_{ij} is the cost per unit shipped from region i to region j. The c_{ij} obey triangle inequality, that is, $c_{ij} \leq c_{ih} + c_{hj}$ for all i, j, h. The decision variables in the problem are: p_i = equilibrium price in the ith region, y_i = net imports into the ith region, x_{ij} = actual exports from region i to region j.

If $p_i > a_i$ then supply locally exceeds demand in the ith region, the difference being available for export. y_i is not restricted in sign; negative values of y_i are interpreted as exports. The following linear relation holds.

$$p_i = a_i - b_i y_i \tag{5.49}$$

Interregional trade equilibrium conditions are

$$p_i + c_{ij} \gtreqqless p_j, \quad \text{for all } i, j \tag{5.50}$$

$$(p_i + c_{ij} - p_j)x_{ij} = 0, \quad \text{for all } i, j \tag{5.51}$$

If (5.50) does not hold, exports from i to j will increase until the elasticity effects in markets i and j rise, and prices will adjust so that additional profit for export no longer exists. If $x_{ij} > 0$, we must have $p_i + c_{ij} = p_j$; hence the complementary slackness conditions (5.51) hold. The y_i and x_{ij} are linked through the flow conservation equations.

$$y_i - \sum_{j=1}^{N} x_{ji} + \sum_{j=1}^{N} x_{ij} = 0, i = 1 \text{ to } N. \tag{5.52}$$

The problem is to develop a procedure for computing the equilibrium prices p_i and flows x_{ij}, given the data. Consider the quadratic program (QP):

$$\text{maximize } z = \sum_{i=1}^{N} (\quad a_i y_i \quad -\frac{1}{2}b_i y_i^2 - \sum_{j=1}^{N} c_{ij} x_{ij}) \tag{5.53}$$

$$\text{subject to} \quad x_{ij} \overset{\geq}{=} 0, \text{ and } (5.52)$$

The objective function z in (5.53) can be interpreted as a net social payoff function. Show that the KKT optimality conditions for (5.53) are the equilibrium conditions (5.49) to (5.52). The dual of the QP (5.53) is the following problem, and an optimum λ for it is an equilibrium price vector p.

$$\text{minimize } \quad \sum_{j=1}^{N} \frac{(a_i - \lambda_i)^2}{2b_i} \tag{5.54}$$

$$\text{subject to } \lambda_j - \quad \lambda_i \quad \overset{\leq}{=} c_{ij}, \text{ for all } i, j$$

(i) Prove that there exists an equilibrium solution in which the trade routes of positive flows from a forest.

(ii) Using (i) show that there exists an equilibrium solution in which the set of regions is partitioned into trading coalitions. The members of each coalition trade only with each other and the set of trade routes with positive flows within each coalition forms a spanning tree for the coalition. Formally, a coalition is a set C of r nodes and $r - 1$ trade routes with positive flows among them having the following properties

(a) **Internal equilibrium:**   (5.49) to (5.52) are satisfied for all nodes and routes within C, with no flows between the set C and regions outside of C.

(b) **Tree structure:**   The set of trade routes with positive flows inside C form a spanning tree for C.

Prove that any coalition satisfying (a) and (b) has an equilibrium solution in which the following property holds

     (c) Alternating arc orientation: A coalition **C** with $|\mathbf{C}| \gtreqless 2$ satisfying (a) and (b) has an equilibrium solution in which each node is either an exporter or an importer, i.e., no transshipment occurs.

(iii) Develop a tree growing algorithm for finding an equilibrium solution, which builds up a set of coalitions that are, at termination, in equilibrium with each other as well as being in internal equilibrium.

(Glassey [1978]).

    5.18 The Flow Circulation Sharing Problem   Consider a region in which there are p power plants dependent on coal supplies. Over a horizon of w weeks, let d_{ij} = normal amount of coal used by ith plant in jth week. Suppose there is a prolonged strike at some coal mines in the region during the horizon. Let: y_{ij} = amount of coal assigned to ith plant in jth week, $x_{ij} = \sum_{t=1}^{j} y_{it}$ = cumulative amount of coal shipped to plant i through week j, $D_{ij} = \sum_{t=1}^{j} d_{ij}$ = normal demand from plant i through week j. An equitable distribution scheme for coal deliveries in this period of shortage should try to

$$\text{maximize } \{\text{minimum } \{\frac{x_{ij}}{D_{ij}} : i = 1 \text{ to } p, j = 1 \text{ to } w\}\} \qquad (5.55)$$

    Let k_j = total amount of coal available for distribution in week j for all the plants, $j = 1$ to w. Now the problem of finding equitable sharing of coal among the plants can be posed as that of finding a circulation in a directed network, to minimize an objective function of the form in (5.55). For example, the network for such a circulation problem when $p = 2, w = 2$ is given in Figure 5.23.

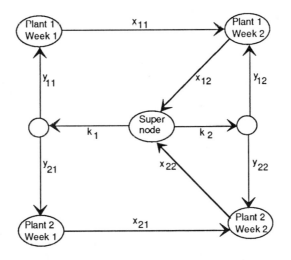

Figure 5.23

Formulate the general problem in the following form: given a single commodity directed network $G = (\mathcal{N}, \mathcal{A}, \ell, k)$ and a subset of arcs $\mathbf{A} \subset \mathcal{A}$ with a positive constant D_{ij} for each arc $(i,j) \in \mathbf{A}$, among feasible circulations $f = (f_{ij})$ in G find one that maximizes minimum $\{f_{ij}/D_{ij} : (i,j) \in \mathbf{A}\}$. Develop a network flow algorithm for this problem (Brown [1983]).

5.19 The "More-for-Less" or "More-for-Nothing" Paradoxes in the Distribution Model   Consider the $m \times n$ balanced transportation problem (5.56) with supply vector $a = (a_1, \dots, a_m) > 0$, demand vector $b = (b_1, \dots, b_n) > 0$, and unit cost matrix $c = (c_{ij}) \geqq 0$.

$$\text{minimize} \quad \sum_{i=1}^{m} \sum_{j=1}^{n} c_{ij} \; x_{ij}$$

$$\text{subject to} \quad \sum_{j=1}^{n} x_{ij} = a_i, \; i = 1 \text{ to } m \qquad (5.56)$$

$$\sum_{i=1}^{m} x_{ij} = b_j, \; j = 1 \text{ to } n$$

$$x_{ij} \geqq 0, \; \text{for all } i, j$$

As usual, we assume that $\sum a_i = \sum b_j$, and denote the optimum objective value in this problem by $g(a, b)$, as a function of the supply-demand vectors a, b, while c is fixed. It is sometimes possible to find vectors $a' = (a'_1, \dots, a'_m) \geq a$, $b' = (b'_1, \dots, b'_n) \geq b$ satisfying $\sum a'_i = \sum b'_j > \sum a_i$, such that $g(a', b') \leq g(a, b)$. If $g(a', b') < g(a, b)$ we are able to ship more total goods for less total cost (even though all $c_{ij} \geqq 0$, and we ship the same amount or more from each source, and to each destination, than in the original problem (5.56)). Hence this is known as the *more for less paradox*. If $g(a', b') = g(a, b)$ under the same conditions, it is called the *more for nothing paradox*. As an example, consider the problem with the following data.

| to destination $j =$ | | cost of shipping ($/unit) | | | | supply |
		1	2	3	4	a_i
from source $i =$	1	1	6	3	5	20
	2	7	3	1	6	10
	3	9	4	5	4	25
demand	b_j	11	13	17	14	

(a) In this numerical example, verify that the optimum objective value strictly decreases if a_2 and b_1 are increased by the same positive quantity δ up to 9.

(b) Assuming that a primal optimum BFS for this problem is nondegenerate, give an explanation for these paradoxes based on marginal analysis, by deriving the marginal value associated with such a change from the optimum dual solution.

(c) Show that an optimum solution for (5.56) may not be optimal to (5.57) even if $c > 0$

$$\text{minimize} \quad \sum_{i=1}^{m} \sum_{j=1}^{n} c_{ij} \; x_{ij}$$

$$\text{subject to} \quad \sum_{j=1}^{n} x_{ij} \geqq a_i, \; i = 1 \text{ to } m \qquad (5.57)$$

$$\sum_{i=1}^{m} x_{ij} \geqq b_j, \; j = 1 \text{ to } n$$

$$x_{ij} \geqq 0, \; \text{for all } i, j$$

(d) Let $\bar{x} = (\bar{x}_{ij})$ be a nondegenerate optimal BFS for (5.56), and $(\bar{u}, \bar{v})$ the corresponding dual optimum solution. If i, j are such that $\bar{u}_i + \bar{v}_j \leq 0$, both a_i and b_j can be increased by the same positive quantity δ while decreasing (if $\bar{u}_i + \bar{v}_j < 0$) or not changing (if $\bar{u}_i + \bar{v}_j = 0$) the optimum objective value. Give δ the maximum possible value which keeps the basis associated with the BFS $\bar{x}$ primal feasible. Repeat with other pairs i, j satisfying the same condition. So we are increasing the supplies and demands maximally to where the "more for less (nothing) paradox" just stops. Let a', b' be the resulting supply, demand vectors obtained at the end of this operation. Prove that every optimum BFS for the balanced transportation problem (5.56) with a', b' replacing a, b, must be degenerate.

(Charnes and Klingman [1971], Szwarc [1971], Charnes, Duffuaa and Ryan [1980])

5.20 Consider the single commodity flow network $G = (\mathcal{N}, \mathcal{A}, 0, k, c, \check{s}, \check{t})$, with $k > 0$ and finite. Let the "parametric value problem" refer to the problem of finding a minimum cost flow of value v in G, for each possible value v. Let $\bar{v}$ be the maximum flow value in G from $\check{s}$ to $\check{t}$. For $0 \leqq v \leqq \bar{v}$, let $g(v)$ be the minimum cost for a flow value of v in G. Then $g(v)$ is a piecewise linear convex function defined over the interval $0 \leqq v \leqq \bar{v}$. A breakpoint for $g(v)$ is a v where the slope of $g(v)$ changes.

Let x, y be real variables. For any convex function $p(x) : \mathbb{R}^1 \to \mathbb{R}^1 \cup \{\infty\}$ the conjugate of $p(x)$ is defined to be

$$p^*(y) = \text{minimum } \{p(x) - yx\} \text{ over } x \in \mathbb{R}^1.$$

The function $p^*(y)$ is called the conjugate of $p(x)$. It can be shown that $p^*(y)$ is a well-defined convex function and that $p^{**} = p$. Also, if $p(x)$ is piecewise linear and $\{x : p(x) < \infty\}$ is bounded, then $p^*(y)$ is also piecewise linear and the number of breakpoints of $p^*(y)$ is one less than the number of breakpoints of $p(x)$. In fact, each breakpoint of $p^*(y)$ corresponds to an interval of constant slope for $p(x)$ and vice versa. See Rockafellar [1970].

Let $[\mathbf{X}, \bar{\mathbf{X}}]$ be any cut separating $\breve{s}$ and $\breve{t}$ in G with $\breve{s} \in \mathbf{X}, \breve{t} \in \bar{\mathbf{X}}$. Introduce one new arc $(\breve{s}, \breve{t})$ with 0 lower bound, 0 cost, and infinite capacity and let G' denote the resulting network. Define parametric cost-coefficients $c_{ij}(\lambda)$ on G' by the following

$$
c_{ij}(\lambda) = \begin{cases}
c_{ij} - \lambda & \text{if } (i,j) \in (\mathbf{X}, \bar{\mathbf{X}}). \\
c_{ij} + \lambda & \text{if } (i,j) \in (\bar{\mathbf{X}}, \mathbf{X}). \\
c_{ij} & \text{otherwise.}
\end{cases}
$$

Let parametric cost problem refer to the problem of finding a feasible flow vector $f = (f_{ij})$ of value $\bar{v}$ in G' that minimizes $\sum (c_{ij}(\lambda) f_{ij} : \text{ over } (i,j) \in \mathcal{A}')$. Let $h(\lambda)$ denote the minimum objective value in this parametric cost problem as a function of the parameter λ.

Prove that $h(\lambda) = -g^*(\lambda)$, and therefore that $h(\lambda)$ is a piecewise-linear concave function in which the number of breakpoints is one less than that for $g(v)$ (Carstensen [1983], Zadeh [1973]).

5.21 Minimum Cost Flow Problems in Series- Parallel Networks   A two-terminal series-parallel network is a directed network with exactly one source and one sink, which is generated recursively as follows:

i) a single arc together with its tail and head as the source and sink nodes is a series parallel network.

ii) If S_1 and S_2 are series-parallel networks, so is the network obtained by either of the following operations.

    a) parallel composition: identify (i.e., make into a single node) the source of S_1 with the source of S_2, and the sink of S_1 with the sink of S_2.

    b) Series composition: identify the sink of S_1 with the source of S_2.

Let $G = (\mathcal{N}, \mathcal{A}, 0, k, c, \breve{s}, \breve{t})$ be a directed acyclic single commodity flow network. Let $\bar{v}$ denote the maximum value of flow from $\breve{s}$ to $\breve{t}$ in G. Prove that G is a two terminal series-parallel network iff for every arbitrary nonnegative capacity vector k and arbitrary cost vector c and $0 \overset{\le}{=} v \overset{\le}{=} \bar{v}$, a minimum cost flow of value v in G can be obtained by the scheme discussed in Section 5.2 beginning with $f = 0$ for value 0, and augmenting flow along a shortest chain from $\breve{s}$ to $\breve{t}$ consisting of arcs in G with positive residual capacity at that stage until the required value is reached, reverse arcs are never used and flow is never reduced on any arc (Bein, Brucker and Tamir [1985]).

5.22 Let $G = (\mathcal{N}, \mathcal{A})$ be a connected directed single commodity flow network with $V = (V_i)$ as the vector of exogenous flow amounts at the nodes satisfying $\sum_{i \in \mathcal{N}} V_i = 0$. $\mathbb{T}$ is a spanning tree in G with a node, say 1, selected as the root node. Fix the flow amounts on all the out-of-tree arcs at 0, and let $\bar{f}$ denote the resulting basic flow vector corresponding to $\mathbb{T}$. For each $i \in \mathcal{N}, i \neq 1$, let $P(i)$ denote the immediate predecessor of i in $\mathbb{T}$ and let $\mathbf{H}(\mathbb{T}, i)$ be the family of node i in $\mathbb{T}$. For each $i \neq 1$, prove that the flow amount in $\bar{f}$ on the in-tree arc joining i and $P(i)$ is $\sum(\alpha_j V_j :$ over $j \in \mathbf{H}(\mathbb{T}, i))$ where $\alpha_j = +1$ or -1 for all $j \in \mathbf{H}(\mathbb{T}, i)$. Discuss how to determine whether $\alpha_j = +1$ or -1 for each j in $\mathbf{H}(\mathbb{T}, i)$.

5.23 Consider the ordinary uncapacitated transportation problem

$$\text{minimize} \quad \sum_{i=1}^{m} \sum_{j=1}^{n} c_{ij} \ x_{ij}$$

$$\text{subject to} \quad \sum_{j=1}^{n} x_{ij} = a_i, \ i = 1 \text{ to } m \qquad (5.58)$$

$$\sum_{i=1}^{m} x_{ij} = b_j, \ j = 1 \text{ to } n$$

$$x_{ij} \geqq 0, \ \text{for all } i, j$$

where all the cost coefficients satisfy $L \leq c_{ij} \leq 2L$ for some positive L and $a_i > 0, b_j > 0$ for all i, j and $\sum a_i = \sum b_j$. Now consider the following optimization problem, with the same data, in which the variables are unrestricted in sign.

$$\text{minimize} \quad \sum_{i=1}^{m} \sum_{j=1}^{n} c_{ij} \ |x_{ij}|$$

$$\text{subject to} \quad \sum_{j=1}^{n} x_{ij} = a_i, \ i = 1 \text{ to } m \qquad (5.59)$$

$$\sum_{i=1}^{m} x_{ij} = b_j, \ j = 1 \text{ to } n$$

(i) Prove that both these problems have the same optimal solutions.

(ii) Prove that if $\bar{x} = (\bar{x}_{ij})$ is feasible to (5.59), it is optimal iff there exists $u = (u_i), v = (v_j)$ satisfying the following for all $i = 1$ to m, $j = 1$ to n.

$$|u_i + v_j| \leq c_{ij}$$

$$\bar{x}_{ij} > 0 \text{ implies } c_{ij} - u_i - v_j = 0 \qquad (5.60)$$

$$\bar{x}_{ij} < 0 \text{ implies } c_{ij} + u_i + v_j = 0$$

(iii) Consider the following algorithm for solving (5.59). It moves along a path of basic solutions of (5.59); the solutions are not necessarily positive. It starts with any basic solution of (5.59) (which may not be $\geqq 0$) and generates a sequence of basic solutions using operations (a) and (b) given below, until the termination criterion is satisfied.

(a) Let β be the set of basic cells associated with the current solution $\bar{x}$ of (5.59). Define for $(p, q) \in \beta$

$$
c^*_{p,q} = \begin{cases} c_{pq}, & \text{if } \bar{x}_{pq} \geq 0 \\ -c_{pq}, & \text{if } \bar{x}_{pq} < 0. \end{cases}
$$

Determine $u = (u_i), v = (v_j)$ to satisfy $u_p + v_q = c^*_{pq}$ for $(p, q) \in \beta$, and the additional condition $v_n = 0$, say, to take care of the redundancy in the constraints in (5.59).

Termination condition: If $|u_i + v_j| \leq c_{ij}$ for all i, j, then the present solution $\bar{x}$ is optimal for (5.59); terminate.

Go to (b) if termination condition is not satisfied.

(b) Select a cell (h, g) such that $|u_h + v_g| > c_{hg}$ as the entering cell into the basic set β. To determine the dropping basic cell, find the unique cycle in $\beta \cup \{(h, g)\}$. Make x_{hg} a small positive quantity and determine the modified values of x_{ij} for (i, j) on the cycle. Let Δ^+ be the set of cells in the cycle that decrease their absolute value in this operation, and Δ^- the remaining cells in the cycle.

1. Prove that $\Delta^+ \neq \emptyset$ if $u_h + v_g > c_{hg}$, and $\Delta^- \neq \emptyset$ if $u_h + v_g < -c_{hg}$. If $u_h + v_g > c_{hg}$, select the dropping cell from β to be (p, q) corresponding to the smallest $|\bar{x}_{ij}|$ among cells in Δ^+. If $u_h + v_g < -c_{hg}$, select the dropping cell from β to be (p, q) corresponding to the smallest $|\bar{x}_{ij}|$ among cells in Δ^-.

2. Find the basic solution of (5.59) corresponding to the new basic set, and repeat the whole process with it. Prove that the change in the objective value of the solution for (5.59) in this step is $|\bar{x}_{pq}|(c_{hg} - u_h - v_g)$, if $u_h + v_g > c_{hg}$, or $|\bar{x}_{pq}|(c_{hg} + u_h + v_g)$, if $u_h + v_g < -c_{hg}$.

(iv) Discuss a method for solving (5.58), through (5.59) using this method.

(Finke and Ahrens [1978])

5.24 The Bottleneck Transportation Problem Bottleneck transportation problems usually arise in shipping perishable commodities which have to be distributed quickly. Suppose there are m sources and n destinations. Let $a_i > 0$

$$\text{minimize } \left(\text{maximum } \{t_{ij} : \ x_{ij} > 0\}\right)$$

$$\text{subject to } \sum_{j=1}^{n} x_{ij} = a_i, \ i = 1 \text{ to } m \qquad (5.61)$$

$$\sum_{i=1}^{m} x_{ij} = b_j, \ j = 1 \text{ to } n$$

$$0 \leqq x_{ij} \leq k_{ij}, \text{ for all } i,j$$

be the units available at the ith source, and $b_j > 0$ the units required at the jth destination, satisfying $\sum a_i = \sum b_j$. Let $k_{ij} > 0$ be the capacity of the arc (i,j) joining source i and destination j, and $t_{ij} \geq 0$ be the time necessary to carry out the shipment on this arc. Then the balanced capacitated bottleneck transportation problem with this data is (5.61). Given a feasible solution $\bar{x} = (\bar{x}_{ij})$, develop necessary and sufficient optimality conditions for it to be optimal for (5.61). Also develop an algorithm for solving (5.61). Apply the algorithm to solve the problem with data, $m = n = 5$, $k_{ij} = 24$ for all (i,j), and

		t_{ij} in days					a_i
$j =$		1	2	3	4	5	a_i
$i =$	1	13	16	3	12	28	15
	2	22	28	15	12	20	15
	3	21	29	6	18	14	12
	4	17	29	3	25	29	32
	5	4	14	4	15	2	26
b_j		3	39	27	7	24	

The transportation paradox which occurs in the linear transportation problem can also occur in the bottleneck problems. For example, consider the following version of (5.61) with overshipments allowed at both the sources and destinations.

$$\text{minimize } \left(\text{maximum } \{t_{ij} : \ x_{ij} > 0\}\right)$$

$$\text{subject to } \sum_{j=1}^{n} x_{ij} \geq a_i, \ i = 1 \text{ to } m \qquad (5.62)$$

$$\sum_{i=1}^{m} x_{ij} \geq b_j, \ j = 1 \text{ to } n$$

$$0 \leq x_{ij} \leq k_{ij}$$

Develop conditions for the occurrence of the paradox (i.e., the optimum objective value in (5.62) being strictly less than that in (5.61)), and give an explanation for the paradox in terms of dual solutions and marginal analysis.

Show that the paradox occurs in the numerical problem with the data given above (the optimum objective value strictly decreases as a_1 and b_1 are both increased by the same positive quantity, keeping all other data unchanged).

Let BT(OS) denote the bottleneck time (optimum objective value in (5.62)) as a function of overshipment amount $\sum_i \sum_j x_{ij} - \sum_i a_i$. Develop an algorithm to draw the complete curve of BT(OS). Apply this algorithm on the numerical example discussed above, and draw the BT(OS) curve for it (Finke [1983]).

5.25 Applications in Portfolio Management   Consider the problem of selecting a dynamic portfolio of securities in order to maximize total return over a fixed planning horizon. This problem can be modeled as a network flow problem in which each arc represents a security and each node (other than a terminal sink node representing the end of the planning horizon) signifies the beginning of an investment period.

For example, if the investor has the option of purchasing one-year bonds that return 4% per annum, two-year bonds at 6% per annum, or four-year bonds at 5% per annum, the network corresponding to a four year planning horizon is given in Figure 5.24.

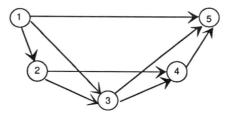

Figures 5.24

(i)   Show that the network for this model is acyclic. Show that the problem of finding the best investment policy over an n period planning horizon can be formulated as that of finding a longest chain from node 1 to node $n+1$ in this network. Using this, find the best investment policy in the example cited above for a 4 year planning horizon.

(ii) Assume that the accumulated interest is considered as profit and not reinvested. Consider the problem of maximizing this total profit over the entire planning horizon, where the original amount can be reinvested as many times as necessary until the end of the planning horizon. Also, suppose we are given lower and upper bounds on the amount to be invested in each possible security at each time point (these bounds may vary with time for each security). Formulate this problem as a maximum profit network flow problem.

Construct this model for the three bond examples discussed above over a six year planning horizon. Assume that the money available to be invested is $100, and that the bounds are as in the table given below. Find an optimum investment policy in this numerical example.

| | First two years | | Next three years | | Final year |
Security	Lower	Upper	Lower	Lower	
1-year bonds	15	35	15	40	0-100
2-year bonds	30	60	40	70	x
4-year bonds	20	40	0	25	x

(Golden and Keating [1982])

5.26 Consider the ordinary transportation problem

$$\text{minimize } \sum_{i=1}^{m} \sum_{j=1}^{n} c_{ij} \; x_{ij}$$

$$\text{subject to } \sum_{j=1}^{n} x_{ij} \; \le a_i, \; i = 1 \text{ to } m$$

$$\sum_{i=1}^{m} x_{ij} \; \ge b_j, \; j = 1 \text{ to } n$$

$$x_{ij} \; \ge 0, \text{ for all } i, j$$

with the usual assumptions that $a_i > 0, b_j > 0$ for all i, j and $\sum a_i \ge \sum b_j$. Suppose there is the additional condition that each demand center j must be entirely supplied from a single source. This requirement appears in many practical applications. It appears frequently in the supplying of supermarket orders from a network of central warehouses. It is commonly required in military applications as troops going in the same mission usually leave from the same staging area, etc. Discuss approaches for solving the transportation problem with this additional condition (Nagelhout and Thompson [1980]).

5.27 Consider a stationary dynamic maximum value flow problem on the network $G = (\mathcal{N}, \mathcal{A}, 0, k, \check{s}, \check{t})$ with $a = (a_{ij})$ as the vector of arc traversal times, over a T-period horizon. Let $v_1, \dots, v_T$ represent the flow value reaching the sink in periods $1, \dots, T$. Prove that any feasible flow of value $v = v_1 + \dots + v_T$ over the horizon which satisfies either of conditions $(a), (b)$ below, also satisfies the other two.

(a) Maximize $\sum_{\tau=1}^{p} v_\tau$ for $p = 1, \dots, T$ (i.e., maximize the output for the first p periods for all p).

(b) Minimize $\sum_{\tau=1}^{T} c_\tau v_\tau$, where $c_1 < c_2 < \dots < c_\tau$ (i.e., minimize the weighted sum of flow values in the various time periods, where the weights are increasing with time).

(c) Minimize p such that $v_{p+\tau} = 0$ for $\tau = 1, \ldots, T - p$ (i.e., minimize the number of time periods required to send a flow of v from $\check{s}$ to $\check{t}$).

(Jarvis and Ratliff [1982]).

5.28 Warehousing Problem   In this problem, we are required to plan the operations of a warehouse over an n-period planning horizon. The warehouse purchases, stores and sells in each period, a single commodity that is subject to known fluctuations in selling prices and purchasing costs. The warehouse has a fixed capacity (γ units) and all new purchases and hold-overs from previous periods are stored there before selling. In each period any amount can be purchased, stored and sold at the known specified prices, subject to the warehouse capacity. We have the additional data and variables listed below.

b_i = warehousing cost per unit in period i (this applies to z_i defined below)
c_i = purchase cost per unit in period i, $i = 1$ to n
d_i = selling price per unit in period i, $i = 1$ to n
s_o = initial stock in warehouse, given
s_n = specified final stock desired in warehouse after selling in period n
x_i = amount purchased in period i, $i = 1$ to n
s_i = amount held in warehouse after selling in period i, $i = 1$ to n
      (s_n is specified in the data given above)
$z_i = s_{i-1} + x_i$ = amount in warehouse after purchasing (but before selling) in
      period i, this has to be $\stackrel{<}{=}$ capacity γ for $i = 1$ to n
y_i = amount sold in period i (this is $\stackrel{<}{=} z_i$), $i = 1$ to n

It is required to determine a purchasing and selling plan which will maximize the total net profit. Formulate this as a network flow problem. Prove that the maximum total net profit is a multiple of the warehouse capacity γ.

5.29 Consider an m-source, n-sink, p-commodity transportation problem denoted by MCTP(m, n, p): it is to find $(x_{ij}^r : i = 1$ to m, $j = 1$ to n, $r = 1$ to $p)$ to

$$\text{Minimize} \sum_i \sum_j \sum_r c_{ij}^r x_{ij}^r$$

$$\text{subject to} \sum_j x_{ij}^r \quad = a_i^r, \text{ for all } i, r$$

$$\sum_i x_{ij}^r \quad = b_j^r, \text{ for all } j, r \qquad (5.63)$$

$$\sum_r x_{ij}^r \quad + s_{ij} = k_{ij}, \text{ for all } i, j$$

$$x_{ij}^r, s_{ij} \quad \stackrel{\geq}{=} 0, \text{ for all } i, j, r$$

where the data satisfy $\sum_i a_i^r = \sum_j b_j^r$, for all r. Prove that the constraint coefficient matrix of MCTP$(m, n, 2)$ is totally unimodular if all capacitated arcs are incident to a common node.

Prove that the necessary and sufficient condition for the constraint coefficient matrix of a capacitated MCTP(m, n, p) with $p \geqq 2$, and k finite, to be unimodular is that either m or n to be $\leqq 2$.

Show that every MCTP$(m, 2, p)$, or MCTP$(2, n, p)$ can be transformed into a one commodity capacitated transportation problem (Rebman [1974]; Evans, Jarvis, and Duke [1977]; Evans [1976]).

Comment 5.2   The minimum cost flow problem is a classical problem. Historically it is among the first linear programming problems to be modeled and studied. The problem was posed and a rudimentary algorithm discussed for it in Kantorovitch [1939]. A computational method for the transportation model that would now be called "primal simplex" has been proposed by Hitchcock [1941] where he shows that an optimum solution will be at an extreme point, and develops methods to iteratively construct better extreme point solutions. Later Koopmans [1949] developed node potentials and the optimality criterion, and showed that an extreme point is based on a spanning tree. Dantzig [1951] developed the special variant of the primal simplex algorithm for the transportation problem. He observed the spanning tree property of the basis, the integrality properties of the optimum solutions, and the Dantzig property. Orden [1956] showed that these results extend to the transshipment problem.

The first combinatorial algorithms based on the primal-dual approach were developed by Ford and Fulkerson [1957]. This work was motivated by the Hungarian method of Kuhn [1955 of Chapter 3] for the assignment problem.

The shortest augmenting path (or the build-up) method based on successive shortest chain computations was developed independently in Jewell [1958], Iri [1960], and Busacker and Gowan [1961]. Originally, all these approaches were based on solving the minimum cost flow problem through a sequence of shortest chain computations on residual networks which may have arbitrary arc lengths. Later, Tomizawa [1971] pointed out that the approach can be implemented using node potentials in such a way that all shortest chain computations are on networks with nonnegative arc lengths.

Minty [1960] and Fulkerson [1961] independently developed the out-of-kilter method for minimum cost flow problems. Edmonds and Karp [1972 of Chapter 2] showed that the out-of-kilter method can be converted into a polynomially bounded algorithm by implementing it using a special capacity scaling technique developed by them. This technique applies the out-of-kilter method on a series of problems which provide successively closer approximations to the original problem. It is the first polynomial time algorithm for the minimum cost flow problem, and the first polynomial time algorithm for a significant special class of linear programs. The practical significance of scaling techniques is not very clear, but it has great theoretical value.

The original negative cycle canceling algorithm is due to Klein [1967]. This simple, classical algorithm may not even terminate if the capacities are irrational. On problems with integer data it is a finite algorithm, but the number of iterations can grow exponentially with the size of the data in the worst case. Recently Goldberg and Tarjan [1989] have shown that this algorithm can be converted into a strongly polynomial algorithm by an appropriate choice of the cycle to cancel in each iteration (a minimum mean residual cycle).

Zadeh [1973a, b] has presented examples of network flow problems on which the classical methods, the primal simplex algorithm with Dantzig's pivot rule, the dual simplex algorithm, the original negative cycle canceling algorithm, the shortest augmenting path algorithm, and the out-of-kilter algorithm, all require an amount of effort growing exponentially with the size of the problem.

The fact that these algorithms displayed exponential growth on one specially constructed pathological class of problems does not mean that they won't work well on network models arising in applications. Computational experience with these methods has always turned out to be much better than their bleak performance on these pathological examples. In the 1950s and 1960s, primal-dual and then the out-of-kilter were popular methods for solving network models. In those days data structures for generating efficient implementations of the primal simplex method for network flow problems were not available. The first tree label data structures for manipulating trees were suggested by Scions [1964] and Johnson [1966]. Early implementations based on them for the primal simplex method for minimum cost flow problems, by Srinivasan and Thompson [1972]; Glover, Karney and Klingman [1974]; and Glover, Karney, Klingman, and Napier [1974] performed much better than implementations of the primal-dual and out-of-kilter methods available at that time. They not only showed that the primal implementations are faster, but also that they require less storage and are most suitable when using secondary storage devices, and are more compatible as embedded parts of more general optimization systems. Improved data structures were developed in Glover, Karney, and Klingman [1972]; Glover, Klingman, and Stutz [1974]; Bradley, Brown, and Graves [1977]; and Barr, Glover, and Klingman [1979]; and numerous others. With these breakthroughs in coding, the primal simplex method has replaced the other methods as the computational workhorse of commercial network software in the 1970s.

The primal simplex method and efficient ways of implementing it to solve minimum cost flow problems have been the subject of very rigorous investigations by many groups in the 1970s and 1980s. There is a gulf of difference between the mathematical statement of an algorithm and its implementation using appropriate data structures, and methods for updating them from one step to the next. Empirical computational tests comparing different algorithms or different implementations of the same algorithm on representative classes of problems of varying sizes, sparsity and other characteristics, are an important research activity. Substantial reductions in computational effort can typically be achieved through appropriate implementation of an algorithm. Data structure design must go hand in hand with pivot selection strategy in implementing the primal simplex method. Any change in pivot strategy requires a commensurate change in data structures to be effective. Mulvey [1978] describes the results of an organized experimental design and a detailed series of empirical tests to compare various pivot strategies, methods for attaining feasibility (big-M, Phase I, etc.) and a variety of other features. He found that the internal tactics of the code are very important. His most successful implementation uses a candidate list of nonbasic arcs (about 40) to select the entering arc from, and re-forms it periodically (every 20 pivots or so) to improve performance. Some of the other papers dealing with computational studies and implementation issues are Aashtiani and Magnanti [1976]; Ali, Helgason, Kennington, and Lall [1978]; Clausen [1968]; Gibby, Glover, Klingman, and Mead [1983]; Glover and Klingman [1982]; Grigoriadis [1986]; Grigoriadis and Hsu [1980]; Hatch [1975]; and Helgason and Kennington [1977].

The supremacy of the primal simplex method for solving minimum cost flow problems has continued into the 1980s. Recently, however, a new class of relaxation algorithms (Section 5.6) were proposed by Bertsekas and his associates, and investigations reported in Bertsekas [1985]; Bertsekas, Hosein, and Tseng [1987]; and Bertsekas and Tseng [1988] indicate that these algorithms exhibit nice empirical behavior, and are competitive or better than the primal simplex method.

Public domain computer codes for the minimum cost flow problems are available from Bertsekas and Tseng [1988]; Grigoriadis and Hsu [1980]; Kennington and Helgason [1980 of Chapter 1]; and Simeone, Toth, Gallo, Maffioli, and Pallotino [1988].

Armacost and Mehrotra [1991] compared an implementation of the network primal simplex method with an implementation of the recently developed interior point method (the dual affine scaling method) on sparse minimum cost flow problems. They found that the network simplex method outperforms the affine scaling method by a substantial margin on these problems; also its advantage seems to grow with the number of nodes and the density of the network.

The first strongly polynomial time minimum cost flow algorithm is due to Tardos [1985]. This method is based on the repeated use of a concept of approximate optimality. Theoretically, the simplest strongly polynomial time algorithm for minimum cost flow problems is perhaps that of Goldberg and Tarjan [1989]. Many other strongly polynomial time algorithms are now available for minimum cost flow problems, among them are Fujishige [1986]; Galil and Tardos [1988]; Orlin [1984, 1988]; and Tardos [1986]. The corresponding question in general LP (is there a strongly polynomial algorithm for it?) still remains an open question.

Many different approaches for obtaining exact or approximate optimum solutions for multicommodity network flow problems have been suggested over the years. In this chapter we discussed an approach suggested by Ford and Fulkerson [1958] based on column generation for the problem in arc-chain formulation. This approach subsequently inspired Dantzig and Wolfe [1960] to develop the decomposition principle for LP. For other approaches for multicommodity flow problems based on partitioning or decomposition, see the survey papers of Ali, Helgason,

Kennington and Lall [1980]; Assad [1978]; Geoffrion and Graves [1974]; Grigoriadis and White [1972]; Hartman and Lasdon [1972]; Kennington [1978]; Kennington and Shalaby [1977]; Kleitman [1971]; Swoveland [1973]; Tomlin [1966]; and Kennington and Helgason [1980 of Chapter 1]. Since 1984, the highly publicized success of interior point methods based on barrier functions, or method of centers, or other approaches of nonlinear programming specialized to LP, has resulted in many researchers shifting their attention to these new methods for solving large scale LPs and multicommodity flow problems in particular. See Adler, Resende, Veiga, and Karmarkar [1989], and Kapoor and Vaidya [1985].

5.13   References

H. A. AASHTIANI and T. L. MAGNANTI, June 1976, "Implementing Primal-Dual Network Flow Algorithms," OR Center report 055-76, MIT, Cambridge, Mass.

I. ADLER, M. G. C. RESENDE, G. VEIGA and N. KARMARKAR, 1989, "An Implementation of Karmarkar's Algorithm for Linear Programming," *MP A*, 44, no. 3(297-335).

R. K. AHUJA, J. BATRA, and S. K. GUPTA, May 1984, "A Parametric Algorithm for Convex Cost Network Flow and Related Problems," *EJOR*, 16, no. 2(222-235).

A. ALI, R. HELGASON, J. KENNINGTON, and H. LALL, 1978, "Primal Simplex Network Codes: State of the Art Implementation Technology," *Networks* 8(315-339).

A. ALI, R. HELGASON, J. KENNINGTON, and H. LALL, 1980, "Computational Comparison Among Three Multicommodity Network Flow Algorithms," *OR*, 28, no. (995-1000).

A. ARMACOST and S. MEHROTRA, 1991, "A Computational Comparison of the Network Simplex Method with the Dual Affine Scaling Method," *Opsearch*, to appear.

A. A. ASSAD, 1978, "Multicommodity Network Flows, A Survey," *Networks*, 8(37-91).

F. BARAHONA and E. TARDOS, 1989, "Note on Weintraub's Minimum Cost Circulation Algorithm," *SIAM Journal on Computing* , 18(579-583).

R. BARR, F. GLOVER, and D. KLINGMAN, Feb. 1979, "Enhancements of Spanning Tree Labeling Procedures for Network Optimization," *INFOR* 17, no. 1(16-34).

M. S. BAZARAA and H. D. SHERALI, 1982, "A Property Regarding Degenerate Pivots for Linear Assignment Networks," *Networks*, 12, no. 4(469-474).

W. W. BEIN, P. BRUCKER, and A. TAMIR, 1985, "Minimum Cost Flow Algorithms for Series-Parallel Networks," *DAM* 10,no. 2(117-124).

D. P. BERTSEKAS, 1985, "A Unified Framework for Primal-Dual Methods in Minimum Cost Network Flow Problems," *MP*, 32(125-145).

D. P. BERTSEKAS, P. HOSEIN, and P. TSENG, 1987, "Relaxation Methods for Network Flow Problems With Convex Arc Costs," *SIAM J. on Control and Optimization*, 25(1219-1243).

D. P. BERTSEKAS and P. TSENG, Jan.-Feb. 1988, "Relaxation Methods for Minimum Cost Ordinary and Generalized Network Flow Problems," *OR* 36, no. 1(93-114).

R. G. BLAND and D. L. JENSEN, 1985, "On the Computational Behavior of a Polynomial Time Network Flow Algorithm," Tech. report 661, SORIE, Cornell University, Ithaca, N.Y.

G. H. BRADLEY, G. G. BROWN and G. W. GRAVES, 1977, "Design and Implementation of Large Scale Transshipment Algorithms," *MS*, 24(1-34).

J. R. BROWN, 1983, "The Flow Circulation Sharing Problem," *MP*, 25, no. 2(199-227).

R. G. BUSACKER and P. J. GOWAN, 1961, "A Procedure for Determining a Family of Minimum-Cost Network Flow Patterns," ORO Tech. report 15, John Hopkins University.

P. J. CARSTENSEN, 1983, "Complexity of Some Parametric Integer and Network Programming Problems," *MP*, 26, no. 1(64-75).

A. CHARNES and D. KLINGMAN, 1971, "The More-for-Less Paradox in the Distribution Model," *Cahiers de Centre d'Etudes Recherche Operationelle*, 13, no. 1(11-22).

A. CHARNES, S. DUFFUAA, and M. RYAN, 1980, "Degeneracy and the More-for-Less Paradox," *Journal of Information and Optimization Sciences*, 1, no. 1(52-56).

R. J. CLAUSEN, March 1968, "The Numerical Solution of Network Problems Using the Out-of-Kilter Algorithm," RAND Corp. memo RM-5456-PR, Santa Monica, Calif.

W. H. CUNNINGHAM, 1976, "A Network Simplex Method," *MP*, 11(105-116).

W. H. CUNNINGHAM, 1979, "Theoretical Properties of the Network Simplex Method," *MOR*, 4(196-208).

W. H. CUNNINGHAM and J. G. KLINCEWICZ, 1983, "On Cycling in the Network Simplex Method," *MP*, 26(182-189).

G. B. DANTZIG, 1951, "Application of the Simplex Method to a Transportation Problem," in T. C. Koopmans(ed.) *Activity Analysis of Production and Allocation*, Wiley, N.Y.

G. B. DANTZIG and P. WOLFE, 1960, "Decomposition Principle for Linear Programs," *OR*, 8(101-111).

J. R. EVANS, 1976, "A Combinatorial Equivalence Between a Class of Multicommodity Flow Problems and the Capacitated Transportation Problem," *MP*, 10(401-404).

J. R. EVANS, J. J. JARVIS, and R. A. DUKE, 1977, "Graphic Matroids and the Multicommodity Transportation Problem," *MP*, 13(323-328).

S. EVEN, A. ITAI, and A. SHAMIR, Dec. 1976, "On the Complexity of Timetable and Multicommodity Flow Problems," *SIAM J. on Computing*, 5, no. 4(691-703).

G. FINKE, May 1983, "Minimizing Overshipments in Bottleneck Transportation Problems," *INFOR*, 21, no. 2(121-135).

G. FINKE and J. H. AHRENS, Feb. 1978, "A Variant of the Primal Transportation Algorithm," *INFOR*,16, no. 1(35-46).

C. O. FONG and V. SRINIVASAN, 1977, "Determining All Nondegenerate Shadow Prices for the Transportation Problem," *TS*, 11, no. 3(199-222).

L. R. FORD and D. R. FULKERSON, 1958, "A Suggested Computation for Maximal Multicommodity Network Flows," *MS*, 5(97-101).

H. FRIESDORF and H. HAMACHER, 1982, "Weighted Minimum Cost Flows," *EJOR*, 11(181-192).

S. FUJISHIGE, 1986, "A Capacity Rounding Algorithm for Minimum Cost Circulation Problem: A Dual Framework of the Tardos Algorithm," *MP*, 35(298-308).

D. R. FULKERSON, 1961, "An Out-of-Kilter Method for Minimum Cost Flow Problems," *SIAM J. on Applied Mathematics*, 9(18-27).

Z. GALIL and E. TARDOS, 1988, "An $O(n^2(m + n \log n) \log n)$ Min. Cost Flow Algorithm," *JACM*, 35(374-386).

A. R. GAMBLE, A. R. CONN, and W. R. PULLEYBLANK, 1988, "A Network Penalty Method," Tech. report, Dept. of Combinatorics and Optimization, University of Waterloo, Waterloo, Ontario, Canada.

B. J. GASSNER, 1964, "Cycling in the Transportation Problem," *NRLQ*, 11(43-58).

B. GAVISH, P. SCHWEITZER, and E. SHLIFER, 1977, "The Zero Pivot Phenomenon in Transportation Problems and its Computational Implications," *MP*, 12(226-240).

A. M. GEOFFRION and G. W. GRAVES, 1974, "Multicommodity Distribution System Design by Benders Decomposition," *MS*, 20(822-844).

D. GIBBY, F. GLOVER, D. KLINGMAN, and M. MEAD, 1983, "A Comparison of Pivot Selection Rules for Primal Simplex Based Network Codes," *OR Letters*, 2(199-202).

C. R. GLASSEY, 1978, "A Quadratic Network Optimization Model for Equilibrium Single Commodity Trade Flows," *MP*, 14(98-107).

S. GLICKSMAN, L. JOHNSON, and L. ESELSON, June 1960, "Coding the Transportation Problem," *NRLQ*, 2(169-184).

F. GLOVER, K. KARNEY, D. KLINGMAN, 1972, "The Augmented Predecessor Index Method for Locating Stepping Stone Paths and Assigning Dual Prices in Distribution Problems," *TS*, 6, no. 1(171-180).

F. GLOVER, D. KARNEY, and D. KLINGMAN, 1974, "Implementation and Computational Comparisons of Primal, Dual and Primal-Dual Computer Codes for Minimum Cost Network Flow Problems," *Networks*, 4(191-212).

F. GLOVER, D. KARNEY, D. KLINGMAN, and A. NAPIER, 1974, "A Computational Study on

Start Procedures, Basis Change Criteria and Solution Algorithms for Transportation Problems," *MS*, 20, no. 5(793-813).

F. GLOVER and D. KLINGMAN, 1977, "Network Applications in Industry and Government," *AIIE Transactions*, 9, no. 4(363-376).

F. GLOVER and D, KLINGMAN, Nov. 1982, "Recent Developments in Computer Implementation Technology for Network Flow Algorithms," *INFOR*, 20, no. 4(433-452).

F. GLOVER, D. KLINGMAN, and A. NAPIER, 1972, "An Efficient Dual Approach to Network Problems," *Opsearch*, 9, no. 1(1-18).

F. GLOVER, D. KLINGMAN, and J. STUTZ, 1974, "Augmented Threaded Index Method for Network Optimization," *INFOR*, 12, no. 3(293-298).

A. V. GOLDBERG and R. E. TARJAN, Oct. 1989, "Finding Minimum-Cost Circulations by Canceling Negative Cycles," *JACM*, 36, no. 4(873-886).

A. V. GOLDBERG and R. E. TARJAN, Aug. 1990, "Finding Minimum-Cost Circulations by Successive Approximation," *MOR*, 15, no. 3(430-466).

B. L. GOLDEN and K. D. KEATING, 1982, "Network Techniques for Solving Asset Diversification Problems in Finance," *COR*, 9, no. 3(173-195).

M. D. GRIGORIADIS, July 1978, "Algorithms for Minimum Cost Single and Multi-Commodity Network Flow Problems," Notes for Summer School in Combinatorial Optimization, SOGESTA, Urbino, Italy.

M. D. GRIGORIADIS, 1986, "An Efficient Implementation of the Network Simplex Method," *MPS*, 26(83-111),

M. D. GRIGORIADIS and T. HSU, 1979, "The Rutgers Minimum Cost Network Flow Subroutines," *SIGMAP Bulletin of the ACM*, 26(17-18).

M. D. GRIGORIADIS and T. HSU, Nov. 1980, "The Rutgers Minimum Cost Network Flow Subroutines," RNET Documentation, Dept. of Computer Science, Rutgers University, New Brunswick, N.J.

M. D. GRIGORIADIS and W. W. WHITE, 1972, "A Partitioning Algorithm for the Multicommodity Network Flow Problem," *MP*, 3(157-177).

H. W. HAMACHER and S. TUFEKCI, 1987, "On the Use of Lexicographic Min Cost Flows in Evacuation Modeling," *NRLQ*, 34(487-503).

J. K. HARTMAN and L. S. LASDON, 1972, "A Generalized Upper Bounding Algorithm for Multicommodity Network Flow Problems," *Networks*, 1(333-354).

R. HASSIN, 1983, "The Minimum Cost Flow Problems, a Unifying Approach to Dual Algorithms and a New Tree Search Algorithm," *MP*, 25(228-239).

R. S. HATCH, 1975, "Bench Marks Comparing Transportation Codes Based on Primal Simplex and Primal-Dual Algorithms," *OR*, 23(1167-1172).

R. V. HELGASON and J. L. KENNINGTON, March 1977, "An Efficient Procedure for Implementing a Dual Simplex Network Flow Algorithm," *AIIE Transactions*, 9(63-68).

F. L. HITCHCOCK, April 1941, "The Distribution of a Product From Several Sources to Numerous Localities," *Journal of Mathematics and Physics* 20, no. 2(224-230).

M. IRI, 1960, "A New Method for Solving Transportation-Network Problems," *J. OR Society of Japan*, 3(27-87).

J. J. JARVIS and H. D. RATLIFF, Jan. 1982, "Some Equivalent Objectives for Dynamic Network Flow Problems," *MS*, 28, no. 1(106-109).

W. S. JEWELL, 1958, "Optimal Flow Through Networks," Tech. report 8, OR Center, MIT, Cambridge, Mass.

E. JOHNSON, 1966, "Networks and Basic Solutions," *OR*, 14(619-623).

L. V. KANTOROVITCH, 1939, "Mathematical Methods in the Organization and Planning of Production," Publication House of the Leningrad University, translated in *MS*, 6, 1960(366-422).

S. KAPOOR and P. M. VAIDYA, 1985, "Fast Algorithms for Convex Quadratic Programming and Multicommodity Flows," *Proc. 18th annual ACM Symp. on Theory of Computing*, (147-159).

S. KAPOOR and P. M. VAIDYA, 1988, "Speeding Up Karmarkar's Algorithm for Multicommodity Flows," Tech. report, Dept. of Computer Science, University of Illinois at Urbana-Champaign, Il.

R. M. KARP, 1978, "A Characterization of the Minimum Cycle Mean in a Digraph," *Discrete*

Mathematics, 23(309-311).

J. L. KENNINGTON, Mar. -Apr. 1978, "A Survey of Linear Cost Multicommodity Network Flow," *OR,* 26, no. 2(209-236).

J. L. KENNINGTON and M. SHALABY, 1977, "An Effective Subgradient Procedure for Minimal Cost Multicommodity Flow Problems," *MS,* 23(994-1004).

M. KLEIN, Nov. 1967, "A Primal Method for Minimum Cost Flows With Applications to Assignment and Transportation Problems," *MS,* 14, no. 3(205-220).

D. L. KLEITMAN, 1971, "An Algorithm for Certain Multicommodity Flow Problems," *Networks,* 1(75-90).

T. C. KOOPMANS, 1949, "Optimum Utilization of the Transportation System," *Proceedings of the International Statistical Conference,* Vol. 5.

E. MINIEKA, 1974, "Dynamic Network Flows with Arc Changes," *Networks,* 4(255-265).

G. J. MINTY, 1960, "Monotone Networks," *Proceedings of the Royal Society of London,* 257A(196-212).

J. M. MULVEY, 1978, "Pivot Strategies for Primal Simplex Network Codes," *JACM,* 25(266-270).

J. M. MULVEY, 1978, "Testing of a Large Scale Network Optimization Program," *MP,* 15(291-314).

R. V. NAGELHOUT and G. L. THOMPSON, 1980, "A Single Source Transportation Algorithm," *COR,* 7, no. 3(185-198).

M. NANIWADA, 1969, "Multicommodity Flows in a Communication Network," *Electronics Commun.* , Japan, 52-A(34-41).

A. ORDEN, April 1956, "The Transshipment Problem," *MS,* 2, no. 3(276-285).

J. B. ORLIN, 1984, "Genuinely Polynomial Simplex and Non-Simplex Algorithms for the Minimum Cost Flow Problem," OR 1615-84, Sloan School of Mgt. MIT, Cambridge, Mass.

J. B. ORLIN, 1988, "A Faster Strongly Polynomial Minimum Cost Flow Algorithm," *Proc. 20th ACM Symp. Theory of Computing,* (377-387).

M. H. PARTOVI, 1984, "A Study of Degeneracy in the Simplex Algorithm for Linear Programming and Network Flow Problems," Ph. D. dissertation, IOE Dept. University of Michigan, Ann Arbor, Mich.

K. R. REBMAN, 1974, "Total Unimodularity and the Transportation Problem: A Generalization," *Linear Algebra and its Applications,* 8(11-24).

H. ROCK, 1980, "Scaling Techniques for Minimal Cost Network Flows," in U. Page (ed.) *Discrete Structures and Algorithms,* Carl Hanser, Munchen,(181-191).

R. T. ROCKAFELLAR, 1970, *Convex Analysis,* Princeton University Press, Princeton, N.J.

R. SAIGAL, 1967, "Multicommodity Flows in Directed Networks," ORC 67-38, University of California, Berkeley, Calif.

S. R. SCHMIDT, P. A. JENSEN, and J. W. BARNES, 1982, "An Advanced Dual Incremental Network Algorithm," *Networks,* 12, no. 4(475-492).

H. I. SCOINS, 1964, "The Compact Representation of a Rooted Tree and the Transportation Problem," International Symposium on Mathematical Programming, London.

B. SIMEONE, P. TOTH, G. GALLO, F. MAFFIOLI, and S. PALLOTINO (eds.), *Fortran Codes for Network Optimization, AOR,* 13.

S. SINGH, July 1986, "Improved Methods for Storing and Updating Information in the Out-of-Kilter Algorithm," *JACM,* 33, no. 3(551-567).

V. SRINIVASAN and G. L. THOMPSON, 1972, "Accelerated Algorithms for Labeling and Relabeling of Trees With Applications to Distribution Problems," *JACM,* 19, no. 4(712-726).

V. SRINIVASAN and G. L. THOMPSON, 1973, "Benefit Cost Analysis of Coding Techniques for the Primal Transportation Algorithm," *JACM,* 20(194-213).

C. SWOVELAND, 1973, "A Two-Stage Decomposition Algorithm for a Generalized Multicommodity Flow Problem," *INFOR,* 11(232-244).

W. SZWARC, 1971, "The Transportation Paradox," *NRLQ,* 18(185-202).

E. TARDOS, 1985, "A Strongly Polynomial Minimum Cost Circulation Algorithm," *Combinatorica,* 5(247-256).

N. TOMIZAWA, 1971, "On Some Techniques Useful for Solution of Transportation Network Problems," *Networks,* 1(173-194).

J. A. TOMLIN, 1966, "Minimum Cost Multicommodity Network Flows," *OR*, 14(45-51).

R. R. VEMUGANTI and M. BELLMORE, 1973, "On Multicommodity Maximal Dynamic Flows," *OR*, 21, no. 1(10-21).

A. WEINTRAUB, 1974, "A Primal Algorithm to Solve Network Flow Problems with Convex Costs," *MS*, 21(87-97).

W. W. WHITE, 1972, "Dynamic Transshipment Networks: An Algorithm and its Application to the Distribution of Empty Containers," *Networks*, 2(211-236).

N. ZADEH, 1973a, "A Bad Network Flow Problem for the Simplex Method and Other Minimum Cost Flow Algorithms," *MP*, 5(255-266).

N. ZADEH, 1973b, "More Pathological Examples for Network Flow Problems," *MP*, 5(217-224).

Chapter 6

Single Commodity Flows with Additional Linear Constraints

Here we consider the single commodity flow problem (5.3) in the directed connected pure network $G = (\mathcal{N}, \mathcal{A}, \ell, k, c, V)$ with $|\mathcal{N}| = n, |\mathcal{A}| = m$, in which the flow vector is required to satisfy additional linear constraints. Select a node, say n, and fix the root node at it, and eliminate the flow conservation equation at it as the redundant constraint. After this our problem is of the form: find $f = (f_{ij} : (i,j) \in \mathcal{A})$ to

$$\text{Minimize } \sum (c_{ij}f_{ij} : \text{ over } (i,j) \in \mathcal{A})$$

$$\text{Subject to } -f(i, \mathcal{N}) + f(\mathcal{N}, i) = -V_i, \text{ for each } i \neq n \qquad (6.1)$$

$$\sum_{(i,j)} a_{ijr}f_{ij} = (\text{ or } \leqq) b_r, \text{ for } r = 1 \text{ to } \rho \qquad (6.2)$$

$$\ell_{ij} \leqq f_{ij} \leqq k_{ij}, \qquad \text{for all } (i,j) \in \mathcal{A} \qquad (6.3)$$

Because of (6.2) this problem cannot be solved by network methods alone. To solve it by the bounded variable simplex method, we have to deal with bases of order $n - 1 + \rho$ which could be very large even if ρ is small. We discuss an efficient special implementation (belonging to the area of *structured LP*) of the bounded variable primal simplex method for this problem due to Chen and Saigal [1977] that exploits the special structure of this problem. It uses spanning trees in G to handle the flow conservation constraints (6.1), and maintains the inverse of a working basis of order ρ (as opposed to that of maintaining the inverse of bases of order $(n-1+\rho)$ in the usual simplex algorithm) to handle the additional linear constraints (6.2). It is very convenient to solve this problem as long as ρ is not very large. If ρ is large, it may be better to solve this problem directly by large scale implementations of interior point methods mentioned in Section 5.11.

We denote the arcs in $\mathcal{A}$ by $e_1, \ldots, e_m$; and by ℓ_t, k_t, c_t, f_t, the lower bound, capacity, unit cost coefficient, and flow amount, respectively, on e_t for $t = 1$ to m. ρ_0, ρ_1 are the numbers of equality, inequality constraints in (6.2). Introduce the nonnegative slack variables $f_{m+1}, \ldots, f_{m+\rho_1}$ and convert all the inequality constraints in (6.2) into equations. For $t = 1$ to ρ_1, let $A_{.t}$ (a column vector of the unit matrix of order ρ) denote the column vector of f_{m+t} in the system (6.2) after it is transformed into a system of equations. Let $H_{.t}$ be the column vector of f_t in (6.1) for $t = 1$ to m, and let $H_{.t} = 0 \in \mathbb{R}^{n-1}$ for $t = m+1$ to $m+\rho_1$. Define $\ell_t = 0, k_t = \infty, c_t = 0$ for $t = m+1$ to $m+\rho_1$. Let ℓ, k, c, f denote the vectors in $\mathbb{R}^{m+\rho_1}$ consisting of the associated quantities, $\gamma = (-V_1, \ldots, -V_{n-1})^T, b = (b_1, \ldots, b_\rho)^T, H = (H_{.1} \ldots H_{.m+\rho_1}), A = (A_{.1} \ldots A_{.m+\rho_1})$. Then the problem (6.1), to (6.3) becomes the following LP

$$
\begin{array}{lll}
\text{Minimize} & cf & \\
\text{Subject to} & Hf = \gamma & (6.4) \\
& Af = b & (6.5) \\
& \ell \leqq f \leqq k & (6.6)
\end{array}
$$

By the results in Chapter 1 any basis for just the system of conservation equations (6. 4) corresponds to a spanning tree in G and vice versa. If $\{e_{p_1}, \ldots, e_{p_{n-1}}\}$ are the in-tree arcs in a spanning tree $\mathbb{T}$ of G arranged in this order, $H^1 = (H_{.p_1} \ldots H_{.p_{n-1}})$ is the basis for (6.4) associated with $\mathbb{T}$.

For $t = m+1$ to $m+\rho_1$, since $H_{.t} = 0, (H^1)^{-1} H_{.t} = 0$. For $1 \leqq t \leqq m$, let e_t be an out-of-tree arc wrt $\mathbb{T}$, and let $(\lambda_{1t}, \ldots, \lambda_{n-1,t})^T$ be the in-tree arc-fundamental cycle incidence vector associated with it. Then from Chapter 1 we know that $(\lambda_{1t}, \ldots, \lambda_{n-1,t})^T = (H^1)^{-1} H_{.t}$, i.e., $H_{.t} = \sum_{u=1}^{n-1} \lambda_{u,t} H_{.p_u}$. As an example, the in-tree arc-fundamental cycle incidence vector of the out-of-tree arc e_8 wrt the spanning tree $\mathbb{T}$ with in-tree arcs e_1, e_2, e_3, e_4 arranged in this order, in Figure 6.1 is $(1, -1, -1, 1)^T$. $H_{.t}$ is the column vector of f_t among the top 4 rows in Tableau 6.1. It can be verified that $H_{.8} = H_{.1} - H_{.2} - H_{.3} + H_{.4}$.

We will assume that the constraints in (6.4), (6.5) are linearly independent. Hence if B is a basis for (6.4), (6.5), it has a natural row partition as in (6.7). Since the submatrix consisting of the first $(n - 1)$ rows of B is of full row rank, the nonzero columns in it (i.e., those associated with the flow variables, and not the slack variables discussed earlier) correspond to arcs in a connected subnetwork which contains a spanning tree, say $\mathbb{T}$, of G (in general, there may be several, choose one and call it $\mathbb{T}$). Once the choice of $\mathbb{T}$ is made, basic variables associated with arcs in $\mathbb{T}$ are called *key basic variables* and the corresponding columns in B are called *key basic columns*; the remaining ρ basic variables are called *nonkey basic variables* and their columns in B are called *nonkey basic columns*. The spanning tree $\mathbb{T}$ itself is called the *key tree* in this step. Notice that the choice of key basic variables in a basic vector for (6.4), (6.5) may not be unique, it depends on the key tree selected in the subnetwork corresponding to basic flow variables. All key basic variables are always flow variables, the slack variables in the additional linear constraints will either be nonkey basic variables or nonbasic variables in every step. With this, B has been partitioned as below.

$$
\begin{array}{cc}
n-1 \text{ key} & \rho \text{ nonkey} \\
\text{basic cols.} & \text{basic cols.}
\end{array}
$$

$$
B = \left(\begin{array}{ccc} H^1 & \vdots & H^2 \\ \cdots & & \cdots \\ A^1 & \vdots & A^2 \end{array} \right) \begin{array}{l} n-1 \text{ rows from (6.4)} \\ \\ \rho \text{ rows from (6.5)} \end{array} \tag{6.7}
$$

Let the key and nonkey basic variables be $(f_{p_1}, \ldots, f_{p_{n-1}})$ and $(f_{p_n}, \ldots, f_{p_{n-1+\rho}})$, arranged in some order. When ordered this way, f_{p_r} is known as the rth *key basic variable*, for $r = 1$ to $n-1$; and f_{p_g+n-1} is known as the gth *nonkey basic variable* for $g = 1$ to ρ. Let $\lambda_{.u} = (\lambda_{1,u}, \ldots, \lambda_{n-1,u})^T$ be 0 if f_{p_u+n-1} is a slack variable, or the in-tree arc-fundamental cycle incidence vector of e_{p_u+n-1} wrt $\mathbb{T}$ otherwise. Then $\lambda = (\lambda_{.1} \ldots \lambda_{.\rho}) = (H^1)^{-1} H^2$ is known as the $\lambda-matrix$ corresponding to the present basis B, and the key, nonkey choice. So, the rth row $\lambda_{r.}$ of λ is associated with the key basic arc e_{p_r}, and its gth column $\lambda_{.g}$ is associated with the nonkey basic variable f_{p_g+n-1}. As discussed above we have

$$
H_{.p_g+n-1} = \lambda_{1,g} H_{.p_1} + \ldots + \lambda_{n-1,g} H_{.p_{n-1}} \tag{6.8}
$$

All entries in the λ- matrix are 0, or ± 1, so it can be stored very compactly (it is only necessary to store the two sets of cells in which the entries are $+1$, -1 respectively).

For carrying out the computations in the simplex algorithm in the step in which B is the present basis, we will make use of the following square upper triangular matrix M of order $(n-1+\rho)$ known as the *transformation matrix*. Here, for any t, I_t is the unit matrix of order t. Also, verify that BM has the form given below.

$$
M = \left(\begin{array}{ccc} I_{n-1} & \vdots & -\lambda \\ \cdots & & \cdots \\ 0 & \vdots & I_\rho \end{array} \right) \tag{6.9}
$$

$$
BM = \left(\begin{array}{ccc} H^1 & \vdots & 0 \\ \cdots & & \cdots \\ A^1 & \vdots & A^2 - A^1\lambda \end{array} \right) \tag{6.10}
$$

The square matrix in the lower right corner of BM, $W = A^2 - A^1\lambda$ of order ρ is called the *working basis* in this step. It depends on the key, nonkey choice. W must be nonsingular, as otherwise the set of last ρ columns in (6.10) will be a linearly dependent set, a contradiction since B and M are both nonsingular. The special implementation of the bounded variable primal simplex method discussed in this chapter for solving (6.4) to (6.6), carries out all the computations in the step when B is the present basis, using the key tree, and the inverse of the working basis W (either the explicit inverse, or the inverse in some convenient product or factorization form). After each pivot step it updates the tree labels for the key tree, the λ-matrix, and the inverse of the working basis, very efficiently.

EXAMPLE 6.1

Tableau 6.1

f_1	f_2	f_3	f_4	f_5	f_6	f_7	f_8	f_9	f_{10}		
0	0	0	−1	−1	0	0	−1	−1	0	=	$-V_1$
0	0	1	1	0	0	−1	0	0	0	=	$-V_2$
−1	0	−1	0	1	1	0	0	0	0	=	$-V_3$
0	−1	0	0	0	−1	1	1	0	0	=	$-V_4$
1	2	3	−1	1	−1	0	2	1	0	=	13
0	1	2	−2	2	0	1	1	2	0	=	7
1	−1	−1	3	−1	1	−1	2	2	1	=	35

$$
B = \begin{pmatrix}
\begin{array}{cccc|ccc}
 & \text{key} & & & \text{nonkey} & & \\
0 & 0 & 0 & -1 & -1 & -1 & 0 \\
0 & 0 & 1 & 1 & 0 & 0 & 0 \\
-1 & 0 & -1 & 0 & 1 & 0 & 0 \\
0 & -1 & 0 & 0 & 0 & 1 & 0 \\
\hline
1 & 2 & 3 & -1 & 1 & 2 & 0 \\
0 & 1 & 2 & -2 & 2 & 1 & 0 \\
1 & -1 & -1 & 3 & -1 & 2 & 1
\end{array}
\end{pmatrix}
= \left(\begin{array}{c|c} H^1 & H^2 \\ \hline A^1 & A^2 \end{array} \right)
$$

We will illustrate the derivation of the working basis in this example. Only data relevant to this illustration are given in this example. Consider the network in Figure 6.1, with arcs numbered e_1 to e_9. f_r denotes the flow amount on arc e_r for $r = 1$ to 9. Node 5 is the root node. There are 3 additional linear constraints on the variables in this problem, two equations; and one inequality, the slack variable associated with which is called f_{10}. The conservation equation corresponding to the root node 5 is omitted as a redundant constraint, the remaining conservation equations, and the additional linear constraints are given in Tableau 6.1.

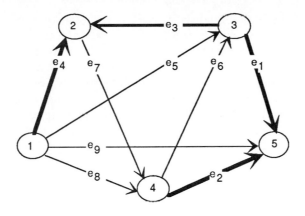

Figure 6.1

Let $(f_1, f_2, f_3, f_4, f_5, f_8, f_{10})$ be the basic vector under consideration for this system. Choose (f_1, f_2, f_3, f_4) as the key basic variables, and (f_5, f_8, f_{10}) as the nonkey basic variables in that order. The associated basis B, partitioned as in (6.7) is given above.

The subnetwork corresponding to the basic flow variables in this basic vector consists of the arcs $e_1, e_2, e_3, e_4, e_5, e_8$. The key tree is marked with thick arcs in Figure 6.1. In this example we have

$$\lambda = (H^1)^{-1}H^2 = \begin{pmatrix} 0 & 1 & 0 \\ 0 & -1 & 0 \\ -1 & -1 & 0 \\ 1 & 1 & 0 \end{pmatrix}$$

$$W = A^2 - A^1\lambda = \begin{pmatrix} 5 & 7 & 0 \\ 6 & 6 & 0 \\ -5 & -4 & 1 \end{pmatrix}, \quad W^{-1} = \begin{pmatrix} -1/2 & 7/12 & 0 \\ 1/2 & -5/12 & 0 \\ -1/2 & 5/4 & 1 \end{pmatrix}$$

W is the working basis in this example, and its inverse, W^{-1}, is given above.

To solve (6.4) to (6.6) the bounded variable primal simplex algorithm deals with partitions of the variables in this problem of the form $(\mathbf{B}, \mathbf{L}, \mathbf{U})$ where $\mathbf{B}$ is the vector of basic variables associated with a basis B; say, $\mathbf{L}$ is the vector of all nonbasic variables whose values are fixed equal to their lower bound in (6.6) in the current basic solution, and $\mathbf{U}$ is the vector of all nonbasic variables for which the capacity is finite, and the values of these variables are fixed equal to their capacity. Given the partition $(\mathbf{B}, \mathbf{L}, \mathbf{U})$, the primal basic solution associated with it is obtained by fixing the values of all the variables in $\mathbf{L}, \mathbf{U}$ at the respective bounds as mentioned above, in (6.4), (6.5); and then solving the remaining system

for the values of the basic variables in **B** (this will lead to a system of the form
(6.12) given below). The partition is a *feasible partition* if the values of all the basic
variables in **B** in the associated basic solution are within their bounds. The primal
simplex algorithm for (6.4) to (6.6) is initiated with a feasible partition (**B**, **L**, **U**)
which can be generated, if the problem is feasible, by applying the same algorithm
on a Phase I problem set up as discussed later on. In the algorithm, after each
pivot step, the new primal BFS is obtained by updating the one from the previous
step. We will now discuss how all the computations in a step of this algorithm can
be carried out efficiently using a key tree selected in the basic subnetwork and the
associated working basis inverse.

How to Carry Out All the Computations in a Pivot Step
Using the Key Tree and the Working Basis Inverse

Let **B**, **L**, **U** be the present feasible partition for the problem, where **B** is the
vector of basic variables associated with the basis B for (6.4), (6.5). Let $\mathbb{T}$ be
the key tree selected in the basic subnetwork as discussed above, resulting in the
partition of the basis B as in (6.7), and the λ- matrix λ and working basis W. To
carry out the computations in this step, we need to solve systems of equations of
the following forms.

$$\pi B \;=\; c_B \tag{6.11}$$

$$By \;=\; d \tag{6.12}$$

c_B is the row vector of the original cost coefficients of the basic variables, and
the solution π of (6.11) is the dual basic solution associated with the basis B. Once
π is obtained, the relative cost coefficient of the nonbasic variable f_t wrt B is

$$\bar{c}_t = c_t - \pi \begin{pmatrix} H_{.t} \\ \cdots \\ A_{.t} \end{pmatrix}$$

The present feasible partition (**B**, **L**, **U**) and the associated BFS are optimal
if $\bar{c}_t \gtreqless 0$ for all t such that f_t is in **L**, and $\bar{c}_t \lesseqgtr 0$ for all t such that f_t is in **U**.
If this optimality criterion is violated, any nonbasic variable violating it is *eligible
to be the entering variable* in this step, one of them is selected as the entering
variable for a pivot step in the present partition. Then, we need to compute its
updated column vector, which is the solution y of (6.12), where d is its original
column vector. This updated column vector is the pivot column for the pivot step.
Using the pivot column, and the values of the basic variables in the current BFS,
the minimum ratio test of the bounded variable primal simplex algorithm is carried
out to determine the dropping variable in this pivot step. If the dropping variable

is the same as the entering variable, it moves from $\mathbf{L}$ or $\mathbf{U}$ where it is contained in the present partition, to the other set. In this case these nonbasic sets are revised appropriately, the primal BFS is updated, and the algorithm moves to the next step with the same basis. Since there is no change in $\mathbf{B}$ in this case, the key tree, working basis inverse, and the dual basic solution all remain unchanged. On the other hand, if the dropping variable is not the same as the entering variable, it will be a basic variable in $\mathbf{B}$, which should be replaced by the entering variable in this pivot step. In this case we need to update the key tree, the λ-matrix, and the working basis inverse, to get the corresponding things for the next partition. We will now discuss efficient procedures for doing all this work.

To Solve (6.11)

Let $\pi^1 = (\pi_1, \ldots, \pi_{n-1})$, $\pi^2 = (\pi_n, \ldots, \pi_{n-1+\rho})$. $\pi = (\pi^1, \pi^2)$ is the vector of dual variables corresponding to (6.4), (6.5) in that order. Let M be the transformation matrix discussed in (6.9). Compute $c_B M = \nu = (\nu_1, \ldots, \nu_{n-1+\rho})$. Let $\nu^1 = (\nu_1, \ldots, \nu_{n-1})$, $\nu^2 = (\nu_n, \ldots, \nu_{n-1+\rho})$. Multiplying both sides of (6.11) by the nonsingular transformation matrix M leads to $\pi BM = c_B M = \nu$. Using (6.10) this reduces to

$$\pi^1 H^1 + \pi^2 A^1 = \nu^1 \tag{6.13}$$
$$\pi^2 W = \nu^2 \tag{6.14}$$

Hence $\pi^2 = \nu^2 W^{-1}$, it can be computed directly using ν^2 and the available W^{-1}. Let $h = \nu^1 - \pi^2 A^1$. Then from (6.13), $\pi^1 H^1 = h$. H^1 is the coefficient matrix of the flow variables on in-tree arcs in $\mathbf{T}$ in the flow conservation equations (6.4). So, π^1 is the vector of node prices of non-root nodes in $\mathbb{T}$, determined as in Section 5.4, with h as the vector of cost coefficients for in-tree arcs in $\mathbf{T}$, and 0 as the node price for the root node. So, π^1 can be determined by back substitution, beginning at the root node n, and going down in increasing order of level in the rooted tree $\mathbf{T}$ as described in Section 5.4.

As an example consider the basis B associated with the key basic vector (f_1, f_2, f_3, f_4) and the nonkey basic vector (f_5, f_8, f_{10}) discussed in Example 6.1. Suppose $c_B = (c_1, c_2, c_3, c_4, c_5, c_8, c_{10}) = (3, -4, 7, 5, 6, -2, 0)$. In this example, we have $\nu = c_B M = (3, -4, 7, 5, 4, 3, 0)$ using the λ derived in Example 6.1. So, $\nu^1 = (3, -4, 7, 5)$ and $\nu^2 = (4, 3, 0)$. Hence $\pi^2 = (\pi_5, \pi_6, \pi_7) = (4, 3, 0)W^{-1} = (-1/2, 13/12, 0)$. So, $h = \nu^1 - \pi^2 A^1 = (7/2, -49/12, 19/3, 20/3)$. We show the key tree with h as the arc cost coefficients in Figure 6.2.

Fixing the price of the root node 5 to 0, the equations to solve to get $\pi^1 = (\pi_1, \pi_2, \pi_3, \pi_4)$ are:

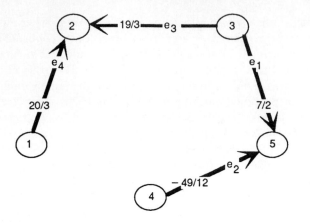

Figure 6.2

$$0 - \pi_4 = -49/12$$
$$0 - \pi_3 = 7/2$$
$$\pi_2 - \pi_3 = 19/3$$
$$\pi_2 - \pi_1 = 20/3$$

yielding the solution $\pi^1 = (-23/6, 17/6, -7/2, 49/12)$. So, the solution for (6.11) in this example is $\pi = (-23/6, 17/6, -7/2, 49/12, -1/2, 13/12, 0)$.

To Solve (6.12)

Define $\xi = (\xi_1, \ldots, \xi_{n-1+\rho})^T$ as a column vector of new variables, with $\xi^1 = (\xi_1, \ldots, \xi_{n-1})^T$, $\xi^2 = (\xi_n, \ldots, \xi_{n-1+\rho})^T$. Define $d^1 = (d_1, \ldots, d_{n-1})^T$, $d^2 = (d_n, \ldots, d_{n-1+\rho})^T$; (d^1, d^2) is a partition of d corresponding to the partition (ξ^1, ξ^2) of ξ. We will first solve the system of equations $(BM)\xi = d$, i.e.,

$$H^1 \xi^1 = d^1 \qquad\qquad (6.15)$$

$$W \xi^2 = d^2 - A^1 \xi^1 \qquad\qquad (6.16)$$

H^1 is the coefficient matrix of the flow variables on the in-tree arcs in T in the flow conservation equations (6.4). Hence, the solution ξ^1 for (6.15) is the vector of flows on in-tree arcs in T to satisfy exogenous flow amounts of $p_i = d_i$ at non-root nodes $i = 1$ to $n - 1$, and $-(d_1 + \ldots + d_{n-1})$ at the root node n. Since H^1 is a triangular matrix, the solution ξ^1 of (6.15) can be obtained by back substitution as discussed in Section 5.4, in $(n - 1)$ steps. In each step we find an in-tree arc e

incident to a non-root terminal node i of the remaining tree at this stage. Make the flow amount on e equal to $+p'_i$ or $-p'_i$ depending on whether i is the tail or head node of e, where p'_i is the present updated exogenous flow amount at node i. If j is the other node on e, update the exogenous flow amount of j by adding (subtracting) the flow amount on e to (from) it depending on whether e is directed into (or out of) j. Now delete e from the tree and go to the next step if any in-tree arcs remain.

Once the solution ξ^1 for (6.15) is obtained, we obtain ξ^2 from (6.16) to be $W^{-1}(d^2 - A^1\xi^1)$, which is computed using the available W^{-1}.

Once ξ^1, ξ^2 are both determined, we compute the solution y for (6.12) from $y = M\xi$.

As an example, consider the basis B associated with the key basic vector (f_1, f_2, f_3, f_4) and the nonkey basic vector (f_5, f_8, f_{10}) discussed in Example 6.1. Suppose we need to solve (6.12) with $d = (0, -1, 0, 1, 0, 1, -1)^T$. Here $d^1 = (0, -1, 0, 1)^T$ and $d^2 = (0, 1, -1)^T$. We show the key tree with exogenous flow amounts of d^1 at the non-root nodes, and exogenous flow amount of $-(d_1 + d_2 + d_3 + d_4) = 0$ at root node 5 in Figure 6.3.

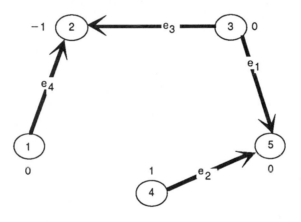

Figure 6.3

Following the procedure discussed above, we first select the non-root terminal node 1 and make $\xi_4 = 0$, and delete e_4. Next, we select node 2, make $\xi_3 = 1$, delete e_3, and revise the exogenous flow amount at node 3 to -1. Next we select node 3 and make $\xi_1 = -1$, delete e_1, and revise the exogenous flow amount at the root node 5 to -1. Next we select node 4, and make $\xi_2 = 1$. So, in this example, the solution for (6.15) is $\xi^1 = (-1, 1, 1, 0)^T$

We have $d^2 - A^1\xi^1 = (4, 4, -4)^T$. So, $\xi^2 = W^{-1}(4, 4, -4)^T = (1/3, 1/3, -1)^T$.

Hence, the solution for (6.15), (6.16) in this example is $\xi = (1, -1, -1, 0, 1/3, 1/3, -1)^T$. Therefore, the solution for (6.12) in this example is $y = M\xi = (2/3, -2/3, -1/3, -2/3, 1/3, 1/3, -1)^T$.

Updating the Key Tree, λ-Matrix, and the Working Basis Inverse in a Pivot Step

Let B be the basis associated with the basic vector $(f_{p_1}, \ldots, f_{p_{n-1+\rho}})$ at the beginning of a pivot step, with $(f_{p_1}, \ldots, f_{p_{n-1}})$ as the key basic vector, the key tree $\mathbb{T}$, the λ-matrix λ, and the working basis W. Suppose the key, nonkey partition of B, is the one given in (6.7). Let f_s be the entering variable into this basic vector in this pivot step, with $(H_{.s}, A_{.s})$ as its original column vector in (6.4), (6.5). For updating the key tree, the λ-matrix, and the working basis inverse in this pivot step, we consider three cases depending on which present basic variable is the dropping variable.

Case 1 : The Dropping Variable Is a Nonkey Basic Variable

Let the dropping variable be the gth nonkey basic variable, $f_{p_{g+n-1}}$. The entering variable f_s replaces $f_{p_{g+n-1}}$ as the gth nonkey basic variable in this pivot step. By definition, the present $W_{.g} = A_{.p_{g+n-1}} - A^1 \lambda_{.g}$. Compute $\delta = A_{.s} - A^1(H^1)^{-1} H_{.s}$. Let $\bar{\delta} = W^{-1}\delta$.

To update the λ-matrix in this case, replace $\lambda_{.g}$ by $(H^1)^{-1} H_{.s}$. To update the working basis, replace its gth column by δ. To update the working basis inverse, put the column $\bar{\delta}$ by the side of the present W^{-1}, and perform a pivot step with the gth row as the pivot row and $\bar{\delta}$ as the pivot column. This pivot step updates W^{-1} into the new working basis inverse. Because there is no change in the key basic variables, the key tree remains unchanged.

As an example, consider the basic vector $(f_1, f_2, f_3, \ f_4, f_5, f_8, \ f_{10})$ with (f_1, f_2, f_3, f_4) as the key part, for the problem discussed in Example 6.1. Suppose f_7 is the entering variable into this basic vector, replacing the second nonkey basic variable, f_8, from it. The column vector δ defined above, is $A_{.7} - A^1(H^1)^{-1} H_{.7} = (4, 4, -4)^T$ in this example, and $\bar{\delta} = W^{-1}\delta = (1/3, 1/3, -1)^T$. The new working basis is obtained by replacing $W_{.2}$ in the present W by δ. So, it is

$$\text{New working basis} = \begin{pmatrix} 5 & 4 & 0 \\ 6 & 4 & 0 \\ -5 & -4 & 1 \end{pmatrix}$$

For updating the working basis inverse, the pivot column $\bar{\delta}$ is entered on the right-hand side following the present working basis inverse in the tableau given below. The pivot row is row 2, and the pivot element is in a box. So, the new working basis inverse is the matrix on the left-hand side bottom of the tableau given below. To update λ, replace $\lambda_{.2}$ from the present λ by $(H^1)^{-1}h_{.7} = (1, -1, -1, 0)^T$.

Working basis inverse			Pivot col. $\overline{\delta}$
$-1/2$	$7/12$	0	$1/3$
$1/2$	$-5/12$	0	$\boxed{1/3}$
$-1/2$	$5/4$	1	-1
-1	1	0	0
$3/2$	$-5/4$	0	1
1	0	1	0

The new λ-matrix is

$$\begin{pmatrix} 0 & 1 & 0 \\ 0 & -1 & 0 \\ -1 & -1 & 0 \\ 1 & 0 & 0 \end{pmatrix}$$

Case 2 : The Dropping Variable Is an Essential Key Basic Variable

A key basic variable is said to be *essential* if none of the present nonkey basic variables can replace it as a key basic variable in the present basic vector, *inessential* otherwise.

Consider the uth key basic variable f_{p_u} associated with the in-tree arc e_{p_u} in the key tree $\mathbb{T}$. Let $[\mathbf{X}, \overline{\mathbf{X}}]$ be the fundamental cutset corresponding to e_{p_u} in $\mathbb{T}$. e_{p-u} can be replaced by any of the out-of-tree arcs in the cutset $[\mathbf{X}, \overline{\mathbf{X}}]$ to yield another spanning tree in G. So, e_{p_u} is an essential key basic arc (and f_{p_u} is an essential key basic variable) iff none of the present nonkey basic variables $f_{p_n}, \dots, f_{p_{n-1+\rho}}$ is the flow variable on an arc in the cutset $[\mathbf{X}, \overline{\mathbf{X}}]$, which happens iff $\lambda_{u.} = 0$.

Here we consider the case where the dropping variable in this pivot step is an essential key basic variable, say f_{p_u}. So in the present λ-matrix, $\lambda_{u.} = 0$. The entering variable f_s must therefore be a flow variable corresponding to an arc in the fundamental cutset of e_{p_u}. Since $\lambda_{u.} = 0$, this change in the basic vector leaves the λ-matrix unchanged. The working basis remains unchanged and so does its inverse. So the only updating to be done in this case is to update the tree labels for replacing the in-tree arc e_{p_u} by e_s, which is carried out as described in Section 5.4.

As an example, consider basic vector $(f_1, f_2, f_3, f_4, f_5, f_9, f_{10})$ for the problem given in Example 6.1, with (f_1, f_2, f_3, f_4) as the key basic variables, and (f_5, f_9, f_{10}) as the nonkey basic variables. The λ-matrix in this case is

$$\begin{pmatrix} 0 & 1 & 0 \\ 0 & 0 & 0 \\ -1 & -1 & 0 \\ 1 & 1 & 0 \end{pmatrix}$$

The key tree is marked with thick lines in Figure 6.1. Suppose f_8 is the entering variable into this basic vector, and the second key basic variable f_2 is the dropping variable. Since $\lambda_{2.} = 0$ in this case, f_2 is an essential key basic variable. f_8 replaces f_2 as the second key basic variable in this pivot step, but there is no change in λ or W.

Case 3 : The Dropping Variable Is an Inessential Key Basic Variable

Here we consider the case where the dropping variable is an inessential key basic variable f_{p_r}. So, $\lambda_{r.} \neq 0$. Let $[\mathbf{X}, \overline{\mathbf{X}}]$ be the fundamental cutset of the in-tree arc e_{p_r} in the key tree $\mathbb{T}$. In this case the updating will be done in two stages.

In Stage 1, a nonkey basic variable which is a flow variable for an arc in the cutset $[\mathbf{X}, \overline{\mathbf{X}}]$ is selected to replace f_{p_r} as the rth key basic variable. The gth nonkey basic variable $f_{p_{g+n-1}}$ is eligible to be selected for this if $\lambda_{rg} \neq 0$. Suppose the nonkey basic variable $f_{p_{u+n-1}}$ has been selected for this. This operation just rearranges the basic vector as $(f_{p_1}, \ldots, f_{p_{r-1}}, f_{p_{u+n-1}}, f_{p_{r+1}}, \ldots, f_{p_{u-2+n}}, f_{p_r}, f_{p_{u+n}}, \ldots, f_{p_{\rho+n-1}})$. In this order the vector corresponds to the basis $\hat{B}$ that is obtained from B by interchanging its rth and $(u+n-1)$th column vectors. Even though $\hat{B}$ is just the same as B except for the rearrangement of two of its columns, since the key basic variables are different, there will be a change in the key tree, the λ-matrix and the working basis.

In the original basis B, each nonkey column in H^2 can be expressed as a linear combination of the key columns in H^1, with the coefficients coming from the λ-matrix. This relationship is expressed in the first tableau given below. The left hand part of this tableau is $-\lambda^T$, followed on the right by the unit matrix of order ρ. To get the λ-matrix corresponding to the new key basic vector, we need to perform a pivot step in this tableau with the column vector under $H_{.P_r}$ as the pivot column, and the uth row as the pivot row. This leads to the second tableau at the bottom. The matrix under the columns headed with $H_{.p_1}, \ldots, H_{.p_{r-1}}, H_{.p_{u+n-1}}, H_{.p_{r+1}}, \ldots, H_{.p_{n-1}}$, in that order in this second tableau is $-(\hat{\lambda})^T$ where $\hat{\lambda}$ is the new λ-matrix. Since all the $\lambda_{ij}, \hat{\lambda}_{ij}$ are 0, ± 1, this pivot step can be carried out very efficiently.

Let $W_{.g}, \hat{W}_{.g}$ denote the gth column of the original and new working bases respectively, for $1 \leqq g \leqq \rho$. Using the formulas for $\hat{\lambda}$ in terms of λ, and the definition of the working bases, we get the formula for $\hat{W}_{.g}$ given below.

$H_{.p_1}$	$\cdots$	$H_{.p_r}$	$\cdots$	$H_{.p_{n-1}}$	$H_{.p_n}$	$\cdots$	$H_{.p_{u+n-1}}$	$\cdots$	$H_{.p_{\rho+n-1}}$	
$-\lambda_{11}$	$\cdots$	$-\lambda_{r1}$	$\cdots$	$-\lambda_{n-1,1}$	1	$\cdots$	0	$\cdots$	0	0
$\vdots$		$\vdots$		$\vdots$	$\vdots$		$\vdots$		$\vdots$	$\vdots$
$-\lambda_{1u}$	$\cdots$	$\boxed{-\lambda_{ru}}$	$\cdots$	$-\lambda_{n-1,u}$	0	$\cdots$	1	$\cdots$	0	0
$\vdots$		$\vdots$		$\vdots$	$\vdots$		$\vdots$		$\vdots$	$\vdots$
$-\lambda_{1\rho}$	$\cdots$	$-\lambda_{r\rho}$	$\cdots$	$-\lambda_{n-1,\rho}$	0	$\cdots$	0	$\cdots$	1	0
$-\hat\lambda_{11}$	$\cdots$	0	$\cdots$	$-\hat\lambda_{n-1,1}$	1	$\cdots$	$-\hat\lambda_{r1}$	$\cdots$	0	0
$\vdots$		$\vdots$		$\vdots$	$\vdots$		$\vdots$		$\vdots$	$\vdots$
$-\hat\lambda_{1u}$	$\cdots$	1	$\cdots$	$-\hat\lambda_{n-1,u}$	0	$\cdots$	$-\hat\lambda_{ru}$	$\cdots$	0	0
$\vdots$		$\vdots$		$\vdots$	$\vdots$		$\vdots$		$\vdots$	$\vdots$
$-\hat\lambda_{1\rho}$	$\cdots$	0	$\cdots$	$-\hat\lambda_{n-1,\rho}$	0	$\cdots$	$-\hat\lambda_{r\rho}$	$\cdots$	1	0

$$\hat{W}_{.g} = \begin{cases} W_{.g} - (\lambda_{rg}/\lambda_{ru})W_{.u}, & \text{for } g \neq u \\[2mm] (-1/\lambda_{ru})W_{.u}, & \text{for } g = u \end{cases}$$

From this it can be verified that $(\hat{W})^{-1} = Q^{-1}W^{-1}$, where Q^{-1} is the elementary matrix of order $\rho \times \rho$ that differs from the unit matrix of order ρ in just its uth row, given below. Notice that the λs in Q^{-1} are those from the λ-matrix corresponding to the original partition of the basis as in B. Hence, updating the inverse of the working basis in this Stage 1 consists of multiplying the present working basis inverse on the left by Q^{-1}.

$$Q^{-1} = \begin{pmatrix} 1 & 0 & \cdots & 0 & 0 & 0 & \cdots & 0 \\ 0 & 1 & \cdots & 0 & 0 & 0 & \cdots & 0 \\ \vdots & \vdots & \ddots & \vdots & \vdots & \vdots & \ddots & \vdots \\ 0 & 0 & \cdots & 1 & 0 & 0 & \cdots & 0 \\ -\lambda_{r1} & -\lambda_{r2} & \cdots & -\lambda_{r,u-1} & -\lambda_{ru} & -\lambda_{r,u+1} & \cdots & -\lambda_{r\rho} \\ 0 & 0 & \cdots & 0 & 0 & 1 & \cdots & 0 \\ \vdots & \vdots & \ddots & \vdots & \vdots & \vdots & \ddots & \vdots \\ 0 & 0 & \cdots & 0 & 0 & 0 & \cdots & 1 \end{pmatrix} \quad \text{row } u$$

$$\text{col. } u$$

As an example for this Stage 1, consider the basis for the problem in Example 6.1 with the key basic vector (f_1, f_2, f_3, f_4), and the nonkey basic vector (f_5, f_8, f_{10}). Suppose we have to replace the third key basic variable f_3 by a nonkey basic variable. In the above notation, $r = 3$ (since f_3 is the third key basic variable here), and we verify that the first two entries in $\lambda_{3.}$ are nonzero. Either the first or the second nonkey basic variable can replace f_3 as a key basic variable, suppose we

to significant gains in computational efficiency and reduction in memory requirements. It is useful for handling problems in which the number of additional linear constraints is small (up to a few hundreds). It is based on the GUB techniques of Dantzig and Van Slyke [1967]. Other methods for handling additional linear constraints in network models are discussed by Barr, Farhangian, and Kennington [1986], and McBride [1985].

If the number of additional linear constraints is itself large, this implementation may not offer any particular advantage for solving the problem. In this case one may consider solving the overall problem as an LP using some of the recently developed interior point methods.

6.2 References

R. BARR, K. FARHANGIAN, and J. KENNINGTON, 1986, "Networks with Side Constraints: An LU Factorization Update," *The Annals of the Society of Logistics Engineers* , 1 (66-85).

S. CHEN and R. SAIGAL, 1977, "A Primal Algorithm for Solving a Capacitated Network Flow Problem with Additional Linear Constraints," *Networks*, 7 (59-79).

W. CHOI, H. W. HAMACHER, and S. TUFEKCI, 1988, "Modeling of Building Evacuation Problems by Network Flows with Side Constraints," *EJOR*, 35 (98-110).

G. B. DANTZIG and R. M. VAN SLYKE, Oct. 1967, "Generalized Upper Bounding Techniques," *Journal of Computer and System Sciences*, 1, no. 3 (213-226).

F. GLOVER, D. KARNEY, D. KLINGMAN, and R. RUSSELL, Nov. 1978, "Solving Singly Constrained Transshipment Problems," *TS*, 12, no. 4 (277-297).

R. D. McBRIDE, July 1985, "Solving Embedded Generalized Network Problems," *EJOR*, 21, no. 1 (82-92).

Chapter 7

Critical Path Methods in Project Networks

Critical path methods (CPM in abbreviation) deal with the application of shortest chain and minimum cost flow algorithms to schedule the jobs in a project along the time axis. Large civil engineering projects (construction of a skyscraper, a highway, etc.); projects that involve the manufacture of large items like ships, generators; projects that involve the development, planning, and launching of new products; large scale research and development projects; etc.; consist of a collection of many individual *jobs* or *activities* with a *partial ordering* defined among them, which arises from technological constraints that require certain jobs to be completed before others can be started (for example, the job "painting the walls" can only be started after the job "erecting the walls" is completed). The words *job, activity* are used synonymously in this chapter. We assume that each job in the project can be started and completed independently of the others within the technological sequence defined by the partial ordering among the jobs. The partial ordering among the jobs defines a directed network known as the *project network*. The simplest CPM derives the project duration and a schedule for the various jobs to achieve this duration, given the project network and the time needed to complete each job. Another critical path model takes as input the cost of applying more workers or other resources to shorten the job durations, and determines which jobs need to be expedited to achieve a specified project duration at minimum cost. In this sense, CPM are concerned with obtaining the trade-off between cost and duration of the project. CPM are most useful in projects such as construction projects, for which there has been considerable experience, and a data base is available to derive reliable cost estimates. CPM are among the most commonly used optimization techniques. Software based on CPM for project planning, analysis, scheduling, and control, is one of the biggest money-makers among all OR software.

If job 2 cannot be started until after job 1 has been completed, then job 1 is known as a *predecessor* or *ancestor* of job 2; and job 2 is known as a *successor* or *descendent* of job 1. If job 1 is a predecessor of job 2, and there is no other job which is a successor of job 1 and predecessor of job 2, then job 1 is known as an *immediate predecessor* of job 2, and job 2 is known as an *immediate successor* of job 1. A job may have several immediate predecessors, it can be started as soon as all its immediate predecessors have been completed. If a job has two or more immediate predecessors, by definition every pair of them must be unrelated in the sense that neither of them is a predecessor of the other.

If 1 is a predecessor of 2, and 2 is a predecessor of 3, then obviously 1 is a predecessor of 3. This property of precedence relationships is called *transitivity*. Given the set of immediate predecessors of each job, it is possible to determine the set of predecessors, or the set of successors of any job, by recursive procedures using transitivity. The predecessor relationships are inconsistent if they require that a job has to be completed before it can be started, so, no job can be a predecessor of itself.

Because of these properties, the precedence relationships define an ordering among the jobs in a project called a *partial ordering* in mathematics. The planning phase of the project involves the breaking up of the project into various jobs using practical considerations, identifying the immediate predecessors of each job based on engineering and technological considerations, and estimating the time required to complete each job.

Inconsistencies may appear in the predecessor lists due to human error. The predecessor data is said to be *inconsistent* if it leads to the conclusion that a job precedes itself, by the transitivity property. Inconsistency implies the existence of a circuit in the predecessor data, i.e., a subset of jobs $1, \ldots, r$, such that j is listed as a predecessor of $j + 1$ for $j = 1$ to $r - 1$, and r is listed as a predecessor of 1. Such a circuit represents a logical error and at least one link in this circuit must be wrong. As it represents a logical error, inconsistency is a serious problem.

Also, in the process of generating the immediate predecessors for an activity, an engineer may put down more than necessary and show as immediate predecessors some jobs that are in reality more distant predecessors. When this happens, the predecessor data is said to contain *redundancy*. Redundancy poses no theoretical or logical problems, but it unnecessarily increases the complexity of the network used to represent the predecessor relationships. Given the list of immediate predecessors of each job, one must always check it for any inconsistency, and redundancy, and make appropriate corrections.

As an example, we give below the precedence relationships among jobs in the project: *building a hydroelectric power station*. In this example, we have not gone into very fine detail in breaking up the project into jobs. In practice, a job like 11 (dam building) will itself be divided into many individual jobs involved in dam building. The job duration is the estimated number of months needed to complete the job. This is followed by a discussion of two different ways of representing the precedence relationships among the jobs in a project as a directed network.

Hydroelectric Power Station Building Project

No.	Job Description	Immediate Predecessors	Job Duration
1.	Ecological survey of dam site		6.2
2.	File environmental impact report and get EPA approval	1	9.1
3.	Economic feasibility study	1	7.3
4.	Preliminary design and cost estimation	3	4.2
5.	Project approval and commitment of funds	2, 4	10.2
6.	Call quotations for electrical equipment (turbines, generators, ...)	5	4.3
7.	Select suppliers for electrical equipment	6	3.1
8.	Final design of project	5	6.5
9.	Select construction contractors	5	2.7
10.	Arrange construction materials supply	8, 9	5.2
11.	Dam building	10	24.8
12.	Power station building	10	18.4
13.	Power lines erection	7, 8	20.3
14.	Electrical equipment installation	7, 12	6.8
15.	Build up reservoir water level	11	2.1
16.	Commission the generators	14, 15	1.2
17.	Start supplying power	13, 16	1.1

Activity on Node (AON) Diagram of the Project

As the name implies, each job is represented by a node in this network. Let node i represent job i, $i = 1$ to $n =$ number of jobs. Include arc (i, j) in the network iff job i is an immediate predecessor of job j. The resulting directed network called the *Activity on Node (AON) diagram*, is very simple to draw, but not too convenient for project scheduling, so we will not use it in the sequel. The AON diagram of the hydroelectric power station building project is given in Figure 7.1.

Arrow Diagram of the Project

The *Arrow diagram* or the *Activity on Arc (AOA) diagram* represents jobs by arcs in the network. We refer to the job corresponding to arc (i, j) in this network, as job (i, j) itself. Nodes in the arrow diagram represent *events* over time. Node i represents the event that all jobs corresponding to arcs incident into node i have been completed, and after this event any job corresponding to an arc incident out

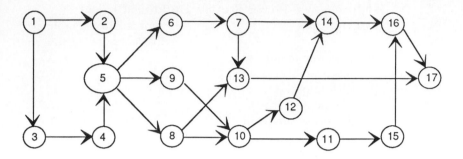

Figure 7.1 AON diagram for the hydroelectric power station building project.

of node i can be started. The arrow diagram is drawn so as to satisfy the following property.

Property 1 If (i,j), (p,q) are two jobs, job (i,j) is a predecessor of job (p,q) iff there is a chain from node j to node p in the arrow diagram.

In order to represent the predecessor relationships through Property 1, it may be necessary to introduce *dummy arcs* which correspond to *dummy jobs*. The need for dummy jobs is explained with illustrative examples later on. In drawing the arrow diagram, the following Property 2 must also be satisfied.

Property 2 If (i,j), (p,q) are two jobs, job (i,j) is an immediate predecessor of job (p,q) iff either $j = p$, or there exists a chain from node j to node p in the arrow diagram consisting of dummy arcs only.

In drawing the arrow diagram, we start with an initial node called the *start node* representing the event of starting the project, and represent each job that has no predecessor, by an arc incident out of it. In the same way, at the end we represent jobs that have no successors by arcs incident into a single final node called the *finish node* representing the event of the completion of the project.

A dummy job is needed whenever the project contains a subset $\mathcal{A}_1$ of two or more jobs which have some, but not all, of their immediate predecessors in common. In this case we let the arcs corresponding to common immediate predecessors of jobs in $\mathcal{A}_1$ to have the same head node and then add dummy arcs from that node to the tail node of each of the arcs corresponding to jobs in $\mathcal{A}_1$. As an example consider the following project, the arrow diagram corresponding to which is given in Figure 7.2.

Job	Immediate predecessors
e_1	
e_2	
e_3	
e_4	e_1, e_2
e_5	e_3, e_2

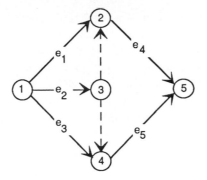

Figure 7.2 Arrow diagram. Dashed arcs represent dummy jobs.

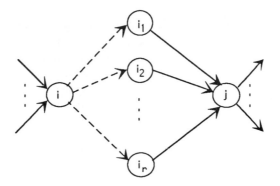

Figure 7.3 Representing jobs with identical sets of immediate predecessors and immediate successors. Arc (i_h, j) represents job h, for $h = 1$ to r. The dashed arcs represent dummy jobs.

Suppose there are $r \ (\geqq 2)$ jobs, say $1, \ldots, r$, all of which have the same set $\mathcal{A}_1$ of immediate predecessors and the same set $\mathcal{A}_2$ of immediate successors; and there are no other immediate successors for any of the jobs in $\mathcal{A}_1$, or immediate predecessors for any of the jobs in $\mathcal{A}_2$. Then, all jobs in the set $\mathcal{A}_1$ can be represented by arcs incident into a common node, i, say; and all jobs in the set $\mathcal{A}_2$ can be represented by arcs incident out of a common node j, say. Then the jobs $1, \ldots, r$, can be represented by r parallel arcs joining nodes i, j. However project engineers do not usually like to deal with parallel arcs, so we introduce additional nodes $i_1, \ldots, i_r$ and represent job h by the arc $(i_h, j), h = 1$ to r; and include dummy arcs (i, i_h) for each $h = 1$ to r. See Figure 7.3.

If a job b has a single immediate predecessor a, then b can be represented by an arc incident out of the head node of the arc representing a.

If job b has more than one immediate predecessor, let $p_1, \ldots, p_r$ be the head nodes of all the arcs representing its immediate predecessors. If no other job has the same set of immediate predecessors, see if it is possible to represent b by an arc incident out of one of the nodes $p_1, \ldots, p_r$ with dummy arcs emanating from the other nodes in this set into that node. If this is not possible, or if there are other jobs which have identically the same set of immediate predecessors as b, introduce a new node q and represent b and each of these jobs by an arc incident out of q, and include dummy arcs $(p_1, q), \ldots, (p_r, q)$.

If some jobs have identical sets of immediate successors, make the head node of the arcs representing these jobs the same.

We continue this way, at each stage identifying the common immediate predecessors of two or more jobs, and representing these immediate predecessors by arcs with the same head node, and letting dummy arcs issue out of this node if necessary. In introducing dummy arcs, one should always watch out to see that precedence relationships not implied by the original data are not introduced, and those in the original specification are not omitted.

After the arrow diagram is completed this way, one can review and see whether any of the dummy arcs can be deleted by merging the two nodes on it into a single node, while still representing the predecessor relationships correctly. For example, if there is a node with a single arc incident out of it, or a single arc incident into it, and this arc is a dummy arc, then the two nodes on that dummy arc can be merged and that dummy arc eliminated. Other simple rules like these can be developed and used to remove unnecessary dummy arcs.

In this way it is possible to draw an arrow diagram for a project using simple heuristic rules. There are usually many different ways of selecting the nodes and dummy arcs for drawing the arrow diagram to portray the specified precedence relationships through Properties 1,2. Any of these that leads to an arrow diagram satisfying Properties 1,2 correctly and completely is suitable for project planning and scheduling computations. For example, a procedure is described in Exercise 7. 2 for converting the AON diagram into an arrow diagram. However, the resulting arrow diagram has too many nodes and dummy arcs and hence it is not efficient to use it. One would prefer an arrow diagram with as few nodes and dummy arcs as possible. But the problem of constructing an arrow diagram with the minimum number of dummy arcs is in general a hard problem (see Krishnamoorthy and Deo [1979], and Exercise 7.11). In practice, it is not very critical whether the number of dummy arcs is the smallest that it can be or not. Any arrow diagram obtained using the simple rules discussed above is quite reasonable and satisfactory.

As an example, the arrow diagram for the hydroelectric power plant building project discussed above is given in Figure 7.4.

Since dummy arcs have been introduced just to represent the predecessor relationships through Properties 1,2, they correspond to dummy jobs, and the time and cost required to complete any dummy job are always taken to be 0.

The transitive character of the precedence relationships, and the fact no job can precede itself, imply that an arrow diagram cannot contain any circuits (i.e.,

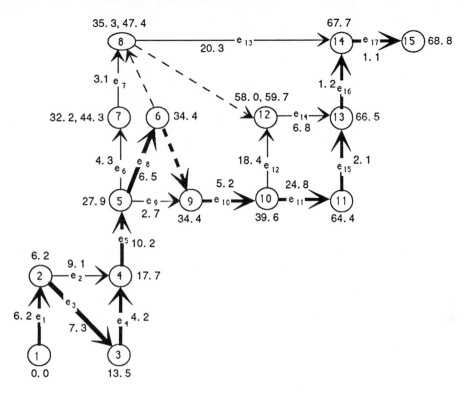

Figure 7.4   Arrow diagram for the hydroelectric power plant building problem. Numbers by the side of job arcs are job durations. The critical path defined in Section 7.1 is marked with thick arcs. If earliest and latest occurrence times for a node are the same, that value is entered by the side of the node, otherwise they are both entered in this order.

it is acyclic). By the results in Chapter 1, an acyclic numbering of nodes in the arrow diagram is possible, i.e., a numbering such that if (i,j) is an arc in the network, then $i < j$. In the sequel we assume that the nodes in the arrow diagram are numbered this way.

Exercises

7.1 Given the predecessor data for a project, develop efficient procedures for checking the data for consistency and for removing redundancies in the specified immediate predecessor lists if the data is consistent.

7.2 Let G be the AON diagram for a project. Replace each node i in G by an arc of the form (i', i''). Let G' be the resulting network. In G' let the arc (i', i'')

represent the same job that node i represented in G. Also, let all arcs in G' which correspond to arcs in G be dummy arcs. Show that G' is an arrow diagram for the project.

7.3 Write a practically efficient computer program to derive an arrow diagram for a project, given the list of immediate predecessors of each job. Include in your program simple rules to try to keep the number of nodes and the number of dummy arcs as small as possible.

7.1   Project Scheduling

Let $G = (\mathcal{N}, \mathcal{A})$, with $|\mathcal{N}| = n$, be the arrow diagram for a project with an acyclic numbering for its nodes, and nodes 1, n as the start, finish nodes, respectively. Given the job durations, project scheduling deals with the problem of laying out the jobs along the time axis with the aim of minimizing the project duration. It is concerned with temporal considerations such as (1) how early would the event corresponding to each node materialize, (2) how far can an activity be delayed without causing a delay in project completion time, etc. For $(i, j) \in \mathcal{A}$ let $t_{ij} \geqq 0$ be the time duration required for completing job (i, j) ($t_{ij} = 0$ if (i, j) is a dummy arc), make t_{ij} the length of arc (i, j) in G. The minimum time needed to complete the project, known as the *minimum project duration*, is obviously the length of the longest chain from 1 to n in G, a longest chain like that is known as a *critical path* in the arrow diagram. There may be alternate critical paths in G. Any arc which lies on a critical path is called a *critical arc*, it represents a *critical job* or *critical activity*. Jobs which are not on any critical path are known as *slack jobs* in the arrow diagram.

For each $i \in \mathcal{N}$ let t_i denote the length of a longest chain from start node 1 to node i in G. t_n, the length of a critical path in G, is the minimum time duration required to complete the project. The quantity t_i is the earliest occurrence time of the event associated with node i assuming that the project has commenced at time 0. For each arc (i, j) incident out of node i, t_i is the earliest point of time at which job (i, j) can be started after the project has commenced, hence it is known as the *early start time of job* (i, j) and denoted by $\mathrm{ES}(i, j)$. For all arcs (i, j) incident out of node i, $\mathrm{ES}(i, j)$ is the same, and $t_i + t_{ij}$ is the earliest point of time that job (i, j) can be completed. This time is known as the *early finish time of job* (i, j), and denoted by $\mathrm{EF}(i, j)$.

Since G is acyclic, the t_is can be computed by applying the algorithm discussed in Section 4.4, with appropriate modifications to find the longest instead of the shortest chain, on G, this process is called *the forward pass* through the arrow diagram. Once the forward pass has been completed, one schedule that gets the project completed in minimum time is to start each job at its early start time. If the duration of any critical job increases by ϵ while all the other data remain unchanged, the project duration also increases by ϵ. If it is required to complete the project in less than t_n units of time, it is necessary to reduce the time required to complete at least one job on every critical path. For this it is helpful if all the critical arcs can be identified. The forward pass identifies only one critical path, it does not identify all the critical arcs.

Slack jobs can be delayed to a limited extent without causing any delay in the whole project. It is interesting to know how late the starting and completion of a job (i, j) can be delayed without affecting the project completion time. We define the *late start time of job* (i, j) to be the latest time that this job can be started without affecting the project completion in minimum time and denote it by LS(i, j). The *late finish time of job* (i, j), denoted by LF(i, j) is LS$(i, j) + t_{ij}$. To compute the late finish times, we begin at the finish node at time point t_n and work backwards, this process is known as the *backward pass* through the arrow diagram. An arc (i, j) is a critical arc iff ES$(i, j) = $ LS(i, j). Hence when both forward and backward passes have been completed, all the critical and slack arcs in the arrow diagram can be identified easily. The combined algorithm comprising the forward and backward passes is described below. In these passes t_{ij} are given data. In the forward pass, node i acquires the *forward label* (L_i, t_i) where t_i is the quantity defined above, it is the earliest event time associated with node i ; and L_i is the predecessor of node i on a longest chain from 1 to i. In the forward pass nodes are labeled in serial order from 1 to n. In the backward pass node i acquires the *backward label* denoted by μ_i; it is the latest event time associated with node i so that the project completion will still occur in minimum time. In the backward pass, nodes are labeled in decreasing serial order beginning with node n.

FORWARD PASS

Step 1 Label the start node, node 1, with the forward label $(\emptyset, 0)$.

General step r , $r = 2$ to n At this stage, all the nodes $1, \ldots, r - 1$ would have been forward labeled, let these forward labels be (L_i, t_i) on node $i = 1$ to $r - 1$. Find

$$t_r = \text{Maximum } \{t_i + t_{ir} : i \in \mathbf{B}_r\} \tag{7.1}$$

where $\mathbf{B}_r$ is the set of tail nodes on arcs incident into node r. Let L_r be any of the i that attains the maximum in (7.1). Label node r with the forward label (L_r, t_r). If $r = n$ go to the backward pass, otherwise go to the next step in the forward pass.

BACKWARD PASS

Step 1 Label the finish node, node n, with $\mu_n = t_n$.

General Step r , $r = 2$ to n At this stage all the nodes $n, n-1, \ldots, n-r+2$ would have received backward labels, let these be $\mu_n, \ldots, \mu_{n-r+2}$, respectively. Find

$$\mu_{n-r+1} = \text{Minimum } \{\mu_j - t_{n-r+1,j} : j \in \mathbf{A}_{n-r+1}\} \qquad (7.2)$$

where $\mathbf{A}_{n-r+1}$ is the set of head nodes on arcs incident out of $n - r + 1$. If $r = n$ terminate; otherwise go to the next step in the backward pass.

Discussion

The fact that t_i in the forward label on node i is the length of the longest chain from node 1 to i follows from the results in Section 4.4.

We will now show that the backward label, μ_i, on node i is the latest point of time at which the event associated with node i has to occur if the project is to be completed in minimum time. This is clearly true for $i = n$. Set up an induction hypothesis that this statement is true for $i \geqq n - r + 2$ for some r between 2 and n. Suppose the minimum in (7.2) is attained by $j = p$. So, $(n - r + 1, p) \in \mathcal{A}$ (this implies that $p \geqq n - r + 2$), and μ_{n-r+1} defined in (7.2) satisfies $\mu_{n-r+1} + t_{n-r+1,p} = \mu_p$. Thus, if the event associated with node $(n - r + 1)$ does not occur before μ_{n-r+1}, then the event associated with node p cannot occur before μ_p, and since $p \geqq n - r + 2$, by the induction hypothesis this implies that the project will not be completed in minimum time. Also, from (7.2) it is clear that if the event associated with node $(n - r + 1)$ occurs at time μ_{n-r+1}, then the events associated with nodes $i \geqq n - r + 1$ can occur by time μ_i. All these facts together imply that under the induction hypothesis, the statement made at the beginning must also be true for $i = n - r + 1$. By induction, it is true for all i.

Hence, for any $(i, j) \in \mathcal{A}$, $\text{LF}(i, j) = \mu_j$, and so $\text{LS}(i, j) = \mu_i - t_{ij}$. We have already seen that $\text{ES}(i, j) = t_i$. The difference $\text{LS}(i, j) - \text{ES}(i, j) = \mu_j - t_{ij} - t_i$ is known as *the total slack* or *the total float* of job (i, j) and denoted by $\text{TS}(i, j)$. Also, the following activity floats can be similarly interpreted.

$$\mu_j - \mu_i - t_{ij} = \textit{Safety float} \text{ of job } (i, j)$$

$$t_j - t_i - t_{ij} = \textit{free float or free slack} \text{ of job } (i, j)$$

Job (i, j) is a *critical job* iff $\text{TS}(i, j) = 0$. Hence, after the forward and backward passes, all the critical jobs are easily identified. Any chain from node 1 to n on which all the arcs are critical arcs is a *critical path*. In particular, the chain from node 1 to n traced by the forward pass labels is a critical path. Critical jobs have to start exactly at their early start times if the project has to be completed in minimum time. However, slack jobs can be started any time within the interval between their early and late start times, allowing the scheduler some freedom in choosing their starting times. One should remember that if the start time of a slack

The ES, EF, LF, LS, and TS
for Jobs in the Hydroelectric
Dam Building Project

Job	ES	EF	LF	LS	TS
1	0.0	6.2	6.2	0.0	0.0
2	6.2	15.3	17.7	8.6	2.4
3	6.2	13.5	13.5	6.2	0.0
4	13.5	17.7	17.7	13.5	0.0
5	17.7	27.9	27.9	17.7	0.0
6	27.9	32.2	44.3	40.0	12.1
7	32.2	35.3	47.4	44.3	12.1
8	27.9	34.4	34.4	27.9	0.0
9	27.9	30.6	34.4	31.7	3.8
10	34.4	39.6	39.6	34.4	0.0
11	39.6	64.4	64.4	39.6	0.0
12	39.6	58.0	59.7	41.3	1.6
13	35.3	55.6	67.7	47.4	12.1
14	58.0	64.8	66.5	59.7	1.7
15	64.4	66.5	66.5	64.4	0.0
16	66.5	67.7	67.7	66.5	0.0
17	67.7	68.8	68.8	67.7	0.0

job is delayed beyond its early start time, the start times of all its successor jobs are delayed too, and this may affect their remaining total slacks.

Free slack can be used effectively in project scheduling. For example, if a job has positive free slack, and its start is delayed by any amount $\leqq$ its free slack, this delay will not affect the start times or slack of succeeding jobs.

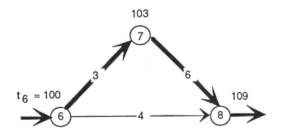

Figure 7.5 An illustration of a job (6, 8) with free slack. Thick arcs are on the critical path.

A node i is on a critical path iff $t_i = \mu_i$. Two nodes i, j may both be on a critical path, and yet the arc joining them (i, j) may not be a critical arc. An example is given in Figure 7.5. Here, the numbers on the arcs are the job durations, the numbers by the side of the nodes are the t_is, and critical arcs are thick. Even though both nodes 6, 8 are on the critical path, job (6, 8) is not a critical job, and its free slack is $109 - 100 - 4 = 5$. Job $(6, 8)$ has positive float even though both the nodes on it have zero slack. The start time of job $(6, 8)$ can be anywhere between 100 to 105 time units after project start, this delay in job $(6, 8)$ has absolutely no effect on the start times or slack of any of its successors.

Consider the arrow diagram for the hydroelectric dam building project in Figure 7.4. The critical path identified by the forward labels is marked with thick lines. For each i, if $t_i = \mu_i$ we entered their common value by the side of node i or entered the pair t_i, μ_i in that order if they are not equal. Minimum project duration is 68.8 months. The critical path in this example is unique, as all the nodes not on it satisfy $t_i > \mu_i$. The ES, EF, LF, LS, TS of all the jobs listed under the project (i.e., not the dummy jobs) are given above. It can be verified that job 2 has positive free slack of 2.4 in this example.

7.2   The Project Shortening Cost Curve

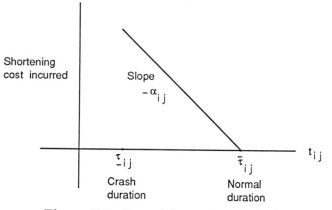

Figure 7.6 Cost of shortening job (i, j)

Let $G = (\mathcal{N}, \mathcal{A})$ be the arrow diagram for a project with $1, n$ as the start and finish nodes. Suppose it is required to complete a project before the minimum completion time, t_n, computed as discussed above, to meet a desired project due date. Then, ways have to be found to shorten the project duration by expediting one or more jobs. In practice, most job durations can be reduced by devoting additional resources (such as more workers, machines, overtime, etc.). This operation of expediting is known as *job shortening* or *crashing*, and the expense incurred on it is known as the *job shortening cost*. We consider here the problem of determining which subset of jobs to shorten, and each by how much, in order to complete the project by the given due date at minimum shortening cost. When a job is shortened, the job shortening cost is incurred in addition to the normal cost of carrying out the job. The normal cost is incurred anyway whether the job is shortened or not, hence we ignore it here, and try to minimize the total job shortening cost. We assume that the following data is available for each job (i, j) in the project.

$\overline{\tau}_{ij}$ = normal time duration for completing job (i, j)
$\underline{\tau}_{ij}$ = the minimum, or crash time duration for completing job (i, j)
α_{ij} = shortening cost in \$/unit time

$\alpha_{ij} \geq 0$ is the cost for reducing the time required for completing job (i, j) below the normal time $\overline{\tau}_{ij}$, per unit. See Figure 7.6. For all dummy jobs (i, j) we always have $\overline{\tau}_{ij} = \underline{\tau}_{ij} = 0$ and $\alpha_{ij} = 0$.

The time allowed for completing job (i, j), t_{ij}, is itself a variable in this problem, subject to the bounds $\underline{\tau}_{ij} \leq t_{ij} \leq \overline{\tau}_{ij}$, and the shortening cost associated with t_{ij} is $(\overline{\tau}_{ij} - t_{ij})\alpha_{ij}$. $\overline{\tau}_{ij} = \underline{\tau}_{ij}$ implies that job (i, j) cannot be shortened, in this case $t_{ij} = \overline{\tau}_{ij} = \underline{\tau}_{ij}$, and we take $\alpha_{ij} = 0$. If $\overline{\tau}_{ij} > \underline{\tau}_{ij}$ and $\alpha_{ij} = 0$ we will obviously select $t_{ij} = \underline{\tau}_{ij}$ as this is likely to reduce project duration at no additional cost. So, we assume that if $\overline{\tau}_{ij} > \underline{\tau}_{ij}$ then $\alpha_{ij} > 0$.

Let t_i be the clock time at which the event corresponding to node i occurs. The t_i, t_{ij} variables have to satisfy $t_j - t_i \geq t_{ij}$ for each $(i, j) \in \mathcal{A}$ in this problem. Given the job durations, t_{ij}s, the total shortening cost is $\sum((\overline{\tau}_{ij} - t_{ij})\alpha_{ij}$: over $(i, j) \in \mathcal{A})$, and since $\sum(\overline{\tau}_{ij}\alpha_{ij}$: over $(i, j) \in \mathcal{A})$ is a known constant, minimizing the total shortening cost is equivalent to maximizing $\sum(\alpha_{ij}t_{ij}$: over $(i, j) \in \mathcal{A})$. Hence, if λ is the specified project duration, the problem of meeting this deadline with minimum shortening cost is equivalent to : find $(t_{ij} : (i, j) \in \mathcal{A}), (t_i : i \in \mathcal{N})$ that solve (7.3) to (7.6).

$$Q(\lambda) = \text{Maximum} \sum(\alpha_{ij}t_{ij} \quad : \quad \text{over } (i, j) \in \mathcal{A})$$

$$\text{Subject to} \quad t_{ij} - (t_j - t_i) \;\leq\; 0, \text{ for all } (i, j) \in \mathcal{A} \qquad (7.3)$$

$$t_{ij} \;\leq\; \overline{\tau}_{ij}, \text{ for all } (i, j) \in \mathcal{A} \qquad (7.4)$$

$$-t_{ij} \;\leq\; -\underline{\tau}_{ij}, \text{ for all } (i, j) \in \mathcal{A} \qquad (7.5)$$

$$t_n - t_1 \;\leq\; \lambda \qquad (7.6)$$

The optimum project shortening cost is $P(\lambda) = \sum(\overline{\tau}_{ij}\alpha_{ij}$: over $(i, j) \in \mathcal{A})$ $-Q(\lambda)$. Associating the dual variables f_{ij}, g_{ij}, h_{ij} to the constraints in (7.3), (7.4), (7.5), respectively, and the dual variable v to the constraint in (7.6), we see that the dual of the above problem is (7.7). The dual problem (7.7) has the structure of a minimum cost flow problem. From the complementary slackness optimality conditions for this primal, dual pair of LPs, it can be seen that if (7.7) has an optimum solution, then it has an optimum solution in which at least one of the two variables g_{ij} or h_{ij} is 0 for each arc $(i, j) \in \mathcal{A}$. The constraints in (7.7) imply that in such an optimum dual solution $g_{ij} = (\alpha_{ij} - f_{ij})^+ = \max. \{0, \alpha_{ij} - f_{ij}\}$,

$$\text{Min. } W(\lambda, v, f, g, h) = \lambda v \; + \; \sum_{(i,j)\in A}(\overline{\tau}_{ij}g_{ij}) - \sum_{(i,j)\in A}(\underline{\tau}_{ij}h_{ij})$$

$$\text{Subject to } f_{ij} + g_{ij} - h_{ij} \; = \; \alpha_{ij}, \text{ for all } (i,j) \in A$$

$$f(i, \mathcal{N}) - f(\mathcal{N}, i) \; = \; \left\{ \begin{array}{l} 0, \text{ for all } i \neq 1 \text{ or } n \\ -v, \text{ for } i = n \end{array} \right. \tag{7.7}$$

$$f, g, h, v \; \gtreqqless \; 0$$

and $h_{ij} = (\alpha_{ij} - f_{ij})^- = | \min. \{0, \alpha_{ij} - f_{ij}\}|$. So, in such a solution we have $W(\lambda, v, f, g, h) = \lambda v + \sum(\omega_{ij}(f_{ij}) : \text{over } (i,j) \in A)$, where $\omega_{ij}(f_{ij})$ is a piecewise linear convex function defined below (it is convex because $\overline{\tau}_{ij} \gtreqqless \underline{\tau}_{ij}$). See Figure 7.7.

$$\omega_{ij}(f_{ij}) = \left\{ \begin{array}{l} -\overline{\tau}_{ij}(f_{ij} - \alpha_{ij}) \text{ if } f_{ij} \lesseqqgtr \alpha_{ij} \\ -\underline{\tau}_{ij}(f_{ij} - \alpha_{ij}) \text{ if } f_{ij} > \alpha_{ij} \end{array} \right.$$

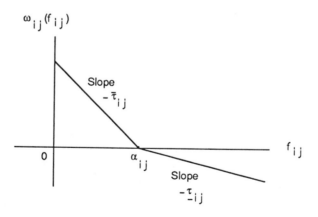

Figure 7.7

By these arguments, the dual (7.7) is equivalent to the minimum cost flow problem (7. 8) in the network G, with a piecewise linear convex objective function. We can of course solve the original project shortening cost problem (7.3) to (7.6) as an LP, but as its dual is a minimum cost flow problem for which there are very efficient special algorithms, it turns out to be much more convenient to solve the dual problem using these algorithms, and then obtain an optimum solution of the primal problem by the complementary slackness conditions for optimality.

If $(i, j) \in A$ is such that $\overline{\tau}_{ij} = \underline{\tau}_{ij}$, then $t_{ij} = \overline{\tau}_{ij} = \underline{\tau}_{ij}$ is known and fixed, in this case $\omega_{ij}(f_{ij})$ defined above is in fact linear. $\omega_{ij}(f_{ij})$ is piecewise linear convex, and not linear, only if $\overline{\tau}_{ij} > \underline{\tau}_{ij}$. (7.8) can be transformed into a linear minimum

$$\text{Minimize } \lambda v + \sum (\omega_{ij}(f_{ij})) \quad : \quad \text{over } (i,j) \in \mathcal{A}$$

$$\text{Subject to } f(i, \mathcal{N}) - f(\mathcal{N}, i) = \begin{cases} 0, & \text{for all } i \neq 1 \text{ or } n \\ -v, & \text{for } i = n \end{cases} \tag{7.8}$$

$$f, v \; \geqq \; 0$$

cost flow problem as in Section 5.9. For this we replace each arc (i, j) on which ω_{ij} is not linear, by the pair of arcs $(i,j)_1, (i,j)_2$ called Type 1 and Type 2 arcs, with the following data (see Figure 7.8):

lower bounds on all the arcs $= 0$
capacity of $(i,j)_1$ is $k(i,j,1) = \infty$
capacity of $(i,j)_2$ is $k(i,j,2) = \alpha_{ij}$
unit cost coefficient on $(i,j)_1 = c(i,j,1) = -\underline{\tau}_{ij}$
unit cost coefficient on $(i,j)_2 = c(i,j,2) = -\overline{\tau}_{ij}$.

If arc (i, j) is such that $\overline{\tau}_{ij} = \underline{\tau}_{ij}$, it is treated as a Type 1 arc itself, and no Type 2 arc corresponding to it is introduced (since $\omega_{ij}(f_{ij})$ is linear for such arcs). See Figure 7.8.

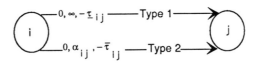

Figure 7.8 If arc (i, j) is such that $\overline{\tau}_{ij} > \underline{\tau}_{ij}$, it corresponds to the pair of Type 1, 2 arcs with this data (lower bound, capacity, unit cost) in $\mathcal{A}'$. If $\overline{\tau}_{ij} = \underline{\tau}_{ij}$, there will be no Type 2 arc.

Let $G' = (\mathcal{N}, \mathcal{A}')$ be the augmented network with the data on arcs as defined above. Let $f(i,j,r)$ denote the flow amount on the Type r arc $(i,j)_r \in \mathcal{A}'$ for $r = 1$, 2, and let $f' = (f(i,j,r) : (i,j,r) \in \mathcal{A}')$. Given a feasible flow vector $f' = (f(i,j,r))$ in G', the corresponding flow vector $f = (f_{ij})$ feasible to (7.8) in the original network G is obtained from

$$f_{ij} = \begin{cases} f(i,j,1) + f(i,j,2) & \text{if both Type 1,2 arcs corresponding to } (i,j) \text{ exist in } G' \\ f(i,j,1) & \text{otherwise} \end{cases}$$

(7.8) is equivalent to the problem of minimizing $\lambda v + \sum (c(i,j,r)f(i,j,r)$: over $(i,j)_r \in \mathcal{A}', r = 1, 2)$ in G'. The flow vector f in G corresponding to an optimum flow vector f' in G', is an optimum flow vector for (7.8).

Let $\overline{\lambda}, \underline{\lambda}$ be the lengths of the critical paths in G with $(\overline{\tau}_{ij}), (\underline{\tau}_{ij})$ as the arc length vectors respectively. When $\lambda = \overline{\lambda}$, the optimum $t_{ij} = \overline{\tau}_{ij}$ for each $(i,j) \in \mathcal{A}$, and the optimum shortening cost is 0. $\underline{\lambda}$ is at the other end, it is the minimum project

duration. Since the dual problem is (7.8) is always feasible, the objective function
must be unbounded below in it (and the same thing happens in G') whenever $\lambda < \underline{\lambda}$,
and vice versa.

We would like to solve the project shortening cost minimization problem para-
metrically in λ as it decreases from $\bar{\lambda}$ to $\underline{\lambda}$. For this, we need to solve the minimum
cost flow problem of minimizing $\lambda v + \sum (c(i,j,r)f(i,j,r)$: over $(i,j)_r \in \mathcal{A}', r = 1, 2)$
in G', treating λ as a parameter. This problem is in the same form as the parametric maximum profit flow problem discussed in Section 5.3, with the exception
that the objective function here is to be minimized instead of being maximized,
so it can be solved by an appropriate modification of that algorithm. The node
price vector $\pi = (\pi_i)$ in G' in this algorithm turns out to be the vector of early
occurrence times, (t_i), for nodes i in G, associated with the present λ. If π is the
node price vector in G', the relative cost coefficients wrt it are

$$
\begin{aligned}
\bar{c}(i,j,1) &= c(i,j,1) + (\pi_j - \pi_i) = -\underline{\tau}_{ij} + (\pi_j - \pi_i), \text{ for Type 1 arcs } (i,j)_1 \in \mathcal{A}' \\
\bar{c}(i,j,2) &= c(i,j,2) + (\pi_j - \pi_i) = -\bar{\tau}_{ij} + (\pi_j - \pi_i), \text{ for Type 2 arcs } (i,j)_2 \in \mathcal{A}'
\end{aligned}
$$

The signs of these relative cost coefficients are opposite to those in Section
5.3 since we discussed the maximum profit flow problem there, but we have a
minimum cost flow problem in G'. The feasible flow vector, node price vector pair
$(f' = (f(i,j,r)), \pi = (\pi_i))$ is optimal in G' for a given value of λ if the following
optimality conditions hold.

$$
\begin{aligned}
\bar{c}(i,j,1) &\gtreqless 0 \text{ for all } (i,j)_1 \in \mathcal{A}' \\
\bar{c}(i,j,1) &> 0 \text{ implies } f(i,j,1) = 0 \\
\text{if } (i,j)_2 \in \mathcal{A}' \quad, \quad &\text{then} \begin{cases} \bar{c}(i,j,2) > 0 \text{ implies } f(i,j,2) = 0 \\ \bar{c}(i,j,2) < 0 \text{ implies } f(i,j,2) = k(i,j,2) \end{cases} \quad (7.9) \\
\text{and } \pi_n - \pi_1 &= \lambda
\end{aligned}
$$

Let $(f'^p, \pi^p = (\pi^p))$ be a feasible flow vector, node price vector pair with the
value of f'^p being v^p, satisfying (7.9) in G' for $\lambda = \lambda_p$. To find optimum flow
vectors in G' for $\lambda < \lambda_p$, the parametric algorithm applies the labeling algorithm
to increase the flow value while continuing to satisfy the optimality conditions (7.9)
keeping $\lambda = \lambda_p$. Only arcs $(i,j)_r \in \mathcal{A}'$ for which $\bar{c}(i,j,r) = 0$ are *admissible* for flow
change in this step. Labeling is carried out in two stages in this step. In Stage 1
we check whether the flow value can be increased to ∞. If this is possible, it would
imply that if λ is decreased from its present value λ_p, the objective value in G'
becomes unbounded below, which in turn implies that the original problem (7.3) to
(7.6) becomes infeasible, i.e., $\lambda_p = \underline{\lambda}$, the minimum possible project duration under
crashing. Another explanation for this is the following. The flow value can be
increased to ∞ making flow changes on admissible arcs only, iff there exists a chain
from node 1 to node n, say $\mathcal{C}$, consisting of admissible Type 1 arcs only, since only

they have infinite capacity. Hence $\bar{c}(i,j,1) = -\mathcal{T}_{ij} + \pi_j^p - \pi_i^p = 0$, or $\pi_j^p - \pi_i^p = \mathcal{T}_{ij}$, for each (i,j) on $\mathcal{C}$. Hence $\mathcal{C}$ is a critical path, and each job on it on it is at its crash time, which implies that $\pi_n^p = \lambda_p = \underline{\lambda}$, and hence λ cannot be reduced below its current value.

In Stage 1, if it has been verified that the flow value cannot be increased to ∞, in Stage 2 we obtain the maximum flow value making flow changes on admissible arcs of both Types 1 and 2.

During the labeling process, if an FAP from node 1 to node n consisting of admissible Type 1 arcs only has been identified (which only happens if n is labeled in Stage 1) we say that an *infinite breakthrough* has occurred, this is a signal that the current value of λ is $\underline{\lambda}$. Any FAP identified during Stage 2 will consist of some Type 2 arcs which have finite capacities, and we refer to its occurrence as a *finite breakthrough*.

The label on a node j in this algorithm will be of the form $(i, \pm, r = 1 \text{ or } 2)$. If it is $(i, +, r)$, it means that $(i,j)_r$ is a forward arc on the FAP from 1 to j. If it is $(i, -, r)$ it means that $(j,i)_r$ is a reverse arc on the same FAP.

THE PARAMETRIC SHORTENING COST MINIMIZATION ALGORITHM

Initial Step   Find the longest chains from node 1 to all the other nodes in G, using $\bar{\tau}_{ij}$s, normal durations, as the lengths of arcs $(i,j) \in \mathcal{A}$, by the forward pass routine discussed earlier. For $i \in \mathcal{N}$ let (L_i, t_i^1) be the forward pass label on i. Define $\pi^1 = (\pi_i^1)$, where $\pi_i^1 = t_i^1$ for each $i \in \mathcal{N}$. These node prices satisfy $\pi_j^1 - \pi_i^1 \geqq \bar{\tau}_{ij}$, which is opposite to those in Chapter 4, since we are finding the longest chains here. Define $f'^1 = 0$. It can be verified that (f'^1, π^1) satisfies (7.9) when $\lambda = \bar{\lambda} = \lambda_1$, and hence is an optimum pair in G' for λ_1. When $\lambda = \lambda_1$, $l_{ij} = \bar{\tau}_{ij}$ for each $(i,j) \in \mathcal{A}$, and $t_i = \pi_i^1$ is the earliest occurrence time for the event associated with node i, $i \in \mathcal{N}$. Enter the labeling routine.

Stage 1 of the Labeling Routine

Labeling Step 1   Label the start node 1 with $(\emptyset, +)$. List $= \{ 1 \}$.

Labeling Step 2 Select a node for Stage 1 scanning   Let λ_p be the current value of λ, and (f'^p, π^p) the present optimum pair with v^p as the value of f'^p. If list $= \emptyset$, go to Stage 2 of the labeling routine. Otherwise select the node from the top of the list to scan, delete it from the list, and go to Labeling Step 3.

Labeling Step 3 Stage 1 scanning   Let i be the node to be scanned. Label all unlabeled nodes j such that $(i,j,1)$ is admissible for flow change, with $(i,+,1)$, and include them at the bottom of the list as they are labeled.

If node n is now labeled, there is an infinite breakthrough. This implies that $\lambda_p = \underline{\lambda}$, terminate the algorithm. If n is not yet labeled, go back to Labeling Step 2.

Stage 2 of the Labeling Routine   Make the list = set of all labeled nodes at this time

> **Labeling Step 4 Select a node for Stage 2 scanning**   If list $= \emptyset$, go to the node price change step. Otherwise select the node from the top of the list to scan, delete it from the list, and go to Labeling Step 5.
>
> **Labeling Step 5 Stage 2 scanning**   Let i be the node to scan.
>
>> **Forward labeling**   Identify all unlabeled nodes j such that (i,j,r) is admissible for flow change, and $f^p(i,j,r) < k(i,j,r)$ for $r = 1$ or 2 or both, then label j with $(i,+,r)$ with any of the r satisfying the above condition and include j at the bottom of the list as it is labeled.
>>
>> **Reverse labeling**   Identify all unlabeled nodes j such that (j,i,r) is admissible for flow change, and $f^p(j,i,r) > 0$ for $r = 1$ or 2 or both, then label j with $(i,-,r)$ with any of the r satisfying the above condition, and include j at the bottom of the list as it is labeled.
>
> If node n is now labeled, there is a finite breakthrough, go to the flow augmentation step. Otherwise go back to Labeling Step 4.

Flow augmentation   Trace the admissible FAP from node 1 to node n using the node labels, and carry out flow augmentation using it. Erase the labels on all the nodes and go back to Labeling Step 1.

Node price change   Let $\mathbf{X}$, $\overline{\mathbf{X}}$ be the current sets of labeled, unlabeled nodes respectively. Define

$$
\begin{aligned}
\mathbf{A}^1 &= \{(i,j)_r : r = 1 \text{ or } 2, (i,j)_r \in (\mathbf{X},\overline{\mathbf{X}}), \text{ and current } \bar{c}(i,j,r) > 0\} \\
\mathbf{A}^2 &= \{(i,j)_r : r = 1 \text{ or } 2, (i,j)_r \in (\overline{\mathbf{X}},\mathbf{X}), \text{ and current } \bar{c}(i,j,r) < 0\} \\
\delta_1 &= \text{min. } \{\bar{c}(i,j,r) : (i,j)_r \in \mathbf{A}^1\} \\
\delta_2 &= \text{min. } \{-\bar{c}(i,j,r) : (i,j)_r \in \mathbf{A}^2\} \\
\delta &= \text{min. } \{\delta_1, \delta_2\} \\
\pi_i^p(\nu) &= \begin{cases} \pi_i^p, & \text{if } i \in \mathbf{X} \\ \pi_i^p - \nu, & \text{if } i \in \overline{\mathbf{X}} \end{cases}
\end{aligned}
$$

for $0 \leqq \nu \leqq \delta$. Let $\pi^p(\nu) = (\pi_i^p(\nu) : i \in \mathcal{N})$. It can be verified that the feasible pair $(f'^p, \pi^p(\nu))$ satisfies the optimality conditions (7.9) when $\lambda = \pi_n^p - \nu$ for all $0 \leqq \nu \leqq \delta$. So, if we define

$$t_{ij}^p(\nu) = Min.\{\bar{\tau}_{ij}, \pi_j^p(\nu) - \pi_i^p(\nu)\}, \text{ for } (i,j) \in \mathcal{A}$$
$$t_i^p(\nu) = \pi_i^p(\nu), \text{ for } i \in \mathcal{N}$$

then $(t_{ij}^p(\nu) : (i,j) \in \mathcal{A}), (t_i^p(\nu) : i \in \mathcal{N})$ is an optimum solution for the original project shortening cost minimization problem (7.7) when $\lambda = \lambda_p - \nu = \pi_n^p - \nu$, for $0 \leqq \nu \leqq \delta$. Since there is no change in the flow vector, the optimum objective value in G' (which is equal to $Q(\lambda)$ by the duality theorem of LP) decreases with slope v^p as λ decreases in this interval from λ_p to $\lambda_{p+1} = \lambda_p - \delta$. Hence, the minimum job shortening cost $P(\lambda)$ increases with slope v^p as λ decreases in this interval. In other words, $P(\lambda)$ is linear in this interval with slope $-v^p$.

Define $\pi^{p+1} = \pi^p(\delta), \lambda_{p+1} = \lambda_p - \delta$. Take (f'^p, π^{p+1}) as the new pair in G', λ_{p+1} as the new value for λ, include all the labeled nodes in the list, and resume labeling by going back to Labeling Step 2 in Stage 1.

Discussion

In the set $\mathbf{A}^2$ defined in a node price change step, all arcs are always saturated Type 2 arcs. Likewise, all arcs in the set $\mathbf{A}^1$ have zero flow.

The set $\mathbf{A}^1$ is always nonempty in a node price change step. The reason for this is the following. Let node q be the unlabeled node with the smallest number in G'. The only node in G' which has no arcs incident into it is 1 and it is labeled. So, there exists an arc of the form $(i, q) \in \mathcal{A}'$. By the acyclic numbering of the nodes in G, $i < q$, and from the definition of q, i must be in $\mathbf{X}$. If (i, q) is currently admissible, node q would have been labeled when node i is scanned, a contradiction. So, (i, q) is inadmissible, and hence by (7.9), the current value of $\bar{c}(i, q, 1) > 0$. So, $(i, q)_1 \in \mathbf{A}^1$. Thus $\mathbf{A}^1 \neq \emptyset$. Hence the quantity δ in a node price change step in this algorithm is always positive and finite.

The algorithm obtains the optimum job durations, and the earliest occurrence times for the events associated with the nodes in the network, $(t_{ij}), (t_i)$, corresponding to each value of λ in its range. For any λ, the latest occurrence times associated with the nodes, can be obtained using $t_n = \lambda$ and the job duration values (t_{ij}) for that λ, by applying the backward pass routine. These provide all the necessary information to the scheduler to identify all the critical jobs, to compute the total slack of each job, and to schedule the jobs over time for that value of λ.

As mentioned above, the optimum job shortening cost, $P(\lambda)$, increases as λ decreases. Whenever flow augmentation occurs, the slope of $P(\lambda)$ as λ decreases, increases. We have already seen that $P(\lambda)$ is piecewise linear. Hence, $P(\lambda)$ is a piecewise linear convex function.

As an example consider the project consisting of six jobs denoted by e_1 to e_6, with data given in the following table. The arrow diagram for the project is given in Figure 7.9. Numbers on the arcs there are the normal durations, and the entries by the side of the nodes are the forward pass labels corresponding to these normal durations.

Job	Immediate predecessors	Normal duration	Crash duration	Unit shortening cost
e_1		3	1	3
e_2		4	2	4
e_3	e_1	4	2	1
e_4	e_1, e_2	6	1	3
e_5	e_1	5	5	0
e_6	e_3, e_4	4	4	0

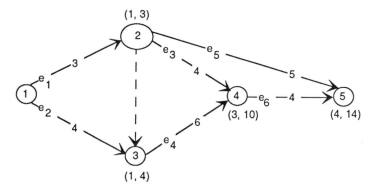

Figure 7.9

In Figure 7.10 we show the augmented network G'. The entries on arc (i, j, r) in this figure are the capacity $k(i, j, r)$ and the cost coefficient $c(i, j, r)$. The Type 1 arcs are all the arcs with ∞ capacity. All lower bounds are 0. The initial node prices in π^1, from the earliest occurrence times associated with the nodes, under normal job durations, are entered by the side of the nodes. The arcs admissible for flow change in Figure 7.10 are $(4, 5)_1, (3, 4)_2, (1, 2)_2$ and $(1, 3)_2$. After labeling and flow augmentation, we get the flow vector of value $v^1 = 3$ (this flow vector is marked in Figure 7.11 with nonzero flow amounts entered inside little boxes by the side of the arcs) and reach the node price change step with the cut $(\mathbf{X}, \overline{\mathbf{X}}) = (\{1, 2, 3\}, \{4, 5\})$. So, $\mathbf{A}^1 = \{(2, 4)_2, (2, 4)_1, (3, 4)_1, (2, 5)_1,\}, \mathbf{A}^2 = \emptyset$, and $\delta = \min.$ $\{3, 5, 5, 6\} = 3$. Hence for project duration $\pi_5^1 - \nu = 14 - \nu$, the earliest occurrence times associated with the nodes 1 to 5 in that order are $(0, 3, 4, 10 - \nu, 14 - \nu)$, for $0 \leqq \nu \leqq 3$. So, when the project duration is $14 - \nu$, the optimum job durations for e_1 to e_6 in that order are $(3, 4, 4, 6 - \nu, 5, 4)$, for $0 \leqq \nu \leqq 3$, and the optimum shortening cost increases with slope $v^1 = 3$ as project duration decreases from 14 to 11. Making $\nu = \delta = 3$, we get the optimum flow vector, node price vector pair marked in Figure 7.11.

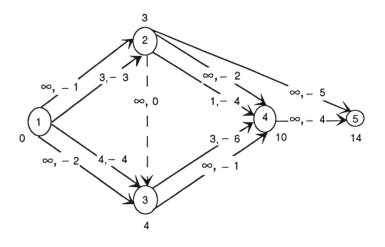

Figure 7.10 All lower bounds are zero. Data on the arcs is capacity, unit cost.

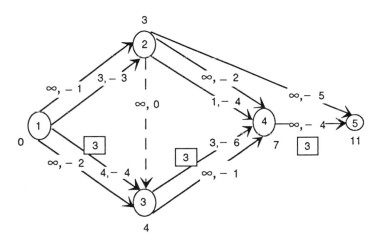

Figure 7.11 Data on the arcs is capacity, unit cost. Nonzero flow amounts are entered in little boxes by the side of the nodes. Present node prices are entered by the side of the nodes.

The algorithm can be continued in the same manner. It terminates with an infinite breakthrough when the project duration reaches 7. The project shortening cost curve for this project is shown in Figure 7. 12.

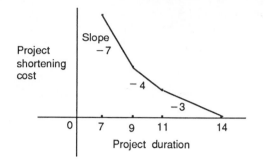

Figure 7.12

7.3   Resource Constrained Project Scheduling Problems

In the CPM models discussed so far, we assumed that the only constraints in scheduling jobs over time are those imposed by the predecessor relationships among the jobs. To carry out jobs in practical project scheduling problems, we require resources such as a crane, or other piece of equipment, or trained personnel, etc. Two or more jobs may require the same resources, and it may not be possible to carry them out simultaneously because of limited supply of resources, even though the precedence constraints do not prevent them from being scheduled simultaneously. The limited availability of resources imposes a new set of constraints. Before starting a job, the project scheduler now has to make sure that all its predecessors have been completed, and also that the resources required to carry it out are available. Problems of this type are known as *resource constrained project scheduling problems*. See Exercise 7.6, for a problem of this type. Usually, solving resource constrained project scheduling cannot be accomplished purely by network techniques alone; typically, they require combinatorial optimization methods. Also, practical resource constrained project scheduling normally leads to very large combinatorial optimization problems, for which efficient exact algorithms are not available at the moment. Hence, a variety of heuristic algorithms have been developed for resource constrained project scheduling, see Battersby [1967], Burman [1972], Elmaghraby [1977], Weist and Levy [1977], and Willis and Hastings [1976].

7.4   PERT

In the CPM models discussed so far, the job durations are assumed to be either known constants, or deterministic variables whose values can be selected from known intervals by spending a deterministic sum of money. In real world project scheduling, job durations may not be known with certainty, in fact there may be quite a bit of uncertainty or random variation in them. Uncertainty in job durations appears most often in research and development projects, in projects dealing with designing or launching a new product, etc. The PERT (Program Evaluation and Review Technique) model deals with project scheduling under such uncertainty.

The PERT model usually assumes that the various job durations are mutually independent random variables. Replacing each of these random variables by their expected values leads to a deterministic problem which can be analyzed using the CPM models already discussed. This often leads to an optimistic estimate of the expected project duration. In engineering problems, the expected value of a job duration can itself be approximated by a convex combination of the most probable job duration, an optimistic job duration, and a pessimistic job duration, all guessed by qualified project engineers; PERT normally uses this approach to estimate the expected job durations.

If the job durations can be assumed to be random variables with known probability distributions, then a simulation can be run by selecting values for job durations from these distributions. Once the job durations are known, the critical path and the scheduling information can be obtained using the CPM methods. The procedure can be repeated many times by selecting different sets of values for the random variables. From these simulation runs, information like the average project duration and its standard deviation, probabilities for the various jobs being critical, etc., can be computed. Statistical analysis of the data from these simulation runs can give the scheduler very useful information. Because of the random nature of job durations, it is not possible to lay down a rigid time schedule for the jobs at the beginning of the project, and expect to stick with it. A rough time schedule is prepared using the information from CPM based on expected job durations, and the simulation runs. As the event corresponding to each node in the arrow diagram materializes, a review is made, and the time schedule for the remaining jobs is revised. See Burman [1972], Elmaghraby [1977], Weist and Levy [1977].

7.5 Exercises

7.4 Let $G = (\mathcal{N}, \mathcal{A})$ be the arrow diagram for a project. λ is the project completion time, and d_1 denotes a target value for λ. For each job p, $\underline{\tau}_p, \bar{\tau}_p, \alpha_p$ have the same meaning as in Section 7.2. For each unit of time the project is completed before the target time d_1, there is a profit of δ\$. If $\lambda > d_1$, a penalty is incurred, this penalty, denoted by $f(\lambda)$, is 0 if $\lambda \leqq d_1$, and a positive piecewise linear function of λ in the range $\lambda > d_1$, with slopes given below.

Interval	Slope of $f(\lambda)$
$d_1 < \lambda \leqq d_2$	g_1
$d_2 < \lambda \leqq d_3$	g_2
$\vdots$	$\vdots$
$d_u < \lambda$	g_u

where $d_1 < d_2 < \ldots < d_u$ and $g_1 < g_2 < \ldots < g_u$. Define the net cost of the project to be the cost of job shortening $+ f(\lambda) -$ any profit due to project completion before target completion date d_1. Formulate the problem of finding an optimum project duration to minimize the net cost as an LP, and show how it can be solved using a network flow approach. As a numerical example consider the following project.

Project: Setting Up a Fossil Fuel Power Plant

No. p	Job	IPs	$\overline{\tau}_p$	$\underline{\tau}_p$ if $< \overline{\tau}_p$	α_p
1.	Land acquisition		6		
2.	Identi. trained personnel	1	3		
3.	Land dev. & infrastructure	1	2		
4.	Control room eng.	1	12		
5.	Lag in turbine civil works	1	8		
6.	Delivery of TG	1	12		
7.	Delivery of boiler	1	10		
8.	Joining time for personnel	2	3		
9.	Boiler prel. civil works	3	2	1	6
10.	Control room civil works	4	5	2	3
11.	TG civil works	5	9	7	15
12.	Training	8	6		
13.	Boiler final civil works	9	9	8	15
14.	Erection of control room	10	8	7	5
15.	Erection of TG	6, 11	10	8	20
16.	Boiler erection	7, 13	12	11	35
17.	Hydraulic test	16	2		
18.	Boiler light up	14, 17	1.5	0.5	7.5
19.	Box up of turbine	15	3	2	15
20.	Steam blowing, safety valve floating	18, 19	2.5	1.5	10
21.	Turbine rolling	20	1.5	1	20
22.	Trial run	21	1	0.5	25
23.	Synchronization	22	1	0.5	20

IP = Immediate predecessors, $\overline{\tau}_p, \underline{\tau}_p$ in months, α_p in \$mil.

The plant is scheduled for erection and commissioning in $d_1 = 36$ months from land acquisition. There is a profit of \$35 million/month if the plant is completed before 36 months. If the erection division fails to hand over the plant to the customer at the end of 36 months, there will be a penalty with increasing slopes of \$30, 35, 40, and 55 million beyond 36, 37, 38, and 39 months, respectively. Solve this problem and obtain an optimum project schedule. (Kanda and Singh[1988])

7.5 How does the formulation in Exercise 7.4 change if the profit for early completion is a constant, \$ ξ, irrespective of what the value of λ is $< d_1$, but everything else remains unchanged.

7.6 Coke Depot Project A depot is to be built to store coke and to load and dispatch trucks. There will be three storage hoppers (SH. in abbreviation), a block of bunkers (B. in abbreviation), interconnecting conveyors (abbreviated as C.), and weigh bridges (called WB.). Around the bunkers there will be an area of hard-standing and an access road will have to be laid to the site. There are six resources required for construction, their availability is limited to the quantities given below during the construction.

Resource	Symbol	Available quantity
Laborers	R_1	10
Steel men	R_2	5
Concrete men	R_3	4
Bricklayers	R_4	2
Cranes	R_5	1
Dumpers	R_6	2

Data on this project, the duration (in weeks) and resource requirements of each job are given below. Draw an arrow diagram for this project, and determine the earliest and latest start and finish times, and the total float of each job. Schedule the jobs so that the project is finished as quickly as possible without the resource availabilities being exceeded using an appropriate heuristic approach. (Willis and Hastings[1976])

Job	IPs	Duration	R_1	R_2	R_3	R_4	R_5	R_6
1. B. piling		5	2	1				
2. Clear site for SH.		8	6					1
3. B. excavation for cols.	1	4	4					1
4. SH. excavations for C.	2, 3	4	4					1
5. Concrete tops of piles for B.	3	3	2		2			
6. Place cols for B.	5	4	3	1			1	
7. Excavate access road	5	4	3					1
8. Put in B.	6	3		2				
9. Stairways inside B.	6	1		2				
10. Excavate pit for WB.	4	6	1					1
11. Concrete for SH.	4	12	2		4			
12. Main C. foundation	4	4	1					
13. Brick walls for B.	8, 9	3	2			2		
14. Clad in steel for B.	8, 9	1		2				
15. Install internal equip. in B.	8, 9	6	2	1		2		
16. Erect gantry for main C.	12, 6	1	2	1			1	
17. Install C. under hoppers	11	1	2	2				
18. Concrete pit for WB.	11, 10	2	1		2			
19. Excavate for hard-standing	7	9	4					1
20. Lay access roadway	7	9	4					
21. Install outloading equip. for B.	15	2	2	3				
22. Line B.	13, 14	1	1	1				
23. Install main C.	16	1	2	2				
24. Build weighhouse	18	4	1			2		
25. Erect perimeter fence	19	4	2					
26. Install C. to SH.	17, 23	1	2	2				
27. Install WB.	24	1	2	2			1	
28. Lay hard-standing	19, 18	6	4					
29. Commission hoppers	26	1						

IP = Immediate predecessors

7.7 Draw an arrow diagram for each of the following projects. For the values of project duration in its feasible range, obtain the optimum job durations and the earliest occurrence times of events associated with nodes in the network, and draw the project shortening cost curve as a function of the project completion time in each case. (R. Visweswara Rao).

(a) Data Process and Collection System Design for a Power Plant

No. p	Activity	IPs	Duration (days)	α_p
1.	Prel. Syst. description		40	
2.	Develop specs.	1	100	
3.	Client approval & place order	2	50	
4.	Develop I/O summary	2	40 - 60	15
5.	Develop alarm list	4	40	
6.	Develop log formats	3, 5	40	
7.	Software def.	3	35	
8.	Hardware requirements	3	35	
9.	Finalize I/O summary	5, 6	50 - 60	18
10.	Anal. performance calculation	9	50 - 70	20
11.	Auto. turbine startup anal.	9	60	
12.	Boiler guides anal.	9	30	
13.	Fabricate & ship	10, 11, 12	400	
14.	Software preparation	7, 10, 11	60 - 80	22
15.	Install & check	13, 14	100 - 130	30
16.	Termination & wiring lists	9	30	
17.	Schematic wiring lists	16	60	
18.	Pulling & term. of cables	15, 17	60	
19.	Operational test	18	80 - 125	30
20.	First firing	19	1	

IP = Immediate predecessors

α_p = shortening cost of job p/day shortened

(b) Electrical Auxiliary System Design for a Nuclear Plant

No. p	Activity	IPs	Duration (days)	α_p
1.	Aux. load list		100 - 120	15
2.	13.8 switchgear load ident.	1	140 - 190	12
3.	4.16kv & 480 v. switchgear load ident.	1	45	
4.	Vital AC load determination	1	200 - 300	14
5.	DC load determ.	1	110 - 165	18
6.	Voltage drop study	2	84	
7.	Diesel gen. sizing	3	77	
8.	Inventer sizing	4	20	
9.	Battery sizing	5, 8	40	
10.	DC fault study	9	80	
11.	Prel. AC fault current study	6, 7	20	
12.	Power transformer sizing	2, 11	80	
13.	Composite oneline diagram	2,3	72	
14.	Safety (class 1E) system design	13	150 - 200	25
15.	Non-class 1E system design	13	160 - 190	20
16.	Relaying oneline & metering dia.	13	80	
17.	3-line diagram	14, 15, 16	150	
18.	Synchronizing & phasing diagrams	17	100	
19.	Client review	10, 18	25	
20.	Equipment purchase & installation	19	800	

(c) Sewer and Waste System Design for a Power Plant

No. p	Activity	IPs	Duration (days)	α_p
1.	Collection system outline		25 - 40	10
2..	Final design & approval	1	30	
3.	Issue construction drawings	2	23 - 30	8
4.	Get sewer pipe & manholes	1	145	
5.	Fabricate & ship	3,4	45	
6.	Treat. system drawings & approval		50 - 70	15
7.	Issue treat. system construction drawings	6	30	
8.	Award contract	7	60	
9.	Final construction	8, 5	200 - 300	30

IP = Immediate predecessors

α_p = shortening cost of job p/day shortened

7.8 In the job crashing model discussed in Section 7. 2, intuitively it seems correct to assume that if a job is crashed in an optimum schedule for a project duration, then that job will stay crashed in optimum schedules as the project duration decreases further. Show that this could be wrong, using the following example.

Job no. r	IPs	Duration		α_r
		$\underline{\tau}_r$	$\overline{\tau}_r$	
1.		1	3	3
2.		2	4	1
3.	1	0	2	1
4.	1	2	5	1
5.	2, 3	1	6	3

IP = Immediate predecessors

α_r = shortening cost of job r/ unit time shortened

(Ford and Fulkerson [1962 of Chapter 1])

7.9 Consider the project with precedence relationships given below. Draw the arrow diagram for it using the smallest number of nodes. Let G be this arrow diagram. Show that it is possible to decrease the number of dummy arcs in G by increasing the number of nodes. Using this example show that it may not be possible to minimize the number of arcs and the number of nodes in the arrow diagram for a project simultaneously even if there are no parallel activities in it.

Jobs	Immediate Predecessors
a, b, b', c, g, p, l	None
e	a, b, b', p
f	b, b', c, l
d	g, b, b', c
h	a
i	b
j	c
m	b'

(Syslo[1984])

7.10 Develop an efficient algorithm to check whether a given project can be represented by an arrow diagram using no dummy arcs at all. (Syslo[1984])

7.11 Let H = (N, A) be a graph with $N = \{ 1, \ldots, n \}$, $A = \{ e_1, \ldots, e_m \}$, in which each node has degree 2 or 3. A *node cover* for H is a subset of nodes in N which covers all the edges in A.

Define a project with $n + m + 1$ activities numbered 1 to $n + m + 1$, related to the nodes and edges in H with precedence relations among them defined by the data in H as follows: for $i = 1$ to n activity i in the project corresponds to node

i in H; for $p = 1$ to m activity $n + p$ in the project corresponds to edge e_p in H; activity $n + m + 1$ is an additional activity; for $p = 1$ to m and $i = 1$ to n, activity $n + p$ is a predecessor of activity i if e_p is incident to i in H; and for all $p = 1$ to m, activity $n + p$ is a predecessor of activity $n + m + 1$.

In drawing the arrow diagram for this project, since each of the activities $n + 1$ to $n + m$ have no predecessors, they can be represented by arcs with the same tail node (this is the *start* node). Similarly activities $1, \ldots, n, n + m + 1$ have no successors, so they can be represented by arcs with the same head node (this is the *finish* node) in the arrow diagram. When drawn in this way, show that the minimum number of dummy activities needed to represent this project is precisely the minimum number of nodes that cover all the edges in H. Since the problem of finding a minimum cardinality node cover in H is known to be NP-hard, this establishes that the problem of drawing an arrow diagram for a project using the smallest number of dummy arcs is also NP-hard. (Krishnamoorthy and Deo [1979])

7.12 Let $G = (\mathcal{N}, \mathcal{A})$ be the arrow diagram for a project. Consider the project shortening cost minimization problem on G. In addition to the features discussed in Section 7.2, suppose a subset of nodes $P \subset \mathcal{N}$ called *penalty nodes* is specified, with a due date of d_i for $i \in P$. Nodes in P correspond to *key events*, and for each i in it, a penalty of $p_i\$$ is levied per unit time delay of event associated with it beyond its due date d_i. It is required to minimize the total cost of shortening the activities and the penalties of violating the target dates of key events, treating the project duration λ as a parameter. Develop a modification of the algorithm discussed in Section 7.2 to solve this problem. Apply this algorithm on the project network given in Figure 7.13. (Kanda and Rao [1984])

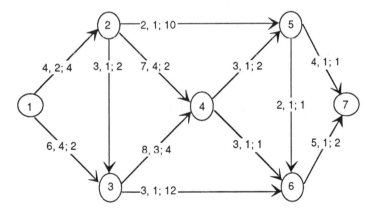

Figure 7.13 Data on arc (i, j) is $\bar{\tau}_{ij}, \underline{\tau}_{ij}.\alpha_{ij}$, in that order. Penalty nodes are 4 ($d_4 = 12$, $p_4 = 10$) and 7 ($d_7 = 18$, $p_7 = 20$).

7.13 Let $G = (\mathcal{N}, \mathcal{A})$ be the AON network for a project. Each job takes exactly one day to process. On each day, any number of jobs can be processed provided they are all unrelated and each of their predecessors has all been processed already, i.e., if job i is processed on day t and $(i, j) \in \mathcal{A}$, then job j can be processed on the $t + 1$th day or any later day. c_{it} is the cost of doing job i on day t. All c_{it} are given and they are all > 0.

Define decision variables $x_{it} = 1$ if job i is processed on the tth day or before, 0 otherwise. Formulate the problem of completing the project at minimum cost using these decision variables. Show how this problem can be solved using the algorithms discussed in Chapter 2. (G. Chang and J. Edmonds)

7.14 The Payment Scheduling Problem　　Let $G = (\mathcal{N}, \mathcal{A})$ be the arrow diagram for a project with nodes 1, n as the start and finish nodes, and $(t_{ij} : (i, j) \in \mathcal{A})$ as the vector of job durations. Define $t_1 = 0$, t_i for $i = 2$ to n to be the time when the event corresponding to node i is realized. In this problem, the variables are t_2 to t_n, the vector $t = (t_2, \ldots, t_n)$ is called the *schedule*. The realization of the event corresponding to any node i is usually the occasion for a transaction (either the contractor doing the project is given a stage payment for reaching this milestone in this project, or he may have to pay a subcontractor whom he hired to do part of the work), let c_i (> 0 for receipts, < 0 for payments) denote the cash reward to the contractor at this event. The payment scheduling problem is concerned with finding a feasible schedule that maximizes the present value (at the commencement of the project) of all the cash transactions. Assuming that the discount rate for money per unit per unit time is β, this is the problem of finding a schedule t that maximizes $P(t) = \sum (c_i \exp(-\beta t_i) :$ over $i = 2$ to $n)$, subject to $t_j - t_i \overset{\geq}{} t_{ij}$ for each $(i, j) \in \mathcal{A}$ and $t_n - t_1 \overset{\leq}{} \lambda =$ upper bound on project completion time. Show that this problem can be transformed into an LP by transforming the variables using $t_i = -(1/\beta) \log(y_i), i = 2$ to n.

Show that every extreme point of this LP corresponds to a spanning tree in G' $= (\mathcal{N}, \mathcal{A}')$ where $\mathcal{A}' = \mathcal{A} \cup \{(n, 1)\}$, and vice versa. Hence the search for optimal schedules can be restricted to feasible trees in G'. Using standard complementary slackness results for checking the optimality of feasible trees, develop a simplex -like procedure for this problem that moves from one feasible tree to an adjacent one obtained by changing the tree by one arc, improving the objective function value in each move, until an optimum schedule is obtained. (Grinold [1972], Russel [1970]. See Elmaghraby and Herroelen [1990] for a critique of this model.)

Comment 7.1　　The first paper to discuss the problem of computing the cost curves for a project composed of many individual jobs is that by Kelly and Walker [1959]. In this chapter we discussed methods for computing the project cost curve by network flow methods due to Fulkerson [1961] and Kelly [1961]. Since the appearance of these papers, the network-based CPM has become a part of the language of project management, and has been used extensively in planning, scheduling and controlling large projects. The glamorous successes claimed for their initial applications, and the adoption of these models as standard requirement in contracts by many governments, have added to their importance. Computer packages for these network based techniques specialized to the needs of a variety of industries continue to be the best sellers of all OR software. In the chapter's exercises, we have included some modeling problems taken from simplified real world applications.

The literature on network techniques for project management is enormous. Battersby [1967], Burman [1972], Elmaghraby [1977], Weist and Levy [1977] are some of the books devoted exclusively to this area.

Krishnamoorthy and Deo [1979] are the first to show that the problem of generating an arrow diagram for a project using the smallest number of dummy arcs is NP-hard. The papers of Syslo [1981, 1984] explore some other complexity issues associated with arrow diagrams. Dimsdale [1963], Fisher, Liebman, and Nemhauser [1968], discuss practical techniques for generating arrow diagrams.

The papers of Russel [1970], and Grinold [1972] discuss the payment scheduling problem. Elmaghraby and Herroelen [1990] give a critique of this model.

7.6 References

A. BATTERSBY, 1967, *Network Analysis for Planning and Scheduling*, Macmillan & Co., London.

P. J. BURMAN, 1972, *Precedence Networks for Project Planning and Control* , McGraw-Hill, London.

D. G. CORNEIL, C. C. GOTLIEB and Y. M. LEE, 1973, "Minimal Event-Node Network of Project Precedence relations," *CACM*, 16(296-298).

D. DIMSDALE, March 1963, "Computer Construction of Minimal Project Network," *IBM Systems Journal*, 2(24-36).

S. E. ELMAGHRABY, 1977, *Activity Networks*, Wiley, NY.

S. E. ELMAGHRABY and W. S. HERROELEN, 1990, "The Scheduling of Activities to Maximize the Net Present Value of Projects," *EJOR*, 49(35-49).

A. C. FISHER, D. S. LIEBMAN, and G. L. NEMHAUSER, July 1968, "Computer Construction of Project Networks," *CACM*, 11(493-497).

D. R. FULKERSON, Jan. 1961, "A Network Flow Computation for Project Cost Curve," *MS*, 7, no. 2 (167-178).

R. C. GRINOLD, 1972, "The Payment Scheduling Problem," *NRLQ*, 19, no. 1(123-136).

A. KANDA and V. R. K. RAO, May 1984, "A Network Flow Procedure for Project Crashing with Penalty Nodes," *EJOR*, 16, n0. 2(123 -136).

A. KANDA and N. SINGH, July 1988, "Project Crashing with Variations in Reward and Penalty Functions: Some Mathematical Programming Formulations," *Engineering Optimization*, 13, no. 4(307-315).

J. E. KELLY, Jr., 1961, "Critical Path Planning and Scheduling: Mathematical Basis," *OR*, 9(296-320).

J. E. KELLY, Jr. and M. R. WALKER, Dec. 1959, "Critical Path Planning and Scheduling," *Proceedings of the Eastern Joint Computer Conference*, Boston, MA.

M. S. KRISIINAMOORTHY and N. DEO, 1979, "Complexity of the Minimum Dummy Activities Problem in a PERT Network," *Networks*, 9(189-194).

A. H. RUSSEL, Jan. 1970, "Cash Flows in Networks," *MS*, 16, no. 5(357-373).

M. M. SYSLO, 1981, "Optimal Constructions of Event-Node Networks," *RAIRO Recherche Operationelle*, 15(241-260).

M. M. SYSLO, 1984, "On the Computational Complexity of the Minimum-Dummy-Activities Problem in a PERT Network," *Networks*, 14(37-45).

J. D. WEIST and F. K. LEVY, 1977, *A Management Guide to PERT/CPM*, Prentice-Hall, Englewood Cliffs, NJ, 2nd Ed.

R. J. WILLIS and N. A. J. HASTINGS, 1976, "Project Scheduling With Resource Constraints Using Branch and Bound Methods," *ORQ*, 27, no. 2, i(341-349).

Chapter 8

Generalized Network Flows

In flow problems discussed so far, we assumed that if f_{ij} units of a commodity enter an arc (i, j) at its tail node i and travel across the arc, then exactly the same f_{ij} units will reach its head node j. This assumption may not hold in some flow models. For example, in a water distribution network, if some quantity of water is shipped across an open canal linking two nodes, some is lost due to evaporation and seepage during transit, and the amount reaching the destination will only be a fraction of the amount that left the origin. The same phenomenon takes place in the transmission of electric power through high voltage transmission lines, because of transmission losses. On the other hand, if we transmit money from one period to the next by holding it in a bank account (cash flow), because of the interest earned, the amount reaching the destination will be more than the amount that left the origin. In all these examples there exists a positive *multiplier* p_{ij} associated with arc (i, j) such that if a packet of f_{ij} units of the commodity enter the arc (i, j) at node i, and travels through the arc, then by the time it reaches node j, the packet contains $p_{ij}f_{ij}$ units of the commodity. If $0 < p_{ij} < 1$, arc (i, j) is said to be *lossy*; and if $1 < p_{ij} < \infty$, it is said to be *gainy*. In pure networks studied so far, $p_{ij} = 1$ for all arcs (i, j), flow problems on them have been called *pure network flow problems*. If $p_{ij} \neq 1$ for at least one arc, the network is called a *generalized network,* or a *network with multipliers,* or a *network with gains or losses,* and flow problems on it are called *generalized network flow problems*.

We assume that the flow variable f_{ij} associated with an arc (i, j) in a generalized network always refers to the amount of material entering this arc at its tail node i for transit to node j. We also assume that all the data on this arc (lower bound, capacity, cost coefficient, multiplier) applies to this variable. Let $G = (\mathcal{N}, \mathcal{A}, \ell = (\ell_{ij}), k = (k_{ij}), p = (p_{ij}), c = (c_{ij}), \check{s}, \check{t})$ be a connected directed generalized network with $|\mathcal{N}| = n \geqq 2, |\mathcal{A}| = m, p$ as the vector of multipliers associated with the arcs in $\mathcal{A}$, and the usual meaning for the other symbols. Let $v_{\check{s}}, v_{\check{t}}$ denote the amounts of material leaving the source node $\check{s}$, and arriving at the sink node $\check{t}$ respectively. Then the flow vector $f = (f_{ij})$ is feasible in G if it satisfies

$$- \sum_{j \in \mathbf{A}(i)} f_{ij} + \sum_{j \in \mathbf{B}(i)} p_{ji} f_{ji} \;=\; \left\{ \begin{array}{ll} -v_{\breve{s}} & \text{if } i = \breve{s} \\[2ex] 0 & \text{if } i \neq \breve{s} \text{ or } \breve{t} \\[2ex] v_{\breve{t}} & \text{if } i = \breve{t} \end{array} \right. \tag{8.1}$$

$$\ell_{ij} \leqq f_{ij} \leqq k_{ij} \quad , \qquad \text{for all } (i,j) \in \mathcal{A}$$

where $\mathbf{A}(i)$, $\mathbf{B}(i)$ are the after i, and before i sets in G. Because of the multipliers, $v_{\breve{t}}$ may not be equal to $v_{\breve{s}}$ in (8.1). In the coefficient matrix of the equality constraints in (8.1), each column has exactly 2 nonzero entries, one of them a "-1," and the other the positive multiplier associated with the arc in G corresponding to this column. We may be interested in maximizing $v_{\breve{t}}$ given $v_{\breve{s}}$ subject to (8.1), this problem is known as the *minimum loss problem*. A feasible flow vector which maximizes $v_{\breve{t}}$ is known as a *maximum feasible flow vector*. Among all maximum feasible flow vectors, the one which is associated with the smallest value of $v_{\breve{s}}$ is known as an *optimum maximum feasible flow vector*. The minimum cost flow problem in G deals with minimizing $\sum (c_{ij} f_{ij} : \text{over } (i,j) \in \mathcal{A})$ subject to (8.1), given $v_{\breve{t}}$ or $v_{\breve{s}}$. This is the general problem that we will consider.

Sometimes there may be gains or losses occurring at the nodes, on the material passing through them. On such networks, replace each node i associated with a gain or loss factor, by an arc (i_1, i_2) with the new nodes i_1, i_2 representing the receiving and departing ends of the original node i as discussed in Section 2.1, and make the multiplier of this arc (i_1, i_2) equal to the gain or loss factor at i. In the modified network only arcs have multipliers.

THEOREM 8.1 *Let A be the $n \times m$ coefficient matrix of the system of equality constraints in (8.1). Each row (column) is associated with a node in $\mathcal{N}$ (arc in $\mathcal{A}$). If G is connected, the rank of A is $n - 1$ or n.*

Proof   Let $\mathbb{T}$ be a spanning tree in G. Draw $\mathbb{T}$ as a rooted tree with any arbitrary node selected as the root node. Let B be the square matrix of order $n - 1$ consisting of the columns in A associated with the arcs in $\mathbb{T}$, with the row corresponding to the root node struck off. We will now prove that the determinant of B is nonzero. Since $n \geqq 2$, there exists at least one nonroot terminal node in $\mathbb{T}$, say i_1. So, in the row corresponding to node i_1 there exists a single nonzero entry in B. Hence the determinant of B is a nonzero multiple of the determinant of the matrix obtained by striking off the row associated with node i_1, and the column of the nonzero entry in it, from B. The resulting matrix is of order $n - 2$, and it is associated with the tree obtained by deleting node i_1 and the unique arc incident at it, from $\mathbb{T}$. The same process can be repeated on this matrix, and continued. After $n - 2$ repetitions it leads to the conclusion that the determinant of B is nonzero. Hence the rank of A is $n - 1$ or n. ∎

THEOREM 8.2 *If* G *is connected, and the rank of A, the coefficient matrix of the system of equality constraints in (8.1), is $n - 1$, then (8.1) can be transformed into a pure network flow system.*

Proof Suppose the rank of A is $n - 1$. So, there exists $\alpha = (\alpha_1, \ldots, \alpha_n) \neq 0$ such that

$$\alpha_1 A_1. + \ldots + \alpha_n A_n. = 0 \tag{8.2}$$

We will now show that in any α satisfying (8.2), $\alpha_i \neq 0$ for all i, by contradiction. Suppose $\alpha_n = 0$. Select a spanning tree, $\mathbb{T}$, in G and draw it as a rooted tree with n as the root node. Let B be the square matrix of order $n - 1$, consisting of columns in A corresponding to arcs in $\mathbb{T}$, with the row corresponding to the root node n struck off. Since $\alpha_n = 0$, from (8.2), we have $\alpha'B = 0$, where $\alpha' = (\alpha_1, \ldots, \alpha_{n-1})$. The column in B corresponding to any in-tree arc incident at node n contains only a single nonzero entry (it lies in the row corresponding to the other node on that arc). This, and $\alpha'B = 0$ together imply that $\alpha_i = 0$ for all nodes i which are immediate successors of n in $\mathbb{T}$. In the same manner, going down $\mathbb{T}$ one level at a time, we conclude that α_i must be 0 for all i, a contradiction. Thus α_n could not have been 0. Similarly, $\alpha_i \neq 0$ for all i.

For $i = 1$ to n, multiply the equation in (8.1) corresponding to node i by α_i on both sides, and let A' be the matrix of coefficients of the modified system. From (8.2), the sum of the row vectors of A' is 0. Since there are exactly two nonzero entries in each column of A', this implies that the two nonzero entries in any column of A' must have the same absolute value, and opposite signs. Let γ_{ij} denote the absolute value of the nonzero entries in the column of A' corresponding to the arc $(i, j) \in \mathcal{A}$. Transform the variables using the linear transformation $f_{ij} = f'_{ij}/\gamma_{ij}$, for each $(i, j) \in \mathcal{A}$. So, if A'' is the coefficient matrix of the equality constraints after these transformations, each of its columns contains exactly two nonzero entries, a "-1" and a "$+1$." Thus A'' is the node-arc incidence of G, and therefore these transformations have converted (8.1) into a pure network flow system. ∎

Therefore, in the sequel, we will assume that the rank of A is n. Let $\mathbb{C}$ be an oriented simple cycle in G. The quantity

$$\frac{\text{Product of multipliers associated with reverse arcs in } \mathbb{C}}{\text{Product of multipliers associated with forward arcs in } \mathbb{C}} \tag{8.3}$$

is known as the *loop factor* of the cycle $\mathbb{C}$ under this orientation. If the orientation of $\mathbb{C}$ is reversed its loop factor gets inverted. As an example, consider the cycle in the subnetwork on the left in Figure 8.1 oriented so that (1, 2) is a forward arc. The numbers on the cycle arcs are the multipliers. Arcs (1, 2), (3, 4), (4, 5) are the forward arcs; and arcs (3, 2), (1, 5) are the reverse arcs. The loop factor of this cycle is $(5(4))/(2(1/2)1) = 20$. If the orientation of this cycle is reversed, its loop factor becomes 1/20.

$e_1, \ldots, e_r$, and the sequence of nodes is 1, 2, $\ldots$, r. Γ is linearly dependent iff there exists $(\alpha_1, \ldots, \alpha_r) \neq 0$ such that $\sum_{t=1}^{r} \alpha_t (\text{ column of } e_t \text{ in (8.8)}) = 0$. This equation holds iff when we make the flow amount on e_t equal to α_t for $t = 1$ to r, conservation holds at all the nodes $1, \ldots, r$. To check whether we can find a nonzero flow vector on $\mathbb{C}$ that maintains flow conservation at all the nodes, we select the initial arc, $e_1 = (1, 2)$, and fix the flow amount on it to be 1. This brings the amount p_{12} to node 2. If the next arc e_2 is a forward arc (i.e., $e_2 = (2, 3)$), for conservation to hold at node 2, the flow on it must be equal to p_{12}, and this brings $p_{12}p_{23}$ to node 3. On the other hand, if e_2 is the reverse arc $(3, 2)$, for conservation to hold at node 2, the flow on it must be $-p_{12}/p_{32}$, this leaves an amount of p_{12}/p_{32} at node 3 which has to be shipped out of the other arc incident at it for conservation to hold. Continuing this procedure, we can determine the flows on all the arcs on $\mathbb{C}$. When this procedure is completed by going around $\mathbb{C}$ once, and we come back to the initial node 1, we determine the flow on the initial arc $(1, 2)$ for conservation to hold at node 1; that flow amount, β, say, must turn out to be 1, the same quantity that we fixed it to be at the beginning of this procedure. Otherwise there is no nonzero flow on $\mathbb{C}$ that maintains conservation at all the nodes on $\mathbb{C}$.

Let $\mathcal{P}_{i1}$ denote the path from 1 to i as we travel along the oriented cycle $\mathbb{C}$ from node 1 to node i.

Let e_t be any arc on $\mathbb{C}$, i its tail node, and let $\gamma_t = 1/(\text{path factor of } \mathcal{P}_{i1})$. It can be verified that the flow on e_t obtained in this procedure is $-\gamma_t$ if e_t is a reverse arc on $\mathbb{C}$, or γ_t if e_t is a forward arc on $\mathbb{C}$. So, $\beta = 1/\Delta$. Hence, a nonzero flow vector maintaining conservation at all the nodes exists in $\mathbb{C}$ iff $1/\Delta = 1$, i.e., iff $\Delta = 1$. Therefore, Γ is linearly dependent iff $\Delta = 1$. ∎

There are algorithms for solving the maximum flow problem in generalized networks, or the optimum maximum flow problem, based on FAPs obtained by using a labeling method or the shortest chain method, but we will not discuss these special methods. Instead we discuss an implementation of the primal simplex algorithm to solve (8.8) using tree labels without the need for computing the basis inverse in any step. This algorithm is general enough to handle any of these generalized network flow problems.

8.1 The Primal Simplex Method for Generalized Network Flow Problems

A *basis* for (8.8) is a square nonsingular submatrix of order n of the coefficient matrix of the system of equality constraints in it. The *basis network* corresponding to a basis B, denoted by G_B, is the subnetwork of G consisting of the arcs corresponding to columns of B. Arcs in the basis network are called *basic arcs*, those not in the basis network are called *nonbasic arcs*.

THEOREM 8.6 *A basis network* G_B *for (8.8) may consist of several connected components. Each connected component of* G_B *consists of a tree plus an additional* *arc which may be a self loop, and hence contains a unique loop or cycle.*

Proof   There are n columns in B, so if G_B is connected it is a connected network consisting of a spanning tree with one extra arc, and hence the statement of the theorem holds in this case.

Suppose G_B consists of t connected components, with the gth one consisting of r_g nodes for $g = 1$ to t. Since the gth component is connected, it must contain at least $r_g - 1$ arcs. The set of row vectors, $\mathbf{S}$, of the basis B corresponding to the r_g nodes in this component, must be linearly independent, as B is a basis. If this component contains only $r_g - 1$ arcs, there are only $r_g - 1$ columns of B which contain nonzero entries in rows of this set $\mathbf{S}$, this implies that the rank of $\mathbf{S}$ is $\leqq r_g - 1$, a contradiction. So, each connected component of G_B must contain at least as many arcs as nodes. Since B is square, the number of nodes and arcs in G_B are equal. These facts imply that each connected component of G_B has the same number of arcs as nodes, and hence it is a spanning tree in these nodes with one additional arc. This additional arc may either be a self loop or an out-of-tree arc. Hence each connected component of G_B contains a unique loop or cycle.   ∎

Rearrange the rows of the basis B so that the rows corresponding to nodes in each connected component of G_B appear consecutively. Now rearrange the columns of B so that the columns associated with arcs in each connected component of G_B appear consecutively; and in the same order in which these components appear among the rows. Then clearly B takes the following block diagonal form.

$$
B = \begin{pmatrix} B_1 & & & \\ & B_2 & & \\ & & \ddots & \\ & & & B_t \end{pmatrix}
$$

$$
B_g = \begin{pmatrix}
a_1 & & & & & \\
-1 & a_2 & & & & \\
& -1 & & & & \\
& & \ddots & & & \\
& & & -1 & a_{r_g-1} & \\
& & & & -1 & \boxed{a_{r_g}} \\
& & & & & \text{self loop} \\
& & & & & \text{portion}
\end{pmatrix}
$$

Here B_g is a square nonsingular matrix of order r_g for $g = 1$ to t, and all elements in B outside these diagonal blocks are 0. B_g is itself a triangular matrix if the loop in the connected component of G_B corresponding to it is a self loop. In this case, the rows and columns of B_g can be rearranged so that it has the structure given above, where a_{r_g} is either $+1$ or -1, and is the unique nonzero entry in the column in B corresponding to this self loop. $a_1, \ldots, a_{r_g-1}$ are the multipliers associated with the other arcs in this connected component.

If the loop in the gth connected component of G_B is a cycle which is not a self loop, the rows and columns of B_g can be rearranged so that it has the following structure.

$$
B_g = \begin{pmatrix}
a_1 & & & & & & & & \\
-1 & a_2 & & & & & & & \\
& -1 & & & & & & & \\
& & \ddots & & & & & & \\
& & & a_{w-1} & & & & & \\
& & & -1 & a_w & & & & -1 \\
& & & & -1 & a_{w+1} & & & \\
& & & & & -1 & & & \\
& & & & & & \ddots & & \\
& & & & & & & -1 & a_{r_g}
\end{pmatrix}
\tag{8.9}
$$

$$
\text{The cycle portion}
$$

Columns 1 to $w-1$ in (8.9) correspond to arcs in the gth connected component of G_B which are not on the cycle, the remaining columns w to r_g correspond to arcs on the cycle. If the "-1" entry in row w and the last column is made 0 in (8.9), then B_g becomes triangular. Hence in this case B_g is said to be *near triangular*.

A *quasitree* in G is a partial subnetwork $(\hat{\mathcal{N}}, \hat{\mathcal{A}})$ with $\hat{\mathcal{N}} \subseteq \mathcal{N}$, and $\hat{\mathcal{A}}$ consisting of the arcs in a tree spanning the nodes in $\hat{\mathcal{N}}$, plus either a self loop at one of the nodes in $\hat{\mathcal{N}}$, or an arc joining two nodes in $\hat{\mathcal{N}}$. By Theorem 8.6 every connected component in the basis network G_B is a quasitree. Thus a basis network for the generalized network flow problem (8.8) is not a spanning tree as in pure network flow problems, but a set of disjoint quasitrees. We will now discuss how quasitrees can be stored and manipulated using node labels just as trees were. Node labels for storing each quasitree are derived separately as described below.

When all the cycle arcs are deleted from a quasitree leaving all the nodes as they are, we are left with a set of *tributary trees*, some of which may consist of a single node on the cycle. See Figures 8.1, 8.2. For each tributary tree choose the node in it that belongs to the cycle as the root, and generate the predecessor, successor, and brother labels for all the nonroot nodes exactly as in Section 1.2. If the cycle is a self loop, at node i say, make the predecessor index of i to be $+i$ or $-i$ depending on whether the nonzero entry in the column corresponding to it is $+1$ or -1. If the cycle is not a self loop, let the nodes on it be $i_1, \ldots, i_h$ in this order when the cycle is written as a path from some node i_1 back to itself with some orientation. Then for $r = 1$ to h make the predecessor index of node i_r either $+i_{r-1}$ or $-i_{r-1}$ depending on whether the arc joining them is a forward or reverse arc under the orientation chosen for the cycle, where $i_0 = i_h$. Thus each node on the cycle is its own ancestor. The predecessor indices of nodes on the cycle can be used to trace these nodes uniformly in a clockwise or counterclockwise direction. Also, each node in the cycle has an equal status as an ancestor of all nodes in the quasitree. For each $r = 1$ to h, list i_r as the youngest son of i_{r-1} (so, i_r is younger to every immediate successor of i_{r-1} in the tributary tree rooted at it in the quasitree).

These labels are said to define the *rooted loop labeling* of the quasitree. In this labeling, every node has a predecessor, and hence the predecessor index of no node is empty. As an example consider the basis network given in Figure 8.1. There are two quasitrees in it. In the left hand quasitree there is a cycle consisting of nodes 1, 2, 3, 4, 5, and of these only 1, 3 have tributary trees with one or more arcs (see Figure 8.2). The second quasitree on the right has a self loop at node 17 which is a slack self loop. The node labels corresponding to this basis are given in the following table (empty labels are left blank in the table).

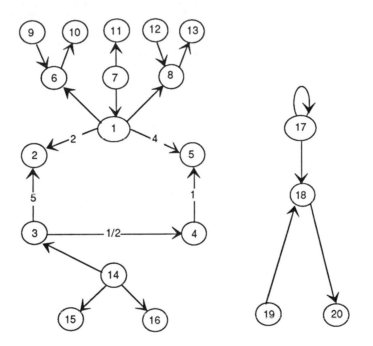

Figure 8.1 A basis network with two quasitrees. The numbers on the cycle arcs are the arc multipliers.

To find the cycle or the loop in any quasitree using the predecessor indices, select any node, say i, in this quasitree and trace its predecessor path. Keep tracing this path until at some stage a node on this path, say i_1, appears for a second time (this will always happen). The cycle in this quasitree consists of all the arcs and

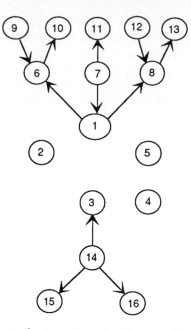

Figure 8.2   The tributary trees in the quasitree on the left of Figure 8.1.

Node	1	2	3	4	5	6	7	8	9	10
PI	− 5	1	− 2	3	4	1	− 1	1	− 6	6
SI	− 8	3	14	− 5	1	−10	−11	−13		
EBI		6		14		7	8		10	
YBI						2	6	7		9

Node	11	12	13	14	15	16	17	18	19	20
PI	7	− 8	8	− 3	14	14	17	17	−18	18
SI				−15			−18	19		
EBI		13				15	18			19
YBI			12	4	16			17	20	

nodes obtained in this process after i_1 has been obtained for the first time. The entire path traced in this process beginning with node i is known as the *predecessor path* of i, it duplicates no arcs, and it contains the cycle or loop in this quasitree. As an example, the predecessor path of node 12 in the quasitree on the left hand side in Figure 8.1 is: 12, (12, 8), 8, (1, 8), 1, (1, 5), 5, (4, 5), 4, (3, 4), 3, (3, 2), 2, (1, 2), 1, and this path contains the unique cycle in this quasitree.

Thus predecessor paths in a basis network for a generalized network flow problem are elementary but not simple paths.

Node labels other than the predecessor, successor, and brother indices are sometimes used in generalized network codes to store and manipulate quasitrees. Among these are the distance or depth label, thread label, etc. The distance or depth label is defined to be 0 for every node in the cycle or self loop in a quasitree. The depth label for a non-cycle node is exactly its depth in the tributary tree in which it is contained as defined in Section 1.2.2. So, the depth label of any node in a quasitree is the number of non-cycle arcs in its predecessor path. The thread label is defined exactly as in Section 1.2.2. As an example, consider the quasitree in Figure 8.3 (arc orientations are not shown for simplicity), the depth and thread labels for nodes on it are shown in the table following the figure. With the thread label defined this way, it can be verified that the set of descendents of any node can be obtained by recursive applications of the maps $t^r(i)$ exactly as discussed in Section 1.2.2.

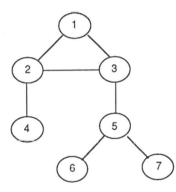

Figure 8.3

Node	1	2	3	4	5	6	7
Predecessor label	3	1	2	2	3	5	5
Depth label	0	0	0	1	1	2	2
Thread label	2	4	5	3	6	7	1

Let A be the coefficient matrix of the system of equality constraints in (8.8). Since G is a connected generalized network with $|\mathcal{N}| = n$, rank of A is n. A basic solution for (8.8) corresponds to a partition of the arcs in G into $(\mathbf{Q}, \mathbf{L}, \mathbf{U})$ where $\mathbf{Q}$ is the set of basic arcs, and $\mathbf{L}, \mathbf{U}$ are the sets of nonbasic arcs on which the flows are fixed equal to the lower, upper bounds respectively. In any partition $(\mathbf{Q}, \mathbf{L}, \mathbf{U})$, $|\mathbf{Q}| = n$, the matrix B consisting of the column vectors in A corresponding to arcs in $\mathbf{Q}$ is a basis for A, and every arc in $\mathbf{U}$ has finite capacity. By Theorem 8.5, the loop factors of all the cycles in the basis network $(\mathcal{N}, \mathbf{Q})$ which are not self loops, are different from 1.

**How To Compute The Primal Basic Solution
Corresponding To A Given Partition (Q, L, U)**

Let B be the basis for (8.8) corresponding to $\mathbf{Q}$, and let $\bar{f} = (\bar{f}_{ij})$ denote the basic solution of (8.8) corresponding to the partition $(\mathbf{Q}, \mathbf{L}, \mathbf{U})$. By definition, $\bar{f}_{ij} = \ell_{ij}$ if $(i,j) \in \mathbf{L}$, and $= k_{ij}$ if $(i,j) \in \mathbf{U}$. For each $(i,j) \in \mathbf{L} \cup \mathbf{U}$, multiply the column vector corresponding to it in (8.8) by $\bar{f}_{ij}$ and transfer it to the right hand side, and suppose this changes the right hand side constants vector in (8.8) to b'. b' is the vector of remaining requirements at the nodes after the flows on arcs in $\mathbf{L}$, $\mathbf{U}$ are fixed at their lower bounds, capacities respectively. Now only the flow amounts on the basic arcs remain to be computed in order to satisfy these remaining requirements. These are obtained by solving

$$B(\bar{f}_{ij} : (i,j) \in \mathbf{Q}) = b' \tag{8.10}$$

Since B has a block diagonal structure as discussed earlier, we can solve (8.10) by finding the flow amounts on arcs in each quasitree in the basis network $(\mathcal{N}, \mathbf{Q})$ separately.

Consider the gth quasitree in $(\mathcal{N}, \mathbf{Q})$. Computing the flow amounts on arcs in this quasitree can be conveniently carried out in two stages. In Stage 1 flows on arcs on the tributary trees in this quasitree are determined using back substitution as discussed in Section 5.4. The Stage 1 procedure processes each non-cycle node on the tributary tree exactly once. Once a node is processed, the flow amounts on all the arcs incident at it are known. At any time in Stage 1, $\mathbf{Y}$ denotes the set of processed nodes at that time. This procedure also maintains another set $\mathbf{X}$ of non-cycle nodes in the quasitree satisfying the property that for each node in $\mathbf{X}$, the flow amounts on all the arcs incident at it are already known at this time except on the arc joining it to its immediate predecessor. After the flow amount on each arc is obtained, the requirements at the nodes on it are updated to cancel the requirements fulfilled by it, and obtain the remaining requirements still to be fulfilled, denoted by (b'_i) itself.

PROCEDURE FOR DETERMINING THE FLOW AMOUNTS ON BASIC ARCS IN A QUASITREE IN $(\mathcal{N}, \mathbf{Q})$ IN THE BASIC SOLUTION ASSOCIATED WITH THE PARTITION $(\mathbf{Q}, \mathbf{L}, \mathbf{U})$

Stage 1   Initiate with $\mathbf{X} =$ set of non-cycle leaf nodes in the tributary trees in this quasitree, $\mathbf{Y} = \emptyset$.

In a general step, select and delete a node from $\mathbf{X}$, say i. Let j be the immediate predecessor of i. Let b'_i, b'_j be the remaining requirements at nodes i, j at this time. If the basic arc joining i and j is (i,j), make $\bar{f}_{ij} = -b'_i$ and change the remaining requirement at j to $b'_j - \bar{f}_{ij}p_{ij}$. If the basic arc joining i, j is (j,i), make $\bar{f}_{ji} = b'_i/p_{ji}$ and change the remaining requirement at j to $b'_j + b'_i/p_{ji}$. Include i in $\mathbf{Y}$. If j is a non-cycle node, and if all the brothers of i in its tributary tree are already in $\mathbf{Y}$, include j in $\mathbf{X}$.

If $\mathbf{X} \neq \emptyset$ repeat the general step above. If $\mathbf{X} = \emptyset$, go to Stage 2.

Stage 2   If the cycle in the quasitree is a self loop, suppose it is (i, i). Let b'_i be the remaining requirement at node i at this time. Then $\bar{f}_{ii} = -b'_i$ or $+b'_i$ depending on whether the self loop at i is a surplus or slack self loop.

Suppose the cycle is not a self loop. Suppose the nodes in it are $1, 2, \ldots, t$ in this order when it is oriented with an arc incident at 1, say $e_1 = (1, 2)$, as a forward arc. Let the remaining arcs on the cycle in this order be $e_2, \ldots, e_t$; so, e_r is either $(r, r+1)$ or $(r+1, r)$ depending on whether it is a forward or a reverse arc. Let p_r be the multiplier associated with e_r, and let a_r be p_r or $1/p_r$ depending on whether e_r is a forward or reverse arc, for $r = 1$ to t. Then $\Delta = a_1 a_2 \ldots a_t$ is the inverse of the loop factor of this cycle under the present orientation, $\Delta \neq 1$ by Theorem 8.5. Denote the flow amount on e_r in the solution we are computing by $\bar{f}_r$. Let b'_r be the remaining requirement at node r at this time, for $r = 1$ to t. Then from (8.9) we see that $(\bar{f}_r : r = 1$ to $t)$ is the solution of a system of equations of the form

$$
\begin{pmatrix}
a_{11} & a_{12} & & & & \\
 & a_{22} & a_{23} & & & \\
 & & a_{33} & & & \\
 & & & \ddots & & \\
 & & & & a_{t-1,t-1} & a_{t-1,t} \\
a_{t1} & & & & & a_{tt}
\end{pmatrix}
\begin{pmatrix}
\bar{f}_1 \\ \bar{f}_2 \\ \bar{f}_3 \\ \vdots \\ \bar{f}_{t-1} \\ \bar{f}_t
\end{pmatrix}
=
\begin{pmatrix}
b'_1 \\ b'_2 \\ b'_3 \\ \vdots \\ b'_{t-1} \\ b'_t
\end{pmatrix}
\tag{8.11}
$$

where for each r, $\{a_{rr}, a_{r,r+1}\} = \{-1, p_{r+1}\}$ in (8.11). This system is *near triangular*, in the sense that if the flow amount on one of the arcs in the cycle is given, say α, the flow amounts on the other arcs can be computed by back substitution from (8.11) as affine functions of α. Also, by going around the cycle once we get an equation for α from which α can be uniquely determined.

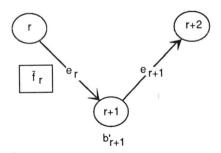

Figure 8.4 Flow amount on e_r is $\bar{f}_r$.

If $\bar{f}_r$ is given, to determine the flow amount on the next arc in the cycle, $\bar{f}_{r+1}$, four cases arise corresponding to the orientations of the two arcs e_r, e_{r+1}. If both e_r, e_{r+1} are forward arcs (see Figure 8.4), for conservation to hold at node $r + 1$, we clearly have $\bar{f}_{r+1} = -b'_{r+1} + p_r f_r$.

Similarly, if e_r is a forward arc and e_{r+1} is a reverse arc, $\bar{f}_{r+1} = (b'_{r+1} - p_r \bar{f}_r)/p_{r+1}$. If e_r is a reverse arc and e_{r+1} is a forward arc, $\bar{f}_{r+1} = -b'_{r+1} - \bar{f}_r$. If both e_r, e_{r+1} are reverse arcs, $\bar{f}_{r+1} = (b'_{r+1} + \bar{f}_r)/p_{r+1}$.

Assume that the flow amount on e_1, $\bar{f}_1$, is α. By using the formulas discussed above, all $\bar{f}_r$ can be obtained as functions of α. By going around the cycle once completely we get an expression for α in terms of itself, which leads to

$$\alpha = (-b'_1 - a_t b'_t - a_t a_{t-1} b'_{t-1} - \ldots - a_t a_{t-1} \ldots a_2 b'_2)/(1 - \Delta) \qquad (8.12)$$

Get $\alpha = \bar{f}_1$ from (8.12). Then, compute the flow amounts on the other arcs in the cycle using the formulas developed above.

A partition $(\mathbf{Q}, \mathbf{L}, \mathbf{U})$ for (8.8) is said to be *primal feasible* if the primal basic solution corresponding to it, $\bar{f} = (\bar{f}_{ij})$, is feasible to (8.8), i.e., if $\ell_{ij} \leqq \bar{f}_{ij} \leqq k_{ij}$ for all $(i, j) \in \mathbf{Q}$; otherwise it is said to be *primal infeasible*.

How to Compute the Dual Basic Solution Associated with a Partition

Let $(\mathbf{Q}, \mathbf{L}, \mathbf{U})$ be a given partition for (8.8). The dual basic solution associated with this partition is the node price vector $\pi = (\pi_i)$ obtained from

$$
\begin{aligned}
p_{ij}\pi_j - \pi_i &= c_{ij}, \text{ for each } i \neq j, (i, j) \in \mathbf{Q} \\
a_{ii}\pi_i &= c_{ii}, \text{ for each self loop } (i, i) \in \mathbf{Q}
\end{aligned}
\qquad (8.13)
$$

where $a_{ii} = \pm 1$ is the single nonzero entry in the column of f_{ii} in (8.8), and c_{ii} the cost coefficient of f_{ii}. The node price computation in each quasitree of the basis network can be carried out separately.

If a quasitree contains a self loop, at node i say, then from (8.13), $\pi_i = c_{ii}/a_{ii}$. Using this value of π_i, the node prices for the other nodes in this quasitree can be computed from the remaining equations in (8.13) by back substitution, going down this quasitree beginning at i level by level.

Suppose a quasitree contains a cycle consisting of t nodes; $1, \ldots, t$ say; in this order, when it is oriented with an arc incident at 1, $e_1 = (1, 2)$ as a forward arc. Let the remaining arcs on the cycle in this order be $e_2, \ldots, e_t$; so, e_r is either $(r, r + 1)$ or $(r + 1, r)$ depending on whether it is a forward or a reverse arc. Let c_r, p_r be the cost coefficient, multiplier associated with e_r, $r = 1$ to t. For $r = 1$ to t define $a_r = p_r$, $\gamma_r = 1/p_r$ if e_r is a forward arc on the cycle; otherwise $a_r = 1/p_r, \gamma_r = -1$. $\Delta = a_1 a_2 \ldots a_t$ is the inverse of the loop factor of this cycle under the present orientation, $\Delta \neq 1$ by Theorem 8.5. Applying (8.13) to e_r on the cycle we get

$$\pi_{r+1} = \begin{cases} \frac{\pi_r}{p_r} + \frac{c_r}{p_r}, & \text{if } e_r \text{ is a forward arc} \\[2mm] p_r\pi_r - c_r, & \text{if } e_r \text{ is a reverse arc} \end{cases} \tag{8.14}$$

By applying (8.14) once around the cycle we get

$$\pi_1 = \left(\frac{\Delta}{\Delta - 1}\right)\left(c_t\gamma_t + \frac{c_{t-1}\gamma_{t-1}}{a_t} + \frac{c_{t-2}\gamma_{t-2}}{a_t a_{t-1}} + \cdots + \frac{c_2\gamma_2}{a_t \cdots a_3} + \frac{c_1\gamma_1}{a_t \cdots a_2}\right)$$

Having obtained π_1, the value of π_j for each node j on the cycle can be computed using (8.14) by going around the cycle. Once node prices for all nodes on the cycle are known, the node prices of all the non-cycle nodes in this quasitree can be computed using (8.13) by going down each tributary tree beginning with the cycle node on it level by level.

The Primal Simplex Optimality Criterion

Given a primal feasible partition $(\mathbf{Q}, \mathbf{L}, \mathbf{U})$ for (8.8) associated with the BFS $\bar{f}$, and the dual basic solution π, the relative cost coefficient of arc $(i,j) \in \mathcal{A}$ wrt this partition is $\bar{c}_{ij} = c_{ij} - (p_{ij}\pi_j - \pi_i)$ if $i \neq j$, or $= c_{ii} - a_{ii}\pi_i$ if $i = j$, where a_{ii} is the single nonzero entry in the column associated with f_{ii} in (8. 8). $(\mathbf{Q}, \mathbf{L}, \mathbf{U})$ is an optimum partition for (8.8), and $\bar{f}$ is an optimum solution for it if

$$\begin{aligned} \bar{c}_{ij} > 0 \quad &\text{implies} \quad \bar{f}_{ij} = \ell_{ij}, \text{ i. e.,} (i,j) \in \mathbf{L} \\ \bar{c}_{ij} < 0 \quad &\text{implies} \quad k_{ij} \text{ is finite and } \bar{f}_{ij} = k_{ij}, \text{ i.e.,} (i,j) \in \mathbf{U} \end{aligned} \tag{8.15}$$

If (8.15) does not hold, define

$$\mathbf{E} = \{(i,j) : (i,j) \in \mathbf{L} \cup \mathbf{U} \text{ and violates (8.15)} \} \tag{8.16}$$

$\mathbf{E}$ is the set of nonbasic arcs eligible to be an entering arc for a pivot step in the present primal feasible partition to continue the primal simplex algorithm.

Basis Representation of an Entering Arc

Let $\mathbf{Q}, \mathbf{L}, \mathbf{U}$ be the present primal feasible partition for (8.8), and suppose the nonbasic arc (g, h) has been selected as the entering arc for a pivot step in it. To carry out the computations in the ensuing pivot step we need the updated column of the entering variable f_{gh} wrt the present basic set $\mathbf{Q}$, which is the pivot column for the pivot step. Let $H(g, h)$ denote the original column of f_{gh} in (8.8). $H(g, h) = -I_{.g} + p_{gh}I_{.h}$ (here I is the unit matrix of order n) if $g \neq h$, or $= a_{gg}I_{.g}$ if $g = h$ ($a_{gg} = \pm 1$ is the unique nonzero entry in the column of f_{gg} in (8.8)). The updated column of f_{gh} wrt $\mathbf{Q}$ is the vector of coefficients in the representation of $H(g, h)$ as a linear combination of the basic columns, hence it is called the *basis representation of the entering arc* wrt the basic set $\mathbf{Q}$. If B is the present basis (the matrix consisting of the columns corresponding to the basic arcs in $\mathbf{Q}$ in (8.8)), this representation, denoted by $\bar{y}$ is the solution of

$$B\bar{y} = H(g, h) \tag{8.17}$$

So, $\bar{y}$ is the vector of flow values on the basic arcs in $\mathbf{Q}$ to meet a requirement of -1 unit at node g, and p_{gh} units at node h, while the flow amounts on all the nonbasic arcs are 0. So, $\bar{y}$ can be computed directly using the procedure described earlier for computing the flows on basic arcs to meet specified requirements at the nodes. For any $i \in \mathcal{N}$, let $\mathcal{P}_i(\mathbf{Q})$ denote both the predecessor path of i in its quasitree in $\mathbf{Q}$, or the set of arcs on this path. Obviously, nonzero entries in the basis representation $\bar{y}$ of (g, h) correspond to arcs in $\mathcal{P}_g(\mathbf{Q}) \cup \mathcal{P}_h(\mathbf{Q})$.

Selection of the Dropping Arc and Updating the Primal Solution in a Pivot Step

Let $(\mathbf{Q}, \mathbf{L}, \mathbf{U})$ be the present primal feasible partition for (8.8) associated with the BFS $\bar{f} = (\bar{f}_{ij})$ in which the optimality criterion (8.15) is violated. Suppose the nonbasic arc (g, h) has been selected as the entering arc in this partition for a primal simplex pivot step. Let $\bar{y} = (\bar{y}_{ij} : (i, j) \in \mathbf{Q})$ be the updated column of the entering variable f_{gh}. Let $\mathcal{P}_g(\mathbf{Q}), \mathcal{P}_h(\mathbf{Q})$ denote the predecessor paths (or the set of arcs on these paths) of g, h in their quasitrees in $\mathbf{Q}$. In this pivot step we keep the flow amounts on all the nonbasic arcs other than (g, h) at their present values, change the flow amount on the entering arc (g, h) from its present $\bar{f}_{gh}$ to $\bar{f}_{gh} + \delta\lambda$, where $\delta = +1$ if $(g, h) \in \mathbf{L}$, or -1 if $(g, h) \in \mathbf{U}$; and reevaluate the flow amounts on the basic arcs in $\mathbf{Q}$ as functions of λ. This leads to the solution $\hat{f}(\lambda) = (\hat{f}_{ij}(\lambda))$ where

$$\hat{f}_{ij}(\lambda) = \begin{cases} \bar{f}_{ij} & \text{for } (i, j) \neq (g, h) \text{ in } \mathbf{L} \cup \mathbf{U} \\ \bar{f}_{ij} + \delta\lambda & \text{for } (i, j) = (g, h) \\ \bar{f}_{ij} - \delta\lambda\bar{y}_{ij} & \text{for } (i, j) \in \mathbf{Q} \end{cases} \tag{8.18}$$

Thus $\hat{f}_{ij}(\lambda)$ differs from $\bar{f}_{ij}$ only if $(i, j) \in \{(g, h)\} \cup (\mathcal{P}_g(\mathbf{Q}) \cup \mathcal{P}_h(\mathbf{Q}))$. The minimum ratio in this pivot step is the maximum value for λ that keeps $\hat{f}(\lambda)$ primal feasible, it is

$$\theta = \text{Min. } \{k_{gh} - \ell_{gh}\} \cup \left\{ \frac{k_{ij} - \bar{f}_{ij}}{-\delta\bar{y}_{ij}} : (i, j) \in \mathbf{Q} \text{ and } \delta\bar{y}_{ij} < 0 \right\} \cup$$
$$\left\{ \frac{\bar{f}_{ij} - \ell_{ij}}{\delta\bar{y}_{ij}} : (i, j) \in \mathbf{Q} \text{ and } \delta\bar{y}_{ij} > 0 \right\} \tag{8.19}$$

In computing θ we adopt the convention that the minimum in the empty set is ∞. θ is always nonnegative. If $\theta = \infty$, the unboundedness criterion is satisfied, $\hat{f}(\lambda)$ remains feasible for all $\lambda \geqq 0$ and its objective value $\rightarrow -\infty$ as $\lambda \rightarrow +\infty$, we terminate the algorithm. If θ is finite, let

$$\mathbf{D} = \{(i,j) : (i,j) \in \mathbf{Q} \cup \{(g,h)\} \text{ ties for the minimum in (8.19)}\} \qquad (8.20)$$

$\mathbf{D} \subset \{(g,h)\} \cup (\mathcal{P}_g(\mathbf{Q}) \cup \mathcal{P}_h(\mathbf{Q}))$ is the set of *blocking arcs*, i.e., arcs which are among those first driven to an upper or lower bound by an attempted change in the value of f_{gh} away from the bound it currently equals. Arcs in $\mathbf{D}$ are those which are eligible to be the *dropping* or *leaving arc* in this pivot step. Let $(r,s) \in \mathbf{D}$ denote the arc chosen as the dropping arc in this pivot step.

This pivot step is *degenerate* if $\theta = 0$, *nondegenerate* if $\theta > 0$. $\hat{f}(\theta)$ is the new BFS obtained after this pivot step. Notice that there is no change in the BFS in a degenerate pivot step.

If $(r,s) = (g,h)$, (g,h) is moved from $\mathbf{L}$ or $\mathbf{U}$ where it is presently contained, into the other set. There is no change in $\mathbf{Q}$. This gives the new partition. Since there is no change in $\mathbf{Q}$, there are no changes in the node labels or the dual solution π.

If $(r,s) \neq (g,h)$, delete (g,h) from $\mathbf{L}$ or $\mathbf{U}$ which contains it presently. Drop (r,s) from $\mathbf{Q}$, and include (g,h) in $\mathbf{Q}$ in its place. Include (r,s) in $\mathbf{L}$ or $\mathbf{U}$ depending on whether $\hat{f}_{rs}(\theta) = \ell_{rs}$ or k_{rs} respectively. Since there is a change in the basic set of arcs, the node labels and the dual basic solution have to be updated as described below.

After all the necessary updatings are carried out, the primal simplex algorithm moves to the next step with the new partition.

Updating the Node Labels and Node Prices in a Pivot Step

Let $(\mathbf{Q}, \mathbf{L}, \mathbf{U})$ be the present partition, (g,h) the entering arc, and $(r,s) \neq (g,h)$ the leaving arc. Let $\mathcal{P}_g(\mathbf{Q})$, $\mathcal{P}_h(\mathbf{Q})$ be the predecessor paths (or the set of arcs on these paths) of g, h in their quasitrees in $\mathbf{Q}$. $\hat{\mathbf{Q}} = \mathbf{Q} \cup \{(g,h)\} \backslash \{(r,s)\}$ is the new basic set of arcs in the next partition. $\hat{\mathbf{Q}}$ contains a new cycle that $\mathbf{Q}$ does not contain iff $(r,s) \in \mathcal{P}_g(\mathbf{Q}) \cap \mathcal{P}_h(\mathbf{Q})$.

If $\mathcal{P}_g(\mathbf{Q}) \cap \mathcal{P}_h(\mathbf{Q}) = \emptyset$, nodes g, h belong to different quasitrees in $\mathbf{Q}$. If $\mathcal{P}_g(\mathbf{Q}) \cap \mathcal{P}_h(\mathbf{Q}) \neq \emptyset$, nodes g, h belong to the same quasitree, and both $\mathcal{P}_g(\mathbf{Q})$, $\mathcal{P}_h(\mathbf{Q})$ contain the cycle in this quasitree.

For each $i \in \mathcal{N}$ define $\mathcal{P}_{ig}(\mathbf{Q})$ to be the empty path if i is not an ancestor of g, or the portion of the predecessor path of g up to node i the first time that i appears on the path if i is an ancestor of g. In the latter case $\mathcal{P}_{ig}(\mathbf{Q})$ is a simple path beginning with g and ending with i. We will also use the same symbol $\mathcal{P}_{ig}(\mathbf{Q})$ to denote the set of arcs on this path.

Let $\pi = (\pi_i), \hat{\pi} = (\hat{\pi}_i)$ denote the dual basic solutions wrt $\mathbf{Q}$, $\hat{\mathbf{Q}}$ respectively. Removing the outgoing arc (r, s) from $\mathbf{Q}$ without adding the incoming arc (g, h) creates a unique tree in $\mathcal{P}_g(\mathbf{Q}) \cup \mathcal{P}_h(\mathbf{Q})$, which we denote by $\mathbb{T}$. It is possible that $\mathbb{T}$ may consist of a single node only. See Figures 8.5, 8.6, 8.7. $\hat{\pi}_i = \pi_i$ for all nodes i except those on $\mathbb{T}$, as can be verified from the manner in which the dual solution is computed by the procedure discussed above. So, it is necessary to update the node prices only for the nodes in $\mathbb{T}$ in this step.

When $\mathcal{P}_g(\mathbf{Q}) \cap \mathcal{P}_h(\mathbf{Q}) \neq \emptyset$, we define

$$x \quad : \quad \text{the node on } \mathcal{P}_g(\mathbf{Q}) \text{ such that } \mathcal{P}_{xg}(\mathbf{Q}) = \mathcal{P}_g(\mathbf{Q}) \setminus \mathcal{P}_h(\mathbf{Q})$$

$$(8.21)$$

$$y \quad : \quad \text{the node on } \mathcal{P}_h(\mathbf{Q}) \text{ such that } \mathcal{P}_{yh}(\mathbf{Q}) = \mathcal{P}_h(\mathbf{Q}) \setminus \mathcal{P}_g(\mathbf{Q})$$

If g and h belong to the same tributary tree, then $x = y$, and this is the first common node on $\mathcal{P}_g(\mathbf{Q})$ and $\mathcal{P}_h(\mathbf{Q})$. See Figure 8.5. If g and h belong to different tributary trees of the same quasitree, then x, y are the roots (i.e., cycle nodes) of these quasitrees. See Figure 8.6. We now discuss the procedures for updating the node labels and node prices under several cases separately.

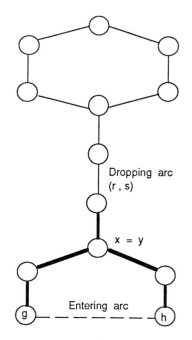

Figure 8.5 The tree $\mathbb{T}$ is marked with thick lines.

In Figures 8.5 to 8.9 orientations of the arcs are not shown. On each non-cycle arc, the node at the top is the predecessor of the node at the bottom. Cycles are drawn in such a way that as you move in the clockwise direction, you move from a node to its predecessor.

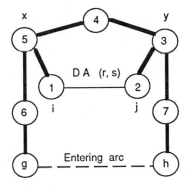

Figure 8.6 The dropping arc, DA, is on $\mathcal{P}_{xy}(\mathbf{Q})$. The tree $\mathbb{T}$ is marked with thick lines. Position after this pivot step is indicated in Figure 8.9.

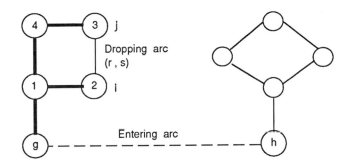

Figure 8.7 Illustration of Case 1. The tree $\mathbb{T}$ is marked with thick arcs. Position after the pivot step is indicated in Figure 8.8.

CASE 1 : $\mathcal{P}_g(\mathbf{Q}) \cap \mathcal{P}_h(\mathbf{Q}) = \emptyset$, AND THE DROPPING ARC $(r, s) \in \mathcal{P}_g(\mathbf{Q})$.

Let i, j be the predecessor, successor nodes respectively, among r, s. So, PI(j) $= \pm i$ and (r, s) is either (i, j) or (j, i). Make PI(g) $= -h$, SI(h) $= g$, SI(i) $= $ SI(j) $= \emptyset$. Reverse the predecessor, successor relationships on each arc along the path $\mathcal{P}_{jg}(\mathbf{Q})$. Make corresponding changes in the successor index and brother indices of nodes along this path, and their brothers, as in Section 5.4. See Figures 8.7, 8.8. The updating of other node labels such as the depth label, thread label, etc. can be carried out in a similar way. We leave these to the reader, in this and in the following cases.

Node prices change only for nodes in T. After updating all the node labels, the values of $\hat{\pi}_i$ for $i \in \mathsf{T}$ are computed using (8.13) beginning with the known value of $\hat{\pi}_h = \pi_h$.

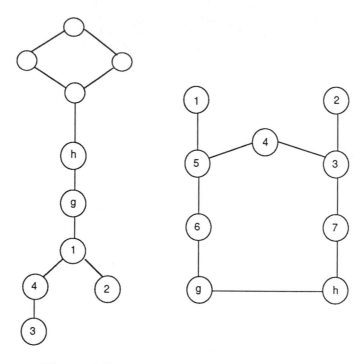

Figure 8.8　　　　　　　　　Figure 8.9

CASE 2 : $\mathcal{P}_g(\mathbf{Q}) \cap \mathcal{P}_h(\mathbf{Q}) = \emptyset$, AND THE DROPPING ARC $(r, s) \in \mathcal{P}_h(\mathbf{Q})$.

Define nodes i, j as in Case 1. Make $\mathrm{PI}(h) = g$, $\mathrm{SI}(g) = -h$, $\mathrm{SI}(i) = \mathrm{SI}(j) = \emptyset$. Reverse the predecessor, successor relationships on each arc along the path $\mathcal{P}_{jh}(\mathbf{Q})$. Make corresponding changes in the successor index and brother indices of nodes along this path, and their brothers, as in Section 5.4.

In this case node prices are updated for nodes on T using the known value of $\hat{\pi}_g = \pi_g$ as mentioned in Case 1.

CASE 3 :　$\mathcal{P}_g(\mathbf{Q}) \cap \mathcal{P}_h(\mathbf{Q}) \neq \emptyset$, AND THE DROPPING
　　　　　　ARC (r, s) IS NOT IN $\mathcal{P}_g(\mathbf{Q}) \cap \mathcal{P}_h(\mathbf{Q})$.

If $(r, s) \in \mathcal{P}_g(\mathbf{Q}) \backslash \mathcal{P}_h(\mathbf{Q})$, update the node labels and node prices as in Case 1. If $(r, s) \in \mathcal{P}_h(\mathbf{Q}) \backslash \mathcal{P}_g(\mathbf{Q})$ update the node labels and node prices as in Case 2.

CASE 4 :   THE DROPPING ARC $(r, s) \in \mathcal{P}_g(\mathbf{Q}) \cap \mathcal{P}_h(\mathbf{Q})$,
BUT NOT ON THE CYCLE IN THIS QUASITREE.

See Figure 8.5. In this case nodes g, h belong to the same tributary tree in $\mathbf{Q}$ and nodes x, y defined in (8.21) are the same. $\mathsf{T} \cup \{(g, h)\}$ forms a new quasitree in the new basic set $\hat{\mathbf{Q}}$. The cycle in this new quasitree is a new cycle that is created, and it contains the entering arc (g, h). Update the node labels as in Case 1. Compute the new node prices $\hat{\pi}_i$ for nodes i in this new quasitree by the procedure discussed earlier. Node prices at all other nodes remain unchanged.

CASE 5 : THE DROPPING ARC (r, s) IS ON THE CYCLE IN $\mathcal{P}_g(\mathbf{Q}) \cap \mathcal{P}_h(\mathbf{Q})$.

See Figure 8.6. In this case $\mathsf{T} \cup \{(g, h)\}$ forms a new quasitree in $\hat{\mathbf{Q}}$. Define nodes i, j as in Case 1. If $(r, s) \in \mathcal{P}_{xy}(\mathbf{Q})$, make PI$(h) = g$, SI$(g) = -h$; and reverse the predecessor, successor relationships on each arc along the path $\mathcal{P}_{jh}(\mathbf{Q})$. See Figures 8.6, 8.9. If $(r, s) \in \mathcal{P}_{yx}(\mathbf{Q})$, make PI$(g) = -h$, SI$(h) = g$; and reverse the predecessor, successor relationships on each arc along the path $\mathcal{P}_{jg}(\mathbf{Q})$. Make corresponding changes in the successor index and brother indices of nodes along these paths, and their brothers, as in Section 5.4.

Compute the new node prices $\hat{\pi}_i$ for nodes i in this new quasitree by the procedure discussed earlier. Node prices at all other nodes remain unchanged.

The Primal Simplex Method for (8.8)

If a primal feasible partition is known, (8.8) can be solved by initiating the primal simplex algorithm with it. Each primal feasible partition obtained in this algorithm is checked for optimality. If the optimality criterion (8.15) is satisfied, the present BFS is optimal and the algorithm terminates. Otherwise, an entering arc is selected among those nonbasic arcs eligible to enter, and a pivot step is carried out using the procedures discussed above. The same process is then repeated with the new partition.

If a primal feasible partition for (8.8) is not known, select a spanning tree in G, and make it into a basic set $\mathbf{Q}$ for (8.8) by including an additional arc in this spanning tree. Partition the nonbasic arcs into $\mathbf{L}$, $\mathbf{U}$ arbitrarily and generate the initial partition $(\mathbf{Q}, \mathbf{L}, \mathbf{U})$. If it is primal feasible, use it to initiate the primal simplex algorithm. If the initial partition is primal infeasible, alter the lower bounds and capacities on basic arcs and construct a Phase I problem exactly as discussed in Section 5.4 for the pure network flow problem. Solve the Phase I problem by the primal simplex algorithm beginning with this initial feasible partition for it. At Phase I termination, we will either obtain a feasible partition for the original problem (8.8), or conclude that it is infeasible.

8.2   Resolution of Cycling Under Degeneracy in the Primal Simplex Algorithm for Generalized Network Flows

The potential theoretical consequences of degeneracy (cycling and stalling) in the primal simplex method have already been explained in Chapter 5 and possible remedies for them in the minimum cost pure network flow problem have been discussed. Similar techniques are needed in the generalized network flow problem too under degeneracy. In this section we discuss a method developed by Elam, Glover, and Klingman [1979] for resolving cycling under degeneracy in the primal simplex algorithm for the minimum cost generalized network flow problem (8.8) when the multiplier vector $(p_{ij}) > 0$. It is called the *method of strongly convergent partitions* or the *strongly convergent primal simplex algorithm* or the *SC method* in short. It can be viewed as an extension of the method of strongly feasible partitions for the minimum cost pure network flow problem discussed in Section 5.4, to the generalized network case.

The main idea behind the SC method is the following. A subset of feasible partitions for (8.8) called *strongly convergent feasible partitions* or *SC partitions* is defined. It is shown that (8.8) always has an SC partition when it is feasible. The method is initiated with an SC partition. A dropping arc choice rule which determines the dropping arc uniquely and unambiguously, and maintains the SC-character in the next partition, is developed. So, when the primal simplex algorithm is initiated with an SC partition and executed using this special dropping arc choice rule, it is restricted to SC partitions throughout. Finally it is shown that under this restriction it cannot cycle, and hence has to terminate in a finite number of pivot steps, and it can only do so by satisfying either the unboundedness or the optimality criterion.

Let $(\mathbf{Q}, \mathbf{L}, \mathbf{U})$ be a feasible partition for (8.8). In this partition, a basic arc $(i, j) \in \mathbf{Q}$ is said to be a *forward basic arc* if node i is a predecessor of node j, or if $i = j$ and the arc is a slack self loop; a *reverse basic arc* otherwise. For any subset of basic arcs $\mathbf{A} \subset \mathbf{Q}$, FOR($\mathbf{A}$), REV($\mathbf{A}$) denote the sets of forward and reverse basic arcs in $\mathbf{A}$ respectively. We have already defined loop factors of cycles, and path factors of simple paths, which depend on the orientation selected for those objects. Extending this concept, here we define a *collective gain/loss factor* for any subset of basic arcs $\mathbf{A} \subset \mathbf{Q}$ to be $\psi(\mathbf{A})$ given by

$$\psi(\mathbf{A}) = \frac{\text{Product of multipliers associated with arcs in REV}(\mathbf{A})}{\text{Product of multipliers associated with arcs in FOR}(\mathbf{A})}$$

where the product of multipliers in a set is defined to be 1 if that set is $\emptyset$.

In a feasible flow vector $\bar{f} = (\bar{f}_{ij})$ for (8.8), an arc (i, j) is said to have *lower leeway* if $\bar{f}_{ij} > \ell_{ij}$, *upper leeway* if $\bar{f}_{ij} < k_{ij}$, *both lower and upper leeways* or *double leeway* if $\ell_{ij} < \bar{f}_{ij} < k_{ij}$.

In a primal feasible partition $(\mathbf{Q}, \mathbf{L}, \mathbf{U})$ for (8.8), a basic arc $(i, j) \in \mathbf{Q}$ is said to be an *SC arc* or *to have SC leeway* if it is a forward basic arc with a lower leeway or a reverse basic arc with an upper leeway. The primal feasible partition $(\mathbf{Q}, \mathbf{L}, \mathbf{U})$ for (8.8) is said to be a *strongly convergent partition* or *SC partition* if: (1) all basic arcs have SC leeway in it, and (2) the collective gain/loss factor of every cycle in $\mathbf{Q}$ that is not a self loop is < 1. It is always possible to construct an initial SC partition which may involve some artificial variables, since a basic set consisting of slack or surplus self loops, possibly artificial, satisfies the condition for being an SC partition for the appropriately set Phase I problem.

In a pivot step of the primal simplex algorithm in a feasible partition $(\mathbf{Q}, \mathbf{L}, \mathbf{U})$ for (8.8), let (g, h) be the nonbasic arc selected as the entering arc, and let $(\overline{y}_{ij} : (i, j) \in \mathbf{Q})$ be its basis representation. Define

$$
\begin{aligned}
\mathbf{B}^+ &= \{(i, j) : (i, j) \in \mathbf{Q} \text{ and } \overline{y}_{ij} > 0\} \\
\mathbf{B}^- &= \{(i, j) : (i, j) \in \mathbf{Q} \text{ and } \overline{y}_{ij} < 0\} \\
\mathbf{B} &= \mathbf{B}^+ \cup \mathbf{B}^- \\
\mathbf{H} &= \{(i, j) : \text{ either } (i, j) \in \text{FOR}(\mathbf{B}) \text{ and } \overline{y}_{ij} < 0, \\
&\quad \text{or } (i, j) \in \text{REV}(\mathbf{B}) \text{ and } \overline{y}_{ij} > 0\} \\
\mathbf{J} &= \{(i, j) : \text{ either } (i, j) \in \text{FOR}(\mathbf{B}) \text{ and } \overline{y}_{ij} > 0, \\
&\quad \text{or } (i, j) \in \text{REV}(\mathbf{B}) \text{ and } \overline{y}_{ij} < 0\}
\end{aligned}
\tag{8.22}
$$

Hence $\mathbf{H}$ is the set of all forward basic arcs whose flow change has the same sign, and all reverse basic arcs whose flow change has the opposite sign, to that of the entering arc (g, h) in this pivot step. A similar symmetric interpretation holds for $\mathbf{J}$. Elam, Glover, and Klingman [1979] have proved that $\mathbf{H} = \mathcal{P}_g(\mathbf{Q}), \mathbf{J} = \mathcal{P}_h(\mathbf{Q})$, if $\mathcal{P}_g(\mathbf{Q}) \cap \mathcal{P}_h(\mathbf{Q}) = \emptyset$; and that in general even when $\mathcal{P}_g(\mathbf{Q}) \cap \mathcal{P}_h(\mathbf{Q}) \neq \emptyset$, $\mathbf{H}$ consists of consecutive arcs in $\mathcal{P}_g(\mathbf{Q})$ and $\mathbf{J}$ consists of consecutive arcs in $\mathcal{P}_h(\mathbf{Q})$. Index the arcs in $\mathbf{H}, \mathbf{J}$ in ascending order, in the same sequence as they are encountered by a trace of the predecessor paths of g and h starting at these nodes. Thus if $\mathbf{H} \neq \emptyset$, the arc joining g and its predecessor $\text{PI}(g)$ is the first in $\mathbf{H}$; and if $\mathbf{J} \neq \emptyset$, the arc joining h and its predecessor $\text{PI}(h)$ is the first in $\mathbf{J}$, etc. In addition, include (g, h) in $\mathbf{H}$ at the beginning of this set.

The set of arcs eligible to be the dropping arc in this pivot step is the set $\mathbf{D}$ defined in (8.20). The *SC dropping arc selection rule* or the *SC pivot rule* specifies that the dropping arc (r, s) in this pivot step should be selected from $\mathbf{D}$ according to the following:

if $(g, h) \in \mathbf{L}$, (r, s) is the last arc on $\mathbf{J}$ in the set $\mathbf{J} \cap \mathbf{D}$ if this set is nonempty, otherwise it is the first arc in $\{(g, h)\} \cup \mathbf{H}$ that is in $\mathbf{D}$

if $(g, h) \in \mathbf{U}$, (r, s) is the last arc in the set $(\{(g, h)\} \cup \mathbf{H}) \cap \mathbf{D}$ if this set is nonempty, otherwise it is the first arc in $\mathbf{J}$ that is in $\mathbf{D}$

Elam, Glover, and Klingman [1979] have shown that if the feasible partition (**Q, L, U**) at the beginning of this pivot step is an SC partition, and if the dropping arc in this pivot step is selected by this special rule, then the next partition will also be an SC partition. Furthermore, any choice of the dropping arc other than that specified by the above rule destroys the SC property in the next partition.

The method of strongly convergent partitions is the primal simplex algorithm initiated with an SC Partition, and executed using the special dropping arc choice rule given above, in each pivot step. In each pivot step the entering arc can be selected by any rule, or even arbitrarily from the set of those eligible (this is the set **E** of (8.16)). All partitions obtained in the algorithm will be SC partitions. It has also been shown that in any sequence of consecutive degenerate pivot steps in this SC algorithm, any node price that changes always increases, i.e., they always change in a uniform direction throughout the entire sequence of degenerate pivot steps. This is the reason for the phrase *strongly convergent* in the name. A similar property was shown to hold in Section 5.4 in the method of strongly feasible partitions for pure network flow problems. This property guarantees that the SC algorithm cannot cycle, and hence must terminate in a finite number of pivot steps by satisfying either the unboundedness or the optimality criteria. Orlin [1985] has shown that the SC pivot rule is equivalent to lexicography in its choice of the exiting arc. Thus the SC algorithm extends the method of strong feasibility to generalized network flow problems with positive multipliers.

A Phase I problem, which is in the same form but with a readily available artificial SC partition, can be formulated for (8.8). When the SC algorithm is applied to solve this Phase I problem beginning with this initial SC partition, it will terminate finitely with either a proof of infeasibility of the original problem, or with an SC partition for it. If, feasible, beginning with the SC partition provided by Phase I, the original problem can be solved finitely by the SC algorithm in Phase II.

8.3   The Most General Generalized Network Flow Problem

The generalized network flow model discussed so far can accommodate any LP in which the coefficient matrix has the property that each column has at most two nonzero entries, and if there are two nonzero entries in a column they have opposite signs. This sign condition guarantees that the multiplier associated with every arc in the corresponding generalized network is always a positive number.

The most general problem of this type is any LP in which every column in the coefficient matrix has at most two nonzero entries, without any restriction on the sign of these nonzero entries in any column. Given such an LP, scale each column with two nonzero entries so that one of these entries becomes equal to -1. After scaling, associate a node with each row of the coefficient matrix. If a column has a single nonzero entry, say in row i, associate that column with a self loop at node i. If a column has two nonzero entries, say a "-1" in row i and an entry of $p \neq 0$ in row j, associate it with arc (i, j) and make p the multiplier of that arc. Here p may be positive or negative. The LP is equivalent to the generalized network flow problem of the form (8.8) on this network, but some of the multipliers may be negative. The results at the beginning of this chapter continue to hold for this problem, and it can be solved by the implementation of the primal simplex algorithm discussed in Section 8.1. However, the SC algorithm of Section 8.2 and all the special convergence properties mentioned there are based on the assumption that all the arc multipliers are positive, and these properties and results may not hold for the general problem with some negative multipliers. But recently, Arantes, Birge, and Murty [1992] developed a special method for resolving the problem of cycling under degeneracy in the primal simplex algorithm applied to this general generalized network flow problem. This method is also a specialization of lexicography using the special structure of this problem.

8.4 Exercises

8.1 Let $(\mathbf{Q}, \mathbf{L}, \mathbf{U})$ be an SC partition for (8.8). Prove that the pivot step in this partition with (g, h) as the incoming arc is nondegenerate iff either the dropping arc $(r, s) = (g, h)$, or $(r, s) \in \mathbf{J}$ if $(g, h) \in \mathbf{L}$, or $(r, s) \in \mathbf{H}$ if $(g, h) \in \mathbf{U}$; where $\mathbf{J}, \mathbf{H}$ are the sets defined in Section 8.2.

8.2 In Section 3.3 we discussed a method for transforming a single commodity minimum cost flow problem in a pure non-bipartite network, into a similar problem on a pure bipartite network. Show that the minimum cost flow problem (8.8) in a non-bipartite generalized network G can similarly be transformed into a problem on a bipartite generalized network (Malek-Zavarei and Aggarwal [1972]).

8.3 Consider the mixed 0-1 integer programming problem of minimizing $z(x) = cx$ subject to, $d \overset{\leq}{=} Ax \overset{\leq}{=} b$, $x_j \in \{0, 1\}$ for all $j \in \mathbf{N}_1$, $0 \overset{\leq}{=} x_j \overset{\leq}{=} u_j$ for all $j \in \mathbf{N} \backslash \mathbf{N}_1$; where A is of order $m \times n$, $\mathbf{N} = \{1, \cdots, n\}$, $\mathbf{N}_1 \subset \mathbf{N}$, and $A_{\cdot j}$ has at most two nonzero entries for each $j \in \mathbf{N} \backslash \mathbf{N}_1$. Show that this problem can be transformed into a mixed 0-1 generalized network flow problem (i.e., a generalized network flow problem in which the flow on a specified subset of arcs is required to be 0 or 1). In addition, for each $j \in \mathbf{N} \backslash \mathbf{N}_1$ suppose the following holds: If $A_{\cdot j}$ contains a single nonzero entry, it is $+1$ or -1; if it contains two nonzero entries, one is $+1$ and the other is -1. In this case show that this problem can be transformed into a mixed $0 - U$ pure network flow problem (i.e., one in which the flow on a specified subset of arcs is required to be either the lower bound 0 or the capacity on that arc).

8.4 Let $G = (\mathcal{N}, \mathcal{A}, \ell, k, p)$ be a generalized network with p as the multiplier vector. Prove that a necessary and sufficient condition for the existence of a feasible circulation in G is that the following condition holds for every tree (N, A) in G, and any node price vector $\pi = (\pi_i)$ in G satisfying: $\pi_i = 0$ for $i \in \mathcal{N} \setminus$ N, and $\pi_i - \pi_j p_{ij} = 0$ for all $(i, j) \in$ A.

$$\sum_{(i,j)\in\mathcal{A}} k_{ij}(\, \text{max. } \{0, \pi_i - p_{ij}\pi_j\}) \geqq \sum_{(i,j)\in\mathcal{A}} \ell_{ij}(\, \text{max. } \{0, p_{ij}\pi_j - \pi_i\})$$

Verify that no feasible circulation exists in the network in Figure 8.10, where the data on arc (i, j) is $\ell_{ij}, k_{ij}, p_{ij}$ in that order; and that the above condition fails to hold for the tree with arcs (1, 2), (3, 1) and the vector $\pi = (1, 1/2, 1)$.

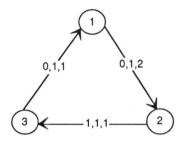

Figure 8.10

(Hassin[1981])

8.5 Cash Flow Management Model   Consider a discrete time deterministic cash flow problem. There are n time periods in the planning horizon. The supply and demand for cash in each time period is known in advance. Spare cash can be saved from one period to the next in a savings account at fraction yield of α per period, or invested in a bond portfolio at fraction yield of β per period. All transactions occur at the beginning of a period, and returns are obtained at the end of a period. c_1, c_2 are the per unit fraction cost of converting cash into bond portfolio, a bond back into cash, respectively. Conversions are instantaneous and can be carried out either way in any quantity. $S(t)$ is the new supply (incoming) of cash in addition to any return from investments, and $D(t)$ is the demand for cash (to be paid out) at the beginning of period $t = 1$ to n. There is an initial inventory in bond portfolio of y_0 at the start.

(i)   It is required to determine the amounts to be invested in each asset in each period so as to maximize the return from the final planning period. Formulate this as a generalized network flow problem on a network with $2n + 3$ nodes. Construct this network model for the numerical problem in which $n = 3$, $(S(1), S(2), S(3)) = (12, 8, 5)$, $(D(1), D(2), D(3)) = (4, 10, 10)$ in million\$, $\alpha = .05, \beta = .08, c_1 = c_2 = .02$ and $y_0 = 1$ in million\$ units; and obtain the optimum solution.

(ii) Consider the same problem with two additional features. One is a minimum cash balance specification, i.e., the firm is required by policy to put down a minimum of L units in cash in the savings account from each period to the next. The other is a borrowing capability at a fraction interest of γ per period up to an upper limit of U units in any period. Discuss the modifications to be made in the network model of (i) to incorporate these features. Construct this model for the numerical problem in which $n = 4$, $(S(1) \text{ to } S(4)) = (5, 10, 20, 15)$, $(D(1) \text{ to } D(4)) = (16, 12, 8, 8)$, $\alpha = .05$, $\beta = .08$, $\gamma = .10$, $L = 5$, $U = 10$, $y_0 = 10$. The cash unit is one million \$. Find the optimum cash flow policy. Solve the same numerical problem with all the data the same, except $\gamma = .09$, $.1025$, $.12$ respectively. Verify that as γ increases from $.09$ to $.10$, and from $.1025$ to $.12$, the optimum policy actually borrows more, a counterintuitive phenomenon. Explain.

(iii) Assume $\beta > \alpha$. N^* is said to be a forecast horizon for period t if demands in periods beyond N^* have no effect on amounts converted from cash into bonds or vice versa at period t in an optimum policy. If N' is the smallest positive integer N satisfying $(1 - c_1)(1 - c_2)(1 + \beta)^N \geqq (1 + \alpha)^N$, then prove that min. $\{n, t + N' - 1\}$ is a forecast horizon for period t.

(iv) Instead of a single bond-portfolio, suppose there are r different assets $(r \geqq 2)$ in which cash can be invested in each period. Given corresponding data for each asset, discuss the modifications to be made in the network model to handle this change.

(Golden, Liberatore, and Lieberman [1979])

8.6 Consider the minimum cost flow problem (8.8) on the generalized network $G = (\mathcal{N}, \mathcal{A})$, with the multiplier vector $(p_{ij} : (i, j) \in \mathcal{A}) > 0$, and $|\mathcal{N}| = n$. Let $(\mathbf{Q}, \mathbf{L}, \mathbf{U})$ be a feasible partition for this problem with the arcs in $\mathbf{Q}$ arranged and numbered as $e_1, \ldots, e_n$. Let B be the basis for (8.8) corresponding to $\mathbf{Q}$. Let e_{n+1} be a nonbasic arc with $(\overline{y}_t : t = 1 \text{ to } n)$ as its basis representation wrt $\mathbf{Q}$. Let $\beta = B^{-1}$. For each $t = 1$ to n such that $\overline{y}_t \neq 0$ define $S_{t.} = (\beta_{t.})/\overline{y}_t$, and let S be the matrix with all these rows $S_{t.}$. So, if $r = |\{t : \overline{y}_t \neq 0\}|$, then $S = (s_{tj})$ is of order $r \times n$. Prove the following:

(i) For each $t = 1$ to n, all the nonzero elements of $\beta_{t.}$ have the same sign. This shows that the unisign property discussed in Theorem 1.8 also holds in generalized networks in which all multipliers are positive.

(ii) For some t_1, t_2, if $S_{t_1.}$ is lexicographically greater than $S_{t_2.}$, then $s_{t_1 j} \geqq s_{t_2 j}$ for all $j = 1$ to n.

(iii) Suppose the partition $(\mathbf{Q}, \mathbf{L}, \mathbf{U})$ was obtained in the process of solving (8.8) by the primal simplex algorithm, and that e_{n+1} is the arc selected to be the entering arc into it. Suppose the dropping arc in this pivot step has been selected by the special dropping arc choice rule specified by the SC algorithm, and it is e_r. Then show that S_r is the lexico maximum row among the rows of S.

Comment 8.1   Bhaumik [1973], Glover, Hultz, Klingman, and Stutz [1978], Golden, Liberatore, and Lieberman [1979], Hamacher and Tufekci [1987], are some references discussing applications of generalized network flow models.

In general, any LP in which the coefficient matrix of constraints other than individual bound restrictions on single variables contains at most two nonzero entries per column, can be treated as a generalized network flow problem. In this chapter we discussed an implementation of the primal simplex method based on rooted loop labelings of quasitrees, for the generalized minimum cost flow problem. Although total unimodularity is not there, this implementation makes it possible to execute the primal simplex method on this problem in a purely combinatorial mode without the use of basis inverses. Brown and Mcbride [1984], Elam, Glover, and Klingman [1979], Glover and Stutz [1973], Nulty and Trick [1988] are some references on this topic. Elam, Glover, and Klingman [1979] report that this implementation gave excellent performance in computational tests. For solving randomly generated generalized minimum cost flow problems it was far superior than simplex based general purpose LP codes. However it takes about 3 times the computer time than a corresponding code does to solve a pure minimum cost flow problem of comparable size.

Several other algorithms are available for this problem. One of the earliest is a primal-dual algorithm due to Jewell [1962]. Jensen and Bhaumik [1977] describe a shortest augmenting path type approach for this problem. Bertsekas and Tseng[1988] have extended their relaxation methods (discussed in Section 5.6 for pure minimum cost flow problems) to handle minimum cost flow problems in generalized networks.

8.5   References

J. C. ARANTES, 1991, "Resolution of Degeneracy in Generalized Networks and Penalty Methods for Linear Programming," Ph. D. dissertation, University of Michigan, Ann Arbor, MI.

J. C. ARANTES, J. R. BIRGE, and K. G. MURTY, 1992, "Studies of Lexicography in the Generalized Network Simplex Method," Tech. report, University of Cincinnati, Cincinnati, Ohio.

D. BERTSEKAS and P. TSENG, Jan. - Feb. 1988, "Relaxation Methods for Minimum Cost Ordinary and Generalized Network Flow Problems," *OR*, 36, n0. 1(93-114).

G. BHAUMIK, 1973, "Optimum Operating Policies of a Water Distribution System with Losses," Ph. D. dissertation, University of Texas at Austin, TX.

G. G. BROWN and R. MCBRIDE, 1984, "Solving Generalized Networks," *MS*, 30(1497-1523).

V. CHANDRU, C. R. COULLARD, and D. K. WAGNER, July 1985, "On the Complexity of Recognizing a Class of Generalized Networks," *OR Letters*, 4, n0. 2(75-78).

J. ELAM, F. GLOVER, and D. KLINGMAN, Feb. 1979, "A Strongly Convergent Primal Simplex Algorithm for Generalized Networks," *MOR*, 4, no. 1(39-59).

F. GLOVER, J. HULTZ, D. KLINGMAN, and J. STUTZ, Aug. 1978, "Generalized Networks: A Fundamental Computer Based Planning Tool," *MS*, 24, no. 12(1209-1220).

F. GLOVER and D. KLINGMAN, 1973, "On the Equivalence of Some Generalized Network Problems to Pure Network Problems," *MP*, 4(369-378).

F. GLOVER and J. MULVEY, May-June 1980, "Equivalence of the 0-1 Integer Programming Problem to Discrete Generalized and Pure Networks," *OR*, 28, no. 3(829-836).

F. GLOVER and J. STUTZ, 1973, "Extensions of the Augmented Predecessor Index Method to Generalized Network Problems," *TS*, 7(377-384).

B. GOLDEN, M. LIBERATORE, and C. LIEBERMAN, 1979, "Models and Solution Techniques for Cash Flow Management," *COR*, 6, no. 1(13-20).

H. W. HAMACHER and S. TUFEKCI, 1987, "On the Use of Lexicographic Minimum Cost Flows in Evacuation Modeling," *NRLQ*, 34(487-503).

R. HASSIN, 1981, "Generalizations of Hoffman's Existence Theorem for Circulations," *Networks*, 11, no. 3(243-254).

P. A. JENSEN and G. BHAUMIK, 1977, "A Flow Augmentation Approach to the Network with Gains Minimal Cost Flow Problem," *MS*, 23(631-643).

W, S, JEWELL, 1962, "Optimal Flow Through Networks with Gains," *OR*, 10(476-499).

M. MALEK-ZAVAREI and J. K. AGGARWAL, 1972, "Optimal Flow in Networks with Gains and Costs," *Networks*, 1(355-365).

J. MAURRAS, 1972, "Optimization of the Flow Through Networks with Gains," *MP*, 3(135-144).

R. D. MCBRIDE, July 1985, "Solving Embedded Generalized Network Problems," *EJOR*, 21, n0. 1(82-92).

W. G. NULTY and M. A. TRICK, April 1988, "GNO/PC Generalized Network Optimization System," *OR Letters*, 7, no. 2(101-102).

J. B. ORLIN, 1985, "On the Simplex Algorithm for Networks and Generalized Networks," *MPS*, 24(166-178).

K. TRUEMPER, 1977, "On Max Flow with Gains and Pure Min-Cost Flows," *SIAM J. Applied Mathematics*, 32(450-456).

Chapter 9

The Minimum Cost Spanning Tree Problem

Given a set of points on the two-dimensional plane, the problem of finding the shortest connecting network that connects all the points, using only lines joining pairs of points from the given set arises in many applications. If the length between every pair of points is positive, the shortest connecting network will obviously be a spanning tree, it will be a *minimum length spanning tree.*

The minimum length (or cost) spanning tree problem is one of the nicest and simplest problems in network optimization, and it has a wide variety of applications. The input in this problem is a connected undirected network $G = (\mathcal{N}, \mathcal{A})$ with $c = (c_{ij} : (i; j) \in \mathcal{A})$ as the vector of edge costs or lengths. Define the cost (or length) of a tree in G to be the sum of the costs (or lengths) of edges in it. The problem is to find a minimum cost (or length) spanning tree in G. Applications include the design of various types of distribution networks in which the nodes represent cities, centers etc.; and edges represent communication links (fiber glass phone lines, data transmission lines, cable TV lines, etc.), high voltage power transmission lines, natural gas or crude oil pipelines, water pipelines, highways, etc. The objective is to design a network that connects all the nodes using the minimum length of cable or pipe or other resource. The minimum cost spanning tree problem also appears as a subproblem in algorithms for many routing problems such as the traveling salesman problem.

As an example consider the network G with 6 nodes representing 6 Michigan cities, and edges joining every pair of these nodes, given in Figure 9.1. The number on each edge is its length in miles. The minimum length spanning tree in this network is marked with thick lines.

The network G may in fact be a directed or mixed network. The problem that we consider in this chapter is that of finding a minimum cost spanning tree in G

466

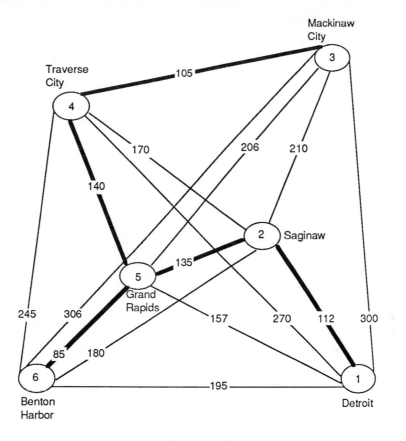

Figure 9.1

without paying any regard to the orientations of the arcs in G. So, if there are any arcs in the input network, we ignore their orientations and treat every line as an edge. We also assume that the network G has no self loops. If there are r parallel edges joining a pair of nodes in G, any tree can contain at most one of them, and if a minimum cost spanning tree contains any of them it will be a least cost edge among them. So, we identify one least cost edge among these r parallel edges and keep it, but delete all the rest. Hence, in the sequel, we assume that there is at most one edge joining any pair of nodes in the input network $G = (\mathcal{N}, \mathcal{A})$. Let $|\mathcal{N}| = n, |\mathcal{A}| = m$. So, $m \overset{\leq}{=} n(n-1)/2 < n^2/2$.

What we are considering is the *unconstrained minimum cost spanning tree problem*. The *constrained minimum cost spanning problem* with constraints on the type of spanning tree obtained is discussed briefly in Section 9.4.

A Method That Does Not Work

Let the input network be $G = (\mathcal{N}, \mathcal{A}, c)$ with $c > 0$. Consider the following approach. Select any node, say 1, and find the shortest paths from it to all the other nodes in G. This yields a shortest path tree $\mathbb{T}_0$ rooted at 1. For every node $j \neq 1$, the path between 1 and j in $\mathbb{T}_0$ is a shortest path between these nodes in G. Because of this one may be tempted to conclude that $\mathbb{T}_0$ is a minimum length spanning tree in G. This may not be the case as illustrated in Figure 9.2. Here $\mathbb{T}_0$, consisting of the dashed arcs; has length 11. It is not a minimum length spanning tree; since the spanning tree consisting of the thick edges (dashed or continuous) has length only 9.

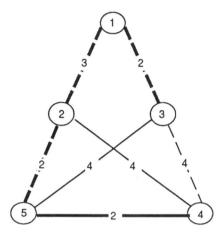

Figure 9.2 The shortest path tree rooted at a node may not be a minimum length spanning tree.

9.1   Some Results on Minimum Cost Spanning Trees

Let the input network be G $= (\mathcal{N}, \mathcal{A}, c)$ with $|\mathcal{N}| = n$. No assumptions are made on the entries in c, each of them could be positive, 0, or negative.

THEOREM 9.1 *Let* $\mathbb{T}_0$ *be a minimum cost spanning tree in G. Then every in-tree edge in* $\mathbb{T}_0$ *must be a minimum cost edge in the fundamental cutset associated with it.*

Proof   Let $(\mathbf{X}; \overline{\mathbf{X}})$ be the fundamental cutset associated with the in-tree edge $(i;j)$ in $\mathbb{T}_0$. So, $(i;j)$ is the only in-tree edge in $(\mathbf{X}; \overline{\mathbf{X}})$, and deletion of $(i;j)$ from $\mathbb{T}_0$ disconnects it into two smaller trees, one spanning the nodes in $\mathbf{X}$, and the other spanning the nodes in $\overline{\mathbf{X}}$. Suppose a minimum cost edge in the cutset $(\mathbf{X}; \overline{\mathbf{X}})$ is an out-of-tree edge $(p;q)$ and not $(i;j)$, i.e., $c_{pq} < c_{ij}$. Then replacing $(i;j)$ in $\mathbb{T}_0$ by $(p;q)$ leads to a spanning tree $\mathbb{T}_1$ with cost $=$ cost of $\mathbb{T}_0 - (c_{ij} - c_{pq}) <$ cost of $\mathbb{T}_0$, contradicting the optimality of $\mathbb{T}_0$. Hence $(i;j)$ must be a minimum cost edge in the fundamental cutset $(\mathbf{X}; \overline{\mathbf{X}})$ associated with it in $\mathbb{T}_0$. ∎

THEOREM 9.2 *Let $\mathbb{T}_0$ be a minimum cost spanning tree in* G. *Then every out-of-tree edge must be a maximum cost edge in the fundamental cycle associated with it.*

Proof Let $(p;q)$ be an out-of-tree edge. If there is an in-tree edge $(i;j)$ on the fundamental cycle of $(p;q)$ satisfying $c_{pq} < c_{ij}$, replacing $(i;j)$ in $\mathbb{T}_0$ by $(p;q)$ leads to a new spanning tree with cost = cost of $\mathbb{T}_0 + (c_{pq} - c_{ij}) <$ cost of $\mathbb{T}_0$, contradicting the optimality of $\mathbb{T}_0$. So, c_{pq} must be $\geqq$ the cost of every in-tree edge in the fundamental cycle of $(p;q)$. ∎

THEOREM 9.3 *Any spanning tree in* G *satisfying*

$$\textit{Every out-of-tree edge has maximum cost in} \atop \textit{its fundamental cycle wrt that tree.} \qquad (9.1)$$

is a minimum cost spanning tree.

Proof Let $\mathbb{T}_1$, $\mathbb{T}_2$ be two distinct spanning trees in G satisfying (9.1). We will now prove that the costs of $\mathbb{T}_1$ and $\mathbb{T}_2$ must be equal.

Classify the edges in $\mathbb{T}_1 \cup \mathbb{T}_2$ into 3 classes. The $\mathbb{T}_1$-edges are those which are in $\mathbb{T}_1$ but not in $\mathbb{T}_2$. The $\mathbb{T}_2$-edges are those in $\mathbb{T}_2$ but not in $\mathbb{T}_1$. The $\mathbb{T}_1\mathbb{T}_2$-edges are those which are in both $\mathbb{T}_1$ and $\mathbb{T}_2$. Since $\mathbb{T}_1$ and $\mathbb{T}_2$ are distinct, both the sets of $\mathbb{T}_1-$, $\mathbb{T}_2$-edges are nonempty.

Select any $\mathbb{T}_2$-edge, say $(r;s)$. Let $\mathcal{P}_1$ be the unique path in $\mathbb{T}_1$ between r, s. Then $\mathbb{C}_1 = \{(r;s)\} \cup \mathcal{P}_1$ is the fundamental cycle of $(r;s)$ wrt $\mathbb{T}_1$. All edges on $\mathcal{P}_1$ are in $\mathbb{T}_1$, and at least one of them is not in $\mathbb{T}_2$, as otherwise $\mathbb{T}_2$ contains the entire simple cycle $\mathbb{C}_1$, a contradiction since $\mathbb{T}_2$ is a tree. So, there are some $\mathbb{T}_1$-edges on $\mathcal{P}_1$. Since $\mathbb{T}_1$ satisfies (9.1), all these $\mathbb{T}_1$ edges on $\mathcal{P}_1$ have cost coefficients $\leqq c_{rs}$.

Claim: At least one of the $\mathbb{T}_1$-edges on $\mathcal{P}_1$ has cost coefficient $= c_{rs}$.

We will prove the claim by contradiction. Suppose each $\mathbb{T}_1$-edge on $\mathcal{P}_1$ has cost $< c_{rs}$. Let $(p;q)$ be a $\mathbb{T}_1$-edge on $\mathcal{P}_1$. So, $c_{pq} < c_{rs}$. $(p;q)$ is an out-of-tree edge for $\mathbb{T}_2$, let $\mathcal{S}_{pq}$ be the unique path between p and q in $\mathbb{T}_2$. Since $\mathbb{T}_2$ satisfies (9.1), each edge $(i;j)$ on $\mathcal{S}_{pq}$ satisfies $c_{ij} \leqq c_{pq} < c_{rs}$. Hence $\mathcal{S}_{pq}$ does not contain the edge $(r;s)$. Replace the $\mathbb{T}_1$-edge $(p;q)$ on $\mathcal{P}_1$ by the path $\mathcal{S}_{pq}$, but continue denoting it by the same symbol $\mathcal{P}_1$. If there are other $\mathbb{T}_1$-edges remaining on $\mathcal{P}_1$, carry out the same procedure for them. When this process is completed, the path $\mathcal{P}_1$ gets converted into a path from r to s, not containing the edge $(r;s)$, but completely contained in $\mathbb{T}_2$, call it $\mathcal{P}$ at this stage. So, $\mathbb{T}_2$ contains the edge $(r;s)$ and the path $\mathcal{P}$ from r to s not containing the edge $(r;s)$, a contradiction since $\mathbb{T}_2$ is a tree. So, the claim must be true.

So, at least one of the $\mathbb{T}_2$-edges on $\mathcal{P}_1$ has cost $= c_{rs}$, suppose it is $(j_1;j_2)$. Replace $(j_1;j_2)$ from $\mathbb{T}_1$ by $(r;s)$. This leads to a new spanning tree $\mathbb{T}'_1$ with the same cost as $\mathbb{T}_1$, but it has one more edge in common with $\mathbb{T}_2$, and it can be verified that it also satisfies (9.1). If $\mathbb{T}'_1$ and $\mathbb{T}_2$ are distinct, repeat this process again with them. After each repetition we get another spanning tree satisfying (9.1) and having the same cost as $\mathbb{T}_1$, but containing one more edge in common with $\mathbb{T}_2$. So, after at most $n-1$ repetitions of this process it must lead to $\mathbb{T}_2$. Hence $\mathbb{T}_1$ and $\mathbb{T}_2$ have the same cost.

There are only a finite number of spanning trees in G, and hence a minimum cost spanning tree exists in G. If $\mathbb{T}_0$ is a minimum cost spanning tree, it satisfies (9.1) by Theorem 9.2. But we have just proved that any pair of spanning trees satisfying (9.1) have the same cost. Hence every spanning tree in G satisfying (9.1) is a minimum cost spanning tree. ∎

THEOREM 9.4 *Let* **F** *be the set (possibly empty) of edges in a forest in* G. *Let* $(\mathbf{X}; \overline{\mathbf{X}})$ *be a cut in* G *containing none of the edges from* **F**. $(p; q)$ *is a minimum cost edge in the cut* $(\mathbf{X}; \overline{\mathbf{X}})$. *Consider the problem of finding a minimum cost tree among the spanning trees of* G *containing all the edges in* **F**. *There is a minimum cost spanning tree for this problem which contains the edge* $(p; q)$.

Proof Since G is connected and **F** is the set of edges in a forest, there exist spanning trees of G which contain all the edges in **F**; let $\mathbb{T}_0$ be a minimum cost one among them. If $\mathbb{T}_0$ contains $(p; q)$ we are done. Suppose $(p; q) \notin \mathbb{T}_0$. Let $\mathbb{C}$ be the fundamental cycle of $(p; q)$ wrt $\mathbb{T}_0$. Since $\mathbb{C}$ is a cycle and it contains the edge $(p; q)$ from the cut $(\mathbf{X}; \overline{\mathbf{X}})$, it must contain at least another edge, say $(r; s)$ from $(\mathbf{X}; \overline{\mathbf{X}})$. By the hypothesis $(\mathbf{X}; \overline{\mathbf{X}})$ contains no edges from **F**, so $(r; s) \notin \mathbf{F}$ and since $(p; q)$ is a minimum cost edge in $(\mathbf{X}; \overline{\mathbf{X}})$, $c_{pq} \stackrel{\leq}{=} c_{rs}$. Let $\mathbb{T}_1$ be the spanning tree obtained by replacing $(r; s)$ in $\mathbb{T}_0$ with $(p; q)$. Since $(r; s) \notin \mathbf{F}$, $\mathbb{T}_1$ contains all the edges in **F**, and also the edge $(p; q)$. If $c_{pq} < c_{rs}$, the cost of $\mathbb{T}_1$ would be $<$ cost of $\mathbb{T}_0$, contradicting the definition of $\mathbb{T}_0$. So, c_{pq} must equal c_{rs}, and $\mathbb{T}_1$ is also a minimum cost one among the spanning trees of G containing all the edges in **F**, and it contains $(p; q)$, proving the theorem. ∎

COROLLARY 9.1 *Let* $(p; q)$ *be a minimum cost edge in a cut in* G. *Then there is a minimum cost spanning tree containing* $(p; q)$.

Proof Follows from Theorem 9.4 with $\mathbf{F} = \emptyset$. ∎

COROLLARY 9.2 *Suppose* $\mathbf{F} \subset \mathcal{A}$ *satisfies the property that there exists a minimum cost spanning tree in* G *containing all the edges in* **F**. *Let* $(p; q)$ *be a minimum cost edge in a cut in* G *which contains no edges from* **F**. *Then, there exists a minimum cost spanning tree in* G *containing all the edges in* $\mathbf{F} \cup \{(p; q)\}$.

Proof Follows from Theorem 9.4. ∎

THEOREM 9.5 *Let* $\mathbf{F} \subset \mathcal{A}$ *be the set of edges in a forest in a connected undirected network* G. *Let* $\mathbb{C}$ *be a simple cycle in* G, *and* $(r; s)$ *a maximum cost edge among those edges in* $\mathbb{C}$ *not in* **F**. *Consider the problem of finding a minimum cost spanning tree among those spanning trees of* G *containing all the edges in* **F**. *There is a minimum cost spanning tree for this problem which does not contain* $(r; s)$.

Proof Let $\mathbb{T}_0$ be a minimum cost spanning tree in G among those containing all the edges in **F**. If $(r; s) \notin \mathbb{T}_0$ we are done. Suppose $\mathbb{T}_0$ contains $(r; s)$. Let $(\mathbf{X}; \overline{\mathbf{X}})$ be the fundamental cutset of $(r; s)$ wrt $\mathbb{T}_0$. So, $(r; s)$ is the only in-tree arc in $(\mathbf{X}; \overline{\mathbf{X}})$, and hence this cut contains no arcs from **F**. Since $(\mathbf{X}; \overline{\mathbf{X}})$ contains $(r; s)$ from $\mathbb{C}$, it must contain at least one other edge from $\mathbb{C}$, suppose it is $(p; q)$. So, $(p; q)$ is an out-of-tree edge on $\mathbb{C}$, and since $(p; q) \notin \mathbf{F}$ we have $c_{rs} \overset{\geq}{=} c_{pq}$. Let $\mathbb{T}_1$ be the spanning tree obtained by replacing $(r; s)$ from $\mathbb{T}_0$ by $(p; q)$. $\mathbb{T}_1$ contains all the edges in **F** and does not contain $(r; s)$. If $c_{rs} > c_{pq}$, cost of $\mathbb{T}_1 <$ cost of $\mathbb{T}_0$, contradicting the definition of $\mathbb{T}_0$. So, we must have $c_{rs} = c_{pq}$, and $\mathbb{T}_1$ is a minimum cost spanning tree among those containing all the edges in **F**, and does not contain $(r; s)$. ∎

Exercises

9.1 Prove the following converse of Theorem 9.1. If $\mathbb{T}$ is a spanning tree in G satisfying

$$\text{each edge in } \mathbb{T} \text{ is a min. cost edge in its fundamental cutset wrt } \mathbb{T} \qquad (9.2)$$

then $\mathbb{T}$ is a minimum cost spanning tree in G.

9.2 Let $G = (\mathcal{N}, \mathcal{A})$ be an undirected connected network with $|\mathcal{N}| = n$. With each spanning tree $\mathbb{T}$ in G associate its 0-1 incidence vector defined on $\mathcal{A}$ to be the vector $(a_{ij} : (i; j) \in \mathcal{A})$ where $a_{ij} = 1$ if $(i; j)$ is an in-tree edge, 0 otherwise. Consider the following system of constraints in variables $(x_{ij} : (i; j) \in \mathcal{A})$

$$\sum (\quad x_{ij} \quad : \text{ over } (i; j) \in \mathcal{A} \qquad = n - 1$$

$$\sum (\quad x_{ij} \quad : \text{ over } (i; j) \in \mathcal{A}; i, j \in \mathbf{J}) \overset{\leq}{=} |\mathbf{J}| - 1,$$

$$\text{for all } \mathbf{J} \subset \mathcal{N} \text{ with } |\mathbf{J}| \overset{\geq}{=} 2 \qquad (9.3)$$

$$x_{ij} \quad \overset{\geq}{=} 0, \text{ for all } (i; j) \in \mathcal{A}$$

Prove that **K**, the set of feasible solutions of (9.3), is the convex hull of the 0-1 incidence vectors of all the spanning trees in G. Also prove that the 0-1 incidence vector of any spanning tree in G is an extreme point of **K** and conversely.

Given two extreme points of **K**, derive necessary and sufficient conditions for them to be adjacent, based on the properties of spanning trees corresponding to them.

9.2 Algorithms for the Minimum Cost Spanning Tree Problem

All efficient algorithms known for this problem are based on an incremental approach which builds up a minimum cost spanning tree, edge by edge, called the *greedy method*. It is so named because at each stage it selects the next edge to be the cheapest that can be selected at that time. The greedy method is very naive, once a decision to include an edge is made, that decision is never reversed (so, no backtracking), and the selection at each stage is based on the situation at that time without any features of look-ahead etc. Hence, the method is also known as a *myopic method*. The fact that such a simple approach solves the minimum cost spanning tree problem indicates the very nice structure and the simplicity of the problem.

All the algorithms begin with the trivial spanning forest in G consisting of isolated nodes: $(\{1\}, \emptyset)$, ..., $(\{n\}, \emptyset)$. In each step they select one or more edges merging the forest components. Finally, after $(n-1)$ edges are selected, all the forest components merge into a single spanning tree, which will be a minimum cost spanning tree. We will denote the set of trees in the forest in a general stage by $\mathbf{F}_1 = (\mathcal{N}_1, \mathcal{A}_1), \ldots, \mathbf{F}_l = (\mathcal{N}_l, \mathcal{A}_l)$, where each of $\mathbf{F}_1, \ldots, \mathbf{F}_l$ is a tree; and $\mathcal{N}_1, \ldots, \mathcal{N}_l$ is a partition of $\mathcal{N}$. $\mathbf{F}_1, \ldots, \mathbf{F}_l$ are the connected components in the forest at this stage.

The first algorithm that we discuss is known as *Prim's algorithm* (Jarnik [1930], Prim [1957], Dijkstra [1959 of Chapter 4]). It grows only one component in the forest as a connected tree, leaving all the other components as isolated nodes. So, we will call the component being grown as the "tree." Select a node arbitrarily, say node 1, and make it the root node of the tree. Since the network is undirected, we maintain the predecessor indices of the nodes in the tree without any + or − signs. Nodes in the tree are labeled with predecessor indices, which are called *permanent labels*. Each out-of-tree node j carries a *temporary label* (which can change from step to step until this node joins the tree) of the form (P_j, d_j), where $d_j = +\infty$ and $P_j = \emptyset$ if there is no edge joining j with an in-tree node so far; and $d_j = \min.$ $\{c_{ij}:$ over i an in-tree node such that $(i; j) \in \mathcal{A} \}$ and P_j is an i that attains this minimum, otherwise. So, for an out-of-tree node j with temporary label (P_j, d_j), if $d_j < \infty$, P_j is a "nearest" in-tree node to j at this stage. We will now describe Prim's algorithm.

PRIM'S ALGORITHM

Initialization   Permanently label the root node 1 with the label $\emptyset$. For each node j such that $(1; j) \in \mathcal{A}$ $[(1; j) \notin \mathcal{A}]$ temporary label j with $(1, c_{1j})$ $[(\emptyset, \infty)]$.

General step   Among all out-of-tree nodes j at this stage, find one with the smallest d_j, suppose it is r with temporary label (P_r, d_r). Delete this temporary label on r and give it the permanent label P_r (its predecessor index in the tree), i.e., make (P_r, r) an in-tree edge and r an in-tree node. If there are no out-of-tree nodes left, terminate, the spanning tree defined by the permanent labels is a minimum cost spanning tree in G. If there are some out-of-tree nodes left update their temporary labels as follows: for each out-of-tree node j with temporary label (P_j, d_j), change it to (r, c_{rj}) only if $(r; j) \in \mathcal{A}$ and $c_{rj} < d_j$; leave it unchanged otherwise. Then go to the next step.

Discussion

At some stage in the algorithm let $\mathbf{X}$ be the set of in-tree nodes and $\overline{\mathbf{X}}$ its complement. So, there are no in-tree edges in the cut $(\mathbf{X}; \overline{\mathbf{X}})$ at this stage. The edge selected at this stage for being the next in-tree edge is a minimum cost edge in the cut $(\mathbf{X}, \overline{\mathbf{X}})$ by the manner in which it is selected. This is a *greedy selection*, that's what makes this a greedy method. Applying Corollaries 9.1, 9.2 in each step, we conclude that at every stage there is a minimum cost spanning tree in G containing all the in-tree edges at that stage. Hence when the tree becomes a spanning tree in G, it is a minimum cost spanning tree in G.

The computational effort in each step is clearly $O(n)$ and there are exactly $n-1$ steps. Hence the overall computational complexity of this algorithm is $O(n^2)$.

As an example we apply this algorithm on the network in Figure 9.1. We select node 1 as the root node. We show the node labels, and the in-tree edge selected in each step in the following table. Permanent labels are shown in bold face. Non-bold face labels are temporary.

Step	1	2	3	4	5	6	In-tree node and in-tree edge selected in step
			Label on node at end of step				
Initial	$\emptyset$	(1,112)	(1,300)	(1,270)	(1,157)	(1,195)	
1	$\emptyset$	**1**	(2,210)	(2,170)	(2,135)	(2,180)	2, (1; 2)
2	$\emptyset$	**1**	(5,206)	(5,140)	**2**	(5,85)	5, (2; 5)
3	$\emptyset$	**1**	(5,206)	(5,140)	**2**	**5**	6, (5; 6)
4	$\emptyset$	**1**	(4,105)	**5**	**2**	**5**	4, (5; 4)
5	$\emptyset$	**1**	**4**	**5**	**2**	**5**	3, (4; 3)

There can be at most $\binom{n}{2}$ edges in G, the network is dense if their number is of this order. Since the cost of each edge has to be examined at least once, for dense networks $O(n^2)$ is the best complexity we can expect for this problem. Hence Prim's algorithm attains the best complexity for this problem on dense networks. However, on sparse networks, i.e., when $|\mathcal{A}| = m$ is much smaller than n^2, it is possible to construct better algorithms. The next algorithm that we discuss is *Kruskal's algorithm*. It begins with the trivial spanning forest $(\{1\}, \emptyset), \ldots, (\{n\}, \emptyset)$, and it examines the edges in $\mathcal{A}$ in increasing order of cost, one after the other, and either discards it or includes it in the forest. The forest develops in several disconnected components until all of them are connected in the end. The various components may be of different sizes during the algorithm. It stores and updates each component as a tree, and maintains the set of nodes in each component at each stage.

KRUSKAL'S ALGORITHM

Initialization   Order the edges in $\mathcal{A}$ in increasing order of cost. Begin with the trivial spanning forest $(\{1\}, \emptyset), \ldots, (\{n\}, \emptyset)$.

General step   Get the next least cost remaining edge in $\mathcal{A}$. Suppose it is $(i; j)$. If both the nodes i, j on this edge lie in the same component of the forest at this stage, discard this edge and go to the next step. If i, j lie in different components of the forest at this stage, include $(i; j)$ as an in-forest edge and merge the two components that it connects into a single component. If there is only one component now, it is a minimum cost spanning tree, terminate. Otherwise go to the next step.

Discussion

Let $(i; j)$ be the edge that has come up for examination at some stage of this algorithm. Suppose i, j both appear in a component, $\mathbf{F}_r = (\mathcal{N}_r, \mathcal{A}_r)$ say, at this stage. All the arcs in $\mathcal{A}_r$ were examined earlier, so the cost of every edge in $\mathcal{A}_r$ is $\leqq c_{ij}$. So, if $\mathbb{C}_1$ is the fundamental cycle of $(i; j)$ wrt $\mathbf{F}_r$, then $(i; j)$ is a maximum cost edge on $\mathbb{C}_1$, and by Theorem 9.5 there exists a minimum cost spanning tree not containing edge $(i; j)$, among all the spanning trees containing all the edges in the forest at this stage.

Suppose i, j appear in different components, $\mathbf{F}_r = (\mathcal{N}_r, \mathcal{A}_r)$, and $\mathbf{F}_s = (\mathcal{N}_s, \mathcal{A}_s)$, say. Then $(i; j)$ is a least cost edge in the cut $(\mathcal{N}_r; \mathcal{N} \backslash \mathcal{N}_r)$, and this cut contains none of the forest edges at this stage. So, by Theorem 9.4, there exists a minimum cost spanning tree containing edge $(i; j)$ among the spanning trees containing all the edges in the forest at this stage.

Applying these arguments in each step from the beginning, we conclude that the spanning tree obtained at the termination of this algorithm is a minimum cost spanning tree in G. It can also be verified that this spanning tree satisfies (9.1).

Computationally, the most expensive operation in this algorithm is that of arranging the edges in $\mathcal{A}$ in ascending order of cost at the beginning, which requires $O(m \log m)$ effort, where $m = |\mathcal{A}|$. Also since $m < n^2$, this is the same as $O(m \log n)$, where $n = |\mathcal{N}|$. This is the worst case computational complexity of this method.

It is not actually necessary to order all the edges in $\mathcal{A}$ in ascending order of cost, since we will actually select only $(n - 1)$ of the edges for the final spanning tree. Any partial quick sort scheme which produces the least cost remaining edge would be adequate. An ideal scheme for this is a multipass sorting routine in which each pass produces the next edge in the ordered sequence very efficiently. Even with all these refinements, this method is clearly suitable to solve minimum cost spanning tree problems in relatively sparse networks.

When applied to find a minimum cost spanning tree in the network in Figure 9.1, this algorithm examines the edges in the order (5; 6), (3; 4), (1; 2), (2; 5), (4; 5), includes each of them in the forest and then terminates with the spanning tree consisting of the thick edges.

The next algorithm for this problem that we discuss is *Boruvka's algorithm*, it dates back to 1926, and seems to be one of the earliest algorithms for a network optimization problem. One of the main features of this algorithm is that it grows the forest, with all its components growing simultaneously in every step. This algorithm does not need the edges ordered in any specific order. The operations in this algorithm can be carried out unambiguously if all the edge cost coefficients are distinct. However, if some edge cost coefficients are equal, we need a unique tie breaking rule for the minimum in every pair of edge cost coefficients, that holds throughout the algorithm; otherwise there is ambiguity in the selection of the edges to be included in the forest, and this may result in cycles being formed. So, we adopt the following simple tie breaking strategy. Let $e_1, \ldots, e_m$ be the edges in the order in which they are recorded in the input data, with their cost coefficients $c_1, \ldots, c_m$ respectively. Brand the edges with these numbers. Among a pair of edges, say e_r, e_s; the edge e_r is considered to be the least cost one iff either $c_r < c_s$, or $c_r = c_s$ and $r < s$ (conceptually, this strategy is equivalent to replacing c_p by $c_p + p\epsilon$, for $p = 1$ to m, where ϵ is treated as an arbitrarily small positive parameter without being given any specific value).

BORUVKA'S ALGORITHM

Initialization   Begin with the trivial spanning forest $(\{1\}, \emptyset), \ldots, (\{n\}, \emptyset)$.

General step   Let $\mathbf{F}_1 = (\mathcal{N}_1, \mathcal{A}_1), \ldots, \mathbf{F}_l = (\mathcal{N}_l, \mathcal{A}_l)$, be the connected components in the forest at this stage. Find a least cost edge in the cut $(\mathcal{N}_h, \mathcal{N} \backslash \mathcal{N}_h)$, for each $h = 1$ to l. This can be carried out in a single operation of examining all the edges once. Add all these edges to the forest. Determine the connected components in the new forest. If there is only one, it is a minimum cost spanning tree in G, terminate. Otherwise, go to the next step.

Discussion

We need to show that the addition of all the new edges in a step does not create a cycle. Suppose at the beginning of some step, the trees in the forest are $\mathbf{F}_1 = (\mathcal{N}_1, \mathcal{A}_1), \ldots, \mathbf{F}_l = (\mathcal{N}_l, \mathcal{A}_l)$. Suppose a cycle is created when all the new edges selected in this step are added to this forest. Each of the selected edges joins a node in one of these trees, to a node outside this tree. So, a cycle can only be created if a subset of 2 or more selected edges, $e_{p_1}, \ldots, e_{p_t}$, say, are such that : e_{p_r}, the least cost edge in the cut $(\mathcal{N}_r, \mathcal{N} \backslash \mathcal{N}_r)$, joins $\mathbf{F}_r$ and another tree, $\mathbf{F}_{r+1}$ say, for

$r = 1$ to t, with $t + 1$ being 1. So, the trees that these selected edges join are as in Figure 9.3.

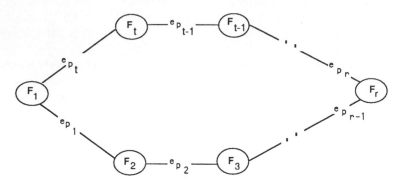

Figure 9.3 A situation for the selected edges that could lead to a cycle in Boruvka's algorithm.

Both edges $e_{p_{r-1}}$ and e_{p_r} are in the cut $(\mathcal{N}_r, \mathcal{N} \backslash \mathcal{N}_r)$, but e_{p_r} has been selected as the least cost edge in this step. So, by the arrangement made at the beginning of this algorithm, $c_{p_r} \leqq c_{p_{r-1}}$, and if $c_{p_r} = c_{p_{r-1}}$ then $p_r < p_{r-1}$ for $r = 1$ to t with $t + 1$ being 1. So, $c_{p_t} \leqq c_{p_{t-1}} \leqq \ldots \leqq c_{p_1} \leqq c_{p_t}$ which implies that $c_{p_t} = c_{p_{t-1}} = \ldots = c_{p_1} = c_{p_t}$. Hence we must have $p_t < p_{t-1} < \ldots < p_1 < p_t$, which is impossible. Hence cycles can never be created in this algorithm.

The validity of this algorithm follows from Theorem 9.4. In this method, the number of connected components in the forest is reduced by a factor of 2 in each step. So, the method terminates after at most $\log n$ steps, and the worst case computational complexity of the method is $O(m \log n)$.

When applied to find the minimum spanning tree in the network in Figure 9.1, this algorithm selects the edges (1; 2), (3; 4), (5; 6) in the first step after initialization. At this stage there are three trees in the forest, each one consisting of a single edge. In the next step it selects the edges (2; 5), (4; 5) to be added to the forest. With this addition, we get the spanning tree marked by the thick edges in Figure 9.1, and the method terminates.

There are several other algorithms for the minimum cost spanning tree problem based on the same idea of growing a forest using the greedy principle, and on the results in Section 9.1, but using very sophisticated data structures. Some of the best algorithms have the worst case computational complexity of $O(m \log \log n)$. The interested reader is referred to Yao [1975] and Tarjan [1983 of Chapter 1].

Other Problems Solvable by the Greedy Approach

The greedy method applies to a variety of problems besides the minimum cost spanning tree problem, in the general setting of matroidal optimization, see Lawler [1976 of Chapter 1]. However, the real and fundamental importance of the greedy approach is not mainly due to its ability to solve a few theoretical problems in matroidal optimization exactly; but because of its amenability in constructing heuristic algorithms to obtain reasonable solutions to many hard problems for which efficient exact algorithms are not known. There are a large number of combinatorial optimization problems which do not seem to have the basic structural properties to guarantee that an optimum solution for them can be found efficiently by any known approach. The greedy approach is the basis for the most commonly used heuristic algorithms to attack such problems that arise in applications. Thus the practical importance of the greedy approach is enormous. The greedy approach has 3 characteristic features.

The Incremental Feature   It poses the problem in such a way that the solution can be viewed as a set of elements. The approach builds up this solution set one element at a time.

The No-backtracking Feature   Once an element is selected for inclusion in the solution set, it is never taken back or replaced by some other element.

The Greedy Selection Feature   Each additional element selected for inclusion in the solution set is the best among those available for selection at that stage.

Several different criteria could be used to characterize the "best" when making the greedy selection, depending on the problem being solved. When used as a base for developing heuristic algorithms, the success of the greedy approach depends critically on the choice of this criterion.

A second approach that is commonly used in designing heuristic algorithms is the *interchange approach*. This approach obtains a solution set for the problem first, by using a greedy or any other approach. Then it selects a small positive integer, r, and searches to see if a better solution set can be obtained by exchanging r or less elements from the present solution set with those outside it. If an improvement is obtained, the whole process is repeated with the new solution set. If no improvement is obtained by the search, the method terminates with the present solution. Since the computational effort for the search grows rapidly with r, it is usually taken to be a small positive integer. The greedy approach coupled with an interchange approach is the basis for the most commonly used methods by practitioners to handle hard combinatorial optimization problems.

The Worst Case Performance of the Greedy Approach

On some problems like the minimum cost spanning tree problem, the greedy method has been shown to yield an optimum solution. However, it is frequently used on a variety of other combinatorial optimization problems on which it is not guaranteed to find an optimum solution. How does it perform on such problems? Here we summarize the disturbing results of Jenkyns [1986], who has constructed examples of assignment and traveling salesman problems on which the greedy method produces the worst possible solution.

Consider the assignment problem of order n in which the objective function is required to be *maximized* (the algorithms for the assignment problem in Chapter 3 are described in terms of minimization). Let $C = (c_{ij})$ be the objective coefficient matrix. The most direct version of the greedy method identifies a cell (p, q) with the maximum coefficient in C, puts an allocation in that cell, strikes off row p and column q, and continues in the same manner with the remaining matrix until n allocations are made. Let $n = 6$, consider this problem with the following matrix $C = (c_{ij})$.

	1	2	3	4	5	6	$\bar{u}_i$
$j =$ c_{ij}							
$i = 1$	1	2	6	10	17	29	0
2	3	4	8	11	20	30	2
3	5	7	9	12	22	33	4
4	13	14	15	16	23	34	9
5	18	19	21	24	25	35	16
6	26	27	28	31	32	36	24
$\bar{v}_j$	1	2	5	7	9	12	

It can be verified that the greedy method produces the unit matrix with all allocations along the main diagonal, as the optimum assignment. However, from the $\bar{u}, \bar{v}$ vectors given in the above tableau, it can be verified that $\bar{u}_i + \bar{v}_j \leqq c_{ij}$ for all i, j; and $\bar{u}_i + \bar{v}_i = c_{ii}$ for $i = 1$ to 6; from the results in Chapter 3 this establishes that the unit matrix is the minimum objective value assignment in this problem. Hence the unit matrix is the worst candidate for the problem of maximizing the objective function considered here. So, on this problem, any random selection is a better answer than that produced by the greedy algorithm.

A square matrix C of order n is defined to be *gullible* (to indicate that it will be an easy victim of a shady lady) if the greedy algorithm produces the worst possible answer for the objective maximizing assignment problem with C as the objective coefficient matrix. Let $\hat{u} = (\hat{u}_1, \ldots, \hat{u}_n), \hat{v} = (\hat{v}_1, \ldots, \hat{v}_n)$ be any pair of vectors in which the components are nondecreasing left to right. Select c_{ij} to satisfy $\hat{u}_i + \hat{v}_j \leqq c_{ij} \leqq \hat{u}_m + \hat{v}_m$ where $m = \max. \ \{i, j\}$, for all i, j. Then it can be verified that the greedy method produces the unit matrix of order n as the solution to the objective maximizing assignment problem with C as the objective coefficient matrix, and this is actually a minimum objective solution in this problem. Hence such matrices are gullible. Jenkyns [1986] has given many other procedures to generate gullible matrices and shows that they are not intricately contrived pathological cases.

At least for the assignment problem there are very efficient exact algorithms available, and no one would have to resort to heuristics to solve it. However, the traveling salesman problem (TSP) is a well known NP-hard problem, large scale versions of which are commonly solved using greedy type methods in many applications. Jenkyns [1986] has constructed examples of TSPs in n cities for all

n, and showed that a simple greedy heuristic produces the worst possible tour on each of them.

These results show that it is not wise to rely entirely on a greedy heuristic by itself, to tackle hard combinatorial optimization problems without some sort of error analysis.

How to Find Maximum Length Spanning Trees

If a maximum length spanning tree in $G = (\mathcal{N}, \mathcal{A})$ with c as the edge length vector is required, it can be obtained by finding a minimum length spanning tree in G with $-c$ as the edge length vector. The algorithms discussed above for the minimum cost spanning tree problem work for arbitrary cost coefficients.

9.3 An Algorithm for Ranking the Spanning Trees in Ascending Order of Cost

Let $G = (\mathcal{N}, \mathcal{A}, c)$ be a connected undirected network with c as the edge cost vector, and $|\mathcal{N}| = n, |\mathcal{A}| = m$. In Section 9.2 we discussed several algorithms for finding an unconstrained minimum cost spanning tree in G. In some applications, a minimum cost spanning tree satisfying some constraints may be required. One possible approach for finding such a tree is to rank the spanning trees in G in increasing order of cost until one satisfying the constraints is found. Here we present an algorithm for this ranking from Murty, Saigal, and Suurballe [1974] based on the application of the partitioning technique discussed in Sections 3.6 and 4.7. We denote the ranked sequence of spanning trees in G by $\mathbb{T}_1, \mathbb{T}_2, \ldots$ where $\mathbb{T}_1$ is a minimum cost spanning tree in G, and for $u \geqq 1$, $\mathbb{T}_{u+1}$ is a minimum cost spanning tree in G excluding the trees $\mathbb{T}_1, \ldots, \mathbb{T}_u$.

The algorithm deals with subsets of spanning trees in G which will be denoted using a special notation defined below. Here $a_1, \ldots, a_r, b_1, \ldots, b_s$ are edges in $\mathcal{A}$.

$$\tau_0 = [a_1, \ldots, a_r; \bar{b}_1, \ldots, \bar{b}_s] \quad = \quad \begin{array}{l} \text{Set of all spanning trees in G} \\ \text{containing each of the edges} \\ a_1, \ldots, a_r; \text{ and none of the} \\ \text{edges } b_1, \ldots, b_s. \end{array} \quad (9.4)$$

The edges $a_1, \ldots, a_r$ are said to be the *included edges* in τ_0; and the edges $b_1, \ldots, b_s$ are said to be the *excluded edges* from τ_0.

Let $\tau = [a_1, \ldots, a_r; \bar{b}_1, \ldots, \bar{b}_s]$ be a set of spanning trees in G, and let $\mathbb{T}$ consisting of edges $a_1, \ldots, a_r, e_{r+1}, \ldots, e_{n-1}$; be a minimum cost spanning tree among those in τ. Define the following new subsets.

$$\tau_1 \quad = \quad [a_1, \ldots, a_r; \bar{b}_1, \ldots, \bar{b}_s, \bar{e}_{r+1}]$$

$$\tau_2 = [a_1, \ldots, a_r, e_{r+1}; \bar{b}_1, \ldots, \bar{b}_s, \bar{e}_{r+2}]$$
$$\tau_3 = [a_1, \ldots, a_r, e_{r+1}, e_{r+2}; \bar{b}_1, \ldots, \bar{b}_s, \bar{e}_{r+3}]$$
$$\vdots$$
$$\tau_{n-r-1} = [a_1, \ldots, a_r, e_{r+1}, \ldots, e_{n-2}; \bar{b}_1, \ldots, \bar{b}_s, \bar{e}_{n-1}]$$

For $j = 1$ to $n - r - 1$, it is easy to check whether $\tau_j \neq \emptyset$, and if so to find a minimum cost spanning tree in it. Find the fundamental cutset of the in-tree arc e_{r+j} wrt $\mathbb{T}$, suppose it is $(\mathbf{X}; \overline{\mathbf{X}})$. The edge e_{r+j} is the unique in-tree edge in $\mathbb{T}$ in the cutset $(\mathbf{X}; \overline{\mathbf{X}})$, and it is a minimum cost edge among edges of this cutset not in $\{b_1, \ldots, b_s\}$ since $\mathbb{T}$ is a minimum cost spanning tree among those in τ. The subset τ_j is empty iff the cutset $(\mathbf{X}; \overline{\mathbf{X}}) \subset \{b_1, \ldots, b_s, e_{r+j}\}$, the set of excluded edges in τ_j. If $(\mathbf{X}; \overline{\mathbf{X}}) \not\subset \{b_1, \ldots, b_s, e_{r+j}\}$, let e'_{r+j} be a minimum cost edge in the cutset $(\mathbf{X}; \overline{\mathbf{X}})$ that is not contained in $\{b_1, \ldots, b_s, e_{r+j}\}$. Then the spanning tree $\mathbb{T}^j$ obtained by replacing e_{r+j} in $\mathbb{T}$ by e'_{r+j}; is a minimum cost spanning tree among those in τ_j. Clearly, the computational effort needed to check whether $\tau_j = \emptyset$ and finding a minimum cost spanning tree $\mathbb{T}^j$ in it if it is nonempty is at most $O(m)$ by this approach.

Each of the sets $\tau_1, \ldots, \tau_{n-r-1}$ is again in the same form as in (9.4), and clearly they are mutually disjoint, and their union is $\tau \backslash \{\mathbb{T}\}$. These sets are said to be the *new sets generated when τ is partitioned using the minimum spanning tree $\mathbb{T}$ in it.* The number of these sets is $n - 1 -$(number of included edges in τ), and some of these sets may be empty. Clearly, a minimum cost spanning tree among those in $\tau \backslash \{\mathbb{T}\}$ is the best among $\{\mathbb{T}^j : j = 1$ to $n - r - 1$ such that $\tau_j \neq \emptyset \}$.

The ranking algorithm generates various sets of spanning trees in G of the form defined in (9.4). Each nonempty set among these is stored in a list, together with a minimum cost spanning tree in it, in increasing order of the cost of this tree, going from top to bottom. Each step of the algorithm generates one additional spanning tree in the ranked sequence, and adds at most $n - 1$ new sets to the list. The computational effort of each step is at most $O(mn)$, and the method can be terminated any time, after enough spanning trees in the ranked sequence have been obtained.

RANKING ALGORITHM

Initial step   Find a minimum cost spanning tree in G and call it $\mathbb{T}_1$, the first tree in the ranked sequence. Suppose the edges in it are $e_1, \ldots, e_{n-1}$. Define the subsets: $\tau_1 = [\bar{e}_1]$, $\tau_2 = [e_1; \bar{e}_2]$, $\ldots$, $\tau_{n-1} = [e_1, \ldots, e_{n-2}; \bar{e}_{n-1}]$. Find a minimum cost spanning tree in each of these subsets, and arrange the nonempty subsets among them in increasing order of cost of the minimum cost spanning tree in them (as you go from top to bottom) in the list.

General step   Suppose the trees $\mathbb{T}_1, \ldots, \mathbb{T}_u$ in the ranked sequence have already been obtained. Pull out the set from the top of the list now, suppose it is τ, and delete it from the list. $\mathbb{T}_{u+1}$ is the stored minimum cost spanning tree in τ. If no more elements in the ranked sequence are needed, terminate. Otherwise, partition τ using T_{u+1}. Find the minimum cost spanning tree in each of the new sets generated at this partition, as described above, and add the nonempty sets among them, together with the minimum cost spanning tree in it in the appropriate place in the list, and go to the next step.

Discussion

If only a certain number, α say, of the spanning trees in the ranked sequence are required, only the top α sets in the list need be kept, and the rest discarded. If the aim is to obtain all spanning trees whose cost is $\leqq$ some specified β, any set generated during the algorithm is stored only if the cost of the minimum spanning tree in it is $\leqq \beta$.

9.4 Other Types of Minimum Cost Spanning Tree Problems

So far we have dealt with the unconstrained minimum cost spanning tree problem in networks, ignoring any orientations of the lines. Here we will discuss other types of minimum spanning tree problems briefly.

Constrained Minimum Cost Spanning Tree Problems

If we are required to find a minimum spanning tree subject to degree constraints at one or more nodes (i.e., those of the form: at certain nodes i, the tree can have no more than a specified number d_i of incident edges, or should have exactly a specified number of incident edges, etc.), we have a *degree constrained minimum cost spanning tree problem*. Problems with a degree constraint at only one node can be transformed into unconstrained problems (Gabow and Tarjan [1984]). Problems with degree constraints at nodes no pair of which are adjacent in G can be formulated as weighted matroid intersection problems and solved by efficient matroidal algorithms (Gabow and Tarjan [1984], Lawler [1976 of Chapter 1]).

But if degree constraints appear at all the nodes in G, the problem becomes a hard combinatorial optimization problem. As an example consider the case where the degree in the tree at every node is required to be $\leqq 2$. Any feasible spanning tree to this problem is a node covering simple path and vice versa, such paths are called *Hamiltonian paths*. To check whether a Hamiltonian path exists in an undirected network is NP-complete, and to find a minimum cost Hamiltonian path is NP-hard.

The Steiner Tree Problem in Networks

So far we have considered problems of finding optimum trees that span all the nodes in the network. In the *Steiner tree problem*, we are given a connected undirected network $G = (\mathcal{N}, \mathcal{A}, c)$ and a subset of nodes $\mathbf{X} \subset \mathcal{N}$. The problem is to find a minimum cost tree $\mathbb{T}$ in G subject to the constraint that $\mathbb{T}$ must include all the nodes in $\mathbf{X}$, but may or may not include the nodes in $\mathcal{N} \setminus \mathbf{X}$. So, the problem is equivalent to finding a subset of nodes $\mathbf{Y}$ satisfying $\mathbf{X} \subseteq \mathbf{Y} \subseteq \mathcal{N}$ such that the cost of a minimum spanning tree in the subnetwork of G induced by $\mathbf{Y}$ is the least subject to this constraint. Nodes in the optimal $\mathbf{Y}$ which are not in $\mathbf{X}$ are called *Steiner points* for this problem. So, the problem boils down to the optimal choice of Steiner points from $\mathcal{N} \setminus \mathbf{X}$.

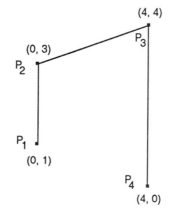

Figure 9.4 Shortest connecting network with no outside points.

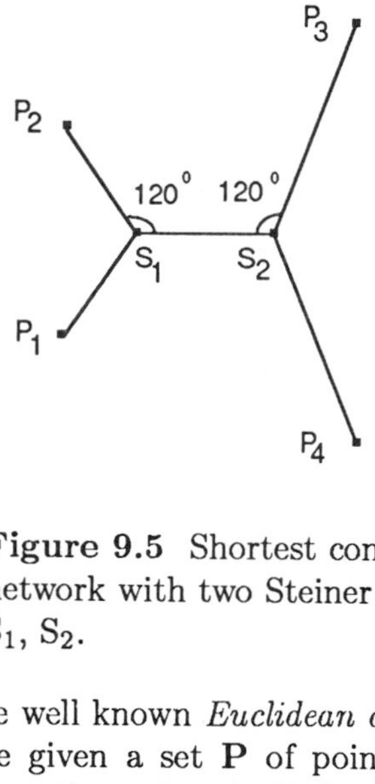

Figure 9.5 Shortest connecting network with two Steiner points S_1, S_2.

This problem is the network version of the well known *Euclidean or rectilinear Steiner problem in geometry*. There, we are given a set $\mathbf{P}$ of points in the 2-dimensional plane. The problem is to connect the points in $\mathbf{P}$ by straight line segments so that the total length (Euclidean or rectilinear) of the line segments drawn is a minimum. Some of these lines could meet at points in the plane which are not in the set $\mathbf{P}$, such points are the *Steiner points* in the solution of this problem. As an example consider the problem in which $\mathbf{P} = \{P_1, \text{to } P_4\}$ as shown in Figure 9.4. If no outside points are allowed, the shortest connecting network has Euclidean length of 10.123. By introducing 2 Steiner points S_1, S_2 as shown in Figure 9.5, the Euclidean length of the connecting network reduces to 9.196.

The Steiner tree problem in networks is a hard combinatorial optimization problem. Exact enumerative algorithms are available for solving it, but the computational effort required by these algorithms grows exponentially with $|\mathcal{N} \setminus \mathbf{X}|$.

Directed Spanning Tree Problems

So far we have discussed spanning tree problems in undirected networks. Consider now a directed network $G = (\mathcal{N}, \mathcal{A}, c)$ with $|\mathcal{N}| = n, |\mathcal{A}| = m$. Suppose we are given a designated node, say node 1, as the root node. A *branching* rooted at 1 is a spanning outtree rooted at 1. The problem of finding a minimum cost branching rooted at 1 in G has a very efficient algorithm of time complexity $O(nm)$. But it does not seem to be of much practical importance, so we will not discuss it. See references given at the end of this chapter.

9.5   Exercises

9.3 Construct a directed network with positive arc lengths to show that a shortest chain tree rooted at a node may not be a minimum cost branching rooted at that node (try the network in Figure 9.6, with node 1 as the root node. Arc lengths are entered on the arcs).

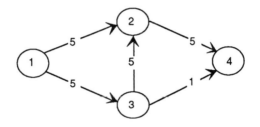

Figure 9.6

9.4 Given a minimum cost spanning tree in an undirected network, develop an efficient procedure for doing cost ranging on it, i.e., to determine the range of values of each cost coefficient which does not affect the minimality of the tree (Tarjan [1982]).

9.5 We are given a minimum cost spanning tree T in an undirected connected network. Develop an efficient procedure to update T into a minimum cost spanning tree when a new node and incident edges are added to the network (Spira and Pan [1975]).

9.6 Develop good heuristic algorithms based on the greedy approach for the following combinatorial optimization problems.

(a) The Knapsack Problem   In this problem we are given n objects with w_i, v_i as the weight, value of the ith object, $i = 1$ to n. A knapsack with weight capacity w is provided. It is required to find an optimal subset of objects to be loaded into the knapsack, so as to maximize its value, subject to the knapsack's weight capacity. All data are positive integers. Mathematically, it is to find $x = (x_i : i = 1$ to $n)$ to

$$\text{maximize} \quad v_1 x_1 + \ldots + v_n x_n$$
$$\text{subject to} \quad w_1 x_1 + \ldots + w_n x_n \leq w$$
$$\text{and } x_i = 0, \text{ or } 1, \quad i = 1 \text{ to } n$$

Apply your heuristic algorithm for obtaining a solution to the following journal subscription problem faced by a librarian. The librarian has to decide an optimum subset among 8 journals to renew the subscription, with only 500\$ available in the new budget for these journals. Other data is given below. The journals selected by the librarian should satisfy the maximum number of users subject to the budget constraint.

Journal i	w_i, annual subscription in \$	v_i, no. of users per year
1	80	7840
2	95	6175
3	115	8510
4	165	15015
5	125	7375
6	78	1794
7	69	897
8	99	8315

(b) The Set Covering Problem This is a 0-1 integer programming problem of the following form

$$\text{minimize} \quad cx$$
$$\text{subject to} \quad Ax \geq e$$
$$x \text{ is a } 0 - 1 \quad \text{vector}$$

where e on the right hand side is a column vector of all 1's in $\mathbb{R}^m$, $A = (a_{ij})$ is a given 0-1 matrix of order $m \times n$, and c is a positive row vector in $\mathbb{R}^m$. Apply your heuristic algorithm on the following numerical example. Here $m = 10$, $n = 13$, c is the row vector of all 1's in $\mathbb{R}^{13}$, and the sets $\Gamma_i = \{j : j = 1$ to 13 such that $a_{ij} = 1\}$ are: $\Gamma_1 = \{7, 9, 10, 13\}$, $\Gamma_2 = \{2, 8, 9, 13\}$, $\Gamma_3 = \{3, 9, 10, 12\}$, $\Gamma_4 = \{4, 5, 8, 9\}$, $\Gamma_5 = \{3, 6, 8, 11\}$, $\Gamma_6 = \{3, 6, 7, 10\}$, $\Gamma_7 = \{2, 4, 5, 12\}$, $\Gamma_8 = \{4, 5, 6, 13\}$, $\Gamma_9 = \{1, 2, 4, 11\}$, $\Gamma_{10} = \{1, 5, 7, 12\}$.

9.7 Let $G = (\mathcal{N}, \mathcal{A}, c)$ be an undirected connected network. Let d_1 be the degree on node 1 in G. Let $\mathbf{D}_r$ denote the set of spanning trees in G containing exactly r edges incident at 1. Show that there exists a $u \overset{\geq}{=} 1$ such that the set of all r for which $\mathbf{D}_r \neq \emptyset$ is $u \overset{\leq}{=} r \overset{\leq}{=} d_1$.

Suppose $\mathbf{T}$ is a minimum cost spanning tree in $\mathbf{D}_r$ for some r. Then prove that

(a) if $\mathbf{D}_{r-1} \neq \emptyset$, there are edges e_1 of $\mathbf{T}$ incident at 1, and $e_2 \in \mathcal{A}$ not incident at 1 and not in $\mathbf{T}$; such that replacing e_1 in $\mathbf{T}$ by e_2 yields a minimum cost tree in $\mathbf{D}_{r-1}$

(b) if $\mathbf{D}_{r+1} \neq \emptyset$, there are edges e_1 not in $\mathbf{T}$ and incident at 1, and e_2 in $\mathbf{T}$ not incident at 1; such that replacing e_2 in $\mathbf{T}$ by e_1 yields a minimum cost tree in $\mathbf{D}_{r+1}$.

Use these results to develop an efficient algorithm for finding a minimum cost spanning tree in G satisfying a degree constraint at node 1, as this specified degree varies parametrically (Gabow [1978]).

9.8 Minimum Ratio Spanning Trees Let $G = (\mathcal{N}, \mathcal{A})$ be a connected undirected network with two vectors $a = (a_{ij})$, $b = (b_{ij})$ of edge cost coefficients defined over $\mathcal{A}$. For any tree $\mathbf{T}$ in G, define $z(\mathbf{T}) = (\sum(a_{ij} : \text{over } (i; j) \in \mathbf{T}) / (\sum(b_{ij} : \text{over } (i; j) \in \mathbf{T})$. Assume that the vector b is such, that the denominator of $z(\mathbf{T})$ has the same sign for every spanning tree in G, without any loss of generality we assume that this sign is positive. Consider the problem of finding a spanning tree $\mathbf{T}$ in G that minimizes $z(\mathbf{T})$. Let z^* denote the unknown optimum objective value for this problem

Consider the following greedy approach for this problem. It grows a connected tree by adding one selected edge at a time, making sure that each edge added does not form a cycle with those already selected, and among such edges it selects the one which increases the objective function by the smallest quantity. Use the network in Figure 9.7 to show that this approach may not lead to an optimum solution of this problem.

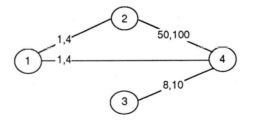

Figure 9.7 Entries on the edges are a_{ij}, b_{ij} in that order.

Let α be a real valued parameter. Define for each $(i; j) \in \mathcal{A}$, $c_{ij}(\alpha) = a_{ij} - \alpha b_{ij}$. Let $\hat{\mathbb{T}}(\alpha)$ denote a minimum cost spanning tree in G with $c(\alpha) = (c_{ij}(\alpha))$ as the edge vector, and let $h(\alpha)$ be its cost. Prove that

$$h(\alpha) \begin{cases} > 0 \text{ iff } z^* > \alpha \\ < 0 \text{ iff } z^* < \alpha \\ = 0 \text{ iff } z^* = \alpha \end{cases}$$

Let e, g denote any pair of edges in G. When $a_e - \alpha b_e, a_g - \alpha b_g$ are plotted against α on the 2-dimensional Cartesian plane, we get two straight lines which may either be identical, or nonintersecting parallel lines, or two lines intersecting at a unique point. For every pair of edges e, g in G for which the two lines given above intersect at a unique point, define $\alpha(e, g) = (a_e - a_g)/(b_e - b_g)$, and let Δ be the set of all such $\alpha(e, g)$. So, $|\Delta| \overset{\leq}{=} |\mathcal{A}|^2$. Suppose there are r distinct elements in Δ, order them in increasing order, and then suppose they are $-\infty < \alpha_1 < \alpha_2 < \cdots < \alpha_r < +\infty$. Prove that for any α in any one of the open intervals $(-\infty, \alpha_1), (\alpha_1, \alpha_2), \ldots, (\alpha_r, +\infty)$, the minimum cost spanning tree $\hat{T}(\alpha)$ is actually a minimum cost spanning tree for every α in that interval. Hence $h(\alpha)$ is linear in each of these intervals. Using this result, develop an efficient algorithm for finding a spanning tree T in G that minimizes $z(T)$, and determine its computational complexity (Chandrasekaran [1977]).

9.9 In applying Kruskal's algorithm to find a minimum cost spanning tree in G, if r is the number of forest edges in a step, prove that the forest at that stage is a minimum cost forest of cardinality r in G.

9.10 It is required to find a single tree in G of minimum cost containing r edges. Develop an efficient approach for solving this problem.

9.11 $G = (\mathcal{N}, \mathcal{A}, c)$ is a connected undirected network with $c > 0$. Develop efficient methods for finding spanning trees in G that minimize : (*a*) the product of the cost coefficients of in-tree edges, (*b*) the cost of the costliest in-tree edge, (*c*) any symmetric cost function of edge cost coefficients.

9.12 Let $G = (\mathcal{N}, \mathcal{A})$ be a connected undirected network with $|\mathcal{N}| = n, |\mathcal{A}| = m$. The characterization of the convex hull of spanning tree incidence vectors in G described in Exercise 9.2 has about 2^n constraints. Here we discuss a constraint formulation of the convex hull of spanning trees in G, in a higher dimensional space, with only $O(mn)$ constraints. Additional variables called *auxiliary variables* are introduced. In this formulation, a particular node, say 1, is selected as a source node. Each of the other nodes $r \neq 1$ is associated with a distinct commodity for which it is the sink node. So, there are $(n - 1)$ different commodities in this formulation, and node 1 has one unit of each of these commodities available for shipment.

For each $(i; j) \in \mathcal{A}$ with both $i, j \neq 1$ create two arcs (i, j) and (j, i). For each edge $(1; j)$ in G create an arc $(1, j)$. Let **A** denote the set of all these created arcs. So, $|\mathbf{A}| = 2m -$ degree of node 1 in G. Define $\mathbf{A}_i = \{j : (i, j) \in \mathbf{A}\}$, $\mathbf{B}_i = \{j : (j, i) \in \mathbf{A} \}$, for each $i \in \mathcal{N}$.

Introduce a variable $x_{i;j}$ for each $(i;j) \in \mathcal{A}$; it is the spanning tree-edge incidence variable for that edge; and auxiliary variables y_{ij}^r for each arc $(i,j) \in \mathbf{A}$ and $r = 2$ to n. y_{ij}^r denotes the flow of commodity r on arc (i,j). Let $x = (x_{i;j}), y = (y_{ij}^r)$. Consider the following system of constraints.

$$
\begin{aligned}
\sum (x_{i;j} : \text{ over } (i;j) \in \mathcal{A}) &= n-1 \\
\sum (y_{1i}^r : \text{ over } i \in \mathbf{A}_1) &= 1, \quad r = 2 \text{ to } n \\
\sum (y_{ij}^r : \text{ over } i \in \mathbf{B}_j) & \\
-\sum (y_{ji}^r : \text{ over } i \in \mathbf{A}_j) &= 0, \quad j = 2 \text{ to } n, j \neq r; \\
& \qquad\qquad r = 2 \text{ to } n \\
\sum (y_{ir}^r : \text{ over } i \in \mathbf{B}_r) &= 1, \quad r = 2 \text{ to } n \quad\quad (9.5) \\
y_{1j}^r - x_{1;j} &\leqq 0, \quad j \in \mathbf{A}_1, r = 2 \text{ to } n \\
y_{ij}^r + y_{ji}^p - x_{i;j} &\leqq 0, \quad i, j \text{ both } \neq 1, (i;j) \in \mathcal{A}; \\
& \qquad\qquad r, p = 2 \text{ to } n \\
0 \leqq x_{i;j} \leqq 1, & \qquad (i;j) \in \mathcal{A} \\
y_{ij}^r \geqq 0, & \qquad (i,j) \in \mathbf{A}, r = 2 \text{ to } n
\end{aligned}
$$

Prove that the projection of the feasible solution set of (9.5) in the x-space, is the convex hull of spanning tree incidence vectors in G (Kipp Martin [1986], Wong [1984]).

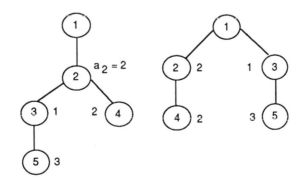

Figure 9.8 Capacity of every edge is 5. a_i is entered by the side of node i. The tree at the left is infeasible, and the one on the right is feasible, for the capacitated problem.

9.13 The Capacitated Minimum Spanning Tree Problem $G = (\mathcal{N}, \mathcal{A}, c, k)$ is a connected undirected network with c as the vector of edge cost coefficients, and

k as the vector of edge flow capacities. One of the nodes in G, 1, is the sink node; and all other nodes are source nodes, with $i \neq 1$ required to ship a_i units of a commodity from it to node 1. The problem is to find a spanning tree in G along the simple paths in which all these shipments will be carried out. A spanning tree $\mathbb{T}$ is said to be feasible to this problem if all these flows can be carried out on it without violating the edge capacity on any in-tree edge, infeasible otherwise. For example, the spanning tree on the left in Figure 9.8 is infeasible for the capacitated problem, whereas the one on the right is feasible (in the figure, capacities of all the edges are 5, and for each $i \neq 1$, a_i is entered by the side of node i, the data is the same in both the trees). Develop either an exact method, or a good heuristic method, for finding a minimum cost feasible spanning tree for the capacitated problem (Chandy and Lo [1973]).

9.14 A Constrained Minimum Spanning Tree Problem   Node 1 is the center that will be the root of a spanning tree in G $= (\mathcal{N}, \mathcal{A}, c)$. Each node $i \neq 1$ is associated with a load of a_i. It is required to find a minimum cost spanning tree in G satisfying: there should be no more than n_0 nodes connected to the center by an in-tree edge, there should be no more than l edges in tandem in a path connecting a node to the center, the total load on any edge (the sum of the loads of all nodes it connects to the center) cannot be greater than a specified d, and that no node other than the center have degree greater than a specified number r. Constrained minimum spanning tree problems of this type arise in designing centralized telecommunication and data communication networks using a single speed line. Develop a heuristic algorithm based on Kruskal's algorithm to generate high quality solutions to this problem (Kershenbaum [1974]).

9.15 G $= (\mathcal{N}, \mathcal{A}, c)$ is a connected undirected network with $c \geqq 0$. Develop 0-1 integer programming formulations for the following constrained minimum cost spanning tree problems in G: a) for each node in a specified subset of nodes, an upper bound on the number of in-tree edges incident at that node is specified, b) the number of descendents of each son node of a specified root node is required to be $\leqq$ a specified positive integer α (Gavish [1982]).

9.16 For any subset $\mathbf{N} \subseteq \mathcal{N}$ of nodes in an undirected network G $= (\mathcal{N}, \mathcal{A})$, let $\mathbf{E(N)}$ denote the set of edges with both their nodes in $\mathbf{N}$. Prove that the set of feasible solutions $x = (x_e : e \in \mathcal{A})$ of the following system, is the convex hull of forest-edge incidence vectors in G.

$$\sum(x_e : \text{ over } e \in \mathbf{E(N)}) \quad \leqq \quad |\mathbf{N}| - 1, \text{ for each } \mathbf{N} \subseteq \mathcal{N} \text{ with } |\mathbf{N}| \geqq 2$$

$$x_e \quad \geqq \quad 0, \text{ for all } e \in \mathcal{A}$$

(Edmonds [1969]).

9.17 Let l_{ij} denote the number of lines on the simple path between nodes i, j in a tree $\mathbb{T}$. For each i in $\mathbb{T}$, define $\theta_i = \max. \{l_{ij} : \text{over nodes } j \text{ in } \mathbb{T}\}$. A *center* of $\mathbb{T}$ is defined to be a node in $\mathbb{T}$ that minimizes θ_i. Show that the following algorithm finds a center for $\mathbb{T}$. Also, find the center of the tree in Figure 9.9 using this algorithm.

Step 1 If the number of nodes in the tree is ≤ 2, any node in the tree is a center, terminate. Otherwise go to Step 2.

Step 2 Identify all the terminal nodes in the tree, and delete them and the lines incident at them from the tree. Go to Step 3.

Step 3 If the number of remaining nodes is ≤ 2, any of these is a center for the original tree, terminate. Otherwise return to Step 2 with the remaining tree.

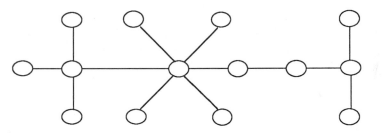

Figure 9.9

9.18 Let $\mathbb{T}$ be a tree spanning n nodes. For nodes i, j in $\mathbb{T}$ define l_{ij} as in Exercise 9.17. For a node i in $\mathbb{T}$ let $j_1, \ldots, j_p$ be all its adjacent nodes in $\mathbb{T}$, and let $\mathbb{T}_{j_r}$ be the subtree containing node j_r when the line joining i and j_r is deleted from $\mathbb{T}$, $r = 1$ to p. Let n_r be the number of nodes in $\mathbb{T}_{j_r}, r = 1$ to p. Node i in $\mathbb{T}$ is said to be a *centroid* of $\mathbb{T}$ iff $n_r \leq n/2$ for all $r = 1$ to p. Prove that $\sigma_i = \sum(l_{ij} : \text{over } j \text{ in } \mathbb{T})$ is minimized iff i is a centroid of $\mathbb{T}$. Develop an efficient algorithm for finding a centroid of $\mathbb{T}$. Apply your algorithm to find a centroid of the tree given in Figure 9.10 (Kang, Lee, Chang, and Chang [1977], Kang and Ault [1975]).

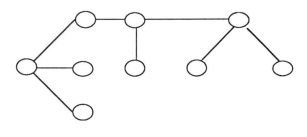

Figure 9.10

9.19 Shallow Minimum Spanning Trees $G = (\mathcal{N}, \mathcal{A}, c)$ is a connected undirected network. Among all minimum cost spanning trees in G, it is required to find one which is as shallow as possible (i.e., has the smallest number of levels when it is drawn as a rooted tree). Show that the following variant of Prim's algorithm finds such a tree. In this algorithm, **X** denotes the set of in-tree nodes, which grows by one in each step.

Step 0   Select any node i, let $\mathbf{X} = \{i\}$. Go to Step 1.

Step 1   For each $j \in \mathcal{N}\backslash\mathbf{X}$, define $d_j = \min. \{c_{ij}:$ over $i \in \mathbf{X}\}$. Find a node $q \in \mathcal{N}\backslash\mathbf{X}$ which minimizes d_j over j in this set. Let **Y** be the set of all i in **X** which tie for the minimum in the definition of d_q. If **Y** is a singleton set, let p be the node in it, otherwise define p to be the node in **Y** which has the maximum number of in-tree edges incident at it at this stage. Select $(p;q)$ as the next in-tree edge, and transfer q into **X** from $\mathcal{N}\backslash\mathbf{X}$. If **X** is now $\mathcal{N}$, terminate. Otherwise, repeat this step.

Develop appropriate data structures to obtain an efficient implementation of this algorithm (Kang, Lee, Chang, and Chang [1977]).

9.20 A min-max spanning tree in G is one which minimizes the maximum cost among all in-tree edges. Prove that every minimum cost spanning tree is also a min-max spanning tree, but the converse may not be true. Develop an O(m) complexity algorithm to find a min-max spanning tree in G (Camerini [1978]).

9.21 $\mathbb{T}$ is a minimum cost spanning tree in the connected undirected network $G = (\mathcal{N}, \mathcal{A}, c)$. $G' = (\mathcal{N}', \mathcal{A}', c')$ is the network obtained when a node i and all the edges incident at it are deleted from G. Assume that G' is connected. Let $\mathbb{T}_i = (\mathcal{N}', \mathrm{E}_i)$ be the forest remaining when node i and all the in-tree edges incident at it are deleted from $\mathbb{T}$.

If i is a leaf node in $\mathbb{T}$, show that $\mathbb{T}_i$ is a minimum cost spanning tree in G'.

Suppose i is not a leaf node in $\mathbb{T}$. Let $G'' = (\mathcal{N}'', \mathcal{A}'')$ be an undirected network whose vertex set is the set of connected components of $\mathbb{T}_i$, and every edge $e = (a,b)$ in $\mathcal{A}''$ corresponds to an edge e' in $\mathcal{A}'$ such that e' has the smallest cost among all the edges connecting the components a and b of $\mathbb{T}_i$ in G'. Shoe that G'' is connected, and that if $\mathbb{T}'' = (\mathcal{N}'', \mathrm{E}'')$ is a minimum cost spanning tree in G'' then $\mathrm{E}' = \mathrm{E}_i \cup \mathrm{E}''$ is such that $\mathbb{T}' = (\mathcal{N}', \mathrm{E}')$ is a minimum cost spanning tree in G'.

Using these results, develop an efficient algorithm to update the minimum cost spanning tree when a vertex is deleted from the network (Chin and Houck [1978], Tsin [1988]).

9.22 Minimum Cost Spanning Tree Problen as a Min-Max Path Problem   Let $G = (\mathcal{N} = \{1, 2, \ldots, n\}, \mathcal{A}, c)$ be a connected undirected network. Assume that all the entries in c are distinct, otherwise the perturbation discussed under Bourvka's can be used to achieve this. Define the cost of any path in G to be the max. $\{c_{ij} : (i;j)$ is an edge on the path$\}$. Define $l_{ij}^0 = \infty$ if $(i;j) \notin \mathcal{A}$, c_{ij} if $(i;j) \in \mathcal{A}$. For all $i, j, r = 1$ to n define

$$l_{ij}^r = \text{cost, as defined above, of the least cost path from } i \text{ to } j$$
$$\text{that passes through only vertices in the set } \{1, 2, \ldots, r\}.$$

For any $i, j \in \mathcal{N}$, the least cost path between i and j with no intermediate vertex higher than r either passes through r (in this case $l_{ij}^r = \text{max. } \{l_{ir}^r, l_{rj}^r\}$) or does not (in this case $l_{ij}^r = l_{ij}^{r-1}$). Using this, prove that $l_{ij}^r = \text{min. } \{l_{ij}^{r-1}, \text{max. } \{l_{ir}^{r-1}, l_{rj}^{r-1}\}\}$.

Prove that the unique minimum cost spanning tree in G in this case can be recovered from the costs of the least cost paths using the rule: the edge $(i;j) \in \mathcal{A}$ is in the minimum cost spanning tree in G iff $l_{ij}^0 = l_{ij}^n$ (Maggs and Plotkin [1988]).

9.23 Develop an efficient algorithm for finding a minimum cost spanning tree in G, subject to the constraint that it contain a specified number, r, of edges from a given subset $\mathbf{S} \subset \mathcal{A}$ (Gabow and Tarjan [1984]).

9.24 Optimum Face of the Minimum Cost Spanning Tree Problem Let $G = (\mathcal{N}, \mathcal{A}, c)$ be a connected undirected network with $c = (c_{ij})$ as the vector of edge cost coefficients. There may be several spanning trees in G, which tie for being a minimum cost spanning tree in G. Let $\mathbf{E}$ denote the set of all edges in $\mathcal{A}$, each of which is contained on at least one minimum cost spanning tree in G.

1. Is the statement "every spanning tree in the partial subnetwork $(\mathcal{N}, \mathbf{E})$ is a minimum cost spanning tree in G" correct? Either prove it, or construct a counterexample.

2. If all the entries in the cost vector c are distinct, prove that there is a unique minimum cost spanning tree in G.

3. Suppose that some of the edge cost coefficients are equal to each other. Arrange the distinct values among $\{c_{ij} : (i;j) \in \mathcal{A}\}$ in increasing order, and let them be $\alpha_1, \ldots, \alpha_p$. For $t = 1$ to p, let $\mathbf{S}_t = \{(i;j) : c_{ij} = \alpha_t\}$. Then $\mathbf{S}_1$ is the set of least cost edges in G, $\mathbf{S}_2$ is the set of edges with the second best cost, etc.

   (a)   If after deleting all the edges in $\mathbf{S}_p$ (the maximum cost edges) the network remains connected, then prove that none of the edges in $\mathbf{S}_p$ is contained on any minimum cost spanning tree in G.

   (b)   Prove that every minimum cost spanning tree in G must contain at least one of the edges in $\mathbf{S}_1$ as an in-tree edge.

   (c)   Let $\mathbf{T}_0$ be any minimum cost spanning tree in G, and let r_t be the number of edges from $\mathbf{S}_t$ in $\mathbf{T}_0$, for $t = 1$ to p. Then prove that every spanning tree in G containing exactly r_t edges from $\mathbf{S}_t$ for each $t = 1$ to p is a minimum cost spanning tree in G and vice versa.

In this case, prove that the convex hull of minimum cost spanning tree incidence vectors in G is the set of feasible solutions of (9.3) and the additional constraints

$$\sum(x_{ij} : \quad \text{over} \quad (i;j) \in \mathbf{S}_t) = r_t, \quad t = 1 \ \text{to} \ p$$

Define a *second best valued spanning tree* in G to be a minimum cost spanning tree among those with cost $>$ cost of $\mathbb{T}_0$. In this case, using the algorithm mentioned in Exercise 9.23, develop an O($pm \log n$) algorithm to find a second best valued spanning tree in G.

(Murty, Hamacher, and Maffioli [1991]).

Comment 9.1   The minimum cost spanning tree problem (MCST) is a classical problem concerned with the design of optimum networks. For example, it arises when several points in a region have to be linked via cable at minimum cost. It is the first network optimization problem for which an efficient algorithm has been developed. Graham and Hell [1985] survey the history of the problem.

Boruvka [1926] seems to have been the first to develop an algorithm for this problem. Choquet [1938] has rediscovered the same algorithm. We discussed this algorithm, as well as those of Kruskal [1956] and Prim [1957]. All these methods are based on the incremental technique of building the optimum tree, edge by edge, using the greedy selection rule. The greedy method solves a variety of other problems besides the MCST. The MCST has a powerful generalization in the setting of matroid theory, it is the problem of finding a minimum cost base of a matroid (the cost of a base is the sum of the costs associated with the elements in it, see Lawler [1976 of Chapter 1] for a discussion of matroid theory relevant to the area of optimization). The MCST is a special case of this problem corresponding to a graphic matroid. The greedy method can be used to find a minimum cost base of a matroid.

Prim's algorithm has been developed earlier by Jarnik [1930], and improved by Dijkstra [1959 of Chapter 4].

A variety of techniques have been developed to improve the worst computational complexity of these algorithms on sparse networks. These techniques involve sophisticated data structures. See Cheriton and Tarjan [1976], Johnson [1975], and Yao [1975]. Tarjan [1983 of Chapter 1] contains surveys on the MCST and variants of it. Parallel algorithms for the MCST are discussed by Akl [1986], Lavallee [1984], Lavallee and Roucairol [1986], and Nath and Maheswari [1982]. Some applications of the MCST are discussed by Ali and Kennington [1986], Gavish [1982], Held and Karp [1970], Hu[1961], Kang, Lee, Chang, and Chang [1977], and Magnanti and Wong [1984]. Sensitivity analysis and updating an optimum MCST under changes in the data are discussed in Chin and Houck [1978], Spira and Pan [1975], Tarjan [1982], and Tsin [1988].

The related theoretical problem of characterizing $\mathbf{K}$ = convex hull of the spanning tree incidence vectors in $G = (\mathcal{N}, \mathcal{A})$ with n nodes and m edges, through a system of linear constraints has attracted a good deal of attention. If we use only the edge incidence variables, this characterization requires about 2^n constraints in them (Exercise 9.2). Recently Kipp Martin [1986] and Wong [1984] have shown that by introducing additional variables called auxiliary variables, a convex polyhedron Γ can be defined through a system of $O(nm)$ linear constraints only, such that $\mathbf{K}$ is the projection of Γ in the subspace of the edge incidence variables.

The greedy algorithm has been used for a long time as a heuristic approach to obtain solutions for many hard combinatorial optimization problems arising in real world applications, even though the method is not guaranteed to produce an optimum or even near optimum solution on such problems. Recently Jenkins [1986] has constructed simple assignment and traveling salesman problems on which he shows that the greedy method leads to the worst possible solution for them (i.e., a solution actually minimizing the objective function which is required to be maximized). These examples indicate the need for extreme caution while using such heuristic methods to solve problems on which the methods performance is unknown.

Most of the constrained minimum cost spanning problems tend to be hard problems. See Chandy and Lo [1973], Gabow [1978], and Kershenbaum [1974] for a discussion of some constrained spanning tree problems.

9.6   References

S. AKL, 1986, "An Adaptive and Cost-Optimal Parallel Algorithm for Minimum Spanning Trees," *Computing*, 36(271-277).

I. ALI and J. KENNINGTON, 1986, "The Asymmetric M-Traveling Salesman Problem: A Duality Based Branch and Bound Algorithm," *DAM*, 13(259-276).

O. BORUVKA, 1926, "O Jistém Problému Minimalnim," *Pracá Moravské Přirodovĕdecké Společnosti*, 3(37-58), in Czech.

P. M. CAMERINI, Jan. 1978, "The Min-Max Spanning Tree Problem and Some Extensions," *IPL*, 7, no. 1(10-14).

P. M. CAMERINI, L. FRATTA, and F. MAFFIOLI, 1979, "A Note on Finding Optimum Branchings," *Networks*, 9(309-312).

R. CHANDRASEKARAN, 1977, "Minimal Ratio Spanning Trees," *Networks*, 7, no. 4(335-342).

K. M. CHANDY and T. LO, 1973, "The Capacitated Minimum Spanning Tree," *Networks*, 3, no. 2(173-181).

D. CHERITON and R. E. TARJAN, 1976, "Finding Minimum Spanning Trees," *SIAM Journal on Computing*, 5(724-742).

F. CHIN and D. HOUCK, 1978, "Algorithm for Updating Minimum Spanning Trees," *Journal of Computer and System Science*, 16(333-344).

G. CHOQUET, 1938, "Etude de Certains Reseaux de Routes," *C. R. Acad. Sci. Paris*, 206(310-313).

J. EDMONDS, 1970, "Submodular Functions, Matroids and Certain Polyhedra," PP 69-87 in R. Guy(ed.), *Combinatorial Structures and Their Applications, Proceedings of the Calgary International Conference*, Gordon Breach, N.Y.

H. N. GABOW, 1978, "A Good Algorithm for Smallest Spanning Tree With a Degree Constraint," *Networks*, 8(201-208).

H. N. GABOW and R. E. TARJAN, Mar. 1984, "Efficient Algorithms for a Family of Matroid Intersection Problems," *Journal of Algorithms*, 5, no. 1(80-131).

R. G. GALLAGER, P. A. HUMBLET, and P. M. SPIRA, 1983, "A Distributed Algorithm for Minimum-Weight Spanning Trees," *ACM Transactions on Programming Languages and Systems*, 5(66-77).

B. GAVISH, 1982, "Topological Design of Centralized Computer Networks - Formulations and Algorithms," *Networks*, 12(355-377).

R. L. GRAHAM and P. HELL, Jan. 1985, "On the History of the Minimum Spanning Tree Problem," *Ann. History of Computing*, 7, no. 1(43-57).

M. HELD and R. KARP, 1970, "The Traveling Salesman Problem and Minimum Spanning Trees," *OR*, 18(1138-1162).

T. C. HU, 1961, "The Maximum Capacity Route Problem," *OR*, 9(898-900).

V. JARNIK, 1930, "O Jistém Problému Minimalnim," *Pracá Moravské Přirodovědecké Společnosti*, 6(57-63), in Czech.

T. A. JENKYNS, 1986, "The Greedy Algorithm is a Shady Lady," *Congressus Numerantium*, 51(209-215).

D. B. JOHNSON, 1975, "Priority Queues with Update and Finding Minimum Spanning Trees," *IPL*, 4(53-57).

A. KANG and A. AULT, Sept. 1975, "Some Properties of a Centroid of a True Tree," *IPL*, 4, no. 1(18-20).

A. N. C. KANG, R. C. T. LEE, C. CHANG, and S. K. CHANG, May 1977, "Storage Reduction Through Minimal Spanning Trees and Spanning Forests," *IEEE Transactions on Computers*, C-26, no. 5(425-434).

A. KERSHENBAUM, 1974, "Computing Capacitated Minimal Spanning Trees Efficiently," *Networks*, 4, no. 4(299-310).

R. KIPP MARTIN, May 1986, "A Sharp Polynomial Size Linear Programming Formulation of the Minimum Spanning Tree Problem," Graduate school of Business, U. of Chicago.

J. B. KRUSKAL, 1956, "On the Shortest Spanning Subtree of a Graph and the Traveling Salesman Problem ," *Proceedings of the AMS*, 7(48-50).

I. LAVALLEE, 1984, "An Efficient Parallel Algorithm for Computing a Minimum Spanning Tree," *Parallel Computing*, 83(259-262).

I. LAVALLEE and G. ROUCAIROL, 1986, "A Fully Distributed (Minimal) Spanning Tree Algorithm," *IPL*, 23(55-62).

B. M. MAGGS and S. A. PLOTKIN, Jan. 1988, "Minimum-cost Spanning as a Path-Finding Problem," *IPL*, 26, no. 6(291-293).

T. L. MAGNANTI and R. T. WONG, 1984, "Network Design and Transportation Planning: Models and Algorithms," *TS*, 18(1-55).

N. MEGIDDO, 1979, "Combinatorial Optimization with Rational Objective Functions," *MOR*, 4(414-424).

K. G. MURTY, R. SAIGAL, and J. SUURBALLE, 1974, "Ranking Algorithm for Spanning Trees," Bell labs. memo.

K. G. MURTY, H. HAMACHER, and F. MAFFIOLI, 1991, "Characterization of the optimum face of the minimum cost spanning tree problem, and applications," Tech. report, IOE Dept., University of Michigan, Ann Arbor, Mich.

D. NATH and S. MAHESWARI, 1982, "Parallel Algorithms for the Connected Components and Minimal Spanning Tree Problems," *IPL*, 14, no. 1(7-11).

R. C. PRIM, 1957, "Shortest Connection Networks and Some Generalizations," *Bell System Tech. J.*, 36(1389-1401).

P. M. SPIRA and A. PAN, 1975, "On Finding and Updating Spanning Trees and Shortest Paths," *SIAM J. on Computing*, 4(375-380).

R. E. TARJAN, 1982, "Sensitivity Analysis of Minimum Spanning Trees and Shortest Path Trees," *IPL*, 14(30-33).

Y. H. TSIN, April 1988, "On Handling Vertex Deletion in Updating Minimum Spanning Trees," *IPL*, 27, no. 4(167-168).

A. YAO, 1975, "An $O(|E|\log \log |V|)$ Algorithm for Finding Minimum Spanning Trees," *IPL*, 4(21-23).

References on Steiner Tree Problems in Networks

Y. P. ANEJA, 1980, "An Integer Linear Programming Approach to the Steiner Problem in Graphs," *Networks*, 10(167-178).

S. E. DREYFUS and R. A. WAGNER, 1972, "The Steiner Problem in Graphs," *Networks*, 1(195-207).

E. N. GILBERT and H. O. POLLOCK, Jan. 1968, "Steiner Minimal Trees," *SIAM J. Applied Math.*, 16, no. 1(1-29).

S. L. HAKIMI, 1971, "Steiner's Problem in Graphs and Its Implications," *Networks*, 1(113-133).

M. HANAN, Mar. 1966, "On Steiner's Problem with Rectilinear Distance," *SIAM J. Applied Math.*, 14, no. 2(255-265).

Z. A. MELZAK, 1961, "On the Problem of Steiner," *Canadian Mathematical Bulletin*, 4(143-148).

R. T. WONG, 1984, "A Dual Ascent Approach for Steiner Tree Problems on a Directed Graph," *MP*, 28(271-287).

References on Optimum Branchings

F. BOCK, 1971, "An Algorithm to Construct a Minimum Directed Spanning Tree in a Directed Network," PP 29-44 in *Developments in Operations Research*, Gordon and Breach, N.Y.

Y. J. CHU and T. H. LIU, 1965, "On the Shortest Arborescence of a Directed Graph," *Sin. Sinica*, 14(1396-1400).

J. EDMONDS, 1967, "Optimum Branchings," *J. Res. Nat. Bur. Standards*, Section B, 71(233-240).

R. M. KARP, 1971, "A Simple Derivation of Edmonds' Algorithm for Optimum Branchings," *Networks*, 1(265-272).

R. E. TARJAN, 1977, "Finding Optimum Branchings," *Networks*, 7(25-35).

Chapter 10

Blossom Algorithms for 1-Matching/Edge Covering Problems in Undirected Networks

In this chapter we consider only undirected networks. Let $G = (\mathcal{N}, \mathcal{A}, c = (c_{ij}))$ be a given undirected network, with c as the vector of edge weights or edge cost coefficients, and $|\mathcal{N}| = n, |\mathcal{A}| = m$. There may be parallel edges, but we assume that there are no self loops or isolated nodes in G. Each c_{ij} could be positive, 0, or negative. As defined in Section 1.2.2, a *matching* (or *1-matching* to be specific, also called *an independent set of edges*) in G is a subset of edges $M \subset \mathcal{A}$ that contains at most one edge incident at node i, for each $i \in \mathcal{N}$. The "1" in the name "1-matching" refers to the fact that the degree in M of every node is required to be $\leqq 1$. Hence, every pair of distinct edges in a matching are node disjoint. In Figure 10.1, the set of thick edges is a matching. The cardinality of a maximum cardinality matching in G is known as the *edge independence number* of G, clearly this is $\leqq n/2$. The cost of a matching M is defined to be $\sum(c_{ij} : \text{over } (i; j) \in M)$.

A *perfect matching* (or *perfect 1-matching* to be specific, also called a *one factor*) in G is a matching M that contains exactly one edge incident at each $i \in \mathcal{N}$. If a perfect matching M exists in G, n must be even, and $|M| = n/2$. The network in Figure 10.1 has 12 nodes, and it has the wavy perfect matching.

When referring to a matching M, edges in it are called *matching edges*, and those not in it are called *nonmatching edges*. A node i is said to be a *matched node* in M if M contains an edge incident at i, or an *unmatched node* wrt M otherwise. So, M is a perfect matching iff there are no unmatched nodes wrt it. If, $(i; j) \in M$, nodes i, j are called *mates of each other* in M.

496

The algorithm for finding a minimum length postman's route (an ECR) in an undirected network G discussed in Section 1.3.8 required a minimum cost perfect matching on the complete undirected network H defined on the set of odd degree nodes in G. This is a very important application of a matching problem in routing. The minimum cost perfect matching problem is also used in an algorithm for finding a near optimum tour whose cost is guaranteed to be no more than 50% greater than the cost of an optimum tour in the Euclidean distance traveling salesman problem, see Christofides [1976]. For other applications of matching problems see Fujii, Kasami, and Ninomiya [1969], and Montreuil, Ratliff, and Goetschalckx [1987].

We will discuss efficient primal-dual algorithms called *blossom algorithms* for several different types of matching problems, the *maximum cardinality matching problem*, the *minimum cost perfect matching problem*, the problem of finding a *minimum cost matching* among all matchings, and the *parametric matching problem* of finding a minimum cost matching among all matchings of cardinality γ treating γ as an integer parameter varying between 1 and $\lfloor n/2 \rfloor$. To solve a *maximum weight matching problem* wrt an edge weight vector, define the edge cost vector to be the negative of the edge weight vector, and then find the minimum cost matching subject to the same conditions. Both problems have the same set of optimum matchings, and the minimum cost matching algorithms are shown to work for arbitrary cost vectors.

A subset of edges $\mathbf{E} \subset \mathcal{A}$ is said to *cover the nodes* in G if each node in $\mathcal{N}$ is incident to at least one edge in $\mathbf{E}$, and such a set is called an *edge cover* or *1-edge cover* to be specific, or a *dominating edge set* in G because the edges in $\mathbf{E}$ cover or dominate all the nodes. For the network in Figure 10.1, the set of dashed edges is an edge cover. A matching is an edge cover iff it is a perfect matching. Since we assumed that there are no isolated nodes in G, $\mathcal{A}$ itself is an edge cover in G. We will discuss efficient algorithms to find minimum cardinality edge covers, and minimum cost edge covers.

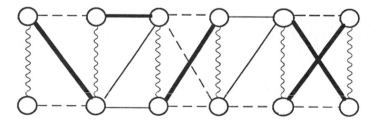

Figure 10.1 The set of thick edges is a matching, the set of wavy edges is a perfect matching and also an edge cover, and the set of dashed edges is an edge cover.

Another type of covers usually discussed in graph theory are *node covers*. These are subsets of nodes $\mathbf{X} \subset \mathcal{N}$ that cover all the edges, i.e., every edge in $\mathcal{A}$ is incident to at least one node in $\mathbf{X}$. From Exercise 1.28 we know that the node-edge incidence matrix of an undirected network is totally unimodular iff the network is bipartite. Because of this property, we can solve minimum cost node cover problems in G by linear programming methods if G is bipartite. However, if G is not bipartite, i.e., if it contains odd cycles, to find either a minimum cardinality or minimum cost node cover, are difficult problems for which no polynomially bounded algorithms are known. In contrast, in a general undirected network a minimum cost edge cover can be found with a computational effort of at most $O(n^3)$ using the blossom algorithms discussed in this chapter. We do not discuss any algorithms for node covers.

1-Matching/Edge Coverings

Let $(\mathcal{N}^{\leq}, \mathcal{N}^{=}, \mathcal{N}^{\geq}, \mathcal{N}^0)$ be a given partition of the node set $\mathcal{N}$, where some of these subsets may be empty. A subset of edges $\mathbf{E} \subset \mathcal{A}$ is called a *1-matching/edge covering* or *1-M/EC* in short, wrt the partition $(\mathcal{N}^{\leq}, \mathcal{N}^{=}, \mathcal{N}^{\geq}, \mathcal{N}^0)$ of $\mathcal{N}$ if

every node $i \in \mathcal{N}^{\leq}$ is incident to at most one edge in $\mathbf{E}$

every node $i \in \mathcal{N}^{=}$ is incident to exactly one edge in $\mathbf{E}$

every node $i \in \mathcal{N}^{\geq}$ is incident to at least one edge in $\mathbf{E}$.

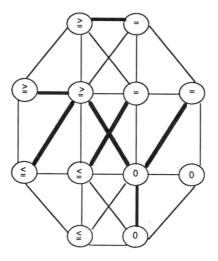

Figure 10.2 The symbols $\geq, \leq, =, 0$ inside the nodes identify the subset $\mathcal{N}^{\geq}, \mathcal{N}^{\leq}, \mathcal{N}^{=}, \mathcal{N}^0$ to which they belong. The set of thick edges is a 1-M/EC.

There are no constraints on the degrees of nodes in $\mathcal{N}^0$ in $\mathbf{E}$. The notation for the subsets in the partition of $\mathcal{N}$ is very suggestive. See Figure 10.2. One can easily see that 1-matching corresponds to $\mathcal{N} = \mathcal{N}^{\leq}, \mathcal{N}^{=} = \mathcal{N}^{\geq} = \mathcal{N}^0 = \emptyset$; 1-perfect matching corresponds to $\mathcal{N} = \mathcal{N}^{=}, \mathcal{N}^{\leq} = \mathcal{N}^{\geq} = \mathcal{N}^0 = \emptyset$; and 1-edge covering corresponds to $\mathcal{N} = \mathcal{N}^{\geq}, \mathcal{N}^{\leq} = \mathcal{N}^{=} = \mathcal{N}^0 = \emptyset$. Hence all these are special cases of 1-M/EC. We discuss efficient primal-dual blossom algorithms for the minimum cost 1-M/EC problem.

Let $\mathbf{E} \subset \mathcal{A}$ be an arbitrary subset of edges in G. Its incidence vector is the 0-1 vector $x^{\mathbf{E}} = (x_{ij}^{\mathbf{E}})$ defined on $\mathcal{A}$, where $x_{ij}^{\mathbf{E}} = 1$ if $(i; j) \in \mathbf{E}$, 0 otherwise. Conversely, given a 0-1 vector $x = (x_{ij})$ defined on $\mathcal{A}$, it is the incidence vector of (and is therefore said to correspond to) the subset of edges $\mathbf{E}^x = \{(i; j) : x_{ij} = 1\}$.

A 0-1 vector $x = (x_{ij})$ defined on $\mathcal{A}$ is said to be a *matching vector, perfect matching vector, edge covering vector,* or *1-M/EC vector,* if it is the incidence vector of a matching, perfect matching, edge covering, or 1-M/EC, respectively.

In this chapter an *edge vector* in G always refers to a 0-1 vector $x = (x_{ij})$ defined on $\mathcal{A}$. The variable x_{ij} associated with edge $(i; j)$ is known as the *decision variable associated with that edge* in the vector x. Given the edge vector x, we define for each $i \in \mathcal{N}$

$$x(i) = \sum (x_{ij} : \text{ over } j \text{ such that } (i; j) \in \mathcal{A}) \tag{10.1}$$

Thus an edge vector x is

a matching vector   iff $x(i) \overset{\leq}{=} 1$, for each $i \in \mathcal{N}$

a perfect matching vector   iff $x(i) = 1$, for each $i \in \mathcal{N}$

an edge covering vector   iff $x(i) \overset{\geq}{=} 1$, for each $i \in \mathcal{N}$ (10.2)

a 1-M/EC vector wrt partition $(\mathcal{N}^{\leq}, \mathcal{N}^{=}, \mathcal{N}^{\geq}, \mathcal{N}^0)$ iff $x(i) \begin{cases} \overset{\leq}{=} 1, & \text{for } i \in \mathcal{N}^{\leq} \\ = 1, & \text{for } i \in \mathcal{N}^{=} \\ \overset{\geq}{=} 1, & \text{for } i \in \mathcal{N}^{\geq} \end{cases}$

An Application of the Minimum Cost 1-M/EC Problem in Integer Programming

Consider the following pure 0-1 integer programming problem

$$\begin{array}{ll} \text{Minimize} & cy \\ \text{subject to} & Ay \quad \square \quad \mathbf{e} \\ & y_j = 0 \text{ or } 1 \quad \text{for all } j \end{array} \tag{10.3}$$

where $A = (a_{ij})$ is a 0-1 matrix of order $p \times q$, $\mathbf{e}$ is the column vector of all 1's in $\mathbb{R}^q$, and $\square$ denotes the vector of either $\overset{\leq}{=}, =,$ or $\overset{\geq}{=}$ for each constraint. The well known set covering problem with many applications in airline crew scheduling, facility location, vehicle routing, scheduling, etc., is a special case of (10.3) when $\square$ consists of all $\overset{\geq}{=}$ symbols. (10.3) has many applications.

First consider the case in which there are at most two nonzero entries in each column of A. In this case, (10.3) is a 1-M/EC problem. To see this, let $\mathcal{N} = \{1, \ldots, p\}$ if each column of A contains exactly two nonzero entries, $\mathcal{N} = \{1, \ldots, p, p+1\}$ if some columns of A contain only one nonzero entry. For $i = 1$ to p, node i in $\mathcal{N}$ is associated with constraint i in the problem. Associate each column of A with an edge. If $A._j$ contains only one nonzero entry, in row i say, associate it with the edge $(i; p+1)$ and make the cost of this edge equal to c_j. If $A._j$ contains two nonzero entries; in rows h and w say; associate it with the edge $(h; w)$ and make the cost of this edge equal to c_j. Let $\mathcal{A}$ be the set of q edges associated with the columns in A. Define $\mathcal{N}^{\leq}, \mathcal{N}^{=}, \mathcal{N}^{\geq}$ to be the set of rows $i = 1$ to p in which the entry in $\square$

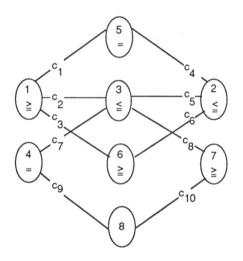

Figure 10.3 Node i is associated with constraint i. The symbol $\overset{\geq}{=}, =, \overset{\leq}{=}$ inside a node represents the type of constraint associated with it, and the subset $\mathcal{N}^{\geq}, \mathcal{N}^{=}, \mathcal{N}^{\leq}$ to which it belongs.

is $\overset{\leq}{=}, =, \overset{\geq}{=}$ respectively. Define $\mathcal{N}^0 = \emptyset$ if every column of A contains two nonzero entries, or $= \{p+1\}$ if there are some columns of A with a single nonzero entry. Let $G = (\mathcal{N}, \mathcal{A})$. Then, it can be verified that a minimum cost 1-M/EC vector in G wrt the partition $(\mathcal{N}^{\leq}, \mathcal{N}^{=}, \mathcal{N}^{\geq}, \mathcal{N}^0)$ of $\mathcal{N}$ is an optimum solution of (10.3) and vice versa, in this case. As an example consider the following instance of (10.3). Each column of the coefficient matrix contains at most two nonzero entries among the constraints. So, this instance is equivalent to a 1-M/EC problem. The corresponding network, constructed as mentioned above, is given in Figure 10.3. The variable y_j in the problem corresponds to the edge with c_j marked on it in

Figure 10.3, and c_j is the cost of this edge in the 1-M/EC problem.

Constraint	y_1	y_2	y_3	y_4	y_5	y_6	y_7	y_8	y_9	y_{10}		
1	1	1	1								$\geq$	1
2				1	1	1					$\leq$	1
3		1			1		1	1			$\leq$	1
4							1		1		$=$	1
5	1			1							$=$	1
6			1			1					$\geq$	1
7								1		1	$\geq$	1
	c_1	c_2	c_3	c_4	c_5	c_6	c_7	c_8	c_9	c_{10}	Min.	

Now consider the general case where there are columns in A with 3 or more nonzero entries. The best known methods for solving (10.3) in this case are partial enumeration methods known as branch and bound methods (see Murty [1976 of Chapter 1]). A basic operation in this method is the bounding operation whose main function is to compute a lower bound for the minimum objective value in the problem. There is a lower bounding strategy for (10.3) based on the minimum cost 1-M/EC problem. It consists of identifying a subset $\mathbf{I} \subset \{1,\ldots,p\}$, so that every column of A contains at most two nonzero entries among rows in the subset $\mathbf{I}$. There may be many subsets $\mathbf{I}$ with this property, the best among these is one with as large a cardinality as possible. In branch and bound methods, a subset $\mathbf{I}$ like this is initially selected by inspection, maximizing its cardinality to the extent possible. Once $\mathbf{I}$ is selected, constraints in it are said to be *included constraints*, and those outside it are called *unincluded constraints*. In the candidate problems generated after one or more branching operations, the corresponding set of included constraints is obtained automatically through the branching strategy.

The lower bounding strategy relaxes the unincluded constraints using Lagrangian relaxation. Let $\mathbf{\bar{I}} = \{1,\ldots,p\} \setminus \mathbf{I}$; $A_{\mathbf{I}}, A_{\mathbf{\bar{I}}}$ the submatrices of A consisting of all the rows in $\mathbf{I}$, $\mathbf{\bar{I}}$ respectively; and $\mathbf{e}_{\mathbf{I}}, \mathbf{e}_{\mathbf{\bar{I}}}$ the subvectors of $\mathbf{e}$ corresponding to $\mathbf{I}$, $\mathbf{\bar{I}}$ respectively. Let $u_{\mathbf{\bar{I}}} = (u_i : i \in \mathbf{\bar{I}})$ be a vector of Lagrange multipliers associated with the unincluded constraints satisfying

$$u_i \begin{cases} \leq 0, & \text{if the } i\text{th constraint is } \leq \\ \geq 0, & \text{if it is } \geq \\ \text{unrestricted}, & \text{if it is } = \end{cases} \tag{10.4}$$

Then the relaxed problem is

$$\begin{aligned} \text{Minimize} \quad & L(y, u_{\mathbf{\bar{I}}}) = cy - u_{\mathbf{\bar{I}}}(A_{\mathbf{\bar{I}}}y - \mathbf{e}_{\mathbf{\bar{I}}}) \\ \text{subject to} \quad & A_{\mathbf{I}}y \;\boxed{\mathbf{I}}\; \mathbf{e}_{\mathbf{I}} \\ & y \text{ is a } 0, 1 \text{ vector} \end{aligned} \tag{10.5}$$

where $\boxed{\mathbf{I}}$ is the subvector of the original $\square$ corresponding to the constraints in I. In (10.5), the u_i for $i \in \bar{\mathbf{I}}$ are given constants satisfying (10.4), and only the y are the decision variables. Since each column in $A_{\mathbf{I}}$ has at most two nonzero entries, (10.5) can be solved directly as a 1-M/EC problem as discussed above. It can be shown that the optimum objective value in (10.5) is a lower bound for that in (10.3) for any $u_{\bar{\mathbf{I}}}$ satisfying (10.4). In the branch and bound method one uses a subgradient optimization procedure to select a $u_{\bar{\mathbf{I}}}$ satisfying (10.4) to make the optimum objective value in (10.5) as large as possible, in order to get a high quality lower bound for the optimum objective value in (10.3), see Chapter 15 in Murty [1976 of Chapter 1].

Eliminating Parallel Edges

In the 1-M/EC problem associated with a partition of nodes $(\mathcal{N}^{\leqq}, \mathcal{N}^=, \mathcal{N}^{\geqq}, \mathcal{N}^0)$ of $\mathcal{N}$, an optimum solution set of edges can contain at most one edge joining a pair of nodes i, j with $i \in \mathcal{N}^{\leqq} \cup \mathcal{N}^=$ and $j \in \mathcal{N}$. Hence for each such pair of nodes, if there are parallel edges joining them, keep only one of these parallel edges corresponding to the minimum cost and delete all the rest. If $\mathcal{N}^{\geqq} \cup \mathcal{N}^0 \neq \emptyset$, consider a pair of nodes i, j in this set. If there are parallel edges joining this pair, define $\mathbf{U}(i; j)$ to be the subset of these parallel edges consisting of all of them with cost coefficient $\leqq 0$ if some of them belong to this category, or the one associated with the minimum cost among these parallel edges if all of them have cost coefficients > 0. Replace all the parallel edges with a single edge $(i; j)$ representing the set of parallel edges $\mathbf{U}(i; j)$ in the original network, and make its cost coefficient equal to the sum of the original cost coefficients of edges in $\mathbf{U}(i; j)$. If this edge $(i; j)$ is contained in an optimum 1-M/EC in the transformed network, replace it with the set of parallel edges $\mathbf{U}(i; j)$ to get an optimum 1-M/EC in the original network. So, in the sequel we assume that the network G on which our problems are defined contains no parallel edges.

The Minimum Cost 1-M/EC Problem

The minimum cost 1-M/EC problem in the network $G = (\mathcal{N}, \mathcal{A}, c)$ with the partition $(\mathcal{N}^{\leqq}, \mathcal{N}^=, \mathcal{N}^{\geqq}, \mathcal{N}^0)$ of $\mathcal{N}$, is the following problem (10.6), (10.7); and all the matching, edge covering problems that we consider are special cases of this problem.

The coefficient matrix of the variables in (10.6) is a submatrix of the node-edge incidence matrix of G. First consider the case where G is bipartite. In this case, the coefficient matrix in (10.6) is totally unimodular (Exercise 1.28) and all the BFSs of (10.6) are themselves integer vectors without even the integer constraint (10.7). Hence we can ignore the integer restriction (10.7) and solve (10.6) as a continuous variable LP by any algorithm that is guaranteed to terminate with an optimum BFS when feasible solutions exist, and this provides an optimum solution satisfying (10.7) automatically. Thus all these problems can be solved directly by linear programming algorithms (for example, variants of the primal-dual Hungarian method of Chapter 3) in the case when G is bipartite.

$$\text{Minimize} \quad \sum (c_{ij} x_{ij} \quad : \quad \text{over } (i;j) \in \mathcal{A})$$

$$\text{Subject to} \quad x(i) \quad \begin{cases} \leqq 1, & \text{for } i \in \mathcal{N}^{\leqq} \\ = 1, & \text{for } i \in \mathcal{N}^{=} \\ \geqq 1, & \text{for } i \in \mathcal{N}^{\geqq} \end{cases} \tag{10.6}$$

$$0 \leqq x_{ij} \leqq 1 \quad \text{for all } (i;j) \in \mathcal{A}$$

$$\text{and} \quad x_{ij} \text{ integer} \quad \text{for all } (i;j) \in \mathcal{A} \tag{10.7}$$

Now consider the case where G is not bipartite, i.e., it contains some odd cycles. In this case (10.6) may have extreme point solutions which violate (10.7); so (10.6), (10.7) is a genuine integer programming problem. As an example, consider (10.8), (10.9), the minimum cost perfect matching problem in the network in Figure 10.4, with all edge cost coefficients equal to −1.

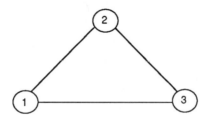

Figure 10.4

$$\begin{array}{llll}
\text{Minimize} & -x_{12} - x_{13} - x_{23} & & \\
\text{subject to} & x_{12} + x_{13} & = & 1 \\
& x_{13} + x_{23} & = & 1 \\
& x_{12} \quad\;\; + x_{23} & = & 1 \\
& x_{12}, x_{13}, x_{23} & \geqq & 0
\end{array} \tag{10.8}$$

$$x_{12}, x_{13}, x_{23} \qquad \text{integer} \tag{10.9}$$

Ignoring (10.9), the unique optimum solution of (10.8) is $\bar{x} = (\bar{x}_{12}, \bar{x}_{13}, \bar{x}_{23})^T = (1/2, 1/2, 1/2)^T$, which is a noninteger BFS of (10.8). Actually, (10.8), (10.9) together have no feasible solution.

Again consider the minimum cost edge covering problem in the network in Figure 10. 4, with a cost coefficient of 1 for each edge. This problem is

$$
\begin{aligned}
\text{Minimize}\quad & x_{12} + x_{13} + x_{23} \\
\text{subject to}\quad & x_{12} + x_{13} && \geqq 1 \\
& \qquad\quad x_{13} + x_{23} && \geqq 1 \\
& x_{12} \qquad\quad + x_{23} && \geqq 1 \\
& 0 \leqq x_{12}, x_{13}, x_{23} && \leqq 1
\end{aligned}
\tag{10.10}
$$

$$
x_{12}, x_{13}, x_{23} \qquad \text{integer}
\tag{10.11}
$$

It can be verified that $\bar{x} = (1/2, 1/2, 1/2)^T$ with an objective value of $3/2$, is again the unique optimum solution of (10.10) without (10.11). The optimum objective value in (10.10), (10.11) together is 2. These examples indicate that matching and edge covering problems in nonbipartite networks are nontrivial integer programs. The thing that makes them nontrivial is the presence of odd cycles in nonbipartite networks.

The approach taken to solve these integer programs is to develop additional linear constraints (usually called *blossom constraints*) such that every extreme point solution of the system consisting of (10.6) and the blossom constraints, is a feasible solution of (10.6), (10.7), and vice versa. One then relaxes the integer restrictions (10.7), and solves the LP (10.6) with the additional blossom constraints, by primal-dual or primal methods leading to extreme point optimum solutions when the problem is feasible. The primal-dual algorithms of this type are called *blossom algorithms*.

The Symmetric Assignment Problem

Consider the minimum cost perfect matching problem in the undirected network $G = (\mathcal{N}, \mathcal{A}, c)$ with $|\mathcal{N}| = n$ even. If G is a bipartite network with $(\mathcal{N}_1, \mathcal{N}_2)$ as its bipartition, where $|\mathcal{N}_1| = |\mathcal{N}_2| = n/2$, we have seen in Chapter 3 that the minimum cost perfect matching problem in G is an assignment problem of order $n/2$. Even when G is nonbipartite, the minimum cost perfect matching problem in G can still be posed as an assignment problem, but with additional constraints. In this case the formulation leads to an assignment problem of order n (in comparison to the bipartite case where we had an assignment problem of order $n/2$), and the additional constraints are symmetry constraints, hence it is called a *symmetric assignment problem*.

Set up an $n \times n$ transportation array. In this array associate both row i and column i with node i in G, for $i = 1$ to n. For $i \neq j$, associate the edge $(i; j)$ in G with the pair of cells (i, j) and (j, i) in the array. Hence, in this formulation, the cell (i, j) in the array is never individually considered by itself, it is always considered together with the cell (j, i). If the edge $(i; j)$ is included in the matching, both the cells (i, j) and (j, i) will be cells with allocations in the corresponding assignment, and vice versa. With this convention any perfect matching in G corresponds to an assignment $y = (y_{ij})$ of order n, in which $y_{ij} = y_{ji}$ for all i, j. These are the symmetry constraints, and assignments satisfying them are called *symmetric assignments*. Every perfect matching in G corresponds to a symmetric assignment y with no allocations along the main diagonal (i.e., $y_{ii} = 0$ for all $i = 1$ to n) and vice versa. Let d_{ij} denote the cost coefficient in cell (i, j) in the array. To guarantee that the cost of any perfect matching in G will be the same as that of the symmetric assignment corresponding to it in the array, we need to make sure that $d_{ij} + d_{ji} = c_{ij}$ for each edge $(i; j) \in \mathcal{A}$. In practice one may prefer to take $d_{ij} = d_{ji} = (1/2)c_{ij}$ for every $(i; j)$. With this cost matrix $d = (d_{ij})$ defined on the array, the minimum cost perfect matching problem in G is equivalent to the symmetric assignment problem in the array.

From the results in Chapter 1, we know that the matrix of coefficients of the usual assignment constraints (for example, see (3.1) in Chapter 3) is totally unimodular. Unfortunately, when the symmetry constraints are added to this system, this total unimodularity property is lost. Because of this, there is no guarantee that all the BFSs of the symmetric assignment problem will be integer vectors. Hence in the symmetric assignment problem, the integer requirements on the variables have to be taken into consideration and cannot be ignored. Thus, the formulation of the minimum cost perfect matching problem as a symmetric assignment problem, does not lead to any direct linear programming approaches to solve it.

10.1 The 1-Matching Problems

In this section we discuss results on the 1-matching problems on $G = (\mathcal{N}, \mathcal{A}, c)$ with $|\mathcal{N}| = n, |\mathcal{A}| = m$, and efficient algorithms of worst case computational complexity $O(n^3)$ for solving them.

Alternating and Augmenting Paths, APs, Alternating Trees

In Chapter 3 we have seen that alternating paths are a key ingredient in the Hungarian method for the assignment problem, which is the minimum cost perfect matching problem in a bipartite network. By the 1950's it was realized that alternating paths play a key role in algorithms for matching problems in nonbipartite networks as well. Let M be a given matching in G. A simple path $\mathcal{P}$ in G is said to be an *alternating path* wrt M, or an M-*alternating path* if it satisfies the following conditions (10.12). Usually the matching M is understood from the context, then we simply refer to the path as being alternating. By definition, all nodes on an alternating path, except possibly the end nodes, are matched nodes.

Edges in $\mathcal{P}$ are alternately in $\mathbf{M}$ and not in $\mathbf{M}$. If any of the two end nodes (these are nodes incident to only one edge of $\mathcal{P}$) is a matched node, $\mathcal{P}$ contains the matching edge incident at it.   (10.12)

Let $\mathcal{P}$ be an alternating path wrt a matching $\mathbf{M}$. Let $\mathbf{A}_1, \mathbf{A}_2$ be the sets of matching, nonmatching edges on $\mathcal{P}$. Let $\mathbf{M}_1 = \mathbf{A}_2 \cup (\mathbf{M} \backslash \mathbf{A}_1)$. Clearly, $\mathbf{M}_1$ is obtained by making all the edges in $\mathbf{A}_1$ (the original matching edges on $\mathcal{P}$) into nonmatching edges, and all the edges in $\mathbf{A}_2$ (the original nonmatching edges on $\mathcal{P}$) into matching edges; and it is another matching in G. We say that $\mathbf{M}_1$ is obtained from $\mathbf{M}$ by *changing the polarity of matching and nonmatching edges on* $\mathcal{P}$, or just by *rematching* $\mathbf{M}$ *using* $\mathcal{P}$.

An *augmenting path* wrt a matching $\mathbf{M}$ is an alternating path both of whose end nodes are unmatched nodes. An augmenting path always contains an even number of nodes and an odd number of edges. If it contains $2r + 1$ edges, r are matching edges, and the remaining $r + 1$ are nonmatching edges. Thus the operation of rematching using an augmenting path, increases the cardinality of the matching by one, and hence is called an *augmentation step*. The name augmenting path stems from this fact.

As an example, consider the paths $\mathcal{P}_1 = 2, (2, 5), 5, (5, 8), 8, (8, 9), 9$; $\mathcal{P}_2 = 1, (1, 2), 2, (2, 5), 5$; $\mathcal{P}_3 = 1, (1, 3), 3, (3, 4), 4, (4, 6), 6$ in the network in Figure 10.6. All these are alternating paths wrt the wavy matching. Rematching the wavy matching using $\mathcal{P}_1, \mathcal{P}_2, \mathcal{P}_3$ leads to the matchings: $\mathbf{M}_1 = \{(5, 8), (7, 10), (3, 4)\}$, $\mathbf{M}_2 = \{(1, 2), (8, 9), (7, 10), (3, 4)\}$, $\mathbf{M}_3 = \{(1, 3), (4, 6), (7, 10), (8, 9), (2, 5)\}$ respectively, in the network in Figure 10.6. It can be verified that among these three alternating paths, only $\mathcal{P}_3$ is an augmenting path wrt the wavy matching in Figure 10.6, and rematching using it increases the cardinality of the matching from 4 to 5.

Verify that there is no augmenting path wrt the wavy matching in the network in Figure 10.14.

An augmenting path begins at an unmatched node, and by moving alternately through nonmatching and matching edges, it reaches another unmatched node. To trace an augmenting path, it is convenient to label the points along it alternately as *outer* (labeled with a +) or *inner* (labeled with a −) nodes, beginning with an outer label for the initial unmatched node. The last node will be an unmatched node labeled as an inner node. All nodes which are reached after passing through an odd (even) number of edges are inner (outer) nodes. That's why in some books outer, inner nodes are called even, odd nodes respectively. We will now discuss a theorem due to Berge [1957] and Norman and Rabin [1959] that shows the importance of augmenting paths in the maximum cardinality matching problem.

THEOREM 10.1 *A matching* $\mathbf{M}$ *in G is a maximum cardinality matching iff there exists no augmenting path* wrt *it.*

Proof If there exists an augmenting path wrt **M**, augmentation using it leads to a matching of cardinality $1 + |M|$, and hence **M** is not a maximum cardinality matching.

Now suppose **M** is a matching in G that is not of maximum cardinality. Let **M*** be a maximum cardinality matching in G. So, $|M^*| \geq 1 + |M|$. Let $A = (M \cup M^*)\setminus(M \cap M^*)$. Since both **M**, **M*** are matchings, each of them contains at most one edge incident at any node. Hence **A** is the disjoint union of some simple paths and simple cycles. Each simple cycle in **A** consists alternately of an edge in **M** and an edge in **M***, and hence contains an equal number of edges from **M** and from **M***. Each simple path in **A** consists alternately of an edge in **M** and an edge in **M***. Since $|M^*| > |M|$, these facts imply that there must exist a simple path in **A** which contains more edges from **M*** than from **M**, and clearly that path will be an augmenting path wrt **M**. So, if **M** is not a maximum cardinality matching, there exists an augmenting path wrt it. ∎

As an example consider the network in Figure 10.5 and the two matchings M(thick), M*(wavy) in it, with one common edge (17; 18). The set $A = (M \cup M^*)\setminus(M \cap M^*)$ consists of an even alternating cycle containing the edges (11; 16), (14; 15), (12; 13) from **M**, and the edges (11; 12), (13; 14), (15; 16) from **M***; and three augmenting paths wrt **M**. They are the single edge augmenting path (8, 9); the second 1, (1, 2), 2, (2, 3), 3, (3, 4), 4; and the third 5, (5, 6), 6, (6, 7), 7, (7, 10), 10.

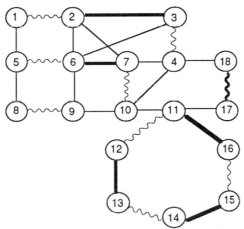

Figure 10.5 M(M*) is the matching of thick (wavy) edges. (17; 18) is the only edge that is in both matchings.

Hence the maximum cardinality matching problem in G can be solved by starting with an arbitrary matching (for example the empty matching), searching for an augmenting path wrt it; and if one is found, carrying out augmentation using it, and repeating the whole process with the new matching, until we reach a matching wrt which there is no, augmenting path. To implement this approach, we need an efficient method for finding an augmenting path if one exists, or determining that no augmenting path exists. The labeling routine discussed in Section 3.1 provided a very efficient method for doing this in bipartite networks. When the same labeling routine is applied on a nonbipartite network, we will see that odd cycles in the network pose obstacles in discovering augmenting paths.

Often, we have to deal with simple paths which have the property that edges in it are alternately from a matching M and not from it, we will use the symbol "**AP**" to denote such paths. So, an AP between nodes i, j wrt a matching M is a simple path with its edges alternately in M and out of M. An AP may not be an alternating path as defined in (10.12) because it may not satisfy the second property in (10.12). For example the path 2, (2, 3), 3, (3, 4), 4, (4, 6), 6 in Figure 10.6 is an AP wrt the wavy matching, but it is not an alternating path as defined in (10.12) because it does not contain the matching edge (2; 5) incident at its end node 2. While we are developing alternating paths in matching algorithms, in intermediate steps they will actually be APs because they may not satisfy the second condition in (10.12) at that time. But any path with which we carry rematching operation will always be an alternating path as defined in (10.12).

An *alternating tree* wrt a matching M in G is a tree rooted at an unmatched node and satisfying the property that the predecessor path of every nonroot node in it is an AP (in the sense that it consists alternately of matching and nonmatching edges). The labeling routine of Section 3.1 to find an augmenting path, is equivalent to rooting an alternating tree at an unmatched node, and growing it. In this procedure list always refers to the set of labeled and unscanned nodes. The procedure is initiated by selecting an unmatched node, r say, and making it the root of an alternating tree by labeling it as an *outer node* with the label $(\emptyset, +)$. At this stage the list $= \{r\}$. The tree is grown by selecting nodes from the list in some order and scanning them using the following rules.

(i) **Scanning an Outer Node** i Find all unlabeled nodes j satisfying $(i; j) \in \mathcal{A}, (i; j) \notin M$, label them as *inner nodes* with the label $(i, -)$, and include them in the list. Node i is their *immediate predecessor* or *predecessor index*, and all these nodes are the *immediate successors* of i. Now delete i from the list.

(ii) **Scanning an Inner Node** i If i is unmatched, an augmenting path has been found, and the tree is said to have become an *augmenting tree*. The augmenting path is the predecessor path of i. Terminate labeling.

If i is matched, and its mate j is unlabeled, label it with $(i, +)$ and include it in the list. In this case i is the immediate predecessor or predecessor index of j, and j is the only successor of i. Now delete i from the list.

At any stage, the set of in-tree edges is the set of edges joining the labeled nodes to their immediate predecessors. Alternating trees can be verified to satisfy the following properties: The predecessor path of any outer (inner) node consists of an even (odd) number of edges. Each inner node has at most one immediate successor, and if it has one, the edge joining them is a matching edge. If a node has two or more immediate successors, that node must be an outer node. Each inner node is incident to at most two in-tree edges, one of which must be a nonmatching edge. Each outer node is incident to any number of nonmatching in-tree edges, and one matching edge, excepting the root which meets no matching edge. If i is a non-root outer node with label $(P(i), +)$, then $P(i)$ must be its mate and an inner node. And the in-tree simple path from any outer node to a descendent inner node has an odd number of edges in it.

As an example, consider the network in Figure 10.6 with the wavy matching in it. An alternating tree is rooted at the unmatched node 1 in this network and grown. Nodes are scanned in the order 1, 2, 3, 4, and node labels are entered by the side of the nodes. When node 4 is scanned, the unmatched node 6 became labeled and the tree became an augmenting tree. The augmenting path is the predecessor path of node 6, it is 6, (6, 4), 4, (4, 3), 3, (3, 1), 1. All the properties mentioned above can be verified to hold in this tree.

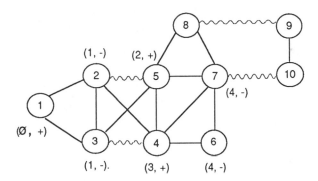

Figure 10.6 Matching edges are wavy. The tree became augmenting when unmatched node 6 is labeled.

In Section 3.1, 3.2, this scheme was applied on bipartite networks. There, outer nodes correspond to rows of the transportation array, and inner nodes to columns. Hence, the possibility of a pair of outer nodes, or a pair of inner nodes, being joined by an edge in the network never arises there. However when this labeling scheme is applied on a nonbipartite network, it may happen that there exists an edge joining a pair of outer nodes, or a pair of inner nodes (which indicates the existence of odd cycles in the network) and this may prevent this simple labeling scheme from discovering augmenting paths even when they exist. As an example of this consider the network in Figure 10.7 with the wavy matching in it. An alternating tree is rooted at the unmatched node 1 and grown by scanning the nodes in the list in serial order. Node labels are entered by the side of the nodes. The path at the bottom of the network is an augmenting path, but the simple labeling scheme has been unable to discover it because the unmatched node 12 could not be labeled.

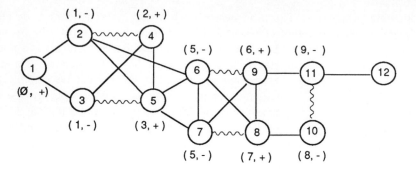

Figure 10.7 Matching edges are wavy. The augmenting path between 1 and 12 is not discovered when nodes are scanned in serial order.

Simple Blossoms

The labeling scheme failed to discover the augmenting path in the network in Figure 10.7 because of the odd alternating cycle encountered when nodes 10, 11 on the matching edge (10; 11) are both labeled as inner nodes. This is the cycle marked with thick solid lines in the following Figure 10.8.

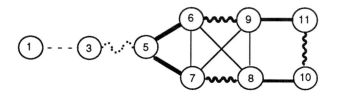

Figure 10.8 Matching edges are wavy. Simple blossom with base 5 is in solid lines. Odd cycle is thick. Dotted portion is the stem.

In general, an *odd alternating cycle* is encountered whenever a pair of nodes on a matching (nonmatching) edge are both labeled as inner (outer) nodes. Such a pair is called the *identifying pair of nodes* for the odd alternating cycle. Let i, j be the identifying pair of nodes. The first common node on the predecessor paths of i, j is known as the *base node* for the odd alternating cycle, suppose it is t. Node t has at least two successors, one on each of the predecessor paths of i and j, and possibly others not on these paths. So, t must be an outer node. Hence, the two edges incident at t on the predecessor paths of i and j must be nonmatching edges. If i, j are both inner (outer) nodes, the in-tree paths from t to i, and from t to j, are edge disjoint paths, each containing an odd (even) number of edges. Hence in either case, the in-tree paths from t to i, and from t to j, together with edge $(i; j)$ form an odd alternating cycle. Let $\mathcal{N}_B$ be the set of nodes on this odd cycle, and

$A_B = \{(i;j) : (i;j) \in A; i,j \in N_B\}$. $B = (N_B, A_B)$ is the partial network of G determined by N_B. B has the following properties.

> B is a partial network determined by an odd subset of nodes of cardinality $\geqq 3$.

> There is exactly one node in B, defined to be its *base node* which is not incident to a matching edge within B. All other nodes in B are incident to a matching edge within B. The base may be incident to a matching edge which is not an edge in B.   (10.13)

> There is a simple spanning cycle in B containing all the nodes in B, in which the edges incident at the base are nonmatching edges, and the others are alternately matching and nonmatching edges. This is the spanning odd alternating cycle in B.

A partial network B of G satisfying (10.13) is called a *simple blossom* wrt the present matching **M**. B also satisfies the following additional property.

> The base t of B is either an unmatched node (if it is the root node); or there exists an alternating path from t beginning with a matching edge, to an unmatched node (the root node), with all its edges outside B.   (10.14)

If the base node t is matched, its predecessor path meets the requirement in (10.14), and is called the *stem* of the simple blossom B. If the base node t is unmatched, it is the root node itself, and the stem is empty (i.e., contains no edges), and in this case B is called a *rooted simple blossom*. Since the base node is always an outer node, it can be verified that the stem always contains an even number (which could be 0) of edges. As an example, consider the alternating tree in the network in Figure 10.7. The two inner nodes 11, 10 joined by a matching edge are the identifying pair of nodes for a simple blossom whose base node is 5. The set of nodes on this simple blossom is $N_B = \{5, 6, 9, 11, 10, 8, 7\}$. See Figure 10.8.

An odd cycle is identified whenever the pair of nodes on an edge, $(h_1; h_2)$ say, are both labeled as outer or inner nodes. Suppose h_1, h_2 are both labeled as outer nodes. If $(h_1; h_2)$ is a matching edge, both h_1, h_2 are nonroot nodes since the root is unmatched, and if $P(h_1)$, $P(h_2)$ are the immediate predecessors of h_1, h_2 respectively, both $P(h_1)$, $P(h_2)$ must be inner nodes, so $P(h_1) \neq h_2$, and $(P(h_1); h_1)$ must be a matching edge, a contradiction since the two edges $(h_1; h_2)$, $(P(h_1); h_1)$ incident at h_1 cannot both be matching edges. So, $(h_1; h_2)$ must be a nonmatching edge, and this is a signal that a simple blossom has been identified.

Suppose h_1, h_2 are both labeled as inner nodes. In this case, if $(h_1; h_2)$ is a nonmatching edge, an odd cycle in the network is of course identified, but this odd cycle does not satisfy the last two properties in (10.13) to qualify as a simple blossom. For example, in the alternating tree in Figure 10.7, nodes 6, 7 on the nonmatching edge are both inner nodes. The odd cycle in the union of the edge (6; 7), and the predecessor paths of nodes 6, 7 is 6, (6, 5), 5, (5, 7), 7, (7, 6), 6, and this cycle does not satisfy the last two properties in (10.13). Such odd cycles do not pose obstacles in the way of the labeling scheme from identifying augmenting paths, only simple blossoms do.

Whenever a simple blossom is identified in the process of growing an alternating tree, we say that the tree has *blossomed*. By recognizing simple blossoms when they occur, and developing methods for handling them, the labeling scheme can be modified into a method that is guaranteed to find an augmenting path if one exists. The method of handling simple blossoms involves an operation called *shrinking them*, and later on expanding them when necessary.

We will now describe some of the properties of simple blossoms. Let $B = (\mathcal{N}_B, \mathcal{A}_B)$ be a simple blossom wrt the matching $\mathbf{M}$ in G, with node t as the base node. Let $\mathbf{M}_B = \mathbf{M} \cap \mathcal{A}_B$. Then $\mathbf{M}_B$ is a maximum cardinality matching in B, since all the nodes in B other than the base node t are matched in it. There can be at most one matching edge joining a node outside a simple blossom to a node inside it, and if there is such a matching edge, it must be incident at the base node.

If i is an inner node in B, the portion of the predecessor path of i up to the base node t, $\mathcal{P}_1$, is an AP between i and t, consisting of an odd number of edges, and beginning with the nonmatching edge incident at i on the odd cycle in B. By deleting all the edges on $\mathcal{P}_1$ from the odd alternating cycle in B, we are left with another AP between i and t, this one begins with the matching edge incident at i and has an even number of edges.

If i is a non-base outer node in B, the portion of the predecessor path of i up to t, $\mathcal{P}_2$, is an AP between i and t, consisting of an even number of edges and beginning with the matching edge incident at i. The AP between i and t consisting of an odd number of edges is obtained by deleting all the edges on $\mathcal{P}_2$ from the odd alternating cycle, this path begins with the nonmatching edge incident at i.

As an example, consider the non-base node 9 in the simple blossom in Figure 10.8. The AP between 9 and the base node 5 consisting of an even number of edges is 9, (9, 6), 6, (6, 5), 5; and the AP consisting of an odd number of edges is 9, (9, 11), 11, (11, 10), 10, (10, 8), 8, (8, 7), 7, (7, 5), 5.

A simple blossom has an empty stem iff its base is the root of the alternating tree, i.e., it is a rooted simple blossom. The simple blossom in Figure 10.7, with 4, 5 as the identifying pair of nodes is a rooted simple blossom.

Shrinking a Simple Blossom into a Pseudonode

Let $B = (\mathcal{N}_B, \mathcal{A}_B)$ be a simple blossom wrt a matching $\mathbf{M}$ in G, discovered during the growth of an alternating tree. At this stage we store this simple blossom (its base node, its identifying pair of nodes, and all the nodes on it together with the present labels on them). This stored information is sufficient to recognize all the current matching edges in this simple blossom (for example, if $i \in \mathcal{N}_B$ has label $(g, +)[\ (g, -)\ ]$, then $(g; i)$ is a current matching [nonmatching] edge on the odd alternating cycle in B); and the APs between any non-base node i and the base node, beginning with the matching or nonmatching edge incident at i on the odd alternating cycle in B. Then we perform an operation called *shrinking the simple blossom B* into a single new node known as a *pseudonode* (to distinguish it from the original nodes), which we denote by the same symbol B. This operation is essentially that of replacing all the present nodes in the simple blossom B by the single pseudonode B. This operation is also called *contraction* in graph theory.

Introduce the pseudonode B. For each $p \in \mathcal{N} \backslash \mathcal{N}_B$, if p is joined to one or more nodes in $\mathcal{N}_B$ by edges in $\mathcal{A}$ of which there is a matching edge, introduce a new edge $(p; B)$ and make it a matching edge. If all the edges in $\mathcal{A}$ joining p to nodes in $\mathcal{N}_B$ are nonmatching edges, make the new edge $(p; B)$ into a nonmatching edge.

Then eliminate all the nodes in $\mathcal{N}_B$, and all the edges in $\mathcal{A}$ which are incident to at least one node in $\mathcal{N}_B$. In the resulting network, give the pseudonode B the same label as that on the base of the simple blossom B, so, it will be an outer node. For each $p \in \mathcal{N} \backslash \mathcal{N}_B$ which was an immediate successor of a node in the simple blossom B before the shrinking, change the immediate predecessor of p into the pseudonode B, but leave its inner, outer status unchanged. Now the shrinking operation is completed. Let G^1 denote the resulting network, and M^1 the remaining matching in it. G^1 is said *to have been obtained by shrinking the simple blossom B into the pseudonode B*. M^1 is a matching in G^1, and it contains an edge incident at B iff the base in the simple blossom B was a matched node (if t was the base node of the simple blossom B, and $(t; g)$ was the matching edge incident at it in M, M^1 contains $(B; g)$). Define the base of the pseudonode B to be the base of the simple blossom B. The labels on nodes in G^1 define an alternating tree wrt M^1. Thus, after the shrinking of a simple blossom, a matching and an alternating tree are available in the resulting network. G^1 is known as the *current network*, and M^1, the *current matching* in it.

The pseudonode B is an unmatched node in G^1, iff the base of the simple blossom B was the root, in this case pseudonode B is called a *rooted pseudonode*, it becomes the root node in the alternating tree in G^1. This shows that the number of unmatched nodes in G^1 wrt M^1, is exactly the same as the number of unmatched nodes in G wrt M. Also, the number of nodes in G^1 is $|\mathcal{N}| - |\mathcal{N}_B| + 1$, and since $|\mathcal{N}_B| \geqq 3$ and odd, this number is $\leqq |\mathcal{N}| - 2$. Thus the operation of shrinking a simple blossom into a pseudonode reduces the total number of nodes in the network by an even number $\geqq 2$, but leaves the total number of unmatched nodes unchanged.

As examples, when the simple blossom defined by the subset of nodes {11, 10, 8, 7, 5, 6, 9} in Figure 10.7 is shrunk into the pseudonode B_1, the network in Figure 10.9 is obtained. In the same network in Figure 10.7, if the simple blossom defined by the subset of nodes {4, 2, 1, 3, 5} is shrunk into the pseudonode B_2, we get the network in Figure 10.10.

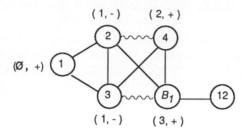

Figure 10.9 Network after shrinking the simple blossom in Figure 10.8 in the original network in Figure 10.7, into the pseudonode B_1.

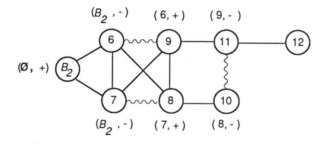

Figure 10.10   Network after shrinking the simple blossom defined by $\{4,2,1,3,5\}$ in the original network in Figure 10.7, into the pseudonode B_2.

The importance of simple blossoms and the process of shrinking them, arises from the following procedure, and Theorem 10.2 that comes later.

Procedure for Deriving an Augmenting Path $\mathcal{P}$
in G from an Augmenting Path $\mathcal{P}^1$ in G^1

Let $\mathbf{M}$ be a matching in the original network $G = (\mathcal{N}, \mathcal{A})$, and $B = (\mathcal{N}_B, \mathcal{A}_B)$ a simple blossom in G wrt $\mathbf{M}$ with base node t, which is shrunk into the pseudonode B resulting in the current network G^1 and current matching $\mathbf{M}^1$. Let $\mathcal{P}^1$ be an augmenting path in G^1 wrt $\mathbf{M}^1$. Here we discuss a procedure for obtaining an augmenting path $\mathcal{P}$ wrt $\mathbf{M}$ in G from $\mathcal{P}^1$.

If $\mathcal{P}^1$ does not contain the pseudonode B, all the nodes and edges on $\mathcal{P}^1$ are original nodes and edges in G itself, and hence $\mathcal{P}^1$ itself is an augmenting path wrt

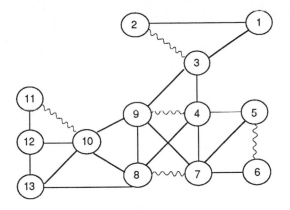

Figure 10.11 Network G. M is the wavy matching. B the partial network of $\mathcal{N}_B = \{5,4,9,10,8,7,6\}$, is a simple blossom.

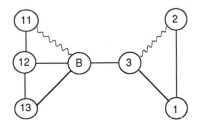

Figure 10.12 Network G^1 obtained after shrinking B in Figure 10.11 into the pseudonode B.

M in G. Now suppose $\mathcal{P}^1$ contains the pseudonode B. We consider two cases.

Case 1 B is an end node on $\mathcal{P}^1$. Since $\mathcal{P}^1$ is an augmenting path in G^1, B must be unmatched in this case. Let $(q; B)$ be the edge in $\mathcal{P}^1$ incident at B. There must exist an $i \in \mathcal{N}_B$ such that $(q; i) \in \mathcal{A}$, and since B is unmatched, $(q; i) \notin M$. If $i = t$ let $\mathcal{P}'$ be the edge $(q; t)$. If $i \neq t$, let $\mathcal{P}'$ be the path between q and t consisting of $(q; i)$, and the AP from i to t beginning with the matching edge incident at i in the odd cycle in the simple blossom corresponding to B. Replace the edge $(q; B)$ in $\mathcal{P}^1$ by the path $\mathcal{P}'$, this leads to $\mathcal{P}$, an augmenting path in G wrt M.

Case 2 B is an intermediate node on $\mathcal{P}^1$. In this case there exists an edge $(q_1; B) \in M^1$, and an edge $(B; q_2) \notin M^1$ incident at B on $\mathcal{P}^1$. So, $(q_1; t) \in M$, and there exists an $i \in \mathcal{N}_B$ such that $(i; q_2) \in \mathcal{A} \backslash M$. If $i = t$ let $\mathcal{P}'$ be $q_1, (q_1, t), t, (t, q_2), q_2$. If $i \neq t$ let $\mathcal{P}'$ be the AP between q_1 and q_2 beginning with the matching edge $(q_1; t)$, then going from t to i using the AP on the odd cycle in the simple blossom corresponding to B ending with the matching edge incident at i, and then finally the nonmatching edge $(i; q_2)$. Replace the pair of edges $(q_1; B), (B; q_2)$ on $\mathcal{P}^1$ by the path $\mathcal{P}'$, this converts it into an augmenting path $\mathcal{P}$ in G wrt M.

In both cases the augmenting path $\mathcal{P}$ in G is said to have been obtained by *expanding the pseudonode* B on $\mathcal{P}^1$. This operation of expanding the pseudonode B on $\mathcal{P}^1$ uses the stored information about the simple blossom corresponding to B.

As an example consider the current network in Figure 10.10 obtained by shrinking the simple blossom determined by the set of nodes $\{4, 2, 1, 3, 5\}$ in the original network in Figure 10.7 into the pseudonode B_2. B_2, $(B_2, 7)$, $(7, 8)$, 8, $(8, 10)$, $(10, 11)$, 11, $(11, 12)$, 12 is an augmenting path wrt the wavy matching in the network in Figure 10.10. By expanding the pseudonode B_2 on it as described in Case 1 above, we get the augmenting path between nodes 1 to 12 at the bottom of the original network in Figure 10.7.

For another example consider the original network G in Figure 10.11, and the current network G^1 obtained after shrinking the simple blossom defined by the subset of nodes $\{5, 4, 9, 10, 8, 7, 6\}$ in G into the pseudonode B. $\mathcal{P}^1 = 1$, $(1, 2)$, 2, $(2, 3)$, 3, $(3, B)$, B, $(B, 11)$, 11, $(11, 12)$, 12 is an augmenting path in G^1. By expanding the pseudonode B on it as described in Case 2 above, we obtain the augmenting path 1, $(1, 2)$, 2, $(2, 3)$, 3, $(3, 4)$, 4, $(4, 9)$, $(9, 10)$, 10, $(10, 11)$, 11, $(11, 12)$, 12 in G.

THEOREM 10.2 *Let* M *be a matching in* $G = (\mathcal{N}, \mathcal{A})$. *Let* $B = (\mathcal{N}_B, \mathcal{A}_B)$ *be a simple blossom in* G *wrt* M, *which is shrunk into the pseudonode* B, *resulting in the current network* G^1 *with the current matching* M^1.

(i) *If there exists an augmenting path in* G^1 *wrt* M^1, *then there exists an augmenting path in* G *wrt* M.

(ii) *If* B *satisfies (10.14), and if there exists an augmenting path in* G *wrt* M, *then there exists an augmenting path in* G^1 *wrt* M^1.

Proof   (i) follows from the expansion procedure discussed above. To prove (ii), let $\mathcal{P}$ be an augmenting path in G wrt M. Let t be the base node of B. If $\mathcal{P}$ contains no nodes from $\mathcal{N}_B$, $\mathcal{P}$ itself is an augmenting path in G^1, we are done.

Now suppose $\mathcal{P}$ contains one or more nodes from $\mathcal{N}_B$. We consider two cases.

Case 1   One of the End Nodes of $\mathcal{P}$ is in $\mathcal{N}_B$ Since $\mathcal{P}$ is an augmenting path, its end nodes are unmatched. So, B must be an unmatched node in G^1, and t must be an end node of $\mathcal{P}$. The path $\mathcal{P}$ may go in and out of the simple blossom B several times, but let i be the successor of the last node in $\mathcal{N}_B$ on $\mathcal{P}$ as you travel along it beginning at t. Replace the entire portion of the path $\mathcal{P}$ from t to i by the single edge $(B; i)$, this converts $\mathcal{P}$ into an augmenting path $\mathcal{P}^1$ wrt M^1 in G^1.

Case 2   The Only Nodes from $\mathcal{N}_B$ on $\mathcal{P}$ are Intermediate Nodes Let p_1, p_2 be the end nodes of $\mathcal{P}$. As you travel from p_1 to p_2 along $\mathcal{P}$, let q_1, q_2 be the first and last nodes of $\mathcal{N}_B$ encountered, with $(s_1; q_1)$ as the edge leading to q_1, and $(q_2; s_2)$ as the edge leading out of q_2.

If t is unmatched, replace the entire portion of the path $\mathcal{P}$ from s_1 to p_2 by the single edge $(s_1; B)$ and let the resulting path be called $\mathcal{P}^1$.

If t is a matched node and one of the two edges $(s_1; q_1), (q_2; s_2)$ is a matching edge, say $(s_1; q_1)$, replace the portion of the path $\mathcal{P}$ from s_1 to s_2 with the pair of edges $(s_1; B), (B; s_2)$ and let the resulting path be called $\mathcal{P}^1$.

Suppose t is a matched node and both the edges $(s_1; q_1), (q_2; s_2)$ are nonmatching edges. By hypothesis, B satisfies (10.14). Let $(t; i_1)$ be the matching edge incident at t. Define the following paths in G.

$$\mathcal{P}_1 = \text{portion of the path from } p_1 \text{ to } s_1 \text{ on } \mathcal{P}$$
$$\mathcal{P}_2 = \text{portion of the path from } s_2 \text{ to } p_2 \text{ on } \mathcal{P}$$
$$\mathcal{P}_3 = \text{stem of } B, \text{ beginning with } (t; i_1), \text{ to an unmatched node}, i_u, \text{ say.}$$
$$\mathcal{P}_4 = \text{AP from } i_1 \text{ to } i_u \text{ obtained by deleting } (t; i_1) \text{ from } \mathcal{P}_3$$

If $\mathcal{P}_1$ and $\mathcal{P}_3$ have no common nodes, let $\mathcal{P}^1$ be the path $\mathcal{P}_1 \cup \{(s_1; B), (B; i_1)\} \cup \mathcal{P}_4$.

If $\mathcal{P}_1$ and $\mathcal{P}_3$ have common nodes, but $\mathcal{P}_2$ and $\mathcal{P}_3$ have no common nodes, let $\mathcal{P}^1$ be the path $(\mathcal{P}_4 \text{ in reverse direction from } i_u \text{ to } i_1) \cup \{(i_1; B), (B; s_2)\} \cup \mathcal{P}_2$.

Now consider the case where both $\mathcal{P}_1$ and $\mathcal{P}_2$ have nodes in common with the stem $\mathcal{P}_3$. As you travel on $\mathcal{P}_2$ in reverse direction (i.e., from p_2 to s_2) let $\tilde{i}$ be the first node encountered on $\mathcal{P}_3$. Likewise let $\hat{i}$ be the first node encountered on $\mathcal{P}_3$ as you travel on $\mathcal{P}_1$ from p_1 to s_1. Since $\mathcal{P}_1$ and $\mathcal{P}_2$ are node disjoint (because $\mathcal{P}$ is a simple path) $\tilde{i} \neq \hat{i}$. As you travel from i_u to t along $\mathcal{P}_3$ suppose $\tilde{i}$ comes after $\hat{i}$ (the case where $\hat{i}$ comes after $\tilde{i}$ is handled in a symmetric way). Define

$$\mathcal{P}_5 = \text{portion of the path from } \tilde{i} \text{ to } p_2 \text{ on } \mathcal{P}_2$$
$$\mathcal{P}_6 = \text{portion of the path from } i_u \text{ to } \tilde{i} \text{ on } \mathcal{P}_3$$
$$\mathcal{P}_7 = \text{portion of the path from } \tilde{i} \text{ to } i_1 \text{ on } \mathcal{P}_3$$

Verify that $\mathcal{P}_5$ and $\mathcal{P}_3$ are disjoint and $\mathcal{P}_7$ is disjoint with both $\mathcal{P}_1$ and $\mathcal{P}_2$. Also the edge on $\mathcal{P}_5$ incident at $\tilde{i}$ will be a nonmatching edge. If the edge incident at $\tilde{i}$ on $\mathcal{P}_6$ is a matching edge define $\mathcal{P}^1$ to be $\mathcal{P}_5 \cup \mathcal{P}_6$. If the edge incident at $\tilde{i}$ on $\mathcal{P}_6$ is a nonmatching edge define $\mathcal{P}^1$ to be $\mathcal{P}_1 \cup \{(s_1; B), (B; i_1)\} \cup \mathcal{P}_7 \cup \mathcal{P}_5$.

It can be verified that $\mathcal{P}^1$, so constructed in each case discussed above, is an augmenting path in G^1. This completes the proof of (ii). ■

The result in part (ii) of Theorem 10.2 depends critically on the hypothesis that the simple blossom B satisfies (10.14). If B does not satisfy (10.14), there may not exist an augmenting path in G^1 wrt M^1, even though there exists an augmenting path in G wrt M. Consider the network G in Figure 10.13 with the wavy matching M. The partial network determined by the subset of nodes {4, 5, 6, 7, 11} is a simple blossom in G satisfying (10.13) but not (10.14). Shrinking this simple blossom into the pseudonode B leads to the current network G^1 in Figure 10.14 with the wavy matching M^1. The path on the top of G is an augmenting path wrt **M**. It can be verified that M^1 is a maximum cardinality matching in G^1, and hence no augmenting path exists in G^1 wrt M^1.

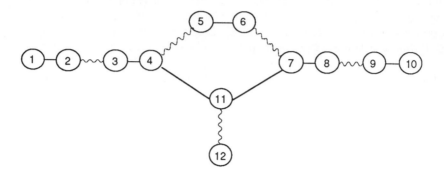

Figure 10.13 Network G. M is the wavy matching. The path at top is an augmenting path wrt **M**.

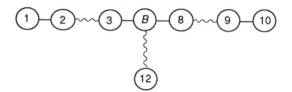

Figure 10.14 Network G^1. M^1 is the wavy matching, it has maximum cardinality in G^1.

All simple blossoms that we encounter in the blossom algorithm for the maximum cardinality matching problem discussed later will always satisfy (10.14). However, in the blossom algorithm for the minimum cost matching problem, we sometimes encounter simple blossoms not satisfying (10.14), that algorithm has special routines for handling such simple blossoms.

Nodes and edges in the current network G^1 are called *current nodes, current edges* respectively. A current node may either be an original node or a pseudonode. As mentioned earlier, G^1 now has a rooted alternating tree wrt the current matching M^1. Continue growing the alternating tree in G^1 by resuming labeling, using the same scheme discussed earlier.

Simple Blossoms in the Current Network, Blossoms

When labeling is resumed in G^1, a new simple blossom, say B_2, satisfying (10.13) wrt the current matching M^1 in it, may be identified. The signal for this is the same as before, i.e., when two nodes on a current matching (nonmatching) edge are labeled as inner (outer) nodes. In this case, the alternating tree being grown is said to have blossomed again. B_2 is identified exactly as before, its base node is the first common node on the predecessor paths of the identifying pair of nodes; and the odd alternating cycle in it consists of the current edge joining the identifying pair of nodes, and the portions of their predecessor paths up to the base node. If this simple blossom does not contain the pseudonode formed earlier, say B_1, it is itself a simple blossom in G wrt M. Otherwise, B_2 is a simple blossom in the current network containing the pseudonode B_1 as a node on it.

Let $\mathcal{N}_{B_2}$ be the set of original nodes which are either current nodes in B_2, or those contained within a pseudonode which is a current node in B_2. Let $\mathcal{A}_{B_2} = \{(i;j) : (i;j) \in \mathcal{A}, i, j \in \mathcal{N}_{B_2}\}$. Then the partial network $(\mathcal{N}_{B_2}, \mathcal{A}_{B_2})$ is called a *blossom* in the original network G wrt M, and B_2 in the current network G^1 is the simple blossom corresponding to it.

In general, a blossom in G wrt the matching M is a partial network $B = (\mathcal{N}_B, \mathcal{A}_B)$ satisfying the following properties.

$|\mathcal{N}_B|$ is an odd number ≥ 3

$M_B = M \cap \mathcal{A}_B$ is a maximum cardinality matching in B that leaves exactly one node unmatched. This unique unmatched node, t say, is called the *apex* of B. Notice that t may be a matched node in G, with the matching edge incident at it not in $\mathcal{A}_B$. (10.15)

There is a simple blossom corresponding to B in the current network at the stage that B is discovered. Some of the nodes on this simple blossom may be pseudonodes formed in earlier stages. It satisfies (10.13) in the current network and current matching at the stage that it is discovered. $\mathcal{N}_B$ is the set of original nodes which are either nodes on this simple blossom, or contained within pseudonodes on it.

Whenever a new blossom is discovered, the simple blossom corresponding to it is stored by storing all the current nodes on it together with the labels on them, its identifying pair of nodes and its base node; and it is then shrunk into a new pseudonode, p say, exactly as before. Then p gets the same label as on the base node on the simple blossom, and the predecessor index of any immediate successor of that base node is changed into p. The new pseudonode p and its descendents are put in the list of labeled and unscanned nodes, and the current alternating tree is grown again by resuming labeling. In matching algorithms, blossoms may be found and shrunk repeatedly.

The *base* of a blossom B is defined to be the base node of the simple blossom corresponding to it. If the base is an original node, it is the same as the apex of the blossom. If the base is a pseudonode, the apex of the blossom is the apex of its base node. Blossoms B discovered in the maximum cardinality matching algorithm also satisfy the following additional property.

> The simple blossom corresponding to it is either a rooted simple blossom (i.e., the base is the root of the alternating tree in the current network when it is discovered), or there exists an alternating path beginning with the current matching edge incident at the base, to an unmatched node in the current network, such that none of the nodes on this path other than the base are in the simple blossom. This alternating path is the stem of this simple blossom.   (10.16)

As an example consider the network in Figure 10.15 with the wavy matching in it. An alternating tree is rooted at the unmatched node 1 and grown by scanning the nodes in serial order. We show the detection and shrinking of two simple blossoms, at the end of which we have the current network shown in Figure 10.17.

Let $G^1 = (\mathcal{N}^1, \mathcal{A}^1)$ be the current network at some stage after some blossoms have been shrunk. Nodes in $\mathcal{N}^1$ are called *current nodes*, they may be either pseudonodes, or original nodes in $\mathcal{N}$ not contained in any blossoms shrunk so far. Let $p \in \mathcal{N}^1$ be a current node. We will say that an original node i is *inside* p or that p contains i *inside it*, if either $i = p$, or if i is a node in the blossom corresponding to p. An original edge $(i; j) \in \mathcal{A}$ is said to be inside p if both i and j are inside p. Another pseudonode q which is not a current node is said to be inside p if the set of original nodes inside q is a subset of the set of original nodes inside p.

When a pseudonode is just formed, it is a current node in the current network at that stage. Afterwards, it might get absorbed inside another pseudonode, at which stage it is said to have become *dormant*. We will say that a pseudonode is a *current pseudonode* or *outermost pseudonode* if it is a current node at that stage, or a *dormant pseudonode* if it is contained inside another pseudonode.

Levels of Pseudonodes and Blossoms, and Their Nesting Properties

Original nodes refer to nodes in $\mathcal{N}$ in the original network G. They are defined to be *level 0 nodes*. For the sake of consistency, the base node and the apex node corresponding to a level 0 node is defined to be that node itself. A simple blossom, all the nodes on which are level 0 nodes is said to be a *level 1 simple blossom*, and the pseudonode into which it is shrunk is called a *level 1 pseudonode*. In general, for any $s \geqq 1$, a *level $(s+1)$ simple blossom* is one, at least one node on which is a level s pseudonode, and all other nodes on which are either pseudonodes of level $\leqq s$, or original nodes. For any $s \geqq 1$, the blossom (pseudonode) corresponding to a level s simple blossom is called a level s blossom (pseudonode). Outermost pseudonodes may be of any level, and at each stage there may be outermost pseudonodes of various levels in the current network. We will use the term *labeled node* to always mean a current node in the current alternating tree labeled as an outer or inner node, which may be either an original node not contained in any pseudonode, or an outermost pseudonode.

The base node of a simple blossom, pseudonode, or blossom, is the node on the corresponding simple blossom that is not contained on any matching edge within that simple blossom. Its apex node is the original node in the corresponding blossom, that is not contained on any matching edge inside that blossom. So, for any simple blossom, pseudonode, or blossom, the apex node is the same as the apex of its base node.

After some blossom shrinkings, let G^1 denote the current network at some stage. Let u be the total number of blossoms that have been shrunk into pseudonodes up to this stage. Let these blossoms, or the pseudonodes corresponding to them, be $B_1, \ldots, B_u$. For $v = 1$ to u let $\mathcal{N}_{B_v}$ be the set of original nodes contained within B_v. We will now discuss some of the properties of the blossoms at this stage.

THEOREM 10.3 *Each shrinking operation leads to a new current network in which the number of current nodes is reduced by an even number $\overset{\geq}{=} 2$.*

Proof   Each simple blossom consists of an odd number of current nodes $\overset{\geq}{=} 3$. When it is shrunk, all the nodes in the simple blossom are eliminated and just one new pseudonode introduced. This leads to the result stated in the theorem. ∎

THEOREM 10.4 *The class of subsets $\{\mathcal{N}_{B_1}, \ldots, \mathcal{N}_{B_u}\}$ has the property that for v, w, if $\mathcal{N}_{B_v} \cap \mathcal{N}_{B_w} \neq \emptyset$, then one of them is a subset of the other.*

Proof   From the manner in which blossoms are identified and shrunk, it is clear that if $\mathcal{N}_{B_v} \cap \mathcal{N}_{B_w} \neq \emptyset$, one of B_v or B_w is contained within a pseudonode on the simple blossom corresponding to the other. The theorem follows from this. ∎

A class of subsets satisfying the property discussed in Theorem 10.4 is known as a *nested class of subsets*. Thus the class of subsets of original nodes contained within the pseudonodes at any stage will always be a nested class of subsets.

THEOREM 10.5 *If $\mathcal{N}_{B_v} \supset \mathcal{N}_{B_w}$ then $|\mathcal{N}_{B_v}| \overset{\geq}{=} 2 + |\mathcal{N}_{B_w}|$.*

Proof   Since $\mathcal{N}_{B_v} \supset \mathcal{N}_{B_w}$, B_w must be contained within a pseudonode on the simple blossom corresponding to B_v. This simple blossom must have at least two other nodes besides the pseudonode containing B_w. These facts imply the theorem. ∎

THEOREM 10.6 *The total number of pseudonodes of all levels, which are either current nodes, or contained within other outermost pseudonodes is always $\overset{\leq}{=} (n/2)$.*

Proof   This theorem can be proved very easily and directly from Theorem 10.3. However, we provide here a somewhat lengthy proof giving information on bounds for the number of pseudonodes of various levels. We consider all the pseudonodes at this stage in the current network G^1, let r_t be the number of pseudonodes of level $t = 1, 2, \ldots$ among these. So, we have to prove that $\sum_t r_t \overset{\leq}{=} (n/2)$. Each level 1 pseudonode is obtained by shrinking a simple blossom in G, which contains at least 3 nodes. So $r_1 \overset{\leq}{=} (n/3)$. Let S_1 denote the set of all level 1 pseudonodes, and original nodes of G which are not contained in any level 1 pseudonode. Nodes in S_1 are the nodes on the simple blossom corresponding to any level 2 pseudonode. We have $|S_1| \overset{\leq}{=} r_1 + n - 3r_1 = n - 2r_1$. Each simple blossom corrresponding to a level 2 pseudonode consists of at least 3 nodes from S_1, so $r_2 \overset{\leq}{=} (n - 2r_1)/3$. So, $r_1 + r_2 \overset{\leq}{=} ((n - 2r_1)/3) + r_1 = (n/3) + (r_1/3) \overset{\leq}{=} (n/3) + (n/3^2)$.

In a similar manner, for $t \overset{\geq}{=} 2$, define S_t to be the set of all pseudonodes of level t, and all original nodes and pseudonodes of levels $\overset{\leq}{=} (t - 1)$ which are not contained within any pseudonode of level $\overset{\leq}{=} t$. The nodes in the simple blossom corresponding to any level $(t + 1)$ pseudonode are those from the set S_t, each of these simple blossoms contains at least three nodes from S_t. Using the same arguments as above, and induction, it can be seen that $|S_t| \overset{\leq}{=} r_t + |S_{t-1}| - 3r_t \overset{\leq}{=} n - 2r_1 - 2r_2 - \ldots - 2r_t$. So, $r_{t+1} \overset{\leq}{=} (n - 2r_1 - \ldots - 2r_t)/3$. So, $r_{t+1} + r_t + \ldots + r_1 \overset{\leq}{=} ((n - 2r_1 - \ldots - 2r_t)/3) + r_t + \ldots + r_1 = (n/3) + (r_1 + \ldots + r_t)/3 \overset{\leq}{=} (n/3) + (n/3^2) + \ldots + (n/3^{t+1})$, for any $t \overset{\geq}{=} 2$. Hence

$$\sum_t r_t \overset{\leq}{=} \sum_{t=1}^{\infty} \frac{n}{3^t} = \frac{n}{3}(1 + \frac{1}{3} + \frac{1}{3^2} + \ldots)$$

$$= \frac{n}{3}(1 - \frac{1}{3})^{-1} = \frac{n}{3} \times \frac{3}{2} = \frac{n}{2}$$

This proves the theorem. ∎

THEOREM 10.7 *The number of unmatched nodes in the current network* G^1 wrt *the current matching* M^1 *is always the same as the number of unmatched nodes in the original network* G wrt *the matching* M *in it.*

Proof   Consider the operation of shrinking one simple blossom into a pseudonode. All the nodes on the simple blossom, with the possible exception of its base node, are matched nodes. If the base node is a matched (the unmatched root) node, the pseudonode into which this simple blossom is shrunk will be a matched (the new unmatched root) node after it is formed. Thus each shrinking operation leaves the total number of unmatched nodes unchanged, which implies the theorem. ∎

THEOREM 10.8 *Let B be a blossom in* G wrt *the matching* M. *Let t be the apex node of B, and* $j \neq t$ *an original node inside B. There exist two edge disjoint APs in B from j to t, one beginning with the matching edge incident at j, and the other beginning with a nonmatching edge incident at j.*

Proof Let B_1 be the simple blossom corresponding to the blossom B. Suppose the base node of B_1 is q, and let p be the node in B_1 that contains j inside it. First suppose $p \neq q$. Since B_1 is a simple blossom, there exist two edge disjoint APs from p to q on its alternating cycle, one beginning with the matching edge incident at p on it, and the other beginning with the nonmatching edge incident at p on it. Let these paths be $\mathcal{P}_1, \mathcal{P}_2$ respectively. Expand the pseudonodes on these paths, always expanding those of the highest level among the remaining pseudonodes first, using the procedure described earlier. This leads eventually to the alternating paths from j to t in the statement of the theorem. If $p = q$, go into the pseudonode q and repeat the same argument there. ■

As an example consider the simple blossom in Figure 10.16 corresponding to the pseudonode B_2 in the current network in Figure 10.17, or the blossom in the original network in Figure 10.15 defined by the subset of original nodes {4, 5, 8, 9, 7, 12, 11, 10, 6} corresponding to B_2. The base node of B_2 is B_1, and its apex node is 4. 11 is an original node inside B_2. 11 is a node on this simple blossom and the AP in it beginning with the matching edge incident at 11 to the base B_1 is 11, (11, 12), 12, (12, B_1), B_1. We now expand the pseudonode B_1 on this path. The simple blossom corresponding to B_1 can be seen from Figure 10.15, node 12 is connected only to node 5 in this simple blossom by a nonmatching edge. So, expanding the pseudonode B_1 on the above path leads to the AP 11, (11, 12), 12, (12, 5), 5, (5, 8), 8, (8, 9), 9, (9, 7), 7, (7, 4), 4 in the original network in Figure 10.15, from 11 to 4.

THEOREM 10.9 *Let* $B = (\mathcal{N}_B, \mathcal{A}_B)$ *be a blossom in* G wrt *the matching* M, *with* t *as its apex node. If* B *satisfies (10.16), either* t *is unmatched in* M, *or* t *is a matched node and there exists an alternating path wrt* M *from* t *beginning with the matching edge incident at it, to an unmatched node such that none of the edges on this path are from* $\mathcal{A}_B$.

Proof If the stem of the simple blossom corresponding to B is empty, t is the unmatched root node, otherwise t is a matched node and the required path is obtained by expanding this stem. ■

THEOREM 10.10 *(i) If there exists an augmenting path in the current network* G^1 wrt *the current matching* M^1, *then there exists an augmenting path in the original network* G wrt *the matching* M *in it.*
(ii) If all the pseudonodes in G^1 *correspond to blossoms which satisfy (10.16), and if an augmenting path exists in* G wrt M, *then there exists an augmenting path in* G^1 wrt M^1.

Proof The current network and the current matching in it change after each blossom shrinking. The result here follows by applying the results in Theorems 10.2 and 10.9 after each successive blossom shrinking. ■

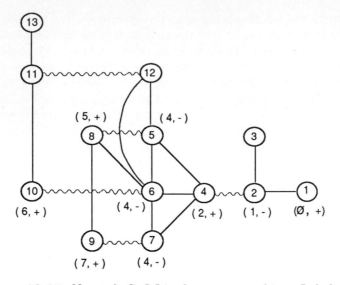

Figure 10.15 Network G. M is the wavy matching. Labels are recorded by the side of the nodes.

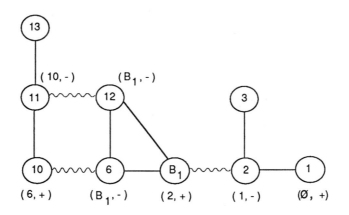

Figure 10.16 Labeling is continued after shrinking B_1.

Once an alternating tree is rooted at an unmatched node r, another unmatched node can only join it as an inner labeled node, and if that happens the tree is said to have become an *augmenting tree* . If the unmatched node labeled is i, its predecessor path $\mathcal{P}^1$ in the current network is an augmenting path. The corresponding augmenting path $\mathcal{P}$ in G between the apex of i and r is obtained by expanding the pseudonodes on $\mathcal{P}^1$, always expanding the highest level pseudonode, one at a time. Hence pseudonodes are expanded in reverse order to the one in which they are formed.

By Theorem 10.6, if there exists an augmenting path in G wrt the matching M, containing the root node r, then the alternating tree grown will become an augmenting tree after at most $(n/2)$ blossom shrinking operations. If an augmenting path is discovered, a rematching operation is carried out. If there are still some unmatched nodes, the whole procedure can be repeated with the new matching.

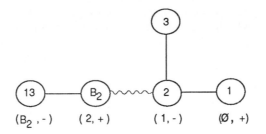

Figure 10.17 Labeling continued. Unmatched node 13 is labeled.

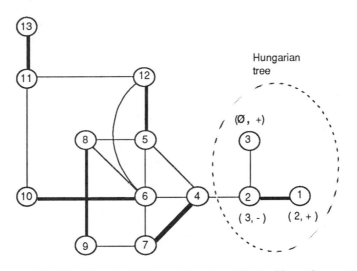

Figure 10.18 New matching edges are thick. New alternating tree rooted at 3, and grown.

As an example, consider the alternating tree rooted at the unmatched node 1 and grown in the network G in Figure 10.15 with the wavy matching M. We adopt the rule of selecting the node with the least serial number among those in the list for scanning. Nodes 8, 9 are the identifying pair for the simple blossom defined by the subset of nodes $\{9, 8, 5, 4, 7\}$, which is shrunk into the pseudonode B_1 resulting in the current network in Figure 10.16. Labeling is continued after shrinking B_1.

Nodes 11, 12 are the identifying pair of nodes for the simple blossom defined by the subset of nodes $\{11, 12, B_1, 6, 10\}$ which is shrunk into the pseudonode B_2

resulting in the current network in Figure 10.17. When tree growth is resumed, the unmatched node 13 is labeled, and the alternating tree in Figure 10.17 has become an augmenting tree with the horizontal path between nodes 1 and 13 as an augmenting path. We derive the corresponding augmenting path in the original network by expanding the pseudonode on it, and rematch. This leads to the new matching marked with thick edges in Figure 10.18. A new alternating tree is rooted at the unmatched node 3 and grown. After nodes 3, 2, 1 are labeled, we are unable to label any more, even though the tree has not become an augmenting tree, and has not blossomed. We discuss the implications of this next (in this example, the original network has 13 nodes, and the thick matching in Figure 10.18 has 6 edges, so clearly this matching is a maximum cardinality matching).

Hungarian Trees and Their Properties

At some stage it may happen that the current alternating tree has not become an augmenting tree, it has no blossoms to be shrunk, and it cannot grow any further (i.e., labeling cannot be continued any further). At this stage we say that the alternating tree has become *hungarian*, and the tree itself is called a *Hungarian tree*. By the above results, this can only happen if there exists no augmenting path wrt the present matching **M**, beginning with the unmatched root node r in G. Clearly the only unmatched node in a Hungarian tree is its root node.

An example of a Hungarian tree is the tree consisting of nodes 3, 2, 1 rooted at the unmatched node in Figure 10.18.

THEOREM 10.11 *A Hungarian tree contains an odd number of current nodes, all but the root node in which are matched nodes contained on in-tree current matching edges. The total number of original nodes contained within the current nodes in a Hungarian tree is odd, and all but the apex of the root node among these is a matched node.*

Proof Since a Hungarian tree has not become an augmenting tree, all nonroot nodes in it must be matched nodes lying on current in-tree matching edges. This also implies that the number of current nodes in the Hungarian tree is one plus twice the number of current matching edges in it, which is an odd number. Since every current node contains an odd number of original nodes inside it, the total number of original nodes inside current nodes in this Hungarian tree is also an odd number. As all the current nodes in the Hungarian tree other than the root node are matched, all the original nodes inside current nodes in it are matched except the apex of the root node. ∎

Let $\mathcal{N}_H$ be the set of original nodes contained within nodes on a Hungarian tree H in the current network. Let $G_H = (\mathcal{N}_H, \mathcal{A}_H)$ be the partial network of the original network determined by $\mathcal{N}_H$, and M_H the set of edges in the present matching **M** in G contained within G_H. Theorem 10.11 implies that all the nodes in $\mathcal{N}_H$ excepting the apex of the root node in the Hungarian tree H, are matched by the matching edges in $\mathbf{M}_H$. So, $|\mathcal{N}_H|$ is odd, and $\mathbf{M}_H$ is a maximum cardinality matching in $\mathbf{G}_H$.

THEOREM 10.12 *All leaf nodes in a Hungarian tree must be outer nodes.*

Proof   Let $j \neq$ root be a leaf node in a Hungarian tree. If j is an inner node, and it is unmatched, the tree is an augmenting tree contrary to the hypothesis that it is hungarian. If j is matched, let its mate be p. If p is unlabeled, it can be labeled, a contradiction. If p is already labeled, it is not the root since it is matched, and it cannot be an outer node because for every nonroot outer node its predecessor must be its mate. So, if p is an outer node, j must be its predecessor, contradiction to the fact that j is a leaf node. Hence p must be an inner node. But since $(j;p)$ is a current matching edge, if j, p are both inner nodes, they are the identifying pair for a blossom that can be shrunk, contradiction to the hungarianness of the tree. Hence j must be an outer node. ∎

THEOREM 10.13 *If $(i;j)$ is a current edge with i in a Hungarian tree and j not in it, then i must be an inner node.*

Proof   On the contrary, suppose i is an outer node. If $(i;j)$ is a nonmatching edge, then j could have been labeled, contradicting the fact that the tree has become hungarian. If $(i;j)$ is a matching edge, then j must be the predecessor of i, contradicting the fact that j is not an in-tree node. So, i must be an inner node. ∎

COROLLARY 10.1 *Suppose the current network contains a Hungarian tree. If an edge in this network is incident at node i which is an outer node in the Hungarian tree, then that edge must contain an inner node in this Hungarian tree at its other end.*

Proof   Follows directly from Theorem 10.13. ∎

THEOREM 10.14 $G = (\mathcal{N}, \mathcal{A})$ *is the original network with the matching* $\mathbf{M}$ *in it. An alternating tree is rooted at an unmatched node r and grown. It becomes the Hungarian tree* H *at the stage when* $G^1 = (\mathcal{N}^1, \mathcal{A}^1)$ *is the current network.* $\mathcal{N}_H$ *is the set of original nodes contained within current nodes on* H, *and* $\mathcal{N}_{\overline{H}} = \mathcal{N} \backslash \mathcal{N}_H$. $G_H = (\mathcal{N}_H, \mathcal{A}_H)$, $G_{\overline{H}} = (\mathcal{N}_{\overline{H}}, \mathcal{A}_{\overline{H}})$ *are the partial networks of* G *determined by* $\mathcal{N}_H, \mathcal{N}_{\overline{H}}$ *respectively.* $\mathbf{M}_H = \mathbf{M} \cap \mathcal{A}_H$, *and* $\overline{\mathbf{M}_{\overline{H}}}$ *is a maximum cardinality matching in* $G_{\overline{H}}$. *Then* $\mathbf{M}_H \cup \overline{\mathbf{M}_{\overline{H}}}$ *is a maximum cardinality matching in* G.

Proof   Let $\mathbf{M}^1$ be the current matching in G^1. Let $\mathbf{A} = \{(i;j) : i \in \mathcal{N}_H, j \in \mathcal{N}_{\overline{H}}, (i;j) \in \mathcal{A}\}$. Since all nodes in $\mathcal{N}_H$ other than r are matched by matching edges within $\mathcal{A}_H$, all edges in $\mathbf{A}$ are nonmatching edges, so $\mathbf{A} \cap \mathbf{M} = \emptyset$. Let $\hat{\mathcal{A}} = \{(i;j) : (i;j) \in \mathcal{A}, \text{at least one of } i \text{ or } j \text{ is in } \mathcal{N}_H\} = \mathcal{A}_H \cup \mathbf{A}$. So, $\hat{\mathbf{M}} = \mathbf{M} \cap \hat{\mathcal{A}} = \mathbf{M} \cap \mathcal{A}_H = \mathbf{M}_H$. Let $\hat{\mathcal{N}}$ be the set of nodes on edges in $\hat{\mathcal{A}}$, and $\hat{G} = (\hat{\mathcal{N}}, \hat{\mathcal{A}})$. We will now prove that $\mathbf{M}_H$ is a maximum cardinality matching in $\hat{G}$. Define

$$\mathcal{N}_H^1 \;=\; \text{Set of all current nodes in H}$$

$$\hat{\mathcal{N}}^1 \;=\; \mathcal{N}_H^1 \cup (\hat{\mathcal{N}} \backslash \mathcal{N}_H) = \text{Set of all current nodes}$$
$$\text{corresponding to original nodes in } \hat{\mathcal{N}}$$

$$\hat{\mathcal{A}}^1 \;=\; \text{Set of current edges with at least one node in } \mathcal{N}_H^1$$

$$\hat{G}^1 \;=\; (\hat{\mathcal{N}}^1, \hat{\mathcal{A}}^1) = \text{Current network corresponding to } \hat{G}$$

$$\hat{M}^1 \;=\; M^1 \cap \hat{\mathcal{A}}^1$$

$$\mathbf{A}^1 \;=\; \{(p;q) : p \in \mathcal{N}_H^1, q \in \hat{\mathcal{N}} \backslash \mathcal{N}_H, (p;q) \in \mathcal{A}^1\}$$

By previous arguments, $\mathbf{A}^1 \cap \hat{M}^1 = \emptyset$ and the only nodes in $\hat{G}^1$ which are unmatched in M^1 are those in $\hat{\mathcal{N}} \backslash \mathcal{N}_H$, and the root node of H. So any augmenting path wrt M^1 in $\hat{G}^1$ must contain at least one node in $\hat{\mathcal{N}} \backslash \mathcal{N}_H$ as a terminal node.

Suppose $\mathcal{P}$ is an augmenting path in $\hat{G}^1$ wrt $\hat{M}^1$ beginning with an unmatched node $p \in \hat{\mathcal{N}} \backslash \mathcal{N}_H$, to another unmatched node $q \in \hat{\mathcal{N}}^1$. Give the nodes on this path alternately *outer and inner designations* (the reader is cautioned not to confuse these outer and inner designations with the outer and inner labels already existing on nodes in H) beginning with an outer designation for p. If $\mathcal{P}$ contains any nodes from $\mathcal{N}_H^1$, the fact that all the edges in $\mathbf{A}^1$ are nonmatching edges, and Theorem 10.13 together imply that any node from $\mathcal{N}_H^1$ in $\mathcal{P}$ bears an inner or outer designation iff it is an inner or outer labeled node respectively in H. Since $\mathcal{P}$ is an augmenting path, its terminal node q must bear an inner designation, and since all nodes labeled as inner nodes in H are matched nodes, q cannot be a node in H, and hence q must also be a node in $\hat{\mathcal{N}} \backslash \mathcal{N}_H$.

Hence an alternating path in $\hat{G}^1$, starting from an unmatched node in $\hat{\mathcal{N}}^1$ outside the Hungarian tree H, can only be an augmenting path if it terminates at another node outside H. But an augmenting path always consists of an odd number of edges. So, if the first edge in an augmenting path in $\hat{G}^1$ is a current edge joining a node in $\hat{\mathcal{N}} \backslash \mathcal{N}_H$ to a current inner labeled node in H, then the last edge on this path must be a current edge joining an outer labeled node in H with an unmatched node. This is impossible by Theorem 10.13. Hence there exists no augmenting path wrt the current matching $\hat{M}^1$ in $\hat{G}^1$. By Theorem 10.10, this implies that there exists no augmenting path in $\hat{G}$ wrt $\hat{M}$. Hence $\hat{M} = M_H$ is a maximum cardinality matching in $\hat{G}$.

Now let $\mathbf{N}$ be any other matching in G. Let $\hat{\mathbf{N}} = \mathbf{N} \cap \hat{\mathcal{A}}$, and $\mathbf{N}_{\overline{H}} = \{(i;j) : (i;j) \in \mathbf{N}, \text{ and } i \text{ and } j \text{ are both in } \mathcal{N}_{\overline{H}}\}$. Then $\mathbf{N} = \hat{\mathbf{N}} \cup \mathbf{N}_{\overline{H}}$. By definition, $|\overline{M}_{\overline{H}}| \geqq |\mathbf{N}_{\overline{H}}|$. Since $\hat{M} = M_H$ is a maximum cardinality matching in $\hat{G}$, we have $|M_H| \geqq |\hat{\mathbf{N}}|$. These facts imply that $|M_H \cup \overline{M}_{\overline{H}}| \geqq |\mathbf{N}|$, so $M_H \cup \overline{M}_{\overline{H}}$ is a maximum cardinality matching in G. ∎

So, to find a maximum cardinality matching in G, start with an arbitrary matching, root an alternating tree at an unmatched node and grow it. If the tree blossoms, shrink the blossom. If the tree becomes augmenting, trace the augmenting path, rematch using it, and repeat the procedure with the new matching. If the tree becomes hungarian, identify the set $\mathcal{N}_H$ of all the original nodes contained within that Hungarian tree. Let $\mathbf{M}_H$ be the set of all matching edges in the present matching, both of whose incident nodes are in $\mathcal{N}_H$. Delete all the nodes in $\mathcal{N}_H$ and all the edges which contain at least one node from $\mathcal{N}_H$, let the remaining network be $\mathbf{G}_{\overline{H}}$. The union of $\mathbf{M}_H$ and a maximum cardinality matching in $\mathbf{G}_{\overline{H}}$, is a maximum cardinality matching in G. The problem of finding a maximum cardinality matching in the smaller network $\mathbf{G}_{\overline{H}}$ remains. For that, root an alternating tree at an unmatched node in $\mathbf{G}_{\overline{H}}$ and repeat this process with it.

A different procedure is to root an alternating tree at each unmatched node in G and grow them all simultaneously. This is called a *planted forest*. In general the forest will consist of disjoint alternating trees each rooted at a separate unmatched node. This forest growth procedure leads to a more efficient algorithm than the procedure of growing only one alternating tree at a time. When growing a forest, it is necessary to maintain the identity of the rooted tree to which each labeled node belongs. The current network $\mathbf{G}^1 = (\mathcal{N}^1, \mathcal{A}^1)$ changes each time a simple blossom is shrunk in this forest growth process. Edges in the present $\mathbf{G}^1$ can be generated as they are needed, and we outline methods for doing it efficiently here.

The original network is stored by storing $\mathcal{A}$. Each new pseudonode created is given a distinct identification number when it is formed; and we store the numbers of the current nodes on the simple blossom corresponding to it, together with the present labels on them; its identifying pair of nodes, base and apex nodes; and the set of original nodes contained inside it. We store the present matching M, and update it whenever it changes in augmentation steps. We store $\mathcal{N}^1$, the set of numbers of all the current nodes, and update it whenever it changes.

If i is an original node which is a current node, the set of current edges incident at i is $\{(i;p) : p \in \mathcal{N}^1$ and p contains a j inside it such that $(i;j) \in \mathcal{A}\}$. This set can be easily generated from the stored data whenever needed. If i is a matched node, let its mate be j, the current matching edge incident at i is $(i;p)$ where p is the unique current node containing j inside it.

If p is a pseudonode which is a current node, the set of current edges incident at p is $\{(p;q) : q$ is a current node such that there exists an original node i inside p, and an original node j inside q, with $(i;j) \in \mathcal{A}\}$. There can exist at most one original node i inside p such that it is joined to an original node j outside p by a matching edge. If no such node i exists, p is unmatched. If such a node i exists, find its mate j, and the current node q containing j inside it, then $(p;q)$ is the current matching edge incident at p. All this can be generated efficiently from the stored information whenever needed.

In practice, it seems to be convenient to also maintain and update the set of current matching edges, but generate all other data about the current network from the stored information as needed. We also maintain a record of all the pseudonodes in which each node lies, in the order of outermost first, this information is needed for expanding the pseudonodes on a path containing the node.

10.1.1   Blossom Algorithm for the Maximum Cardinality Matching Problem

Based on the ideas discussed so far, we provide an augmenting path method called the *blossom algorithm* for finding a maximum cardinality matching in an undirected network $G = (\mathcal{N}, \mathcal{A})$. It grows an alternating forest with an alternating tree planted at each unmatched node. In this algorithm, an augmenting path is identified when a pair of trees become augmenting trees together, the augmenting path will then be the alternating path between the root nodes of these trees. M, $G^1 = (\mathcal{N}^1, \mathcal{A}^1)$, M^1 always denote the present matching in G, the present current network, and the present current matching respectively. In the algorithm, node labels have 3 entries, the predecessor index, the symbol $+$ (for outer nodes) or $-$ (for inner nodes), and the root index, in that order. List always refers to the present set of labeled and unscanned nodes.

BLOSSOM ALGORITHM FOR THE
MAXIMUM CARDINALITY MATCHING PROBLEM

Step 1 Initialization   Choose an initial matching (it could be $\emptyset$) in G.

Step 2 Rooting an Alternating Forest   If there are no unmatched nodes, go to Step 7. Otherwise root an alternating tree at each unmatched node i, by labeling it with $(\emptyset, +, i)$. List now consists of all these root nodes.

Step 3 Select A Node to be Scanned If list $= \emptyset$, go to Step 7. Otherwise select one node from the list to scan and delete it from the list.

Step 4 Scanning Let the node to be scanned be the current node i with label $(P(i), \pm, r)$.

Scanning an Outer Node If i is an outer node, for each $j \neq P(i)$ such that $(i; j)$ is a current edge (all these will be nonmatching edges) do the following.

If j is an already labeled outer node associated with a root $\neq r$, an augmenting path has been found, go to Step 5.

If j is an already labeled outer node with the same root r, the alternating tree containing i and j has blossomed, go to Step 6.

If j is an already labeled inner node, continue.

If j is unlabeled, label it with $(i, -, r)$ and include it in the list.

Scanning an Inner Node If i is an inner node, it cannot be unmatched since all unmatched nodes are outer root nodes, let $(i; j)$ be the current matching edge incident at i.

If j is already labeled, it must be an inner node too. If the root indices of i and j are the same, the tree containing them has blossomed, go to Step 6. If the root indices of i, j are different, an augmenting path has been found, go to Step 5.

If j is unlabeled, label it with $(i, +, r)$ and include it in the list.

Go back to Step 3.

Step 5 Augmentation   We come to this step when scanning has revealed a pair of adjacent current nodes i, j associated with different root nodes $r(i) \neq r(j)$, such that either i, j are both outer nodes and $(i; j) \notin \mathbf{M}^1$, or i, j are both inner nodes and $(i; j) \in \mathbf{M}^1$. Combine the predecessor paths of i, j together with edge $(i; j)$, leading to the path $\mathcal{P}^1$. $\mathcal{P}^1$ is an augmenting path between $r(i)$ and $r(j)$. Let t_1, t_2 be the apex nodes of $r(i), r(j)$. Find the corresponding augmenting path $\mathcal{P}$ between t_1 and t_2 in G by expanding all the pseudonodes on $\mathcal{P}^1$. Erase all the labels on the nodes in the two trees containing i, j, and throw away all the blossoms in them (this operation is called *dismembering the two trees*). Rematch using $\mathcal{P}$, and revise the current matching accordingly. Nodes t_1, t_2 are matched in the new matching. If there are no trees left in the forest go to Step 7, otherwise put all the outer current nodes in the list and go back to Step 3.

Step 6 Blossom Shrinking   We come to this step when scanning has revealed a pair of adjacent current nodes i, j associated with the same root node, such that either i, j are both outer nodes and $(i; j) \notin \mathbf{M}^1$, or i, j are both inner nodes and $(i; j) \in \mathbf{M}^1$. i, j are the identifying pair of nodes for a simple blossom in the current network, whose base node t is the first common node on their predecessor paths. The nodes on this simple blossom are the base node, and those remaining in these paths after eliminating the common nodes on them; store these nodes together with the predecessor index and the $+$ or $-$ indicator for outer, inner status, in the labels on them. Shrink this simple blossom into a pseudonode, say B_p. Change the predecessor index of all the nodes outside this simple blossom which are immediate successors of nodes on this simple blossom, into B_p. Give B_p the same label as on the base node t. If t is unmatched, B_p becomes the new root node of the tree containing it after the shrinking, change the root index of all the nodes on the tree to B_p. Include B_p in the list and go back to Step 3.

Step 7 Termination   When we come to this step either we have a perfect matching in G, or all the planted trees have become Hungarian trees. In the latter case we say that the labeling has become hungarian, and the forest has become a *Hungarian forest*. The present matching in G is a maximum cardinality matching, terminate.

Discussion

Each augmentation step increases the cardinality of the matching by 1. So, Step 5 is carried at most $n/2$ times in the algorithm. Between any two consecutive occurrences of Step 5, Step 6 can be carried out at most $n/2$ times by Theorem 10.6, and Step 3 will be carried out at most $O(n)$ times. Each execution of Step 3 requires at most $O(n)$ effort. Each execution of Steps 5 or 6 requires tracing the predecessor paths, which requires at most $O(n)$ effort. Thus the overall work between two consecutive occurrences of Step 5 requires at most $O(n^2)$ effort. So, the overall computational effort in the algorithm is bounded above by $O(n^3)$.

Comment 10.1   The pioneering work on the blossom algorithm is due to Edmonds [1965a, b], and the original version of this algorithm is due to him.

10.1.2   The Minimum Cost Perfect Matching Problem

We consider the problem of finding a minimum cost perfect matching in $G = (\mathcal{N}, \mathcal{A}, c = (c_{ij}))$ with $|\mathcal{N}| = n$, $|\mathcal{A}| = m$ and c as the vector of edge cost coefficients. We assume that n is even, otherwise there is no perfect matching in G. The problem is to find $x = (x_{ij} : (i;j) \in \mathcal{A})$ that

$$
\begin{aligned}
\text{Minimizes} \quad z(x) &= \sum (c_{ij}x_{ij} : \text{ over } (i;j) \in \mathcal{A}) \\
\text{subject to} \quad x(i) &= 1, \text{ for all } i \in \mathcal{N} \qquad\qquad (10.17)\\
x_{ij} &\geqq 0 \text{ for all } (i;j) \in \mathcal{A}
\end{aligned}
$$

$$
x_{ij} \qquad \text{integer for all } (i;j) \qquad\qquad (10.18)
$$

where $x(i)$ is defined in (10.1). It is not necessary to include the constraint $x_{ij} \leqq 1$ in this problem, as the first constraint in (10.17) implies it automatically. Let $\mathbf{Y} \subset \mathcal{N}$ with $|\mathbf{Y}|$ odd and $\geqq 3$. Define $\mathbf{Y}^-(x)$ as below. If x is a matching vector, (10.20) must hold.

$$
\mathbf{Y}^-(x) = \sum (x_{ij} : \text{ over } i, j \text{ both } \in \mathbf{Y} \text{ and } (i;j) \in \mathcal{A}) \qquad (10.19)
$$

$$
\mathbf{Y}^-(x) \leqq (|\mathbf{Y}| - 1)/2 \qquad\qquad (10.20)
$$

The constraint (10.20) is known as the *matching blossom inequality* or *matching blossom constraint corresponding to* $\mathbf{Y}$ for (10.17), (10.18). Each subset of $\mathcal{N}$ of odd cardinality $\geqq 3$ leads to a blossom inequality, and every matching vector satisfies all matching blossom inequalities. Let $\{\mathbf{Y}_1, \dots, \mathbf{Y}_L\}$ be the set of all distinct subsets of $\mathcal{N}$ of odd cardinality $\geqq 3$. Now consider the following LP.

$$\text{Minimize} \quad z(x) \;=\; \sum (c_{ij}x_{ij} : \text{ over } (i;j) \in \mathcal{A})$$

$$\text{subject to} \quad x(i) \;=\; 1, \text{ for all } i \in \mathcal{N}$$

$$\mathbf{Y}_\sigma^-(x) \;\leqq\; (|\mathbf{Y}_\sigma| - 1)/2, \sigma = 1 \text{ to } L \qquad (10.21)$$

$$x_{ij} \;\geqq\; 0 \text{ for all } (i;j) \in \mathcal{A}$$

Every perfect matching vector in G is feasible to (10.21) and every integer feasible vector for (10.21) is a perfect matching vector in G. So, if an optimum solution of (10.21) is an integer vector, it is a minimum cost perfect matching vector in G. We will discuss a primal-dual algorithm for solving (10.21) known as the *blossom algorithm for the minimum cost perfect matching problem* and show that if a perfect matching exists in G, then this algorithm terminates with an optimum solution of (10.21) which is an integer vector, and this integer vector is therefore a minimum cost perfect matching vector in G.

To write the dual of (10.21), associate a dual variable π_i with the constraint corresponding to node i in (10.21), and a dual variable μ_σ with the blossom inequality corresponding to the odd subset $\mathbf{Y}_\sigma, \sigma = 1$ to L. The dual variables π_i are only associated with original nodes in G, and hence are also called *original node prices*. The dual variables μ_σ are known as *pseudonode prices*, the reason for this name will become clear later. Let $\pi = (\pi_i), \mu = (\mu_\sigma)$. Since the number of odd subsets of nodes, L, grows exponentially with n, the vector μ has a lot of entries. Given the dual solution (π, μ), define for each $(i;j) \in \mathcal{A}$

$$\mu^-(i;j) \;=\; \sum (\mu_\sigma : \text{ over } \sigma \text{ s.t. } \mathbf{Y}_\sigma \text{ contains both } i \text{ and } j) \qquad (10.22)$$

$$d_{ij}(\pi, \mu) \;=\; \pi_i + \pi_j - \mu^-(i;j) \qquad (10.23)$$

The dual of (10.21) is

$$\text{Maximize} \quad W(\pi, \mu) \;=\; \sum_{i \in \mathcal{N}} \pi_i - \sum_{\sigma=1}^{L} (|\mathbf{Y}_\sigma| - 1)(\mu_\sigma)/2$$

$$\text{subject to} \quad d_{ij}(\pi, \mu) \;\leqq\; c_{ij}, \text{ for each } (i;j) \in \mathcal{A} \qquad (10.24)$$

$$\mu \;\geqq\; 0$$

Given a dual feasible solution (π, μ), define $\mathcal{A}_*(\pi, \mu)$ as in (10.25). Since (π, μ) satisfies the first constraint in (10.24) as an equation for each edge in $\mathcal{A}_*(\pi, \mu)$, they are called *equality edges* wrt (π, μ), and the subnetwork $G_*(\pi, \mu) = (\mathcal{N}, \mathcal{A}_*(\pi, \mu))$ is known as the *equality subnetwork* wrt (π, μ) for (10.21). The complementary slackness conditions for optimality in the primal, dual pair (10.21), (10.24) are (10.26), (10.27) given below.

$$\mathcal{A}_*(\pi, \mu) = \{(i; j) : (i; j) \in \mathcal{A}, \text{ and } d_{ij}(\pi, \mu) = c_{ij}\} \qquad (10.25)$$

$$x_{ij} > 0 \text{ implies } d_{ij}(\pi, \mu) = c_{ij}, \text{ for each } (i; j) \in \mathcal{A} \qquad (10.26)$$

$$\mu_\sigma > 0 \text{ implies } \mathbf{Y}_\sigma^-(x) = (|\mathbf{Y}_\sigma| - 1)/2, \text{ for } \sigma = 1 \text{ to } L \qquad (10.27)$$

The blossom algorithm is initiated with an initial dual feasible solution (π^0, μ^0) in which $\mu^0 = 0$. If we can find a perfect matching in the equality subnetwork $\mathbf{G}_*(\pi^0, \mu^0)$, it is a minimum cost perfect matching in $\mathbf{G}$, since the corresponding perfect matching vector x satisfies the complementary slackness optimality conditions (10.26), (10.27) together with (π^0, μ^0). For this we apply the maximum cardinality matching algorithm of Section 10.1.1 on $\mathbf{G}_*(\pi^0, \mu^0)$ beginning with the empty matching. Suppose this leads to the matching $\mathbf{M}_1$ in $\mathbf{G}_*(\pi^0, \mu^0)$, corresponding to the matching vector x^1. So, all matching edges in $\mathbf{M}_1$ are equality edges in $\mathcal{A}_*(\pi^0, \mu^0)$ and (10.26), (10.27) hold for $x^1, (\pi^0, \mu^0)$. In the process of applying this algorithm blossoms may have been discovered and shrunk into pseudonodes, leading to *current equality subnetworks* $\mathbf{G}_*^1(\pi^0, \mu^0)$ of the form $(\mathcal{N}^1, \mathcal{A}_*^1(\pi^0, \mu^0))$ where $\mathcal{N}^1$ is the set of current nodes, and $\mathcal{A}_*^1(\pi^0, \mu^0)$ is the set of current equality edges.

If $\mathbf{M}_1$ is not a perfect matching, we are left with a Hungarian forest in the equality subnetwork $\mathbf{G}_*^1(\pi^0, \mu^0)$, with a Hungarian tree rooted at each unmatched current node in it. Each pseudonode at this stage corresponds to a shrunken blossom wrt the present matching $\mathbf{M}_1$, the set of original nodes inside it has odd cardinality ≥ 3, and hence there is a blossom inequality and a dual variable μ_σ associated with it. By the second property in (10.15) of blossoms, we verify that the present matching vector x^1 satisfies the blossom inequality corresponding to every existing pseudonode (whether it is an outermost pseudonode or not) as an equation.

To get things moving, the approach now goes to a dual solution change step. The purpose of this is to obtain a new dual feasible solution, the equality subnetwork corresponding to which allows a matching of higher cardinality than $\mathbf{M}_1$. This dual solution change step is designed to satisfy the following properties.

(i)   All present matching edges (which are equality edges now) remain equality edges after the change, so that the present matching vector x^1 continues to satisfy (10.26) together with the new dual solution.

(ii)   The equality, nonequality status of all original edges contained within any existing pseudonode remains unchanged, and all in-tree current equality edges remain equality edges after the dual solution change. So, all the existing alternating trees in the current equality subnetwork, are also contained in the new current equality subnetwork after the dual solution change.

(iii) In the new current equality subnetwork obtained after the dual solution change, the existing forest is not hungarian, i.e., at least one of the trees can grow, or there is at least one augmenting path, etc. This makes it possible to repeat the application of the maximum cardinality matching algorithm in the new current equality subnetwork, beginning with the existing forest, and make some movement in the algorithm.

(iv) In the new dual solution, some of the dual variables μ_σ may be given positive values, but if a μ_σ is positive, the corresponding subset of original nodes $\mathbf{Y}_\sigma$ will always be the set of original nodes inside an existing pseudonode (outermost or not). We have already seen that the present matching vector satisfies the blossom inequality corresponding to every existing pseudonode as an equation, so this guarantees that the complementary slackness condition (10.27) continues to hold.

The application of the maximum cardinality matching algorithm is continued with the present alternating forest in the new current equality subnetwork, and this process is repeated. The following summarizes the properties maintained by the algorithm.

> It maintains a matching vector x, and (π, μ) always feasible to (10.24). It alternates between changing the matching vector x, keeping (π, μ) constant, using the maximum cardinality matching algorithm in the equality subnetwork $G_*(\pi, \mu)$; or changing (π, μ) keeping x constant.
>
> $x_{ij} = 1$ implies that $(i; j) \in \mathcal{A}_*(\pi, \mu)$, so, (10.26) holds always.   (10.28)
>
> $\mu_\sigma > 0$ always implies that the associated $\mathbf{Y}_\sigma$ is the set of original nodes inside an existing pseudonode, so (10.27) holds always. This also guarantees that even though the dual vector $\mu = (\mu_\sigma)$ has a large number of entries, all but at most $(n/2)$ of them will be 0 at every stage. Hence, it is only necessary to store values of μ_σ associated with each pseudonode at that stage of the algorithm. That's why μ_σ are known as *pseudonode prices*.

Changes in Blossoms after an Augmentation Step

Let x, (π, μ) be the solution pair at some stage during the matching change phase of the algorithm. If an augmenting path $\mathcal{P}^1$ is discovered, we would augment. In the maximum cardinality matching algorithm we then discard all the existing blossoms on the two trees containing nodes along $\mathcal{P}^1$. However, some of these blossoms may correspond to pseudonodes associated with a positive μ_σ in the present dual solution (π, μ) and discarding these blossoms will violate the third property in (10.28). So, in this algorithm, all the blossoms along the augmenting path are retained after augmentation, but changes have to be made in the stored data on the simple blossoms corresponding to them, to reflect the change in the matching caused by the augmentation step. This is called the operation of *revising all the blossoms along* $\mathcal{P}^1$.

We describe this operation now for a pseudonode B either on $\mathcal{P}^1$ or inside some node on $\mathcal{P}^1$. Let $\mathcal{P}$ be the augmenting path in $G_*(\pi, \mu)$ corresponding to $\mathcal{P}^1$. Let i_1 be the base node (which may itself be another pseudonode) of B. When all the pseudonodes along $\mathcal{P}^1$ containing B within them are expanded, we will get the portion of the augmenting path passing through the simple blossom corresponding to B. We will denote this portion of the path by $\mathcal{P}^0$. We consider two cases.

Case 1 Apex of Pseudonode B Is an Intermediate Node On $\mathcal{P}$ In this case the augmenting path $\mathcal{P}$ passes through the matching edge incident at the apex node of B, and leaves through either the apex node or some nonapex node within B. We consider two subcases.

> **Subcase 1 $\mathcal{P}$ Passes Only Through the Apex Node of B**   In this subcase $\mathcal{P}^0$ appears as in Figure 10.19. After the augmentation is carried out, $(q_1; i_1)$ becomes a nonmatching edge and $(i_1; q_2)$ becomes a matching edge. The only change needed in the stored data on this simple blossom, is to change the label on its base node i_1 as in Figure 10.20.

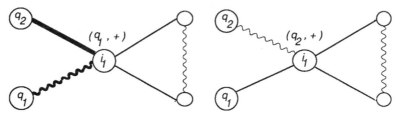

Figure 10.19   **Figure 10.20**

> Before augmentation (Figure 10.19), and after (Figure 10.20). Matching edges are wavy. In Figure 10.19, $\mathcal{P}^0$ is thick. q_1, q_2 are not on the simple blossom. Stored label on i_1 is entered by its side.

> **Subcase 2 $\mathcal{P}$ Contains Some Nonapex Nodes Contained in B**   In this subcase $\mathcal{P}^0$ appears as in Figure 10.21. The i-nodes are nodes on the simple blossom corresponding to B, and q_1, q_2 are nodes outside B. The odd cycle in this simple blossom remains the same, but augmentation changes the matching edges along $\mathcal{P}^0$ into nonmatching edges and vice versa. After augmentation (see Figure 10.22) $(i_1; i_2)$ is a matching edge, and i_1 is no longer the base node, i_g becomes the new base node of this simple blossom, and its apex will be the revised apex of i_g (i.e., after the corresponding change due to augmentation is carried out for i_g). Change

the label on i_g in the stored data to $(q_2, +)$. Change the labels on both the neighbor nodes of i_g on the odd cycle, i_{g+1} and i_{g-1}, to $(i_g, -)$. Change the labels on i_{g+2}, i_{g-2} to $(i_{g+1}, +)$ and $(i_{g-2}, +)$ respectively. Keep on changing the labels along the odd cycle this way until all the node labels are changed. The last pair of nodes to be relabeled in this manner are the new identifying pair of nodes for this simple blossom. It is clear that the odd alternating cycle in this simple blossom, the new matching and nonmatching edges in it, can all be retrieved by the procedures discussed earlier using the revised labels on the nodes.

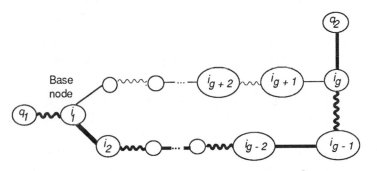

Figure 10.21 Matching edges are wavy. $\mathcal{P}^0$, the thick path, contains more than one node on the simple blossom corresponding to B. See Figure 10.22 for position after augmentation.

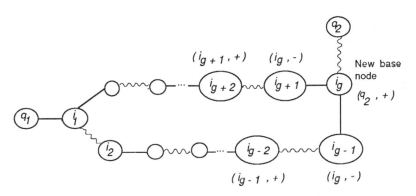

Figure 10.22 Position after augmentation in the simple blossom in Figure 10.21. New base node and new node labels are indicated.

Case 2 Apex of Pseudonode B Is a Terminal Node of $\mathcal{P}$ In this case i_1, the base node of B is an unmatched node. We again consider two subcases.

Subcase 1 $\mathcal{P}$ Contains Only the Apex Node of B In this subcase $\mathcal{P}^0$ appears as in Figure 10.23. After the augmentation $(q_1; i_1)$ becomes a matched edge, change the stored label on i_1 as in Figure 10.24.

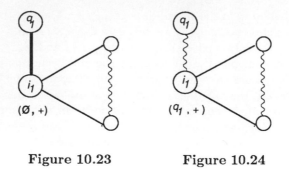

Figure 10.23   **Figure 10.24**

Before augmentation (Figure 10.23), and after (Figure 10.24). $\mathcal{P}^0$ is thick in Figure 10.23, and contains only unmatched base node i_1. q_1 is not on the simple blossom of B. Stored label on i_1 is entered by its side.

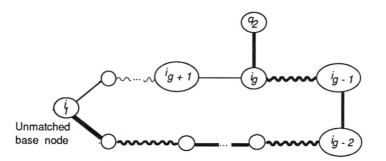

Figure 10.25 Matching edges are wavy. $\mathcal{P}^0$ is thick, it contains more than one node on the simple blossom of B. See Figure 10.26 for position after augmentation.

Subcase 2 $\mathcal{P}$ **Contains Some Nonapex Nodes Contained in** B In this subcase $\mathcal{P}^0$ appears as in Figure 10.25. In this subcase i_g becomes the new base node after augmentation, and the revision is carried out as in Subcase 2 of Case 1. See Figure 10.26.

After the augmentation, the existing two alternating trees containing nodes on the path used are no longer rooted at unmatched nodes, in fact all the nodes on these two trees are matched. So, we erase these two trees by erasing the present labels on the current nodes in them, and leaving all these current nodes as unlabeled

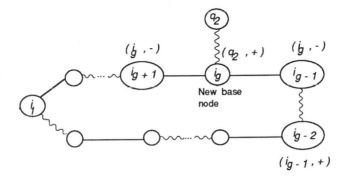

Figure 10.26 New base node, new labels after augmentation are indicated.

nodes. This operation is called *chopping down the trees*. Notice the difference between the operation of dismembering the trees (in Step 5 of the algorithm in Section 10.1.1, all the blossoms on the trees are thrown away), and that of chopping down the trees discussed here (only the tree structures are eliminated, but all the current nodes on them are left as unlabeled nodes in the current network).

Each of the out-of-tree (i.e., unlabeled) current nodes which are pseudonodes, are matched nodes, and the blossoms corresponding to them satisfy the conditions (10.15) wrt the present matching, but they may not satisfy (10.16). This is another difference between the algorithm in Section 10.1.1, and the one to be discussed here.

When the growth of the remaining alternating trees in the present $G_*^1(\pi, \mu)$ is resumed, it is possible that some of the unlabeled pseudonodes get labeled. In this process they may get labeled either as outer or inner nodes. In Section 10.1.1, pseudonodes were always outer nodes, and carried the same label as their base nodes. Here these labels may be quite different. Also, pseudonodes may be labeled as inner nodes.

In the algorithm to be discussed in this section a freshly shrunk pseudonode always gets labeled as an outer node, its label at that time will be the same as that on the base node of the corresponding simple blossom before it was shrunk. So, any inner labeled pseudonode in the current network must have received that label after remaining as an unlabeled node for some time.

THEOREM 10.15 *Let pseudonode p be an outer labeled current node, then the blossom corresponding to p satisfies (10.16).*

Proof   If p is unmatched, it is the root of its alternating tree and (10.16) is satisfied trivially. Suppose p is matched. The root of its alternating tree is unmatched. In this case, the alternating path required in condition (10.16) can be obtained from the predecessor path of p. ∎

Hence in the algorithm to be discussed in this section, the only pseudonodes whose blossoms may not satisfy (10.16) are inner labeled pseudonodes. When there are pseudonodes in $G_*^1(\pi, \mu)$ violating (10.16), there may exist augmenting paths wrt the present matching in $G_*(\pi, \mu)$, and yet there may not exist any augmenting path in $G_*^1(\pi, \mu)$ wrt the current matching. Hence these pseudonodes may prevent us from discovering augmenting paths in $G_*(\pi, \mu)$ through operations on $G_*^1(\pi, \mu)$. The only reason for keeping such pseudonodes is to satisfy the third property in (10.28), if the dual variables μ_σ corresponding to them are strictly positive in the present dual solution. Therefore, whenever there is a pseudonode which is an inner current node associated with $\mu_\sigma = 0$ in the present dual solution, we unshrink that pseudonode into the simple blossom corresponding to it. This *unshrinking operation* is discussed next.

Unshrinking an Inner Labeled Pseudonode Associated with $\mu_\sigma = 0$

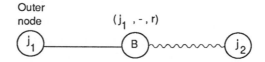

Figure 10.27 Wavy edge is a current matching edge. Inner labeled pseudonode B associated with $\mu_\sigma = 0$ to be unshrunk. Label of B is by its side.

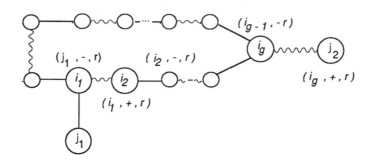

Figure 10.28 Connecting the tree through the alternating path in the simple blossom, after a pseudonode is unshrunk.

This operation is carried out only on pseudonodes which are inner labeled current nodes associated with $\mu_\sigma = 0$. Let B with label $(j_1, -, r)$ be such a pseudonode. So, j_1, is an outer node and $(j_1; B)$ is a current nonmatching edge. By earlier discussion such a pseudonode will always be matched, let $(B; j_2)$ be the current matching edge incident at it. See Figure 10.27. Let $\overline{G}_B = (\overline{\mathcal{N}}_B, \overline{\mathcal{A}}_B)$ be the simple blossom corresponding to B. Since $(j_1; B)$ is a current in-tree nonmatching edge, there must exist an $i_1 \in \overline{\mathcal{N}}_B$ such that $(j_1; i_1)$ is a current equality edge at the stage that pseudonode B was formed. Either i_1 is the base node of B, or there exists an

AP in $\overline{G}_B$ from i_1 beginning with the matching edge incident at i_1, to its base node. Let this path be $i_1, (i_1, i_2), i_2, \ldots, (i_{g-1}, i_g), i_g$, with i_g being the base node of $\overline{G}_B$ (if i_1 is itself the base node, $g = 1$, and this path is the empty path containing no edges). So, $(i_g; j_2)$ is a current matching edge at the stage that pseudonode B was formed. See Figure 10.28.

Unshrinking of B replaces it with the simple blossom $\overline{G}_B$ with all its nodes and edges on it. Each of the current edges of the form $(p; B)$ are replaced by edges of the form $(p; i)$ for $i \in \overline{\mathcal{N}}_B$. After the unshrinking, all the nodes in $\overline{\mathcal{N}}_B$ become current nodes. Update the current matching by replacing the single current matching edge $(B; j_2)$ from it by $(i_g; j_2)$ and all the matching edges within the simple blossom $\overline{G}_B$. The alternating tree that contained B before unshrinking, is connected again by labeling the newly introduced nodes along the path $j_1, (j_1, i_1), i_1, \ldots, i_g, (i_g, j_2), j_2$ alternately as inner and outer nodes as indicated in Figure 10.28.

In the algorithm discussed below, the dual solution change routine has the effect of reducing the value of μ_σ corresponding to pseudonodes which are inner labeled current nodes. So, after each dual solution change step, we always check for possibilities of carrying out unshrinking.

The operation of unshrinking is not needed in the blossom algorithm for the maximum cardinality matching problem, it is needed in the blossom algorithms for minimum cost matching problems. The reader is cautioned not to confuse the two operations of expanding pseudonodes along an alternating path, and unshrinking pseudonodes.

A *dormant period* for a pseudonode B begins when it is just absorbed inside another newly formed pseudonode (i.e., when it just became dormant), and it ends when the pseudonode B becomes a current node again due to all the pseudonodes containing it inside becoming unshrunk.

In the algorithm described below, the matching change phase stops when we reach a current equality subnetwork containing a Hungarian forest. This state is characterized by the following conditions.

> No alternating tree can grow any further (i.e., the list of labeled and unscanned nodes is empty). There are no augmenting paths in the current equality subnetwork wrt the current matching. There are no blossoms which can be shrunk. There are no inner labeled pseudonodes associated with pseudonode price $\mu_\sigma = 0$.

These are the *Hungarian forest conditions* for the blossom algorithm for minimum cost matching problems, and when they are satisfied we say that *the labeling has become hungarian*. The algorithm then goes to the dual change phase, which obtains a new dual feasible solution satisfying the following properties: All the present matching edges, and the present alternating trees lie in the new equality subnetwork, but the Hungarian forest conditions are not satisfied in it anymore. So, in the new equality subnetwork, at least one of tree growth, augmentation, blossom shrinking, or pseudonode unshrinking steps can be carried out when labeling is resumed. And the method continues.

BLOSSOM ALGORITHM FOR THE MINIMUM COST PERFECT MATCHING PROBLEM

Step 1 Initialization   An initial dual feasible solution is $(\pi^0 = (\pi_i^0), \mu^0)$ where $\mu^0 = 0$ and $\pi_i^0 = (1/2)(\min. \{c_{ij} : (i;j) \in \mathcal{A}\})$ for each $i \in \mathcal{N}$. Choose an initial matching (could be empty) in the equality subnetwork $G_*(\pi^0, \mu^0)$.

Step 2 Rooting an Alternating Forest   If there are no unmatched nodes, go to Step 10. Otherwise root an alternating tree at each unmatched node i, by labeling it with $(\emptyset, +, i)$. List now consists of all these root nodes.

Step 3 Select a Node to be Scanned   If list $= \emptyset$, go to Step 8. Otherwise select one node from the list to scan and delete it from the list.

Step 4 Scanning   Let the node to be scanned be the current node i with label $(P(i), \pm, r)$.

Scanning an Outer Node   If i is an outer node, for each $j \neq P(i)$ such that $(i;j)$ is a current equality edge (all these will be nonmatching edges) do the following.

If j is an already labeled outer node associated with a root $\neq r$, an augmenting path has been found, go to Step 5.

If j is an already labeled outer node with the same root r, the alternating tree containing i and j has blossomed, go to Step 6.

If j is an already labeled inner node, continue.

If j is unlabeled, label it with $(i, -, r)$ and include it in the list.

Scanning an Inner Node   If i is an inner node, it cannot be unmatched since all unmatched nodes are outer root nodes, let $(i;j)$ be the current matching edge incident at i.

If j is already labeled, it must be an inner node too. If the root indices of i and j are the same, the tree containing them has blossomed, go to Step 6. If the root indices of i, j are different, an augmenting path has been found, go to Step 5.

If j is unlabeled, label it with $(i, +, r)$ and include it in the list.

Go back to Step 3.

Step 5 Augmentation   We come to this step when scanning has revealed a pair of adjacent current nodes i, j associated with different root nodes $r(i) \neq r(j)$, contained on an augmenting path. Combine the predecessor paths of i, j together with edge $(i;j)$, leading to the path $\mathcal{P}^1$. $\mathcal{P}^1$ is an augmenting path between $r(i)$ and $r(j)$ in the current equality subnetwork. Let t_1, t_2 be the apex nodes of $r(i), r(j)$. Find the corresponding augmenting path $\mathcal{P}$ between t_1 and t_2 in the present equality subnetwork in G by expanding all the pseudonodes on $\mathcal{P}^1$. Rematch using $\mathcal{P}$, and revise all the blossoms along $\mathcal{P}^1$ as discussed above. Chop down the two trees containing nodes on $\mathcal{P}^1$. If there are no unmatched nodes, go to Step 10, otherwise put all the outer current nodes in the list and go back to Step 3.

Step 6 Blossom Shrinking   We come to this step when scanning has revealed a pair of adjacent current nodes i, j associated with the same root node, which are the identifying pair for a simple blossom in the current equality subnetwork. Identify this simple blossom, store all the necessary data on it, and shrink it into a new pseudonode B_p, label B_p, revise the root index of all the current nodes in the tree, revise the labels on the immediate successors of nodes in this simple blossom, revise $\mathcal{N}^1$ and $\mathbf{M}^1$, exactly as in Step 6 of the algorithm in Section 10.1.1. Include B_p in the list and go back to Step 3

Step 7 Pseudonode Unshrinking   Unshrink all pseudonodes that are current inner nodes with the associated pseudonode price $\mu_\sigma = 0$ in the present dual solution. Revise the set of current matching edges, and the set of current nodes accordingly. Include in the list all the new outer labeled nodes in the simple blossoms corresponding to the unshrunk pseudonodes. Repeat this procedure again if necessary, until there are no pseudonodes that are inner labeled current nodes associated with $\mu_\sigma = 0$. Go back to Step 3.

Step 8 Dual Solution Change   We reach this step if we do not yet have a perfect matching, and the Hungarian forest conditions are satisfied. This implies that the present matching is a maximum cardinality matching in the current equality subnetwork. Let (π, μ) be the present dual feasible solution. Compute the following using the convention that the minimum in the empty set is $+\infty$.

$$\delta_1 \;=\; \text{Min.} \; \{c_{ij} - d_{ij}(\pi, \mu) : (i; j) \in \mathcal{A}, i[j] \text{ is inside an outer [an unlabeled] current node}\}$$

$$\delta_2 \;=\; \text{Min.} \; \{\tfrac{1}{2}(c_{ij} - d_{ij}(\pi, \mu)) : (i; j) \in \mathcal{A}, i \text{ and } j \text{ are inside distinct outer current nodes}\}$$

$$\delta_3 \;=\; \text{Min.} \; \{\tfrac{1}{2}\mu_\sigma : \sigma \text{ s. t. } \mathbf{Y}_\sigma \text{ is the set of original nodes inside a current inner labeled pseudonode}\}$$

$$\delta \;=\; \text{Min.} \; \{\delta_1, \delta_2, \delta_3\}$$

If $\delta = +\infty$, go to Step 9. If δ is finite, it will be positive (this is proved below), define the new dual solution to be $\hat{\pi} = (\hat{\pi}_i), \hat{\mu} = (\hat{\mu}_\sigma)$ where

$$\hat{\pi}_i \;=\; \begin{cases} \pi_i + \delta & \text{for all } i \text{ inside outer current nodes} \\ \pi_i - \delta & \text{for all } i \text{ inside inner current nodes} \\ \pi_i & \text{for all } i \text{ inside unlabeled nodes} \end{cases}$$

$$\hat{\mu}_\sigma \;=\; \begin{cases} \mu_\sigma + 2\delta & \text{if } \mathbf{Y}_\sigma \text{ is the set of original nodes in} \\ & \text{a current outer labeled pseudonode} \\ \mu_\sigma - 2\delta & \text{if } \mathbf{Y}_\sigma \text{ is the set of original nodes in} \\ & \text{a current inner labeled pseudonode} \\ \mu_\sigma & \text{otherwise} \end{cases}$$

Find $G_*(\hat{\pi}, \hat{\mu})$. Include all outer current nodes in the list. If $\delta = \delta_3$ go to Step 7. If $\delta < \delta_3$ go to Step 3.

Step 9 Infeasibility   We come to this step if $\delta = +\infty$ in a dual solution change step. In this case there exists no perfect matching in G (this is proved below).

Step 10 Optimality   We come to this step if the matching in the present $G_*(\pi, \mu)$ is a perfect matching, it is a minimum cost perfect matching in G, and the corresponding perfect matching vector is an optimum solution of (10.17), (10.18), or (10.21). Terminate.

Validity of the Algorithm and Its Computational Complexity

THEOREM 10.16 *If the present (π, μ) is dual feasible, in Step 8, δ will either be finite and > 0, or $+\infty$.*

Proof   We execute Step 8 only when the Hungarian forest conditions are satisfied. This and the dual feasibility of (π, μ) implies that each of the sets of which $\delta_1, \delta_2, \delta_3$ are minima, is either empty or consists only of strictly positive entries. So, each of $\delta_1, \delta_2, \delta_3$ is either finite and > 0, or $+\infty$, hence the same thing holds for δ. ■

THEOREM 10.17 *If (π, μ) is dual feasible just before executing Step 8, the vector $(\hat{\pi}, \hat{\mu})$ obtained after completing the execution of Step 8 is also dual feasible.*

Proof   By Theorem 10.16, $\delta > 0$, and since $\delta \leqq \delta_3$ and $\mu \geqq 0$, we have $\hat{\mu} \geqq 0$.

Let $(i; j) \in \mathcal{A}$ with i and j both contained within a current node in the present $G_*^1(\pi, \mu)$ which is a pseudonode. Whether this pseudonode is unlabeled or labeled, it can be verified that in this case $d_{ij}(\hat{\pi}, \hat{\mu}) = d_{ij}(\pi, \mu) \leqq c_{ij}$, the last inequality because (π, μ) is dual feasible. So, $(\hat{\pi}, \hat{\mu})$ satisfies the dual constraint corresponding to this edge $(i; j)$. This also points out that if $d_{ij}(\pi, \mu) = c_{ij}$, then $d_{ij}(\hat{\pi}, \hat{\mu}) = c_{ij}$, so the equality, nonequality status of any edge contained within a pseudonode is unaffected by a dual solution change step.

Now consider an edge $(i; j) \in \mathcal{A}$ with nodes i, j contained within distinct current nodes, i in p, and j in q. Hence, in this case there exists no pseudonode which contains both i and j. If p is outer labeled and q is inner labeled, $d_{ij}(\hat{\pi}, \hat{\mu}) = \hat{\pi}_i + \hat{\pi}_j = \pi_i + \pi_j = d_{ij}(\pi, \mu) \leqq c_{ij}$. If both p, q are outer labeled, $d_{ij}(\hat{\pi}, \hat{\mu}) = \hat{\pi}_i + \hat{\pi}_j = \pi_i + \pi_j + 2\delta = d_{ij}(\pi, \mu) + 2\delta \leqq c_{ij}$ since $\delta \leqq \delta_2$ and (π, μ) is dual feasible. If p is outer labeled and q is unlabeled, $d_{ij}(\hat{\pi}, \hat{\mu}) = \hat{\pi}_i + \hat{\pi}_j = \pi_i + \pi_j + \delta = d_{ij}(\pi, \mu) + \delta \leqq c_{ij}$

since $\delta \leqq \delta_1$ and (π, μ) is dual feasible. If p, q are either inner labeled or unlabeled each, $d_{ij}(\hat{\pi}, \hat{\mu}) = \hat{\pi}_i + \hat{\pi}_j$, and this is either $\pi_i + \pi_j$ or $\pi_i + \pi_j - \delta$ or $\pi_i + \pi_j - 2\delta$, and $\pi_i + \pi_j = d_{ij}(\pi, \mu) \leqq c_{ij}$, this and $\delta > 0$ implies that $d_{ij}(\hat{\pi}, \hat{\mu}) \leqq c_{ij}$. Hence $(\hat{\pi}, \hat{\mu})$ satisfies the dual constraint corresponding to this edge $(i; j)$ in all these cases. This completes the proof that $(\hat{\pi}, \hat{\mu})$ is dual feasible.

Since $d_{ij}(\hat{\pi}, \hat{\mu}) = d_{ij}(\pi, \mu)$ for every $(i; j) \in \mathcal{A}$ with i inside an outer labeled node, and j inside an inner labeled node, we see that the equality, nonequality status of such edges is unaffected by a dual solution change step. ∎

The initial price vector (π^0, μ^0) is clearly dual feasible. By repeated application of the result in Theorem 10.17 after each dual solution change step in the algorithm, we conclude that all price vectors obtained in this algorithm are dual feasible.

THEOREM 10.18 *In this algorithm, all matching edges are always equality edges. Also, after each dual solution change step, in-tree edges remain equality edges.*

Proof The initial matching $\mathbf{M}_0 = \emptyset$, so the property that all matching edges are equality edges is trivially satisfied initially. Every augmenting path discovered in the algorithm is a path in the equality subnetwork at that stage, and new matching edges are only created during the augmentation step (Step 5), so they are all equality edges when they are created. If $(i; j)$ is a matching edge at the beginning of execution of Step 8, we must have, either (i) both i and j are contained inside a current node which is a pseudonode (labeled or unlabeled), or (ii) one of i, j is inside an outer labeled node and the other is within an inner labeled node in the same tree, or (iii) both i, j are inside distinct unlabeled nodes. The arguments in the proof of Theorem 10.17 imply that any such equality edge remains an equality edge wrt the new dual feasible solution obtained at the end of this Step 8. By repeated application of these results we conclude that all matching edges are always equality edges.

Each new in-tree edge created during scanning (Step 4) is always an equality edge when it joins the tree. Each in-tree edge always joins an outer to an inner node, and by the argument in the proof of Theorem 10.17, the equality, nonequality status of such edges remains unaffected during a dual solution change step. By repeated application of this result, we conclude that all in-tree edges are always equality edges. ∎

Thus, after each dual solution change step, the existing alternating forest is contained in the new current equality subnetwork, and its growth can be resumed. If δ was equal to δ_1 in that step, the edges which produced the value for δ_1 lead to new current equality edges, using which the trees containing the labeled nodes on them can be grown further, or a discovery of an augmenting path made. If δ was equal to δ_2 in that step, the edges which produced the value for δ_2 lead to new current equality edges, using which the trees containing the labeled nodes on them can be grown further, or a discovery of a new blossom made. If δ was equal to δ_3 in that step, all the current inner labeled pseudonodes which produced the value for δ_3 will now have zero pseudonode price, and these can now be unshrunk. Thus after an execution of Step 8, the existing alternating forest is one that is no longer hungarian in the new current equality subnetwork, and scanning can be resumed, and at least one tree growth step or Steps 5, or 6, or 7 can be carried out.

Initially $\mu^0 = 0$. A μ_σ can become positive only during Step 8, and there μ_σ is only made positive if the associated set Y_σ is the subset of original nodes corresponding to an outermost pseudonode. In Step 7 pseudonodes are unshrunk, but only if the corresponding pseudonode price μ_σ is 0. Hence the third property in (10.28) holds throughout the algorithm. The first property in (10.28) holds by the nature of the algorithm, and the second holds by the results in Theorem 10.17. Hence all the properties in (10.28) hold throughout the algorithm.

THEOREM 10.19 *Let* M, (π, μ), $G^1_*(\pi, \mu)$, *be the present matching in* G, *present dual feasible solution, present current equality subnetwork respectively, when the algorithm arrives at Step 8 at some stage. If* $\delta = +\infty$ *in that Step 8, there exists no perfect matching in* G, *in fact the present* M *is a maximum cardinality matching in* G.

Proof   Suppose $\delta = +\infty$ in that Step 8. So, all of $\delta_1, \delta_2, \delta_3$ are $+\infty$. $\delta_3 = +\infty$ implies that there are no current nodes which are inner labeled pseudonodes. So, all the inner labeled current nodes are original nodes at this stage, and all pseudonodes that are current nodes are either unlabeled or outer labeled. Define $(\pi(\lambda) = (\pi_i(\lambda)), \mu(\lambda) = (\mu_\sigma(\lambda)))$ by

$$
\pi_i(\lambda) = \begin{cases}
\pi_i + \lambda & \text{for all } i \text{ inside outer current nodes} \\
\pi_i - \lambda & \text{for all } i \text{ inside inner current nodes} \\
\pi_i & \text{for all } i \text{ inside unlabeled nodes}
\end{cases}
$$

$$
\mu_\sigma(\lambda) = \begin{cases}
\mu_\sigma + 2\lambda & \text{if } Y_\sigma \text{ is the set of original nodes in} \\
& \quad \text{a current outer labeled pseudonode} \\
\mu_\sigma & \text{otherwise}
\end{cases}
$$

The dual objective function $W(\pi, \mu)$ in (10.24) at this stage can be written as $\sum(f_p(\pi, \mu) : \text{over current nodes } p)$, where

$$
f_p(\pi, \mu) = \sum(\pi_i : \text{over } i \text{ inside } p) - \sum \quad (\quad (|Y_\sigma| - 1)\mu_\sigma/2 : \text{over } \sigma \text{ s. t. } Y_\sigma
$$
$$
\text{is set of original nodes inside}
$$
$$
\text{a pseudonode within } p)
$$

If p is an unlabeled node it can be verified that $f_p(\pi(\lambda), \mu(\lambda)) = f_p(\pi, \mu)$ for all λ. Also, if p is an inner labeled node, it must be an original node as mentioned above, and hence $f_p(\pi(\lambda), \mu(\lambda)) = f_p(\pi, \mu) - \lambda$.

Suppose p is an outer labeled current node. If p is an original node, then $f_p(\pi(\lambda), \mu(\lambda)) = f_p(\pi, \mu) + \lambda$, because in this case $\pi_p(\lambda) = \pi_p + \lambda$. If p is a pseudonode, let $\mathbf{Y}_{\sigma_1}$ be the set of original nodes inside it. Let $\mathbf{S} = \{\sigma : \mathbf{Y}_\sigma \neq \mathbf{Y}_{\sigma_1}$ is the set of original nodes inside a pseudonode strictly within $p\}$. Then $\mu_\sigma(\lambda) = \mu_\sigma$ for all $\sigma \in \mathbf{S}$ because the associated pseudonode is not a current node. So

$$
\begin{aligned}
f_p(\pi(\lambda), \mu(\lambda)) &= \sum(\pi_i(\lambda) : \text{ over } i \text{ inside } p) \\
&\quad -(|\mathbf{Y}_{\sigma_1}| - 1)\mu_{\sigma_1}(\lambda)/2 - \sum((|\mathbf{Y}_\sigma| - 1)\mu_\sigma/2 : \text{ over } \sigma \in \mathbf{S}) \\
&= \sum(\pi_i : \text{ over } i \text{ inside } p) + |\mathbf{Y}_{\sigma_1}|\lambda \\
&\quad -(|\mathbf{Y}_{\sigma_1}| - 1)\mu_{\sigma_1} + 2\lambda)/2 - \sum((|\mathbf{Y}_\sigma| - 1)\mu_\sigma/2 : \text{ over } \sigma \in \mathbf{S}) \\
&= f_p(\pi, \mu) + \lambda
\end{aligned}
$$

So, if p is an outer labeled current node, whether p is an original or pseudonode, $f_p(\pi(\lambda), \mu(\lambda)) = f_p(\pi, \mu) + \lambda$.

All the nodes in a Hungarian tree in $\mathbf{G}_*^1(\pi, \mu)$ are matched nodes with the exception of the root node which is an outer labeled current unmatched node. Each current matching edge $(p; q)$ in a Hungarian tree contains one inner labeled node and one outer labeled node, and hence from the above facts we have $f_p(\pi(\lambda), \mu(\lambda)) + f_q(\pi(\lambda), \mu(\lambda)) = f_p(\pi, \mu) + f_q(\pi, \mu)$. If r is the root node of a Hungarian tree, since it is an outer labeled node we have $f_r(\pi(\lambda), \mu(\lambda)) = f_r(\pi, \mu) + \lambda$. Since matching edges are node disjoint, all the nonroot nodes in a Hungarian tree can be partitioned into pairs, each pair being the two nodes on a current matching edge in that tree. So, from the above, $\sum(f_p(\pi(\lambda), \mu(\lambda)) : \text{ over current nodes } p \text{ in a Hungarian tree}) = \sum(f_p(\pi, \mu) : \text{ over current nodes } p \text{ in that Hungarian tree}) + \lambda$. Summing over all the Hungarian trees in $\mathbf{G}_*^1(\pi, \mu)$ and over all the unlabeled current nodes, we have

$$
W(\pi(\lambda), \mu(\lambda)) = W(\pi, \mu) + \lambda l
$$

where l is the number of distinct Hungarian trees (same as the number of unmatched nodes at this stage). Since $\delta = +\infty$, $(\pi(\lambda), \mu(\lambda))$ remains dual feasible for all $\lambda \geqq 0$ by Theorem 10.17, and $W(\pi(\lambda), \mu(\lambda)) = W(\pi, \mu) + l\lambda \to +\infty$ as $\lambda \to +\infty$. Thus the dual problem (10.24) is feasible and the dual objective function is unbounded above on it. By the duality theorem of LP the primal problem (10.21) is infeasible in this case. We have discussed earlier that every perfect matching vector is feasible to (10.21), however, since (10.21) is infeasible in this case, there exists no perfect matching in G.

We will now provide an alternate proof of this theorem which does not need the duality theorem of LP. Let $\mathbf{G}^1$ denote the current network consisting of all of the equality and nonequality edges. Suppose $\delta = +\infty$ in Step 8 at some stage. So, all of $\delta_1, \delta_2, \delta_3$ are $+\infty$. $\delta_3 = +\infty$ implies that there are no inner labeled current nodes in $\mathbf{G}^1$ that are pseudonodes. So, at this stage all the current nodes that are pseudonodes are either outer labeled in-tree nodes or matched unlabeled nodes. Throw away (or dismember as discussed in Section 10.1.1) all the unlabeled pseudonodes and replace that part of the network with the corresponding original network, leaving the matching, nonmatching status of each edge unchanged. This changes the current network G^1 into $\overline{G}^1$ say. $\overline{G}^1$ contains all the alternating trees in G^1, but all the nodes in $\overline{G}^1$ outside the alternating trees are matched original nodes. The fact that δ_1, δ_2 are $+\infty$ implies that the present forest is a Hungarian forest in $\overline{G}^1$. Since $\overline{G}^1$ is a current network corresponding to G containing a Hungarian forest, this implies that the present matching (which is not a perfect matching) is a maximum cardinality matching in G. So, there exists no perfect matching in G in this case. ∎

Let $(\pi, \mu), (\hat{\pi}, \hat{\mu})$ be the dual feasible solutions at the beginning and end of a Step 8 in which δ is finite, and l the number of alternating trees in the Hungarian forest (or the number of unmatched nodes in G) at that stage. If M is the matching at that stage, $l = n - 2|M|$. From the arguments in the proof of Theorem 10.19, we have $W(\hat{\pi}, \hat{\mu}) = W(\pi, \mu) + l\delta$. So the dual objective value strictly increases (by δ times the number of unmatched nodes at that stage) each time Step 8 is carried out.

Suppose the algorithm terminates in Step 10. Let $x, (\pi, \mu)$ be the perfect matching vector, dual feasible solution at that stage. $x, (\pi, \mu)$ together satisfy primal feasibility (10.21), dual feasibility (10.24), and the complementary slackness conditions for optimality (10.26), (10.27). So, by duality theory of LP, x is optimal to (10.21), and since it is a perfect matching vector, it is a minimum cost perfect matching vector in G.

We will now analyze the worst case computational complexity of this algorithm. Define an *iteration* in this algorithm to begin just after an augmentation step has been completed (the *first iteration* of course begins at the start), and end when the next augmentation is completed (the final iteration may also end with the infeasibility conclusion). Thus augmentation is carried out exactly once in each iteration, at the end of that iteration (if the problem is infeasible, no augmentation occurs in the final iteration). The algorithm goes through at most $(n/2)$ iterations.

Consider the ρth iteration in this algorithm. A newly formed blossom in this iteration is shrunk into a pseudonode, that is labeled as an outer node when it is formed. This pseudonode can become an unlabeled node only at the end of the iteration, only if it lies on the augmenting path discovered in this iteration. Until an augmentation step occurs after it is formed, this pseudonode either remains an outer current node, or gets absorbed inside another pseudonode which is an outer current node. So, by Theorem 10.6, the blossom shrinking step (Step 6) can occur at most $(n/2)$ times during this iteration.

Only pseudonodes which are inner current nodes are unshrunk during the algorithm. This implies that any pseudonode which is unshrunk in this ρth iteration must be a pseudonode formed in earlier iterations, which is an existing pseudonode at the beginning of this iteration. Again by Theorem 10.6, this implies that the pseudonode unshrinking step (Step 7) can occur at most $(n/2)$ times during this iteration.

Let Γ = set of all original nodes contained within outer labeled current nodes, $\mathbf{I}$ = set of inner labeled current nodes. So, if $i \in \Gamma$, then i is not contained within any current node in $\mathbf{I}$. So, $|\Gamma| + |\mathbf{I}| \stackrel{\leq}{=} n$.

After each execution of Step 8 we either terminate (if $\delta = +\infty$), or δ is finite and equal to δ_1, δ_2, or δ_3.

If δ is finite and $= \delta_2$, at least one new simple blossom will be shrunk into a new outer labeled pseudonode when labeling is resumed. All the pseudonodes which were inner labeled nodes on this simple blossom before its shrinking, will be lost from the set $\mathbf{I}$ because of this shrinking, but each original node contained within any such pseudonode will get included in the set Γ after this blossom shrinking. So, $|\Gamma| + |\mathbf{I}|$ either stays the same or increases in this shrinking step. Also, the maximum number of times that this blossom shrinking step can occur in this iteration is $(n/2)$. Hence, the maximum number of dual solution change steps in this iteration in which δ turns out to be finite and $-\delta_2$ is $(n/2)$.

If δ is finite and $= \delta_3$, at least one pseudonode which is an inner current node will be unshrunk when labeling is resumed. After the unshrinking this pseudonode will be lost from the set $\mathbf{I}$. However, at least 3 nodes on the simple blossom replacing this pseudonode will get labeled, at least 2 of them as inner nodes, and at least one as an outer node. Hence, after the unshrinking, $|\mathbf{I}|$ increases by at least 1, and $|\Gamma|$ increases by at least 1, i.e., $|\Gamma| + |\mathbf{I}|$ increases by at least 2.

If δ is finite and $= \delta_1$, at least one unlabeled current node gets labeled as an inner node immediately after this dual solution change step, as a result $|\mathbf{I}|$ increases by at least one, while $|\Gamma|$ either stays the same (if no more nodes are labeled) or increases (if some more nodes are labeled). Hence, each dual solution change step in this iteration in which δ is finite and equal to δ_1 or δ_3 has the effect of increasing $|\Gamma| + |\mathbf{I}|$ by at least 1, and hence the total number of such steps cannot exceed n in this iteration.

Hence the total number of dual solution change steps in this iteration cannot exceed $(3n/2)$.

The computational effort in a dual solution change step, a pseudonode unshrinking step, a blossom shrinking step, an augmentation step, or a tree growth step, is clearly bounded above by $O(m)$. Hence the total computational effort in one iteration of this algorithm is bounded above by $O(nm)$. Hence the overall computational effort in this algorithm is bounded above by $O(n^2 m)$.

In a straightforward implementation of this algorithm, the values of $d_{ij}(\pi, \mu)$ will be revised for each $(i; j) \in \mathcal{A}$ after each dual solution change step, and this itself involves $O(m)$ effort right away in this step, and makes the overall complexity of this algorithm $O(n^2 m)$. As it was done for the Hungarian method for the bipartite matching problem in Section 3.1, it is possible to implement this algorithm so that for any $(i; j) \in \mathcal{A}$, $d_{ij}(\pi, \mu)$ is computed at most twice per iteration. Because of this, the improved implementation has an overall worst case computational complexity of $O(n^3)$ (Lawler [1976 of Chapter 1]). We discuss this efficient implementation next.

**An $O(n^3)$ Implementation on the Blossom Algorithm
for the Minimum Cost Perfect Matching Problem**

As before, we define an *iteration* in this blossom algorithm to begin either
at initialization, or just after an augmentation step (Step 5) is carried out, and
to finish when the next augmentation step is carried out. So there are at most
$(n/2)$ iterations in the algorithm. In this implementation, we will use a quantity
denoted by $\alpha_t(\pi, \mu)$, and subsets of original edges $\Gamma_t(\pi, \mu), \Delta_{tv}(\pi, \mu)$, defined for
each current node t and current edge $(t; v)$ wrt the present dual solution (π, μ) as
below.

$$
\begin{aligned}
\alpha_t(\pi, \mu) \;&=\; \text{Min}.\{c_{ij} - d_{ij}(\pi, \mu) : (i; j) \in \mathcal{A}, i \text{ inside } t, \\
&\qquad\qquad j \text{ inside an outer current node } \neq t\} \\
\Gamma_t(\pi, \mu) \;&=\; \{(i; j) : (i; j) \in \mathcal{A} \text{ attains the min.} \\
&\qquad\qquad \text{in the definition of } \alpha_t(\pi, \mu)\} \\
\Delta_{tv}(\pi, \mu) \;&=\; \{(i; j) : (i; j) \in \mathcal{A} \text{ attains the min.} \\
&\qquad\qquad \text{in min. } \{c_{pq} - d_{pq}(\pi, \mu) : (p; q) \in \mathcal{A}, p \in t, q \in v\}\}
\end{aligned}
$$

We will now describe the work in one iteration and how it is executed. Let $G^1 =
(\mathcal{N}^1, \mathcal{A}^1), (\overline{\pi}, \overline{\mu})$ be the current network, and the dual feasible solution at the begin-
ning of the iteration. Compute $d_{ij}(\overline{\pi}, \overline{\mu})$ for every $(i; j) \in \mathcal{A}$, and $\alpha_t(\overline{\pi}, \overline{\mu}), \Gamma_t(\overline{\pi}, \overline{\mu})$
for all $t \in \mathcal{N}^1$, and $\Delta_{tv}(\overline{\pi}, \overline{\mu})$ for all $(t; v) \in \mathcal{A}^1$. In subsequent steps in this itera-
tion, we compute $d_{ij}(\pi, \mu)$ only for those original edges $(i; j)$ needed for updating
$\alpha_t(\pi, \mu), \Gamma_t(\pi, \mu), \Delta_{tv}(\pi, \mu)$ at that stage; and we show that this needs to be done
at most once for any edge.

At some stage during this iteration, let $G^1 = (\mathcal{N}^1, \mathcal{A}^1), (\pi, \mu)$ be the current
network, and dual feasible solution. We will now describe how the various tasks in
the algorithm are carried out at this stage.

TO SCAN AN OUTER NODE   For scanning an outer node v, look for unlabeled
current nodes t such that $\alpha_t(\pi, \mu) = 0$ and $\Gamma_t(\pi, \mu)$ contains an edge incident to
an original node inside v; these are all the unlabeled nodes which can be labeled
from v. If $\alpha_v(\pi, \mu) > 0$, labeling the above nodes is all you can do when scanning
v. If, on the other hand, $\alpha_v(\pi, \mu) = 0$, look for current outer labeled nodes t
containing inside it an original node j incident to an edge in $\Gamma_v(\pi, \mu)$. If one of
these outer nodes t has a root index different from that of v, the predecessor paths
of t and v together with the current edge $(t; v)$ is an augmenting path, go to the
augmentation step (Step 5) and after this step terminate this iteration and go to
the next iteration. If no augmenting path is found, using one of the above nodes t
with the same root index as v, a simple blossom is identified in the current equality
subnetwork, go to the blossom shrinking step (Step 6).

UPDATING OF $\alpha_t(\pi,\mu), \Gamma_t(\pi,\mu), \Delta_{tv}(\pi,\mu)$ WHEN AN UNLABELED NODE IS LABELED AS AN INNER NODE   If an unlabeled current node gets labeled as an inner node, there is no change in any of these quantities or sets. Continue.

UPDATING WHEN AN UNLABELED NODE IS LABELED AS AN OUTER NODE   Suppose an unlabeled current node u becomes labeled as an outer node. For every other current node w such that $(u;w) \in \mathcal{A}^1$ do the following: Find $\beta_w = c_{ij} - d_{ij}(\pi,\mu)$ for some $(i;j) \in \Delta_{uw}(\pi,\mu)$. If

$$\alpha_w(\pi,\mu) \quad < \quad \beta_w, \text{ no change in } \alpha_w(\pi,\mu) \text{ or } \Gamma_w(\pi,\mu)$$
$$\alpha_w(\pi,\mu) \quad = \quad \beta_w, \text{ replace } \Gamma_w(\pi,\mu) \text{ by } \Gamma_w(\pi,\mu) \cup \Delta_{uw}(\pi,\mu)$$
$$\alpha_w(\pi,\mu) \quad > \quad \beta_w, \text{ change value of } \alpha_w(\pi,\mu) \text{ to } \beta_w$$
$$\text{and replace } \Gamma_w(\pi,\mu) \text{ by } \Delta_{uw}(\pi,\mu)$$

No change in the other quantities or sets.

UPDATING WHEN A BLOSSOM IS SHRUNK   Suppose a new pseudonode, t_0 has just been formed. So, at this stage t_0 is an outer labeled node. Find out all current nodes t such that $(t;t_0)$ is a current edge. For each such t, do the following: let $\mathbf{D}$ be the set of nodes t' on the simple blossom corresponding to t_0 such that $(t;t')$ was a current edge before t_0 was formed. For each $t' \in \mathbf{D}$ let $\nu_{tt'} = c_{ij} - d_{ij}(\pi,\mu)$ for any $(i;j) \in \Delta_{tt'}(\pi,\mu)$. Let $\beta_{tt_0} = \min.\ \{\nu_{tt'} : t' \in \mathbf{D}\}$. Let $\mathbf{X}_t = $ union of $\Delta_{tt'}(\pi,\mu)$ over $t' \in \mathbf{D}$ satisfying $\nu_{tt'} = \beta_{tt_0}$. If

$$\alpha_t(\pi,\mu) \quad < \quad \beta_{tt_0}, \text{ no change in } \alpha_t(\pi,\mu) \text{ or } \Gamma_t(\pi,\mu)$$
$$\alpha_t(\pi,\mu) \quad = \quad \beta_{tt_0}, \text{ replace } \Gamma_t(\pi,\mu) \text{ by } \Gamma_t(\pi,\mu) \cup \mathbf{X}_t$$
$$\alpha_t(\pi,\mu) \quad > \quad \beta_{tt_0}, \text{ change value of } \alpha_t(\pi,\mu) \text{ to } \beta_{tt_0}$$
$$\text{and replace } \Gamma_t(\pi,\mu) \text{ by } \mathbf{X}_t$$

Also define

$$\Delta_{tt_0}(\pi,\mu) \quad = \quad \mathbf{X}_t$$
$$\alpha_{t_0}(\pi,\mu) \quad = \quad \min.\ \{\beta_{tt_0} : \text{ over outer current nodes}$$
$$t \text{ such that } (t;t_0) \text{ is a current edge } \}$$
$$\Gamma_{t_0}(\pi,\mu) \quad = \quad \text{union of } \mathbf{X}_t \text{ over } t \text{ attaining the minimum}$$
$$\text{in the definition of } \alpha_{t_0}(\pi,\mu)$$

This completes the updating in this case.

UPDATING WHEN A PSEUDONODE IS UNSHRUNK   Suppose a pseudonode v has just been unshrunk. For each node t' on the simple blossom corresponding to v, and current edge $(t;t')$ incident at it after the unshrinking: compute

$\alpha_{t'}(\pi, \mu), \Gamma_{t'}(\pi, \mu), \Delta_{tt'}(\pi, \mu)$ using their definitions. Let $\nu_{tt'} = c_{ij} - d_{ij}(\pi, \mu)$ for any $(i; j) \in \Delta_{tt'}(\pi, \mu)$.

For each node t such that $(t; t')$ is a current edge after the unshrinking for some t' on the simple blossom corresponding to v do the following: Define $\beta_t = \min.$ $\{\nu_{tt'} : t'$ an outer labeled node on the simple blossom corresponding to v, and $(t; t')$ is a current edge after the unshrinking$\}$, $\mathbf{X}_t$ = union of $\Delta_{tt'}(\pi, \mu)$ over all t' attaining the minimum in the definition of β_t. If

$$\begin{aligned} \alpha_t(\pi, \mu) &< \beta_t, \text{ no change in } \alpha_t(\pi, \mu) \text{ or } \Gamma_t(\pi, \mu) \\ \alpha_t(\pi, \mu) &= \beta_t, \text{ replace } \Gamma_t(\pi, \mu) \text{ by } \Gamma_t(\pi, \mu) \cup \mathbf{X}_t \\ \alpha_t(\pi, \mu) &> \beta_t, \text{ change value of } \alpha_t(\pi, \mu) \text{ to } \beta_t \end{aligned}$$

$$\text{and replace } \Gamma_t(\pi, \mu) \text{ by } \mathbf{X}_t$$

TO CARRY OUT A DUAL SOLUTION CHANGE STEP Let (π, μ) denote the present dual feasible solution. Clearly, in this step, $\delta_1 = \min. \{\alpha_t(\pi, \mu) : t$ an unlabeled current node at this stage$\}$, $\delta_2 = \min. \{\frac{1}{2}\alpha_t(\pi, \mu) : t$ an outer labeled current node at this stage$\}$, and δ_3, δ are the same as defined in Step 8 of the algorithm. If the new dual feasible solution is $(\hat{\pi}, \hat{\mu})$, then

$$\alpha_t(\hat{\pi}, \hat{\mu}) = \begin{cases} \alpha_t(\pi, \mu) - \delta & \text{if } t \text{ is an unlabeled current node} \\ \alpha_t(\pi, \mu) - 2\delta & \text{if } t \text{ is an outer labeled current node} \\ \alpha_t(\pi, \mu) & \text{if } t \text{ is an inner labeled current node} \end{cases}$$

and the sets $\Gamma_t(\hat{\pi}, \hat{\mu}) = \Gamma_t(\pi, \mu)$, $\Delta_{tv}(\hat{\pi}, \hat{\mu}) = \Delta_{tv}(\pi, \mu)$.

If carried out this way, it can be verified that for each $(i; j) \in \mathcal{A}$, $d_{ij}(\pi, \mu)$ is computed at most twice during the entire iteration. We will now analyze the computational complexity of this implementation, and show that the effort per iteration is $O(n^2)$. First, consider the computational effort involved in updating after blossom shrinkings during an iteration. Suppose there are s blossom shrinkings in this iteration. None of the resulting pseudonodes are unshrunk during this iteration. Let $g_1, \ldots, g_s$ be the number of nodes on the simple blossoms corresponding to blossoms, in the order in which they are discovered. When the first pseudonode, t_0 say, was formed in this iteration, the updating operations take at most $(n - g_1)g_1$ effort, since there are at most $(n - g_1)g_1$ edges of the form $(t; t')$ discussed above. After this first pseudonode was formed, the number of current nodes is at most $n - g_1 + 1$. So, the effort involved in updating when the second pseudonode is unshrunk in this iteration is at most $g_2(n - g_1 + 1 - g_2)$. After this the number of current nodes is at most $(n - g_1 + 1 - g_2 + 1) = (n - g_1 - g_2 + 2)$. Continuing this argument, we see that the total effort in updating due to blossom shrinking operations in this iteration is at most

$$g_1(n - g_1) + g_2(n - g_1 - g_2 + 1) + \ldots + g_s(n - g_1 - \ldots - g_s + s - 1)$$

$$\stackrel{\leq}{=} n(g_1 + g_2 + \ldots + g_s) \stackrel{\leq}{=} n^2$$

Now consider the computational effort involved in updating after pseudonode unshrinking operations in this iteration. Let u be the number of pseudonodes unshrunk in this iteration, let $a_1, \ldots, a_u$ be the number of nodes on the simple blossoms corresponding to these pseudonodes in the order in which they are unshrunk. Let b_1 be the number of current nodes just before the first pseudonode is unshrunk in this iteration.

Updating after the first unshrinking operation clearly requires at most $b_1 a_1$ effort. After this the number of current nodes is $a_1 + b_1 - 1$. So, the number of current nodes just before the second unshrinking in this iteration is $\leqq a_1 + b_1 - 1$, and by the same argument, the effort involved in updating after it is at most $a_2(a_1 + b_1 - 1)$. Similarly, the number of current nodes just before the third unshrinking in this iteration is $\leqq (a_1 + a_2 + b_1 - 2)$. Continuing the argument in this way, we see that the total effort in updating due to pseudonode unshrinking operations in this iteration is at most

$$a_1 b_1 + a_2(a_1 + b_1 - 1) + a_3(a_1 + a_2 + b_1 - 2) + \ldots +$$

$$a_u(a_1 + \ldots + a_{u-1} + b_1 - (u - 1))$$

But $b_1 + a_1 + \ldots + a_u - (u - 1) \leqq n$. So, the above sum in reverse order is $\leqq (n - a_u)a_u + (n - a_u - a_{u-1} + 1)a_{u-1} + \ldots \leqq O(n^2)$, as before.

The computation of $d_{ij}(\pi, \mu)$ for all $(i; j) \in \mathcal{A}$ at the beginning of this iteration takes $O(m)$ effort. The total effort for tree growth, augmentation, etc., in this iteration is clearly $\leqq O(n^2)$. Summing up, we see that the overall effort in this iteration is $O(n^2)$. Since there are at most $(n/2)$ iterations, the overall effort in this implementation is at most $O(n^3)$.

10.1.3   The Minimum Cost Matching Problem

Here we consider the problem of finding a minimum cost matching in the undirected network $G = (\mathcal{N}, \mathcal{A}, c = (c_{ij}))$ with c as the vector of edge cost coefficients, $|\mathcal{N}| = n$, and $|\mathcal{A}| = m$, without any constraint on the matchings cardinality. This problem is the same as in (10.17), (10.18) except that the constraint $x(i) = 1$ is to be replaced by $x(i) \leqq 1$ for all $i \in \mathcal{N}$; exactly the same change has to be made in (10.21) to get the corresponding LP formulation replacing the integrality requirements on the variables by the same blossom constraints. This is the LP

$$
\begin{aligned}
\text{Minimize} \quad z(x) \ &= \ \sum (c_{ij} x_{ij} : \text{ over } (i; j) \in \mathcal{A}) \\
\text{subject to} \quad x(i) \ &\leqq \ 1, \text{ for all } i \in \mathcal{N} \\
\mathbf{Y}_\sigma^-(x) \ &\leqq \ (|\mathbf{Y}_\sigma| - 1)/2, \sigma = 1 \text{ to } L \\
x_{ij} \ &\geqq \ 0 \text{ for all } (i; j) \in \mathcal{A}
\end{aligned}
\qquad (10.29)
$$

where $\mathbf{Y}_\sigma, \sigma = 1$ to L are all the subsets of $\mathcal{N}$ of odd cardinality $\geqq 3$. Define the dual solution $\pi = (\pi_i : i \in \mathcal{N}), \mu = (\mu_\sigma : \sigma = 1$ to $L)$, and the quantity $\mu^-(i;j)$ for each $(i;j) \in \mathcal{A}$ as in (10.22), exactly as in Section 10.1.2. The dual problem is

$$\text{Maximize} \quad W(\pi, \mu) \;=\; -\sum_{i \in \mathcal{N}} \pi_i - \sum_{\sigma=1}^{L} (|\mathbf{Y}_\sigma| - 1)(\mu_\sigma)/2$$

$$\text{subject to} \quad d_{ij}(\pi, \mu) \;\leqq\; c_{ij}, \text{ for each } (i;j) \in \mathcal{A} \tag{10.30}$$

$$\pi, \mu \;\geqq\; 0$$

where

$$d_{ij}(\pi, \mu) = -\pi_i - \pi_j - \mu^-(i;j) \tag{10.31}$$

This formula for $d_{ij}(\pi, \mu)$ is different from that in (10.23) of Section 10.1.2, because of the difference in the degree constraints here. Also, here π is restricted to be $\geqq 0$ for the same reason. Given a dual feasible solution (π, μ), the set of equality edges wrt it is $\mathcal{A}_*(\pi, \mu)$ defined below in (10.32), and $G_*(\pi, \mu) = (\mathcal{N}, \mathcal{A}_*(\pi, \mu))$ is the equality subnetwork wrt (π, μ). The complementary slackness conditions for optimality in this primal, dual pair of problems are (10.33), (10.34), (10.35) given below.

$$\mathcal{A}_*(\pi, \mu) \;=\; \{(i;j) : (i;j) \in \mathcal{A}, \text{ and } d_{ij}(\pi, \mu) = c_{ij}\} \tag{10.32}$$

$$x_{ij} > 0 \text{ implies } d_{ij}(\pi, \mu) = c_{ij}, \text{ for each } (i;j) \in \mathcal{A} \tag{10.33}$$

$$\mu_\sigma > 0 \text{ implies } \mathbf{Y}_\sigma^-(x) = (|\mathbf{Y}_\sigma| - 1)/2, \text{ for } \sigma = 1 \text{ to } L \tag{10.34}$$

$$\pi_i > 0 \text{ implies } x(i) = 1 \tag{10.35}$$

The blossom algorithm of this section maintains $x, (\pi, \mu)$ satisfying dual feasibility (10.30), and (10.33), (10.34) throughout. In this problem, an unmatched node i does not violate primal feasibility, but if $\pi_i > 0$ for an unmatched node i, then the complementary slackness condition (10.35) is violated. For this reason, in this problem we classify unmatched nodes into two classes: an unmatched node i is said to be an

> *exposed node*   if $\pi_i > 0$ in the present dual solution
>
> *nonexposed node*   otherwise

Since nonexposed nodes do not violate any of the feasibility or optimality (complementary slackness) conditions, the algorithm tries to reduce the number of exposed nodes. This is done by alternating between two phases, just as in the blossom algorithm of Section 10.1.2.

In the *matching change phase*, the dual solution (π, μ) is held constant and the algorithm tries to find alternating paths wrt the present matching, with an exposed node at at least one end of it. This is done by growing an alternating forest with one alternating tree rooted at each exposed node in the equality subnetwork $G_*(\pi, \mu)$. If an augmenting path, i.e., an alternating path between two exposed nodes, is found, we rematch using it, the trees containing nodes on that path are chopped down and the growth of the remaining trees, if any, is resumed. Each augmentation step reduces the number of exposed nodes by 2.

Even if an alternating path joining a matched node t associated with $\pi_t = 0$ beginning with the matching edge incident at it, to an exposed node i is found, we rematch using it. This operation does not change the cardinality of the matching, but converts i into a matched node and t into an unmatched but nonexposed node (since $\pi_t = 0$) and thus reduces the number of exposed nodes by 1.

When such matching changes are not possible any more and a maximum cardinality matching in $G_*(\pi, \mu)$ is obtained, and there are still exposed nodes left, the algorithm moves to a *dual solution change step*. After this step, the process is repeated in the new equality subnetwork.

In all dual feasible solutions obtained during the algorithm, the values of all the original node prices π_i associated with exposed nodes i at that stage will be the same.

If $c \geqq 0$ the empty matching is clearly a minimum cost matching. So, we assume that $c_{ij} < 0$ for at least one $(i; j) \in \mathcal{A}$.

Suppose $x, (\pi, \mu)$ are the present solutions. The matching change phase of the algorithm stops now if the present current equality subnetwork $G_*^1(\pi, \mu)$ contains a Hungarian forest with a Hungarian tree rooted at each exposed node. This state is characterized by the following conditions.

> No alternating tree can grow any further (i.e., the list of labeled and unscanned nodes is empty). There are no augmenting paths in the current equality subnetwork, no blossoms which can be shrunk, and no inner labeled pseudonodes associated with pseudonode price of $\mu_\sigma = 0$. And $\pi_i > 0$ for all original nodes i contained within outer labeled nodes (this guarantees that there are no alternating paths between an exposed node i and a matched node t with $\pi_t = 0$).

When all these conditions are satisfied, we say that the *labeling has become hungarian*, and the alternating forest has become a *Hungarian forest*. Then the algorithm moves to the dual solution change phase.

BLOSSOM ALGORITHM FOR THE MINIMUM COST MATCHING PROBLEM

Step 1 Initialization   Define the initial dual feasible solution to be $(\pi^0 = (\pi_i^0), \mu^0 = (\mu_\sigma^0))$ where $\mu^0 = 0$ and $\pi_i^0 = (1/2)\max.\{0, -c_{pq}, \text{ for all } (p; q) \in \mathcal{A}\}$, for all $i \in \mathcal{N}$. Choose $x^0 = 0$ and the corresponding initial matching $M_0 = \emptyset$.

Step 2 Rooting an Alternating Forest   If there are no exposed nodes, go to Step 9. Otherwise root an alternating tree at each exposed node i, by labeling it with $(\emptyset, +, i)$. List now consists of all these root nodes.

Step 3   Same as Step 3 in the algorithm of Section 10.1.2

Step 4 Scanning   Let the node to be scanned be the current node i with label $(\mathrm{P}(i), \pm, r)$.

Scanning an Outer Node   If i is an outer node, for each $j \neq \mathrm{P}(i)$ such that $(i; j)$ is a current equality edge do the following.

If j is an already labeled outer node associated with a root $\neq r$, an augmenting path joining two exposed nodes has been found, go to Step 5.

If j is an already labeled outer node with the same root r, and there is an original node t inside t such that $\pi_t = 0$, an alternating path joining t to the apex node of r, beginning with the matching edge incident at t has been found, go to Step 5.3.

If j is an already labeled outer node with the same root r, and $\pi_t > 0$ for all original nodes t inside j, the alternating tree containing i and j has blossomed, go to Step 6.

If j is an already labeled inner node, continue.

If j is unlabeled, label it with $(i, -, r)$ and include it in the list.

Scanning an Inner Node   If i is an inner node, and it is an unmatched nonexposed node, go to Step 5.2 (Augmentation 2 step).

If i is a matched inner node, let $(i; j)$ be the current matching edge incident at it. If j is already labeled, it must be an inner node too. If the root indices of i and j are the same, the tree containing them has blossomed, go to Step 6. If the root indices of i, j are different, an augmenting path has been found, go to Step 5.

If j is unlabeled, label it with $(i, +, r)$ and include it in the list.

Go back to Step 3.

Step 5   Same as Step 5 in the algorithm of Section 10.1.2.

Step 5.2 Augmentation 2 Step   We come to this step when scanning has revealed an inner labeled node i which is an unmatched nonexposed node. Let $\mathcal{P}^1$ be the predecessor path of i, it is an augmenting path containing one exposed and one nonexposed unmatched nodes. Find the corresponding augmenting path $\mathcal{P}$ in G between i and the apex of the root node, by expanding the pseudonodes on $\mathcal{P}^1$. Rematch using $\mathcal{P}$. Revise all the blossoms along $\mathcal{P}^1$, as discussed in Section 10.1.2. Chop down the tree containing i. If there are no exposed nodes, go to Step 9. Otherwise include all the outer labeled current nodes in the list and go to Step 3.

Step 5.3 Rematching an Alternating Path between an Exposed and a Matched Node with 0 Node Price   We come to this step when scanning has revealed an outer node i joined to another outer node in the same tree by a current equality edge, such that j is matched and contains inside it an original node t with $\pi_t = 0$. Let the root of this tree be r. Let $\mathcal{P}^1$ be the path consisting of the edge $(j; i)$ and the predecessor path of i. By expanding the pseudonodes on $\mathcal{P}^1$, find the corresponding alternating path $\mathcal{P}$ in G beginning with the matching edge incident at t to the apex node of r. Rematch using $\mathcal{P}$. This makes t unmatched, but since $\pi_t = 0$, it remains nonexposed. The apex of r is now matched. Revise all the blossoms along $\mathcal{P}^1$ as discussed in Section 10.1.2. Chop down the tree containing r. If there are no exposed nodes, go to Step 9. Otherwise, include all the outer nodes in the list and go back to Step 3.

Steps 6, 7   Same as Steps 6, 7 in the algorithm of Section 10.1.2.

Step 8 Dual Solution Change Step   We reach this step only if we still have exposed nodes in the present equality subnetwork and the Hungarian forest conditions are satisfied. This implies that the present matching is a maximum cardinality matching in the current equality subnetwork. Let (π, μ) be the present dual feasible solution. Compute the following using the convention that the minimum in the empty set is $+\infty$.

$$
\begin{aligned}
\delta_1 &= \text{Min. } \{c_{ij} - d_{ij}(\pi, \mu) : (i; j) \in \mathcal{A}, i[j] \text{ is inside an outer [an unlabeled] current node}\} \\
\delta_2 &= \text{Min. } \{\tfrac{1}{2}(c_{ij} - d_{ij}(\pi, \mu)) : (i; j) \in \mathcal{A}, i \text{ and } j \text{ are inside distinct outer current nodes}\} \\
\delta_3 &= \text{Min. } \{\tfrac{1}{2}\mu_\sigma : \sigma \text{ s. t. } \mathbf{Y}_\sigma \text{ is the set of original nodes inside a current inner labeled pseudonode}\} \\
\delta_4 &= \text{Min. } \{\pi_i : i \text{ is inside an outer current node}\} \\
\delta &= \text{Min. } \{\delta_1, \delta_2, \delta_3, \delta_4\}
\end{aligned}
$$

Define the new dual solution to be $\hat{\pi} = (\hat{\pi}_i), \hat{\mu} = (\hat{\mu}_\sigma)$ where

$$
\hat{\pi}_i = \begin{cases} \pi_i - \delta & \text{for all } i \text{ inside outer current nodes} \\ \pi_i + \delta & \text{for all } i \text{ inside inner current nodes} \\ \pi_i & \text{for all } i \text{ inside unlabeled nodes} \end{cases}
$$

$$
\hat{\mu}_\sigma = \begin{cases} \mu_\sigma + 2\delta & \text{if } \mathbf{Y}_\sigma \text{ is the set of original nodes in a current outer labeled pseudonode} \\ \mu_\sigma - 2\delta & \text{if } \mathbf{Y}_\sigma \text{ is the set of original nodes in a current inner labeled pseudonode} \\ \mu_\sigma & \text{otherwise} \end{cases}
$$

Find $G_*(\hat{\pi}, \hat{\mu})$. If $\delta = \delta_4, \hat{\pi}_i = 0$ for all unmatched nodes i, go to Step 9. If $\delta < \delta_4$ but $=\delta_3$ go to Step 7. If $\delta < \delta_4$ and $< \delta_3$ include all outer current nodes in the list and go to Step 3.

Step 9 Optimality We only come to this step when we have a matching M, the associated matching vector x, and dual feasible solution (π, μ), and there are no exposed nodes. M is a minimum cost matching in G. Terminate.

Discussion

By the manner in which scanning is carried out, it can be verified that if p is a root node, it cannot contain inside it an original node t with $\pi_t = 0$. So, whenever we come to Step 5.3, the node j will be a matched node (since it is an outer node and not a root node). Also, by this property and the Hungarian forest conditions, when the algorithm comes to Step 8, then δ_4 is finite and strictly positive. It can be verified that δ will always be finite and positive whenever Step 8 is carried out, and that (π, μ) remains dual feasible throughout the algorithm. Also, verify that all matching edges are all equality edges, and in-tree edges remain equality edges after each dual solution change step. The equality, nonequality status of any edge contained within a pseudonode, or any edge $(i; j) \in \mathcal{A}$ with i inside an outer node, and j inside an inner node, remains unchanged during a dual solution change step. The pair $x, (\pi, \mu)$ always satisfy (10.30), (10.33), (10.34).

By arguments similar to those in the proofs in Section 10.1.2, it can be shown that if $(\pi, \mu), (\hat{\pi}, \hat{\mu})$ are the present and new dual solutions in a Step 8, then $W(\hat{\pi}, \hat{\mu}) = W(\pi, \mu) + l\delta$, where l is the number of alternating trees in the current equality subnetwork at that stage (same as the number of exposed nodes at the beginning of this step). Thus each execution of Step 8 increases the dual objective value strictly.

Exposed nodes are always outer nodes. Every time Step 8 is carried out, the same δ is subtracted from the original node price of exposed nodes, while it may be subtracted or added to the original node price of other nodes. The initial node price is the same for all the nodes. Hence, at every stage, the original node prices of all the exposed nodes are the same, and their value is $\overset{\leq}{=}$ the original node price of any other node.

Finally, when the algorithm arrives at Step 9, there are no exposed nodes, and the complementary slackness condition (10.35) is also satisfied. Since primal feasibility, dual feasibility, and the complementary slackness conditions for optimality are all satisfied at this stage, and the x-vector is an integer vector, it is a minimum cost matching vector in G.

Steps 5 or 5.2 or 5.3 need to be carried at most n times in the algorithm, since each of them reduces the number of exposed nodes by at least 1. Define an *iteration* in this algorithm to begin just after one of Steps 5, 5.2, or 5.3 is completed (the first iteration of course begins at the start) and end when one of these matching change steps is completed next. So, the algorithm goes through at most n iterations.

Using arguments similar to those in Section 10.1.2, it can be verified that the computational effort in one iteration is bounded above by $O(nm)$. Hence, the overall computational effort in the algorithm is bounded above by $O(n^2 m)$, which can be reduced to $O(n^3)$ by implementing it efficiently as discussed in Section 10.1.2.

10.1.4   The Convex Hulls of Matching and Perfect Matching Vectors

Let $\mathbf{K}^{\leq}$, $\mathbf{K}^{=}$ denote the convex hulls of matching, perfect matching vectors respectively in $G = (\mathcal{N}, \mathcal{A})$. Irrespective of what the edge cost vector c may be, we have shown in Section 10.1.3, that the LP (10.29) has an optimum solution that is a matching vector in G, and that the blossom algorithm discussed there will find it. This implies that all the extreme points of the set of feasible solutions of (10.29) are matching vectors in G. Also, clearly every matching vector in G is an extreme point of this set. Hence, $\mathbf{K}^{\leq}$ is the set of feasible solutions of (10.29), i.e., (10.29) provides a linear constraint representation of $\mathbf{K}^{\leq}$.

Using a similar argument, we conclude that (10.21) provides a linear constraint representation of $\mathbf{K}^{=}$.

Here we proved that systems (10.21), (10.29) provide linear constraint representations of $\mathbf{K}^{=}$, $\mathbf{K}^{\leq}$ respectively, using the fact that the blossoms algorithms discussed earlier, based on these systems, find minimum cost extreme points of these polytopes, for all cost vectors c. Direct polyhedral proofs of these results without any recourse to algorithms, have been found recently, see Schrijver [1983].

10.1.5   Other Algorithms for Minimum Cost Matching Problems

An algorithm that begins with a perfect matching in G and finds a minimum cost perfect matching by moving through a sequence of perfect matchings of decreasing cost has been developed by Cunningham and Marsh [1978]. This algorithm has all the features of the primal simplex algorithm for solving LPs, and hence can be thought of as a primal algorithm for the minimum cost perfect matching problem. Its worst case computational complexity is also $O(n^2 m)$, but it is not expected to be computationally competitive with versions of the primal-dual blossom algorithms discussed above. Cunningham and Marsh [1978] also developed methods for doing sensitivity analysis in the minimum cost perfect matching problem using their primal algorithm. Another primal approach for the minimum cost perfect matching problem has been developed based on negative cost alternating cycles. This approach is also initiated with any perfect matching (with artificial edges, if necessary) and it improves this matching successively over negative cost alternating cycles, see Derigs [1986]. Another approach for the minimum cost perfect matching problem is to start with any matching ($M = \emptyset$ for example) satisfying the complementary slackness optimality conditions with a dual feasible solution, and augment it successively using shortest augmenting paths until a perfect matching is obtained, see Derigs [1981, 1991].

10.1.6   Integer Valued Optimum Dual Solutions

Here we will prove that (10.30), the dual of the minimum cost matching problem in $G = (\mathcal{N}, \mathcal{A}, c)$ has an integer optimum solution (π, μ) whenever c is an integer vector.

THEOREM 10.20 *If c is an integer vector, and $\tilde{x}, (\tilde{\pi}, \tilde{\mu})$ are the minimum cost matching vector, and optimum dual solution, obtained by the blossom algorithm of Section 10.1.3 when it is applied on G, then $\tilde{\mu}$ and $2\tilde{\pi}$ are integer vectors.*

Proof   Assume that c is an integer vector. Let (π, μ) be a dual feasible solution obtained during the course of applying the blossom algorithm of Section 10.1.3 on G. We will actually show that (π, μ) satisfy the following properties.

1. μ and 2π are integer vectors.

2.   Let $\mathbf{T} =$ set of all original nodes inside a current in-tree node. Then $\pi_i - \lfloor \pi_i \rfloor$ = the fractional part in π_i has the same value for all $i \in \mathbf{T}$, and this value is either 0 or 1/2.

By the definition of the initial dual feasible solution (π^0, μ^0) in Step 1, it is clear that both properties 1 and 2 hold for it.

Suppose properties 1 and 2 hold for a dual feasible solution (π, μ) obtained at some stage in the algorithm. Since μ is an integer vector, if $(i; j) \in \mathcal{A}$ is an equality edge at this stage such that i is inside a labeled node, and j is inside an unlabeled node, then $d_{ij}(\pi, \mu)$ can only be equal to c_{ij}, an integer, if $\pi_i - \lfloor \pi_i \rfloor = \pi_j - \lfloor \pi_j \rfloor$. This implies that property 2 continues to hold after tree growth steps occur. Properties 1 and 2 imply that in the ensuing dual solution change step, the fractional part in δ_2, if any, is 1/2, and that 2δ is integer. From the dual solution updating formula, these facts imply that properties 1, 2 continue to hold in the new dual feasible solution after a dual solution change step.

Using this repeatedly, we verify that properties 1, 2 hold throughout the algorithm. The theorem follows directly. ∎

THEOREM 10.21 *If $c = (c_{ij})$ is a vector in which all the entries are 0 or -1, (10.30) has an optimum solution $(\pi, \mu = 0)$ which is integer.*

Proof   Let $c_{ij} = 0$ or -1 for all $(i; j) \in \mathcal{A}$. Let $\mathcal{A}_2 = \{(i; j) : c_{ij} = -1\}$. Finding a minimum cost matching in $G = (\mathcal{N}, \mathcal{A}, c)$ in this case is the same problem as finding a maximum cardinality matching in $G_2 = (\mathcal{N}, \mathcal{A}_2)$. Let M_2 be a maximum cardinality matching in G_2, and let x^2 be the matching vector in G corresponding to the matching M_2. Define the 0-1 vector $\pi^2 = (\pi_i^2)$ with π_i^2 equal to 1 for exactly one node on each edge in M_2, and 0 at all other nodes. Verify that $(\pi^2, \mu^2 = 0)$ is feasible to (10.30) in this case, and that $x^2, (\pi^2, \mu^2)$ satisfy the complementary slackness optimality conditions (10.33), (10.34), (10.35) for the primal, dual pair (10.29), (10.30) in this case. So, (π^2, μ^2) is an optimum solution for (10.30) in this case, and it is integer, proving the theorem. ∎

THEOREM 10.22 *Suppose c is an integer vector. Then (10.30) has an integer optimum solution (π, μ).*

Proof   Let $(\tilde{\pi} = (\tilde{\pi}_i), \tilde{\mu})$ be the optimum dual solution obtained when the minimum cost matching problem in G is solved by the blossom algorithm of Section 10.1.3. By Theorem 10.20, $2\tilde{\pi}$ is an integer vector. If $\tilde{\pi}$ is an integer vector, we are done. So, assume that $\tilde{\pi}$ is not an integer vector. By Theorem 10.20, the fractional part of $\tilde{\pi}_i$ is either 0 or 1/2 for each $i \subset \mathcal{N}$. Let $\mathbf{T} = \{i:$ the fractional part of $\tilde{\pi}_i$ is $1/2\}$. For $(i;j) \in \mathcal{A}$ define $c_{ij}' = -1$ if both i and j are in $\mathbf{T}$, 0 otherwise. Let $(\pi', \mu' = 0)$ be an integer dual optimum solution for the minimum cost matching problem in $(\mathcal{N}, \mathcal{A}, c')$ constructed as in the proof of Theorem 10.21. Define $(\pi^* = (\pi_i^* = \lfloor \tilde{\pi}_i \rfloor + \pi_i'), \mu^* = \tilde{\mu})$. Clearly (π^*, μ^*) is an integer feasible solution of (10.30). Also

$$
\begin{aligned}
W(\pi^*, \mu^*) \; &\overset{\geq}{=} \; -\frac{1}{2}|\mathbf{T}| + \left(-\sum_{i \in \mathcal{N}} \lfloor \tilde{\pi}_i \rfloor - \sum_{\sigma=1}^{L} (|\mathbf{Y}_\sigma| - 1)\tilde{\mu}_\sigma/2\right) \\
&= \; -\sum_{i \in \mathcal{N}} \tilde{\pi}_i - \sum_{\sigma=1}^{L} (|\mathbf{Y}_\sigma| - 1)\tilde{\mu}_\sigma/2) \\
&= \; W(\tilde{\pi}, \tilde{\mu})
\end{aligned}
$$

Since, (π^*, μ^*) is feasible to (10.30) and $(\tilde{\pi}, \tilde{\mu})$ is optimal to it, $W(\pi^*, \mu^*) \overset{\geq}{=} W(\tilde{\pi}, \tilde{\mu})$ implies that (π^*, μ^*) is an alternate optimum solution for (10.30). Also, (π^*, μ^*) is an integer vector, so it is an integer optimum solution of (10.30). ∎

Exercises

10.1 Prove that (10.24), the dual of the minimum cost perfect matching problem, always has an integer optimum solution if c is an integer vector, and it has an optimum solution. Develop an algorithm for finding such an integer optimum solution to this problem.

10.2 Show that the various matchings obtained during the successive iterations of the blossom algorithm for the minimum cost matching problem of Section 10.1.3, are minimum cost matchings among all matchings in G having the same cardinality as themselves.

10.3 Let ν be a positive integer. Show that the convex hull of all matching vectors in G whose cardinality is ν, is the set of feasible solutions of the system of constraints in (10.29) and the additional constraint $\sum(x_{ij} : \text{over } (i;j) \in \mathcal{A}) = \nu$.

Comment 10.2   The proof of the integer property in this section has been communicated to me by S. N. Kabadi. For a different proof of this result, see Cunningham and Marsh [1978].

10.2   1-Edge Covering Problems

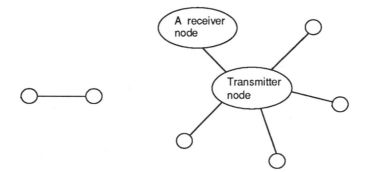

Figure 10.29 Two stars.

Figure 10.30 This network is not a star, as it has two nodes with degree $\geqq 2$.

Let $G = (\mathcal{N}, \mathcal{A})$ be an undirected connected network with $|\mathcal{N}| = n, |\mathcal{A}| = m$. In this section, we discuss algorithms for minimum cardinality and minimum cost edge covers in G. Define a *star* to be a tree spanning at least two nodes, in which at most one node has degree $\geqq 2$. See Figures 10.29, 10.30. In a star, if there is a node of degree $\geqq 2$, it is known as the *transmitter node*; and all other nodes are known as *receiver nodes*.

THEOREM 10.23 *Every minimum cardinality edge cover in G is a star forest.*

Proof Let **E** be a minimum cardinality edge cover in G. **E** cannot contain a cycle, because if it did, an edge from that cycle can be deleted, leading to an edge cover of lower cardinality, a contradiction. So **E** is a forest. If **E** is not a star forest, it must contain a path $\mathcal{P}$ consisting of three or more edges. On this path $\mathcal{P}$ there are two adjacent nodes both of which are of degree ≥ 2 in $\mathbf{E}$. Deleting the edges joining these two nodes in $\mathcal{P}$, leads to an edge cover of lower cardinality than $\mathbf{E}$, a contradiction. Hence, $\mathbf{E}$ is a star forest. ∎

THEOREM 10.24 (i)   *Let $\bar{\mathbf{M}}$ be a maximum cardinality matching in G; and $\bar{\mathbf{E}}$ a subset containing all the edges in $\bar{\mathbf{M}}$, and one arbitrary edge incident at each unmatched node. Then $\bar{\mathbf{E}}$ is a minimum cardinality edge cover in G.*

(ii) *Let $\hat{\mathbf{E}}$ be a minimum cardinality edge cover in G; and $\hat{\mathbf{M}}$ a subset containing exactly one edge from each connected component of $(\mathcal{N}, \hat{\mathbf{E}})$. Then $\hat{\mathbf{M}}$ is a maximum cardinality matching in G.*

Proof   Let $|\bar{\mathbf{M}}| = r$. So, the number of unmatched nodes in G wrt $\bar{\mathbf{M}}$ is $n - 2r$. So, by construction, $|\bar{\mathbf{E}}| = r + (n - 2r) = n - r = n - |\bar{\mathbf{M}}|$, and $\bar{\mathbf{E}}$ is an edge cover.

By hypothesis, $\hat{\mathbf{E}}$ is a minimum cardinality edge cover. So, by Theorem 10.23, $\hat{\mathbf{E}}$ is a star forest, hence each connected component of $(\mathcal{N}, \hat{\mathbf{E}})$ is a star tree. The subset $\hat{\mathbf{M}}$ defined in (ii) is clearly a matching and $|\hat{\mathbf{M}}|$ = number of connected components in $(\mathcal{N}, \hat{\mathbf{E}})$. Take a connected component of $(\mathcal{N}, \hat{\mathbf{E}}), (\mathbf{V}, \mathbf{T})$ say. So, $(\mathbf{V}, \mathbf{T})$ is a star tree. $\hat{\mathbf{M}}$ contains exactly one edge, e, say from $\mathbf{T}$. So, if $(\mathbf{V}, \mathbf{T})$ has a transmitter node, it must lie on e. So, the number of nodes in $(\mathbf{V}, \mathbf{T})$ not on e is exactly $|\mathbf{V}| - 2$, and $|\mathbf{T}\backslash\{e\}| = |\mathbf{V}| - 2$. Summing up the corresponding result over all the connected components of $(\mathcal{N}, \hat{\mathbf{E}})$ and using the above result leads to $|\hat{\mathbf{E}}| = |\hat{\mathbf{M}}| + (n - 2|\hat{\mathbf{M}}|) = n - |\hat{\mathbf{M}}| \geq n - |\bar{\mathbf{M}}|$ (since $\bar{\mathbf{M}}$ is a maximum cardinality matching $|\bar{\mathbf{M}}| \geq |\hat{\mathbf{M}}|$) $= |\bar{\mathbf{E}}|$, by previous results. Since $\hat{\mathbf{E}}$ is a minimum cardinality edge cover and $|\hat{\mathbf{E}}| \geq |\bar{\mathbf{E}}|$, $\bar{\mathbf{E}}$ must also be a minimum cardinality edge cover. Also, from the above, we have $|\hat{\mathbf{M}}| \leq |\bar{\mathbf{M}}|$, and since $\bar{\mathbf{M}}$ is a maximum cardinality matching, $\hat{\mathbf{M}}$ must also be a maximum cardinality matching in G. ∎

As an example, consider the network in Figure 10.18. $\bar{\mathbf{M}} = \{(1; 2),(4; 7),(6; 10),(5; 12);(11; 13)\}$ is a maximum cardinality matching in this network. Node 3 is the only unmatched node wrt $\bar{\mathbf{M}}$. So $\bar{\mathbf{M}} \cup \{(3; 2)\}$ is a minimum cardinality edge cover in this network.

Thus a minimum cardinality edge cover in G can be found by finding a maximum cardinality matching in G first, and then using the procedure described in (i) of Theorem 10.24. The computational effort required by this method is bounded above by $0(n^3)$.

A minimum cardinality edge cover in G can also be found directly by starting with some edge cover (for example, $\mathcal{A}$ can be used as the initial edge cover in G), generating *reducing paths* (these are analogous to augmenting paths in the maximum cardinality matching problem) and using them to move to edge covers of lower cardinality, until an edge cover wrt which no reducing path exists, is obtained. See White [1971] and White and Gillenson [1975].

10.2.1   Minimum Cost Edge Covers

In this section we discuss an algorithm for finding a minimum cost edge cover in $G = (\mathcal{N}, \mathcal{A}, c = (c_{ij}))$. Defining an edge covering vector $x = (x_{ij})$ in G as at the beginning of this chapter, this problem is the integer program

$$
\begin{aligned}
\text{Minimize } z(x) \;&=\; \sum(c_{ij}x_{ij} : \text{ over } (i,j) \in \mathcal{A}) \\
\text{subject to } x(i) \;&\geqq\; 1 \text{ for all } i \in \mathcal{N} \\
0 \leqq x_{ij} \;&\leqq\; 1, \text{ for all } (i,j) \in \mathcal{A}
\end{aligned} \tag{10.36}
$$

$$
x_{ij} \text{ integer} \qquad \text{for all } (i,j) \in \mathcal{A} \tag{10.37}
$$

Let $\mathbf{Y} \subset \mathcal{N}$ with $|\mathbf{Y}|$ odd and $\geqq 3$. Define

$$
\mathbf{Y}^+(x) = \sum (x_{ij} \text{ over } (i,j) \in \mathcal{A} \text{ with at least one of } i \text{ or } j \text{ or both in } \mathbf{Y}) \tag{10.38}
$$

Notice the difference in the definition of $\mathbf{Y}^+(x)$ in (10.38) compared to that of $\mathbf{Y}^-(x)$ in (10.19). If x is an edge covering vector in G, and $|\mathbf{Y}|$ is odd and $\geqq 3$, we must have

$$
\mathbf{Y}^+(x) \geqq (|\mathbf{Y}| + 1)/2. \tag{10.39}
$$

(10.39) is the *covering blossom inequality* or the *covering blossom constraint* corresponding to $\mathbf{Y}$ for the edge covering problem. Also, observe the difference in the matching blossom constraint defined in (10.20) and the covering blossom constraint. Let $\mathbf{Y}_1, \ldots, \mathbf{Y}_L$ be all the distinct subsets of $\mathcal{N}$ of odd cardinality $\geqq 3$. Consider the following LP

$$
\begin{aligned}
\text{Minimize } z(x) \;&=\; \sum(c_{ij}x_{ij} : \text{ over } (i,j) \in \mathcal{A}) \\
\text{subject to } x(i) \;&\geqq\; 1 \text{ for all } i \in \mathcal{N} \\
\mathbf{Y}_\sigma^+(x) \;&\geqq\; (|\mathbf{Y}_\sigma| + 1)/2, \; \sigma = 1 \text{ to } L \\
0 \leqq x_{ij} \leqq 1, \;&\qquad \text{for all } (i,j) \in \mathcal{A}
\end{aligned} \tag{10.40}
$$

Every edge covering vector in G is feasible to (10.40) and every integer feasible solution of (10.40) is an edge covering vector in G. So, if an optimum solution of (10.40) is an integer vector, it is a minimum cost edge covering vector in G. Associate the original node price (dual variable) π_i to the constraint corresponding to node i in (10.40), for each $i \in \mathcal{N}$, and the dual variable (which can be interpreted as the pseudonode price just as in Sections 10.1.2, 10.1.3) μ_σ to the blossom inequality corresponding to $\mathbf{Y}_\sigma, \sigma = 1$ to L. Let $\pi = (\pi_i), \mu = (\mu_\sigma)$. Given (π, μ), for each $(i, j) \in \mathcal{A}$, define

$$\mu^+(i, j) = \sum(\mu_\sigma : \text{over } \sigma \text{ s.t. } \mathbf{Y}_\sigma \text{ contains}$$
$$\text{either } i \text{ or } j \text{ or both}) \tag{10.41}$$

$$d_{ij}(\pi, \mu) = \pi_i + \pi_j + \mu^+(i, j) \tag{10.42}$$

The dual of (10.40) is

$$\text{Maximize} \quad \sum_{i \in \mathcal{N}} \pi_i + \sum_{\sigma=1}^{L}(|\mathbf{Y}_\sigma| + 1)(\mu_\sigma/2) - \sum_{(i,j) \in \mathcal{A}} \xi_{ij}$$
$$\text{subject to} \quad d_{ij}(\pi, \mu) - \xi_{ij} + \nu_{ij} = c_{ij}, \text{ for } (i, j) \in \mathcal{A} \tag{10.43}$$
$$\text{all } \pi_i, \mu_\sigma, \xi_{ij}, \nu_{ij} \geqq 0.$$

Hence the dual feasibility conditions for $(\pi = (\pi_i), \mu = (\mu_\sigma))$ are

$$\pi \geqq 0, \mu \geqq 0. \tag{10.44}$$

The complementary slackness conditions for optimality in the bounded variable primal LP (10.40) and its dual (10.43) are: for each $i \in \mathcal{N}$ and $\sigma = 1$ to L:

$$\pi_i > 0 \quad \text{implies} \quad x(i) = 1 \tag{10.45}$$
$$x(i) > 1 \quad \text{implies} \quad \pi_i = 0 \tag{10.46}$$
$$d_{ij}(\pi, \mu) > c_{ij} \quad \text{implies} \quad x_{ij} = 1 \tag{10.47}$$
$$d_{ij}(\pi, \mu) < c_{ij} \quad \text{implies} \quad x_{ij} = 0 \tag{10.48}$$
$$\mu_\sigma > 0 \quad \text{implies} \quad \mathbf{Y}_\sigma^+(x) = (|\mathbf{Y}_\sigma| + 1)/2 \tag{10.49}$$
$$\mathbf{Y}_\sigma^+(x) > ((|\mathbf{Y}_\sigma| + 1)/2) \quad \text{implies} \quad \mu_\sigma = 0. \tag{10.50}$$

The algorithm for the edge covering problem discussed below, is based on the blossom algorithm for matching problems. At termination, it generates an edge covering vector, $\bar{x}$, and dual feasible solution, $(\bar{\pi}, \bar{\mu})$ which together satisfy (10.45) to (10.50). Hence $\bar{x}$ is optimal to (10.40) and since it is an integer vector, it is a minimum cost edge covering vector.

The algorithm maintains an $\mathbf{E} \subset \mathcal{A}$ called the *solution set of edges*, and the present edge vector x will always be $x^{\mathbf{E}}$. The vector x is also called the *present solution vector*. The solution set of edges $\mathbf{E}$ is partitioned into $\mathbf{M}$ and $\mathbf{A}$. The set $\mathbf{M}$ will always be a matching in G, and $\mathbf{A}$ is called the *set of covering edges*. Throughout the algorithm $\mathbf{M} \cap \mathbf{A}$ will be $\emptyset$; and the set of nodes on edges in $\mathbf{M}$, and the set of nodes on edges in $\mathbf{A}$, will be mutually disjoint.

Original nodes in $\mathcal{N}$ are classified as below during the algorithm. Since $\mathbf{M}$ and $\mathbf{A}$ are always edge and node disjoint, every node is uniquely classifed by this classification.

matched nodes	-	those incident with a matching edge
type 1 nodes	-	those incident with exactly on covering edge
type 2 nodes	-	those incident with two or more covering edges
exposed nodes	-	those incident with no edge from $\mathbf{E} = \mathbf{M} \cup \mathbf{A}$.

Given the dual feasible solution (π, μ), define $d_{ij}(\pi, \mu)$ as in (10. 42) and let

$$\mathcal{A}_*(\pi, \mu) = \{(i, j) : (i, j) \in \mathcal{A}, \text{ and } d_{ij}(\pi, \mu) = c_{ij}\} \qquad (10.51)$$

Edges in $\mathcal{A}_*(\pi, \mu)$ are the *original equality edges* in this problem and $G_*(\pi, \mu) = (\mathcal{N}, \mathcal{A}_*(\pi, \mu))$ is the *equality subnetwork*, wrt the dual feasible solution (π, μ). In the algorithm, the set $\mathbf{M}$ will always be a matching which is a subset of $\mathcal{A}_*(\pi, \mu)$. It tries to obtain matchings of higher cardinality in $G_*(\pi, \mu)$ by growing alternating trees wrt $\mathbf{M}$, rooted at exposed nodes, in it. The alternating trees are grown with the aim of discovering augmenting paths, which are special types of alternating paths that can be used to augment either the matching $\mathbf{M}$, or the set of solution edges $\mathbf{E} = \mathbf{M} \cup \mathbf{A}$.

While growing the alternating trees, simple blossoms may be detected and shrunk into pseudonodes. After some pseudonodes are found, the original network G gets transformed into the current network $G^1 = (\mathcal{N}^1, \mathcal{A}^1)$. $\mathcal{N}^1$ is the set of current nodes at the present stage, these are either the original nodes not contained in any pseudonode, or outermost pseudonodes. If $\mathbf{F} \subset \mathcal{A}$ is a subset of original edges, the set of current edges $\mathbf{F}^1$ corresponding to it is always defined to be

$$\mathbf{F}^1 = \{(p; q) : \ p, q \in \mathcal{N}^1, \text{ there exists } i \in p, j \in q, \text{ such that } (i; j) \in \mathbf{F}\}. \quad (10.52)$$

The sets of current edges corresponding to $\mathcal{A}_*(\pi, \mu), \mathbf{M}, \mathbf{A}, \mathbf{E} = \mathbf{M} \cup \mathbf{A}$ will always be denoted by $\mathcal{A}_*^1(\pi, \mu), \mathbf{M}^1, \mathbf{A}^1, \mathbf{E}^1 = \mathbf{M}^1 \cup \mathbf{A}^1$. Edges in $\mathcal{A}_*^1(\pi, \mu), \mathbf{M}^1, \mathbf{A}^1, \mathbf{E}^1$ are respectively *current equality edges, current matching edges, current covering edges*, and *current solution edges*.

Simple blossoms in this algorithm are partial networks of the current network satisfying the following properties.

It is a partial network of the current network determined by an odd subset with $\geqq 3$ current nodes.

There exists a unique current node in it which is incident with either exactly 2 or 0 current solution edges within it, this node is called the *base node* of the simple blossom. $\qquad$ (10.53)

There exists a simple spanning cycle containing all the current nodes in it, which is an alternating cycle. All solution edges on it other than those incident at the base node, are current matching edges. If a_0 is the base node, and a_1, a_2 are the current nodes adjacent to a_0 on the odd cycle; deleting $(a_0; a_1), (a_0; a_2)$ from the cycle, leaves an AP wrt $\mathbf{M}^1$ (in the sense that edges on it are alternately matching and nonsolution edges). $(a_0; a_1)$ $(a_0; a_2)$ may be nonsolution edges or covering edges. Consequently, every node in the simple blossom other than the base node is incident with exactly one solution edge on the odd alternating cycle in it.

When they are detected, simple blossoms are shrunk into pseudonodes in the algorithm. The *base node* for a pseudonode is the base node on the simple blossom corresponding to it. The *apex node* of a simple blossom or the pseudonode into which it is shrunk, is the original node inside it which is incident exactly to either 2 or 0 solution edges within the simple blossom. For convenience we define the base node and apex node for an original node to be that node itself. Hence the apex node for a pseudonode is the apex node of its base node.

Simple blossoms and the pseudonodes into which they are shrunk are classified in accordance with their configuration. To distinguish between original nodes and pseudonodes we use roman letters to enumerate types of pseudonodes (type A, B, etc.). A pseudonode is said to be:

rooted   -   if the base node is an exposed node (which will be a root node). Correspondingly, the apex of this pseudonode is also an exposed original node.

type A   -   if the base node is a matched node on a matching edge joining the base to a node outside the simple blossom.

type B   -   if the base node is incident to exactly one covering edge joining the base node to a node outside the simple blossom.

type C   -   if the base node is a type 2 original node, incident to exactly two covering edges which are within the simple blossom. The two edges incident at the base node on the odd alternating cycle in the simple blossom, are these covering edges.

type D   -   if the base node is itself a pseudonode which is either a type C or another type D node.

Thus all the edges joining a node in a simple blossom to a node outside are nonsolution edges, with the exception of the unique matching or covering edge

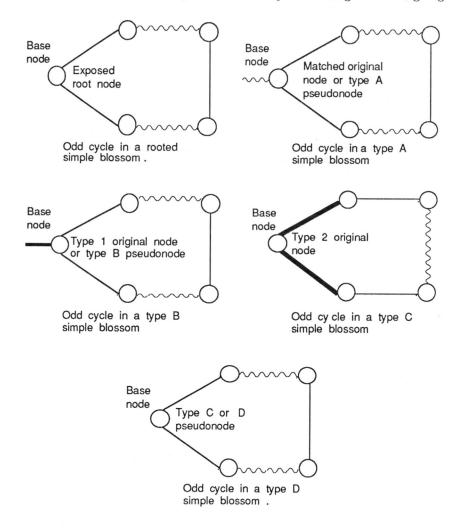

Figure 10.31 Odd cycles in various types of simple blossoms. Wavy (thick) edges are matching (covering) edges.

incident at the base node in type A or B simple blossoms respectively. See Figure 10.31. The base node of a type A (type B) pseudonode is the unique node on its simple blossom incident to the matching (covering) edge joining it to a node outside the simple blossom. The base node of a type C simple blossom is the unique type 2 original node on the two covering edges inside it. The base node of a type D simple blossom is the unique type C or D pseudonode inside it. The apex node of the various types of pseudonodes are:

type A (type B)   -   it is the unique original node inside it joined by an
                         original matching (covering) edge to a node out-
                         side the pseudonode
type C or D   -   it is the unique type 2 original node inside it

A new simple blossom detected during the tree growth process (while scanning some node) will always be either a rooted or a type A simple blossom. Type B, C, D pseudonodes are only created by transformation of an existing rooted or type A pseudonode in a matching and covering augmentation procedure in the algorithm.

If $\mathcal{N}_1$ is the set of original nodes corresponding to an existing pseudonode, since $|\mathcal{N}|$ is odd and $\geqq 3$, there is a covering blossom constraint corresponding to it. If x is the present solution vector, it can be verified that x satisfies the covering blossom constraint corresponding to this pseudonode as an equation if the pseudonode is a type A or B or C or D pseudonode, and violates it if it is a rooted pseudonode. Rooted pseudonodes may exist in intermediate stages of the algorithm but they are eliminated before termination. If any blossoms remain at termination of the algorithm, they will be of types A, B, C or D. Let

$$\mathcal{A}^- = \{(i;j) : (i;j) \in \mathcal{A}, c_{ij} \leqq 0\} \tag{10.54}$$

Obviously $\mathcal{A}^-$ can be assumed to be a subset of a minimum cost edge cover in G. The algorithm maintains $\mathcal{A}^- \subset \mathbf{E}$ always. Solution edges $\mathbf{E}$ are classified into matching and covering edges as follows: matching edges ($\mathbf{M}$) are the edges in the components of $(\mathcal{N}, \mathbf{E})$ consisting of single edges which are not in $\mathcal{A}^-$; the remaining edges of $\mathbf{E}$ (i.e., all edges of $\mathcal{A}^-$, and all edges of $\mathbf{E}$ which are adjacent to at least one other edge in $\mathbf{E}$) are the covering edges.

The algorithm plants a rooted alternating forest with roots (each labeled as an outer node) at each exposed node and grows them. The objective of growing the alternating trees is to detect augmenting paths, which are special APs of solution/nonsolution equality edges that allow augmentations. Every time an augmentation is performed, at least one exposed node becomes nonexposed. Once a node becomes nonexposed it remains nonexposed in the sequel. When all nodes become nonexposed, the solution set of edges at that time will be a minimum cost edge cover, and the algorithm terminates. The following additional properties are always satisfied during the algorithm.

Properties Maintained by the Algorithm

π, μ always satisfy dual feasibility conditions (10.44).

$\mathbf{E} \subset \mathcal{A}^- \cup \mathcal{A}_*(\pi, \mu)$, $\mathbf{M} \subset \mathcal{A}_*(\pi, \mu)$, $\mathbf{A} \supset \mathcal{A}^-$.

$\mathcal{A}^- \subset \mathbf{A}^1 \subset \mathcal{A}_*^1(\pi, \mu) \cup \mathcal{A}^-$.

If $(\mathcal{N}, \mathbf{A})$ or $(\mathcal{N}^1, \mathbf{A}^1)$ has a connected component that is a single edge, that edge is in $\mathcal{A}^-$.

$\pi_i = 0$ if i is a type 2 node or a node on an edge in $\mathcal{A}^-$.

$\mu_\sigma > 0$ implies that $\mathbf{Y}_\sigma$ is the set of original nodes inside an existing pseudonode.

$d_{ij}(\pi, \mu) \leqq c_{ij}$ for all $(i; j) \in \mathcal{A} \backslash \mathcal{A}^-$.

Each in-tree edge is in $\mathcal{A}_*^1(\pi, \mu) \backslash \mathbf{A}^1$.

If node i is an inner labeled current original node, $\pi_i > 0$.

Each in-tree current node that is a pseudonode is either rooted, or a type A labeled outer, or a type A labeled inner associated with pseudonode price $\mu_\sigma > 0$.

Every edge on the odd cycle in the simple blossom corresponding to any pseudonode at any level is an equality edge.

Nodes on edges in $\mathcal{A}^-$ are always current nodes, i.e., these nodes will never be in any pseudonode.

At some stage of the algorithm, if $\mathbf{E}$ is not yet an edge covering in G, no matching and covering augmentations are possible, no blossom shrinking or pseudonode unshrinking steps are possible, and the list of labeled and unscanned nodes is empty (i.e., no tree growth is possible), the labeling at that stage is said to have *become hungarian*, and the alternating trees at that stage are *Hungarian trees* in the current equality subnetwork. The trees constitute a *Hungarian forest* then, and the following conditions will hold. When these conditions are satisfied, the algorithm moves to the dual solution change step in the algorithm.

Hungarian Forest Conditions

Each inner labeled current node is incident to exactly one matching and one nonmatching current in-tree edge.

No node on an edge in $\mathbf{A}$ is in-tree or contained inside an in-tree pseudonode.

If an original node i is inside an outer labeled pseudonode, $\pi_i > 0$.

If $\mathbf{Y}_\sigma$ is the set of original nodes in an inner labeled pseudonode, $\mu_\sigma > 0$.

In the following algorithm the tree growth step is constructed in such a way that only outer nodes are scanned. The list always refers to labeled and unscanned outer current nodes at that stage.

BLOSSOM ALGORITHM FOR THE MINIMUM
COST EDGE COVERING PROBLEM

Step 1 Initialization   Let $M = \emptyset, A = A^-, \pi = (\pi_i) = 0, \mu = (\mu_\sigma) = 0$ initially. If there are no exposed nodes, go to Step 13. Otherwise root an alternating tree at each exposed node i by labeling it with $(\emptyset, +, i)$. List = set of all these root nodes now.

Step 2 Select a Node to Scan   If list $= \emptyset$ go to Step 12. Otherwise, select a node from it to scan, and delete it from the list.

Step 3 Scanning   Let the node to be scanned be p with label $(P(p), +, r)$. Do the following in the order given.

If there is an already labeled outer node j associated with a root node $s \neq r$ such that $(p; j) \in A_*^1(\pi, \mu)$, an augmenting path has been found, go to Step 4.

If there is an already labeled outer node j associated with the same root node r such that $(p; j) \in A_*^1(\pi, \mu)$, the alternating tree containing p, j has blossomed, go to Step 10.

If there is an original current node u such that $(p; u) \in A_*^1(\pi, \mu)$ and $\pi_u = 0$, go to Step 5.

If there is an original current type 1 node or a type B pseudonode v such that $(p; v) \in A_*^1(\pi, \mu)$, go to Step 6.

If there is an unlabeled current node w that is a type C or D pseudonode such that $(p; w) \in A_*^1(\pi, \mu)$, go to Step 7.

Identify all $j \neq P(p)$ such that $(p; j) \in A_*^1(\pi, \mu)$. Since we came to this stage, j cannot be type 1, or 2, or B, C, or D, or exposed, or a rooted pseudonode (in that case we would have already gone to some other step). For each such j do the following: If j is unlabeled, let $(j; t)$ be the current matching edge incident at j, label j with $(p, -, r)$. If t is unlabeled and is a pseudonode inside which there is an original node i associated with node price $\pi_i = 0$ go to Step 8. Otherwise label t with $(j, +, r)$ and include it in the list.

Now p is labeled and scanned. Go to Step 11.

Step 4 Matching Augmentation   We come to this step when scanning or inspection has revealed a pair of current nodes p and j associated with root indices $r(p) \neq r(j)$, contained on an augmenting path $\mathcal{P}^1$ which is obtained by combining the predecessor paths of p and j with the edge $(p; j)$. Let t_1, t_2 be the apex nodes of $r(p), r(j)$. Find the augmenting path $\mathcal{P}$ between t_1 and t_2 by expanding the pseudonodes on $\mathcal{P}^1$. Rematch using $\mathcal{P}$. Revise all the blossoms along $\mathcal{P}^1$ as in Section 10.1.2. Chop down the two trees rooted at $r(p)$ and $r(j)$. If there are no exposed nodes left, go to Step 13. Otherwise change the list into the set of all outer nodes and go back to inspection in

Step 12, or Step 2, depending on whether you arrived here from inspection or Step 3.

Step 5 Matching and Covering Augmentation Procedure 1   We come to this step when scanning or inspection has revealed an outer current node p and an original current node u associated with $\pi_u = 0$, such that $(p; u) \in \mathcal{A}_*^1(\pi, \mu)$. (p, u) must be a nonmatching edge. Let the label on p be $(\mathrm{P}(p), +, r)$. Let j be an original node inside p such that $(j; u) \in \mathcal{A}_*(\pi, \mu)$. Let $\mathcal{P}^1$ be the predecessor path of p. Find $\mathcal{P}$, the alternating path from j to the apex node of r, beginning with the matching edge incident at j if j is a matched node ($\mathcal{P}$ may have no edges if for example $p = r = j$) by expanding the pseudonodes along $\mathcal{P}^1$. Rematch using $\mathcal{P}$. Revise all the blossoms along $\mathcal{P}^1$ as discussed in Section 10.1.2. Chop down the tree rooted at r. If u is a type 1 or 2 node, add the edge $(j; u)$ to the set $\mathbf{A}$. If u is a matched node, let $(u; w) \in \mathbf{M}$, delete $(u; w)$ from $\mathbf{M}$, and add both $(j; u)$ and $(u; w)$ to the set $\mathbf{A}$. So, u becomes a type 2 node as a result of these changes. p becomes a type 1 node if it is an original node; or a type B pseudonode with j as its apex node, if it is a pseudonode.

Go to Step 13 if no exposed nodes are left. Otherwise change the list into the set of all outer nodes at this stage and go back to inspection in Step 12, or Step 2, depending on whether you arrived at this step from inspection or step 3.

Step 6 Matching and Covering Augmentation Procedure 2   We come to this step when scanning has revealed an outer current node p and current type 1 or type B node v such that $(p; v) \in \mathcal{A}_*^1(\pi, \mu)$. Let h be the apex node of v. If v is a type 1 node, π_v must be > 0 (otherwise, by the order in which scanning is done we would have gone to Step 5 instead of coming here) and hence v is not on any edge in $\mathcal{A}^-$. If v is a type B pseudonode, since its apex h is inside a pseudonode, h is not on any edge in $\mathcal{A}^-$. Let $(h; j_1)$ be the unique covering edge incident at h. Since p is an in-tree node, $j_1 \neq p$. So, the connected component of $(\mathcal{N}, \mathbf{A})$ containing $(h; j_1)$ must contain some edge other than $(h; j_1)$, this other edge cannot be incident at h since h is a type 1 node, let $(j_1; j_2)$ be one such edge. So, j_1 is a type 2 node.

Let $\mathcal{P}^1$ be the path consisting of the edge $(v; p)$ and then the predecessor path of p. By expanding the pseudonodes along $\mathcal{P}^1$, obtain the alternating path $\mathcal{P}$ from h to the apex of the root node r corresponding to p. If $(j_1; j_2) \in \mathcal{A}^-$ or if the connected component of $(\mathcal{N}, \mathbf{A})$ that contains $(j_1; j_2)$ has ≥ 3 edges, delete $(h; j_1)$ from $\mathbf{A}$, and then rematch using $\mathcal{P}$. On the other hand, if both $(h; j_1)$ and $(j_1; j_2) \in \mathbf{A} \backslash \mathcal{A}^-$, and the connected component of $(\mathcal{N}, \mathbf{A})$ containing $(h; j_1)$ contains just the two edges $(h; j_1), (h; j_2)$; delete both $(h; j_1)$ and $(j_1; j_2)$ from $\mathbf{A}$, rematch using $\mathcal{P}$ and then include $(j_1; j_2)$ as a new matching edge. Revise all the blossoms along $\mathcal{P}^1$ as discussed in Section 10.1.2. Chop down the tree rooted at r. If there are no exposed nodes left go to Step 13. Otherwise change the list into the set of all outer nodes and go back to Step 2.

Step 7 Matching and Covering Augmentation Procedure 3 We come to this step when scanning or inspection has revealed an outer labeled current node p and a type C or D pseudonode w such that $(p; w) \in \mathcal{A}_*^1(\pi, \mu)$. Let i_1 be the apex of w, and i_2, i_3 the pair of type 1 nodes inside w. So, $(i_1; i_2), (i_1; i_3) \in \mathbf{A}$ and all nodes inside w other than i_1, i_2, i_3 are matched nodes. Let s, j be original nodes inside p, w respectively such that $(s; j) \in \mathcal{A}_*(\pi, \mu)$.

By expanding the pseudonodes along the predecessor paths of p find the alternating path $\mathcal{P}_1$ in G between s and the apex node of the root associated with p, beginning with the matching edge incident at s, if s and the apex of the root node of p are different. By expanding the pseudonodes on the odd cycle in the simple blossom corresponding to w, find the alternating path $\mathcal{P}_2$ in G between i_2 and i_3 that is contained within w. Delete $(i_1; i_2)$ and $(i_1; i_3)$ from $\mathbf{A}$. When the two edges $(i_1; i_2), (i_1; i_3)$ are added to $\mathcal{P}_2$ it becomes a cycle, call it $\mathbb{C}_2$. First make all the edges in $\mathbb{C}_2$ into nonmatching edges. Then traverse $\mathbb{C}_2$ beginning from j, making the edges traveled alternately into nonmatching and matching edges beginning with the first edge incident at j kept as a nonmatching edge, and returning to j at the end.

Then rematch using $\mathcal{P}_1$, and finally make $(s; j)$ into a matching edge. Revise all the blossoms that have nonempty intersections with $\mathcal{P}_1, \mathbb{C}_2$ as in Section 10.1.2. Chop down the tree containing p. Now w becomes a type A pseudonode with j as its apex. If there are no exposed nodes left go to Step 13. Otherwise change the list into the set of all outer labeled current nodes, and go back to Step 2, or the inspection in Step 12, depending on whether you arrived at this step from Step 3 or inspection.

Step 8 Matching and Covering Augmentation Procedure 4 We come to this step when scanning or inspection has identified an inner labeled current node j joined by a current matching edge to a pseudonode t that contains an original node i associated with node price $\pi_i = 0$, inside it. Let h be the apex of t. Let $\mathcal{P}^1$ be the path consisting of the matching edge $(t; j)$ and the predecessor path of j. By expanding the pseudonodes along $\mathcal{P}^1$ obtain the alternating path $\mathcal{P}$ from h to the apex of the root associated with j.

In this procedure we convert t into a type C or D pseudonode with i inside it (associated with $\pi_i = 0$) as its apex node. For this, obtain the alternating cycle $\mathbb{C}$ from h (the present apex of t) to h by expanding the pseudonodes on the odd cycle in the simple blossom corresponding to t. Let $(i; i_1), (i; i_2)$ be the edges on $\mathbb{C}$ incident at i. Let $\mathcal{P}_2$ be the path left between i_1 and i_2 when the edges $(i; i_1), (i; i_2)$ and node i are deleted from $\mathbb{C}$. First make all the edges on $\mathbb{C}$ into nonmatching edges. Then add $(i; i_1), (i; i_2)$ to the set $\mathbf{A}$. Now make the edges along $\mathcal{P}_2$ as you travel from i_1 to i_2, alternately nonmatching and matching edges beginning with the edge incident at i_1 as a nonmatching edge. Revise all the blossoms on the odd cycle in the simple blossom corresponding to t.

Rematch using $\mathcal{P}$. Revise all the blossoms along $\mathcal{P}^1$. If there are no exposed nodes left go to Step 13. Otherwise chop down the tree containing j, change the list into the set of all outer current nodes, and go back to inspection in Step 12, or Step 2 depending on whether you arrived at this step from inspection or Step 3.

Step 9 Matching and Covering Augmentation Procedure 5   We come to this step when inspection has identified an outer current node j which is a rooted pseudonode containing within it an original node i with $\pi_i = 0$. Convert j into a type C or D pseudonode with i as its apex node, as discussed in Step 8. If there are no exposed nodes left, go to Step 13. Otherwise chop down the tree containing j, change the list into the set of all outer current nodes and go back to inspection in Step 12.

Step 10 Blossom Shrinking   We come to this step when scanning or inspection has revealed two outer current nodes p, j with the same root index, such that $(p; j) \in \mathcal{A}^1_*(\pi, \mu)$. Identify the simple blossom (it will either be rooted or type A) and shrink it into a pseudonode as in Step 6 of the algorithm of Section 10.1.1. Include the new pseudonode (now outer labeled) in the list, and go back to inspection in Step 12 or Step 2 depending on whether you arrived here from inspection or Step 3.

Step 11 Type A Pseudonode Unshrinking   Identify all pseudonodes which are current nodes labeled as inner nodes (so they will be type A) associated with pseudonode price $\mu_\sigma = 0$. Unshrink each of them as discussed in Section 10.1.2. Repeat this process as often as possible. Go back to inspection in Step 12, or Step 2, depending on whether you came to this step from inspection or Step 3.

Step 12 Dual Solution Change   We come to this step when the Hungarian forest conditions are satisfied. Let (π, μ) be the present dual feasible solution. The new dual feasible solution is $(\hat{\pi} = (\hat{\pi}_i), \hat{\mu} = (\hat{\mu}_\sigma))$ defined below.

$$\begin{aligned}
\Delta_1 = \quad & \{i : i \in \mathcal{N} \text{ is either an inner current node, or} \\
& \text{contained inside an outer labeled pseudonode}\} \\
\Delta_2 = \quad & \{i : i \in \mathcal{N} \text{ is inside an unlabeled current node}\} \\
\Delta_3 = \quad & \{i : i \in \mathcal{N} \text{ is inside an outer current node}\} \\
\Delta_4 = \quad & \{i : i \in \mathcal{N} \text{ is either an outer current node, or} \\
& \text{contained inside an inner labeled pseudonode}\}
\end{aligned}$$

$$\delta_1 = \quad \text{Min.} \ \{\pi_i : i \in \mathbf{\Delta}_1\}$$

$$\delta_2 = \quad \text{Min.} \ \{c_{ij} - d_{ij}(\pi, \mu) : i \in \mathbf{\Delta}_2, j \in \mathbf{\Delta}_3 \text{ and } (i;j) \in \mathcal{A}\}$$

$$\delta_3 = \quad \text{Min.} \ \{\tfrac{1}{2}(c_{ij} - d_{ij}(\pi, \mu)) : (i;j) \in \mathcal{A}, \text{ and } i, j \in \mathbf{\Delta}_3 \text{ but in different current nodes}\}$$

$$\delta_4 = \quad \text{Min.} \ \{\tfrac{1}{2}\mu_\sigma : \mathbf{Y}_\sigma \text{ is the set of original nodes inside an inner current node that is a pseudonode}\}$$

$$\delta = \quad \text{Min.} \ \{\delta_1, \delta_2, \delta_3, \delta_4\}$$

$$\hat{\pi}_i = \begin{cases} \pi_i - \delta & \text{for all } i \in \mathbf{\Delta}_1 \\ \pi_i + \delta & \text{for all } i \in \mathbf{\Delta}_4 \\ \pi_i & \text{otherwise} \end{cases}$$

$$\hat{\mu}_\sigma = \begin{cases} \mu_\sigma + 2\delta & \text{if } \mathbf{Y}_\sigma \text{ is the set of original nodes in} \\ & \text{a current outer labeled pseudonode} \\ \mu_\sigma - 2\delta & \text{if } \mathbf{Y}_\sigma \text{ is the set of original nodes in} \\ & \text{a current inner labeled pseudonode} \\ \mu_\sigma & \text{otherwise} \end{cases}$$

Inspection

If $\delta = \delta_1$, look for an original node u which is an inner current node associated with $\hat{\pi}_u = 0$. If p is the predecessor of u, with p, u apply Step 5. Repeat as often as possible. Then look for an original node i with $\hat{\pi}_i = 0$ inside an outer current node j that is a pseudonode. If j is rooted (type A) apply Step 9 (Step 8) with it. Repeat as often as possible.

If $\delta = \delta_2$, look for $(i;j) \in \mathcal{A}_*(\hat{\pi}, \hat{\mu})$ with i inside an unlabeled node, and j inside an outer labeled node. Then include the current node containing j in the list.

If $\delta = \delta_3$, look for $(i;j) \in \mathcal{A}_*(\hat{\pi}, \hat{\mu})$ with i, j contained inside different outer nodes. If the root nodes of the current nodes containing i, j are different (the same) apply Step 4 (Step 10) with them. Repeat as often as possible.

If $\delta = \delta_4$ look for inner nodes that are pseudonodes associated with $\hat{\mu}_\sigma = 0$ and go to Step 11 if there are any.

Go to Step 2.

Step 13 Termination We reach this step when there are no exposed nodes left. Let $\overline{\mathbf{E}}, \overline{x}, (\overline{\pi}, \overline{\mu})$ be the present solution set of edges, its incidence vector, and the dual solution at this stage. $\overline{\mathbf{E}}$ is a minimum cost edge cover in G, $\overline{x}$ is the associated minimum cost edge covering vector, and $(\overline{\pi}, \overline{\mu})$ the optimum dual solution (optimum for (10.43)).

Validity of the Algorithm and Its Computational Complexity

1. We will now show that δ is finite and > 0 whenever Step 12 is carried out in the algorithm; and that the conditions :$\pi \geqq 0, \mu \geqq 0, d_{ij}(\pi, \mu) \leqq c_{ij}$ for all $(i; j) \in \mathcal{A} \backslash \mathcal{A}^-$, hold throughout. Clearly, these conditions hold initially, since $\pi = 0, \mu = 0$ at that stage. When the algorithm arrives at Step 12 at some stage suppose (π, μ) is the dual solution and that these conditions hold. Each of δ_1 to δ_4 is the minimum of a set of numbers, and every number in each of these sets is $\geqq 0$ because of these conditions. Also, if any number in any of these sets is 0, it can be verified that it provides an opportunity to carry out one of Steps 4, 5, 8, 9, 10, or 11, or to label a new unlabeled node, contradicting the Hungarian forest conditions which must hold when the algorithm arrives at Step 12. Hence each of δ_1 to δ_4 is either finite and > 0 or $+\infty$ (which happens when the set in which it is the minimum is $\emptyset$).

   So, δ is finite and > 0, or $+\infty$. Suppose $\delta = +\infty$. So, all δ_1 to δ_4 must be $+\infty$. δ_1, δ_4 are both $+\infty$ implies that there are no inner labeled nodes at this stage. Hence, the only labeled nodes must be the root nodes which are outer labeled. Now, δ_2, δ_3 are both $+\infty$ implies that there are no edges joining a pair of outer labeled nodes, and no edges joining an outer node to an unlabeled node. These facts imply that the root nodes are isolated nodes in G, contradicting the hypothesis that G has no isolated nodes. Hence δ cannot be $+\infty$. So, δ is finite and > 0.

   Let $\hat{\pi}, \hat{\mu}$ be the new dual solution obtained in that Step 12. The definition of δ_1 and δ imply that $\hat{\pi} \geqq 0$. The definition of δ_4 and δ imply that $\hat{\mu} \geqq 0$. The definition of $\delta_2, \delta_3, \delta$ imply that $c_{ij} - d_{ij}(\hat{\pi}, \hat{\mu}) \geqq 0$ for all $(i; j) \in \mathcal{A} \backslash \mathcal{A}^-$. Hence if the conditions mentioned above hold at the beginning of a Step 12, they continue to hold after that Step 12 is executed. Repeating the same argument after each occurrence of Step 12, we conclude that these conditions hold always and that $\delta > 0$ and finite in every Step 12 carried out in the algorithm.

2. We will now show that $M \cup (\mathbf{A} \backslash \mathcal{A}^-) \subset \mathcal{A}_*(\pi, \mu)$ throughout the algorithm, and that all in-tree edges are always in $\mathcal{A}_*^1(\pi, \mu)$ at that stage. Also, all edges in the odd cycles of the various simple blossoms corresponding to the various pseudonodes at any stage, are equality edges at that stage.

   Initially $\mathbf{M} = \emptyset$ and $\mathbf{A} = \mathcal{A}^-$, hence these statements hold then trivially. Any new edge added to any alternating tree is always from the $\mathcal{A}_*^1(\pi, \mu)$ at that stage. Also, from the dual solution change formulas it can be verified that all in-tree edges continue to remain in the new current equality subnetwork after a dual solution change step because each in-tree edge joins an outer to an inner node. A newly created matching edge is always from $\mathcal{A}_*(\pi, \mu)$ at that stage, by the manner in which matching and covering augmentations are carried out. Also, during a dual solution change step, a current matching edge is either in-tree, or out-of-tree with both nodes on it unlabeled. From the dual solution change formulas these facts imply that all matching edges continue to remain in the new $\mathcal{A}_*(\pi, \mu)$ after each dual solution change step. Similarly it can be verified that all edges in $\mathbf{A} \setminus \mathcal{A}^-$ are always equality edges.

When a new simple blossom is discovered, it is clear that all the edges on its odd cycle are equality edges. From the manner in which the dual solution is changed, we can verify that these edges remain equality edges as long as this simple blossom remains shrunken as a pseudonode.

Hence all the statements made above hold throughout the algorithm.

3. From these facts we verify that all the properties mentioned before the statement of the algorithm are all maintained in it. Let $\overline{\mathbf{E}}, \overline{x}, (\overline{\pi}, \overline{\mu})$ be the solution set of edges, its incidence vector, and the dual solution at termination. Since all the nodes are covered at that stage $\overline{\mathbf{E}}$ is an edge cover in G. It can be verified that $\overline{x}, (\overline{\pi}, \overline{\mu})$ together satisfy all the conditions (10.45) to (10.50). Hence $\overline{x}$ is an optimum solution of (10.40). Since it is also a 0-1 vector, it is an optimum solution of (10.36), (10.37), and hence an optimum edge covering vector for G, and therefore $\overline{\mathbf{E}}$ is an optimum edge cover.

4. We will now analyze the worst case computational complexity of this algorithm. Each time one of Steps 4 to 9 occur, the number of exposed nodes decreases by 1 or 2. So, Steps 4 to 9 can occur at most n times. Define a *stage* in this algorithm to begin either with the initial Step 1 (the *initial stage*) or after one of Steps 4 to 9 has just been completed, and to end when one of Steps 4 to 9 is completed the next time. So, there are at most n stages in the algorithm, and in each stage, exactly one of Steps 4 to 9 occurs.

A newly shrunken pseudonode always receives an outer label when it is formed, and it either remains outer labeled, or gets absorbed inside another pseudonode till the end of the stage in which it is formed. Thus by the results in Section 10.1, the total number of times that Step 10 (blossom shrinking step) can occur in a stage is at most $(n/2)$.

Only inner labeled pseudonodes are unshrunk in the algorithm. So, any pseudonode which is unshrunk in a stage must have been formed in earlier stages. Hence, by the results in Section 10.1, Step 11 (pseudonode unshrinking step) can occur at most $(n/2)$ times in a stage.

Define $\mathbf{J} = \{i : i \in \mathcal{N}$, and i is inside an outer node$\}$, $\mathbf{I} = \{p : p$ is an inner labeled current node$\}$, $\theta = |\mathbf{J}| + |\mathbf{I}|$. Whenever a blossom shrinking step (Step 10) is carried out, it can be verified that $|\mathbf{J}|$ strictly increases, and that θ does not decrease. Whenever a pseudonode unshrinking step (Step 11) is carried out, it can be verified that θ strictly increases. Also, θ strictly increases whenever a tree growth step occurs. And $\theta \leq n$.

Whenever Step 12 occurs with $\delta = \delta_1$, it leads to Step 5 and then the end of the stage. If Step 12 occurs with $\delta = \delta_2$ or δ_4, a tree growth step, or pseudonode unshrinking step occurs, and since θ increases by at least 1 in each of these steps, Step 12 with $\delta = \delta_2$ or δ_4 can occur at most n times in a stage. Step 12 with $\delta = \delta_3$ leads to Step 4 or Step 10, and hence this can occur at most $1 + (n/2)$ times in a stage. So, in a stage Step 12 can occur at most $1 + n + (n/2)$ times. The computational effort in one of Steps 4 to 12 can be verified to be bounded above by O(m). So, the total effort in a stage of this algorithm is bounded above by O(nm). Since there are at most n stages, the overall computational effort in this algorithm is bounded above by O($n^2 m$).

The Convex Hull of Edge Covering Vectors

Let $\mathbf{K}^{\geq}$ denote the convex hull of edge covering vectors in G. If G has some isolated nodes, it has no edge cover, and hence $\mathbf{K}^{\geq} = \emptyset$, and it can be verified that the set of feasible solutions of the system (10.40) is also empty in this case. Under the assumption that there are no isolated nodes in G, we will use the same fundamental proof technique as in Section 10.1.4 to show that $\mathbf{K}^{\geq}$ is the set of feasible solutions of (10.1.40). We have shown that the LP (10.40) has an optimum solution which is an edge covering vector in G, and that the blossom algorithm discussed above will find it irrespective of what the edge cost vector c may be. This implies that all extreme points of the set of feasible solutions of (10.40) are edge covering vectors in G. Also, every edge covering vector in G is feasible to (10.40), and it is an extreme point of the set of feasible solutions of this system since it is a 0-1 vector, and all the variables in this system are bounded by 0 and 1. Thus the system of constraints in (10.40) provides a linear constraint representation for $\mathbf{K}^{\geq}$, in all cases.

Comment 10.3   The blossom algorithm for the minimum cost edge covering problem, and the results in this subsection are taken from Murty and Perin [1982].

10.3   Minimum Cost 1-Matching/Edge Coverings

Let G $= (\mathcal{N}, \mathcal{A}, c = (c_{ij}))$ be the original undirected network with c as the edge cost vector and $(\mathcal{N}^{\leq}, \mathcal{N}^{=}, \mathcal{N}^{\geq}, \mathcal{N}^0)$ as the partition of $\mathcal{N}$. Here we discuss a primal-dual blossom algorithm for the minimum cost 1-M/EC problem (10.6), (10.7). We will replace the integrality restrictions (10.7) by blossom constraints, thereby transforming the integer program (10.6), (10.7) into an LP for which we develop a primal-dual blossom algorithm. Let $\mathbf{Y} \subset \mathcal{N} \backslash \mathcal{N}^0$ with $|\mathbf{Y}|$ odd and ≥ 3. There are 2 different blossom constraints associated with $\mathbf{Y}$; the *matching blossom constraint* $\mathbf{Y}^-(x) \leq (|\mathbf{Y}| - 1)/2$, and the *covering blossom constraint* $\mathbf{Y}^+(x) \geq (|\mathbf{Y}| + 1)/2$; where $\mathbf{Y}^-(x)$ and $\mathbf{Y}^+(x)$ are defined as in (10.19), (10.38) respectively. From the definitions it is clear that $\mathbf{Y}^-(x) \leq \mathbf{Y}^+(x)$ for every edge vector x in G. Every

1-M/EC vector x satisfies the matching blossom constraint (edge covering blossom constraint) corresponding to $\mathbf{Y}$ if $\mathbf{Y} \subset \mathcal{N}^{\leqq}$ ($\mathbf{Y} \subset \mathcal{N}^{\geqq}$) but may or may not satisfy the covering blossom constraint (matching blossom constraint) corresponding to such a $\mathbf{Y}$. If $\mathbf{Y} \subset \mathcal{N}^{=}$, every 1-M/EC vector x satisfies both the matching and edge covering blossom constraints corresponding to $\mathbf{Y}$. Whenever $\mathbf{Y}$ is any subset of $\mathcal{N} \backslash \mathcal{N}^0$, every 1-M/EC vector satisfies at least one of the matching or covering blossom constraints corresponding to $\mathbf{Y}$. We will introduce either the matching blossom constraint or the edge covering blossom constraint corresponding to $\mathbf{Y}$, to replace the integer requirements (10.7). We use the symbols $\mathbf{MB}$, $\mathbf{CB}$ to denote the class of all these $\mathbf{Y}$ for which the matching blossom constraint, edge covering blossom constraint are introduced respectively.

We will include $\mathbf{Y}$ in $\mathbf{MB}$ if $\mathbf{Y} \subset \mathcal{N}^{\leqq} \cup \mathcal{N}^{=}$, and in $\mathbf{CB}$ if $\mathbf{Y} \subset \mathcal{N}^{\geqq} \cup \mathcal{N}^{=}$ and contains at least one node from $\mathcal{N}^{\geqq}$. If $\mathbf{Y} \subset \mathcal{N}^{\leqq} \cup \mathcal{N}^{=} \cup \mathcal{N}^{\geqq}$ and contains at least one node from $\mathcal{N}^{\leqq}$ and from $\mathcal{N}^{\geqq}$, it will be included either in $\mathbf{MB}$ or in $\mathbf{CB}$ by a rule specified in the algorithm. This rule is based on identifying a specific node in $\mathbf{Y}$ called the BCI (*blossom constraint identifier*) node, using the dual solution at that stage. If the BCI node is from $\mathcal{N}^{\leqq}$ ($\mathcal{N}^{\geqq}$) $\mathbf{Y}$ will be included in $\mathbf{MB}$ ($\mathbf{CB}$). The classes $\mathbf{MB}$, $\mathbf{CB}$ will be fully known only when the algorithm has terminated. The modified LP corresponding to the blossom constraint specification dictated by the choice of the classes $\mathbf{MB}$, $\mathbf{CB}$ is : find $x = (x_{ij})$ to

$$
\text{Minimize} \quad \sum (c_{ij} x_{ij} \quad : \quad \text{over } (i;j) \in \mathcal{A})
$$

$$
\text{Subject to} \quad x(i) \quad
\begin{cases}
\leqq 1, & \text{for } i \in \mathcal{N}^{\leqq} \\
= 1, & \text{for } i \in \mathcal{N}^{=} \\
\geqq 1, & \text{for } i \subset \mathcal{N}^{\geqq}
\end{cases}
\tag{10.55}
$$

$$
\mathbf{Y}^-(x) \leqq (|\mathbf{Y}| - 1)/2 \quad \text{for each } \mathbf{Y} \in \mathbf{MB}
$$

$$
\mathbf{Y}^+(x) \geqq (|\mathbf{Y}| + 1)/2 \quad \text{for each } \mathbf{Y} \in \mathbf{CB}
$$

$$
0 \leqq x_{ij} \leqq 1 \quad \text{for all } (i;j) \in \mathcal{A}
$$

It should be understood that (10.55) is not necessarily equivalent to (10.6), (10.7). However, we develop a blossom algorithm that obtains an optimum solution of (10.6), (10.7) provided it has a feasible solution, using a modified LP of the form (10.55) with the blossom constraint specification classes $\mathbf{MB}$, $\mathbf{CB}$ which are themselves obtained during the algorithm. Associate the original node price π_i for $i \in \mathcal{N}$ to the first set of constraints (node degree constraints) in (10.55); and the pseudonode price μ_σ to the blossom constraint corresponding to a subset of nodes $\mathbf{Y}_\sigma \in \mathbf{MB} \cup \mathbf{CB}$. Given the classes $\mathbf{MB}$, $\mathbf{CB}$, the dual feasibility conditions on $\pi = (\pi_i), \mu = (\mu_\sigma)$ reflecting the structure of the primal problem (10.55), are

$$\pi_i \quad \begin{cases} = 0, & \text{for all } i \in \mathcal{N}^0 \\ \leqq 0, & \text{for all } i \in \mathcal{N}^{\leqq} \\ \geqq 0, & \text{for all } i \in \mathcal{N}^{\geqq} \end{cases}$$

$$\qquad\qquad\qquad (10.56)$$

$$\mu_\sigma \quad \begin{cases} \leqq 0, & \text{for } \sigma \text{ such that } \mathbf{Y}_\sigma \in \mathbf{MB} \\ \geqq 0, & \text{for } \sigma \text{ such that } \mathbf{Y}_\sigma \in \mathbf{CB} \end{cases}$$

The dual of (10.55) is

$$\text{Maximize} \sum (\pi_i \quad : \quad \text{over } i \in \mathcal{N})$$

$$+ \sum (\mu_\sigma(|\mathbf{Y}_\sigma| - 1)/2 : \text{ over } \sigma \text{ s. t. } \mathbf{Y}_\sigma \in \mathbf{MB})$$

$$+ \sum (\mu_\sigma(|\mathbf{Y}_\sigma| + 1)/2 : \text{ over } \sigma \text{ s. t. } \mathbf{Y}_\sigma \in \mathbf{CB})$$

$$+ \sum (v_{ij} : \text{ over } (i;j) \in \mathcal{A})$$

subject to   (10.56) and   (10.57)

$$d_{ij}(\pi, \mu) + v_{ij} \;\leqq\; c_{ij}, \text{ for } (i;j) \in \mathcal{A}$$

$$v_{ij} \;\leqq\; 0, \text{ for } (i;j) \in \mathcal{A}$$

where for $(i;j) \in \mathcal{A}$

$$\mu(i;j) = \sum (\mu_\sigma : \text{ over } \sigma \text{ s. t. } \mathbf{Y}_\sigma \in \mathbf{MB} \text{ contains both } i \text{ and } j\,)$$

$$+ \sum (\mu_\sigma : \text{ over } \sigma \text{ s. t. } \mathbf{Y}_\sigma \in \mathbf{CB} \text{ contains either}$$

$$i \text{ or } j \text{ or both})$$

$$\qquad\qquad\qquad (10.58)$$

$$d_{ij}(\pi, \mu) = \pi_i + \pi_j + \mu(i;j)$$

Notice that the formula for $d_{ij}(\pi, \mu)$ in (10.58) is different from those in earlier sections, since the structure of the LP (10.55) here is different from those in earlier sections. From the structure of (10.57), it is clear that in an optimum solution $(\pi, \mu), v$ for it, we will have for each $(i;j) \in \mathcal{A}$, $v_{ij} = \min. \{0, c_{ij} - d_{ij}(\pi, \mu)\}$. Thus without any loss of generality, we can eliminate the variables v_{ij} from the dual objective function in (10.57) and express it purely in terms of (π, μ), leading to $W(\pi, \mu)$ defined below in (10.59).

$$W(\pi,\mu) \;=\; \sum(\pi_i : \text{ over } i \in \mathcal{N})$$
$$+\; \sum(\mu_\sigma(|\mathbf{Y}_\sigma|-1)/2 : \text{ over } \sigma \text{ s. t. } \mathbf{Y}_\sigma \in \mathbf{MB})$$
$$+\; \sum(\mu_\sigma(|\mathbf{Y}_\sigma|+1)/2 : \text{ over } \sigma \text{ s. t. } \mathbf{Y}_\sigma \in \mathbf{CB}) \qquad (10.59)$$
$$+\; \sum(\min. \{0,\, c_{ij} - d_{ij}(\pi,\mu)\} : \text{ over } (i;j) \in \mathcal{A})$$

The complementary slackness conditions for optimality in the primal, dual pair (10.55). (10.57) are: for all $i \in \mathcal{N}$ and $(i,j) \in \mathcal{A}$

$$\pi_i(x(i)-1) = 0 \qquad (10.60)$$

$$d_{ij}(\pi,\mu) \begin{cases} < c_{ij} \text{ implies } x_{ij} = 0 \\ > c_{ij} \text{ implies } x_{ij} = 1 \end{cases} \qquad (10.61)$$

$$\left.\begin{array}{l} \mu_\sigma((|\mathbf{Y}|-1)/2 - \mathbf{Y}^-(x)) = 0, \text{ for } \sigma \text{ s. t. } \mathbf{Y}_\sigma \in \mathbf{MB} \\ \mu_\sigma((|\mathbf{Y}|+1)/2 - \mathbf{Y}^+(x)) = 0, \text{ for } \sigma \text{ s. t. } \mathbf{Y}_\sigma \in \mathbf{CB} \end{array}\right\} \qquad (10.62)$$

The algorithm maintains a subset of edges $\mathbf{E} \subset \mathcal{A}$ called the *solution set of edges*, and the corresponding edge vector $x = x^{\mathbf{E}}$. Edges not in $\mathbf{E}$ (i.e., those $(i;j)$ with $x_{ij} = 0$) are called *nonsolution edges*. Let

$$\mathcal{A}^- = \{(i;j) : (i;j) \in \mathcal{A}, i \text{ and } j \text{ both in } \mathcal{N}^{\geqq} \cup \mathcal{N}^0, \text{ and } c_{ij} \lesseqgtr 0\} \qquad (10.63)$$

It is clear that if (10.6), (10.7) is feasible, then there exists a minimum cost 1-M/EC which contains $\mathcal{A}^-$ as a subset. The algorithm maintains $\mathbf{E} \supset \mathcal{A}^-$ always. Actually the initial solution set of edges $\mathbf{E}$ will be $\mathcal{A}^-$. The algorithm maintains a dual feasible solution (π,μ) such that $x, (\pi,\mu)$ together satisfy the complementary slackness conditions (10.60). (10.61), (10.62) always. And if $\mu_\sigma \neq 0$, $\mathbf{Y}_\sigma$ will be the set of original nodes inside an existing pseudonode. $\mathbf{E}$ is partitioned into $\mathbf{M}$, $\mathbf{A}$ where $\mathbf{M}$ is a matching ; and $\mathbf{A}$ is called the set of *covering edges* and it always includes $\mathcal{A}^-$, and every solution edge which is adjacent to another solution edge. Every connected component of $(\mathcal{N}, \mathbf{E})$ that consists of a single edge $(i;j)$ satisfying $d_{ij}(\pi,\mu) = c_{ij}$ will be in $\mathbf{M}$. Also, if there is a connected component of $(\mathcal{N}, \mathbf{A})$ that consists of a single edge $(i;j)$, it will satisfy $d_{ij}(\pi,\mu) > c_{ij}$. During the algorithm an original node i is said to be a

Matched node	if it is incident with a matching edge
Type 1 node	if it is incident with exactly one edge from **A**
Type 2 node	if $i \in \mathcal{N}^{\geqq} \cup \mathcal{N}^0$, and it is incident with $\geqq 2$ edges from **A**
Type 0 node	if there is no solution edge incident at it; and either $i \in \mathcal{N}^0$, or $i \in \mathcal{N}^{\leqq}$ and $\pi_i = 0$
Exposed node	if there is no solution edge incident at it; and either $i \in \mathcal{N}^{\geqq} \cup \mathcal{N}^=$, or $i \in \mathcal{N}^{\leqq}$ and $\pi_i < 0$

Original nodes of types 1, 2, 0, or matched nodes are called *nonexposed nodes*. Once a node becomes nonexposed, it remains nonexposed in the sequel, but its classification among matched, types 1, 2, 0 might change from step to step. The algorithm terminates either when there are no more exposed nodes in G, or when it becomes clear that it is impossible to convert any more exposed nodes into nonexposed nodes.

If (π, μ) is the present dual feasible solution, define $\mathcal{A}_*(\pi, \mu) = \{(i;j) : (i;j) \in \mathcal{A}$ and $d_{ij}(\pi, \mu) = c_{ij}\ \}$, where $d_{ij}(\pi, \mu)$ is defined as in (10.58). Edges in $\mathcal{A}_*(\pi, \mu)$ are the *original equality edges*, and $G_*(\pi, \mu) = (\mathcal{N}, \mathcal{A}_*(\pi, \mu))$ is the *equality subnetwork* of G wrt (π, μ). The algorithm plants an alternating tree at each exposed node and grows these trees wrt M in $G_*(\pi, \mu)$. During this process, simple blossoms may be discovered and shrunk into pseudonodes. The nodes in $\mathcal{N}^0$ or any node on an edge in $\mathcal{A}^-$ will never be contained on any simple blossom, and hence will never be inside any pseudonode. Blossoms and pseudonodes into which they are shrunk are classified by the configuration of the simple blossom corresponding to them. The various types of simple blossoms, blossoms, and pseudonodes are

| Types, rooted, A,B,C,D | defined exactly as in Section 10.2.2 |
| Type E | if the base node has no solution edge incident at it, but is either a type 0 node in $\mathcal{N}^{\leqq}$, or another lower level type E pseudonode |

Thus the apex node of a type E pseudonode is always the unique type 0 original node in $\mathcal{N}^{\leqq}$ contained inside it, with no solution edge incident at it. If $\mathcal{N}_B$ is the set of original nodes inside a type E pseudonode, there exists no solution edges joining a node in $\mathcal{N}_B$ to one outside it.

A newly formed pseudonode will always be a type A or a rooted pseudonode when it is formed. Types B,C,D,E pseudonodes are obtained through transformation of a type A or rooted pseudonode during an augmentation step. $G^1 = (\mathcal{N}^1, \mathcal{A}^1)$ denotes the current network. $\mathbf{M}^1, \mathbf{A}^1, \mathbf{E}^1 = \mathbf{M}^1 \cup \mathbf{A}^1$ denote the set of current matching, covering, and solution edges respectively. A type A (B) current pseudonode in G^1 is incident with a current matching edge (exactly one current covering edge). Type C,D,E and rooted current pseudonodes are incident with no current solution edges. Note that there is no pseudonode incident with 2 or more current solution edges.

The BCI Node of a Blossom

BCI (*blossom constraint identifier*) nodes are only defined for blossoms and pseudonodes containing nodes from both $\mathcal{N}^{\leqq}$ and $\mathcal{N}^{\geqq}$ inside them. Let $G_B = (\mathcal{N}_B, \mathcal{A}_B)$ be a blossom identified during the algorithm which has been shrunk into the pseudonode B. If $\mathcal{N}_B \cap \mathcal{N}^{\leqq}$ and $\mathcal{N}_B \cap \mathcal{N}^{\geqq}$ are both nonempty we identify an original node j tying for the minimum in (break ties arbitrarily)

$$\text{Min. } \{|\pi_i| : i \in \mathcal{N}_B \cap (\mathcal{N}^{\leqq} \cup \mathcal{N}^{\geqq})\} \tag{10.64}$$

A node j selected among those attaining the minimum in (10.64) is known as the BCI node for the blossom G_B or the pseudonode B. If the BCI node $j \in \mathcal{N}^{\leqq}$ ($j \in \mathcal{N}^{\geqq}$) $\mathcal{N}_B$ is included in the class **MB** (**CB**) and the matching (covering) blossom constraint corresponding to G_B is introduced. Once a BCI node for a pseudonode B is selected, it never changes in the algorithm as long as B remains, also its BCI node continues to tie for the minimum in (10.64) as long as B remains a current node. If the pseudonode B gets absorbed into another pseudonode, its BCI node may not tie for the minimum in (10.64) during the dormant period for B, but it will start being satisfied again as soon as the dormant period ends and B becomes current again.

If p is a pseudonode containing other pseudonodes inside it, it is possible that p has a BCI node different from the BCI nodes of pseudonodes inside it. Also, p may belong in the class **MB** or **CB**, and contain inside it other pseudonodes which belong in either class or both.

Properties Maintained By the Algorithm

Here we summarize the properties maintained by the algorithm. Labeled nodes are always either exposed nodes or matched nodes. Exposed nodes are the roots of alternating trees and labeled as outer nodes, so all inner nodes will always be matched nodes. All in-tree edges will always be current equality edges from the set $\mathcal{A}^1_*(\pi, \mu) \backslash \mathbf{A}^1$ at that stage. The following properties always hold.

(π, μ) is always dual feasible.

$\mathbf{M} \subset \mathcal{A}_*(\pi, \mu), \mathbf{A} \supset \mathcal{A}^-, \mathbf{E} \subset \mathcal{A}^- \cup \mathcal{A}_*(\pi, \mu)$.

If $(i; j) \notin \mathbf{E}$ then $d_{ij}(\pi, \mu) \overset{\leq}{=} c_{ij}$.

$x(i) > 1$ implies $i \in \mathcal{N}^0 \cup \mathcal{N}^{\overset{\geq}{=}}$ and $\pi_i = 0$.

$x(i) < 1$ implies that either $i \in \mathcal{N}^0 \cup \mathcal{N}^{\overset{\leq}{=}}$ with $\pi_i = 0$, or $i \in \mathcal{N} \backslash \mathcal{N}^0$ and is an exposed node.

$\mu_\sigma < 0$ implies that $\mathbf{Y}_\sigma^-(x) = (|\mathbf{Y}_\sigma| - 1)/2$; $\mu_\sigma > 0$ implies that either $\mathbf{Y}_\sigma^+(x) = (|\mathbf{Y}_\sigma| + 1)/2$, or $\mathbf{Y}_\sigma^+(x) = (|\mathbf{Y}_\sigma| - 1)/2$ with $\mathbf{Y}_\sigma$ being the set of original nodes inside a rooted pseudonode.

Nodes in $\mathcal{N}^0$ and those on edges in $\mathcal{A}^-$ are never contained inside any pseudonode, and $\pi_i = 0$ always for these nodes i.

No type 1,2, or 0 nodes, or type B,C,D, or E pseudonodes will ever be in-tree nodes.

At some stage of the algorithm, if $\mathbf{E}$ is not yet a 1-M/EC; and no augmentation, blossom shrinking, pseudonode unshrinking steps are possible; and the list of labeled and unscanned nodes is empty (i.e., no tree growth is possible), we have a Hungarian forest. The following properties are satisfied at that time, and the algorithm moves to a dual solution change step.

Hungarian Forest Conditions

Each inner node is either matched or type A, incident with a matching edge joining it to an outer node.

There exists no node $i \in \mathcal{N}^{\overset{\leq}{=}}$ which is an outer current node with $\pi_i = 0$.

There exists no inner node $i \in \mathcal{N}^{\overset{\geq}{=}} \cup \mathcal{N}^0$ with $\pi_i = 0$.

If $\mathbf{Y}_\sigma$ is the set of original nodes inside a current inner pseudonode, then $\mu_\sigma \neq 0$.

There exists no current equality edge joining an outer to a non-inner (outer or unlabeled) node.

There exists no outer current pseudonode containing inside it an original node $i \in \mathcal{N}^{\overset{\leq}{=}} \cup \mathcal{N}^{\overset{\geq}{=}}$ with $\pi_i = 0$.

In the blossom algorithm described below, list always refers to the set of labeled and unscanned outer current nodes. The tree growth step in the algorithm is constructed in such a way that only outer nodes are scanned.

BLOSSOM ALGORITHM FOR THE MINIMUM COST 1-M/EC PROBLEM

Step 1 Initialization   Initially let $\mathbf{M} = \emptyset, \mathbf{A} = \mathcal{A}^-, \mu = (\mu_\sigma) = 0$; and $\pi = (\pi_i)$ where $\pi_i = 0$ for all $i \in \mathcal{N}^{\overset{\geq}{=}} \cup \mathcal{N}^0$, and $= $ min. $\{0, c_{jg} : (j; g) \in \mathcal{A} \}$ for all $i \in \mathcal{N}^{\overset{\leq}{=}} \cup \mathcal{N}^=$. If there are no exposed nodes, go to Step 13. Otherwise root an alternating tree at each exposed node i by labeling it with $(\emptyset, +, i)$. List = set of all these root nodes now.

Step 2 Select a Node to Scan   If list $= \emptyset$ go to Step 12. Otherwise, select a node from it to scan, and delete it from the list.

Step 3 Scanning   Let the node to be scanned be p with label $(\mathrm{P}(p), +, r)$. Do the following in the order given.

If there is an already labeled outer node j associated with a root node $s \neq r$ such that $(p; j) \in \mathcal{A}_*^1(\pi, \mu)$, an augmenting path has been found, go to Step 4.

If there is an already labeled outer node j associated with the same root node r such that $(p; j) \in \mathcal{A}_*^1(\pi, \mu)$, the alternating tree containing p, j has blossomed, go to Step 10.

If there is an original current node $u \in \mathcal{N}^0 \cup \mathcal{N}^{\geqq}$ such that $(p; u) \in \mathcal{A}_*^1(\pi, \mu)$ and $\pi_u = 0$, go to Step 5.

If there is an original current type 1 node or a type B pseudonode v such that $(p; v) \in \mathcal{A}_*^1(\pi, \mu)$, go to Step 6.

If there is an unlabeled current node w that is a type C or D or E pseudonode, or a type 0 original node, such that $(p; w) \in \mathcal{A}_*^1(\pi, \mu)$, go to Step 7.

Identify all $j \neq \mathrm{P}(p)$ such that $(p; j) \in \mathcal{A}_*^1(\pi, \mu)$. Since we came to this stage, j cannot be type 0, 1, or 2, or B, C, D, or E, or exposed, or a rooted pseudonode (in that case we would have already gone to some other step). For each such j do the following. If j is unlabeled, let $(j; t)$ be the current matching edge incident at j, label j with $(p, -, r)$. If t is unlabeled and either $t \in \mathcal{N}^0 \cup \mathcal{N}^{\leqq}$ with $\pi_t = 0$, or t is a pseudonode inside which there is an original node $i \in \mathcal{N}^{\leqq} \cup \mathcal{N}^{\geqq}$ associated with node price $\pi_i = 0$ go to Step 8. Otherwise label t with $(j, +, r)$ and include it in the list.

Now p is labeled and scanned. Go to Step 11.

Step 4 Matching Augmentation   Same as Step 4 in the algorithm of Section 10.2.1.

Step 5 Matching and Covering Augmentation Procedure 1   Same as Step 5 in the algorithm of Section 10.2.1.

Step 6 Matching and Covering Augmentation Procedure 2   Same as Step 6 in the algorithm of Section 10.2.1.

Step 7 Matching and Covering Augmentation Procedure 3   We come to this step when scanning or inspection has revealed an outer current node p and a current node w which is either type C,D,E, or 0, and $(p; w) \in \mathcal{A}_*^1(\pi, \mu)$. If w is type C or D, this step is carried out exactly as Step 7 in the algorithm of Section 10.2.1.

Suppose w is either type 0 or E. Let i_1 be the apex node of w. Let $\mathcal{P}^1$ be the path obtained by adding $(p; w)$ to the predecessor path of p. Obtain the augmenting path $\mathcal{P}$ in G connecting i_1 with the apex of the root node associated with p, by expanding the pseudonodes along $\mathcal{P}^1$. Rematch using $\mathcal{P}$. Revise all the blossoms along $\mathcal{P}^1$ as in Section 10.1.2. Chop down the tree containing p. If there are no exposed nodes left, go to Step 13. Otherwise, change the list into the set of all outer labeled current nodes and go back to Step 2 or inspection in Step 12 depending on whether you arrived here from Step 3 or inspection.

Step 8 Matching and Covering Augmentation Procedure 4   We come to this step when scanning or inspection has identified an inner labeled current node j joined by a current matching edge to a current node t, where either $t \in \mathcal{N}^0 \cup \mathcal{N}^{\leq}$ with $\pi_t = 0$, or t is a pseudonode containing inside it an original node $i \in \mathcal{N}^{\leq} \cup \mathcal{N}^{\geq}$ with $\pi_i = 0$.

Let $\mathcal{P}^1$ be the predecessor path of j. Remove the matching edge joining the apex nodes of j and t from $\mathbf{M}$ and remove $(j;t)$ from $\mathbf{M}^1$. Rematch using $\mathcal{P}^1$ and revise all the blossoms along it as in Section 10.1.2. If t is a pseudonode and $i \in \mathcal{N}^{\geq}$ convert t into a type C or D pseudonode with i as its apex, as in Step 8 in the algorithm of Section 10.2.1. If t is a pseudonode and $i \in \mathcal{N}^{\leq}$, rematch within t to convert it into a type E pseudonode with i as its apex.

t gets converted into type 0, C, D, or E. If there are no exposed nodes left, go to Step 13. Otherwise, chop down the tree containing j, change the list into the set of all outer labeled current nodes, and go back to Step 2 or inspection in Step 12 depending on whether you arrived here from Step 3 or inspection.

Step 9 Matching and Covering Augmentation Procedure 5   We come to this step when inspection or blossom shrinking has identified an outer labeled current node t such that either $t \in \mathcal{N}^0 \cup \mathcal{N}^{\leq}$ with $\pi_t = 0$, or t is a pseudonode containing inside it an original node $i \in \mathcal{N}^{\leq} \cup \mathcal{N}^{\geq}$ with $\pi_i = 0$. Rematch the predecessor path of t and revise all the blossoms along it as in Section 10.1.2. If t is an original node, this process converts it into type 0. If t is a pseudonode, either convert it into type C or D with i as its apex if $i \in \mathcal{N}^{\geq}$, or into type E with i as its apex if $i \in \mathcal{N}^{\leq}$. If there are no exposed nodes left, go to Step 13. Otherwise chop down the tree containing t, change the list into the set of all outer labeled current nodes and go back to inspection in Step 12, or Step 10, or Step 2.

Step 10 Blossom Shrinking   We come to this step when scanning or inspection has revealed two outer labeled current nodes p, j with the same root index, such that $(p;j) \in \mathcal{A}^1_*(\pi, \mu)$. This is an indication that the alternating tree containing p, j has blossomed. The simple blossom will either be rooted or type A. Identify the simple blossom and shrink it into a pseudonode as in Step 6 of the algorithm of Section 10.1.1. Include the new pseudonode, say

B, which will now be an outer node, in the list. Let $\mathcal{N}_B$ be the set of original nodes inside this pseudonode. If $\mathcal{N}_B \subset \mathcal{N}^{\leq} \cup \mathcal{N}^{=}$ ($\mathcal{N}_B \subset \mathcal{N}^{\geq} \cup \mathcal{N}^{=}$ and $\mathcal{N}_B \cap \mathcal{N}^{\geq} \neq \emptyset$) include $\mathcal{N}_B$ in the class **MB** (**CB**). In these two cases no BCI node is defined for this pseudonode. If $\mathcal{N}_B \cap \mathcal{N}^{\leq} \neq \emptyset$ and $\mathcal{N}_B \cap \mathcal{N}^{\geq} \neq \emptyset$, then define the BCI node of the pseudonode B to be a node j attaining the minimum in (10.64), breaking ties arbitrarily. If $j \in \mathcal{N}^{\geq}$ ($j \in \mathcal{N}^{\leq}$) include $\mathcal{N}_B$ in the class **CB** (**MB**). If there is an $i \in \mathcal{N}^{=} \cup \mathcal{N}^{\geq}$ inside B with $\pi_i = 0$ carry out Step 9. Return to Step 2 or inspection in Step 12, depending on whether you came here from Step 3 or inspection.

Step 11 Type A Pseudonode Unshrinking   Same as Step 11 in the algorithm of Section 10.2.1.

Step 12 Dual Solution Change   We come to this step when the Hungarian forest conditions are satisfied. Let (π, μ) be the present dual feasible solution. Define (adopt the convention that the minimum in the empty set is $+\infty$)

$$\delta_1 = \text{Min. } \{-\pi_i : i \in \mathcal{N}^{\leq} \text{ is a current outer node}\}$$

$$\delta_2 = \text{Min. } \{\pi_i : i \in \mathcal{N}^{\geq} \text{ is a current inner node}\}$$

$$\delta_3 = \text{Min. } \{-\pi_i : i \in \mathcal{N}^{\leq} \text{ is inside a current outer labeled pseudonode in the class } \mathbf{MB}\}$$

$$\delta_4 = \text{Min. } \{\pi_i : i \in \mathcal{N}^{=} \text{ is inside a current outer labeled pseudonode in the class } \mathbf{CB}\}$$

$$\delta_5 = \text{Min. } \{-\tfrac{1}{2}\mu_\sigma : \mathbf{Y}_\sigma \in \mathbf{MB} \text{ is the set of original nodes in a current inner labeled pseudonode}\}$$

$$\delta_6 = \text{Min. } \{\tfrac{1}{2}\mu_\sigma : \mathbf{Y}_\sigma \in \mathbf{CB} \text{ is the set of original nodes in a current inner labeled pseudonode}\}$$

$$\delta_7 = \text{Min. } \{\tfrac{1}{2}(c_{ij} - d_{ij}(\pi,\mu)) : (i;j) \in \mathcal{A}, i \text{ and } j \text{ are inside distinct outer current nodes}\}$$

$$\delta_8 = \text{Min. } \{(c_{ij} - d_{ij}(\pi,\mu)) : (i;j) \in \mathcal{A}, \text{ one of } i, j \text{ is inside an outer current node, and the other is inside an unlabeled node}\}$$

$$\delta = \text{Min. } \{\delta_1, \ldots, \delta_8\}$$

If $\delta = +\infty$, go to Step 14. If δ is finite, it will be > 0 (proved below), define the new dual solution $\hat{\pi} = (\hat{\pi}_i)$, $\hat{\mu} = (\hat{\mu}_\sigma)$ where

$$\hat{\pi}_i = \begin{cases} \pi_i + \delta & \text{if } i \text{ is an outer current node, or is inside an outer} \\ & \text{current pseudonode in } \mathbf{MB}, \text{ or is inside an inner} \\ & \text{current pseudonode in } \mathbf{CB} \\ \pi_i - \delta & \text{if } i \text{ is an inner current node, or is inside an inner} \\ & \text{current pseudonode in } \mathbf{MB}, \text{ or is inside an outer} \\ & \text{current pseudonode in } \mathbf{CB} \\ \pi_i & \text{if } i \text{ is inside an unlabeled node} \end{cases}$$

$$\hat{\mu}_\sigma = \begin{cases} \mu_\sigma + 2\delta & \text{if } \mathbf{Y}_\sigma \in \mathbf{MB} \text{ forms a current inner labeled} \\ & \text{pseudonode, or if } \mathbf{Y}_\sigma \in \mathbf{CB} \text{ forms a current outer} \\ & \text{labeled pseudonode} \\ \mu_\sigma - 2\delta & \text{if } \mathbf{Y}_\sigma \in \mathbf{MB} \text{ forms a current outer labeled} \\ & \text{pseudonode, or if } \mathbf{Y}_\sigma \in \mathbf{CB} \text{ forms a current in-} \\ & \text{ner labeled pseudonode} \\ \mu_\sigma & \text{otherwise} \end{cases}$$

Inspection

If $\delta = \delta_1$, look for a $t \in \mathcal{N}^{\leq}_{=}$ which is an outer current node associated with $\hat{\pi}_t = 0$, and then apply Step 8.

If $\delta = \delta_2$, look for a $u \in \mathcal{N}^{\geq}_{=}$ which is an inner current node associated with $\hat{\pi}_u = 0$, and then apply Step 5.

If $\delta = \delta_3$, look for an $i \in \mathcal{N}^{\leq}_{=}$ associated with $\hat{\pi}_i = 0$, that is inside an outer current pseudonode t with its predecessor j, and then apply Step 8.

If $\delta = \delta_4$, look for an $i \in \mathcal{N}^{\geq}_{=}$ associated with $\hat{\pi}_i = 0$, that is inside an outer current pseudonode t with its predecessor j, and then apply Step 8.

If $\delta = \delta_5$ or δ_6, look for an inner current pseudonode associated with the set of original nodes $\mathbf{Y}_\sigma$ for which $\hat{\mu}_\sigma = 0$, and unshrink it as in Step 11.

If $\delta = \delta_7$, look for current equality edges $(p; j)$ joining two outer current nodes, and perform Step 4 if p, j have different root indices, or Step 10 if they have the same root index.

If $\delta = \delta_8$, look for outer nodes p and unlabeled nodes j such that $(p; j) \in \mathcal{A}^1_*(\hat{\pi}, \hat{\mu})$, include such nodes p in the list.

Repeat the above checks as many times as possible. Every node or edge whose price or weight led to the value of δ must be checked. Then go to Step 2.

Step 13 Optimality   We reach this step when there are no exposed nodes left. Let $\overline{\mathbf{E}}, \overline{x}, (\overline{\pi}, \overline{\mu})$ be the present solution set of edges, its incidence vector, and the dual solution at this stage. $\overline{\mathbf{E}}$ is a minimum cost 1-M/EC in G, $\overline{x}$ is the associated minimum cost 1-M/EC vector, and $(\overline{\pi}, \overline{\mu})$ the optimum dual solution (optimum for (10.57)). Terminate.

Step 14 Infeasibility   We come to this step if $\delta = +\infty$ in a dual solution change step. In this case there exists no 1-M/EC in G wrt the given partition of the nodes, i.e., (10.6), (10.7) is infeasible. Terminate.

Validity of the Algorithm and Its Computational Complexity

By verifying for each step, we can check that the algorithm maintains all the properties claimed for it, and that the Hungarian forest conditions hold whenever the algorithm arrives at Step 12. These properties also guarantee that $\delta \overset{\geq}{=} 0$ in Step 12. For δ to be 0, at least one among δ_1 to δ_8 must be 0, and if this happens the algorithm would have gone to the appropriate Steps 4 to 11 during scanning and augmented the matching and covering, or shrunk a simple blossom, or unshrunk a type A pseudonode, or labeled an unlabeled node, violating the Hungarian forest conditions at the time of arrival at Step 12. Hence $\delta > 0$ in Step 12 always.

Using exactly the same arguments as in Section 10.2.1, it can be shown that the computational effort required by this algorithm is bounded above by $O(n^2 m)$, which can be reduced to $O(n^3)$ by implementing it with the approach discussed in Section 10.1.2.

If $\overline{x}, (\overline{\pi}, \overline{\mu})$ are the final solution and dual vectors when the algorithm terminates in Step 13, then clearly $\overline{x}$ is optimal to (10.55), and $(\overline{\pi}, \overline{\mu})$ is optimal to its dual (10.57) for the classes **MB, CB** as at the termination of the algorithm. It should be noted that (10.55) depends on the choice of the classes **MB, CB**, and thus is not necessarily equivalent to (10.6), (10.7). We will now prove that $\overline{x}$ is optimal to (10.6), (10.7) even though (10.55) and (10.6), (10.7) are not necessarily equivalent.

THEOREM 10.25 *Let x be an integer feasible solution of (10.6). Let $\mathbf{Y}_\sigma \subset \mathcal{N}$ with $|\mathbf{Y}_\sigma| \overset{\geq}{=} 3$ and odd. Then either $\mathbf{Y}_\sigma^-(x) \overset{\leq}{=} (|\mathbf{Y}_\sigma|-1)/2$, or $\mathbf{Y}_\sigma^+(x) \overset{\geq}{=} (|\mathbf{Y}_\sigma|+1)/2$.*

Proof   Suppose there is an integer feasible solution x of (10.6) which violates the hypothesis of the theorem. Then $\mathbf{Y}_\sigma^-(x) > (|\mathbf{Y}_\sigma|-1)/2$, $\mathbf{Y}_\sigma^+(x) < (|\mathbf{Y}_\sigma|+1)/2$. But $\mathbf{Y}_\sigma^-(x) \overset{\leq}{=} \mathbf{Y}_\sigma^+(x)$. Hence $(|\mathbf{Y}_\sigma| + 1)/2 > \mathbf{Y}_\sigma^+(x) \overset{\geq}{=} \mathbf{Y}_\sigma^-(x) > (|\mathbf{Y}_\sigma| - 1)/2$, a contradiction since all these quantities are integers. ∎

THEOREM 10.26 *Let B be a pseudonode formed during the algorithm, associated with the pseudonode price $\mu(B)$. Suppose this pseudonode B itself gets absorbed inside another pseudonode B_1. Let $(\pi', \mu'), (\tilde{\pi}, \tilde{\mu})$ be the dual solutions at the beginning, and at the end respectively, of this dormant period for pseudonode B. Then $\mu'(B) = \tilde{\mu}(B)$, and $\pi_i' = \tilde{\pi}_i$ for all original nodes i inside B.*

Proof   During this entire dormant period, the pseudonode price of B does not change at all and so it remains equal to $\mu'(B)$. So, $\mu'(B) = \tilde{\mu}(B)$

The pseudonode price of any newly formed pseudonode is 0 just when it is formed. So $\mu'(B_1) = 0$, at the beginning of this dormant period. As long as B_1 remains current, the change in $\mu(B_1)$ is (-2) times the change in π_i for original nodes i contained inside it, in any dual solution change step. So, for any original node i contained inside B, the net effect on π_i of all the dual solution change steps during the time that B_1 remains a current node, is $(-1/2)$ times the net effect on $\mu(B_1)$ in the same steps. And B_1 will not be unshrunk until its pseudonode price becomes 0 again. So, for $i \in B$, the net effect on π_i of all the dual solution change steps during the time that B_1 remains a current node from the time it is formed to the time it is unshrunk again, is 0.

It is possible that B_1 itself gets absorbed inside another pseudonode B_2 before it is unshrunk. As long as B_2 remains current, the value of $\mu(B_1)$ remains unchanged and B_1 will not be unshrunk until B_2 is unshrunk at some stage first. Using the same arguments as above, it can be verified that for all original nodes $i \in B$, the net effect on π_i, of all the dual solution change steps during the time that B_2 remains a current node from the time it is formed to the time it is unshrunk again, is 0. It is possible that B_2 itself goes into dormancy and gets absorbed into another pseudonode B_3. The same argument can be applied again.

The dormant period for pseudonode B ends only when every pseudonode containing B inside it, formed since the beginning of this period, is unshrunk and B becomes a current node again. By repeating the above argument for every pseudonode containing B inside it, formed during this period and unshrunk, we conclude that the net effect of all the dual solution change steps carried out during this dormant period on π_i for $i \in B$ is 0. So, $\pi'_i = \tilde{\pi}_i$ for all $i \in B$ ∎

THEOREM 10.27 *Let $(\pi = (\pi_i), \mu)$ be the dual solution at some stage of the algorithm. Suppose pseudonode B is a current node at this stage, containing some nodes from both the sets $\mathcal{N}^{\leqq}$ and $\mathcal{N}^{\geqq}$. For $j \in \mathcal{N}^{\leqq} \cup \mathcal{N}^{\geqq}$ contained inside B, let f_j denote the node price of j at the time that pseudonode B was just formed for the last time. Also, let p be the BCI node of pseudonode B. Let $\mu(B)$ be the pseudonode price of B in the present dual solution. Then for all $j \in \mathcal{N}^{\leqq} \cup \mathcal{N}^{\geqq}$ inside B, and for p we have*

$$
\begin{array}{rll}
\text{(i)} & f_j & = \quad \pi_j + \tfrac{1}{2}\mu(B) \\
\text{(ii)} & |f_p| & = \quad |\pi_p| + \tfrac{1}{2}|\mu(B)| \\
\text{(iii)} & |f_j| & \geqq \quad \tfrac{1}{2}|\mu(B)| \\
\text{(iv)} & |f_p| & \geqq \quad |\pi_p| \\
\text{(v)} & |\pi_p| & = \quad Min. \ \{|\pi_i| : i \in \mathcal{N}^{\leqq} \cup \mathcal{N}^{\geqq} \ inside \ B\}
\end{array}
$$

Proof While B is a current node, whenever Step 12 is carried out, the change in the pseudonode price of B is (-2) times the change in the node price for original nodes inside B. Also, the pseudonode price associated with B is 0 just when it was formed. Hence (i) holds.

Since p is the BCI node of B, p is inside B and $f_p, \pi_p, \mu(B)$ all have the same sign. From these, and by applying (i) to p, we get (ii).

By the definition of the BCI node, we have $|f_j| \geqq |f_p|$ for all nodes $j \in \mathcal{N}^{\leqq} \cup \mathcal{N}^{\geqq}$ inside B. Using this and (ii) leads to (iii).

(iv) follows from (ii).

From the definition of the BCI node, we have $|f_p| = $ min. $\{|f_i| : i \in \mathcal{N}^{\leqq} \cup \mathcal{N}^{\geqq}$ inside $B\} = $ min. $\{|\pi_i + \frac{1}{2}\mu(B)| : i \in \mathcal{N}^{\leqq} \cup \mathcal{N}^{\geqq}$ inside $B \}$ by (i). So, by (ii), and since π_p and $\mu(B)$ have the same sign, $|\pi_p + \frac{1}{2}\mu(B)| = $ min. $\{|\pi_i + \frac{1}{2}\mu(B)| : i \in \mathcal{N}^{\leqq} \cup \mathcal{N}^{\geqq}$ inside $B \}$. This implies (v). ∎

THEOREM 10.28 *Let (π, μ) be the dual solution at some stage of the algorithm. Suppose pseudonode B containing some original nodes from both $\mathcal{N}^{\leqq}, \mathcal{N}^{\geqq}$, with BCI node p is dormant at this stage. For $j \in \mathcal{N}^{\leqq} \cup \mathcal{N}^{\geqq}$ contained inside B let $f_j [g_j]$ be the node price of j at the time that B was just formed [entered dormancy] for the last time. Let $\mu(B)[\mu'(B)]$ be the pseudonode price associated with B at present [at the beginning of this dormant period]. Then*

$$
\begin{array}{lll}
(i) & \mu(B) &= \mu'(B) \\
(ii) & f_j &= g_j + \frac{1}{2}\mu(B) \text{ for all original nodes } j \text{ inside } B \\
(iii) & |f_p| &= |g_p| + \frac{1}{2}|\mu(B)| \\
(iv) & |f_j| &\geqq \frac{1}{2}|\mu(B)|, \text{ for } j \in \mathcal{N}^{\leqq} \cup \mathcal{N}^{\geqq} \text{ inside } B \\
(v) & |g_p| &= \text{ min. } \{|g_j| : j \in \mathcal{N}^{\leqq} \cup \mathcal{N}^{\geqq} \text{ inside } B\}
\end{array}
$$

Proof (i) follows because the pseudonode price of any pseudonode does not change at all during dormancy.

By Theorem 10.26, and the manner in which the dual solution is updated in Step 12, the net effect of all the dual solution change steps on the node price for original nodes i inside B is $(-1/2)$ times the net effect on the pseudonode price of B in the same steps. Also, the pseudonode price of B was 0 when it was just formed. These facts imply (ii).

(iii) follows by applying (ii) for $j = p$, because $f_p, g_p, \mu(B)$ all have the same sign.

From (iii) we have $|f_p| \geqq \frac{1}{2}|\mu(B)|$. Also, by the definition of the BCI node we have $|f_j| \geqq |f_k|$ for all $j \in \mathcal{N}^{\leqq} \cup \mathcal{N}^{\geqq}$ inside B. These two together imply (iv).

By the definition of the BCI node we have $|f_p| = $ min. $\{ |f_j| : j \in \mathcal{N}^{\leqq} \cup \mathcal{N}^{\geqq}$ and inside $B \}$. Using (ii) in this we get $|g_p + \frac{1}{2}\mu(B)| = $ min. $\{ |g_j + \frac{1}{2}\mu(B)| : j \in \mathcal{N}^{\leqq} \cup \mathcal{N}^{\geqq}$ and inside $B \}$. And from (iv) we have

$$
|g_j + \frac{1}{2}\mu(B)| \begin{cases} = |g_j| + \frac{1}{2}|\mu(B)|, & \text{if } g_j \text{ and } \mu(B) \text{ have the same sign} \\ \leqq |g_j| + \frac{1}{2}|\mu(B)|, & \text{if } g_j \text{ and } \mu(B) \text{ have opposite signs} \end{cases}
$$

Since g_p and $\mu(B)$ have the same sign, we have $|g_p + \frac{1}{2}\mu(B)| = |g_p| + \frac{1}{2}|\mu(B)|$. These facts together imply (v). ∎

THEOREM 10.29 *Let π, μ be the dual solution at some stage of the algorithm. For each dormant original node $i \in \mathcal{N}^{\leq} \cup \mathcal{N}^{\geq}$, let y_i be the node price associated with it at the time that it became a noncurrent node for the last time. Then $|y_i| \geq \frac{1}{2}\sum(|\mu_\sigma| : over\ \sigma\ such\ that\ i \in \mathbf{Y}_\sigma)$.*

Proof Let $B_1, \ldots, B_r$ be the sequence of all the shrunken blossoms containing i inside them, where i is inside B_1, and B_t is inside B_{t+1} for each $t = 1$ to $r - 1$, and B_r is a current node. Let $\mu_1, \ldots, \mu_r$ be the pseudonode prices of $B_1, \ldots, B_r$ at this stage. Let f_i^t be the node price of i when B_t was just shrunk for the last time.

If none of the pseudonodes containing node i inside it, have a BCI node defined for them, the result in this theorem follows directly from the manner in which the dual solution is updated whenever Step 12 is carried out in this algorithm, because the original node price π_i of i, and the pseudonode price μ_σ of any pseudonode containing node i inside it, always have the same sign in this case.

Now consider the case where the lowest level pseudonode containing node i, B_1, itself has a BCI node defined for it. In this case, all the pseudonodes $B_1, \ldots, B_r$ have BCI nodes defined for them. Let j be any original node inside the current pseudonode B_r. The proof in this case is by induction. We now set up an induction hypothesis.

Induction Hypothesis   For some u satisfying $1 < u \leq r$, we have $|f_j^s| \geq \frac{1}{2}(|\mu_r| + |\mu_{r-1}| + \ldots, |\mu_s|)$ for all $u \leq s \leq r$ and for all $j \in \mathcal{N}^{\leq} \cup \mathcal{N}^{\geq}$ inside B_s.

The induction hypothesis holds for $u = r$ by (iii) of Theorem 10.27. We will now prove that under the induction hypothesis, the statement in it must also hold for $s = u - 1$. Let p be the BCI node of B_{u-1}. By (i) and (iii) of Theorem 10.28, we have $|f_p^{u-1}| = |f_p^u| + \frac{1}{2}|\mu_{u-1}|$. Since p is the BCI node of B_{u-1}, we have for all $j \in \mathcal{N}^{\leq} \cup \mathcal{N}^{\geq}$ inside B_{u-1}

$$|f_j^{u-1}| \geq |f_p^{u-1}| \geq |f_p^u| \geq \frac{1}{2}(|\mu_r| + \ldots + |\mu_u|)$$

This shows that if the induction hypothesis holds for u where $1 < u \leq r$, then it also holds for $u - 1$. So, by induction it must hold for $u = 1$ too. But by definition, $y_i = f_i^1$ for all $i \in \mathcal{N}^{\leq} \cup \mathcal{N}^{\geq}$ inside B_1, and so the result in the theorem follows in this case by applying the statement in the induction hypothesis for $u = 1$.

Now consider the case where some of the pseudonodes containing node i inside them do not have a BCI node defined for them, and some of the others do. In this case, the proof of the result in the theorem can be accomplished by combining the arguments in the two cases discussed above. ∎

THEOREM 10.30 *Let $\pi' = (\pi_i'), \mu' = (\mu_\sigma')$ be the dual feasible solution at some stage of the algorithm. For every original node i inside a pseudonode (dormant or* *current) at this stage, let y_i be the original node price of i in the step that node i entered dormancy for the last time in the algorithm. Then $y_i = \pi_i' + \frac{1}{2}\sum(\mu_\sigma' : over \sigma \text{ such that } i \in \mathbf{Y}_\sigma)$.*

Proof For a dormant original node i, whenever Step 12 occurs, the change in the node price π_i is $(-1/2)$ times the change in the pseudonode price of the outermost pseudonode containing i, and the pseudonode price of any other dormant pseudonodes containing i at that stage does not change at all in that step. The theorem follows from this fact. ∎

THEOREM 10.31 *Let $x', (\pi' = (\pi_i'), \mu' = (\mu_\sigma'))$ be feasible solutions to (10.55), (10.57) respectively when the blossom constraint specification classes are chosen as $\mathbf{MB'}, \mathbf{CB'}$. Let B be a pseudonode at this stage, either current or dormant, with $\mathbf{Y}_{\sigma_1}$ as the set of original nodes inside it, with a BCI node. Suppose we move $\mathbf{Y}_{\sigma_1}$ from the class among $\mathbf{MB'}, \mathbf{CB'}$ in which it is contained, into the other class. Let $\mathbf{MB''}, \mathbf{CB''}$ be the resulting blossom constraint specification classes. Define $\pi'' = (\pi_i''), \mu'' = (\mu_\sigma'')$ where $\pi_i'' = \pi_i' + \mu_{\sigma_1}'$ for all $i \in \mathbf{Y}_{\sigma_1}$, and $= \pi_i'$ for all $i \notin \mathbf{Y}_{\sigma_1}$; $\mu_\sigma'' = -\mu_{\sigma_1}'$ for $\sigma = \sigma_1$, and $= \mu_\sigma'$ for $\sigma \neq \sigma_1$. Then (π'', μ'') is feasible to the modified dual (10.57) corresponding to blossom constraint specification classes $\mathbf{MB''}, \mathbf{CB''}$, and $W(\pi', \mu') = W(\pi'', \mu'')$.*

Proof Clearly μ_σ'' satisfies the sign restrictions for dual feasibility for all σ. Original node prices outside $\mathbf{Y}_{\sigma_1}$ keep their values unchanged and hence continue to satisfy the sign restrictions for dual feasibility. It remains to be proved that $\pi_i'' \stackrel{\leq}{=} 0$ for $i \in \mathcal{N}^{\stackrel{\leq}{=}} \cap \mathbf{Y}_{\sigma_1}$ and $\pi_i'' \stackrel{\geq}{=} 0$ for $i \in \mathcal{N}^{\stackrel{\geq}{=}} \cap \mathbf{Y}_{\sigma_1}$. Define y_i for $i \in \mathbf{Y}_{\sigma_1}$ as in Theorem 10.30. Then by Theorems 10.27, 10.28, 10.30, for $i \in \mathbf{Y}_{\sigma_1}$, $y_i = \pi_i' + \frac{1}{2}\sum(\mu_\sigma' : over \sigma \text{ such that } i \in \mathbf{Y}_\sigma) = (\pi_i' + \mu_{\sigma_1}') + \frac{1}{2}(\sum((\mu_\sigma' : over \sigma \text{ such that } i \in \mathbf{Y}_\sigma)) - 2\mu_{\sigma_1}') = \pi_i'' + \frac{1}{2}\sum(\mu_\sigma'' : over \sigma \text{ such that } i \in \mathbf{Y}_\sigma)$. So, $\pi_i'' = y_i - \frac{1}{2}\sum(\mu_\sigma'' : over \sigma \text{ such that } i \in \mathbf{Y}_\sigma)$. Since dual feasibility is maintained during the algorithm, we have $y_i \stackrel{\leq}{=} 0$ for $i \in \mathcal{N}^{\stackrel{\leq}{=}} \cap \mathbf{Y}_{\sigma_1}$ and $y_i \stackrel{\geq}{=} 0$ for $i \in \mathcal{N}^{\stackrel{\geq}{=}} \cap \mathbf{Y}_{\sigma_1}$. This together with the result in Theorem 10. 29, and the formula derived above for π_i'' implies that $\pi_i'' \stackrel{\leq}{=} 0$ for $i \in \mathcal{N}^{\stackrel{\leq}{=}} \cap \mathbf{Y}_{\sigma_1}$ and $\pi_i'' \stackrel{\geq}{=} 0$ for $i \in \mathcal{N}^{\stackrel{\geq}{=}} \cap \mathbf{Y}_{\sigma_1}$, establishing that (π'', μ'') satisfies the sign restrictions for dual feasibility.

It can be verified that $d_{ij}(\pi'', \mu'') = d_{ij}(\pi', \mu')$ for all $(i;j) \in \mathcal{A}$. From this and the above result, it follows that (π'', μ'') is feasible to the modified dual corresponding to the blossom constraint specification classes $\mathbf{MB''}, \mathbf{CB''}$.

From the fact that $d_{ij}(\pi', \mu') = d_{ij}(\pi'', \mu'')$ for all $(i;j) \in \mathcal{A}$, and the definition of π'', μ'', it easily follows that $W(\pi'', \mu'') = W(\pi', \mu')$. ∎

THEOREM 10.32 *If the algorithm discussed above terminates in Step 13, $\bar{x}$, the edge vector at termination, is a minimum cost 1-M/EC vector in G.*

Proof Let $(\bar{\pi}, \bar{\mu})$ be the dual solution at termination. Let $\sigma = 1$ to L_1 correspond to all the blossoms at all levels that are in existence at termination. Of those, let $\sigma = 1$ to L_2 refer to subsets of $\mathcal{N}^{\lessgtr} \cup \mathcal{N}^{=}$ or $\mathcal{N}^{\gtrless} \cup \mathcal{N}^{=}$ for which the type of blossom constraint (matching or covering) is known, and let $\sigma = L_2 + 1$ to L_1 correspond to subsets of nodes containing at least one node from each of $\mathcal{N}^{\lessgtr}$ and $\mathcal{N}^{\gtrless}$.

Let $\mathbf{Y}_\sigma$, $\sigma = L_1 + 1$ to L be all the other odd subsets of $\mathcal{N} \backslash \mathcal{N}^0$ of cardinality $\gtrless 3$, the blossom constraint corresponding to which do not appear at the termination of the algorithm. As discussed earlier, every 1-M/EC vector has to satisfy either the matching or the covering blossom constraint, or possibly both, corresponding to $\mathbf{Y}_\sigma$ for every $\sigma = 1$ to L. Because of this, in formulating the modified problem (10.55), if we eliminate all the blossom constraints (both matching and covering type) corresponding to $\mathbf{Y}_\sigma$ for $\sigma = L_1 + 1$ to L, and yet obtain an integer feasible solution for the resulting modified problem, then that integer solution must be an optimum solution of the original problem (10.6), (10.7).

Let $\overline{\mathbf{MB}}, \overline{\mathbf{CB}}$ be the blossom constraint specification classes at termination. With this specification, $\overline{x}$ is an optimum solution to the modified problem (10.55) as discussed earlier, and it is an integer vector. So, $\overline{x}$ is a 1-M/EC vector and its cost $z(\overline{x}) = W(\overline{\pi}, \overline{\mu})$ by the duality theorem of LP. Also, by applying Theorem 10.31 repeatedly, we see that if the blossom constraint specifications are given by any sets $\mathbf{MB}$, $\mathbf{CB}$ obtained from $\overline{\mathbf{MB}}, \overline{\mathbf{CB}}$ by moving some of the $\mathbf{Y}_\sigma$ for σ between $L_2 + 1$ to L_1 that are in $\overline{\mathbf{MB}}$ into the set $\overline{\mathbf{CB}}$ and vice versa, then the dual of the corresponding modified problem, has a dual feasible solution (π, μ) satisfying $W(\pi, \mu) = W(\overline{\pi}, \overline{\mu})$. This, by the weak duality theorem of LP implies that the optimum objective value in the corresponding modified problem for that blossom constraint specification, is $\gtrless W(\overline{\pi}, \overline{\mu}) = z(\overline{x})$. Hence $\overline{x}$ gives the minimum value for $z(x)$ among all the integer feasible solutions of modified problems given by such blossom constraint specifications. This and the earlier arguments clearly imply that $\overline{x}$ is the optimum solution of (10.6), (10.7), i.e., a minimum cost 1-M/EC vector in G. ∎

The Infeasibility Conclusion

THEOREM 10.33 *At each execution of Step 12 in the blossom algorithm discussed in this section, the dual objective value $W(\pi, \mu)$ increases by δ times the number of exposed nodes at that stage.*

Proof   Let (π, μ) be the dual feasible solution when the algorithm arrives at an occurrence of Step 12, and $(\hat{\pi} = (\hat{\pi}_i), \hat{\mu} = (\hat{\mu}_\sigma))$ the new dual feasible solution at the end of this step. Let $\mathbf{Y}_\sigma$ be the set of original nodes inside a current pseudonode at this stage. Let $a_\sigma = (|\mathbf{Y}_\sigma| - 1)/2$ if $\mathbf{Y}_\sigma \in \mathbf{MB}$, or $(|\mathbf{Y}_\sigma| + 1)/2$ if $\mathbf{Y}_\sigma \in \mathbf{CB}$. Then

$$\left(a_\sigma \hat{\mu}_\sigma + \sum(\hat{\pi}_i : \text{ over } i \in \mathbf{Y}_\sigma)\right) - \left(a_\sigma \mu_\sigma + \sum(\pi_i : \text{ over } i \in \mathbf{Y}_\sigma)\right)$$

$$= a_\sigma(\hat{\mu}_\sigma - \mu_\sigma) + \sum(\hat{\pi}_i - \pi_i : \text{ over } i \in \mathbf{Y}_\sigma)$$

$$= \begin{cases} \delta, & \text{if the pseudonode is outer} \\ -\delta, & \text{if the pseudonode is inner} \\ 0, & \text{if the pseudonode is unlabeled} \end{cases}$$

Since $d_{ij}(\pi, \mu) \lesseqgtr c_{ij}$ for all $(i; j) \notin \mathcal{A} \backslash \mathcal{A}^-$, min. $\{0, c_{ij} - d_{ij}(\pi, \mu)\} \neq 0$ only for edges $(i; j) \in \mathcal{A}^-$, and these edges are always current edges, with $\pi_p = \hat{\pi}_p = 0$ for all nodes p on these edges. Hence the term $\sum(\text{min. } \{0, c_{ij} - d_{ij}(\pi, \mu)\}$: over $(i; j) \in \mathcal{A})$ remains unchanged during any execution of Step 12.

Also $\hat{\mu}_\sigma = \mu_\sigma$ if $\mathbf{Y}_\sigma$ is the set of original nodes in a dormant pseudonode at this stage. The number of outer current nodes at this stage is clearly = the number of exposed nodes + the number of inner current nodes, because of the Hungarian forest conditions holding at the beginning of Step 12. These facts imply the result in this theorem. ∎

THEOREM 10.34 *If the value of δ turns out to be $+\infty$ in a dual solution change step during this algorithm, there exists no 1-M/EC in G with the given partition of $\mathcal{N}$.*

Proof Let (π, μ) be the dual solution at the beginning of that dual solution change step, and **MB, CB** the classes giving the blossom constraint specifications at that stage. Let $\sigma = 1$ to L_1 correspond to all the blossoms at all levels that are in existence at this time. Of these, let $\sigma = 1$ to L_2 refer to subsets of $\mathcal{N}^{\lesseqgtr} \cup \mathcal{N}^=$ or $\mathcal{N}^{\gtreqless} \cup \mathcal{N}^=$ for which the type of blossom constraint (matching or covering) is known, and let $\sigma = L_2 + 1$ to L_1 correspond to subsets of nodes containing at least one node from each of $\mathcal{N}^{\lesseqgtr}$ and $\mathcal{N}^{\gtreqless}$. So, **MB**∪**CB** contain all the $\mathbf{Y}_\sigma$ for $\sigma = 1$ to L_1.

Define $\hat{\pi}(\lambda) = (\hat{\pi}_i(\lambda)), \hat{\mu}(\lambda) = (\hat{\mu}_\sigma(\lambda))$ by

$$\hat{\pi}_i(\lambda) = \begin{cases} \pi_i + \lambda & \text{if } i \text{ is an outer current node, or is inside an outer current pseudonode in } \mathbf{MB}, \text{ or is inside an inner current pseudonode in } \mathbf{CB} \\ \pi_i - \lambda & \text{if } i \text{ is an inner current node, or is inside an inner current pseudonode in } \mathbf{MB}, \text{ or is inside an outer current pseudonode in } \mathbf{CB} \\ \pi_i & \text{if } i \text{ is inside an unlabeled node} \end{cases}$$

$$\hat{\mu}_\sigma(\lambda) = \begin{cases} \mu_\sigma + 2\lambda & \text{if } \mathbf{Y}_\sigma \in \mathbf{MB} \text{ forms a current inner labeled pseudonode, or if } \mathbf{Y}_\sigma \in \mathbf{CB} \text{ forms a current outer labeled pseudonode} \\ \mu_\sigma - 2\lambda & \text{if } \mathbf{Y}_\sigma \in \mathbf{MB} \text{ forms a current outer labeled pseudonode, or if } \mathbf{Y}_\sigma \in \mathbf{CB} \text{ forms a current inner labeled pseudonode} \\ \mu_\sigma & \text{otherwise} \end{cases}$$

Then it can be verified that $(\hat{\pi}(\lambda), \hat{\mu}(\lambda))$ is feasible to (10.57) with the blossom constraint specification classes **MB**, **CB** as at present, for all $\lambda \geqq 0$, and that $W(\hat{\pi}(\lambda), \hat{\mu}(\lambda)) = W(\pi, \mu) + \lambda\gamma$, where $\gamma =$ number of exposed nodes at this stage. Since $\gamma \geqq 1$ (by the Hungarian forest conditions there exists at least one exposed node at this stage) as $\lambda \to +\infty$, $W(\hat{\pi}(\lambda), \hat{\mu}(\lambda)) \to +\infty$. So, by the duality theory of LP (10.55) is infeasible with the present blossom constraint specification classes **MB**, **CB**. Using the arguments in the proof of Theorem 10.31, it can be shown that if $\overline{\textbf{MB}}, \overline{\textbf{CB}}$ is a blossom constraint specification obtained from **MB**, **CB** by moving some of the $\textbf{Y}_\sigma$ for σ between L_2+1 to L_1 that were in **MB** into the set **CB** and vice versa, the same conclusion holds. As discussed earlier, if a 1-M/EC vector exists in G, it must satisfy either the matching or the covering blossom constraint corresponding to every $\sigma = L_2+1$ to L_1, including the other constraints in (10.55). So, the conclusion that (10.55) remains infeasible even when some of the $\textbf{Y}_\sigma$ for σ between $L_2 + 1$ to L_1 that were in **MB** are moved into **CB** or vice versa, implies that there are no 1-M/EC vectors in G. ∎

Exercises

10.4 Specialize the blossom algorithms discussed in Sections 10.2.1, and this section to the case when G is a bipartite network.

10.5 Let $\underline{r}, \overline{r}$ denote the minimum and maximum cardinalities for a 1-M/EC in G $= (\mathcal{N}, \mathcal{A})$ with $(\mathcal{N}^{\leqq}, \mathcal{N}^=, \mathcal{N}^{\geqq}, \mathcal{N}^0)$ as the partition of $\mathcal{N}$. Prove that there exists a 1-M/EC of cardinality r in G for every $\underline{r} \leqq r \leqq \overline{r}$. (Cartensen, Murty, and Perin [1981])

A FORTRAN implementation of the blossom algorithm in this section took on an average 0.333 CPU seconds (on an AMDAHL 470/V6 computer in 1980) to solve minimum cost 1-M/EC problems with randomly generated data in networks with 50 nodes and 250 arcs, and 4.127 seconds in networks with 100 nodes and 3000 arcs (see Perin [1980]).

The Convex Hull of 1-M/EC Vectors

We developed an algorithm for solving the minimum cost 1-M/EC problem in G $= (\mathcal{N}, \mathcal{A}, c)$ with $(\mathcal{N}^{\leqq}, \mathcal{N}^=, \mathcal{N}^{\geqq}, \mathcal{N}^0)$ as the partition of $\mathcal{N}$, but we did not explicitly determine a system of linear constraints whose solution set is the convex hull of all 1-M/EC vectors in G. We will do that now.

Let $\textbf{Y} \subset \mathcal{N} \backslash \mathcal{N}^0$ with $|\textbf{Y}|$ odd and $\geqq 3$, and let $\overline{\textbf{Y}} = \mathcal{N} \backslash \textbf{Y}$. For any edge vector $x = (x_{ij})$ defined on $\mathcal{A}$ define $x(\textbf{Y}; \overline{\textbf{Y}})$ as in Chapter 1 to be the sum of x_{ij} over edges $(i; j)$ joining nodes in $\mathbf{Y}$ to those in $\overline{\mathbf{Y}}$. Verify that $x(\mathbf{Y}; \overline{\mathbf{Y}}) = \mathbf{Y}^{|}(x) - \mathbf{Y}^-(x)$. If $\mathbf{E} = \{(i; j) : x_{ij} = 1\}$, then $x(\mathbf{Y}; \overline{\mathbf{Y}}) =$ the number of edges in $\mathbf{E}$ joining a node in $\mathbf{Y}$ to one outside of $\mathbf{Y}$.

Let $s(i)$ be the slack variable corresponding to node i in the node degree constraints in 1-M/EC problems, i.e., for any 1-M/EC vector x

$$
s(i) = \begin{cases}
1 - x(i) & \text{for } i \in \mathcal{N}^{\leq} \\
0 & \text{for } i \in \mathcal{N}^= \\
x(i) - 1 & \text{for } i \in \mathcal{N}^{\geq} \\
0 & \text{for } i \in \mathcal{N}^0
\end{cases}
$$

Clearly $s(i) \overset{\geq}{=} 0$ whenever x is a 1-M/EC vector.

THEOREM 10.35 *Let* $\mathbf{Y} \subset \mathcal{N} \backslash \mathcal{N}^0$ *with* $|\mathbf{Y}|$ *odd and* $\overset{\geq}{=} 3$, *and let* $\overline{\mathbf{Y}} = \mathcal{N} \backslash \mathbf{Y}$. *Let* $x, s = (s(i))$ *be a 1-M/EC vector and the corresponding vector of slack variables defined above. Then*

$$
x(\mathbf{Y}; \overline{\mathbf{Y}}) + \sum (s(i) : \text{ over } i \in \mathbf{Y}) \overset{\geq}{=} 1 \tag{10.65}
$$

Proof Let $\mathbf{E}$ be the 1-M/EC corresponding to x. Since $x(\mathbf{Y}; \overline{\mathbf{Y}})$ and $s(i)$ are nonnegative, the left hand side of (10.65) is $\overset{\geq}{=} 0$. So, if (10.65) is violated, we must have $x(\mathbf{Y}; \overline{\mathbf{Y}}) = 0$, and $s(i) = 0$ for all $i \in \mathbf{Y}$. $x(\mathbf{Y}; \overline{\mathbf{Y}}) = 0$ implies that $\mathbf{E}$ contains no edges joining a node in $\mathbf{Y}$ to one outside it. $s(i) = 0$ for all $i \in \mathbf{Y} \subset \mathcal{N} \backslash \mathcal{N}^0$ implies that each node in $\mathbf{Y}$ is incident to exactly one edge in $\mathbf{E}$, and by the above, any such edge must join two nodes in $\mathbf{Y}$. Since $|\mathbf{Y}|$ is odd, this is impossible, hence (10.65) must hold. ∎

THEOREM 10.36 *If* x *is the 1-M/EC vector in* G *obtained when the minimum cost 1-M/EC problem is solved by the algorithm discussed above, and* $\mathbf{Y}$ *is the set of original nodes inside a pseudonode at the termination of the algorithm, then (10.65) will hold as an equation for* x *and* $\mathbf{Y}$.

Proof Let $s = (s(i))$ be the slack vector corresponding to x. If the pseudonode is type A or B, $x(\mathbf{Y}; \overline{\mathbf{Y}}) = 1$ and $s(i) = 0$ for all $i \in \mathbf{Y}$, so (10.65) holds. If the pseudonode is type C, D, or E, let p be its apex node, then $x(\mathbf{Y}; \overline{\mathbf{Y}}) = 0, s(p) = 1$, and $s(i) = 0$ for all $i \in \mathbf{Y}, i \neq p$, hence (10.65) holds again. Since every pseudonode at termination will be type A, B, C, D, or E if 1-M/EC vectors exist, this completes the proof of the theorem. ∎

We will show that (10.65) is in fact the blossom inequality for the 1-M/EC problem corresponding to the odd subset of nodes $\mathbf{Y}$ from $\mathcal{N} \backslash \mathcal{N}^0$. Define $a(\mathbf{Y}) = 1 + |\mathbf{Y} \cap \mathcal{N}^{\geq}| - |\mathbf{Y} \cap \mathcal{N}^{\leq}|$, and $g(x, \mathbf{Y}) = x(\mathbf{Y}; \overline{\mathbf{Y}}) + \sum (x(i) : \text{over } i \in \mathbf{Y} \cap \mathcal{N}^{\geq}) -$

$\sum(x(i) :$ over $i \in \mathbf{Y} \cap \mathcal{N}^{\leqq})$. Then after rearranging terms, (10.65) can be written as

$$g(x, \mathbf{Y}) \geqq a(\mathbf{Y}) \tag{10.66}$$

(10.66), or the equivalent (10.65) is the blossom constraint corresponding to the odd subset of nodes $\mathbf{Y}$ for the 1-M/EC problem. In this problem, there is one blossom constraint of this type, for each subset of $\mathcal{N} \backslash \mathcal{N}^0$ of odd cardinality $\geqq 3$. Let $\mathbf{Y}_\sigma, \sigma = 1$ to L denote all the subsets of $\mathcal{N} \backslash \mathcal{N}^0$ of odd cardinality $\geqq 3$. Consider the LP (10.67) given below, obtained from the minimum cost 1-M/EC problem (10.6), (10.7) by replacing the integer requirements on the variables by the blossom inequalities of the form (10.66).

$$\text{Minimize} \quad \sum(c_{ij} x_{ij} \quad : \quad \text{over } (i; j) \in \mathcal{A})$$

$$\text{Subject to} \quad x(i) \quad \begin{cases} \leqq 1, & \text{for } i \in \mathcal{N}^{\leqq} \\ = 1, & \text{for } i \in \mathcal{N}^{=} \\ \geqq 1, & \text{for } i \in \mathcal{N}^{\geqq} \end{cases} \tag{10.67}$$

$$g(x, \mathbf{Y}_\sigma) \quad \geqq \quad a(\mathbf{Y}_\sigma), \sigma = 1 \text{ to } L$$

$$0 \leqq x_{ij} \quad \leqq \quad 1 \text{ for all } (i; j) \in \mathcal{A}$$

To write the dual of (10.67), associate the dual variable ξ_i to the node constraint at node i (with ξ_i defined to be $= 0$ for all $i \in \mathcal{N}^0$ always), η_σ to the blossom constraint corresponding to $\mathbf{Y}_\sigma$, and ω_{ij}, ν_{ij} to the bounds on x_{ij}. Let $\xi = (\xi_i), \eta = (\eta_\sigma)$. The dual of (10.67) is

$$\text{Max.} \quad \sum_{i \in \mathcal{N}} \xi_i \quad + \quad \sum_{\sigma=1}^{L} a(\mathbf{Y}_\sigma) \eta_\sigma + \sum_{(i;j) \in \mathcal{A}} \omega_{ij}$$

$$\text{subject to} \quad \delta_{ij}(\xi, \eta) \quad + \quad \omega_{ij} + \nu_{ij} = c_{ij}, \quad \text{for} \quad (i; j) \in \mathcal{A}$$

$$\xi_i \quad \begin{cases} \leqq 0, & \text{for } i \in \mathcal{N}^{\leqq} \\ \geqq 0, & \text{for } i \in \mathcal{N}^{\geqq} \\ = 0, & \text{for } i \in \mathcal{N}^0 \end{cases} \tag{10.68}$$

$$\eta_\sigma \geqq 0, \sigma = 1 \text{ to } \quad L \quad, \text{ and } \omega_{ij} \leqq 0, \nu_{ij} \geqq 0 \text{ for } (i; j) \in \mathcal{A}$$

where for $(i; j) \in \mathcal{A}$

$$b_{ij}(\mathbf{Y}_\sigma) = \begin{cases} 2 & \text{if both } i,j \in \mathcal{N}^{\geqq} \cap \mathbf{Y}_\sigma, \text{ or if one of} \\ & i \text{ or } j \text{ is } \in \mathcal{N}^{\geqq} \cap \mathbf{Y}_\sigma, \text{ and the other} \\ & \notin \mathbf{Y}_\sigma \\ 1 & \text{if one of } i,j \text{ is } \in \mathcal{N}^= \cap \mathbf{Y}_\sigma, \text{ and the} \\ & \text{other is either } \in \mathcal{N}^{\geqq} \cap \mathbf{Y}_\sigma, \text{ or } \notin \mathbf{Y}_\sigma \\ 0 & \text{if both } i,j \in \mathcal{N}^=\cap\mathbf{Y}_\sigma; \text{ or if both } i,j \notin \\ & \mathbf{Y}_\sigma; \text{ or if one of } i,j \text{ is } \in \mathcal{N}^{\geqq}\cap\mathbf{Y}_\sigma, \text{ and} \\ & \text{the other is } \in \mathcal{N}^{\leqq} \cap \mathbf{Y}_\sigma; \text{ or if one of} \\ & i,j \text{ is } \in \mathcal{N}^=\cap\mathbf{Y}_\sigma, \text{ and the other } \notin \mathbf{Y}_\sigma \\ -1 & \text{if one of } i,j \text{ is } \in \mathcal{N}^{\leqq} \cap \mathbf{Y}_\sigma, \text{ and the} \\ & \text{other is } \in \mathcal{N}^= \cap \mathbf{Y}_\sigma \\ -2 & \text{if both } i,j \in \mathcal{N}^{\leqq} \cap \mathbf{Y}_\sigma \end{cases} \tag{10.69}$$

$$\delta_{ij}(\xi,\eta) = \xi_i + \xi_j + \sum_{\sigma=1}^{L} b_{ij}(\mathbf{Y}_\sigma)\eta_\sigma \tag{10.70}$$

From the structure of the dual problem (10.68), it is clear that at an optimum solution (ξ, η, ω, ν), we will have for each $(i; j) \in \mathcal{A}$

$$\omega_{ij} = \min.\{0, c_{ij} - \delta_{ij}(\xi,\eta)\}, \nu_{ij} = \max.\{0, c_{ij} - \delta_{ij}(\xi,\eta)\} \tag{10.71}$$

Using this, we can express the complementary slackness optimality conditions for the primal-dual pair (10.67), (10.68) in terms of x, ξ, η only. These are, for each $i \in \mathcal{N}, (i;j) \in \mathcal{A}$, and $\sigma = 1$ to L

$$\xi_i(x(i) - 1) = 0$$

$$\delta_{ij}(\xi,\eta) \quad \begin{cases} > c_{ij} \text{ implies } x_{ij} = 1 \\ < c_{ij} \text{ implies } x_{ij} = 0 \end{cases} \tag{10.72}$$

$$\eta_\sigma(g(x, \mathbf{Y}_\sigma) - a(\mathbf{Y}_\sigma)) = 0$$

THEOREM 10.37 *Let* $\tilde{x}, (\tilde{\pi}, \tilde{\mu}), \tilde{\text{MB}}, \tilde{\text{CB}}$ *be the 1-M/EC vector, dual solution, and the blossom constraint specification sets obtained when the 1-M/EC problem in* G *is solved by the algorithm discussed above. So,* $\tilde{\mu}_\sigma = 0$ *if* $\mathbf{Y}_\sigma \notin \tilde{\text{MB}} \cup \tilde{\text{CB}}$. *Define* $\tilde{\xi} = (\tilde{\xi}_i), \tilde{\eta} = (\tilde{\eta}_\sigma)$ *where for* $i \in \mathcal{N}, \sigma = 1$ *to* L

$$\tilde{y}_i = \tilde{\pi}_i + \sum(\frac{1}{2}\tilde{\mu}_\sigma : \text{ over } \sigma \text{ s. t. } \mathbf{Y}_\sigma \in \tilde{\text{MB}} \cup \tilde{\text{CB}} \text{ and contains } i)$$

$$\tilde{\eta}_\sigma = \frac{1}{2}|\tilde{\mu}_\sigma|$$

$$\tilde{\xi}_i = \begin{cases} \tilde{y}_i & \text{for } i \in \mathcal{N}^0 \cup \mathcal{N}^= \\ \tilde{y}_i + \sum(\tilde{\eta}_\sigma : \text{ over } \sigma \text{ s. t. } \mathbf{Y}_\sigma \text{ contains } i) & \text{for } i \in \mathcal{N}^{\leqq} \\ \tilde{y}_i - \sum(\tilde{\eta}_\sigma : \text{ over } \sigma \text{ s. t. } \mathbf{Y}_\sigma \text{ contains } i) & \text{for } i \in \mathcal{N}^{\geqq} \end{cases}$$

Then $\tilde{x}, (\tilde{\xi}, \tilde{\eta})$ are optimal to the primal dual pair of problems (10.67), (10.68); with the corresponding ω, ν given by (10.71).

Proof $\tilde{x}$ is clearly feasible to (10.67). $(\tilde{\xi}, \tilde{\eta})$ satisfy the sign constraints on them in (10.68) because of the results proved in Theorem 10.29. By considering the various cases corresponding to the nodes i, j on an arc $(i; j) \in \mathcal{A}$ lying in various subsets in the partition $(\mathcal{N}^{\leq}, \mathcal{N}^{=}, \mathcal{N}^{\geq}, \mathcal{N}^0)$ of $\mathcal{N}$ separately, it can be verified that $\delta_{ij}(\tilde{\xi}, \tilde{\eta}) = d_{ij}(\tilde{\pi}, \tilde{\mu})$ for all $(i; j) \in \mathcal{A}$.

By the complementary slackness optimality conditions (10.61) satisfied by $\tilde{x}, (\tilde{\pi}, \tilde{\mu})$, this implies that for each $(i; j) \in \mathcal{A}$, $\delta_{ij}(\tilde{\xi}, \tilde{\eta}) > c_{ij}$ implies that $\tilde{x}_{ij} = 1$; and $\delta_{ij}(\tilde{\xi}, \tilde{\eta}) < c_{ij}$ implies that $\tilde{x}_{ij} = 0$. From this, and Theorems 10.36, 10.29, it can be verified that $\tilde{x}, (\tilde{\xi}, \tilde{\eta})$ satisfy the complementary slackness optimality conditions (10.72); and since $\tilde{x}$ is feasible to (10.67), and $(\tilde{\xi}, \tilde{\eta})$ is feasible to its dual (10.68), we conclude that $\tilde{x}, (\tilde{\xi}, \tilde{\eta})$ are optimal to the primal dual pair of problems (10.67), (10.68). ∎

THEOREM 10.38 *The system of constraints in (10.67) provides a linear constraint representation for the convex hull of all the 1-M/EC vectors in G.*

Proof   If 1-M/ECs exist in G, we have shown that for any cost vector c, the algorithm discussed in this section terminates with an $\tilde{x}$ which is a minimum cost 1-M/EC vector, and by Theorem 10.37, this $\tilde{x}$ is an optimum solution of (10.67). Hence, when 1-M/ECs exist in G, for every cost vector c, (10.67) has an optimum solution that is a 1-M/EC vector. And every 1-M/EC vector in G is feasible to (10.67).

When there exist no 1-M/ECs in G, we have shown that (10.57) is unbounded above, using this and the arguments in the proof of Theorem 10.37, it can be shown that the objective function in (10.68) is also unbounded above in this case, which implies that (10.67) is infeasible by the duality theorem of LP.

These facts imply that every extreme point of the set of feasible solutions of (10.67) is a 1-M/EC vector. Also, every 1-M/EC vector in G is a 0-1 vector which is feasible to (10.67), and since all the variables in (10.67) are bounded by 0 and 1, every 0-1 feasible solution for it is an extreme point of its set of feasible solutions. Thus every extreme point of the set of feasible solutions of (10.67) is a 1-M/EC vector in G and vice versa, so this set is the convex hull of all 1-M/EC vectors in G. ∎

Comment 10.4   The algorithm and the results in this section are taken from Perin [1980]. In it, he also derived an out-of-kilter type blossom algorithm for the 1-M/EC problem, and used it to develop efficient techniques for performing sensitivity analysis in the 1-M/EC problem.

10.4 The Minimum Cost 1-M/EC Problem with Specified Cardinality

Consider the undirected network G $= (\mathcal{N}, \mathcal{A}, c)$ with the partition $(\mathcal{N}^{\leq}, \mathcal{N}^{=}, \mathcal{N}^{\geq}, \mathcal{N}^0)$ of $\mathcal{N}$, and $|\mathcal{N}| = n, |\mathcal{A}| = m$. Here we consider the problem of finding a minimum cost 1-M/EC in G having a specified cardinality r. This is the problem (10.6), (10.7), with the additional constraint $\sum(x_{ij}: \text{ over } (i;j) \in \mathcal{A}) = r$. Associate a single Lagrange multiplier λ to this constraint, and include it in the objective function. This partial Lagrangian relaxation leads to the problem of finding a 1-M/EC in G which minimizes $(c - \lambda \mathbf{e}^T)x$, where $\mathbf{e}^T$ is the row vector in $\mathbb{R}^m$ all of whose entries are 1. When λ is given a specific value, this is a 1-M/EC problem that can be solved efficiently by the blossom algorithm of Section 10.3. Let $x(\lambda) = (x_{ij}(\lambda))$ denote an optimum 1-M/EC vector for this problem as a function of λ. Let $r(\lambda) = \sum(x_{ij}(\lambda): \text{ over } (i;j) \in \mathcal{A})$, the cardinality of $x(\lambda)$. Then it can be shown that $x(\lambda)$ is a minimum cost 1-M/EC vector when the specified cardinality $r = r(\lambda)$, and that $r(\lambda)$ increases with λ. Using this result, an algorithm for solving the specified cardinality minimum cost 1-M/EC problem, treating the cardinality, r, as an integer valued parameter, has been developed in Carstensen, Murty, and Perin [1981], based on the blossom algorithm for the 1-M/EC problem of Section 10.3. That algorithm finds $x(\lambda)$ for some specific value of λ first, and then varies λ, treating it as a parameter, through efficient parametric procedures. This produces minimum cost 1-M/ECs of all possible cardinalities, by the above result.

10.5 Degree Constrained Subnetworks, *b*-Matching Problems, and the General Matching Problem

Let G $= (\mathcal{N}, \mathcal{A}, c)$ be an undirected multinetwork (i.e., there may be parallel edges in $\mathcal{A}$). Let $b = (b_i)$ be a vector of specified positive integers defined on $\mathcal{N}$. The problem of finding a minimum cost subnetwork of G, subject to the constraints that for each $i \in \mathcal{N}$ the degree of i in the subnetwork should be equal to (or $\leq$) b_i is known as a *minimum cost degree constrained subnetwork problem,* or a *minimum cost b-matching problem.* Verify that if $b_i = 1$ for all i, this problem becomes a minimum cost 1-matching problem. So, the b-matching problem is a generalization of the 1-matching problems discussed so far. It can be shown that every b-matching problem can be transformed into a 1-matching problem on an enlarged network, but this transformation leads to an inefficient algorithm for it. The 1-matching blossom algorithms have been generalized into blossom algorithms operating on the original network G itself for the b-matching problems, see Edmonds and Johnson [1970], and Edmonds and Pulleyblank [1975]. These algorithms can be used to solve integer programming problems of the following form

$$\text{Minimize} \quad \sum_{j=1}^{n} c_j x_j$$

$$\text{subject to} \quad \sum_{j=1}^{n} a_{ij}x_j \quad \begin{cases} = b_i, i = 1, \ldots, m \\ \leqq b_i, i = m+1, \ldots, m+p \end{cases} \qquad (10.73)$$

$$\ell_j \leqq x_j \quad \leqq u_j, j = 1, \ldots, n$$

$$x_j \quad \text{integer for all } j$$

where all the data is integer, and the a_{ij} satisfy the conditions

$$\sum_{i=1}^{m+p} |a_{ij}| \leqq 2, \text{ for each } j = 1 \text{ to } n$$

An integer program of this form is called a *general matching problem*, and the b-matching blossom algorithms can be used to solve such a problem efficiently.

We have seen that the 1-matching/edge covering problems on general undirected networks are a generalization of the assignment problem on bipartite networks. In the same way, it can be verified that the b-matching problems on general undirected networks, are a generalization of the transportation problem on bipartite networks.

10.6   Exercises

10.6 Let Let G $= (\mathcal{N}, \mathcal{A}, c = (c_{ij}))$ be a connected undirected network with $c \geqq 0$ as the vector of edge lengths, and $\check{s}, \check{t}$ as the specified origin and destination nodes. Make 2 copies of G side by side, with the same edge lengths as above. To distinguish the 2 copies, add a prime, i', to the node numbers in the right hand side copy. On the left hand side copy delete $\check{t}$ and all the edges incident at it. On the right hand side copy delete $\check{s}'$ and all the edges incident at it. For each i introduce the new edge $(i; i')$, and define its length to be 0. Let $\tilde{G}$ denote the resulting network.

(i) Let $\mathcal{P} = \check{s}, (\check{s}; i_1), i_1, (i_1; i_2), i_2, \ldots, (i_{2r-2}; i_{2r-1}), i_{2r-1}, (i_{2r-1}; \check{t}), \check{t}$ be a simple path consisting of an even number, $2r$, edges between $\check{s}$ and $\check{t}$ in G. Define the set of edges, M($\mathcal{P}$), in $\tilde{G}$, corresponding to the simple path $\mathcal{P}$ in G, to be, M($\mathcal{P}$) $= \{(\check{s}; i_1), (i_1'; i_2'), (i_2; i_3), (i_3'; i_4'), \ldots, (i_{2r-2}; i_{2r-1}), (i_{2r-1}'; \check{t}'), \text{ and } (i; i') \text{ for each } i \in \mathcal{N} \text{ not on } \mathcal{P}\}$.

Whenever $\mathcal{P}$ is a simple path between $\check{s}$ and $\check{t}$ in G consisting of an even number of edges, the set M($\mathcal{P}$) constructed as above is a perfect matching in $\tilde{G}$ of the same length as that of the path $\mathcal{P}$ in G. Conversely every perfect matching in $\tilde{G}$ corresponds to a simple path consisting of an even number of edges in G between $\check{s}$ and $\check{t}$, of the same length. Using this develop an algorithm for finding a shortest path between $\check{s}$ and $\check{t}$ in G, subject to the constraint that the path consist of an even number of edges.

(ii) Put 2 copies of G side by side again, as discussed above. Delete nodes $\check{s}'$, $\check{t}'$ and all the edges incident at them from the right hand side copy. For each $i \neq \check{s}, \check{t}$, introduce the edge $(i; i')$ as above. Let the resulting network be called $\hat{G}$. Show that a one to one correspondence, preserving lengths, can be established as above, between perfect matchings in $\hat{G}$ and simple paths in G between $\check{s}$ and $\check{t}$ consisting of an odd number of edges. Using this develop an algorithm for finding a shortest path between $\check{s}$ and $\check{t}$ in G, subject to the constraint that the path consist of an odd number of edges.

10.7 Consider the network in Figure 10.32 with the wavy matching M_1 marked in it. Is it possible to include another edge in the set M_1 and still retain the matching property for it? From this, can you conclude that M_1 is a maximum cardinality matching in this network? Why?

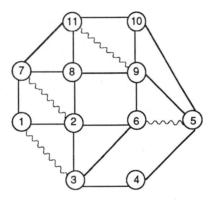

Figure 10.32 Matching edges are wavy.

10.8 M is a perfect matching in the undirected network $G = (\mathcal{N}, \mathcal{A}, c)$ with c as the vector of edge cost coefficients. Given an alternating path or cycle wrt M, $\mathcal{P}$ say, define its cost to be $\sum(c_{ij} : \text{over } (i; j) \in \mathcal{P} \backslash M) - \sum(c_{ij} : \text{over } (i; j) \in \mathcal{P} \cap M)$. Prove that M is a minimum cost perfect matching in G iff there exists no negative cost alternating cycle wrt it.

10.9 M is a matching in the undirected network $G = (\mathcal{N}, \mathcal{A}, c)$ with c as the vector of edge cost coefficients. Let $\mathbb{P}_i(M)$ denote the set of all augmenting paths wrt M beginning with an unmatched node i, and $\mathbb{P}(M)$ the set of all augmenting paths wrt M. Define the cost of any path in $\mathbb{P}(M)$ exactly as in Exercise 10.8. Assume that there exists no negative cost alternating cycle wrt M, and that $\mathcal{P}$ is a least cost augmenting path wrt M in some $\mathbb{P}_i(M)$ or in $\mathbb{P}(M)$. Let M' be the matching obtained by rematching M using $\mathcal{P}$. Prove that M' also satisfies the property that there exists no negative cost alternating cycle wrt it (Derigs [1981]).

10.10 A matching M in an undirected network $G = (\mathcal{N}, \mathcal{A})$ is said to be a *maximal matching* if it is impossible to add another edge to the set M and still keep its matching property. Let $\underline{r}, \bar{r}$ denote the minimum and maximum cardinalities of maximal matchings in G. Prove that $\bar{r} \leqq 2\underline{r}$.

10.11 M is a perfect matching in the undirected network $G = (\mathcal{N}, \mathcal{A}, c)$. Let $(i; j) \in M$. Prove that there exists a negative cost alternating cycle $\mathbb{C}$ wrt M containing edge $(i; j)$ iff the cost of the shortest augmenting path wrt $M' = M \backslash \{(i;j)\}$ is $< c_{ij}$. Hence show that the problem of finding negative cost alternating cycles can be solved by computing shortest augmenting paths (Derigs [1986]).

10.12 A Delivery Problem   A commodity has to be supplied to several users in a geographical region from a central depot by truck. Each truck has a finite capacity and hence can fulfill two users only on a trip. We are given the distance between the depot and each user, and the distance between every possible pair of users. Formulate the problem of fulfilling the user demands at minimum cost (which can be assumed to be proportional to the distance traveled by the trucks) as a matching problem (DeMaio and Roveda [1971]).

10.13 E is a specified subset of edges in an undirected network $G = (\mathcal{N}, \mathcal{A})$. It is required to check whether there exists a perfect matching in G consisting of at most r edges from E. Formulate this problem as a minimum cost maximum cardinality matching problem.

10.14 M_1 is a perfect matching in an undirected network $G = (\mathcal{N}, \mathcal{A})$. Prove that M_1 is not the only perfect matching in G iff there exists an even alternating cycle wrt M in G.

10.15 M_1 is a maximum cardinality matching that is not perfect in an undirected network G. Prove that M_1 is not the only maximum cardinality matching in G iff at least one of the following things exists. (i) an even alternating cycle wrt M_1 in G, (ii) a matching edge $(i; j)$ and an unmatched node k adjacent to i or j in G (Itai, Rodeh, and Tanimoto [1978]).

10.16 M is a matching in a rooted tree $\mathbb{T}$ with node 1 as the root. M is said to be a *proper matching* of $\mathbb{T}$ if it satisfies the following property: if a node $i \neq 1$ is unmatched, then there exists a brother j of i such that $(j; P(j)) \in M$ where $P(j)$ is the parent node of j. Prove that every proper matching in $\mathbb{T}$ must be a maximum cardinality matching. Using this develop an $O(n)$ algorithm (n is the number of nodes in $\mathbb{T}$) for finding a maximum cardinality matching in a tree (Savage [1980]).

10.17 $\mathbb{T}$ is a depth first search spanning tree in an undirected network G. M(G), $M(\mathbb{T})$ are maximum cardinality matchings in G, $\mathbb{T}$ respectively. Prove that $|M(G)|/|M(\mathbb{T})| \leqq 2$ (Savage [1980]).

10.18 A vertex cover or node cover in an undirected network is a subset of nodes satisfying the property that every edge in the network is incident to at least one node in the subset. The problem of finding a minimum cardinality vertex cover is NP-hard in general.

(i)   In a bipartite network, prove that the cardinality of a minimum cardinality vertex cover is the same as the cardinality of a maximum cardinality matching. This result may not be true in nonbipartite networks.

(ii) **M** is a proper matching in a rooted tree $\mathbb{T}$ with node 1 as the root (see Exercise 10.16 for definition). Let **S** be the set of nonroot nodes i such that $(i; P(i)) \in \mathbf{M}$, where $P(i)$ is the parent of i in **T**. Let $\mathbf{F} = \{P(j): j \in \mathbf{S}\}$. Show that any node not in **S**, **F** is unmatched, and that **F** is a minimum cardinality vertex cover for $\mathbb{T}$.

(iii) Let $\mathbb{T}$ be a depth first search spanning tree in an undirected network G (the following results may not be true for arbitrary spanning trees in G). Let $\mathbf{L}(\mathbb{T}), \mathbf{NL}(\mathbb{T})$ be the sets of leaf, non-leaf nodes in $\mathbb{T}$ respectively. Let $\mathbf{M}(\mathbb{T})$ be a proper matching in $\mathbb{T}$ as defined in Exercise 10.16, and $\mathbf{M}(G)$ a maximum cardinality matching in G. Let $\mathbf{C}(G)$ be a minimum cardinality vertex cover in G.

Since $\mathbb{T}$ is a depth first search spanning tree, prove that $\mathbf{NL}(\mathbb{T})$ is a vertex cover for G. Prove the following string of inequalities.

$$|\mathbf{M}(\mathbb{T})| \leqq |\mathbf{M}(G)| \leqq |\mathbf{C}(G)| \leqq |\mathbf{NL}(\mathbb{T})| \leqq 2|\mathbf{M}(\mathbb{T})|$$

From this we get $|\mathbf{M}(G)|/|\mathbf{M}(\mathbb{T})| \leqq 2$ (result in Exercise 10.17) and $|\mathbf{NL}(\mathbb{T})|/|\mathbf{C}(G)| \leqq 2$. So, $\mathbf{NL}(\mathbb{T})$ forms a vertex cover of G guaranteed to be within a factor of 2 of the minimum vertex cover in size. Show that this bound is tight by considering the case in which G is a path of odd length and the depth first search is done from a vertex of degree 1.

(iv) Show that $|\mathbf{C}(G)|/\mathbf{M}(G) \leqq 2$ for any undirected network G.

(v)   Let $\mathbf{M}_1$ be any maximal matching in G (i.e., a matching satisfying the property that any pair of unmatched nodes are nonadjacent in G), and let $\mathbf{V}(\mathbf{M}_1)$ be the set of nodes on edges in $\mathbf{M}_1$. Then $\mathbf{V}(\mathbf{M}_1)$ is a vertex cover for G. Prove that $|\mathbf{C}(G)|/|\mathbf{V}(\mathbf{M}_1)| \leqq 2$.

(Savage [1982]).

10.19 Degree Constrained Subnetwork Problem in Undirected Networks $G = (\mathcal{N}, \mathcal{A})$ is an undirected network with $\mathcal{N} = \{1, \ldots, n\}$. $b_1, \ldots, b_n$ are nonnegative integers. We are required to find a subnetwork $G' = (\mathcal{N}, \mathcal{A}')$ of G such that $|\mathcal{A}'|$ is maximized subject to the constraint that for each $i = 1$ to n, the degree of i in G' is $\leqq b_i$. This is the undirected version of the same problem for directed networks discussed in Exercise 1.57. As stated there, in directed networks the problem easily reduces to a max-flow problem, this is not the case in undirected networks.

Construct an undirected network $H = (\mathcal{N}_H, \mathcal{A}_H)$ from G by the following procedure: for each $i \in \mathcal{N}$ put b_i copies of node i, $i_1, \ldots, i_{b_i}$ say, in $\mathcal{N}_H$. On each edge

$e \in \mathcal{A}$ introduce two new nodes u_e, v_e, and transform the edge e into the subnetwork $H(e)$ as shown in the following Figure 10.33.

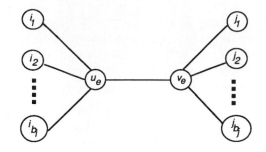

Figure 10.33

Let $\overline{M}$ be a maximum cardinality matching in H and let $\overline{G} = (\mathcal{N}, \overline{\mathcal{A}}')$ be an optimum subnetwork of G to our degree constrained subnetwork problem. Then prove the following.

(a) $|\overline{M}| = 2|\overline{\mathcal{A}}'| + |\mathcal{A} \backslash \overline{\mathcal{A}}'| = |\mathcal{A}| + |\overline{\mathcal{A}}'|$.

(b) Given a maximum cardinality matching $\overline{M}$ in H, the set of edges $\overline{\mathcal{A}}'$ for an optimum subnetwork of G to our problem is obtained by the rule: $e \in \mathcal{A}$ is in $\overline{\mathcal{A}}'$ iff $|H(e) \cap \overline{M}| = 2$, and this satisfies $|\overline{\mathcal{A}}'| + |\mathcal{A}| = |\overline{M}|$.

(c) Conversely, given an optimum subnetwork $\overline{G} = (\mathcal{N}, \overline{\mathcal{A}}')$ for our problem, define a matching $\overline{M}$ in H by the rules : (i) for each $e \in \mathcal{A}$, $(u_e; v_e) \in \overline{M}$ iff $e \notin \overline{\mathcal{A}}'$, (ii) if $e = (i; j) \in \overline{\mathcal{A}}'$, then $(i_r; u_e), (v_e; j_s)$ are in $\overline{M}$ for some $1 \leq r \leq b_i, 1 \leq s \leq b_j$. Since the degree of node i in $\overline{G}'$ is $\leq b_i$, this matching can be accomplished without any conflicts. Then prove that this matching will satisfy $|\overline{M}| = 2|\overline{\mathcal{A}}'| + |\mathcal{A} \backslash \overline{\mathcal{A}}'|$.

(Shiloach [1981]).

10.20 Weighted Degree Constrained Subnetwork Problem in Undirected Networks Consider the degree constrained subnetwork problem discussed in Exercise 10.19, with the exception that the objective function is to maximize $\sum(w(e) :$ over $e \in \mathcal{A}')$ rather than $|\mathcal{A}'|$, where $w(e)$ are specified weights for edges in G. Construct the undirected network H from G as in Exercise 10.19, and define edge weights in H by the rule: every edge in $H(e)$ gets the same weight in H, as that of e in G. Given a feasible subnetwork $G' = (\mathcal{N}, \mathcal{A}')$ of G, define a matching M in H as in (c) of Exercise 10.19. Show that $w(M) = 2w(\mathcal{A}') + w(\mathcal{A} \backslash \mathcal{A}') = w(\mathcal{A}) + w(\mathcal{A}')$. So, maximizing $w(\mathcal{A}')$ is equivalent to maximizing $w(M)$. Using this, show that this problem in G gets transformed into a maximum weight matching problem in H (Shiloach [1981]).

10.21 General Degree Constrained Subnetwork Problems in Undirected Networks Let G = $(\mathcal{N}, \mathcal{A})$ be an undirected network with $\mathcal{N} = \{1, \ldots, n\}$. We are given nonnegative integers $a_1, \ldots, a_n; b_1, \ldots, b_n$ satisfying $a_i \leqq b_i$ for all i. $w(e)$ is the weight of $e \in \mathcal{A}$. It is required to find a subnetwork G' = $(\mathcal{N}, \mathcal{A}')$ of G, such that $d_{G'}(i)$ = degree of node i in G' is between a_i and b_i for all i; and either maximize $|\mathcal{A}'|$ in the cardinality problem, or $\sum(w(e) :$ over $e \in \mathcal{A}')$ in the weighted problem. In Exercises 10.19, 10.20, we considered the cases in which all the a_i are 0.

    (i) Consider the cardinality problem first. Replace all the a_i by 0 and solve the resulting problem which is now in the same form as that in Exercise 10.19, and let G^r be the resulting optimal subnetwork. If G^r is not feasible to our problem, there must be some nodes i whose degree in G^r is $< a_i$, call them deficient nodes. If our problem is feasible, show that we can use alternating paths (not necessarily simple) to transform G^r into an optimum solution of our problem, reducing the total deficiency by one in each step while preserving the cardinality.

    (ii) Consider the weighted problem. Construct the network H as in Exercise 10.19 (this does not use the a_is), and for each $e \in \mathcal{A}$ make the weight of each edge in H(e) in H the same as $w(e)$. Define $\mathcal{N}_1 = \cup_{i=1}^n \{i_1, \ldots, i_{a_i}\}$. Show that our problem has a solution iff there exists a matching in H in which all the nodes in $\mathcal{N}_1$ are matched. Moreover, show that an optimum solution of our problem can be obtained from an optimum solution of the constrained weighted matching problem in H in which it is required to find a maximum weight matching subject to the constraint that every node in $\mathcal{N}_1$ must be matched, using the transformation in (b) of Exercise 10.19.

    Change the data in H as follows: add α (2α) to the weight of each edge incident to exactly one node (two nodes) in $\mathcal{N}_1$, where α is a large positive number. Denote the network with the resulting data as H'. Find a maximum weight matching M' in H'. Show that if some nodes in $\mathcal{N}_1$ are unmatched in M', then the constrained matching problem in H discussed above has no solution. And if all the nodes in $\mathcal{N}_1$ are matched in M', then M' solves the constrained weighted matching problem in H. Using these, discuss an algorithm for solving our weighted subnetwork problem with any maximum weighted 1-matching algorithm (Shiloach [1981]).

10.22 It is required to find a minimum cost matching in an undirected network G = $(\mathcal{N}, \mathcal{A}, c)$ subject to the constraint that all the nodes in a specified subset $\mathcal{N}_1$ must be matched. Discuss a way of transforming this constrained matching problem into an unconstrained minimum cost matching problem.

10.23 The Edge Partitioning Problem For any undirected network L, let $\Delta(L)$ denote the maximum degree among its nodes. Let G = $(\mathcal{N}, \mathcal{A})$ be an undirected network. Let α, β be positive integers such that $\alpha + \beta = \Delta(G)$. It is required to partition $\mathcal{A}$ into disjoint subsets $\mathcal{A}_1, \mathcal{A}_2$ such that if $G_t = (\mathcal{N}, \mathcal{A}_t)$, $t = 1, 2$, then $\Delta(G_1) = \alpha, \Delta(G_2) = \beta$. Formulate this as a special case of the degree constrained subnetwork problem of Exercise 10. 21 (Shiloach [1981]).

10.24 In a connected bipartite network, prove that there exists a matching M, and a node cover N such that every edge in M contains exactly one node in N; and every node in N is contained on exactly one edge in M.

10.25 N_t, $t = 1, 2$ are two subsets of nodes in an undirected network G, satisfying $|N_1| < |N_2|$. M_t is a matching in G covering all the nodes in N_t, $t = 1, 2$. Prove that there must exist a matching in G, that covers all the nodes in N_1 and at least one node in $N_2 \backslash N_1$.

10.26 N is a specified subset of nodes in an undirected network $G = (\mathcal{N}, \mathcal{A}, c)$ with edge cost vector $c \geqq 0$. It is required to find a subset of edges A in G satisfying the constraint that in the subnetwork $(\mathcal{N}, A)$ all nodes in N have odd degree and all nodes not in N have even degree. Formulate the problem of finding a minimum cost subset of edges subject to this constraint, as a matching problem.

Comment 10.5   Matchings have been studied by graph theorists since 1891. At that time it was known that the famous four color conjecture is true if every cubic (i.e., one in which every node has degree 3) planar graph can be factorized into three perfect matchings, this heightened interest in the study of matchings.

Then in the early 1900s matchings in bipartite graphs were studied extensively. Necessary and sufficient conditions for a bipartite graph to have a perfect matching have been derived. These can be explained easily using a marriage interpretation as follows. Think of the two sets of nodes of the bipartite graph as representing men, women respectively. Interpret each edge in the graph as representing a man, woman pair acquainted with each other. A matching in the graph corresponds to a set of man-woman couples (one couple determined by each matching edge) so that each couple is acquainted. To have a perfect matching, clearly the number of men and the number of women should be equal. In this context, the conditions state that if we have n men and n women, a set of n acquainted man-woman couples can be formed iff for each $1 \leqq r \leqq n$, each set of r men collectively are acquainted with at least r women. This result was the forerunner of another fundamental result in bipartite matchings, *Hall's theorem on distinct representatives*. It states that if S_t, $t = 1$ to n are finite sets, there is a set of distinct elements, $x_1, \ldots, x_n$, such that $x_t \in S_t, t = 1$ to n, iff for each $1 \leqq r \leqq n$, the union of any r of the S_ts contains at least r elements. Also around this time, all the theory necessary for finding a maximum cardinality matching in a bipartite graph had been worked out by König and Egerváry.

Then in 1947, a big step in the study of matchings was taken by Tutte. He gave a characterization of general (i.e., nonbipartite) graphs that have a perfect matching, this can be viewed as a generalization of the corresponding result in bipartite graphs. Tutte's characterization states that a connected graph has no perfect matching iff there exists a set S of nodes in it, the deletion of which leaves more than $|S|$ components having an odd number of points each. Tutte's characterization is not algorithmic, it prompted attempts to find efficient algorithms for perfect (or maximum cardinality) matchings in nonbipartite networks. Berge [1957] and Norman and Rabin [1959] have shown that the concepts of alternating and augmenting paths do generalize from the bipartite to the nonbipartite networks, and in fact the augmenting path theorem holds in nonbipartite networks without any change, i.e., that a matching in a nonbipartite network is of maximum cardinality iff it admits no augmenting path. By this time matching problems in bipartite networks became efficiently solvable through the Hungarian method. People believed for a long time that the result of Berge, Norman and Rabin leads to an efficient algorithm for the maximum cardinality matching problem in nonbipartite networks. One begins with the empty matching and repeatedly performs augmentations until there are no augmenting paths. Matters stood there until 1965 when Edmonds showed that the search for augmenting paths in nonbipartite networks is very intricate, and could take exponential time unless special techniques are developed for it. We begin at an unmatched node, and advance along an AP. If this reaches another unmatched node, we have found an augmenting path. This procedure always works nicely in bipartite networks, but in nonbipartite networks, odd cycles create subtle difficulties for it. A node can appear on an AP in either parity (outer or inner); and if we allow two visits to a node, one in each parity, the path is no longer simple; if we don't, we miss the augmenting path.

The algorithmic study of matchings took a giant step with the work of Edmonds [1965a, b]. There, he developed the elegant solution to the problem of finding augmenting paths efficiently by detecting and shrinking blossoms as they appear. In these papers he introduced the characterization of the convex hull of matching incidence vectors through the system of blossom inequalities. Using this characterization, he extended the Hungarian method for the bipartite minimum cost matching problem into the blossom algorithm for minimum cost matching problem in nonbipartite networks, and showed that it is polynomially bounded. In these papers, Edmonds has also proposed *polynomial time* property as the prime characteristic for designating algorithms as *good algorithms*. This has become a fundamental tenet of theoretical computer science ever since.

Another major outgrowth from these papers is the study of other combinatorially defined polyhedra with the aim of characterizing them through a system of linear constraints, this area is now called *polyhedral combinatorics*.

Earlier in 1962, the Chinese mathematician Kwan Mei-Ko posed the important postman's route problem. Hence Edmonds called this the *Chinese postman problem* and developed the efficient procedure for solving it using the blossom algorithm for the minimum cost perfect matching problem on the complete network defined on the set of odd degree nodes in the original network, as described in Section 1.3.8.

With the Hungarian method in bipartite networks, and the blossom algorithm in nonbipartite networks, the study of optimization problems involving matchings has taken off with a vigorous start. Several new algorithms for matching problems, as well as efficient implementations based on appropriate data structures, have been developed. Many applications in a variety of areas have come up. Some recent applications in VLSI chip design etc. lead to matching problems on such large networks, that it is impractical to solve them with the $O(n^3)$ blossom or other exact algorithms. Hence, even though mathematically efficient exact algorithms exist for them, heuristic algorithms of low order complexity are being developed for tackling very large scale matching problems.

For other algorithms for matching problems see Avis [1983], Ball and Derigs [1983], Cunningham and Marsh [1978], Derigs [1981, 1985, 1986, 1988 of Chapter 1, 1991], Even and Kariv [1975], Gabow [1976], and Kameda and Munro [1974].

10.7   References

Y. A. ALYAHYA, 1984, "Matching and Covering Algorithms," Ph. D. dissertation, Dept. of Industrial and Operations Engineering, University of Michigan, Ann Arbor, Mich., USA,

D. AVIS, 1983, "A Survey of Heuristics for the Weighted Matching Problem," *Networks*, 13(475-493).

M. O. BALL and U. DERIGS, 1983, "An Analysis of Alternate Strategies for Implementing Matching Algorithms," *Networks*, 13(517-549).

C. BERGE, 1957, "Two Theorems in Graph Theory," *Proceedings of the National Academy of Sciences, USA*, 43(842-844).

J. BROWN, 1977, "Shortest Alternating Path Algorithms," *Networks*, 4(311-334).

P. CARSTENSEN, K. G. MURTY, and C. PERIN, 1981, "Parametric Specified Cardinality 1-Matching/Edge Covering Problems and Intermediate Feasibility Property," Technical Report 81-6, Dept. of Industrial and Operations Engineering, University of Michigan, Ann Arbor, Mich., USA,

N. CHRISTOFIDES, 1976, "Worst-case Analysis of a New Heuristic for the Traveling Salesman Problem," Technical Report, GSIA, Carnegie-Mellon University, Pittsburgh, PA, USA.

W. H. CUNNINGHAM and A. B. MARSH III, 1978, "A Primal Algorithm for Optimum Matching," *MPS*, 8(50-72).

A. O. DEMAIO and C. A. ROVEDA, 1971, "The Minimal Cost Maximum Matching of a Graph," *Unternehmens-Forschung Operations Research*, 15, no. 3(196-210).

U. DERIGS, 1981, "A Shortest Augmenting Path Method for Solving Minimal Perfect Matching Problems," *Networks*, 11(379-390).

U. DERIGS, Nov. 1985, "An Efficient Dijkstra-like Labeling Method for Computing Shortest Odd/Even Paths," *IPL*, 21, no. 5(253-258).

U. DERIGS, 1986, "Solving Large-Scale Matching Problems Efficiently: A New Primal Matching Approach," *Networks*, 16(1-16).

U. DERIGS and A. METZ, Mar. 1991, "Solving (Large Scale) Matching Problems Combinatorially," *MP*, Series A, 50, no. 1(113-121).

J. EDMONDS, 1962, "Covers and Packings in a Family of Sets," *BAMS*, 68(494-499).

J. EDMONDS, 1965a, "Paths, Trees, and Flowers," *Canadian Journal of Mathematics*, 17(449-467).

J. EDMONDS, 1965b, "Maximum Matching and a Polyhedron With 0, 1 Vertices," *Journal Research NBS*, 69(B)(130-165).

J. EDMONDS, 1965c, "The Chinese Postman Problem," *OR*, 13, Suppl. 1(373).

J. EDMONDS and E. L. JOHNSON, 1970, "Matching: A Well Solved Class of Integer Linear

Programs," PP 89-92 in R. Guy(ed.) *Combinatorial Structures and Their Applications* , Gordon and Breach, N.Y.

J. EDMONDS and W. PULLEYBLANK, 1975, "The Matching Problem and the Blossom Algorithm," Lecture notes, John Hopkins University, Baltimore, MD, USA.

S. EVEN and O. KARIV, 1975, "An $O(n^{2.5})$ algorithm for the Maximum Matching in General Graphs," *Proceedings of the 16th Annual Symposium on Foundations of Computer Science*, IEEE, N.Y. (100-112).

M. FUJII, T. KASAMI, and K. NINOMIYA, 1969, "Optimal Sequencing of Two Equivalent Processors," *SIAM Journal on Applied Mathematics*, 17(784-789), Erratum ibid 20, 1971(141).

H. N. GABOW, 1976, "An Efficient Implementation of Edmond's Algorithm for Maximum Matching on Graphs," *JACM*, 23(221-234).

A. ITAI, M. RODEH, and J. L. TANIMOTO, Oct. 1978, "Some Matching Problems for Bipartite Graphs," *JACM*, 25, no. 4(517-525).

T. KAMEDA and I. MUNRO, 1974, "An $O(|V| \ |E|)$ Algorithm for Maximum Matching of Graphs," *Computing*, 12(91-98).

A. S. LAPAUGH and C. H. PAPADIMITRIOU, 1984, "The Even-Path Problem for Graphs and Digraphs," *Networks*, 14, no. 4(507-513).

KWAN MEI-KO, 1962, "Graphic Programming Using Odd or Even Points," *Chinese Math.* , 1(273-277).

B. MONTREUIL, H. D. RATLIFF, and M. GOETSCHALCKX, Sept. 1987, "Matching Based Interactive Facility Layout," *IIE Transactions*, 19, no. 3(271-279).

K. G. MURTY and C. PERIN, 1982, "A Blossom Type Algorithm for the Minimum Cost Edge Covering Problem," *Networks*, 12(379-391).

R. Z. NORMAN and M. O. RABIN, 1959, "An Algorithm for a Minimum Cover of a Graph," *Proceedings of the American Math. Soc.*, 10(315-319).

M. W. PADBERG and M. R. RAO, 1982, "Odd Minimum Cut-Sets and b-Matchings," *MOR*, 7(67-80).

C. PERIN, 1980, "Matching and Edge Covering Algorithms," Ph. D. dissertation, Dept. of Industrial and Operations Engineering, University of Michigan, Ann Arbor, Mich., USA,

W. R. PULLEYBLANK, 1983, "Polyhedral Combinatorics," PP 312-345 in A. Bachem, M. Grotschel, and B. Korte (eds.), *Mathematical Programming, The State of the Art: Bonn 1982*, Springer-Verlag, Berlin.

C. SAVAGE, 1980, "Maximum Matchings and Trees," *IPL*, 10, no. 4/5(202-205).

C. SAVAGE, July 1982, "Depth First Search and the Vertex Cover Problem," *IPL*, 14, no. 5(233-235).

A. SCHRIJVER, 1983, "Short Proofs of the Matching Polyhedron," *Journal of Combinatorial Theory, (B)*, 34(104-108).

Y. SHILOACH, April 1981, "Another Look at the Degree Constrained Subgraph Problem," *IPL*, 12, no. 2(89-92).

W. T. TUTTE, 1947, "The Factorization of Linear Graphs," *Journal of the London Math. Soc.* , 22(107-111).

L. J. WHITE, 1971, "Minimum Covers of Fixed Cardinality in Weighted Graphs," *SIAM Journal of Applied Math.*, 21(104-113).

L. J. WHITE and M. L. GILLENSON, 1975, "An Efficient Algorithm for Minimum k-Covers in Weighted Graphs," *MP*, 8(20-42).

Index